21st Century Astronomy's new Animations help students master core concepts.

Developed specifically for the Second Edition of *21st Century Astronomy*, these lessons animate 18 core concepts central to astronomy and use interactivity to test students' mastery.

Animations illustrate specific sections of the text

A.	The Celestial Sphere and the Ecliptic	Text Section 2.2
B.	The Earth Spins and Revolves	Text Sections 2.2 and 2.3
C.	The Moon's Orbit: Eclipses and Phases	Text Section 2.4 and 2.5
D.	Kepler's Laws	Text Section 3.2
E.	Newton's Laws and Universal Gravitation	Text Section 3.5
F.	Light as a Wave, Light as a Photon	Text Section 4.2, 4.4
G.	Doppler Effect	Text Section 4.4
H.	Atomic Energy Levels and the Bohr Model	Text Section 4.4
I.	Atomic Energy Levels and Light Emission and Absorption	Text Section 4.4
J.	Geometric Optics and Lenses	Text Section 5.1
K.	Solar System Formation	Text Section 6.1–6.4
L.	Processes That Shape the Planets	Text Section 7.2, 7.3, 7.5, 7.7
M.	Tides and the Moon	Text Section 10.2, 10.3
N.	Solar Spectrum	Text Section 13.3
O.	The H-R Diagram	Text Section 13.5
P.	The Solar Core	Text Section 14.2
Q.	Star Formation	Text Sections 15.3, 15.4
R.	Hubble's Law	Text Sections 20.2, 20.3, 20.4

There are a variety of ways to access the Animations

 Available free from the StudySpace student website at **www.wwnorton.com/astro21**

 Integrated into SmartWork homework assignments

Or ready for classroom presentation as a part of the PowerPoint resources on the Norton Media Library CD-ROM

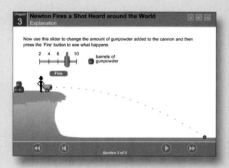

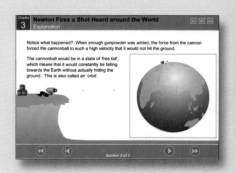

STARS AND GALAXIES

SECOND EDITION

21ST CENTURY
ASTRONOMY

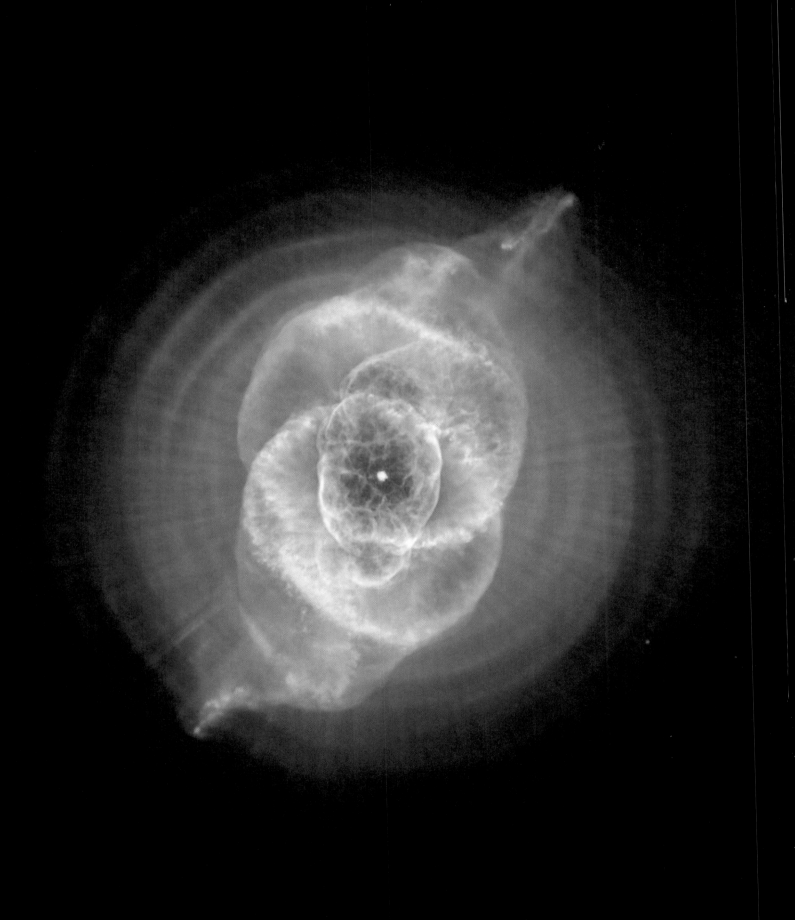

SECOND EDITION

21ST CENTURY

ASTRONOMY

Jeff Hester
ARIZONA STATE UNIVERSITY

David Burstein
ARIZONA STATE UNIVERSITY

George Blumenthal
UNIVERSITY OF CALIFORNIA—
SANTA CRUZ

Ronald Greeley
ARIZONA STATE UNIVERSITY

Bradford Smith
UNIVERSITY OF HAWAII—MANOA

Howard G. Voss
ARIZONA STATE UNIVERSITY

 W. W. NORTON & COMPANY · NEW YORK · LONDON

W. W. Norton & Company has been independent since its founding in 1923, when William Warder Norton and Mary D. Herter Norton first published lectures delivered at the People's Institute, the adult education division of New York City's Cooper Union. The Nortons soon expanded their program beyond the Institute, publishing books by celebrated academics from America and abroad. By mid-century, the two major pillars of Norton's publishing program—trade books and college texts—were firmly established. In the 1950s, the Norton family transferred control of the company to its employees, and today—with a staff of 400 and a comparable number of trade, college, and professional titles published each year—W. W. Norton & Company stands as the largest and oldest publishing house owned wholly by its employees.

Editor: Leo Wiegman
Associate Editor: Sarah England
Editorial Assistant: Lisa Rand
Managing Editor—College: Marian Johnson
Senior Project Editor: Kim Yi
Copy Editor: Meg McDonald
Developmental Editor: Brian Loehr
Science Media Editor: April Lange
Supplements Editor: Karen Misler
Director of Manufacturing—College: Roy Tedoff
Art Director: Rubina Yeh
Photo Researcher: Kelly Mitchell
Illustrations: J/B Woolsey Associates/Penumbra
Composition and layout: Brad Walrod/High Text Graphics, Inc.
Manufacturing: VonHoffmann Company

ISBN-13: 978-0-393-93010-8 (pbk.)
ISBN-10: 0-393-93010-6 (pbk.)

The Library of Congress has cataloged the one-volume edition as follows:
21st century astronomy / Jeff Hester . . . [et al.].—2nd ed.
    p. cm.
  Includes index.
  ISBN-13: 978-0-393-92443-5 (pbk.)
  ISBN-10: 0-393-92443-2 (pbk.)
  1. Astronomy.   I. Hester, John Jeffrey.   II. Title: Twenty-first century astronomy.
  QB45.2.A14 2006
  520—dc22   2006047240

W. W. Norton & Company, Inc., 500 Fifth Avenue, New York, N.Y. 10110
www.wwnorton.com
W. W. Norton & Company Ltd., Castle House, 75/76 Wells Street, London W1T 3QT

1 2 3 4 5 6 7 8 9 0

Jeff Hester dedicates this book to Vicki, to whom he is still married, even after "The Book"; to his children—may they learn about rushing in where the wise fear to tread; to his colleagues and graduate students, who too often were left to pick up the slack as this project became all consuming; and finally to the many students, past and future, who make it all worthwhile.

David Burstein wishes to thank Gail, Jon, and Liz for their love and help.

George Blumenthal gratefully thanks his wife, Kelly Weisberg, and his children, Aaron and Sarah Blumenthal, for their support during this project.

Ronald Greeley dedicates this book to his wife, Cindy.

Bradford Smith dedicates this Second Edition to his patient and understanding wife, Diane.

Howard G. Voss dedicates this book to his wife, Helen Ann, who cheerfully sacrificed much so that he could attend to his teaching and other professional work. Helen Ann passed away at the time of the completion of this edition. She will be greatly missed by Howard, their family, and her many friends. She lived a life of selfless service even despite being ill most of that life.

Brief Contents

CHAPTERS 6–12 ARE NOT INCLUDED IN THIS ALTERNATE EDITION

Contents

PART III Stars and Stellar Evolution

CHAPTER 13 Taking the Measure of Stars

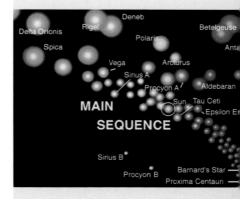

CHAPTER 14 A Run-of-the-Mill G Dwarf: Our Sun

CHAPTER 15   Star Formation and the Interstellar Medium

CHAPTER 16   Stars in the Slow Lane

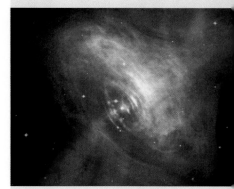

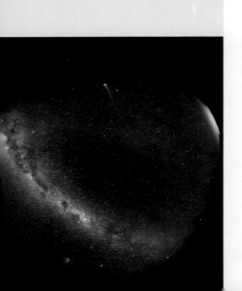

CHAPTER 20 Our Expanding Universe

Preface

What purpose does an introductory course in astronomy serve at today's college and university campuses? Some intrepid students would boldly answer this question, "To satisfy my lifelong curiosity about astronomy, physics, and planetary science!" Thank goodness such students exist. Their intent faces and interested questions get us through those occasional class periods when we think to ourselves, "Boy, am I glad I didn't have to *sit through* the lecture I just gave!"

We know most students take "Stars for Poets" to satisfy a science requirement. When faced with the choices at hand—biology (too squishy), physics (turn the catalog page quickly), chemistry (too smelly), geology (what's geology?)—many students opt for astronomy as the least of the available evils. Looking overhead at night, they see the stars and think, "Big Dipper, Little Dipper, North Star... sure, I can do stars!" It is of course ironic that astronomy turns out to be a mixture of physics, chemistry, and geology, with a bit of math and biology thrown in for good measure.

Students who study introductory astronomy typically come from diverse backgrounds; often their introductory astronomy course is the only formal exposure to science they will have in their lives! When viewed in this way, teaching introductory astronomy seems like a big responsibility, and so it is. As instructors, everything we do should turn on the answer to basic questions such as these: What should students carry away from this course that will still be with them 20 years from now? How should this course change the students who take it? And how can we, as scientists and educators, facilitate those changes within our students?

A thoughtful response to these questions does *not* lead to a course that is primarily about memorization of facts. At best such facts will enter students' short-term memory just long enough to be regurgitated on an exam. At worst students might come to think that science really is nothing more than memorizing information.

Facts are essential to reasoning and understanding in any course. Yet many students arrive at our doors with the idea that *learning* and *memorization* are synonymous. They confuse *knowing* with the ability to call an object by name or recite a few relevant pieces of information about it, as if stars once gathered into constellations and named were of no further interest. If an introductory science course does nothing else, it should encourage students to think of their brains as muscles that grow strong only with exercise and training. An introduction to science should be an introduction to what it means to think deeply about the universe, finding patterns and relationships within the world that go beyond the specifics of a particular object or setting, and applying those patterns and relationships broadly. An introductory science course should first and foremost be about understanding the world through the eyes of a scientist.

One of the authors of this text, Howard Voss, enjoys telling of a time when he was teaching an introductory physics course for nonscientists. One afternoon after a class discussing Newton's laws, a student came up to him in tears. Howard braced for the worst, but instead of a complaint about the difficulty of the course, he was met by a triumphant "I get it!" The student's tears were tears of joy. She had come to see the *sense* of Newton's laws. Not only did she know what they say, but she had come to understand something of what they mean and of how they can be used to make sense of the world.

In that moment of insight into Newton's laws, Howard's student caught a glimpse of what knowledge really is—of what it means to understand something rather than just know about it—and her discovery of that world of understanding was an emotional experience. This is the essence of an introductory science course.

An introductory astronomy course should also show students something of the epistemology of science: Knowledge is arrived at as we constantly challenge and test our ideas. The universe is not capricious; it behaves according to consistent, explicable rules. All scientific knowledge is provisional. How can it be that in science we never really prove things to be true, but instead only fail to show that they are false? Because that is how we give what is *really* true precedence over what we would *like* to be true. It is precisely because science acknowledges lack of certainty that the pinnacle of human knowledge is a well-tested, well-corroborated scientific theory. Here is the key to the difference between science and the myriad pseudoscientific pretenders that would borrow the credibility of science by adopting its trappings but forgo its intellectual rigor. We will have succeeded if a student comes to recognize the difference between science and pseudoscience, and if 20 years from now a former student still chuckles inwardly at the naiveté of the phrase "It's only a theory."

There is a strong practical side to what we teach about the process of science. Thinking about the world as a scientist means learning to sort the information from the opinion or happenstance. Physical scientists tend not to buy lottery tickets, call psychic help lines, or consume the many other forms of modern-day snake oil. Our students are served well if we pass on a healthy dose of our practical skepticism. A democratic society, which in the long run can fare no better than its citizens can think, is served also.

Yet the real reason it matters that students come to see the world through the eyes of a scientist is the sheer beauty of the view. The majesty of the universe in which we live is staggering. Simply staring at a deep image of distant galaxies or the glow of a newly formed star or a picture of Jupiter's swirling Great Red Spot can be a mind-boggling experience. Add a dose of quantum mechanics or the idea of the tortured fabric of spacetime around a black hole, and we are into territory that would awe Aristotle. Science—this marvelous expression of human curiosity and passion and reason—has shattered the mental cage in which we have lived since our ancestors first looked to the sky and wondered about what they saw there. It is extraordinary to face these vistas surrounded not by speculation but by the meticulously constructed edifice of science. Finally, to live during that moment in history when we first recognized our own origins in the Big Bang that occurred some 14 billion years ago—*that* is something to write home about!

Science is a vital, exciting, ongoing expression of our humanity. While an introductory astronomy book should teach students about what science is, showing them what science has revealed and giving them powerful tools for thinking about the world, it should also share with them something of the passion scientists feel for the endeavor to which we have devoted our lives.

The book that meets these goals would have to be very different from most textbooks. Rather than presenting a blocked-out set of facts and descriptions, that book would need to tell a story. On a small scale, it would need to tell the story of specific ideas. Understanding does not come from a recitation of facts. Understanding comes from thinking carefully and critically about things and how they work. Helping a student understand a concept as a scientist means guiding that student through the concept, making heavy use of examples and analogies, and tying the concept back to everyday phenomena and experiences to which the stu-

dent can relate. That is what the text we envisioned would have to do.

So rather than the expository style of traditional textbooks, this book tries for the explanatory style of a good teacher. Rather than stating that "the altitude of the north celestial pole above the horizon equals the observer's latitude" and leaving it at that, we talk about what the sky looks like over the course of the day as viewed from the North Pole and why. Then we invite the student to follow along as we head south, seeing how things change along the way.

On a larger scale, *21st Century Astronomy* tells a web of different stories. Some of these stories have to do with what science is and how it is done. Why did Newton choose the form that he did for his universal law of gravitation? What are the fundamental differences between Kepler's empirical "laws" and Newton's theoretical derivation of the same relationships? And if Einstein was "right," why wasn't Newton "wrong"? Other stories explore the human side of science, such as how science has led us to think beyond the box that evolution built for us. Representing science fairly and honestly even means telling those stories that might make some students uncomfortable, like the story of why our understanding of the evolving universe and our place in it is science, while "creation science" is not.

We tell the story of motion, the story of light, the story of matter and energy, the story of Earth and our planetary system, the story of the Sun and stars, the story of life, and the story of the universe as narratives, with one idea flowing into the next and each idea connected to the whole. That is the way humans learn best. Knowledge and understanding are interwoven, with each idea and insight given meaning by how it fits into the whole. None of the stories in *21st Century Astronomy* exists in isolation. The stories are threads in a grand intellectual tapestry, woven from the recurrent themes of the physics of matter, energy, radiation, and motion.

Our writing style for *21st Century Astronomy* fits this vision of science. The writing you will find here is far less formal than that found in most textbooks. We allow ourselves to be conversational, to pursue the occasional digression, to be irreverent or provocative or idiosyncratic when it suits our taste, and to allow the sense of passion and wonder that we feel for the subject to show through from time to time. Our hope is to catch the feel of a pleasant, stimulating conversation with a student during office hours.

This brings us to the first of many tensions that were with us throughout the writing and editing of *21st Century Astronomy*—the question of who is sitting across the desk, listening to what we have to say. There is a very real pull in the market in the direction of simplification. Books filled cover to cover with the latest images from NASA missions and simple, factual descriptions look pretty and are doubtless much easier to teach out of than more conceptually challenging treatments of the material. The problem is that such texts are so accessible that what they contain sometimes seems hardly worth accessing at all. For the most part, students pass through such courses untouched, the course forgotten the moment they walk out the door.

Plenty of books can be adapted to a course aimed at the lowest common denominator. That is not the slice of the market we chose to target. Instead we hope that *21st Century Astronomy* will appeal to instructors who, like us, have found that a serious nonscience student in today's colleges and universities is capable of thinking more deeply about more conceptually challenging material than is often assumed. Frankly, with the story that *21st Century Astronomy* has to tell—a story that deals head-on with some of the most fundamental, fascinating, and far-reaching questions humans have ever asked themselves—few students can resist being drawn in if approached in the right way.

Another significant tension in the book is between a traditional organization of topics that may be more comfortable for instructors and an organization that more accurately represents the way that scientists think and work today. True to our purpose, we chose an approach that reflects the state of our science. For example, we organized the Solar System section around a theme of comparative planetology rather than the more traditional "this is Tuesday, so it must be Mars" approach. We pulled the discussion of tidal interactions, orbital resonances, chaos, and similar phenomena into a separate chapter (Chapter 10) titled "Gravity Is More Than Kepler's Laws." We posited the existence of a rotating, collapsing interstellar cloud and a rotating protoplanetary disk and discussed the formation of the Solar System in Chapter 6 *before* discussing the planets themselves. This allowed us to look at the planets from the outset within the context of how they formed. Enough examples—you get the idea.

Even so, there is flexibility in the text. For example, we feel the flow of material through the early chapters discussing physical principles works well. We have tried to include enough discussion of the human and social aspects of science to break up the treatment of conceptually difficult material. On the other hand, some instructors may like to show their students a bit more astronomy and planetary science before hitting the physics. Such instructors might want to start with Chapter 6 on the formation of the Solar System and then pull in material about orbits and radiation as it is needed in what follows. Another approach is to start with stars in Chapter 13, pull in radiation and orbits as needed to understand the properties of stars, then insert the Solar System as an extended excursion into the topic of the formation of low-mass stars. While we wrote the book to tell a complete, well-integrated story, we tried to allow for different paths tailored to the needs and tastes of individual instructors.

Astronomy is one of the most rapidly advancing fields in modern science, and as such presents a continuing challenge for textbook authors to keep up with the most recent discoveries, interpretations, and definitions. For example,

in August 2006, shortly before this Second Edition was ready to go to press, the International Astronomical Union (IAU) announced new definitions for a planet. A "classical" planet is now defined as a celestial body that (a) is in orbit around the Sun, (b) has sufficient mass for its self-gravity to overcome rigid body forces so that it assumes a hydrostatic equilibrium (nearly round) shape, and (c) has cleared the neighborhood around its orbit. Definition (c) removes Pluto from the list of classical planets, a distinction it held since its discovery in 1930. Pluto is now a member of a new class of planets called "dwarf planets." A dwarf planet has the general characteristics of a classical planet, except that it has not cleared smaller bodies from the neighboring regions around its orbit. Ceres, formerly called an asteroid, and the recently discovered distant body Eris are also dwarf planets. The IAU defines all other objects orbiting the Sun collectively as "small solar system bodies." This includes most asteroids, comets and most objects orbiting beyond Neptune. These new criteria represent fundamental and historic changes in how we define the major bodies in our Solar System.

We should point out, however, that the decision by the IAU to demote Pluto created a firestorm of controversy. Many professional astronomers and planetary scientists expressed strong disagreement with the action taken by the IAU, and cries of disappointment were heard from the lay public, including disillusioned school children. Even so, it was our decision to accept the ruling by the IAU and the majority astronomers and thereby bring this Second Edition of *21st Century Astronomy* completely up to date. Through the heroic efforts of our editors and production team we have been able to do this.

In closing we should say a few words about who we are. Among the authors of this text you will find a group of accomplished and well-known scientists who have been at the forefront of many of the most exciting and significant events in astronomy and planetary science in the second half of the 20th century. The authors of this text include members of scientific teams that built three of the cameras that have flown on the Hubble Space Telescope, team members and leaders of many of the major planetary missions including the *Voyager* and *Galileo* missions to the giant planets and the *Mars Rover* mission, and a former president of the American Association of Physics Teachers. We have been involved in significant fundamental research on topics ranging from planetary geology, to the origin and evolution of the Solar System, to the formation and evolution of stars, to the structure and dynamics of the interstellar medium, to the nature of galaxies, and to the origin of structure in the universe. There are remarkably few fields discussed in this book to which one or more of the authors have not contributed in some important way over the years. For us, 21st century astronomy is not a textbook. Rather it is the life we live. We hope you find value in our attempt to share with you what *we* see when we look up at the sky at night.

Acknowledgments

Production of a book like this is a far larger enterprise than any of us imagined when we jumped on board. We would like to thank the teachers who reviewed portions of the First Edition manuscript along the way: Jeff Adams, William Andersen, Barbara Anthony-Twarog, Allen Armstrong, Keith Ashman, Edward Baron, David Baum, Dwight Beery, William Bittle, John Blake, Anita Corn, John Cummins, Kathy Eastwood, Terry Ellis, Randy Emmons, Thomas English, Eric Feigelson, Simonetta Frittelli, Martin Gaskell, Billy Graves, Kim Griest, Erick Guerra, Javier Hasbrun, Paul Heckert, Roger Hewins, Paul Hinds, Eric Hintz, David Hufnagel, Andrew Ingersoll, Adam Johnston, Khondkar Karim, Frank Kowalski, Claud Lacy, John Laird, Kenneth LaSota, Irene Little-Marenin, Bruce Margon, Bradley Matson, George McGill, Kenneth McLaughlin, Karen Meech, Zdzislaw Musielak, Robert O'Connell, Aileen O'Donoghue, Charles Peterson, Cynthia Peterson, Randy Phelps, Carlton Pryor, George Rosensteel, Anthony Russo, Carl Rutledge, Stephen Schneider, William Schoenfeld, Michael Sitko, Tim Slater, Larry Smith, Larry Sromovsky, Thomas Statler, Christine Staver, Curtis Struck, Donald Terndrup, Suzanne Willis, Louis Winkler, and George Wolf.

We also thank the following colleagues who served as Second Edition reviewers: Scott Atkins, University of South Dakota; Peter A. Becker, George Mason University; Timothy C. Beers, Michigan State University; Juan Cabanela, Minnesota State University–Moorhead; Judith Cohen, California Institute of Technology; Paul Hintzen, California State University–Long Beach; Paul Hodge, University of Washington; William A. Hollerman, University of Louisiana at Lafayette; Adam Johnston, Weber State University; Laura Kay, Barnard College; Bill Keel, University of Alabama; Leslie Looney, University of Illinois at Urbana–Champaign; Norm Markworth, Stephen F. Austin State University; Scott Miller, Pennsylvania State University; Ata Sarajedini, University of Florida; Paul P. Sipiera, William Rainey Harper College; Donald Terndrup, Ohio State University; Richard Williamon, Emory University.

There are many at W. W. Norton & Company without whom this project would not have come together. The authors would like to thank our editors, Stephen Mosberg, who first looked across the table at us and said, "We will find a way"; John Byram, who stepped into the First Edition midstream and held it together through difficult times; and Leo Wiegman, who spearheaded the Second Edition. Roby Harrington gave us enough rope to hang ourselves, then helped us out of trouble when we tried to do just that. Among those in the trenches were Brian Loehr, Developmental Editor; Lisa Rand, Editorial Assistant; Sarah England, Associate Editor; Kim Yi, Project Editor; Meg McDonald, Copy Editor; April Lange, Science Media Editor; Karen Misler, Supplements Editor; Sally Whiting, Emedia Assistant; Neil Hoos, Manager of Photo Permissions; Kelly Mitchell, Photo Researcher; Marian Johnson, Managing Editor—College Books; Roy Tedoff, Director of Manufacturing—College; Rubina Yeh, Art Director and designer of this book; Brad Walrod, Layout Artist; Science Technologies for the Animations; Sapling Systems for SmartWork; and our ancillary authors, Scott Miller, Ann Schmiedekamp, Don Terndrup, David Wood. Finally, we would like to thank John Woolsey and Kelly Paralis of J/B Woolsey Associates and Penumbra, who became true collaborators on the project rather than simply artists drawing to spec.

Jeff Hester
David Burstein
George Blumenthal
Ronald Greeley
Bradford Smith
Howard G. Voss

The Features of *21st Century Astronomy*

The overall themes and approach of the First Edition remain, emphasizing foundational physical science concepts and stressing scientific literacy for nonmajors by showing the process of scientific discovery. The following features support students as they start chapters, while they read, and as they finish and prepare for the next topic.

> **Key Concepts** boxes at the beginning of each chapter call attention to the main issues and topics that will be discussed in the chapter.

> **Key Idea** statements—one-sentence summaries, appear throughout each chapter to draw students' attention to fundamental concepts as they read. They are also a useful scanning tool for review.

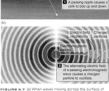

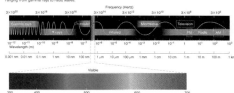

Annotated Figures throughout the book present clear, accurate science in visually compelling ways.

We believe the science must be portrayed accurately, even when it is a little complicated. Hence we kept our use of mathematical expressions within the chapter narrative, where it was in the First Edition. We rarely use more than basic algebra. And we make every effort to apply the equations with real world values, to make the math and the underlying concepts easier to understand.

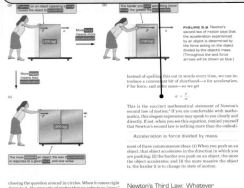

Summaries provide an outline of the key concepts as they were applied in the chapter.

Key Terms lists are ideal for quick review and reference.

Applying the Concepts questions involve problem solving and critical thinking.

Seeing the Forest through the Trees sections conclude each chapter with a thematic synthesis and a glimpse ahead to future chapters.

Thinking about the Concepts questions provide an opportunity for review and emphasize factual recall.

Each chapter ends with a reminder of additional resources available on the StudySpace website for further chapter review. See right for more details.

We retain four types of boxes sprinkled throughout the chapters. These boxes touch on material either because it is somewhat out of the mainstream of our journey or because it deserves to be highlighted.

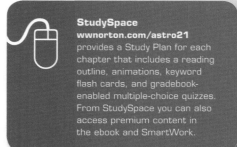

StudySpace
wwnorton.com/astro21
provides a Study Plan for each chapter that includes a reading outline, animations, keyword flash cards, and gradebook-enabled multiple-choice quizzes. From StudySpace you can also access premium content in the ebook and SmartWork.

Foundations boxes discuss the basic science that is central to our physical understanding of the universe.

Connections boxes draw attention to recurring themes—bridges between different parts of our journey. Connections boxes have an "origins" theme in the Second Edition.

Excursions boxes are short but interesting field trips that highlight how scientists apply what they learn to solve problems.

Tools address the technology and techniques that astronomers and planetary scientists use.

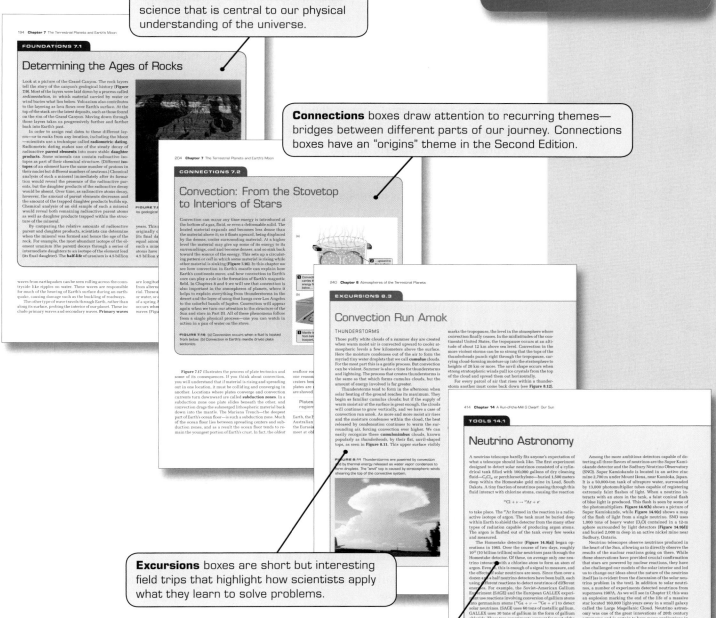

So What's New in the Second Edition?

We were gratified so many instructors found the First Edition of *21st Century Astronomy* a useful teaching tool. This Second Edition is updated, revised, and expanded, but we have kept the "big story" approach to astronomy that was so well-received the first time around.

New Chapter 5 on Telescopes and Instruments The discussion of instrumentation was located principally in the First Edition's Chapter 4. We have consolidated and expanded this material to create a new self-standing Chapter 5, "The Tools of the Astronomer," to close Part I:

- Section 5.1 has expanded coverage of refractors and reflectors, resolution and diffraction, and the limitations imposed by Earth's atmosphere.

- Section 5.2 features new content about optical detectors and instruments, including expanded coverage of charge-coupled devices (CCD).

- Section 5.3 has expanded coverage of radio telescopes and new coverage of radio and optical interferometric arrays.

- Section 5.5 features new coverage of space observatories.

- Section 5.6 has new coverage of flybys, orbiters, rovers, and atmospheric probes.

- Section 5.7 discusses how high-energy colliders help astronomers understand the earliest moments of the universe.

- Section 5.8 has new coverage of high-speed computer use in astronomy.

Updated Science throughout the Second Edition The past few years have been exciting ones in astronomy, with new discoveries revealing themselves all the time. And as mentioned earlier, the IAU announced a new definition for planets in August 2006. We have infused new or updated research findings throughout the book, for example:

- Chapter 6: A new section on and expanded coverage of extrasolar planets.

- Chapter 7: New results from the *Mars Express* and *Mars Odyssey* orbiters and ground discoveries by the *Opportunity* and *Spirit* rovers.

- Chapter 8: An expanded discussion of convection and weather in a new Excursions box about thunderstorms, lightning, and tornadoes.

- Chapter 11: Exciting new discoveries about Saturn's large moon Titan and the geologically active Enceladus from the Cassini/Huygens mission, and a new section on Dwarf Planets, which includes Ceres, Pluto, and Eris.

- Chapter 12: New results from spacecraft visits to Comets Borrelly and Wild 2 and the impact on Comet Tempel 1, and new discoveries of Kuiper Belt objects.

- Chapter 17: New material on gravity waves.

- Chapter 20: New results from the Wilkinson Microwave Anisotropy Probe on cosmic background radiation.

***Origins of Life* Theme** Given the student interest in the origin of life and the growing importance of cosmology, we have added new Connections essay boxes throughout the Second Edition:

- Connections 1.1: Origins—An Introduction

- Connections 6.1: Origins—Choosing the Right Kind of Planet

- Connections 7.1: Origins—Where Have All the Dinosaurs Gone?

- Connections 8.1: Origins—On Atmospheres and Life

- Connections 10.1: Origins—Lunar Tides and Life

- Connections 11.1: Origins—Extreme Environments and an Organic Deep Freeze

- Connections 12.1: Origins—Comets, Asteroids, and Life

- Connections 16.1: Origins—Choosing the Right Kind of Star

- Connections 17.1: Origins—The Chemistry of Life

- Connections 21.2: Origins—Life, the Universe, and Everything

New Pedagogy New Summaries, Key Terms lists, and additional review questions for each chapter round out a robust program in support of the reading.

Many New Photographs and Line Art Figures To keep up with the terrific and pedagogically useful images coming out of the astronomical research community every day, we have added over 100 new photographs In addition, to help illustrate the newly added concepts we have revised almost 100 line art figures and added 30 new ones.

New Design One consequence of adding so much new art to the Second Edition is a new layout of the chapters, so readers can more readily follow the major story within chapters and continue to see the forest for the trees.

New Electronic Resources The StudySpace website is the free and open portal through which students access the new resources that accompany this text.

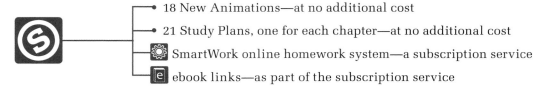

18 New Animations—at no additional cost

21 Study Plans, one for each chapter—at no additional cost

SmartWork online homework system—a subscription service

ebook links—as part of the subscription service

See below for more details.

Student Resources

StudySpace at wwnorton.com/astro21

Ann Schmiedekamp, *Pennsylvania State University–Abbington*

This free and open student Web site offers a sensible study plan that gives students practical assignments to integrate review and assessment resources for each chapter. It includes reading outlines, Keyword flash cards, and Gradebook-enabled concept tests and multiple-choice quizzes. It also features 18 brand new Animations by Science Technologies, which use interactivity to enhance students' understanding of core concepts. From StudySpace students can also access premium content in the ebook and SmartWork (see below for details).

Animations

Brand new for the Second Edition and developed specifically for use with *21st Century Astronomy,* these brief lessons use animation and interactivity to enhance students' understanding of core concepts:

A. The Celestial Sphere and the Ecliptic
B. The Earth Spins and Revolves
C. The Moon's Orbit: Eclipses and Phases
D. Kepler's Laws
E. Newton's Laws and Universal Gravitation
F. Light as a Wave, Light as a Photon
G. Doppler Effect
H. Atomic Energy Levels and the Bohr Model
I. Atomic Energy Levels and Light Emission and Absorption
J. Geometric Optics and Lenses
K. Solar System Formation
L. Processes That Shape the Planets
M. Tides and the Moon
N. Solar Spectrum
O. The H-R Diagram
P. The Solar Core
Q. Star Formation
R. Hubble's Law

Animations are available from the free StudySpace student web site, and are also integrated into assignable SmartWork exercises. Offline versions of the animations for classroom presentation are available from the Norton Media Library instructor's CD-ROM.

SmartWork Online Homework System

SmartWork—Norton's online homework management system—provides ready-made self-grading assignments, including guided problems, simple feedback questions, and animated tutorials—all specifically designed to extend the text's emphasis on critical thinking.

Developed in collaboration with Sapling Systems, SmartWork features an intuitive and easy-to-use interface that offers instructors flexible tools to manage assignments. Helpful and immediate feedback makes it easy for students to assess their understanding of basic concepts. Two types of questions expand on the exposition of concepts in the text:

- **Simple Feedback Problems** anticipate common misconceptions and offer prompts to help them discover the correct answer.

- **Guided Tutorial Problems** address more challenging topics. If a student answers a problem incorrectly, SmartWork guides the student through a series of discrete tutorial steps. Each step is a simple feedback question that the student answers—with hints, if necessary. After completing all of the tutorial steps, the student returns to the original question ready to apply this newly obtained knowledge.

Instructors can easily use these ready-made questions and assignments, customize them to address specific course objectives, or use SmartWork to create their own.

SmartWork and ebook integration SmartWork is available as a standalone purchase or with an integrated ebook version of *21st Century Astronomy*. Links to the ebook make it easy for students to consult the text while completing their homework assignments.

Starry Night Pro 5.0 CD-ROM and Workbook

Workbook by Donald Terndrup, *The Ohio State University*

The remarkably realistic and user-friendly Starry Night Pro 5.0 CD-ROM allows students to explore stars and objects in our cosmic neighborhood and beyond. The accompanying workbook includes 12 observation exercises that guide students' virtual explorations of the night sky and help them apply what they've learned from the text.

Ebook

21st Century Astronomy is also available in a Norton ebook format, a convenient alternative that retains all of the print book's content. The ebook offers a variety of tools for study and review, including sticky notes, highlighters, zoomable images, links to Animations, and a search function.

The ebook is available as a stand-alone item or packaged with SmartWork. The Smart-Work/ebook package makes it easy for students to consult the text when completing their homework assignments. A downloadable PDF version of the ebook is also available from Powells.com. Go to norton**ebooks**.com for more information.

Instructor Resources

Instructor's Manual and Test Bank

Scott Miller, *Pennsylvania State University, University Park*
David Wood, *San Antonio College*

Thoroughly revised and expanded for the Second Edition, this resource includes brief chapter overviews, worked solutions to the end-of-chapter problems, instructor's notes for the Starry Night Workbook activities and approximately 1,000 true-false, multiple-choice, and short answer test questions. The Test Bank is also available in rich-text, ExamView® Assessment Suite, BlackBoard, and WebCT formats.

PowerPoint Lecture Outlines

Donald Terndrup, *The Ohio State University*

These ready-made lecture outlines include selected art from the text, "clicker" questions, and offline and lecture-ready versions of the Animations.

Norton Media Library CD-ROM

This helpful resource includes the PowerPoint lecture outlines with "clicker" questions, offline versions of the Animations, plus selected photographs and all drawn art from the text.

BlackBoard and WebCT Course Cartridges

Course cartridges for BlackBoard and WebCT include access to the Animations, a Study Plan for each chapter, multiple-choice tests, and links to ebook and SmartWork premium content.

Transparencies

Acetates for approximately 200 figures from the text are available to qualified instructors. Contact your local representative for details.

Video Library

A video library will be available to qualified instructors. Contact your Norton representative for details.

About the Authors

Jeff Hester received his Ph. D. from Rice University and is currently a professor of physics and astronomy at Arizona State University. His research interests are the interstellar medium in the Milky Way and external galaxies; structure of the diffuse ISM; interstellar shock waves; supernova remnants; pulsar wind interactions; Herbig-Haro objects; and H II region structure.

David Burstein was born in Englewood, New Jersey on May 19, 1947. After graduating as valedictorian from his high school, he went to Wesleyan University in Middletown, Connecticut for his undergraduate career and University of California–Santa Cruz for his graduate career. After getting his Ph.D. in astronomy and astrophysics, he went to the Department of Terrestrial Magnetism, Carnegie Institution of Washington for 2 years, then on to the National Radio Astronomy Observatory for 3 years. He is now at Arizona State University. Gail and Dave have been married for 35 years as of June 2006, and have two children—Jon who is a reporter for the *South Florida Sun-Sentinel,* and Elizabeth, who is a first grade teacher in the same class in which she was a student.

George Blumenthal is Acting Chancellor at the University of California–Santa Cruz, where he has been a Professor of Astronomy and Astrophysics since 1972. Chancellor Blumenthal received his B.S. degree from the University of Wisconsin–Milwaukee and his Ph.D. in physics from the University of California–San Diego. As a theoretical astrophysicist, Chancellor Blumenthal's research encompasses several broad areas, including the nature of the dark matter that constitutes most of the mass in the universe; the origin of galaxies and other large structures in the universe; the earliest moments in the universe; astrophysical radiation processes; and the structure of active galactic nuclei such as quasars. Besides his teaching and research, Chancellor Blumenthal has served as the chair of the UC Santa Cruz Astronomy and Astrophysics Department, has chaired the Academic Senate for both the UC Santa Cruz campus and the entire University of California system, and has served as the Faculty Representative to the UC Board of Regents.

Ronald Greeley is Regents' Professor in the School of Earth and Space Exploration at Arizona State University. After completing his Ph.D. in geology in 1966, he worked for Standard Oil for a year in exploration and then joined NASA-Ames Research Center, where he remained until 1978 when he went to ASU. While at Ames, he led a consortium of planetary scientists and engineers to design and implement the Mars Surface Wind Tunnel, which he continues to manage as a NASA faculty. Ron was a science team member on *Viking, Galileo-Jupiter, Magellan-Venus, Mars Pathfinder,* and the ill-fated *Soviet Mars 96* orbiter and is currently on the *Mars Exploration Rover* and *Mars Express* orbiter science teams. He has chaired numerous NASA and National Research Council committees charged with formulating plans for solar system exploration.

Bradford Smith has served as an Associate Professor of Astronomy at New Mexico State University, a Professor of Planetary Sciences and Astronomy at the University of Arizona, and a Research Astronomer at the University of Hawaii. Through his interest in Solar System astronomy, he has participated as a team member or imaging team leader on several US and international space missions, including *Mars Mariners* 6, *7* and *9, Viking, Voyager,* and the Soviet *Vega* and *Phobos* missions. More recently, Smith's interests have turned to other planetary systems, working as a team member of the HST NICMOS experiment. He has four times been awarded the NASA Medal for Exceptional Scientific Achievement. Smith is a member of the IAU Working Group for Planetary System Nomenclature and is Chair of the Task Group for Mars Nomenclature. He is now semi-retired but remains affiliated with the Institute for Astronomy at the University of Hawaii, Manoa.

Howard G. Voss is Professor Emeritus of Physics at Arizona State University, where he taught for over four decades, served as chair of the Department of Physics, and received the Distinguished Faculty Award and the Dean's Teaching Award. He was awarded the Melba Phillips Medal by the American Association of Physics Teachers, which he served as president, secretary, a member of the Executive Board, and in other offices. He also served the American Institute of Physics in several positions including as chair of the Publishing Policy Committee and as a member of the Governing Board.

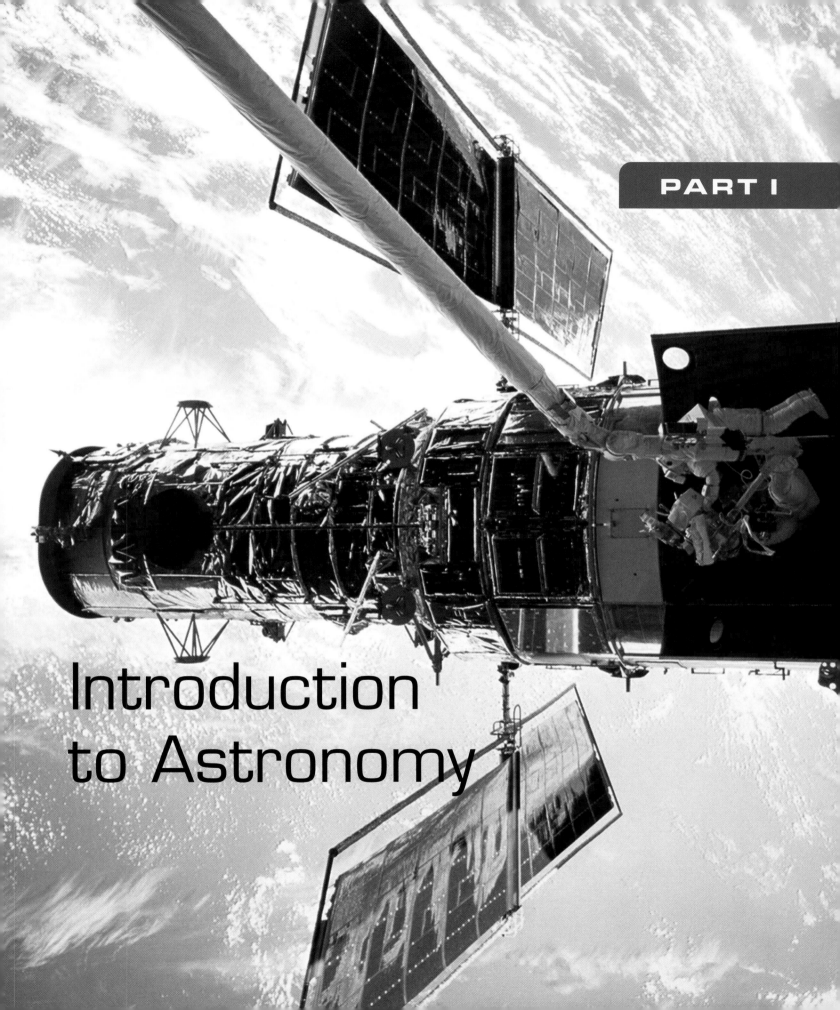

PART I

Introduction to Astronomy

The most beautiful thing we can experience is the mysterious.
It is the source of all true art and all science.
He to whom this emotion is a stranger,
who can no longer pause to wonder and stand rapt in awe,
is as good as dead: his eyes are closed.

ALBERT EINSTEIN (1879–1955)

The moon seen rising over the 2400 year old Greek Temple of Poseidon.

CHAPTER 1

Why Learn Astronomy?

1.1 Starting with a Spark of Interest

Not everyone is fascinated by science, but almost everyone harbors a spark of interest in astronomy. Because you are reading this book, you probably share this spark as well. The spark may have been struck when you were a child looking at the sky and found yourself wondering about what you saw there. What are the Sun and Moon made of? How far away are they? What are the stars? How do they work? Do they have anything to do with me? The prominence of the Sun, Moon, and stars in cave paintings and rock drawings (such as those in **Figure 1.1**) dating back thousands of years tells us that these questions have long occupied the human imagination. Your initial spark of interest in astronomy may have grown over the years as you saw or read news reports about spectacular discoveries made in your lifetime. Some of these discoveries may have sounded so amazing that it was difficult to draw the line between science fact and science fiction.

If you nurture your spark of interest in astronomy as you continue through this book, you may be surprised to find that spark growing into a flame. The title of this book—*21st Century Astronomy*—was chosen to emphasize that this is the most fascinating time in history to be studying this most ancient of sciences. This book will take you to places you never imagined going and will lead you to insights and understandings you never imagined having. To those of you who are reading this book for a course in astronomy at your college or university, we have a special note. The authors of this text have taught introductory astronomy many times over the years. We recognize that you may be

KEY CONCEPTS

Before traveling through unfamiliar terrain, it helps to have some idea of where you are going, what you might see along the way, and what you should pack for the journey. In *21st Century Astronomy*, we will learn not only about the wonders of the universe but also about what it means to look at the world through the eyes of science. In this chapter's overview of what is to come, we will find that

* The universe is vast beyond all human experience, yet it is governed by the same physical laws that shape our daily lives.

* We are a product of that universe; the very atoms of which we are made were formed in stars that died long before the Sun and Earth were formed.

* Science is a creative human activity like art, literature, and music, and it is also a remarkably powerful, successful, and aesthetically beautiful way of viewing the world.

* Understanding comes from thinking carefully and deeply about patterns in the world, not simply from memorization of facts.

* Like climbing a mountain, the journey we are about to make requires effort, but the view from the top is amazing to behold.

FIGURE 1.1 Ancient petroglyphs often include depictions of the Sun, Moon, and stars.

in this course primarily because you need a science credit to graduate. As you flipped through your course catalog, perhaps you were reminded of your interest in astronomy, and that led you to choose astronomy over your other options. (Or perhaps you simply considered astronomy to be the least of the available evils!) Whatever your expectations, the story in *21st Century Astronomy* can fascinate you if you open your mind to it.

The journey of discovery on which we are embarking is not always easy, but few worthwhile journeys are. A hike in the mountains can at times be an easy stroll and at other times a more strenuous climb; but when you arrive, the view from the top is hard to beat. In much the same way, this book will ask you to exercise your mental muscles in different, possibly unaccustomed ways. But as with the hike in the mountains, we feel certain that you will find the rewards worth the investment.

Getting a Feel for the Neighborhood

If you are like many people, your conception of astronomy may not go much beyond learning about the constellations and the names of the stars in them. Loosely translated, the word **astronomy** means "patterns among the stars." But modern astronomy—the astronomy we will talk about in this book—has become far more than looking at the sky and cataloging what is visible there. It may seem something of a contradiction, but a great deal of frontline astronomy is now carried out in physics laboratories like the one shown in **Figure 1.2**. Today astronomers work along with their colleagues in related fields such as **physics**, **chemistry**, **geology**, and planetary science to sharpen our understanding of the physical laws that govern the behavior of **matter** and

energy and to use this understanding to make sense of our observations of the cosmos.

We are confident in our answers to many of the questions that you may have asked yourself as a child when you looked at the sky. We all live on a planet called Earth, which is orbiting under the influence of gravity about a star called the Sun. The Sun is an ordinary, middle-aged star, more massive and luminous than some stars but less massive and

---

Earth exists in the context of the Universe.

---

luminous than others. The Sun is extraordinary only because of its importance within our own Solar System. The Sun is located about halfway out from the center in a flattened collection of approximately 100 billion stars referred to as the **Milky Way Galaxy**. The Milky Way in turn is a member of a small collection of a few dozen galaxies called the **Local Group**, which is part of a vastly larger collection of thousands of galaxies called a **supercluster**. But even this vast structure is part of the *local* universe. The part of the universe that we can see extends outward for the distance that **light** travels in 13.7 billion years, and in this volume we estimate that there are about 100 billion galaxies— roughly as many galaxies as there are stars in the Milky Way!

One of the first conceptual hurdles that we face as we begin to think about the universe is its sheer size. If a hill is big, then a mountain is really big. If a mountain is really big, then Earth is enormous. But where do we go from there? We quickly run out of superlatives as the scale of what we are talking about comes to dwarf our human experience. One technique that can help us develop a sense for the size

FIGURE 1.2 This laboratory, where physicists are studying the properties of atoms, might seem an unlikely place to be doing astronomy. But laboratory astrophysics, studying astronomically important physical processes in a laboratory, has become an important part of astronomy.

of things in the universe is to use a little sleight of hand and move from discussing distance to talking instead about time. If you are driving down the highway at 60 miles per hour, a mile is how far you go in a minute. Sixty miles is how far you go in an hour. Six hundred miles is how far you go in 10 hours. So to get a feeling for the difference in size between 600 miles and 1 mile, you can think about the difference between 10 hours and a single minute.

We can play this same game in astronomy, but the **speed** of a car on the highway is far too low to be useful. Instead we will use the greatest speed in the universe—the speed of light. Light travels at a speed of 300,000 kilometers per second. At that speed, light can circle Earth (a distance of 40,000 kilometers) in just under $\frac{1}{7}$ of a second—about the time it takes you to snap your fingers. Fix that comparison in your mind. The size of Earth is like—*snap!*—a snap of your fingers. Follow along in **Figure 1.3** as we move outward into the universe. We next encounter the Moon, 384,000 kilometers away, or a bit over $1\frac{1}{4}$ seconds when moving at the speed of light. So if the size of Earth is a snap of your fingers, the distance to the

Light travel time helps in understanding size.

Moon is about the time that it takes to turn a page in this book. Continuing on, we find that at this speed the Sun is $8\frac{1}{3}$ minutes away, or the length of a hurried lunch at the student union. Crossing from one side of the orbit of Eris, the outermost planetary body in our Solar System, to the other takes about 19 hours. Think about that for a minute. Let it sink in. Comparing the size of Eris's orbit to the circumference of Earth is like comparing the time of a long plane ride to Australia to a single snap of your fingers.

Yet in crossing Eris's orbit we have only just begun our journey. Many steps remain. It takes us a bit over four years to cover the distance from Earth to the nearest star (other than the Sun), or as much time as you spent in high school. At this point even our analogy using light travel time can no longer bring astronomical distance to a human scale. Light takes about 100,000 years to travel across our galaxy—about the time that modern humans (*Homo sapiens*) have walked the surface of Earth. To reach the nearest large galaxies beyond our own takes several million years, or the time since our australopithecine ancestors appeared on the scene. To reach the limits of the currently observable universe takes light 13.7 billion years—the age of the universe, or about three times the age of Earth.

Look at that comparison again. The size of Earth is to the vast expanse of the universe as a single snap of your fingers is to three times the amount of time that has passed since the Sun and Earth were formed! Here is something to ponder the next time you look up at a star-filled summer sky.

As you read this text, you will occasionally see material set aside in boxes. A topic has been boxed either because

it is somewhat out of the mainstream of our journey or because we wish to highlight it. In particular,

- **Foundations** boxes address material that is central to our physical understanding of the universe.
- **Connections** boxes draw attention to recurring themes —bridges between different parts of our journey.
- **Excursions** boxes are short but interesting side trips.
- **Tools** boxes discuss the technology and techniques that astronomers and planetary scientists use.

Glimpsing Our Place in the Universe

While seeking knowledge about the universe and how it works, modern astronomy and physics have repeatedly come face-to-face with a number of age-old questions long thought to be solely within the domain of philosophers. Issues as seemingly metaphysical as the origin and fate of the universe and the nature of space and time have become the subjects of rigorous scientific investigation. The answers we are finding to these questions are often far more wondrous than our predecessors could have dreamed. They are changing not only our view of the cosmos, but our view of ourselves as well.

Figure 1.4 envisions a traveler raising the veil of the heavens to see what lies there. Throughout most of history, philosophers looked at the universe and saw it as remote and different from Earth—disconnected from our terrestrial existence. When modern astronomers look at the universe, they see instead a network of ongoing processes that we are a part of. Astronomy begins by looking out at the universe, but increasingly that outward gaze turns introspective as we come to appreciate that our very existence is a consequence of those same processes.

The study of the chemical evolution of the universe is such a case. As a result of both observation and theoretical work, we now understand that when the universe was young (13.7 billion years ago), the only chemical elements found in abundance were hydrogen and helium, plus tiny

We are stardust.

amounts of lithium, beryllium, and boron. Yet we are not made exclusively of these lightest elements. Our bodies are built of carbon, nitrogen, oxygen, sodium, phosphorus, and a host of other chemical elements. We live on a planet with a core consisting mostly of iron and nickel, surrounded by a mantle made up of rocks containing large amounts of silicon and other elements. If these more massive elements were not present in the early universe, where did they come from?

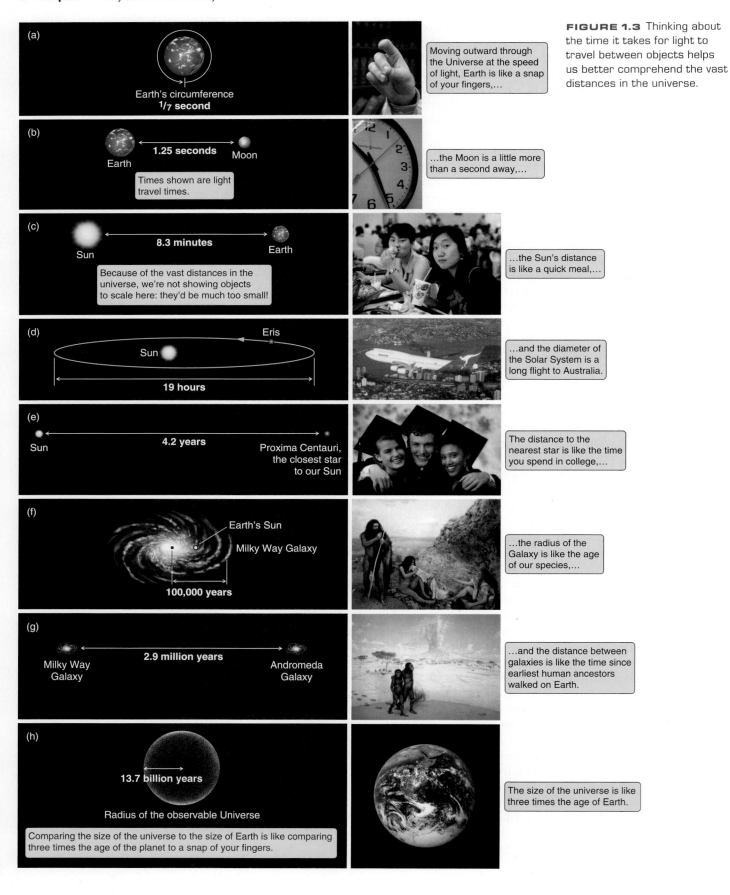

FIGURE 1.3 Thinking about the time it takes for light to travel between objects helps us better comprehend the vast distances in the universe.

FIGURE 1.4 Throughout most of history humans conceived of the rest of the universe as a place apart from us. Here a traveler raises the curtain of the firmament to get a glimpse of what lies beyond.

To answer this question we have but to look at the lights in the night sky. The energy to power stars comes from nuclear fusion reactions that occur deep within their interiors. Fusion reactions in stars take less massive **atoms** like hydrogen and combine them, forming more massive atoms, accomplishing the alchemist's dream of transforming one element into another. When a star exhausts its nuclear fuel and nears the end of its life, it often loses much of its mass—including some of the new atoms formed in its interior—blasting it back into interstellar space. We will talk later about the life and death of stars. For now it is enough to note that our Sun and Solar System formed from a cloud of interstellar gas and dust that had been "polluted" by the chemical effluent from earlier generations of stars. This chemical legacy supplies the building blocks for the interesting chemical processes that go on around us—chemical processes such as life. **Figure 1.5** symbolizes this intimate relationship between the world around us and our heritage in the stars. Look around you. The atoms that make up everything you see were formed in the hearts of stars. Poets sometimes say that "we are stardust," but this is not poetry. It is literal truth.

As humans, we have long speculated about our beginnings. Who or what is responsible for our existence? How were Heaven and Earth created? In the modern world, primitive creation myths have largely given way to scientific theory.[1] And so the topic of *origins* is one that you will en-

---

[1] A few cultures still retain their creation myths. And even within "enlightened" society, certain factions have ignored science in favor of nonscientific "creationism" and "intelligent design."

counter frequently in our journey. We have developed this as a recurring theme that we call (what else?) "Origins." It begins with **Connections 1.1**.

We Live in an Age of Exploration and Discovery

Another reason this is a fascinating time to be learning about astronomy is that we live in an age of exploration. The 1957 launch of *Sputnik,* the first human-made satellite, occurred just one year before the birth of the youngest of the authors of this book. Five decades later, as we begin the 21st century, we have seen humans walk on the Moon (**Figure 1.6**) and have sent unmanned probes to visit all of the classical planets. (We will occasionally refer to eight "classical" planets—Mercury, Venus, Earth, Mars, Jupiter, Saturn, Uranus, and Neptune—to distinguish them from the new "dwarf" class of planets adopted by the International Astronomical Union

---

Space exploration has expanded our view of the universe.

(IAU).) Spacecraft have flown by asteroids, comets, and even the Sun. Our inventions have landed on Mars, Venus, and Titan (Saturn's largest moon) and have plunged into the atmosphere of Jupiter. Most of what we know of the Solar System has been learned over these past five decades as a result of this burst of exploration.

Satellite observatories in orbit around Earth have also given us many new perspectives on the universe. The same atmosphere that shields us from harmful solar radiation also

FIGURE 1.5 You and everything around you, including beautiful waterfalls, are composed of atoms that were forged in the interior of stars that lived and died before the Sun and Earth were formed. The left panel shows a cloud of chemically enriched material that has been ejected from the star Eta Carinae.

CONNECTIONS 1.1

Origins—An Introduction

How and when did the universe begin? What combination of events—some probable, others much less likely—have led to our existence as sentient beings living on a small rocky planet orbiting a typical middle-aged star? Was this a unique happening, or are there others like us scattered throughout the galaxy?

Throughout our journey we will encounter a recurring theme that addresses questions such as these. We call it **origins**. To get a good feeling for this subject we must cast a wide net because origins involves much more than how we humans came to be. We'll look into the genesis of terrestrial life, but we will also examine the possibilities of life elsewhere in our Solar System and beyond, a subject called **astrobiology.** For life to exist around other stars in the galaxy, there must be life-sustaining planets to support it. So our origins theme will include the discovery of extrasolar planets and how they compare with the planets of our own Solar System.

For example, we have learned in this chapter that the early universe was made up almost entirely of only two elements, hydrogen and helium, with just a touch of lithium, beryllium, and boron thrown in. Further along we'll discover why hydrogen and helium were the big winners in a minutes-old universe that had barely cooled down from the Big Bang, and how they were later transformed into more massive elements such as carbon, nitrogen, oxygen, sulfur, and phosphorus, the very atoms that make up the **molecules** of life. We will find that some elements were created in a relatively benign environment in the cores of stars like our Sun, whereas others had their origin in the violence of unimaginable cosmic explosions. Putting it all together, we'll see that we ourselves, the terrestrial life around us, our Earth, the other planets in our Solar System, and those beyond have a common origin: All are made of recycled stardust.

---

blinds us to much of what is going on around us. Space astronomy continues to show us vistas hidden from the gaze of groundbased **telescopes** by the protective but obscuring blanket of our atmosphere. Satellites capable of detecting radiation—ranging from gamma rays and X-rays, to ultraviolet radiation, to infrared radiation to microwaves—have brought surprising discovery after surprising discovery. Each has forever altered our perception of the universe, further expanding the domain of the human mind. At the same time, the closing years of the 20th century also witnessed a renewed vigor in astronomical observations from the surface of Earth. The view of the sky seen by radio telescopes shown in **Figure 1.7** illustrates the new perspectives that have been opened by our growing technological prowess.

Astronomy has also benefited enormously from the computer revolution. The 21st-century astronomer spends far more time peering at a computer screen than peering through the eyepiece of a telescope. Computers are used to do everything from collecting and analyzing data from telescopes, to calculating physical models of the conditions that exist in the hearts of stars, to preparing and disseminating the results of our work.

We truly live in a golden age of exploration and discovery. When we look back at the Renaissance (15th and 16th centuries), we recall few of the concerns that dominated the day-to-day existence of those alive at that time. Instead we

remember the Renaissance as a time of great art, literature, and music. We remember it especially as a time when the spirit of inquiry was reawakened and when much of what we think of as science was born. What might a historian 500 years in the future consider to be of lasting significance about *our* time? It is doubtful that our hypothetical historian of the future will care much about who won the Super Bowl, or which performer was at the top of the pop charts, or which brand of toothpaste tasted best. The media spectacles that often dominate public consciousness will merit little more than an occasional footnote.[2]

Much of what seems so significant to us today will be dust in the wind 500 years from now. However, we can be certain that our future historian *will* remember ours as the time when humankind first stepped beyond the world of our birth and began to reach out with our minds and our science to touch the fabric of the universe itself. It is probably safe to say that few things will have a more lasting impact on our culture than this revolution in our understanding of the universe and our place in it. No history book will ever again be complete without the headline in **Figure 1.8**. What has yet to be determined is whether our future historian

---

[2] That footnote will probably comment on a civilization whose technical ability to spread information had far outstripped its judgment about what information was worth spreading.

(a)

Apollo lunar rover (1971)

(b)

KEY	Space observatories	Lunar and planetary explorers	Historical

Mariner 4 (1964–1965), first images of Mars

Surveyor 1 (1966), lunar lander

Sputnik (1957–1958), USSR, first human-made satellite

Galileo (1989–), first Jupiter orbiter, atmospheric probe

Viking lander 1 (1975–1982), first of two Mars landers

Chandra (1999–), Advanced X-ray Astronomy Observatory (AXAF)

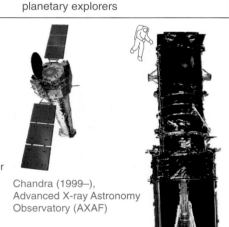

Hubble Space Telescope (1990–), UV, visible, infrared astronomy

FIGURE 1.6 (a) *Apollo 15* astronaut James B. Irwin stands by the lunar rover during an excursion to explore and collect samples from the Moon. (b) Artificial satellites and space probes have progressed a long way since the 1957 launch of *Sputnik I*. These spacecraft are all shown to the same scale. Some are astronomical observatories that view space from Earth's orbit. Others are interplanetary explorers sent to investigate other worlds within our Solar System.

will remember us for reaching out to touch the universe and embracing what we found, or whether we will instead be remembered for a loss of spirit—for stepping back and turning away from the frontier of exploration and discovery. The direction we take from here is a decision in which you will play a part.

1.2 Science Is a Way of Viewing the World

As we look at the universe through the eyes of astronomers, we will also learn something of how **science** works. It is almost impossible to overstate the importance of science in our civilization. One obvious manifestation of science is the fact that almost everything you use in your everyday life is a product of our scientific understanding of the world. The historical "simple life" is often romanticized, but it would be hard for any of us today even to begin to imagine what a truly pretechnological existence was like. Even a "primitive" trip into the backcountry is often accomplished today by eating freeze-dried food, sheltering under ultralight, ultrastrong synthetic fabrics, using a cellular phone to stay minutes

FIGURE 1.7 In the 20th century, new tools opened new windows on the universe. This is the sky as we would see it if our eyes were sensitive to radio waves, shown as a backdrop to the National Radio Astronomy Observatory site in Green Bank, West Virginia.

FIGURE 1.8 There is no doubt that history will remember ours as the time when humankind first stepped beyond our home world and reached out with our minds to embrace the universe.

away from emergency medical assistance, and following maps using a handheld global positioning system (GPS) receiver. Given the visible importance of technology in our lives, you might even be tempted to say that science *is* technology. It is true that science forms the basis of technology and that a mutually supportive relationship exists between science and technology, in which each enables advances in the other. Yet science is much more than technology. Science can no more be reduced to its practical technological application than the accomplishment of an Olympic athlete can be reduced to the utilization of the athlete's Olympic fame to market shoes and other products.

If science is more than technology, you might instead suggest that science is the **scientific method**. When you took science in high school you probably had the scientific method drilled into you—hypothesis and theory, followed by prediction, followed by experiments to test those predictions. There is good reason for placing emphasis on the sci-

entific method. For all practical purposes, it defines what we mean when we use the verb *to know*. It is sometimes said

The scientific method involves trying to falsify ideas.

that the scientific method is how scientists prove things to be true, but actually it is a way of proving things to be *false*. Before scientists accept something as true, they work hard to show that it is false. Only after repeated attempts to disprove an idea have failed do scientists begin to accept its likely validity. This is important! For a theory to be given serious scientific consideration, it *must* be *falsifiable*. It must be capable of being shown to be false. Scientific theories are accepted only as long as they are able to be tested and are not shown to be false.

But science can no more be said to *be* the scientific method than music can be said to *be* the rules for writing down a musical score. The scientific method provides the rules for asking **nature** whether an idea is false, but it offers no insight into where the idea came from in the first place, or how an experiment was designed. If you were to listen to a group of scientists discussing their work, you might be surprised to hear them using words such as *insight, intuition,* and *creativity.* Scientists speak of a beautiful theory in the same way that an artist speaks of a beautiful painting or a musician speaks of a beautiful performance. Yet science is not the same as art or music in one important re-

Nature is the arbiter of science.

spect. Whereas art and music are judged by a human jury alone, in science, it is nature (through the application of the scientific method) that provides the final decisions about which theories can be kept and which theories must be discarded. Nature is completely unconcerned about what we *want* to be true. In the history of science many a beautiful and beloved theory has been abandoned. At the same time, however, **Figure 1.9** makes the point that there is an aesthetic to science that is as human and as profound as any found in the arts.

It is also incorrect to say that science is a body of facts. We do not pretend to have all the answers, and we are constantly having to refine our ideas in response to new data and new insights. (This is, after all, what it means to learn.) The vulnerability of knowledge that is implicit in the sci-

All scientific knowledge is provisional.

entific method may seem like a weakness at first. "Gee, you really don't know anything," the cynical student might say. But this vulnerability is actually science's great strength. It is what keeps us honest. Once an idea is declared to be "truth," then all progress stops. In science even our most cherished ideas about nature remain fair game, subject to challenge by evidence according to the rules of the scien-

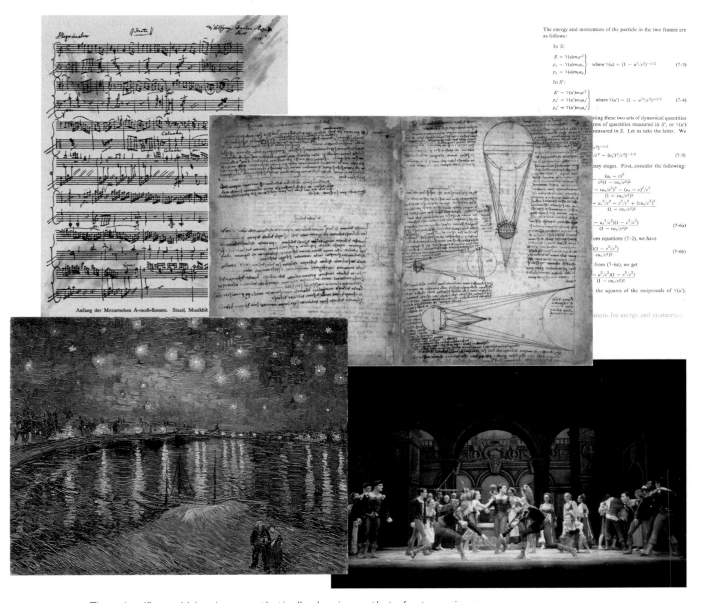

FIGURE 1.9 The scientific worldview is as aesthetically pleasing as that of art, music, or literature; but unlike the arts, nature has the final say about what scientific theories have lasting value.

tific method. Many of history's best scientists earned their place in the forward march of knowledge by successfully goring a sacred cow.

Scientists spend most of their time working within a framework of understanding, extending and refining that framework, and testing its boundaries. Occasionally, however, major shifts occur in the framework of some scientific field itself. Many books have been written about how science progresses, perhaps the most influential being *The Structure of Scientific Revolutions* by Thomas Kuhn. In this work Kuhn emphasizes the constant tension between the scientist's human need to construct a system of beliefs within which to interpret the world, and the occasional

(and likely painful) need to drastically overhaul that system of beliefs.

A scientific revolution is not a trivial thing. We cannot just wish that the universe were one way or another and then expect the universe to oblige. A new theory or way of viewing the world must be able to explain everything that the previous theory could while extending this understanding to new territory into which the earlier theory could not go. If one face can be said to symbolize modern science, it is that of Albert Einstein (**Figure 1.10**). Einstein's theories of special and general relativity replaced the 300-year-old edifice of Newtonian mechanics not by proving Newton wrong, but by showing that Newton's mechanics was a

FIGURE 1.10 Albert Einstein, perhaps the most famous scientist of the 20th century, and *Time* magazine's selection as Person of the Century. Einstein helped to usher in two different scientific revolutions, one of which he himself was never able to accept.

special case of a far more general and powerful set of physical laws. Einstein's new ideas unified our concepts of mass and energy and destroyed our conventional notion of space and time as separate things. Yet scientific revolutions are seldom comfortable for those who live through them, and even the greatest of scientists can be left behind. Einstein actually helped to start two scientific revolutions. He saw the first of these—relativity—through and embraced the world that it opened. Yet Einstein was unable to accept the implications of the second revolution he helped start —quantum mechanics—and went to his grave unwilling to embrace the view of the world it offered.

Science is not simply a body of facts. Science is not simply technology. Science is not simply the scientific method. Perhaps more than anything, science is a way of thinking about the world. It is a way of relating to nature. It is a search for the relationships that make our world what it is. It is a belief that nature is not capricious but instead operates by consistent, explicable, inviolate rules. It is a collection of ideas about how the universe works, coupled with an acceptance of the fact that what is known today may be superseded tomorrow. The scientist's faith is that there is an order in the universe and that the human mind is capable of grasping the essence of the rules underlying that order—or at least of inventing ever better approximations to those rules. The scientist's creed is that nature, through observation and experiment, is the final arbiter of the only thing worthy of the term *objective truth*. Science is an exquisite blend of aesthetics and practicality. And in the final analysis, science has found such a central place in our civilization because *science works*.

It sometimes may seem to you that science is arbitrary. For example, in August 2006 the International Astronomical Union (IAU) redefined what astronomers mean when they call something a planet. Under this new definition Pluto, known to all of us as the "ninth planet" since its discovery in 1930, was unceremoniously stripped of its "planet" status and demoted to "dwarf planet." Ceres, long regarded as the largest asteroid, was also redefined as a "dwarf planet." But it is important to keep in mind that this controversial decision, unsupported by many astronomers, was really more a matter of semantics than science. Pluto and Ceres are still the same scientifically important Solar System bodies that they always were. It is just that "science" has now put new labels on them.

It is beyond the scope of this book for us to provide you with a detailed justification for all that we will say. However, we will try to offer some explanation of where an idea comes from and why we believe it to be valid. We will not present something as fact unless there is a compelling reason to believe it. We will be honest when we are on uncertain, speculative ground, and we will admit it when the truth is that "we really do not know." This book is not a compendium of revealed truth or a font of accepted wisdom. Rather, it is an introduction to a body of knowledge and understanding that was painstakingly built (and sometimes torn down and rebuilt) brick by brick.

For those of us who grew up in a world transformed by science, the scientific worldview might seem anything but subversive. However, for much of history, knowledge was sought in the pronouncements of "authority" rather than through observation of nature. This authoritarian view slowed the advance of knowledge throughout Western Europe for the millennium prior to the European Renaissance, and it was largely the Chinese and Arab cultures that kept the spark of inquiry alive during this time. The greatest scientific revolution of all was the one that overthrew "authority" and replaced it with rational inquiry and the scientific method. Science is not *just* one of many possible worldviews. Science is the most successful worldview in the history of our species. It is worth noting that so far science itself has passed its own test. The foundations of the scientific worldview have withstood centuries of fine minds trying to prove them false.

The Cosmological Principle

The scientific revolution brought about a dramatic shift in our ideas about what knowledge is and how it is sought, but this change alone was not enough to open the universe to our probing gaze. It is likely that every civilization has had some system of beliefs about the relationship between the heavens and Earth. Before the Renaissance most of these included two fundamental tenets. The first was the belief that Earth occupies a special, unique place, usually at the center of the universe. The second was the belief that objects in the heavens are made of a different type of substance than

Earth and behave according to their own rules. The key to our understanding of the universe turned out to be the literal and total negation of both of these beliefs.

At the heart of modern astronomy is a fundamental idea called the **cosmological principle**. The cosmological principle asserts that there is nothing special or unique about Earth, either in our place in the universe or in the rules that govern the behavior of matter and energy here. The cosmological principle states that we are a part of the universe, rather than apart from the universe. The cosmological principle has two important aspects. The first is that

---

There is nothing special about our place in the universe.

---

when we look out around us, what we see is representative of what the universe is generally like. Our location in the universe is what it happens to be by chance—nothing more, nothing less. In a deep image (an image showing very faint objects), even a piece of apparently empty sky is filled with distant galaxies. There are as many galaxies in the observable universe as there are stars in our own Milky Way. The cosmological principle says that nothing sets the Milky Way apart from this group of galaxies. The impression that we get of the universe from our vantage point is representative of the whole. The second aspect of the cosmological principle is the premise that matter and energy obey the same physical laws throughout space and time as they do today on Earth. This means that the same physical laws we learn about in terrestrial laboratories can be used to understand what goes on in the centers of stars or in the hearts of distant galaxies.

The cosmological principle is a theory, and like all scientific theories it is subject to whatever tests our ingenuity and the scientific method can bring to bear. We will not visit a distant star or galaxy in our lifetimes, nor relive the early days of the universe. Yet we can analyze the light that reaches us from events distant in time and space, and ask whether those events follow the same laws that apply today on Earth. Each new success that comes from applying the cosmological principle to observations of the universe around us—each new theory that succeeds in explaining or predicting patterns and relationships among celestial objects—adds to our confidence in the validity of this cornerstone of our worldview.

1.3 Patterns Make Our Lives and Science Possible

Imagine what life would be like if sometimes when you let go of an object it fell up instead of down. What if one day apples were essential nutrition, but when you bit into an

apple the next day you discovered they were deadly poison? What if, unpredictably, one day the Sun rose at noon and set at 1:00 P.M., the next day it rose at 6:00 A.M. and set at 10:00 P.M., and the next day the Sun did not rise at all? In fact, objects do fall toward the ground. Our biochemistry remains stable. The Sun rises, sets, then rises again. Spring turns into summer, summer turns into autumn, autumn turns into winter, and winter turns into spring. The rhythms of nature produce patterns in our lives, and we count on these patterns for our very survival. If nature did not behave according to regular patterns, then our lives—indeed, life itself—would not be possible.

The same patterns that make our lives possible also make science possible. The goal of science is to identify and characterize these patterns and to use them to understand the world around us. Some of the most regular and easily identified patterns in nature are the patterns that we see in the sky. What in the sky will look different or the same a week from now? A month from now? A year from now? Most of us probably lead an indoor and in-town existence, removed from an everyday awareness of the patterns in the sky. However, away from the smog and glare of our cities, the patterns and rhythms of the sky are as easy to see today as they were in ancient times. Patterns in the sky mark the changing of the seasons (**Figure 1.11**), the coming of the rains, the movement of the herds, and the planting and harvesting of

---

Patterns in our lives echo patterns in the sky.

---

the crops. Patterns in the sky share the rhythms of our lives. It is no surprise that astronomy, which is the expression of our human need to understand these patterns, is the oldest of all sciences.

Mathematics Is the Language of Science

There are many kinds of mathematics, most of which deal with more than just numbers. Arithmetic is about counting things. Algebra is about manipulating symbols and the relationships between things. Geometry is about shapes and special relationships. Calculus is about change. Other types of mathematics deal with topics such as topology,

---

Mathematics is the science and language of patterns.

---

the properties of surfaces, or statistics, involving groups of objects and their relationships. What do they have in common? Why do we consider all of them to be part of a single discipline called "mathematics"? All share one thing —they deal with patterns. The best working definition of

mathematics is that "mathematics is the science and language of patterns."

We have seen that science is about patterns—patterns in relationships, patterns in behaviors, and patterns in characteristics. Astronomy is the part of science concerned with patterns related to celestial objects. If patterns are the heart of science, and mathematics is the language of patterns, it should come as no surprise that *mathematics is the language of science.* Trying to study science while avoiding mathematics is the practical equivalent of trying to study Shakespeare while avoiding the written or spoken word.

It quite simply cannot be done, or at least cannot be done meaningfully.

On the other hand, as the authors of this book we understand (and as is humorously pointed out by **Figure 1.12**) that for many of you, *math* is not *a* four-letter word—it is *the* four-letter word. Many people decide early in their education that they cannot "do" math, and from that day forward the mere mention of the word causes their eyes to glaze over and their palms to sweat. A distaste for mathematics is one of the most common obstacles standing between a nonscientist and an appreciation of the beauty and elegance of

FIGURE 1.11 Since ancient times our ancestors recognized that patterns in the sky, such as which stars are overhead at night, change together with the coming and going of the seasons and other patterns that shape our lives.

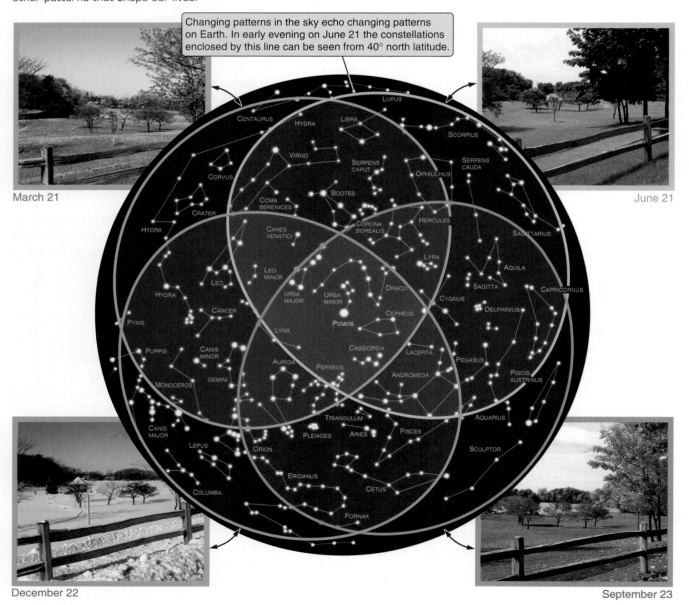

Changing patterns in the sky echo changing patterns on Earth. In early evening on June 21 the constellations enclosed by this line can be seen from 40° north latitude.

March 21

June 21

December 22

September 23

FIGURE 1.12 Mathematics is the science of patterns, which makes mathematics the language of science.

the world as seen through the eyes of a scientist. To move beyond this obstacle, both scientist and nonscientist need to find common ground.

Part of the responsibility for moving beyond this obstacle lies with us, the authors. It is our job to take on the role of translators, using words to express as many concepts as possible, even when these concepts are more concisely and accurately expressed mathematically. When we do use mathematics, we will explain in everyday language what the equations mean and try to show you how equations express concepts that you can connect to the world. We will also limit the mathematics to a few basic tools that all college students should have been exposed to. *Scientific notation* (see **Foundations 1.1**) is needed because of the vast range of sizes of the objects involved. Units are needed to distinguish between time, distance, mass, and energy. A bit of geometry is necessary for understanding the distances, sizes, shapes, and volumes of things. Finally, some algebra —mostly a few ratios and proportionalities—will provide a way of expressing the patterns that relate one physical quantity to another. "Basic" does not necessarily mean easy, but it does mean that we will use the most accessible tools that will make our journey of discovery as comfortable and informative as possible.

Your responsibility is to accept the challenge and make an honest effort to think through the mathematical concepts that we use. Do not concede defeat while still in the starting blocks. It is likely that you know what it means to square a number, or to take its square root, or to raise it to the third power. The mathematics in this book is on a par with what it takes to balance a checkbook, build a

bookshelf that stands up straight, check your gas mileage, estimate how long it will take you to drive to another city, figure your taxes, or buy enough food to feed an extra guest or two at dinner. Foundations 1.1 describes several of the basic mathematical tools we will use throughout this book. (Also see Appendix 1.)

1.4 Bending Your Brain into Shape

This book will likely ask you to think in ways that are different from the ways in which you are accustomed to thinking, and to learn to view the world from new and unfamiliar perspectives. Knowledge and understanding have nothing to do with shoving facts into short-term memory so they can be regurgitated on an exam, and that is certainly not what astronomy is about. Changing the way you think about things takes more effort. It also means studying in ways that may be different from your normal habits. Here are a few practical suggestions for how you might better study this text:

- **Read the text actively.** Think about each section after you have read it. What major concepts were discussed in the section? How are they related to what you have read so far? Have you run into similar concepts elsewhere? Why are the contents of that section important enough to

Mathematical Tools

Mathematics gives scientists many of the tools they need to understand the patterns they see and to communicate that understanding to others. As the authors of this text, we are aware that mathematics is not a friend to many of you taking this course, and so we have worked to keep the math in this text to a minimum. Even so, there are a few tools that we will need:

Scientific notation: Scientific notation is the way that scientists deal with numbers of vastly different sizes. Rather than writing out 7,540,000,000,000,000, 000,000, we write 7.54×10^{21}. Rather than writing out 0.000000000005, we write 5×10^{-12}.

Ratios: Ratios are the most common way that astronomers use to compare things. A star may be 10 times as massive as the Sun or 10,000 times as luminous as the Sun. These are ratios.

Geometry: To describe and understand objects in astronomy and physics, we use concepts such as distance, shape, area, and volume. Apparent separations between objects in the sky are expressed as *angles*. Earth's orbit is an *ellipse* with the Sun at one *focus*. The planets in the Solar System lie close to a *plane*. Geometry provides the tools for working with these concepts.

Algebra: Algebra provides a way of using and manipulating symbols that represent numbers or quantities. We will use algebra to express relationships that are valid not just for a single case, but for many cases. Algebra lets us conveniently express ideas such as "the distance that you travel is equal to the speed at which you are moving times the length of time you go that speed" (in other words, $d = s \times t$, where d is distance, s is speed, and t is time). Algebra also lets us combine these ideas with other ideas to arrive at new relationships.

Proportionality: Often understanding a concept amounts to understanding the *sense* of the relationships that it predicts or describes. "If you have twice as far to go, it will take you twice as long to get here." "If you have half as much money, you will be able to buy only half as much gas." These are examples of proportionality. If you are traveling at a constant speed, then time is proportional to distance. We write $t \propto d$, where $\propto$ means "is proportional to." Proportionalities often involve quantities raised to some power. A circle of radius r has an area A equal to πr^2, so we say that area is proportional to the square of the radius, and write $A \propto r^2$. This means that if you make the radius of a circle three times as large, its area will grow by a factor of 3^2, or 9.

---

be included in the book? Briefly summarize the section and your thoughts about it in your class notes.

- **Draw a picture.** Many physical and mathematical concepts are most easily understood if they are visualized. If you understand a concept well enough to draw a picture that expresses it, you probably understand the concept fairly well. Trying to sketch a picture will also help you better identify what things you understand and what things you do not.

- **Ask yourself "What if?"** when trying to understand a concept. What if Earth were more massive? How would that affect Earth's gravity? What if the Sun were hotter? How would that affect the color of the light from the Sun or the amount of energy that the Sun radiates?

If you cannot "what if" a concept, you probably do not really understand it yet.

- **Try to teach it.** It is often said that you do not really understand something until you try to explain it to someone else. At the end of a reading assignment, talk about the material with a friend or family member. Try to find a partner or a group in your class with whom you can meet, and *take turns teaching each other the material.* Each of you should read the assigned text, then divide up responsibility for who will present which sections. Ask each other lots of questions—the harder the better! (Make a game of playing "stump the instructor.") Even explaining a concept out loud to yourself can help.

- **Share your ideas and insights** with your discussion group. If a particular concept really "clicks" for you—if you really think it is neat—share both your understanding and your enthusiasm with your group.

- **Be honest with yourself** about what things you understand and what things you do not, and try not to avoid concepts you find difficult. Personal growth comes from real accomplishment. Getting the most from this journey will come from facing challenging concepts.

- **Focus your discussion on concepts, relationships, and connections.** You have to know the facts, but the facts are the starting point rather than the end. Use the key concepts, study questions, and other study aids to identify and concentrate your effort on the most important ideas.

- **Do not let discomfort with math keep you from succeeding.** Mathematical formulas are not magical incantations. Rather, they are expressions of logical ideas. We will always present a plain-English discussion of the idea behind any mathematics that we use. Begin by focusing on this discussion. After you grasp the idea, then look at the math. Try to see how the relationships between the quantities in the equation embody the idea. If you need help with basic math skills, work through the appendixes or explore Study Space (wwnorton.com/astro21) for other aids available to you. Your instructor is also there to help.

- **Above all, remember that building understanding is always an *active* process, never passive!**

Our best suggestion for a successful journey through *21st Century Astronomy* is to nurture your spark of interest in astronomy until it grows enough to draw you in. When, as Einstein said in the opening quote, you "pause to wonder and stand rapt in awe"—*then* you will have learned the secret!

1.5 Let the Journey Begin

The journey we are about to begin is sometimes entertaining, sometimes enlightening, often surprising, usually challenging, and always remarkable. We will explore the wonders of the universe, and along the way take a look at what science is, how it is done, and what it means to look at the world as a scientist does. It is a journey that will

- Teach us as much about ourselves as it will about what is "out there."

- Open our minds to new ways of thinking about and experiencing the world.

- Pay tribute to what the human mind and the human spirit are capable of achieving.

Reading the book and taking the journey with us are two different things. This is only a guidebook. It can lead you to the trailhead and tell you something of what you might find along the way, but *you* have to walk the path! If you become an active participant in this adventure, rather than a passive spectator, then what you gain from the journey will remain a part of you long after the final exam is forgotten. And if along the way you find yourself applying your understanding to new situations and new information, and if you learn to combine your understandings and arrive at new insights that are greater than the sum of their parts, then you will have learned something far more than just astronomy.

Summary

- The entire universe is governed by the same physical laws that shape our lives here on Earth.

- There is nothing special about our particular place in the universe.

- The scientific method is a way of trying to *falsify*, not prove, ideas.

- *All* scientific knowledge is provisional.

- Mathematics is the science and language of patterns, and thus it is the language of science.

Seeing the Forest through the Trees

At the end of each chapter of *21st Century Astronomy* you will find a brief narrative about the content of that chapter. The purpose of "Seeing the Forest through the Trees" is not to rehash the entire contents of the chapter, but rather to pick out a few of the high points and put them into a broader context. Building on the analogy of a hike in the mountains, we will spend a lot of time looking in detail at the rocks and the trees, but every so often we need to step back and look around at the forest as a whole.

We live in a world that has been profoundly shaped by the scientific revolution that took hold of Western thought during the Renaissance. That revolution fundamentally altered our way of thinking about the world, as well as our view of the relationship between ourselves and the universe of which we are a part. A new spirit of rational inquiry was turned on the heavens, dislodging Earth and humankind from the center of the cosmos. Observation, experiment, and rigorously applied reason came to replace dogma and authority as the arbiter of knowledge. The heavens became a realm not of mysticism and magic, but instead of physical law—the same physical law that governs the behavior of matter and energy in laboratories here on Earth.

The closing years of the 20th century saw our knowledge of the universe charge ahead at an ever-accelerating pace. This progress comes courtesy of a great many advances both in our technology and in the sophistication of our physical understanding of matter and energy and of space and time themselves. We have seen many fundamental questions about the origin and fate of the universe and the threads that tie our existence to the cosmos move from the realm of philosophical speculation into the realm of rigorous scientific inquiry. The insights that this age of exploration and discovery have brought are often far more profound and startling than dreamed of even a few decades ago. There can be little doubt that this time will be looked on as one of the more significant moments in the intellectual and cultural history of our species.

Like a hike through the mountains, *21st Century Astronomy* will not be an effortless journey. Many travelers will find that they have to flex a few mental muscles in ways they are not used to, and may even have to face an old adversary or two on the trail. But muscles that are sore after the first day of a hike grow comfortable and strong with time, and adversaries can become the best of friends.

In Chapter 2 we begin the journey in earnest, and as with most journeys our starting point is home. What patterns do we see in the skies of our planet Earth, and how do those patterns come to be? This is not an easy or gentle slope on which to begin our trek, but our vistas will change rapidly as we climb.

Key Terms

astronomy, p. 4
matter, p. 4
energy, p. 4
Milky Way Galaxy, p. 4
Local Group, p. 4
supercluster, p. 4
light, p. 4
speed, p. 5
origins, p. 8
astrobiology, p. 8
science, p. 9
scientific method, p. 10
nature, p. 10
cosmological principle, p. 13
mathematics, p. 14
scientific notation, p. 16
algebra, p. 16
proportionality, p. 16

Student Questions

THINKING ABOUT THE CONCEPTS

1. Imagine yourself on a planet orbiting a star in a faraway galaxy. What does the cosmological principle tell you about the way you would perceive the universe from this distant location?

2. It is said that we are made of stardust. Explain why this is a true statement.

3. List patterns in your own life that repeat regularly. How do these patterns affect you? Which patterns are of your own making, which are set by others, and which are determined by nature?

4. The scientific method states that scientific theories must be falsifiable. List some beliefs or views that you conclude are *not* falsifiable.

5. A textbook published in 1945 stated that it takes 800,000 years for light to reach us from the Andromeda Galaxy. In *21st Century Astronomy* we say that it takes 2,900,000 years. What does this tell you about a scientific "fact" and how our knowledge evolves with time?

6. Astrology makes testable predictions. For example, it predicts that the horoscope for your star sign on any day should fit you better than horoscopes for other star signs. Read each of the horoscopes in yesterday's paper without regard to your own sign. How many of them might fit the day that you had yesterday? Repeat the experiment every day for a week and keep records. Was your horoscope consistently the best description of your experiences?

7. A scientist on television states that it is a known fact that life does not exist beyond Earth. Would you consider this scientist reputable? Explain your answer.

8. Some astrologers use elaborate mathematical formulas and procedures to predict the future. Does this show that astrology is a science? Why or why not?

9. You run across an old newspaper with the headline "EINSTEIN PROVES NEWTON WRONG!" Did the newspaper get this story right? Explain your answer.

APPLYING THE CONCEPTS

10. If it takes about 8 minutes for light to travel from the Sun to Earth, and Pluto is 40 times this distance from us, how long does it take light to reach Earth from Pluto? Radio waves travel at the speed of light. What does this imply about the problems you would have if you tried to conduct a two-way conversation between Earth and a spacecraft orbiting Pluto?

11. Imagine the Sun to be the size of a grain of sand and Earth a speck of dust 83 mm away. (On this scale, each light-minute of distance equals 10 mm.) How far would it be from Earth to the Moon on this scale? From the Sun to Pluto? From Earth to the nearest star? To the nearest large galaxies? (Note that $1\,m = 10^3\,mm$ and that $1\,km = 10^3\,m$.) At what point do you lose your "feeling" for these distances?

12. The average distance from Earth to the Moon is 384,000 km. How many days would it take, traveling at 800 km/h (the typical speed of a jet airplane), to reach the Moon?

13. The surface area of a sphere is proportional to the square of its radius. If the Moon has a radius only one-quarter that of Earth, how does the surface area of the Moon compare with that of Earth?

14. Write 86,400 (the number of seconds in a day) and 0.0123 (the Moon's mass compared to Earth's) in scientific notation.

15. Write 1.60934×10^3 (the number of meters in a mile) and 9.154×10^{-3} (Earth's diameter compared to the Sun's) in standard notation.

16. The time (t) it takes for light to reach us from a distant galaxy is equal to the distance (d) of the galaxy divided by the speed of light (c). Use algebra to describe this relationship more simply.

17. If you understand proportionality, then you understand most of the math you need to follow this text. Make a list of five different proportionalities from your daily life. (For example, the price of a bag of apples is proportional to the weight of the bag of apples.) For each proportionality, identify the constant of proportionality (such as the price per pound of apples). How are these constants determined?

18. The circumference of a circle is given by $C = 2\pi r$.
    a. Calculate the approximate circumference of Earth's orbit around the Sun, assuming that the orbit is a circle with a radius of 1.5×10^8 km. You can approximate π as being about equal to 3.
    b. Noting that there are 8,766 hours in a year, how fast, in kilometers per hour, does Earth move in its orbit?
    c. How far along in its orbit does Earth move in one day?

StudySpace
wwnorton.com/astro21
provides a Study Plan for each chapter that includes a reading outline, animations, keyword flash cards, and gradebook-enabled multiple-choice quizzes. From StudySpace you can also access premium content in the ebook and SmartWork.

...marking the conclave of all the night's stars,
those potentates blazing in the heavens
that bring winter and summer to mortal men,
the constellations, when they wane, when they rise.

AESCHYLUS (525–456 B.C.)

The Moon seen rising among the ancient stones of Stonehenge.

CHAPTER 2

Patterns in the Sky—Motions of Earth

2.1 A View from Long Ago

The herds have reached the high meadows where they spend the warm season, and for a time the life of the tribe has settled in as well. The weather is comfortable, and the days are long, with none of the hardships that accompany the time of cold and snow and long, dark nights. It is a time of plenty and a time for telling the age-old stories of the tribe around a fire that guards against the chill of the gathering night. You feel a sense of contentment and are thankful to the gods for this time when life is good. As the embers die down, you gaze upward at the familiar canopy of stars overhead, and as you often do in such moments, you wonder about what you see.

To survive, you must learn the subtle patterns of your world. You must know the ways of the herds, and recognize the gathering of clouds that heralds a coming storm. When you turn your keen eye toward the heavens, you find subtle and changing patterns there as well—patterns that somehow echo those of your life. The spirits of the great animals dwell in the sky; as a child you learned to recognize their pictures there. Above you now are the stars that rule the time of the short nights. These are the stars that bring summer and lead the herds to this pleasant place.

Some of the spirits of the sky can be difficult to please. The mischievous planets wander from place to place, using their fearsome powers to sow chaos through the heavens. The Moon sometimes turns blood red, and the Sun is consumed by an ominous beast. But as long as the tribe remembers them, the gods and spirits of the sky will continue to bring the seasons and send the stars to guide the tribe. This is as your elders taught you when you were young, and this is as

KEY CONCEPTS

In this chapter we begin our journey in earnest, starting out as our ancestors did when they first gazed at the Sun, Moon, and stars, and tried to understand what they saw. With the benefit of knowledge hard-won over the centuries, we will look at patterns present both in the sky and on Earth, and then look beyond appearances to the underlying motions that cause those patterns. Here we will discover

- How the stars appear to move through the sky as Earth rotates on its axis, and how those motions differ when seen from different latitudes on Earth.

- The fundamental concept of a frame of reference, and how Earth's rotating frame of reference affects weather patterns and other terrestrial phenomena.

- How Earth's motion around the Sun and the tilt of Earth's axis relative to the plane of its orbit combine to determine which stars we see at night and the seasons we feel through the year.

- The motion of the Moon in its orbit about Earth, and how that motion, together with the motion of Earth and the Moon around the Sun, shapes the phases of the Moon and the spectacle of eclipses.

you teach the young ones today. So it has always been, and so it shall always be.

Our ancestors lived their lives attuned to the ebb and flow of nature, and the patterns in the sky were a part of that ebb and flow. The coming of night and day, the changing of the seasons, the rising and falling of the tides, the movement of the herds—all of these march in lockstep with the changes that we see in the sky. The repeating patterns of the Sun, Moon, and stars echo the rhythms that have defined the lives of humans since before the beginning of recorded history. By watching the patterns in the sky our ancestors found that they could predict when the seasons would change and the rains would come and the herds would move. Knowledge of the sky offered knowledge of the world, and knowledge of the world was power. It was a small step from here to thinking of the unreachable, untouchable stars as not only

Patterns in the sky have always been important to our species.

a reflection of the patterns in the world, but also as the *cause* of those patterns. The stars found a special place in legend and mythology as the realm of gods and goddesses, holding sway over the lives of humankind. As writing came to replace oral traditions and legends, mythologies of the sky became more elaborate as well. And as humans invented numbers and mathematics to describe and predict and account for things in the world, predictions of the motions of the stars and planets were among their greatest successes. Some of our ancestors came to look upon the orderly and predictable patterns of the sky as the *true* patterns of the world, and our own lives as imperfect reflections of this heavenly reality. They looked for ways to use their knowledge of the sky to find order in the seeming chaos of their everyday lives, and **astrology** was born.

Elements of this same basic history played themselves out many times over and in every part of our globe. From Africa to Asia, from Europe to Central America, from North America to the British Isles, the archeological record holds evidence of early humans who projected ideas from their own cultures onto what they saw in the sky (see **Excursions 2.1**). The connection between the patterns in the sky and the patterns in their lives was simply too compelling to be missed. The idea of the sky as a realm of mysticism and magic is deeply rooted in the traditions and beliefs and history of our species. There is no mystery about the currents of mind that led our ancestors to their belief in astrology and other celestial mythologies. At a time when the causes of things were unknown, and humans existed at the seeming whim of forces they could not comprehend, the sky seemed to offer a window into a mystical and powerful world of spirits, gods, devils, and angels.

What an unwelcome shock it must have been when a few remarkable individuals, with names like Copernicus and Kepler and Galileo and Newton, tugged at the threads of this comfortable and familiar tapestry, *only to discover that it fell apart in their hands!* The stars and other heavenly bodies do not rotate about Earth each day, as humans have thought since they first took notice of the sky. Rather, it is *Earth* that spins on its axis, giving the stars, planets, Sun, and Moon the appearance of following daily paths through the heavens. Nor is Earth at the center of all existence, as

Science shattered ancient mystical views of the heavens.

befits the home of humankind, the pinnacle of all Creation. Earth is just one of eight planets orbiting the Sun. The subtle complexity of the changing patterns we see in the sky results from the motions of planets and moons as they step through their gravitational dance with the Sun. Even the Sun itself, whose radiant energy makes our world what it is, is only one of countless stars, adrift in a universe whose full extent is unknown even today.

The magic of astrology properly belongs to a time long dead, when in the minds of humans Earth rode on the back of a giant sea turtle, and with each passing month the Sun moved from one stellar "house" to the next. Today it is a matter of experimentally verifiable fact that the imaginary patterns seen in the stars hold no more influence over our lives than the random patterns of leaves blowing down the street on an autumn day. The astrologers' quest for deep connections between our lives and the patterns in the sky was both understandable and well placed, but knowledge of the true nature of those connections had to wait for the birth of modern science. Today the sky has become a window of knowledge on the *physical* world. This knowledge has proven worth the wait.

Today saw a long hard climb before you reached the meadow by the river where you now camp, and tomorrow's trek promises to be just as demanding. Even so, the evening is pleasant, and you are content. The embers of your campfire have almost died away when the distant sound of a jetliner interrupts your reverie. Without really thinking you look up to catch sight of the plane high overhead, and are caught off guard by the blazing spectacle of the summer Milky Way and the thousands of pinpoints of light, which seem so close that you can almost reach out and touch them. For a moment the thousands of years separating you from a long-dead tribal nomad vanish as you share the same sense of wonder and awe that has always defined humankind's experience of the universe.

It is here that we begin the journey of *21st Century Astronomy*—with the changing patterns in the sky that captured the attention and imagination of that long-ago nomad and that still beacon overhead on a dark, cloudless night. Yet unlike that nomad, we look on those changing patterns

EXCURSIONS 2.1

Where Are the Constellations?

Where are the **constellations**? The answer may seem obvious: "The constellations are overhead in the sky, for all to see." Yet if you look at the sky, no pictures of winged horses or dragons or chained maidens are painted there. Instead there is only the random pattern of stars —about 6,000 of them visible to the naked eye—spread out across the sky. Constellations exist only within the imagination of the human mind. Constellations are the ideas and pictures that humans imposed on the lights in the sky in an effort to connect our lives on Earth with the workings of the heavens.

As illustrated in **Figure 2.1**, there have been as many different sets of constellations and stories to go with them as there have been cultural traditions in our history. Modern constellations visible from the Northern Hemisphere draw heavily from the list compiled 2,000 years ago by the Alexandrian astronomer Ptolemy. Con-

stellations in the southern sky are drawn from the lists put together by European explorers visiting the Southern Hemisphere during the 17th and 18th centuries. Today astronomers use an officially sanctioned set of constellations as a kind of road map of the sky. The entire sky is broken into 88 different constellations, much as continental landmasses are divided into countries by invisible lines. Every star in the sky lies within the borders of a single constellation, and the names of constellations are used in naming the stars that lie within their boundaries. For example, Sirius, the brightest star in the sky, lies within the boundaries of the constellation Canis Major (meaning the "big dog"). Sirius's official name is therefore Alpha *Canis Majoris* (this is the Latin genitive or "possessive" form—see Appendix 6), indicating that it is the brightest star in that constellation and earning its nickname, the Dog Star.

FIGURE 2.1 The region around what we now call the Big Dipper (Ursa Major, or the "Great Bear") as viewed by three different civilizations. Constellations exist only in the human mind.

Egyptian—1275 B.C.

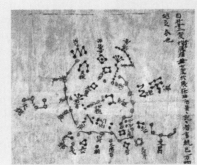

Chinese—A.D. 940

European—A.D. 1540

with the perspective of centuries of hard-won knowledge. We will find that patterns of change in the sky are often the understandable and even unavoidable consequences of the daily rotation of Earth about its axis and Earth's annual trip around the Sun. This is an example of science at its best —the discovery of wonderful variety arising from simple and elegant underlying causes. And just as happened over the history of our species, curiosity about the changing patterns in the sky will show us the way outward into a universe far more vast and awesome than our distant ancestor could have imagined.

2.2 Earth Spins on Its Axis

Despite the apocryphal stories you may have learned in grade school, Columbus did not discover that the world is round. Long before his famous (or possibly infamous) journey to the New World, anyone who had read **Aristotle** or the other Greek philosophers (as had Columbus) knew that Earth is a ball. Far more difficult to accept was the idea that the changes occurring in the sky from day to day and month to month are the result of the motion of Earth rather than the motion of the Sun and stars. The most apparent of these motions is Earth's rotation on its axis, which sets the very rhythm of life on Earth—the passage of day and night. When our remote ancestors first noticed the sky with something approaching human awareness, it was doubtless the daily motion of the Sun in the sky that drew their attention.

As viewed from above Earth's **North Pole**, Earth rotates in a counterclockwise direction (**Figure 2.2**), completing one rotation in a 24-hour period. As the rotating Earth carries us from west to east, objects in the sky *appear* to move in the other direction, from east to west. The path a celes-

---

Earth's rotation is counterclockwise when viewed from above the North Pole.

---

tial body makes across the sky as seen from Earth is called its *apparent daily motion*. The Sun is one such object. When we say "noon," we mean the time of day when our location on Earth faces most directly toward the Sun. By convention, astronomers divide the sky evenly into eastern and

western halves along an imaginary north–south arc called the **meridian**. The meridian runs from due north to due south, passing through the point directly overhead in the sky, called the **zenith**. True *local noon* occurs when the Sun crosses the meridian at our location. Half a day later our spot on Earth comes closest to facing directly away from the Sun. This is *local midnight*.

The View from the Poles

The apparent daily motions of the stars and the Sun witnessed by ancient nomadic tribes would have depended on where on the surface of the planet they happened to live. The apparent daily motions of celestial objects in northern Europe, for example, are quite different from the apparent daily motions seen from a tropical island. (Differences in apparent motions of the Sun are responsible for many of the differences in culture between people living in these two regions.) The daily motions of the stars are easiest to understand when viewed from a place where humans did not set foot until 1909—Earth's North Pole.

Imagine you are standing on the North Pole watching the sky, as shown in **Figure 2.3**. (Ignore the Sun for the moment, and suppose that you can always see stars in the sky.) You are standing where Earth's axis of rotation intersects its surface, which is much the same as standing at the center of a rotating carousel. As Earth rotates, the spot directly above you remains fixed while everything else in the sky appears to revolve in a counterclockwise direction around this spot. (If you are having trouble visualizing this, find a globe and, as you spin it, imagine standing at the pole of the globe.) The direction in which Earth's axis of rotation points, and about which the stars appear to revolve as Earth turns, is called the **north celestial pole**. The greater the angular distance between a celestial object and the north celestial pole, the larger the circular path the object appears to follow. Objects close to the pole appear to follow small circles, whereas the largest circles are followed by objects nearest to the horizon.

You can never see more than half of the sky at any one time, regardless of where you are on the surface of our planet. The other half of the sky is blocked from view by Earth. The boundary between the part of the sky you

---

The same half of the sky is always visible from the North Pole.

---

can see and the part that is blocked by Earth is called the **horizon**. From most locations on Earth, the half of the sky that we can see above the horizon changes constantly as Earth rotates. (The direction in space in which our zenith points at the moment is different now from what it was 12 hours ago, or even 12 seconds ago.) In

FIGURE 2.2 The rotation of Earth and the Moon, the revolution of Earth and the planets about the Sun, and the orbit of the Moon about Earth are counterclockwise as viewed from above Earth's North Pole (not drawn to scale).

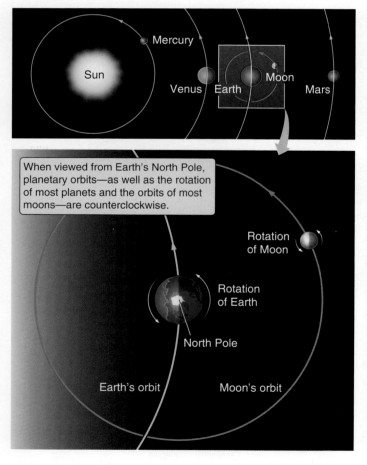

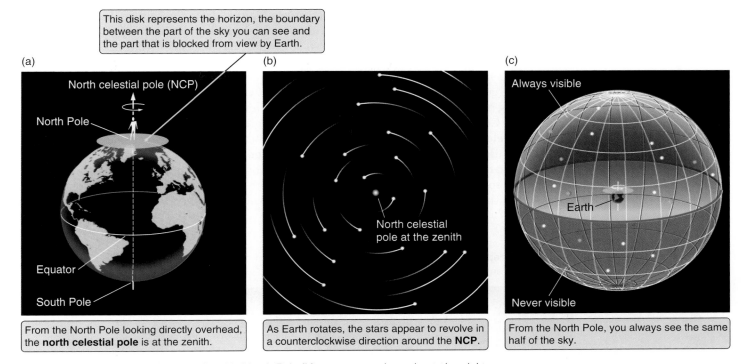

This disk represents the horizon, the boundary between the part of the sky you can see and the part that is blocked from view by Earth.

(a)

North celestial pole (NCP)

North Pole

Equator

South Pole

From the North Pole looking directly overhead, the **north celestial pole** is at the zenith.

(b)

North celestial pole at the zenith

As Earth rotates, the stars appear to revolve in a counterclockwise direction around the **NCP**.

(c)

Always visible

Earth

Never visible

From the North Pole, you always see the same half of the sky.

FIGURE 2.3 (a) As viewed from Earth's North Pole (b) stars move throughout the night on counterclockwise circular paths about the zenith. (c) The same half of the sky is always visible from the North Pole.

contrast, Earth's North Pole points in the *same* direction, hour after hour and day after day. From the North Pole we always see the *same* half of the sky. Nothing rises or sets as Earth turns beneath you. If you look off toward the horizon, you will see that the objects visible there follow circular paths that keep them always the same distance above the horizon.

The view from Earth's **South Pole** is much the same, but with two major differences. First, the South Pole is on the opposite side of Earth from the North Pole, so the half of the sky you see overhead is precisely the half that is hidden from the North Pole. The direction in space that is at the zenith at the South Pole, and about which everything ap-

From the South Pole, the other half of the sky is visible, and stars circle in a clockwise direction.

pears to spin as Earth rotates, is the **south celestial pole**. The second difference is that instead of appearing to move counterclockwise around the sky, stars appear to move *clockwise* around the south celestial pole. (To see this, sit in a swivel chair and spin it around from right to left. As you look at the ceiling things appear to move in a counter-clockwise direction, but as you look at the floor it will appear to be moving clockwise.)

Away from the Poles the Part of the Sky We See Is Constantly Changing

Latitude is a measure of how far north or south we are on the face of Earth. Imagine a line from the center of Earth to your location on the surface of the planet. Now imagine a second line from the center of Earth to the point on the **equator** closest to you. (Refer to **Figure 2.4** for help imagining these lines.) The angle between these two lines is your latitude. The latitude of any point on the equator is 0°. The latitude of the North Pole is 90° north latitude, whereas the South Pole is at 90° south latitude. The latitude of Phoenix, Arizona, is 33.5° north.

Imagine what we see as we leave the North Pole and travel south to lower latitudes. As we follow the curve of Earth, our horizon tilts and our zenith moves away from the north celestial pole. By the time we reach a latitude of 60° north (as shown in Figure 2.4(b)), the north celestial pole has dropped to 60° above the northern horizon. This equality between north latitude and the height of the north celestial pole above the northern horizon holds everywhere. When we reach a latitude of 30° north (as in Figure 2.4(c)), the direction of the north celestial pole has dropped to 30° above the horizon. (A more accurate way to say it is this: The north celestial pole lies in the same direction regardless of where we are on Earth. It is *our horizon* that is tilted at 30° from the north celestial pole when we are at a latitude of 30° north.)

In Figure 2.4(d) we have reached Earth's equator at a latitude of 0°. The north celestial pole is now sitting on the northern horizon. At the same time we get our first look at the south celestial pole, which is sitting opposite the north celestial pole on the southern horizon. Continuing on into the Southern Hemisphere, the south celestial pole is now visible above the southern horizon, while the north celestial pole is hidden from view by the northern horizon. At a latitude of 45° south (Figure 2.4(e)), the south celestial pole lies 45° above the southern horizon. At the South Pole (90°

south latitude—Figure 2.4(f)), the south celestial pole is at the zenith, 90° above the horizon.

Probably the best way to cement your understanding of how the view of the sky changes from one latitude to another is to draw pictures like those in **Figure 2.5(a)** for different latitudes. If you can draw a picture like this for any latitude—filling in the values for each of the angles in the drawing and imagining what the sky looks like from that location—then you will be well on your way to developing a working knowledge of the appearance of the sky. That knowledge will prove useful later when we discuss a variety of phenomena such as the changing of the seasons. When practicing your sketch, however, take care not to make the

FIGURE 2.4 Our perspective on the sky depends on our location on Earth. Here we see how the locations of the celestial poles and celestial equator depend on an observer's latitude.

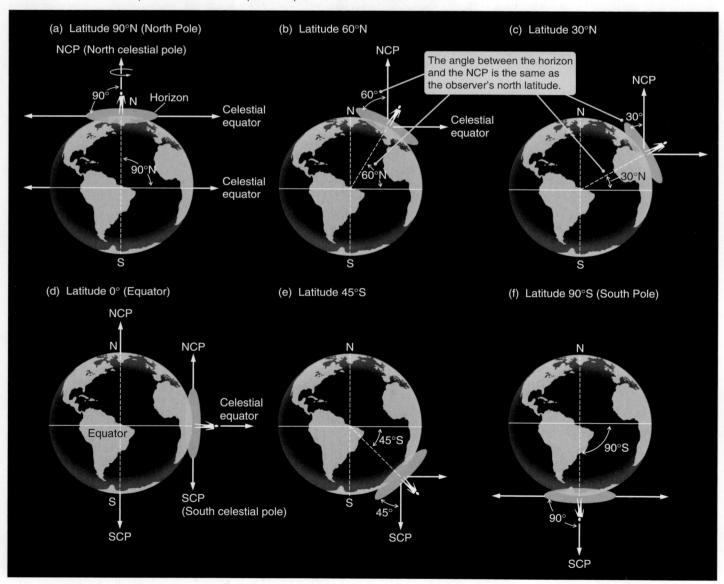

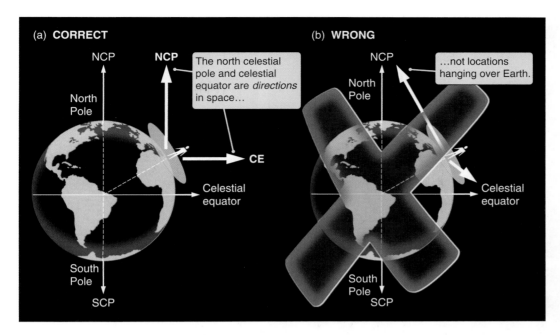

FIGURE 2.5 (a) The celestial poles and the celestial equator are directions in space, not locations hanging above Earth. Do not make the mistake shown in (b).

common mistake shown in **Figure 2.5(b)**. *The north celestial pole is not a location* in space, hovering over Earth's North Pole. Instead, *it is a direction* in space—the direction parallel to Earth's axis of rotation.

For many centuries travelers, including sailors at sea, have used the stars for navigation. Perhaps the simplest of the navigator's techniques is to use the equality between latitude and the altitude of the north (or south) celestial pole. The north or south celestial poles can be found by recognizing the stars that surround them. In the Northern Hemisphere it happens by chance that a moderately bright star is seen within about ¾° of the north celestial pole. This star is called **Polaris**, or more commonly the **north star**. If

The star Polaris marks the north celestial pole.

you can find Polaris in the sky and measure the angle between the north celestial pole and the horizon, then you know your latitude. If you are in Phoenix, Arizona, for example (latitude 33.5° north), you will find the north celestial pole 33.5° above your northern horizon. On the other hand, if you are studying astronomy in Fairbanks, Alaska (latitude 64.6° north), Polaris sits much higher overhead, 64.6° above the horizon in the north.

The location of the north celestial pole in the sky can be used to measure the size of Earth. Suppose we start out in Phoenix, Arizona, and head north. By the time we reach the Grand Canyon, about 290 km (kilometers) later, we notice that the north celestial pole has risen from 33.5° to about 36° above the horizon. This change—2.5°—is 1/144 of the way around a circle. (A circle is 360°, and 2.5°/360° = 1/144.)

This means that we must have traveled 1/144 of the way around the circumference of Earth, so the circumference of Earth must be about 144 × 290 km, or about 42,000 km. The actual circumference of Earth is just a shade over 40,000 km, so our simple measurement was not too bad, given our sloppy measurements of angles and distances. The radius of Earth is this circumference divided by 2π, or about 6,400 km. It was in much this way that the Greek astronomer Eratosthenes (276–194 B.C.) made the first accurate measurements of the size of Earth around 230 B.C. (well before Columbus's time).

The apparent motions of the stars about the celestial poles also differ from latitude to latitude. **Figure 2.6(a)** shows an observer at a point in the Northern Hemisphere other than the North Pole. As Earth rotates, the part of the sky visible to this observer is constantly changing. Of course, from the perspective of the observer it is the horizon that seems to remain fixed, while the stars appear to move past overhead. If we focus our attention on the north celestial pole, from this perspective we still see much the same thing we saw from Earth's North Pole. The north celestial pole remains fixed in the sky, and all of the stars appear to move in daily counterclockwise circular paths around that point. But because the north celestial pole is no longer directly overhead, the apparent circular paths of the stars are now tipped relative to the horizon. (More correctly, our horizon is now tipped relative to the apparent circular paths of the stars.)

Stars located close enough to the north celestial pole are above the horizon 24 hours a day (even if we can't see them in daytime) as they complete their apparent paths around the pole (see **Figures** 2.6(a) and **2.7**). This always-visible re-

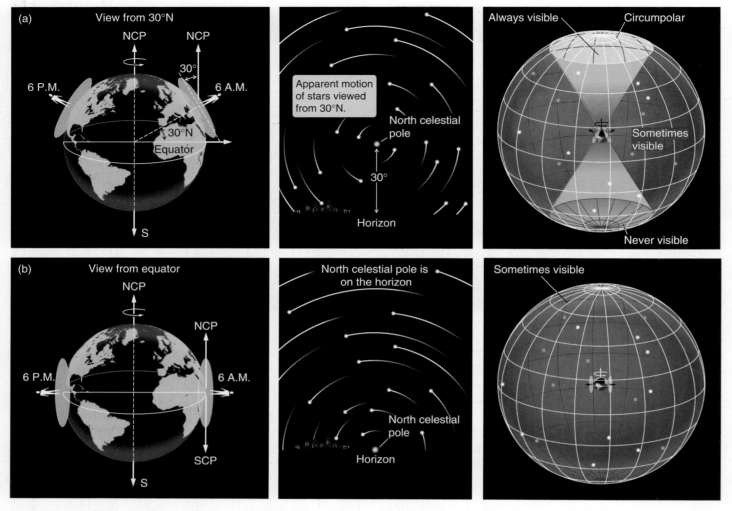

FIGURE 2.6 (a) As viewed from 30° north latitude the north celestial pole is 30° above the northern horizon. Stars appear to move on counterclockwise paths around this point. At this latitude some parts of the sky are always visible, while others are never visible. (b) From the equator the north and south celestial poles are seen on the horizon, and the entire sky is visible over 24 hours.

gion of our sky is referred to as being **circumpolar**, which means "around the pole." There also remains a part of the sky that can *never* be seen from this latitude. This is the part of the sky near the *south* celestial pole that never rises above your horizon. And between this region and the always-visible circumpolar region lies a portion of the sky

> **Circumpolar stars are always above the horizon.**

that can be seen for *part but not all* of each day. Stars in this intermediate region appear to rise above and set below Earth's shifting horizon as Earth turns. The only place on Earth where you can see the entire sky over the course of 24 hours is the equator. From the equator (**Figure 2.6(b)**) the north and south celestial poles sit on the northern and

southern horizons, respectively, and the whole of the heavens passes through the sky each day.

The Celestial Sphere Is a Useful Fiction   It is sometimes useful to think about the sky as if it were a huge sphere with the stars on its surface and Earth at its center. (Our ancient tribesman probably thought this really was the case.) Astronomers refer to this imaginary sphere as the **celestial sphere**.[1] The celestial sphere is a useful concept because it is easy to draw and visualize, but never forget that it is imaginary! Each point on the celestial sphere actually corresponds to a *direction* in space. **Figure 2.8** shows

---

[1] You can find a description of celestial coordinates used with the celestial sphere in Appendix 6.

the celestial sphere as seen by observers at different places on Earth.

We divide the celestial sphere into a northern half and a southern half with an imaginary circle called the **celestial equator**. Just as the north celestial pole is the projection of the direction of Earth's North Pole into the sky, the celestial equator is the projection of the plane of Earth's equator into the sky. If you are on Earth's equator (Figure 2.8(c)), then the celestial equator runs east to west, passing directly overhead through the zenith. As you move north away from

The celestial equator is the projection of Earth's equator into space.

Earth's equator, the celestial equator tips toward the southern horizon by the same amount that the north celestial pole appears to rise above the northern horizon. Just as Earth's North Pole is 90° away from Earth's equator, the north celestial pole is always 90° away from the celestial equator. If you point one arm at a point on the celestial equator and one arm toward the north celestial pole, your arms will always form a right angle. If you are in the Southern Hemisphere, the same holds true there. The angle between the celestial equator and the south celestial pole is 90° as well.

Look at the location of the celestial equator in Figure 2.8. The points where the celestial equator intersects the horizon are always due east and due west. (The only exception to this is at the poles, where the celestial equator is coinci-

dent with the horizon.) An object on the celestial equator rises due east and sets due west. Objects that are north of the celestial equator rise north of east and set north of west. Objects that are south of the celestial equator rise south of east and set south of west.

Also note from Figure 2.8 that regardless of where you are on Earth, half of the celestial equator is always visible above the horizon (again with the exception of the poles). Because half of the celestial equator is always visible, it follows that you can see any object that lies in the direction of the celestial equator half of the time. An object that is in the direction of the celestial equator rises due east, is above the horizon for exactly 12 hours, and sets due west. This is not true for objects that are not on the celestial equator. A look at Figure 2.8(b) shows that from the Northern Hemisphere, you can see more than half of the apparent circular path of any star that is north of the celestial equator. And if you can see more than half of a star's path, then the star is above the horizon for more than half of the time.

As seen from the Northern Hemisphere, stars north of the celestial equator remain above the horizon for more than 12 hours each day. The farther north the star is, the longer it stays up. The circumpolar stars near the north celestial pole are the extreme example of this. They are up 24 hours a day. In contrast, objects south of the celestial equator are above the horizon for less than 12 hours a day, and the farther south you look, the less time a star is visible. Near the south celestial pole there are stars that never rise above our horizon.

From a location in the Canadian woods, the north celestial pole appears high in the sky…

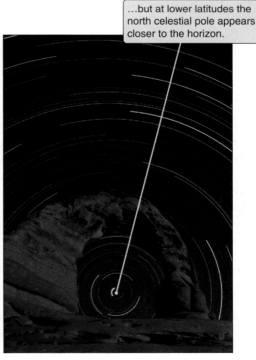

…but at lower latitudes the north celestial pole appears closer to the horizon.

FIGURE 2.7 Time exposures of the sky showing the apparent motions of stars through the night. Note the difference in the circumpolar portion of the sky as seen from the two different latitudes.

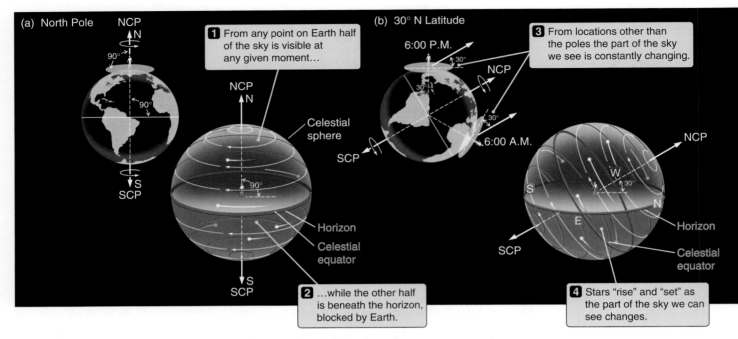

(a) North Pole

1 From any point on Earth half of the sky is visible at any given moment…

2 …while the other half is beneath the horizon, blocked by Earth.

(b) 30° N Latitude

3 From locations other than the poles the part of the sky we see is constantly changing.

4 Stars "rise" and "set" as the part of the sky we can see changes.

NCP · N · Celestial sphere · SCP · Horizon · Celestial equator · 6:00 P.M. · 6:00 A.M. · W · S · E · N · Horizon · Celestial equator

FIGURE 2.8 The celestial sphere is a useful fiction for thinking about the appearance and apparent motion of the stars in the sky. Here the celestial sphere is shown as viewed by observers at four different latitudes. At most latitudes, stars rise and set as the part of the celestial sphere that we see changes during the day.

If you were an observer in the Southern Hemisphere (Figure 2.8(d)), the reverse of the preceding discussion would be true: Objects on the celestial equator would still be up for 12 hours a day, but now objects south of the celestial equator would be up longer than 12 hours, and objects north of the celestial equator would be up less than 12 hours.

A Swinging Pendulum, a Flying Cannonball, and a Swirling Storm Feel the Effect of Earth's Rotation

One reason the ancients did not believe that Earth rotates is that they could not perceive the spinning motion of Earth. Put yourself in their place. As a result of Earth's rotation, the surface of Earth is moving along at a respectable speed—1,674 km/h (kilometers per hour) at the equator (calculated by dividing the circumference of Earth by the period of its rotation). Even so, we do not feel that motion any more than we would "feel" the speed of a car with a perfectly smooth ride cruising down a straight highway. However, Earth's rotation *does* have a number of measurable effects on objects riding along on its surface.

Jean-Bernard-Léon Foucault (pronounced "Foo-coe"; 1819–1868) was the first to carry out an experiment dem-

onstrating that Earth rotates. In 1851 Foucault constructed a **pendulum** by suspending a 28 kg (kilogram) weight at the end of a steel wire 67 meters long from the dome of the Panthéon in Paris, and he set the pendulum swinging. A pendulum swings back and forth because of the combined effects of Earth's gravity and the tension in the wire. A child swinging on a rope is a good example of a pendulum. He falls toward Earth, picking up speed, then slows to a stop as he climbs on the other side of his swing. He then repeats the processes in reverse, returning to his starting point. Left on its own, the laws of physics say that a pendulum will keep swinging back and forth, its motion remaining in the same plane. Yet when Foucault observed his pendulum over several hours, he found that the plane in which the pendulum swung gradually *changed,* rotating in a clockwise direction as viewed from above!

It is easiest to understand the behavior of a **Foucault pendulum** when it is placed at Earth's North or South Pole, as shown in **Figure 2.9**. The pendulum swings back and forth, staying in the same plane, *but Earth is rotating underneath it*. This is obvious to an observer watching from space (Figure 2.9(a)), yet to us riding on the rotating Earth (Figure 2.9(b)), the direction of the pendulum's swing appears to change. After 6 hours, Earth will have rotated 90° on its axis. The pendulum, which is still swinging in the same plane as it always was, appears to us to have changed the direction of its swing by 90° in the opposite direction. (We stress that

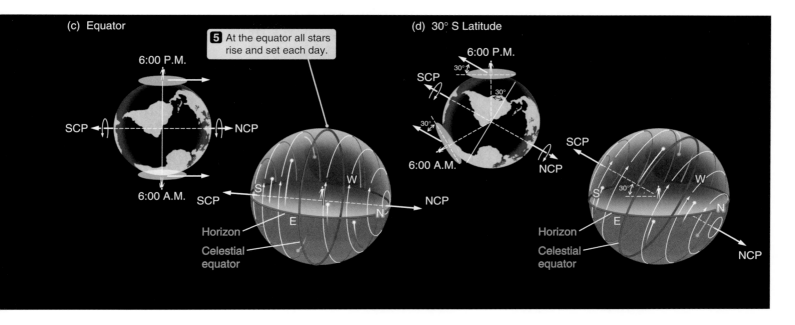

(c) Equator

5 At the equator all stars rise and set each day.

6:00 P.M.

SCP ← → NCP

6:00 A.M. SCP

SCP

NCP

Horizon

Celestial equator

S   W   E   N

(d) 30° S Latitude

6:00 P.M.

30°

SCP

30°

30°

6:00 A.M.

SCP

NCP

NCP

Horizon

Celestial equator

S   30°   W   E   N

NCP

Foucault's pendulum provided the first experimental demonstration of Earth's rotation.

the back-and-forth motion of any pendulum is due to Earth's gravity and the tension in the wire, and *not* to anything having to do with Earth's rotation. It is only the *apparent* change in the *direction* of the back-and-forth swing relative to the ground that is caused by Earth's rotation.)

The plane in which a Foucault pendulum at the North Pole swings appears to make one complete rotation in 24 hours. In contrast, a Foucault pendulum on Earth's equator appears to behave very differently. Imagine that the pendulum is set swinging back and forth along the equator (Figure 2.9(c)). As Earth rotates, the pendulum keeps swinging in the same plane; so from the standpoint of an observer riding along on Earth, the pendulum shows no change in direction. Earth is no longer spinning underneath the pendulum. Instead the pendulum is riding around on a big circle with the surface of Earth.

For latitudes between the pole and the equator, the situation is more difficult to understand in detail, but you can probably guess the answer. If it takes exactly one day for the plane of the swing of a pendulum at the pole to appear to rotate once, and if it takes forever for a pendulum on the equator to change the plane of its swing, then for latitudes

between the two it probably takes longer than a day but less than forever to seem to go around once. This guess would be correct. Foucault's original pendulum, located in Paris at a latitude of 49°, took about 32 hours to complete one rotation. (In science we often start by working out the "easy" or "limiting" cases—the pendulum at the pole or the equator, for example—and then use these to guide our thinking about what happens in more complicated situations.)

Differences in Speed Between Different Latitudes Cause the Coriolis Effect   Foucault may have been the first to conduct an experiment showing that Earth rotates, but he was not the first to experience effects of this rotation. Earth's rotation actually influences things as diverse as the motion of weather patterns on Earth and how an artillery gunner must aim at a distant target. Any object sitting on the surface of Earth follows a circle each day as Earth rotates on its axis. This circle is larger for objects near Earth's equator and smaller for objects nearer to one of Earth's poles; but because Earth is a solid body, all objects must complete their circular motion in exactly one day. Because an object nearer to the equator has farther to go each day than an object nearer a pole, the object nearer the equator must be moving *faster* than the object at a greater latitude. If an object starts out at one latitude and then moves to another, its apparent motion over the surface of Earth is influenced by this difference in speed.

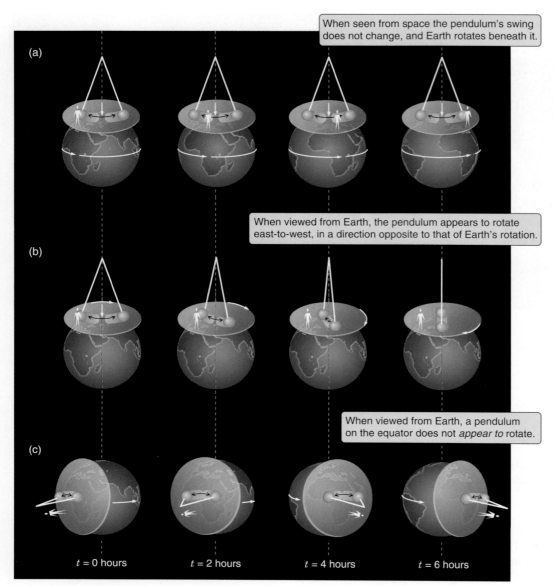

When seen from space the pendulum's swing does not change, and Earth rotates beneath it.

When viewed from Earth, the pendulum appears to rotate east-to-west, in a direction opposite to that of Earth's rotation.

When viewed from Earth, a pendulum on the equator does not *appear to* rotate.

(a)

(b)

(c)

$t = 0$ hours　　　$t = 2$ hours　　　$t = 4$ hours　　　$t = 6$ hours

FIGURE 2.9 (a) A Foucault pendulum at Earth's North Pole swings back and forth in the same plane while Earth rotates underneath it. (b) To an observer on the rotating Earth, however, the swing of the pendulum seems to change directions. (c) A Foucault pendulum on the equator does not appear to rotate.

Imagine that you are riding in a car traveling down a straight section of highway at a constant speed. The windows are blacked over so that you cannot see the scenery go by. How would you tell the difference between this and sitting completely still in your driveway? The fact is that, apart from the vibration due to the roughness of the road,

---

The motion of one object *relative* to another is important.

---

you could not. You might try throwing a ball around in the car. You toss it up and it falls right back in your lap. You can play catch with the person sitting next to you. Even though the car is speeding down the road, the relative mo-

tions between the objects in the car are small, and it is only the **relative motions** that count.

Now imagine that two cars are driving down the road at *different* speeds, as shown in **Figure 2.10**. Ignoring for the moment any real-world complications like wind resistance, if you were to throw a ball from the faster-moving car directly at the slower-moving car as the two cars pass, you would miss. The ball shares the forward motion of the faster car, so the ball outruns the forward motion of the slower car. From your perspective in the faster car (Figure 2.10(b)), the slower car lagged behind the ball. From the slower car's perspective (Figure 2.10(c)), your car and the ball sped on ahead.

Now do the same experiment; but instead of two cars, think about two locations at different latitudes. Suppose you

(a) Frame of reference: Viewer on the street

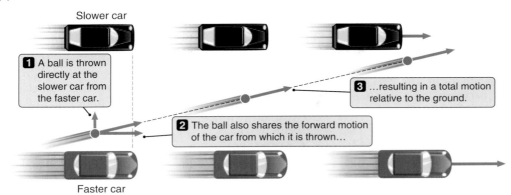

Slower car

1 A ball is thrown directly at the slower car from the faster car.

2 The ball also shares the forward motion of the car from which it is thrown…

3 …resulting in a total motion relative to the ground.

Faster car

FIGURE 2.10 The motion of an object depends on the frame of reference of the observer.

(b) Frame of reference: Viewer in faster car

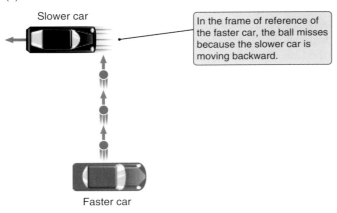

Slower car

In the frame of reference of the faster car, the ball misses because the slower car is moving backward.

Faster car

(c) Frame of reference: Viewer in slower car

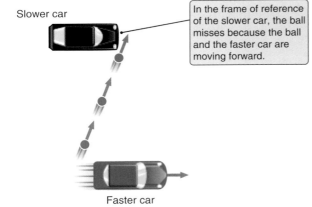

Slower car

In the frame of reference of the slower car, the ball misses because the ball and the faster car are moving forward.

Faster car

fire a cannon directly north from a point in the Northern Hemisphere, as shown in **Figure 2.11(b)**. Because the cannon is located nearer to the equator than its target is, the cannon itself is moving toward the east faster than its target. Even though the cannonball is fired toward the north, it shares in the eastward velocity of the cannon itself. This means that the cannonball is *also* moving toward the east

faster than its target! Recall how the ball thrown from the faster car outpaced the slower-moving car. Similarly, as the cannonball flies north, it finds itself moving toward the east faster than the ground underneath it is. To an observer on the ground, it looks like the cannonball curves toward the east as it outruns the eastward motion of the ground it is crossing. The farther north the cannonball flies, the greater the difference between its eastward velocity and the eastward velocity of the ground. As a result the cannonball follows a path that appears to curve more and more to the east the farther north it goes. If you are located in the Northern Hemisphere and fire a cannonball *south* toward the equator (**Figure 2.11(c)**), the opposite effect will occur. Now the cannon is moving toward the east more slowly than its target. As the cannonball flies toward the south, its eastward motion lags behind that of the ground underneath it, and the cannonball appears to curve toward the west.

This effect of Earth's rotation is called the **Coriolis effect**. In the Northern Hemisphere the Coriolis effect causes a cannonball fired north to drift to the east as seen from the surface of Earth. In other words, the cannonball appears to curve to the right. A cannonball fired south appears to curve to the west, which also gives it the appearance of curving

Deflection caused by the Coriolis effect is to the right in the Northern Hemisphere.

to the right. In the Northern Hemisphere the Coriolis effect seems to deflect things to the *right*. If you think through this example for the Southern Hemisphere, you will see that south of the equator the Coriolis effect seems to deflect things to the *left*. In between, at the equator itself, the Coriolis effect vanishes.

These differences in the Coriolis effect between the Northern and Southern Hemispheres are seen in the rotation of weather systems. As air is pushed from regions of higher pressure toward regions of lower pressure, these motions are influenced by the Coriolis effect. Think about a low-pressure region in the Northern Hemisphere. When air is pushed toward this region of low pressure from the south,

The Coriolis effect causes the counterclockwise rotation of northern hurricanes.

the Coriolis effect deflects this flow of air toward the east. Similarly, air moving toward the region of low pressure from the north is deflected to the west by the Coriolis effect. The net effect is that as air moves toward a region of low pressure in the Northern Hemisphere, the Coriolis effect deflects it into a counterclockwise circulation (**Figure 2.12**). (Think about this carefully. The Coriolis effect deflects objects toward the right as you face north in the Northern Hemisphere, which results in weather patterns that rotate toward the *left*.) The next time you see a television weather map with a low-pressure region, look at the direction of the winds around the region and you will see this counterclockwise flow. The most spectacular example of this **cyclonic motion** is the swirl of wind and clouds around the deep low-pressure area at the eye of a hurricane or typhoon. As the air moves in closer and closer to the central region of low pressure, it rotates faster and faster, giving rise to the winds of hundreds of kilometers per hour that make hurricanes so destructive.

With what you now know about the Coriolis effect, you can show that hurricanes in the Southern Hemisphere rotate in the opposite direction from hurricanes in the Northern Hemisphere. Instead of curving to the right, air moving into

Southern Hemisphere hurricanes rotate clockwise.

a region of low pressure curves to the left, causing a clockwise rotation around the low-pressure region. In crossing from the Northern Hemisphere to the Southern Hemisphere, the Coriolis effect goes away at the equator. The difference in the direction of rotation between the hemispheres and the weakness of the Coriolis effect near the equator mean that a northern hurricane would literally collapse if it tried to cross into the Southern Hemisphere. (Imagine throwing your car into reverse while traveling down the road at 150 km/h.) This is why countries around Earth's equator do not experience the ravages of hurricanes.

As discussed in **Foundations 2.1**, the Coriolis effect is only one example of the type of effect associated with relative motions. On a humorous note, the Coriolis effect has nothing to do with the direction that water swirls in a toilet bowl, as many people imagine. The difference in the speed of Earth's motion between the two sides of a toilet bowl is not enough to matter much. Other effects, such as the direction in which the water flows into the bowl, are much more important. However, the Coriolis effect is enough to deflect a fly ball hit north or south into deep left field

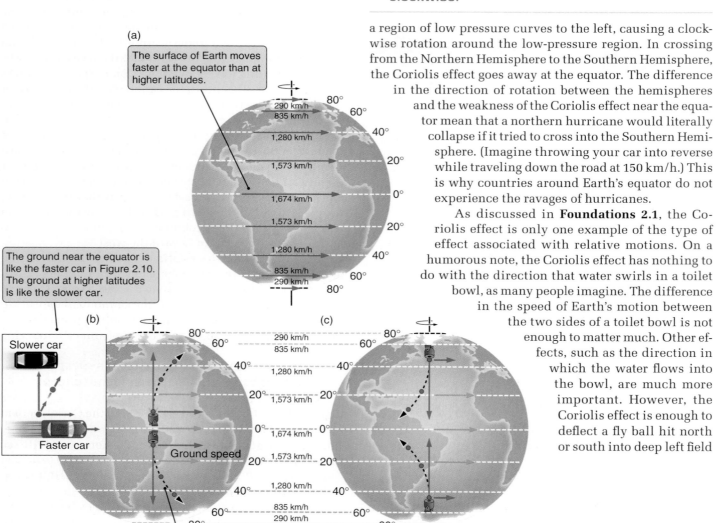

(a)

The surface of Earth moves faster at the equator than at higher latitudes.

290 km/h
835 km/h   80°
                  60°
1,280 km/h   40°
1,573 km/h   20°
1,674 km/h   0°
1,573 km/h   20°
1,280 km/h   40°
835 km/h   60°
290 km/h   80°

The ground near the equator is like the faster car in Figure 2.10. The ground at higher latitudes is like the slower car.

(b)

Slower car

Faster car

80°
60°
40°
1,280 km/h
20° 1,573 km/h
0°   1,674 km/h
Ground speed
20° 1,573 km/h
40° 1,280 km/h
60° 835 km/h
80°

290 km/h
835 km/h

(c)

290 km/h   80°
835 km/h   60°
1,280 km/h   40°
1,573 km/h   20°
1,674 km/h   0°
1,573 km/h   20°
1,280 km/h   40°
835 km/h   60°
290 km/h   80°

A cannonball fired away from the equator outruns the ground it flies over, so its ground track curves to the east.

A cannonball fired toward the equator lags behind the eastward motion of the ground, so its ground track curves to the west.

FIGURE 2.11 The Coriolis effect causes objects to appear to be deflected as they move across the surface of Earth.

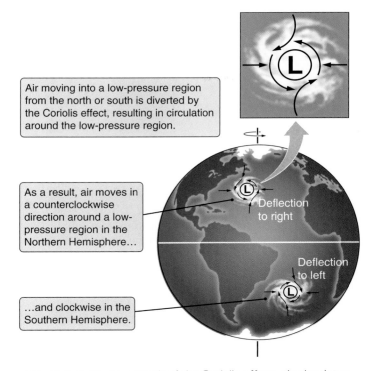

Air moving into a low-pressure region from the north or south is diverted by the Coriolis effect, resulting in circulation around the low-pressure region.

As a result, air moves in a counterclockwise direction around a low-pressure region in the Northern Hemisphere…

…and clockwise in the Southern Hemisphere.

Deflection to right

Deflection to left

FIGURE 2.12 As a result of the Coriolis effect, air circulates around regions of low pressure on the rotating Earth.

in a stadium in the northern United States by about a half a centimeter. At some time or other the Coriolis effect has probably determined the outcome of a ball game.

2.3 Revolution about the Sun Leads to Changes during the Year

The second motion we will discuss is the motion of Earth about the Sun. Earth revolves around the Sun in the same direction Earth spins about its axis—counterclockwise as viewed from above Earth's North Pole. A **year**, by definition, is the time it takes for Earth to complete one revolution around the Sun. The motion of Earth around the Sun is re-

> Earth's orbital motion is counterclockwise as viewed from above Earth's North Pole.

sponsible for many of the patterns of change we see in the sky and on Earth, including changes in which stars we see at night. When you look overhead at midnight, you are looking away from the Sun. As Earth moves around the Sun, this direction changes. Six months from now, Earth will be

on the other side of the Sun, and the stars that we see overhead at midnight will be in nearly the opposite direction from the stars we see near overhead at midnight tonight. The stars that were overhead at midnight six months ago are the stars that are overhead today at noon, but we cannot see those stars today because of the glare of the Sun.

If you could note the position of the Sun relative to the stars each day for a year, you would find it traces out a **great circle** against the background of the stars (**Figure 2.13**). On September 1, the Sun appears to be in the direction of the constellation of Leo. Six months later, on March 1, Earth is on the other side of the Sun, and the Sun appears to be in the direction of the constellation of Aquarius. The apparent path that the Sun follows against the background of the stars is called the **ecliptic**. The constellations that lie along

> The ecliptic is the Sun's apparent yearly path against the background of stars.

the ecliptic and through which the Sun appears to move are called the constellations of the **zodiac**. This is why ancient astrologers assigned special mystical significance to these stars. Actually, the constellations of the zodiac are nothing more than random patterns of distant stars that happen by chance to lie near the plane of Earth's orbit about the Sun.

Earth's Motion through Space Is Measured Using Aberration of Starlight

Just as it is difficult to "feel" the effects of Earth's rotation on its axis, it is even harder to sense the motion of Earth around the Sun. Through most of the history of our species, humans believed that Earth remains stationary while the Sun, the Moon, and the heavens revolve around us. The history of modern astronomy, and to some degree the story of the rise of modern science, can be told as the story of how this view was overthrown during the 17th and 18th centuries. However, the first direct measurement of the effect of Earth's motion was not made until the 18th century. To understand how this measurement was made, we return to our automotive example of a moving frame of reference.

Imagine you are sitting in a car in a windless rainstorm, as shown in **Figure 2.14**. If the car is sitting still and the rain is falling vertically, when you look out your side window you see raindrops falling straight down. That is, if you were to hold a vertical tube out the window, raindrops would fall straight through the tube. When the car is moving forward, however, the situation is different. Between the time a raindrop appears below the top of your window and the time it disappears beneath the bottom of your window, the car has moved forward. The raindrop disappears beneath the window *behind* the point at which it appeared, which means

FOUNDATIONS 2.1

Relative Motions and False Forces

Aside from looking out the window or feeling road vibrations, there is no experiment that you could easily do to tell the difference between riding in a car down a straight section of highway at constant speed and sitting in the car while it is parked in your driveway. Because everything in the car is moving together, the relative motions between objects in the car are all that count. In fact, the only reason you can feel the roughness of the road is that it slightly changes the motion of the car. You feel these brief accelerations as the car's vibration.

The idea that only relative motions count occurs again and again in astronomy and physics. There are numerous examples in this chapter alone. For example, even though Earth is spinning on its axis and flying through space in its orbit about the Sun, the resulting relative motions between objects that are near each other on Earth are small—so small that for most of history humans assumed that Earth was sitting still. Newton's realization that motions are meaningful only when tied to the **frame of reference** of some observer is also at the heart of Einstein's theories of relativity. These theories, which we will return to later, wound up changing the way we think about space and time.

The Coriolis effect, discussed in this chapter, is an example of what can happen if motion is viewed from a frame of reference with a changing velocity. The Coriolis effect is sometimes wrongly called the "Coriolis force" because to someone on the ground it looks like a force has acted on the cannonball, pushing it to the side. In fact, no additional force is acting on the cannonball. It is flying true. The ball *appears* to curve because of the shifting frame of reference of the ground over which it travels. The Coriolis effect and other false forces can be seen in many contexts. When you turn a sharp corner in a car, it seems as though you are thrown against the door by some force. Actually, no force is throwing you against the door at all. Instead, your body is trying to continue moving in a straight line while the frame of reference of the car (and you) is changing. For a playground example of the Coriolis effect and such false forces, return to the same carousel we used to understand the apparent motions of the stars. Sit near the center of the spinning carousel, and try to play catch with someone sitting near the edge. This exercise, if carried out physically and not just in your mind, will help you develop a feeling for how the Coriolis effect works. (It might be good for a few laughs or a queasy stomach as well.)

As we continue on our journey, keep your eyes open for more examples of relative motions. Relative motions among planets, among stars, among galaxies, among atoms—they will all be there.

---

the raindrop *looks as if* it falls at an angle, even though in reality it is falling straight down. For raindrops to fall directly through the tube you are holding out the window now, you would have to tilt the top of the tube forward. The faster you go, the greater the apparent front-to-back motion of the raindrops, and the more tilted their apparent paths become. An observer by the side of the road would say the raindrops are coming from directly overhead, but to you in the moving car they are coming from in front of the car.

This same phenomenon occurs with starlight, as shown in **Figure 2.15**. The light from a distant star arrives at Earth from the direction to this star. Because Earth moves, however, to an observer on Earth the starlight seems to be coming from a slightly different direction, just as the raindrops

appeared to be coming from in front of the car.[2] As the direction of Earth's motion around the Sun continuously changes during the year, the apparent position of a star in the sky moves in a small loop. This shift in apparent position is what we refer to as the **aberration of starlight**.

The speed and direction of Earth's motion and the part of the sky a star is in determine the exact shift in the apparent position of a star. The apparent deflection of a star located at a right angle to Earth's motion is about half of what the typical unaided human eye can detect. This much deflection

---

[2] Actually, in the frame of reference of Earth, the starlight *is* coming from a different direction, and in the frame of reference of the moving car, the raindrops *are* coming from in front of the car.

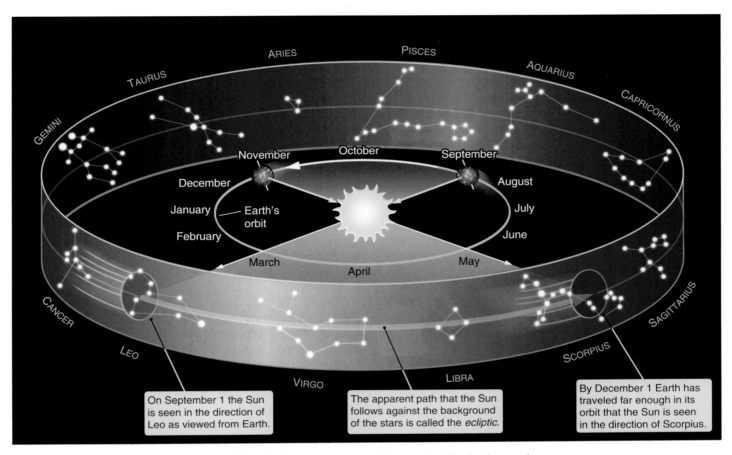

FIGURE 2.13 As Earth orbits about the Sun, the Sun's apparent position against the background of stars changes. The imaginary circle apparently traced by the Sun is called the ecliptic. Constellations along the ecliptic form the zodiac.

Earth's orbital motion around the Sun was first measured from the aberration of starlight.

is easy to measure with a telescope. Aberration of starlight was first detected in the 1720s by two English astronomers, Samuel Molyneux and James Bradley. Measurement of the aberration of starlight shows that Earth is moving on a roughly (but not exactly) circular path about the Sun with an average speed of just under 30 km/s (kilometers per second). Because distance equals speed times time, the distance around this near-circle—its circumference—is just the speed of Earth (29.8 km/s) times the length of one year (3.16×10^7 s). The circumference of Earth's orbit is then

$$\text{Distance} = \text{Speed} \times \text{Time}$$

$$= (29.8 \tfrac{\text{km}}{\text{s}}) \times (3.16 \times 10^7 \text{ s})$$

$$= 9.42 \times 10^8 \text{ km.}$$

The radius of Earth's nearly circular orbit is this circumference divided by 2π, or 1.50×10^8 km, or 150 million kilometers. Astronomers refer to this distance—the average distance between the center of the Sun and the center of Earth[3]—as one **astronomical unit**, abbreviated AU. The astronomical unit is a good unit for measuring

The astronomical unit (AU) is the average distance between the Sun and Earth.

distances within the Solar System. Modern measurements of the size of the astronomical unit are made in very different ways, such as bouncing radar signals off Venus. However, the aberration of starlight provided a simple and compelling demonstration that Earth orbits about the Sun—and a pretty good value for the size of Earth's orbit as well.

---

[3]Distances between celestial objects are almost always taken to be between their centers.

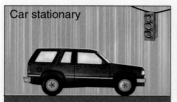

From the vantage point of a person outside, the rain falls vertically, even when the car is moving.

Car stationary

Car moving

From the frame of reference of a person inside the car...

...the rain falls vertically if the car is stationary...

...but at an angle if the car is moving.

FIGURE 2.14 On a windless day the direction in which rain falls depends on the reference frame in which it is viewed. From a stationary car, rain is seen to fall vertically downward. From a moving car, rain is seen to fall at an angle determined by the speed and direction of the car's motion.

Seasons Are Due to the Tilt of Earth's Axis

If you ask most people why it is cold in the winter and warm in the summer, they are likely to tell you that it is because Earth is closer to the Sun in the summer and farther away in the winter. This is a common (and commonsense) idea that *does* have something to do with the seasons on Mars, but has virtually *nothing* to do with the seasons on Earth! Earth's orbit around the Sun is almost a perfect circle centered on the Sun,[4] and so the distance from the Sun changes little during the year. In fact, Earth is slightly closer to the Sun during the northern winter than it is during the northern summer. So far we have discussed the rotation of Earth on its axis and the revolution of Earth about the Sun, and the consequences of each. To understand the changing of the seasons, we need to consider the combined effects of these two motions.

If Earth's spin axis were exactly perpendicular to the plane of Earth's orbit (the ecliptic plane), then the Sun would always appear to lie on the celestial equator. Because the position of the celestial equator is fixed in our sky, the Sun

[4] Planetary orbits are actually elliptical, as will be discussed in Chapter 3.

would follow the same path through the sky day after day, rising due east each morning and setting due west each evening. If the Sun were always on the celestial equator, it would be above the horizon for exactly half the time, and days and nights would always be exactly 12 hours long. In short, if Earth's axis were exactly perpendicular to the plane of Earth's orbit, each day would be just like the last, and there would be no seasons.

However, Earth's axis of rotation is *not* exactly perpendicular to the plane of the ecliptic. Instead it is tilted by 23.5° from the perpendicular. (Astronomers use the term **obliquity** to refer to the angle between a planet's equatorial and orbital planes.) As Earth moves around the Sun, its axis points in almost exactly the same direction through the

---

> Seasons result from the 23.5° tilt of Earth's axis with respect to a line perpendicular to its orbital plane.

---

year and from one year to the next. As a result, sometimes Earth's North Pole is tilted more toward the Sun, and at other times it is pointed more away from the Sun. When Earth's North Pole is tilted toward the Sun, an observer on Earth sees the Sun as lying *north* of the celestial equator. Six months later, when Earth's axis is tilted away from the Sun, the Sun is seen as lying *south* of the celestial equator. If we look at the circle of the Sun's apparent path through the stars—the ecliptic—we see that it is tilted by 23.5° with respect to the celestial equator.

To understand the effect that this has on Earth, begin by looking at **Figure 2.16(a)**. This shows the situation on June 21, the day that Earth's North Pole is tilted most directly toward the Sun.[5] Note first that the Sun is north of the celestial equator. We found earlier in the chapter that from the perspective of an observer in the Northern Hemisphere, an object north of the celestial equator can be seen above the horizon for more than half the time. This is true for the Sun as well as for any other celestial object. Saying that the Sun is above the horizon for more than half the time is just another way of saying that the days are longer than 12 hours. You can see this directly in Figure 2.16(a) by noting that when Earth's North Pole is tilted toward the Sun, over half of Earth's Northern Hemisphere is illuminated by sunlight. These are the long days of the northern summer. Six months later, on December 22, the situation is very different. On December 22 (**Figure 2.16(b)**) Earth's North Pole is tilted away from the Sun, so the Sun appears in the sky south of the celestial equator. Someone in the Northern Hemisphere

[5] We are a bit sloppy with language here. Earth's North Pole tilts in the same direction year-round. On the first day of the northern summer, Earth is on the side of the Sun where the tilt of the North Pole is toward the Sun. On the first day of the northern winter, Earth is on the opposite side of the Sun, so the tilt of the North Pole is away from the Sun.

will see the Sun for less than 12 hours each day. Less than half of the Northern Hemisphere is illuminated by the Sun. It is winter in the north.

Over the course of the year the length of the day changes, courtesy of the 23.5° tilt of Earth's axis relative to the perpendicular to the plane of its orbit. In the preceding paragraph we were careful to specify the length of the day in the *Northern* Hemisphere because things are very different in the Southern Hemisphere. In fact, things in the Southern

Seasons in the Southern Hemisphere are the reverse of those in the Northern Hemisphere.

Hemisphere are exactly reversed from what is going on in the north. Look again at Figure 2.16. On June 21, while the Northern Hemisphere is enjoying long days and short nights, Earth's South Pole is tilted in the direction away from the Sun. Less than half of the Southern Hemisphere is illuminated by the Sun, and days are shorter than 12 hours. Similarly, on December 22 Earth's South Pole is tilted toward the Sun, and the southern days are long.

The differing length of days through the year is part of the explanation for the changing seasons, but we need to consider another important effect. The Sun appears higher in the sky during the summer than it does during the winter, so sunlight strikes the ground *more directly* during the summer than during the winter. To see why this is important, hold a piece of cardboard toward the Sun and look at the size of its shadow. If the cardboard is held so that it is

The angle of sunlight to the ground is closer to perpendicular in summer than in winter, so there is more heating per unit area in summer.

directly face-on to the Sun, then its shadow is large. However, as you turn the cardboard more edge-on, the size of its shadow shrinks. The size of the cardboard's shadow tells you that the cardboard catches less energy from the Sun each second when it is tilted relative to the Sun than it does when it is face-on to the Sun. This is exactly what happens with the changing seasons. During the summer Earth's surface is more nearly face-on to the incoming sunlight, so more energy falls on each square meter of ground each second. During the winter the surface of Earth is more inclined with respect to the sunlight, so less energy falls on each square meter of the ground each second. That is the main

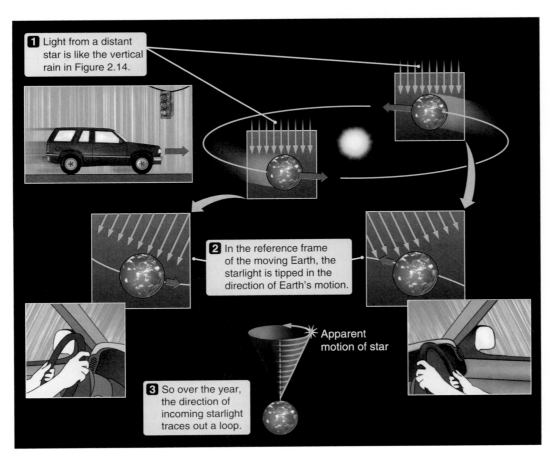

FIGURE 2.15 The apparent positions of stars are deflected slightly toward the direction in which Earth is moving. As Earth orbits the Sun, stars appear to trace out small ellipses in the sky. This effect is called aberration of starlight.

reason why it is hotter in the summer and colder in the winter.

Look at **Figure 2.17**, which shows the direction of incoming sunlight striking Earth at the border between Kansas and Nebraska (40° north latitude). At noon on the first day of summer the Sun is high in the sky—73.5° above the horizon and only 16.5° away from the zenith. Sunlight strikes the ground almost face-on. In contrast, at noon on the first day of winter, the Sun is only 26.5° above the horizon, or 63.5° from the zenith. Sunlight strikes the ground at a rather shallow angle. The difference between the two cases is important. At the Kansas–Nebraska border, over twice as much solar energy falls on each square meter of ground per second at noon on June 21 as falls there at noon on December 22. Together these two effects—the directness of sunlight and the differing length of the day—mean that during the summer there is more heating from the Sun and during the winter there is less heating from the Sun.

We do not have to wait for the seasons to change to see the effect that the height of the Sun in the sky has on terrestrial climate. We need only compare the climates found at different latitudes on Earth. Near the equator the Sun passes high overhead every day, regardless of the season. As a result, the climate is warm throughout the year. At high latitudes, however, the Sun is *never* high in the sky, and the climate can be cold and harsh even during the summer.

Four Special Days Mark the Passage of the Seasons

The apparent path that the Sun follows through the stars each year (the ecliptic) is a great circle that is tilted 23.5° with respect to the celestial equator. Follow along in **Figure 2.18** as we note the four special points on this path that

FIGURE 2.16 (a) On the first day of the northern summer (June 21, the summer solstice), the northern end of Earth's axis is tilted most nearly toward the Sun, while the Southern Hemisphere is tipped away. Seasons are opposite in the Northern and Southern Hemispheres. (b) Six months later, on the first day of the northern winter, the situation is reversed.

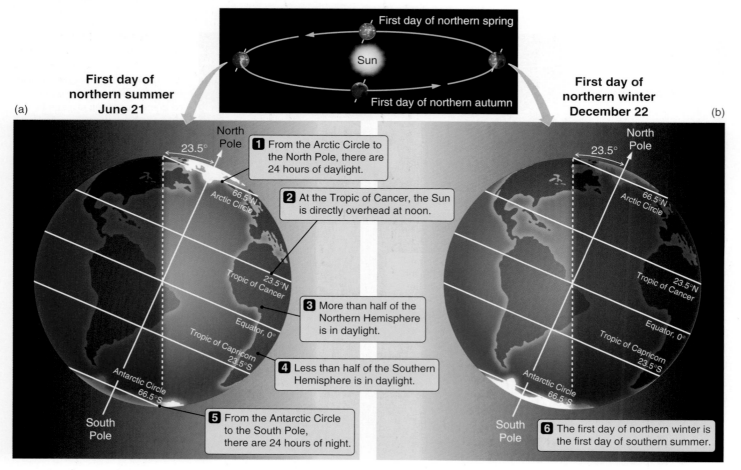

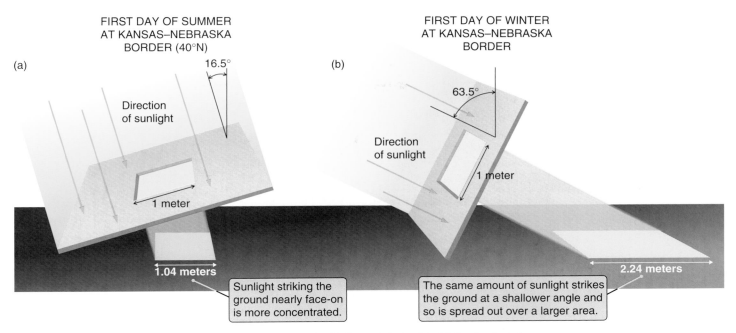

FIGURE 2.17 Local noon at the Kansas–Nebraska border (40° north latitude). (a) On the first day of northern summer, sunlight strikes the ground almost face-on. (b) At local noon on the first day of northern winter, sunlight strikes the ground more obliquely, and less than half as much sunlight falls on each square meter of ground each second.

mark the passage of the seasons. Begin with the point in March when Earth's axis is perpendicular to the direction to the Sun. This is the point where the Sun's apparent motion along the ecliptic crosses the celestial equator moving from the south to the north. That direction in the sky, located in the constellation Pisces, is called the **vernal equinox**. The term *vernal equinox* also refers to the day—about March 21—when the Sun appears at this location. When

The changing seasons are marked by equinoxes and solstices.

the Sun lies on the celestial equator on the vernal equinox, days are 12 hours long. (The term **equinox** means literally "equal night": Everywhere on Earth, night and day are the same length on the days of the equinoxes.) In the Northern Hemisphere, this is the first day of spring.

As Earth continues its journey around the Sun, the direction to the Sun moves into closer alignment with the tilt of the northern end of Earth's axis, and the Sun climbs higher into the northern sky. The Sun is lined up with the tilt of Earth's axis about three months after the vernal equinox. When this happens, the Sun reaches its northernmost point in the sky, located in the constellation Taurus, near its border with Gemini. This day, which marks the longest day of the year and the beginning of summer in the Northern Hemisphere, occurs around June 21. This is the **summer solstice**. (**Solstice** literally means "sun standing still": The

Sun's north–south motion stops as it reverses its direction.) Note that on this same day, the southern end of Earth's axis is tipped directly away from the Sun. This is the shortest day of the year in the Southern Hemisphere, marking the beginning of the southern winter.

Three months later, around September 23, the Sun is again crossing the celestial equator. This point on the Sun's apparent path, located in the constellation Virgo, is called the **autumnal equinox**. *Autumnal equinox* refers both to the location in the sky and the date when this happens. Because the Sun is on the celestial equator, days and nights are exactly 12 hours long. It is the first day of autumn in the Northern Hemisphere and the first day of spring in the Southern Hemisphere.

Around December 22 the Sun reaches its southernmost point in the sky as its apparent path takes it through the constellation Sagittarius. This day is called the **winter solstice**. Earth's North Pole is tipped most directly away from the Sun on this day. In the Northern Hemisphere this is the shortest day of the year—the first day of winter. As the Sun passes the winter solstice and moves on toward the vernal equinox, the northern days begin growing longer again. Almost all cultural traditions in the Northern Hemisphere include some major celebration in late December (**Figure 2.19**). Christmas, for example, is celebrated just three days after the winter solstice. These winter festivals have many different meanings to their various celebrants, but they all share one thing: They celebrate the return of the source of

Motion of Earth around the Sun

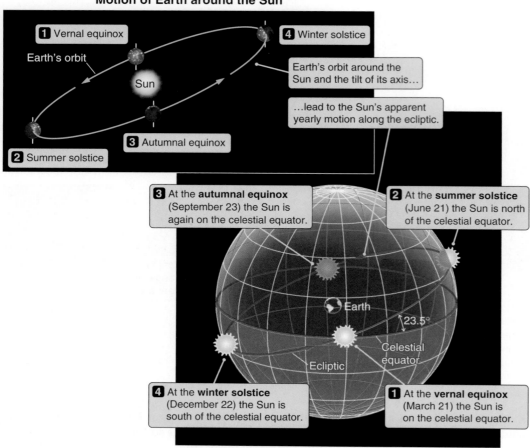

Apparent motion of the Sun seen from Earth

FIGURE 2.18 The motion of Earth about the Sun as seen from the frame of reference of the Sun (upper panel) and Earth (lower panel).

Earth's light and warmth. The days have stopped growing shorter and are beginning to get longer. It is a "new year," and once again the Sun will hold sway over night. Spring will come again.

As an aside, it is interesting to note that there is more to the seasons we feel than the amount of energy we are receiving from the Sun. Just as it takes a pot of water on a stove time to heat up when the burner is turned up and time to cool off when the burner is turned down, it takes Earth time to respond to changes in heating from the Sun. The

Seasonal temperatures lag behind changes in the directness of sunlight.

hottest months of the summer are usually July and August, which come *after* the summer solstice, when the days are growing shorter. Similarly, the coldest months of winter are usually January and February, which occur *after* the winter solstice, when the days are growing longer. The climatic seasons on Earth lag behind changes in the amount of heating we are receiving from the Sun.

Our picture of the seasons must be modified somewhat near Earth's poles. At latitudes north of 66.5° north latitude and south of 66.5° south latitude, the Sun is circumpolar for a part of the year surrounding the first day of summer. These lines of latitude are called the **Arctic Circle** and the **Antarctic Circle**, respectively. When the Sun is circumpolar, it is above the horizon 24 hours a day, earning the arctic regions the nickname "land of the midnight Sun." The Arctic and Antarctic regions pay for these long days, however, with an equally long period surrounding the first day of winter when the Sun never rises and the nights are 24 hours long. The Sun never rises high in the Arctic or Antarctic sky, which means that sunlight is never very direct. This is why even with the long days at the height of summer, the Arctic and Antarctic regions remain relatively cool.

The seasons are also different near the equator. Recall from Figure 2.8(c) that for an observer on the equator, *all* stars are above the horizon 12 hours a day, and the Sun is no exception. On the equator, days and nights are 12 hours long throughout the year. Changes in the directness of sunlight through the year are also different on the equator. Here the

Sun passes directly overhead on the first day of spring and the first day of autumn because these are the days when the Sun is on the celestial equator. Sunlight is most direct on the equator on these days. At the summer solstice the Sun is at its northernmost point along the ecliptic. It is on this day, and on the winter solstice, that the Sun is *farthest* from the zenith at noon, and therefore sunlight is *least* direct. Strictly speaking, the equator experiences only two seasons: summer, when the Sun passes directly overhead, and winter, when the Sun is at its northernmost and southernmost points on the ecliptic. However, summer and winter are not very different. The Sun is always up for 12 hours a day, and the Sun is always so close to being overhead at noon that the directness of sunlight changes by only 8 percent throughout the year.

If you live between the latitudes of 23.5° south and 23.5° north, twice during the year the Sun will be directly overhead at noon. This band is called the **Tropics**. The northern limit of this region is called the Tropic of Cancer. The southern limit is called the Tropic of Capricorn. (As a challenge, think about what the seasons are like at different locations within the Tropics.)

Earth's Axis Wobbles, and the Seasons Shift through the Year

After our discussion of the apparent motion of the Sun along the ecliptic, you might wonder about the names given to the Tropics. The summer solstice is located in the constellation Taurus, so why do we call the northern tropic the tropic of Cancer rather than the tropic of Taurus? Similarly, why is the southern tropic not referred to as the tropic of Sagittarius since the Sun is seen in that constellation on the winter solstice? The answer has to do with the fact that Earth's axis

FIGURE 2.19 Most cultural traditions in the Northern Hemisphere include a major celebration in late December, around the time when days begin to grow longer.

(a)

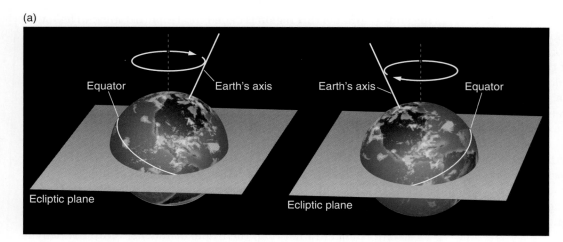

FIGURE 2.20 (a) Earth's rotation axis precesses like a spinning top. (b) Precession causes the projection of the Earth's rotation axis to move in a 47° diameter circle with a period of 25,800 years. The red cross shows the location of the projection of the axis in the early 21st century.

(b)

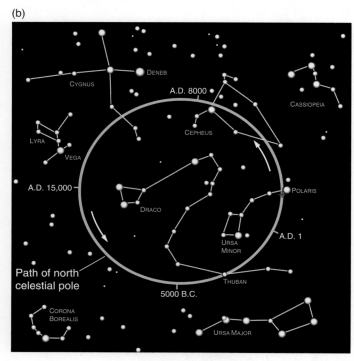

the celestial equator wobbles, the locations where it crosses the ecliptic—the equinoxes—change as well. During each 26,000-year wobble of Earth's axis, the locations of the equinoxes make one complete circuit around the celestial equator. This phenomenon is called the **precession of the equinoxes**. Ptolemy and his cohorts were formalizing their

---

A 26,000-year wobble causes the position of the equinoxes to slowly precess.

---

knowledge of the positions and motions of objects in the sky two thousand years ago. At that time the Sun *was* in the constellation of Cancer on the first day of northern summer and it *was* in the constellation of Capricorn on the first day of northern winter—hence the names of the Tropics.[6]

The tendency of the seasons to shift through the year from century to century has played havoc with human efforts to construct reliable calendars, and history is full of interesting anecdotes related to this difficulty. For example, for much of the 16th, 17th, and 18th centuries the calendars in Protestant Europe lagged behind the calendars in Catholic Europe by first 10, and later 11, days. It was not until 1752 that England and her colonies, including those in America, dropped 11 days from their calendars to bring them into line. This conciliatory step was met by riots among the people, who somehow felt that these 11 days had been stolen from them. Today's calendar (which was not adopted in Russia until the 1917 Bolshevik Revolution) is based on the **tropical year**, which is 365.242199 **solar days** long. The tropical year measures the time from one vernal equinox to the next—from the start of spring to the start of spring. Notice that the tropical year is not an integral number of days long. An elaborate system of **leap years** is used in our calendar

wobbles like the axis of a spinning top (**Figure 2.20**). The wobble is very slow, taking about 26,000 years to complete one cycle. During this time the north celestial pole makes one trip around a large circle through the stars. Polaris is

---

Earth's axis wobbles like the axis of a top.

---

a modern name for the star we see near the north celestial pole. If you could travel several thousand years into the past or future, you would find that the point about which the northern sky appears to rotate is no longer near Polaris.

Recall that the celestial equator is the set of directions in the sky that are perpendicular to Earth's axis. As Earth's axis wobbles, then, so must the celestial equator. And as

---

[6] In fact, due to precession, the location of the summer solstice just passed from Gemini into Taurus in 1990. In about 600 years precession will cause the vernal equinox to pass from Pisces into Aquarius, marking the beginning of the "Age of Aquarius."

EXCURSIONS 2.2

Why Is It Surprising That A.D. 2000 Was a Leap Year?

Everyone knows that years that are divisible by 4 are leap years, and A.D. 2000 was no exception to that rule. Yet A.D. 2000 *was* a special case. To understand why, we need to take a look at the way our calendar is constructed, and the purpose that leap years serve.

It takes Earth 365.242199 days to travel from one vernal equinox to the next. This tropical year is *not* exactly equal to the 365 days we normally think of as making up a year. (And thanks to the precession of the equinoxes, is it also not equal to the time—365.256366 days—it takes for Earth to complete one orbit about the Sun!) For convenience we count years as starting at midnight on the morning of January 1, and ending 365 days later at midnight on the night of December 31. But what about that extra fraction of a day (0.24 plus a bit) that we have not accounted for? Here is where leap years come in. A true tropical year is 0.242199 days—or about ¼ day—longer than a 365-day year, which means that after 4 years these extra parts have added up to make about 1 extra day. So every 4 years we add the extra day back in, and February

gets a 29th day. If we did not add in this extra day, our calendar would slip by ¼ day each year, which means that the seasons would shift by not quite a month each century. If we did not correct for leap years, the first day of summer would come in July, then in August, then in September, and so on.

Even after we add a leap year, the calendar still has problems. A 365-day year is short by 0.242199 days, which is a bit *less* than ¼. If we keep adding an extra day *every* 4 years, then over time we would accumulate an average of 0.007801 extra days per year. In 400 years the calendar would have slipped by about 3 days. To fix this it is necessary to get rid of 3 days every 400 years. This is accomplished by making century years into common 365-day years, *except* for those century years that are divisible by 400—such as A.D. 2000—which remain leap years. With one slight further revision—making years divisible by 4,000 into common 365-day years—the modern **Gregorian calendar** now slips by only about one day in 20,000 years.

to make up for the extra fraction of a day, preventing the seasons from slowly "sliding through" the year (getting increasingly out of sync with the months): winter in December one year, in August in another. **Excursions 2.2** gives further interesting details on this topic.

2.4 The Motions and Phases of the Moon

The second most prominent object in the sky after the Sun is the Moon. Just as Earth orbits about the Sun, the Moon orbits around Earth. (Actually, Earth and the Moon orbit around each other and together they orbit the Sun, as we will see in Chapter 3.) In some respects the appearance of the Moon is constantly changing, but we begin our discus-

sion of the motion of the Moon by talking about an aspect of the Moon's appearance that does *not* change.

We Always See the Same Face of the Moon

The Moon constantly changes its lighted shape and position in the sky, but one thing that does *not* change is the face of the Moon that we see. If we were to go outside next week or next month or 20 years from now or 20,000 *centuries* from

The Moon rotates on its axis once each orbit around Earth.

now, we would still see the same side of the Moon as tonight. Because of this fact, there is a common misconception that the Moon does not rotate. The Moon *does* rotate on its axis—exactly once for each revolution that it makes about Earth.

Imagine walking around the Washington Monument while keeping your face toward the monument at all times (a reasonable thing to do—you want to get a good look at it). By the time you complete one circle around the monument, your head has turned completely around once. (When you were south of the monument, you were facing north; when you were west of the monument, you were facing east; and so on.) But someone looking at you from the monument would never have seen anything other than your face. The Moon does exactly the same thing, rotating on its axis once per revolution around Earth, always keeping the same face toward Earth, as shown in **Figure 2.21**. This phenomenon is referred to as the Moon's **synchronous rotation**. (The Moon's synchronous rotation is not an accident. In Chapter 10 we will find that its cause is related to why we have tides on Earth.)

The Changing Phases of the Moon

Sometimes the Moon appears as a circular disk in the sky. At other times it is nothing more than a thin sliver. At still others its face is dark. The Moon has no light source of its own. Like the planets, including Earth, it shines by reflected sunlight. Like Earth, half of the Moon is always in bright

---

The Moon shines by reflected sunlight.

---

daylight, and half of the Moon is always in darkness. The different **phases** of the Moon result from the fact that the illuminated portion of the Moon that we see is constantly changing. Sometimes (during a new Moon) the side facing away from us is illuminated, and sometimes (during a full

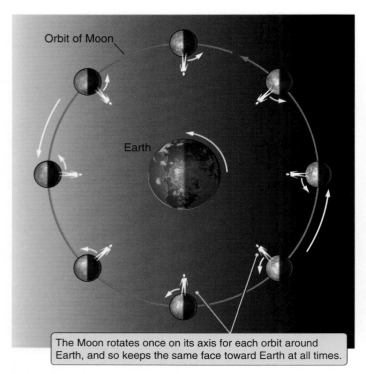

The Moon rotates once on its axis for each orbit around Earth, and so keeps the same face toward Earth at all times.

FIGURE 2.21 The Moon rotates once on its axis for each orbit around Earth, an effect called synchronous rotation.

Moon) the side facing toward us is illuminated. The rest of the time, only part of the illuminated portion can be seen from Earth (see **Excursions 2.3**).

To help you visualize the changing phases of the Moon, go outside at night and have a friend hold up a soccer ball so that it is illuminated from one side by a nearby streetlight (representing the Sun). Have your friend walk around you in a circle, and watch the changes in the ball's appearance.

EXCURSIONS 2.3

The Dark Side of the Moon

Popular culture often refers to the side of the Moon away from Earth as the "dark side of the Moon." This is even the title of a popular piece of 20th-century music. But in fact, there is no dark side of the Moon. At any given time, half of the Moon is in sunlight and half of the Moon is in darkness—just as at any given time, half of Earth is in sunlight and half of Earth is in darkness. The side of the Moon that faces away from Earth, the *far side,* spends just as much time in sunlight as the side of the Moon that faces toward Earth.

On the other hand, "I'll see you on the backside of the Moon" might not have been as wildly successful as a song lyric.

The phase of the Moon is determined by
how much of its bright side you can see.

When you are between the ball and the streetlight, the face
of the ball that is toward you is fully illuminated. The ball
appears to be a bright circular disk. As the ball moves
around its circle, you will see a progression of lighted
shapes, depending on how much of the bright side and how
much of the dark side of the ball you can see. This progression of shapes exactly mimics the changing phases of the
Moon.

Figure 2.22 shows the changing phases of the Moon.
When the Moon is between Earth and the Sun, the illuminated side of the Moon faces away from us, and we see only
its dark side. This is called a **new Moon**. Study Figure 2.22
to see that a new Moon can only be "seen" from the illuminated side of Earth. It appears close to the Sun in the sky,
and so it rises in the east at sunrise, crosses the meridian
near noon, and sets in the west near sunset. A new Moon
is never visible in the nighttime sky.

As the Moon continues on its orbit around Earth, a small
part of the portion being illuminated becomes visible. This
shape is called a **crescent**. Because the Moon appears to be
"filling out" from night to night at this time, the full name
for this phase of the Moon is a waxing crescent Moon. (**Waxing** here means "growing in size and brilliance.") From our
perspective the Moon has also moved away from the Sun
in the sky. Because the Moon travels around Earth in the
same direction in which Earth rotates, we now see the Moon
located to the east of the Sun. A waxing crescent Moon is
visible in the western sky in the evening, near the setting

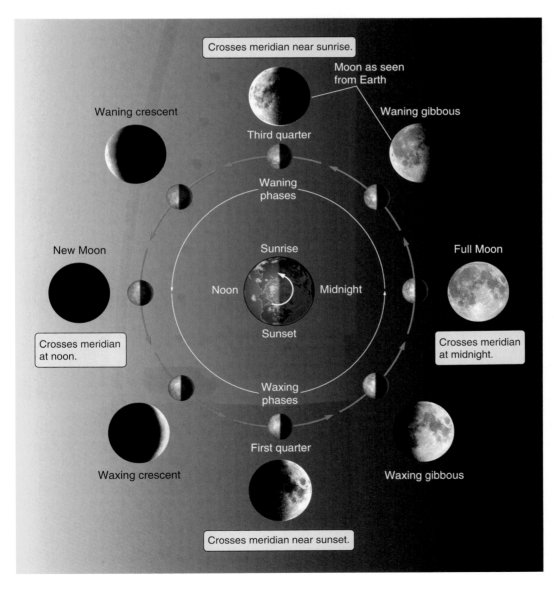

FIGURE 2.22 The inner
circle of images shows the
Moon as it orbits Earth, as
seen by an observer far above
Earth's North Pole. The outer
ring of images shows the
corresponding phases of the
Moon as seen from Earth.

Sun but remaining above the horizon after the Sun sets. The "horns" of the crescent always point away from the Sun.

As the Moon moves farther along in its orbit, more and more of its illuminated side becomes visible each night, so the crescent continues to fill out. At the same time the angular separation in the sky between the Moon and the Sun grows. After about a week the Moon has moved a quarter of the way around Earth. We now see half the Moon illuminated in daylight and half the Moon as dark—a phase that we call **first quarter Moon**. A look at Figure 2.22 shows that the first quarter Moon rises at noon, crosses the meridian at sunset, and sets at midnight. Note that *first quarter* refers not to how much of the face of the Moon that we see illuminated, but rather to the fact that the Moon has completed the first quarter of its cycle from new Moon to new Moon.

As the Moon moves beyond first quarter, we are able to see more than half of its bright side. This phase is called a waxing **gibbous Moon**. The gibbous Moon continues nightly to "grow" until finally Earth is between the Sun and the Moon and we see the entire bright side of the Moon—a **full Moon**. The Sun and the Moon now appear opposite each other in the sky. The full Moon rises as the Sun sets, crosses the meridian at midnight, and sets in the morning as the Sun rises.

The second half of the Moon's orbit proceeds just like the first half but in reverse. The Moon continues in its orbit, again appearing gibbous but now becoming smaller each night. This phase is called a waning gibbous Moon. (**Waning** means "becoming smaller.") A **third quarter Moon** occurs when we once again see half of the illuminated part of the Moon and half of the dark part of the Moon. A third quarter Moon rises at midnight, crosses the meridian near sunrise, and sets at noon. The Moon continues on, visible now as a waning crescent Moon in the morning sky, until the Moon again appears as nothing but a dark circle rising and setting with the Sun, and the cycle begins again.

It takes the Moon 27.32 days to complete one revolution about Earth; this is called its **sidereal** period. However, because of the changing relationship between Earth, the Moon, and the Sun due to Earth's orbital motion, it takes 29.53 days to go from one full Moon to the next; this is called its **synodic** period. (**Figure 2.23** shows how this works.) You can always tell a waxing from a waning Moon because the side that is illuminated is always the side facing the Sun. When the Moon is waxing, it appears in the evening sky, so its western side is illuminated (the right side as viewed from the Northern Hemisphere). Conversely, when the Moon is waning, the eastern side (the left side as viewed from the Northern Hemisphere) appears bright.

Do not try to memorize all of the possible combinations of where the Moon is in the sky at what phase and at what time of day. You do not have to. Instead, work on *understanding* the motion and phases of the Moon, then use your understanding to figure out the specifics of any given case. As a way to study the phases of the Moon, draw a picture like

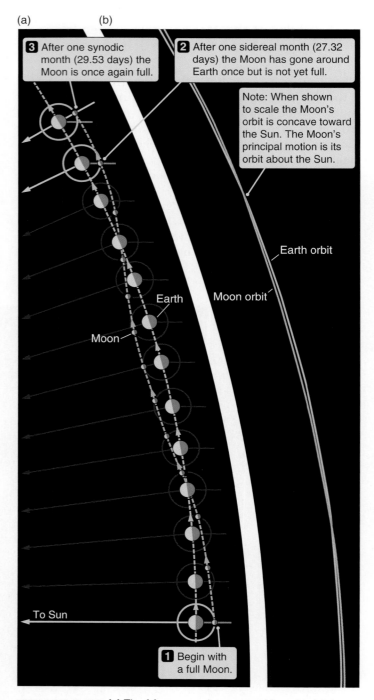

FIGURE 2.23 (a) The Moon completes one sidereal orbit in 27.32 days, but the synodic period (the period between phases seen from Earth) from one full Moon to the next is 29.53 days. (The orange line to the right of the Moon indicates a fixed direction in space.) (b) The orbits of Earth and the Moon are shown here to scale.

Figure 2.22, and use it to follow the Moon around its orbit. Figure out from the drawing what phase you would see and where it would appear in the sky at a given time of day. You might also enjoy thinking about what phase someone on the Moon would see when looking back at Earth.

2.5 Eclipses: Passing through a Shadow

Put yourself in the place of our nomadic friend from the opening section of this chapter. You are finely attuned to the patterns of the sky, and you view these patterns not as the inexorable consequences of physical law but as visible signs from the gods. Can you imagine any celestial event that would strike more terror in your heart than to look up and see the Sun, giver of light and warmth, being eaten away as if by a giant dragon? There is archeological evidence that our ancestors put great effort into trying to find the pattern of eclipses and thereby bring them into the orderly scheme of the heavens. For example, the ancient and massive stone artifact in the English countryside called Stonehenge, pictured in **Figure 2.24**, may have allowed its builders to predict when eclipses might occur. The lives and motivations of the builders of Stonehenge, 4,000 years dead, may be lost in antiquity. Yet how can we doubt their desire to exert some control over the terror of eclipses by learning a few of their secrets—and in the process assuring themselves that an eclipse did not mean that all was lost?

Varieties of Eclipses

The type of eclipse just described, in which Earth moves through the shadow of the Moon, is called a **solar eclipse**.

FIGURE 2.24 Stonehenge is an ancient artifact in the English countryside, used 4,000 years ago to keep track of celestial events.

Three different types of solar eclipses are possible: *total, annular,* and *partial.* To see why, begin by looking at the structure of the shadow of the Sun cast by a round object such as the Moon, as shown in **Figure 2.25**. An observer at point A could see no part of the surface of the Sun. This darkest, inner part of the shadow is called the **umbra**. If a point on Earth passes through the Moon's umbra, the Sun's light is

> There are three types of solar eclipses: partial, annular, and total.

totally blocked by the Moon. This is called a **total solar eclipse** (**Figure 2.26**). Now look instead at points B and C in Figure 2.25. From these points an observer can see one side of the disk of the Sun but not the other. This outer region, which is only partially in shadow, is the **penumbra**. If a point on the surface of Earth passes through the Moon's penumbra, the result is a **partial solar eclipse**, in which the disk of the Moon blocks the light from a portion of the Sun's disk.

In the third type of eclipse, called an **annular solar eclipse**, the Sun appears as a bright ring surrounding the dark

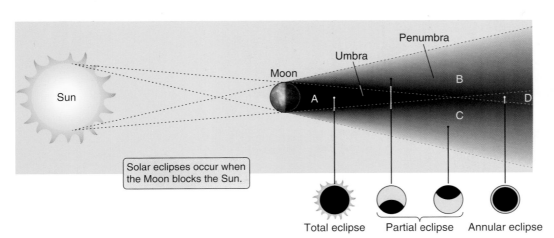

Solar eclipses occur when the Moon blocks the Sun.

Total eclipse   Partial eclipse   Annular eclipse

FIGURE 2.25 Different parts of the Sun are blocked at different places within the Moon's shadow. An observer in the umbra (A) sees a total solar eclipse, observers in the penumbra (B and C) see a partial eclipse, while observers in region D see an annular eclipse.

FIGURE 2.26 The full spectacle of a total eclipse of the Sun.

FIGURE 2.27 An annular eclipse, in which the Moon does not quite cover the Sun.

disk of the Moon (**Figure 2.27**). An observer at point D is far enough from the Moon that the Moon appears to be smaller than the Sun. You may be wondering how one eclipse can be total and another annular. Two things make this possible. One is a fluke of nature: The diameter of the Sun is about 400 times the diameter of the Moon, and the Sun is about 400 times farther away from Earth than the Moon is. As a result, the Moon and Sun have almost exactly the same apparent size in the sky. The other factor is that the Moon's orbit is not a perfect circle. So when the Moon and Earth are a bit closer together than average, the Moon appears larger in the sky than the Sun. An eclipse occurring at that time will be

total. When the Moon and Earth are farther apart than average, the Moon appears smaller than the Sun, so eclipses occurring during this time will be annular. Pictures of total and annular solar eclipses are shown in **Figure 2.28**. Among solar eclipses, one-third are total, one-third are annular, and one-third are seen only as partial eclipses.

Figure 2.29(a) shows a drawing of the geometry of a solar eclipse, with the Moon's shadow falling on the surface of Earth. It is important to realize that figures like this (or like Figures 2.21 or 2.22) are seldom drawn to scale. Instead they show Earth and the Moon much closer together than they are in reality. The reason for distorting figures this way

FIGURE 2.28 Time sequences of images of the Sun taken (a) during a total solar eclipse and (b) during an annular solar eclipse.

(a)

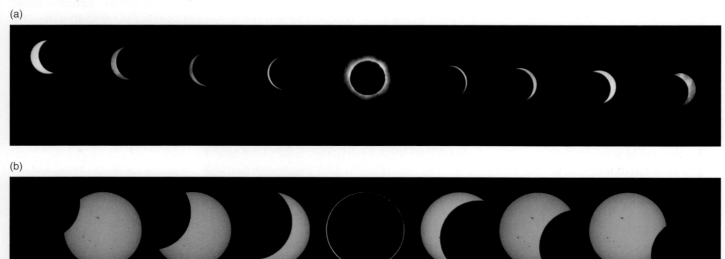

(b)

(a) Solar eclipse geometry (not to scale)

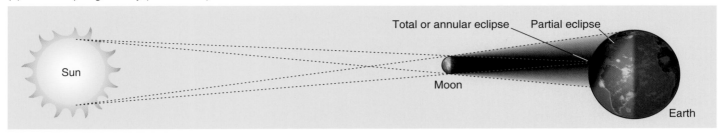

(b) Solar eclipse to scale

(c) Lunar eclipse geometry (not to scale)

(d) Lunar eclipse to scale

FIGURE 2.29 A solar eclipse occurs when the shadow of the Moon falls on the surface of Earth. A lunar eclipse occurs when the Moon passes through Earth's shadow. Note that (b) and (d) are drawn to proper scale.

is simple: There is not enough room on the page to draw them correctly and still keep the smaller details visible. The 384,400 km distance between Earth and the Moon is over 60 times the radius of Earth and over 220 times the radius of the Moon. The relative sizes and distances between Earth and the Moon are roughly like a quarter and a dime held 2 meters apart. **Figure 2.29(b)** shows the geometry of a solar eclipse with Earth, the Moon, and the separation between them drawn to scale. Compare this to Figure 2.29(a), and you will understand why artistic license is normally taken in drawings of Earth and the Moon. If the Sun were drawn to scale in Figure 2.29(b), it would be ⁶⁄₁₀ of a meter across and located almost 64 meters off the left side of the page.

The Moon's penumbra is quite large. In fact, with a bit of thought and a pencil and paper you can convince yourself that the Moon's penumbra, where it hits Earth, must have about twice the diameter of the Moon itself, or almost 7,000 km. This part of the shadow is large enough to cover

a substantial fraction of Earth, so partial solar eclipses are often seen from much of the planet. In contrast, the path along which a total solar eclipse can be seen (**Figure 2.30**) covers only a tiny fraction of the surface of Earth. Earth is so close to the tip of the Moon's umbra that even when the distance between Earth and the Moon is at a minimum, the umbra is only 269 km wide at the surface of Earth. As the Moon moves along in its orbit, this tiny shadow sweeps across the face of Earth at breakneck speed. The Moon moves in its orbit around Earth at a speed of about 3,400 km/h, and its shadow sweeps across the disk of Earth at the same rate. Earth is also rotating on its axis with a velocity of 1,670 km/h at the equator (and less than that at other latitudes). The situation is further complicated by the fact that the Moon's shadow falls on the curved surface of Earth. You may have noticed that the image projected by an overhead projector is distorted when the beam is not perpendicular to the screen. Similarly, the curvature of Earth often causes the region

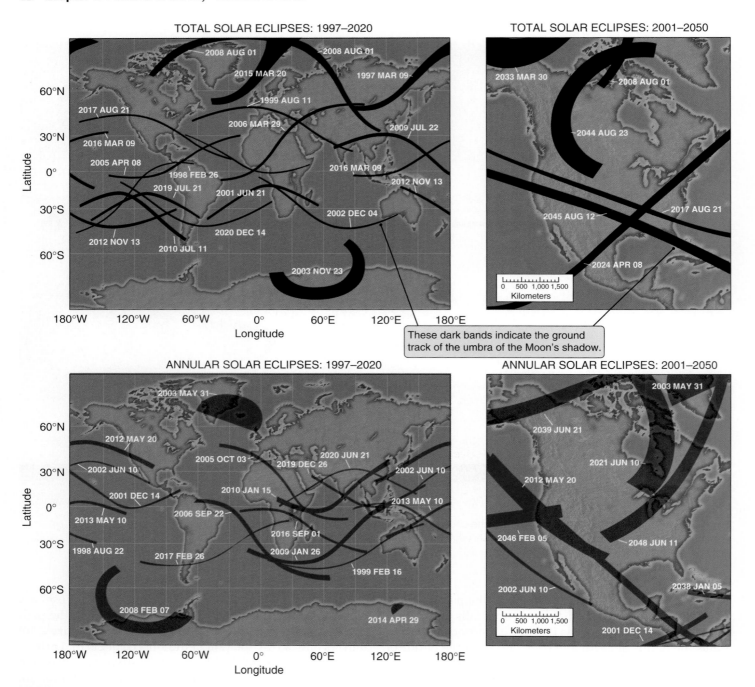

FIGURE 2.30 The paths of total and annular solar eclipses predicted for the late 20th and early 21st centuries.

shaded by the Moon during a solar eclipse to be elongated by differing amounts. The curvature can even cause an eclipse that started out as annular to become total.

When all of these effects are considered, the result is that a total solar eclipse can never last longer than 7½ minutes and is usually significantly shorter. Even so, it is one of the most amazing and awesome sights in nature. People flock from the world over to the most remote corners of Earth to witness the fleeting spectacle of the bright disk of the Sun blotted out of the daytime sky, leaving behind the eerie glow of the Sun's outer atmosphere.

Lunar eclipses are very different in character from solar eclipses. The geometry of a lunar eclipse is shown in **Figure 2.29(c)** (and is shown drawn to scale in **Figure 2.29(d)**). Because Earth is much larger than the Moon, the dark umbra of Earth's shadow at the distance of the Moon is about

Lunar eclipses last much longer than solar eclipses.

9,200 km in diameter, or over 2.5 times the diameter of the Moon. A **total lunar eclipse** is a much more leisurely affair than a total solar eclipse, with the Moon spending as long as 1 hour and 40 minutes in the umbra of Earth's shadow. A **penumbral lunar eclipse** occurs when the Moon passes through the penumbra of Earth's shadow. A penumbral eclipse can be unspectacular: Its appearance from Earth is nothing more than a fading in the brightness of the full Moon. Although the penumbra of Earth is 16,000 km across at the distance of the Moon—over four times the diameter of the Moon—a penumbral eclipse is noticeable only when the Moon passes within about 1,000 km of the umbra.

Eclipse Seasons Occur Roughly Twice Every 11 Months

If the Moon's orbit were in exactly the same plane as the orbit of Earth (imagine Earth, the Moon, and the Sun all sitting on the same flat tabletop), then the Moon would pass directly between Earth and the Sun at every new Moon. The Moon's shadow would pass across the face of Earth, and we would see a solar eclipse. Similarly, Earth would pass directly between the Sun and the Moon every synodic month, and each full Moon would be marked by a lunar eclipse.

Solar and lunar eclipses do *not* happen every month because the Moon's orbit does not lie in exactly the same plane as the orbit of Earth. Look at **Figure 2.31** to see how this works. The plane of the Moon's orbit about Earth is inclined by about 5.2° with respect to the plane of Earth's orbit about the Sun. The line along which the two orbital planes intersect is called the **line of nodes**. For part of the year, the line of nodes passes close to the Sun. During these times, called **eclipse seasons**, a new Moon passes between the Sun and Earth, casting its shadow on Earth's surface and causing a solar eclipse. Similarly, a full Moon occurring during an eclipse season passes through Earth's shadow, and a lunar eclipse results. An eclipse season lasts for only 38 days. That is how long the Sun is close enough to the line of nodes for eclipses to occur. Most of the time, the line of nodes points farther away from the Sun, and Earth, Moon, and Sun cannot line up closely enough for an eclipse to occur. A solar eclipse cannot be seen because the shadow

FIGURE 2.31 Eclipses are only possible when the Sun, Moon, and Earth lie along a line. When the Sun does not lie along the line of nodes, Earth passes under or over the shadow of a new Moon, and a full Moon passes under or over the shadow of Earth.

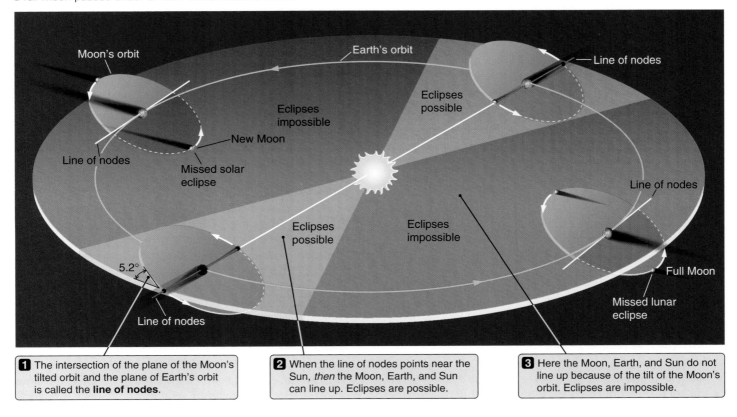

1 The intersection of the plane of the Moon's tilted orbit and the plane of Earth's orbit is called the **line of nodes**.

2 When the line of nodes points near the Sun, *then* the Moon, Earth, and Sun can line up. Eclipses are possible.

3 Here the Moon, Earth, and Sun do not line up because of the tilt of the Moon's orbit. Eclipses are impossible.

of a new Moon passes "above" or "below" Earth. Similarly, no lunar eclipse can be seen because a full Moon passes "above" or "below" the shadow of Earth.

If the plane of the Moon's orbit always had the same orientation, then eclipse seasons would occur twice a year, as suggested by the drawing in Figure 2.31. In actuality, eclipse seasons occur about every 5 months and 20 days. The roughly 10-day difference is due to the fact that the plane of the Moon's orbit slowly wobbles, much like the wobble of a spinning plate balanced on the end of a circus performer's stick. As it does so, the line of nodes changes direction. This wobble rotates in the direction opposite the direction of the motion of the Moon in its orbit. (That is, the line of nodes moves clockwise as viewed from the north.) It takes the Moon's orbit 18.6 years to complete one "wobble" of 360°, so we say that the line of nodes *regresses* by about 360°/18.6 years, or 19.4° per year. This amounts to about a 20-day regression each year. If January 1 marks the middle of an eclipse season, the next eclipse season would be centered around June 20, and the one after that around December 10.

We have come far since our nomadic ancestor looked at the sky and saw there a mystical reflection of the patterns and events that marked the life of the tribe. Yet when we look at the sky today our sense of awe and majesty is no less than that experienced by that long-ago tribesman. Our distant ancestors had to look for patterns in the world to survive. This same human impulse to seek patterns led astrologers to look for connections between ourselves and the heavens, and later led people to seek patterns that gave birth to science. We now know that the patterns of the sky are connected to us much more directly than any mystical link invented by an astrologer. The patterns and changes that we see in the sky are caused by the same forces of nature that bind us to our planet and that cause the wind to blow and the rain to fall. They are the same forces that push the blood through our veins and carry the electric impulses of thought through our brains. So far in our look at the changing patterns of the sky, we have only glimpsed these connections, so it is to these underlying causes that we now turn our attention.

Summary

- Stars appear to move through the sky as Earth rotates on its axis.

- The specific stars we see at night depend on where Earth is in its orbit around the Sun.

- The Coriolis effect causes hurricanes to rotate.

- The tilt of Earth's axis determines the seasons.

- The motion of the Moon in its orbit around Earth shapes the phases of the Moon.

- The phase of the Moon is determined by how much of its bright side you see.

- Special alignments of the Sun, Earth, and Moon result in solar and lunar eclipses.

Seeing the Forest through the Trees

Patterns in the sky change in lockstep with the cycles of life on Earth. It is no surprise that our ancestors sought some link between the stars and the events in their lives, but early astrologers were ultimately limited by their misconceptions about the universe. Earth does not reside at the center of Creation, nor does the Sun move from "house" to "house" along the ecliptic. The eerie spectacle of a solar eclipse blotting out the Sun is no more mystical than the shadow you cast on the sidewalk on a sunny afternoon. The heavens are a place not of magic but of physical law, knowable though observation and experiment and subject to test by the scientific method. The signs of the zodiac that grace the entertainment section of your local newspaper and the covers of supermarket tabloids are anachronisms—pictures born from the human imagination and painted onto the random splash of stars across the night sky.

Yet while astrologers were off the mark in their conclusions, their quest was well motivated. The motions of Earth and the properties of the Sun do, indeed, give rise to the most basic of all of the patterns faced by life on Earth. Earth's rotation is responsible for the coming of night and day. Earth's axial tilt and its passage around the Sun bring the changing of the seasons. Changes in the direction in which sunlight falls on Earth cause dramatic differences in climate from the equator to the poles. These patterns set the stage for the evolution of life on Earth, and they remain with us today, buried deep within the genetic code of our species. Even as humans begin to venture beyond Earth and into space, we carry these patterns with us in everything from the length of our cycle of waking and sleeping, to the temperature we prefer, to the amount of light we need in order to see. Much of human culture also has its roots in the apparent motions of celestial objects. Many of our legends and traditions arose in the ancient view of the sky as a place of gods and spirits. At the same time, patterns of objects in the sky spurred the development of mathematics and, as we will see in Chapter 3, the development of a physical understanding of the world around us. Much of who and what we are as a thinking species has its origins in our experience of the larger universe.

The errors in perception that shaped our views of the universe throughout most of human history are both understandable and forgivable. It is remarkably difficult to directly sense the effects of Earth's motion. As you read this, you are probably (depending on your latitude) moving at more than 1,000 km/h on a circular path around Earth's axis, while Earth itself is moving at over 100,000 km/h in its orbit about the Sun. (And the Sun itself is moving through our galaxy at almost 1 million km/h.) Yet you feel none of this motion. Objects around you share your motion, and so to you are motionless. The telltale signs of your motion are far too subtle to perceive directly. As you watch the Sun and stars cross the sky, it certainly seems as if it is you about which the cosmos revolves. In fact, without the benefit of discussions such as ours, it would be difficult to avoid that impression.

When we replace simple perception with careful experiment and reason, however, this commonsense notion evaporates before our eyes. Minute changes in the direction of starlight through the year provide proof of Earth's motion in its orbit, and they allow us to measure our path around the Sun. The behavior of a simple pendulum hung from the ceiling of a dome in Paris, when carefully considered, provides direct evidence of Earth's rotation about its axis. And as our physical understanding of Earth's motion improves, even the blowing of the wind and the need for a fire on a cold, dark winter night become side effects of Earth's motions through space.

In our journey we will again encounter many of the motions discussed in this chapter, but from a rather different perspective. So far we have concentrated on *describing* the motion of Earth about the Sun and the Moon about Earth. From here we will take the step that separates post-Renaissance science from all that came before by taking up the question of *why* these motions are as they are. The search for understanding will lead us to a discovery that changed our perception of the universe and our place in it—that the same natural forces that dictate the path of a well-hit baseball also govern the clockwork motions of Earth, the Moon, and planets.

Key Terms

constellations, p. 23
meridian, p. 24
zenith, p. 24
celestial poles, north and south, pp. 24, 25
latitude, p. 25
equator, p. 25
Polaris (north star), p. 27
circumpolar, p. 28
celestial sphere, p. 28
celestial equator, p. 29
Foucault pendulum, p. 30
relative motions, p. 32
Coriolis effect, p. 33
cyclonic motion, p. 34
year, p. 35
great circle, p. 35
ecliptic, p. 35
zodiac, p. 35
frame of reference, p. 36
aberration of starlight, p. 36
astronomical unit, p. 37
obliquity, p. 38
equinoxes, vernal and autumnal, p. 41
solstices, summer and winter, p. 41
precession of the equinoxes, p. 44
tropical year, p. 44
solar day, p. 44
leap year, p. 44
synchronous rotation, p. 46
phases, p. 46
sidereal, p. 48
synodic, p. 48
solar eclipse, p. 49
umbra, p. 49
total solar eclipse, p. 49
penumbra, p. 49
partial solar eclipse, p. 49
annular solar eclipse, p. 49
total lunar eclipse, p. 53
penumbral lunar eclipse, p. 53
line of nodes, p. 53
eclipse seasons, p. 53

Student Questions

THINKING ABOUT THE CONCEPTS

1. What is the approximate time of day when you see the full Moon near the meridian? At what time is the first quarter (waxing) Moon on the eastern horizon? Use a sketch to help explain your answers.

2. The Coriolis *effect* is sometimes wrongly called the Coriolis *force* because it can make a cannonball appear to deviate from its projected path. Yet the cannonball truly follows its projected path. Explain the illusion and how it appears to make the projected path deviate.

3. A soldier fires a cannon directly at a distant target toward the east and makes a perfect hit. She then fires a shot directly at another distant target toward the north and finds that her shot has hit to the west of the target. Was the soldier in Australia or Canada? Explain.

4. We tend to associate certain constellations with certain times of year. For example, we see the zodiacal constellation Gemini in the Northern Hemisphere's winter (Southern Hemisphere's summer) and the zodiacal constellation Sagittarius in the Northern Hemisphere's summer. Why do we not see Sagittarius in the Northern Hemisphere's winter (Southern Hemisphere's summer) or Gemini in the Northern Hemisphere's summer?

5. The tilt of Jupiter's rotational axis is 3°. If Earth's axis had this tilt, explain how it would affect our seasons.

6. Why do we not see a lunar eclipse each time the Moon is full or witness a solar eclipse each time the Moon is new?

7. Assume that the Moon's orbit is circular. Suppose you are standing on the side of the Moon that faces Earth. How would Earth appear to move in the sky as the Moon made one revolution around Earth? How would the "phases of Earth" appear to you, as compared to the phases of the Moon as seen from Earth?

8. Sometimes artists paint the horns of the crescent moon pointing toward the horizon. Is this realistic? Explain.

9. The true length of a year is not 365 days, but is actually about 365¼ days. How do we handle this extra quarter day to keep our calendars from getting out of sync?

10. Many cities have main streets laid out in east–west, north–south alignments. Why are there frequent traffic jams on east–west streets during both morning and evening rush hours within a few weeks of the equinoxes? Considering this, if you work in the city during the day, would you rather live east or west of the city?

11. What is the advantage of launching satellites from spaceports located near the equator? Why are satellites never launched in a westerly direction?

12. Does the occurrence of solar and lunar eclipses disprove the notion that the Sun and the Moon both orbit around Earth? Explain your reasoning.

APPLYING THE CONCEPTS

13. The Moon's orbit is tilted by about 5° relative to Earth's orbit around the Sun. What is the highest altitude in the sky that the Moon can reach as seen in Philadelphia (latitude 40°)?

14. Assume that you and your sister can both throw a baseball at a speed of 100 km/h. If you are in an airplane traveling at 800 km/h and play catch with your sister who is near the front of the plane (you being toward the rear), how fast would the ball be traveling as seen by an observer on the ground when thrown (a) by you and (b) by your sister? How fast does it appear to move as seen by you and your sister?

15. You are out in open ocean sailing from the Carolinas to Bermuda (that is, heading due east) and you know there is a hurricane nearby. A strong wind is blowing straight out of the south. Do you continue on to Bermuda or head back? To arrive at your answer, draw a diagram of the Coriolis effect acting on air circulating around the center of a hurricane.

16. Using an atlas, determine the latitude where you live. Draw and label a diagram showing that your latitude is the same as (a) the altitude of the north celestial pole and (b) the angle (along the meridian) between the celestial equator and your local zenith. What is the noontime altitude of the Sun as seen from your home at the times of winter solstice and summer solstice?

17. Solar (and lunar) eclipses can occur only when the Moon is near one of its nodes and is new (or full) during its orbit around Earth. The time between successive passages of the Moon through the same node is 27.2122 days. The period between successive new Moons is 29.5306 days. Show that 223 successive new Moons (223 lunar months) is the same interval (to within one part in 250,000) as 242 successive crossings of the same node, and that this is very close to 18 years in length. It seems likely that the builders of Stonehenge knew about this cycle some 4,000 to 5,000 years ago.

18. Assume that rain is falling at a speed of 5 meters per second (m/s) and you are driving along in the rain at a leisurely 5 m/s (or 18 km/h). Estimate the angle from the vertical at which the rain appears to be falling.

19. Suppose the tilt of Earth's equator relative to its orbit were 10° instead of 23.5°. At what latitudes would the Arctic and Antarctic Circles and the two Tropics be located?

20. Assume Earth is a perfect sphere with a radius of 6,400 km. What is the distance at Earth's equator that corresponds to 1° of longitude? What would be the distance corresponding to 1° of latitude? Estimate the dis-

tance corresponding to 1° of longitude at latitudes of 30° and 60°. What would be the distance of 1° of latitude at these same latitudes?

21. The vernal equinox is now in the zodiacal constellation of Pisces. Wobbling of Earth's axis will eventually cause the vernal equinox to move into Aquarius, beginning the legendary, long-awaited "Age of Aquarius." How long, on average, does the vernal equinox spend in each of the 12 zodiacal constellations?

StudySpace
wwnorton.com/astro21
provides a Study Plan for each chapter that includes a reading outline, animations, keyword flash cards, and gradebook-enabled multiple-choice quizzes. From StudySpace you can also access premium content in the ebook and SmartWork.

The Newtonian principle of gravitation is now more firmly established, on the basis of reason, than it would be were the government to step in, and to make it an article of necessary faith. Reason and experiment have been indulged, and error has fled before them.

THOMAS JEFFERSON (1743–1826)

Sir William Herschel, Sir Isaac Newton, and Johannes Kepler.

CHAPTER 3

Gravity and Orbits— A Celestial Ballet

3.1 Gravity!

Today it is a rare person who does not know that the planets, including Earth, orbit around the Sun. Yet this was not always so. Only 500 years or so have passed since a soft-spoken Polish monk named **Nicholaus Copernicus** (1473–1543) started a revolution when he revived the idea, discarded by the Greeks two thousand years earlier, that the Sun rather than Earth lies at the center of creation. At the time this suggestion seemed outlandish. To think that Copernicus wanted us to believe that humankind—the pinnacle of creation—resides anywhere but at the center of all things! To imagine that we occupy but one of several planets circling the Sun's central fire, that we are nothing but a pebble among all the pebbles on the beach—absurd!

It would be easy from our "modern" perspective to chuckle at the naiveté of our ancestors and their Earth-centered view of the universe, but to do so would be unfair. The previous chapter showed that hard evidence of the motions of Earth was remarkably difficult to come by. To an ancient scholar, educated in Greek and Roman philosophy, Copernicus's view of the universe simply made no sense. Copernicus knew quite well that his ideas flew in the face of authority and would not be welcomed by the powers that be. Wishing to avoid the controversy his theory would certainly cause, Copernicus chose not to publish his ideas until late in his life. His great work *De revolutionibus orbium coelestrium* (On the Revolution of the Celestial Spheres) did not appear until the year of his death. This work pointed the way toward our modern cosmological principle. Copernicus knew his ideas would be unpopular, but he had no way of guessing the consequences of what he had started. In retrospect we see that he knocked the bottom out from

KEY CONCEPTS

In this chapter we follow the story of the birth of modern science as humans discovered regular patterns in the motions of the planets, then went on to explain those patterns with fundamental physical laws. Along the way we will explore

- Empirical rules discovered by Kepler that describe the elliptical orbits of planets around the Sun.

- Theoretical physical laws discovered by Newton and Galileo that govern the motion of all objects.

- The use of proportionality to describe patterns and relationships in nature.

- The nature of scientific theories, the roles of empirical and theoretical science, and the difference between science and pseudoscience.

- How Newton's laws of motion and an inverse square law of gravitation combine to explain an orbit as one body falling freely around another.

- The cosmological principle that grew from the startling realization that the heavens and Earth are governed by the same physical laws.

- How theories lead to new knowledge, such as the way Newton's derivation of Kepler's third law is used to measure the masses of objects from observations of orbital motions.

under the house of cards representing humankind's view of the world around them, and the following centuries would see that house of cards slowly fall apart. The repercussions of Copernicus's insight not only would shape our understanding of the universe around us but would change the direction of the progress of human civilization itself.

Even today we need to avoid too much complacency about what we think we know. People often confuse knowing the name of something with actually *understanding* something. For example, we happily talk about spacecraft in orbit about Earth, or planets in orbit about the Sun, but remarkably few people understand what those words mean. When asked why astronauts float about the cabin of a spacecraft, most educated adults answer, "The spacecraft has escaped Earth's gravity." Yet the gravitational force acting on a space shuttle orbiting Earth is only slightly weaker than when the shuttle is sitting on the launchpad. In fact, were it not for Earth's gravity, the spacecraft would not orbit Earth at all!

Copernicus knew nothing of gravity, but his ideas inevitably led to it. As physicists and astronomers have come to better understand gravity, they have realized that in most respects it is gravity that holds the universe together. Our Solar System is a gravitational symphony. The Sun's gravity shapes the motions of the planets and every other object in its vicinity. These motions range from the almost circular orbits of some planets to the extremely elongated orbits of comets. A comet's orbit may carry it from tens of thousands of astronomical units[1] from the Sun to somewhere inside the orbit of Mercury and back out again. Within this grand symphony, subthemes arise. A chorus of particles orbiting the giant planets give rise to majestic systems of rings, which, in turn, play counterpoint to the gravitational ballet of the planets and their moons. The analogy between the motions of objects in the Solar System and the patterns of music is not new. **Johannes Kepler** (1571–1630), who will figure prominently in our story, titled his great work of 1619 *Harmonice mundi,* or "Harmonies of the World."

As our grasp of the universe has expanded, we have come to realize that our Solar System is but one gravitational opus in a far larger opera. Gravity binds stars into the colossal groups we call galaxies and slows the expansion of the universe. It is gravity that holds the planets and stars together and keeps the thin blanket of air we breathe close to the surface of the planet we live on. It is gravity that caused a vast interstellar cloud of gas and dust to collapse 4.5 billion years ago to form our Sun and Solar System. It is gravity that gives space and time their very shape. As we continue our journey outward through the cosmos, we will come to each of these ideas in turn, and time and time again we will find gravity at the center of our growing understanding.

3.2 An Empirical Beginning: Kepler's Laws Describe the Observed Motions of the Planets

Now that we have extolled the wonders of gravity and the role it plays in the universe, you might expect us immediately to discuss what gravity is and how it works. But the great minds that brought us to our modern understanding did not have the benefit of our 20–20 hindsight. All they could do was watch the motions of the planets over the course of months and years and puzzle over what they saw. With no way even to judge the distances to the planets, it is no wonder that it took humans thousands of years to begin to see the reality behind the celestial patterns before their eyes.

In Chapter 1 we painted a picture of science as a worldview in which nature is governed by physical laws and in which mathematical descriptions of these physical laws are used to explain natural phenomena. But how do scientists go about *discovering* these physical laws? When facing phenomena as complex and puzzling as the motions of the planets in the heavens, where can we find a toehold? Just as the wise sailor settles for any port in a storm, the wise scientist knows that when facing a complex and poorly understood phenomenon, there may be little choice but to turn directly to the information our senses provide. We carefully observe the phenomenon under study, systematically recording as much information as we can as accurately as possible. As our observations start piling up, we look for patterns in those observations and start trying to formulate rules that seem to describe what we have seen.

Imagine that you are a scientist from another world, setting foot for the first time on Earth. You notice right away that many interesting structures are sticking out of the ground on this unexplored planet. As you record your observations, you find that some of these structures are large, some are small, some spread out over the ground, some stick up in the air, and so on. But after a time you realize that almost all these structures are covered with some kind of appendage, and those appendages are green. So you form a descriptive rule about these objects (call them "plants"): Most have green appendages. You decide that green appendages must be fundamental to the nature of plants, so you begin to study what makes these appendages green. After a time you discover that the green appearance always comes

---

[1] Recall from Chapter 2 that an astronomical unit (abbreviated AU) is the average distance between the Sun and Earth.

from the same chemical substance, and when you study that chemical substance you find that it can absorb light and turn water and carbon dioxide into more complex organic molecules. You have discovered photosynthesis, the process responsible for powering the majority of life on Earth. *But you did not start out to discover photosynthesis. You started out noting that plants have green leaves.*

The quest to first note and then accurately describe patterns in nature is called **empirical science**. Empirical science often involves a great deal of creativity and not a small amount of pure (but educated) guesswork. Copernicus's theory that Earth and the planets move in circular orbits about the Sun is an example of empirical science. Copernicus did not understand *why* the planets move about the Sun, but he did realize that his Sun-centered picture provided a much *simpler* description of the observed motions of planets than a model with Earth at its center did. Copernicus's work was made great by the fact that he was able to see beyond the prejudice of his time and to think the unthinkable—that perhaps Earth is "merely" one planet among many.

Copernicus's work paved the way for another great empiricist, Johannes Kepler. Science has often benefited from unlikely and chancy collaborations, and Kepler's is one such story. Kepler, a mathematician who had studied the ideas of Copernicus, worked in 1600 as an assistant to **Tycho Brahe** (1546–1601), a firm believer in an Earth-centered universe. Although Tycho Brahe is described as anything but a pleasant individual, he was also one of the greatest observational astronomers of all time. Toiling away through long nights with primitive equipment, Tycho Brahe amassed a wealth of remarkably accurate observations of the positions of the planets over the course of decades. Kepler, using Tycho's Brahe's data, took the next major step toward understanding the motions of the planets. Working first primarily with Brahe's observations of Mars, Kepler was able to deduce three empirical rules that elegantly and accurately describe the motions of the planets. These three rules are now almost universally referred to as **Kepler's laws**.

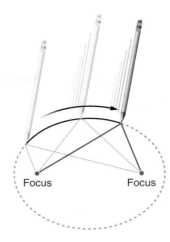

FIGURE 3.1 We can draw an ellipse by attaching a length of string to a piece of paper at two points (called foci), then pulling the string around as shown.

ler discovered that if he replaced Copernicus's circular orbits with elongated elliptical orbits and displaced these orbits to one side so that the Sun was no longer at the center, his calculations and Brahe's observations agreed almost perfectly.

You probably think of an ellipse as an oval shape, but to make sense of Kepler's discovery we need to be a bit more precise about what an ellipse is. The most concrete way to define an **ellipse** is to call it the shape that results when you attach the two ends of a piece of string to a piece of paper, stretch the string tight with the tip of a pencil, and then draw around those two points keeping the string taut (see **Figure 3.1**). Each of the points at which the string is attached is called a **focus** of the ellipse. The closer the two *foci* (plural of focus) are to each other, the more nearly circular the ellipse is. In fact, a circle is just an ellipse with the two foci at the same place. (To see this, just think about the shape you would draw if the two ends of the string were attached at the same spot. In this case, each half of the string would become a radius of the circle.) As the two foci are moved

> The Sun is at one focus of a planet's elliptical orbit.

farther apart, however, the ellipse becomes more and more elongated. Kepler found that the orbit of each planet is an ellipse with the Sun located at one focus. This result is now known as **Kepler's first law** of planetary motion. (You might ask, "If the Sun is located at *one* focus, what is at the other focus?" The answer is "nothing but empty space.")

Figure 3.2 shows a drawing of an ellipse. Half of the length of the long axis of the ellipse is called the **semimajor axis** of the ellipse, often denoted by the letter *A*. The semimajor axis of an orbit turns out to be a handy way to describe the orbit because, apart from being half the longer dimension of the ellipse, it is also the average distance between one focus and the ellipse itself. The average distance between the Sun and Earth, for example, equals the

Kepler's First Law: Planets Move on Elliptical Orbits with the Sun at One Focus

When Kepler used Copernicus's model to calculate where in the sky a planet should be at a particular time, he found quite a lot of disagreement between these predictions and

> Planetary orbits are ellipses.

Brahe's data. He was not the first to notice such discrepancies. Rather than discarding Copernicus's ideas, however, Kepler played with Copernicus's Sun-centered model. Kep-

Kepler's First Law

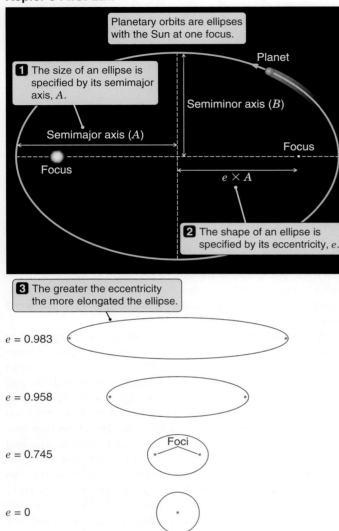

FIGURE 3.2 Planets move on elliptical orbits with the Sun at one focus. Ellipses range from circles to elongated eccentric shapes.

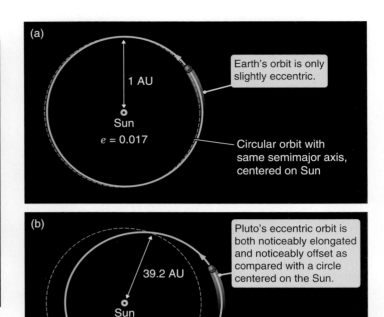

FIGURE 3.3 The shapes of the orbits of Earth and Pluto compared with circles centered on the Sun.

semimajor axis of the Earth's orbit. The same is true for the orbits of all the planets.

In the case of a circular orbit, the semimajor axis is just the radius of the circle. Some ellipses, on the other hand, are very elongated. When describing the shape of an ellipse, we speak of its **eccentricity**. The eccentricity of an ellipse is defined as the separation between the two foci divided by the length of the long axis of the ellipse.[2] A circle has an eccentricity of 0. The more elongated the ellipse becomes,

the closer its eccentricity gets to 1 (see Figure 3.2). Most planets have nearly circular orbits with eccentricities close to 0. The eccentricity of Earth's orbit, for example, is 0.017, which means that the distance between the Sun and Earth departs from its average value by only 1.7 percent. The distance between the two bodies varies from about 0.983 AU to 1.017 AU. It is hard to tell the difference between the orbit of Earth and a circle centered on the Sun (**Figure 3.3(a)**). One of the characteristics that distinguishes the dwarf planet Pluto from its classical cousins is its highly eccentric orbit. With an eccentricity of 0.244, the distance between the Sun and Pluto varies by 24.4 percent from its average value, ranging from 75.6 percent of average to 124.4 percent of average. Pluto's orbit is noticeably oblong, and its center is noticeably displaced from the Sun (**Figure 3.3(b)**).

Kepler's Second Law: Planets Sweep Out Equal Areas in Equal Times

The next empirical rule that Kepler found has to do with how fast planets move at different places on their orbits. A

---

[2] This book tends to use operational definitions. An ellipse, for example, is defined according to how it is drawn. Eccentricity is defined according to how you would calculate its value. Definitions do not mean much in science unless they are connected to how something is actually measured or calculated.

Planets move fastest when they are closest to the sun.

planet moves most rapidly when it is closest to the Sun, and is at its slowest when it is farthest from the Sun. The average speed of Earth in its orbit about the Sun is 29.8 km/s. When Earth is closest to the Sun, it travels at 30.3 km/s. When it is farthest from the Sun, it travels at 29.3 km/s.

Kepler found an elegant way to describe the changing speed of a planet in its orbit about the Sun. Look at **Figure 3.4**, which shows a planet at six different points in its orbit. Imagine a straight line connecting the Sun with this planet. We can think of this line as "sweeping out" an area as it moves with the planet from one point to another. Area *A* (in red) is swept out between times t_1 and t_2, area *B* (in blue) is swept out between times t_3 and t_4, and area *C* (in green) is swept out between times t_5 and t_6. When the planet is closest to the Sun it is moving rapidly, but the distance between the planet and the Sun is small (area *A* in Figure 3.4). Kepler realized that changes in the distance between the Sun and a planet and changes in the speed of a planet work together to produce a surprising result: The area swept out by a planet in the same amount of time is always the same regardless of the location of the planet in its orbit. In Figure 3.4, this means that if the three time intervals are equal ($t_1 \rightarrow t_2 = t_3 \rightarrow t_4 = t_5 \rightarrow t_6$) then the three areas *A*, *B*, and *C* will be equal as well.

This is **Kepler's second law**, which is also referred to as Kepler's **Law of Equal Areas**. It states that the imaginary

A planet "sweeps out" equal areas in equal times.

line connecting a planet to the Sun sweeps out equal areas in equal times, regardless of where the planet is in its orbit. Note that this law applies to only one planet at a time. The area swept out by Earth in a given time is always the same. Likewise, the area swept out by Mars in a given time is always the same. But the area swept out by Earth and the area swept out by Mars in a given time are *not* the same.

Kepler's Third Law: The Harmony of the Worlds

Kepler's first law describes the shapes of planetary orbits, and Kepler's second law describes how the speed of a planet changes as it goes around its orbit. But neither of these laws tells us how long it takes a planet to complete one orbit about the Sun (referred to as the **period** of the orbit). Nor do these laws tell us how this time depends on the distance between the Sun and a planet.

Planets that are closer to the Sun do not have as far to go to complete one orbit as do planets that are farther from the Sun. Jupiter, for example, has an average distance of 5.2 AU from the Sun—5.2 times as far from the Sun as Earth is. That means Jupiter has 5.2 times farther to travel in its orbit about the Sun than Earth does. We might guess, then, that if the

Outer planets have farther to go and move more slowly in their orbits around the sun.

two planets travel at the same speed, Jupiter would complete one orbit in 5.2 years. But such a guess would be wrong. Jupiter takes almost 12 years to complete one orbit. Clearly Jupiter not only has farther to go in its orbit but must be *moving more slowly than Earth* as well. This trend holds true for all the planets. As we go farther out from the Sun, the circumferences of planetary orbits become longer while the speeds at which the planets travel become less. Mercury, at an average distance of 0.387 AU from the Sun, whizzes around its short orbit at an average speed of 47.9 km/s, completing one revolution in only 88 days. At a distance of 30.1 AU from the Sun, Neptune lumbers along at an average speed of 5.48 km/s, taking 163.7 years to make it once around the Sun.

Kepler discovered a simple mathematical relationship between the period of a planet's orbit and its distance from

The square of a planet's orbital period equals the cube of the orbit's semimajor axis.

the Sun. **Kepler's third law** states that the square of the period of a planet's orbit, measured in years, is equal to the cube

FIGURE 3.4 An imaginary line between a planet and the Sun sweeps out an area as the planet orbits. Kepler's second law states that if the three intervals of time shown are equal, then the three areas A, B, and C will be the same.

Kepler's Second Law

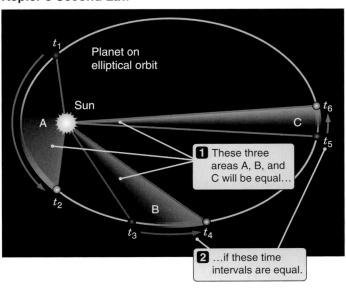

Planet on elliptical orbit

Sun

1 These three areas A, B, and C will be equal...

2 ...if these time intervals are equal.

of the semimajor axis of the planet's orbit, measured in astronomical units. Written as an equation, this says that

$$(P_{years})^2 = (A_{AU})^3,$$

where P_{years} represents the period of the orbit divided by 1 year, and A_{AU} is the semimajor axis of the orbit divided by 1 AU.

This is a case where astronomers use nonstandard units as a matter of convenience. Years are handy units for measuring the periods of orbits, and AU are handy units for measuring the sizes of orbits. When we use years and AU as our units, we get the simple relationship just shown. However, it is important to realize that *our choice of units in no way changes the physical relationship* we are studying. If we instead stayed with standard metric units, this relationship would read $(P_{seconds})^2 = 3 \times 10^{19} (A_{meters})^3$. **Table 3.1** lists the periods and semimajor axes of the orbits of the classical and dwarf planets, together with the values of the ratio P^2 divided by A^3.

Judge for yourself how well Kepler's third law works. These data are also plotted in **Figure 3.5**. This relationship was so beautiful to Kepler that he referred to it as his **harmonic law** or, more poetically, as the "Harmony of the Worlds."

TABLE 3.1

Kepler's Third Law: $P^2 = A^3$

The Orbital Properties of the Classical and Dwarf Planets

Planet	Period P years	Semimajor axis A (AU)	$\frac{P^2}{A^3}$
Mercury	0.241	0.387	$\frac{0.241^2}{0.387^3} = 1.00$
Venus	0.615	0.723	$\frac{0.615^2}{0.723^3} = 1.00$
Earth	1.000	1.000	$\frac{1.000^2}{1.000^3} = 1.00$
Mars	1.881	1.524	$\frac{1.881^2}{1.524^3} = 1.00$
Ceres	4.599	2.765	$\frac{4.559^2}{2.765^3} = 1.00$
Jupiter	11.86	5.204	$\frac{11.86^2}{5.204^3} = 1.00$
Saturn	29.46	9.582	$\frac{29.46^2}{9.582^3} = 0.99*$
Uranus	84.01	19.201	$\frac{84.01^2}{19.201^3} = 1.00$
Neptune	164.79	30.047	$\frac{164.79^2}{30.047^3} = 1.00$
Pluto	247.68	39.236	$\frac{247.68^2}{39.236^3} = 1.02*$
Eris	557.00	67.696	$\frac{557.00^2}{67.696^3} = 1.00$

*These ratios are not exactly 1.00, due to slight perturbations from the gravity of other planets.

3.3 The Rise of Scientific Theory: Newton's Laws Govern the Motion of All Objects

When investigating a newly discovered or poorly understood phenomenon, an empirical approach is often the only available way to proceed. This was certainly the case with Kepler. The development of Kepler's laws of planetary motion was an intellectual accomplishment with few peers. Yet to a modern scientist the empirical rules that Kepler spent his life pursuing are only the first step in the study of a phenomenon. Such empirical laws *describe* some phenomenon and are even useful in predicting what will happen in the future, but they do little to *explain* that behavior. Taken at face value, empirical rules offer little insight into the more fundamental laws describing nature. Kepler was able to characterize the orbits of planets as ellipses, but he did not understand *why* they should be so.

Once the empirical rules that describe some phenomenon have been discovered, a modern scientist will next try to understand those empirical rules in terms of more general physical principles or laws. Beginning with basic physical principles and using the tools of mathematics, the scientist works to *derive* the empirically determined rules. At other times a scientist may start with physical laws and predict relationships, which are then verified empirically. This technique is sometimes referred to as the **theoretical** approach to science. In practice, if the relevant physical laws are already understood, this process is often short-circuited. A scientist may make a theoretical prediction about the behavior of a system, then compare the prediction with experimental data directly to see how well they fit.

Today a great deal of science is done without ever trying to invent an empirical rule. This works only if the relevant physical laws are known ahead of time. If the relevant physical laws are *not* known—as was the case for planetary motion—the empirical rules become a way of *discovering* the physical laws themselves. Can we invent **hypothetical** physical laws that will allow us to derive the empirical rules? If so, what other predictions might we make on the basis of these hypothetical laws? Are these predictions also

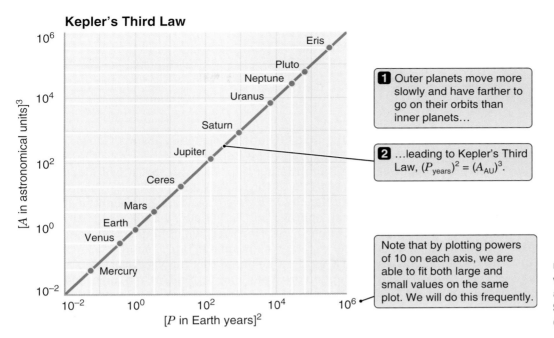

Kepler's Third Law

1 Outer planets move more slowly and have farther to go on their orbits than inner planets…

2 …leading to Kepler's Third Law, $(P_{years})^2 = (A_{AU})^3$.

Note that by plotting powers of 10 on each axis, we are able to fit both large and small values on the same plot. We will do this frequently.

FIGURE 3.5 A plot of A^3 versus P^2 for the eight classical and three dwarf planets in our Solar System shows that they obey Kepler's third law.

borne out by experiment and observation? If so, then we may have discovered something more fundamental about the way the universe works. This is how physical laws are discovered and tested.

One of the earliest great advances in theoretical science was also arguably one of the greatest intellectual accomplishments in the history of our species. In many ways the work of **Sir Isaac Newton** (1642–1727) on the nature of motion set the standard for what we now refer to as *scientific theory* and *physical law*. Building on the work of Kepler and others, Newton proposed three laws that he believed to govern the motions of all objects in the heavens and on Earth.

Newton's laws of motion are the basis of classical mechanics.

Today **Newton's laws** remain the basis for all of what is known as **classical mechanics**. By the time physicists and astronomers complete their formal education, they have spent many hours studying the wondrously subtle and complex consequences of Newton's three laws of motion. (Physics professors pride themselves on their ability to invent truly nasty problems to challenge their graduate students' understanding of Newton's laws.) Even so, Newton's laws themselves are beautifully elegant, and the relationships they describe between such everyday concepts as force, velocity, acceleration, and mass are accessible to all.

However fascinating Newton's laws of motion may be, you might reasonably ask why they are an essential stop on our journey through *21st Century Astronomy*. "After all," you might say, "this is a book about astronomy, not physics." Yet in a very real sense it is with Newton's laws, pub-

lished in 1687, that truly modern astronomy got its start. It was these laws that allowed Newton to look at the motion of a shot fired from a cannon and see instead the motions of the planets on their orbits around the Sun. It was with Newton's laws that the chasm between our thinking about the heavens and about Earth was banished once and for all, and Earth took its true place in the universe.

Newton's First Law: Objects at Rest Stay at Rest; Objects in Motion Stay in Motion

In a strange quirk of history, the physical law almost invariably referred to today as *Newton's first law of motion* did not originate with Newton at all. It was the brainchild of a contemporary of Kepler's by the name of **Galileo Galilei** (1564–1642). Galileo is probably best known to the general public as the first person to use a telescope to make significant discoveries about the heavens and to report those discoveries. In the history of science, however, Galileo's work on the motion of objects is at least as fundamental a contribution as his astronomical observations.

By Galileo's day, Copernicus and others had begun to turn toward the view that knowledge comes from observing nature rather than only from reading the works of classical Greek and Roman philosophers. Yet even in the 16th and 17th centuries the works of one of the greatest of these philosophers, **Aristotle** (384–322 B.C.), who lived almost 2,000 years earlier, still carried the weight of authority. Aristotle

Challenging Authority

It is seldom safe to challenge the entrenched wisdom of your day, and that was especially true at a time when intellectual and religious authority and political power resided in the same hands. In 1600 Giordano Bruno fell victim to the Inquisition and was burned at the stake for his beliefs. These included his support of Copernicus, his belief that the universe is infinite, and his suggestion that Earth is but one of many habitable planets. In 1632 Galileo actually invited the ire of the powers that be when he published his great work, *Dialogo sopra i due massimi sistemi del mondo* (*Dialogue on the Two Great World Systems*). In the *Dialogo* the champion of the Copernican (Sun-centered) view of the universe is a brilliant, witty, and erudite philosopher named Salviati. The *Dialogo*'s defender of Aristotelian authority is named Simplicio and is as much an ignorant buffoon as the name might imply. In Galileo's story the "neutral" but intelligent moderator, Sagredo, is quick to see the truth in Salviati's arguments and dismisses

Simplicio's rebuttals as patently absurd. Galileo's prose is lively and entertaining, and he wrote the *Dialogo* in Italian rather than Latin so that it would be easily accessible to the person on the street. When Galileo published the *Dialogo,* he actually thought he had the tacit approval of the Vatican, which held to the Aristotelian view. However, when he placed a number of the Pope's own arguments into the unflattering mouth of Simplicio, he found that the Vatican's tolerance had limits. Fortunately for Galileo he had more friends in high places than Bruno did, so he spent the closing years of his life under house arrest rather than ending up tied to the stake. To escape a harsher sentence Galileo was forced to publicly recant the Copernican theory that he had supported with such fervor. In one of the great apocryphal stories of the history of astronomy, it is said that as he left the courtroom following his sentencing, he stamped his foot on the ground and muttered, "But it moves!"

believed that the natural state of all objects was to be at rest and that an object in motion would tend toward this natural state. This seemed to be a good empirical rule about how objects in the world around us behaved, and Aristotle had elevated this observation into a fundamental tenet of his philosophy of nature. Aristotle was an extremely sharp individual, and this idea was not easy to refute. A cart rolling down the street coasts to a stop when it is no longer being pulled. A bouncing ball eventually settles to the ground. Even an arrow shot from a bow loses much of its speed before striking its target.

As discussed in **Excursions 3.1**, Galileo was no stranger to controversy. In his writings on motion, Galileo challenged Aristotle's authority by proposing that this apparent tendency of objects to come to rest was a mirage. Galileo argued that in all of the cases just mentioned—indeed in *every* such case—there are hidden reasons why objects come to rest. There is friction as the axle of the cart rubs against its bearing, resisting the motion and eventually bringing it to a halt. Every time a ball bounces, its shape is distorted, and what we might think of as "internal friction" within the ball causes it to bounce less high each time. The resistance of air, which the arrow must push out of the way and which drags against the arrow's shaft, slows the arrow's progress.

Galileo agreed with Aristotle that an object at rest remains at rest unless something causes it to move. But Galileo disagreed with Aristotle by asserting that, *left on its own, an object in motion will remain in motion.* Specifically, Galileo said that *an object in motion will continue moving along a straight line with a constant velocity until an* **unbalanced force** *acts on it to change its state of motion.* Galileo

---

Galileo found that an object left in motion remains in motion.

---

referred to the resistance of an object to changes in its state of motion as **inertia**. Galileo's great insight formed the starting point for Newton's tour de force that was to come. The work of great scientists is always built on the foundation of the great scientists who came before.[3] It is a tribute to Galileo that his law of inertia became the cornerstone of physics as **Newton's first law of motion**.

The idea of inertia has come a long way since the days of Galileo and Newton. In fact, we have already seen a number of very sophisticated applications of this idea. What Galileo

---

[3] Newton himself is credited with the famous quote, "If I have seen further [than you] it is by standing upon the shoulders of giants."

and Newton called *inertia* can actually be viewed as a consequence of what we discovered in Chapter 2's discussions of relative motion and frames of reference. To say that only *relative* motions between objects have meaning is the same as saying that there is *no difference* between an object at rest and an object in uniform motion. What objects are at rest and what objects are in motion, anyway? The object at rest beside you on the front seat of your car as you drive down the highway is moving at 60 miles per hour according to a bystander along the side of the road—but it is moving at 120 mph according to a car in oncoming traffic. All of these perspectives are equally valid.

The connection between inertia and the relative nature of motion is so fundamental that a reference frame moving in a straight line at a constant speed is referred to as an **inertial frame of reference**. Motion is meaningful only when measured relative to an inertial frame of reference, and *all* inertial frames of reference are as good as any other. The realization that the laws of physics are the same in *any* inertial frame of reference is one of the deepest insights ever made into the nature of the universe. When thought of in this way, *of course* an object moving in a straight line at a constant speed remains in motion. As illustrated in **Figure 3.6**, in the frame of reference of that object, *it is already at rest*.

Newton's Second Law: Motion Is Changed by Unbalanced Forces

Newton often gets credit for Galileo's insight about inertia because it was Newton who took the crucial next step. Newton's first law says that in the absence of an unbalanced

---

Unbalanced forces cause changes in motion.

---

force an object's motion does not change; **Newton's second law of motion** goes on to say that *if there is an unbalanced force acting on an object, then the object's motion does*

change. Even more, Newton's second law tells us *how* the object's motion changes in response to that force.

Before going any further, it would be wise to pause to be sure we are all together on this. In the preceding paragraphs we spoke of changes in an object's motion, but what does that phrase really mean? When you are in the driver's seat of a car, a number of controls are at your disposal. On the floor of the car are a gas pedal and a brake. You use these to make the car speed up or slow down. A *change in speed* is one way the motion of an object can change. But also remember the steering wheel beneath your hands. When you are moving down the road and you turn the wheel, your speed does not necessarily change, but the direction of your motion does. A *change in direction* is also a kind of change in motion.

Together the speed and direction of an object's motion are called the object's **velocity**. A change in velocity is called an **acceleration**. Acceleration actually refers to how rapidly the change in velocity happens. If you go from 0 to 60 mph in 4 seconds, you feel the back of your seat shoving your

---

Acceleration measures how quickly a change in motion takes place.

---

body forward, causing you to accelerate along with the car. If you take 2 minutes to get from 0 to 60 mph, on the other hand, the acceleration is so slight that you hardly notice it. To formalize this a bit, your acceleration is determined by how much your velocity changes divided by how long it takes for that change to happen:

$$\text{Acceleration} = \frac{\text{How much velocity changes}}{\text{How long the change takes}}$$

For example, if an object's speed goes from 5 m/s (meters per second) to 15 m/s, then the change in velocity is 10 m/s. If that change happens over the course of 2 seconds, then the acceleration is 10 m/s divided by 2 seconds, which equals 5 meters per second per second. This is the same as saying "5 meters per second squared," which is written 5 m/s^2 or 5 m s^{-2}.

Because the gas pedal on a car is often called the accelerator, some people think *acceleration* means that an object is speeding up. But we need to stress that *any* change in motion is an acceleration. **Figure 3.7** illustrates the point. Slamming on your brakes and going from 60 to 0 mph in 4 seconds is just as much acceleration as going from 0 to 60 mph in 4 seconds. Similarly, the acceleration you experience as you go through a fast, tight turn at a constant speed is every bit as real as the acceleration you feel when you slam your foot on the gas pedal or brake. Faster, slower, turn left, turn right—*if you are not moving in a straight line at a constant speed, you are experiencing an acceleration.*

Newton's second law of motion says that changes in motion—accelerations—are caused by unbalanced forces. The acceleration that an object experiences depends on two

FIGURE 3.6 An object moving in a straight line at a constant speed is at rest in its own inertial reference frame.

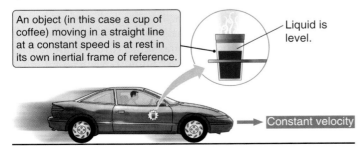

An object (in this case a cup of coffee) moving in a straight line at a constant speed is at rest in its own inertial frame of reference.

Liquid is level.

Constant velocity

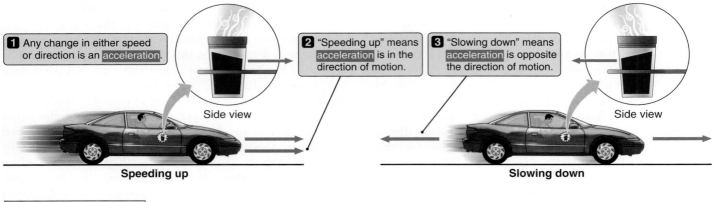

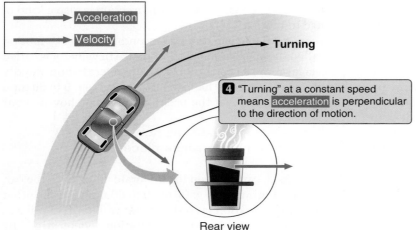

FIGURE 3.7 Any change in the velocity of an object is an acceleration. When driving, for example, any time your speed changes or you follow a curve in the road, you are experiencing an acceleration. (Throughout the text velocity arrows will be shown as red and acceleration arrows will be shown as green.)

things, as shown in **Figure 3.8**. First, it depends on the unbalanced force acting on the object to change its motion. This is a pretty commonsense idea. The stronger the unbalanced force, the greater the acceleration. In fact, the accel-

Greater force means greater acceleration.

eration of an object is *proportional* to the unbalanced force applied. Push on something twice as hard (Figure 3.8(b)) and it experiences twice as much acceleration. Push on something three times as hard and its acceleration is three times as great. (The idea of proportionality, discussed in **Foundations 3.1**, will be used over and over again throughout our journey.) The resulting change in motion occurs in the direction in which the unbalanced force is imposed. Push something forward and it speeds up. Push it to the left and it veers in that direction.

The acceleration that an object experiences also depends on the degree to which the object resists changes in motion (Figure 3.8(c)). Some objects—say a baseball—are easily shoved around by humans. A baseball is thrown at great velocity by the pitcher, only to be hit with a bat and have its motion abruptly changed again. The hard-hit line drive comes suddenly to a stop in the glove of the second base

player. A baseball resists changes in its motion; that is, it has inertia—but not *too* much inertia. Other objects are less obliging. A piece of solid iron the size of a baseball would make a poor substitute in the game. A pitcher would be *very* hard pressed to throw such an iron ball hard enough to get it over home plate; and if he did, being catcher would become an even more dangerous job. A baseball and a ball of

Mass is the property of matter that resists changes in motion.

iron may be the same size, but they are quite different in the degree to which they resist changes in their motion. The property of an object that determines its resistance to changes in motion—the measure of an object's inertia—is referred to as the object's **mass**.

You probably knew this answer intuitively. The iron ball is "heftier" and therefore harder to throw than a baseball. However, you may not have thought much about what we really mean when we say that a ball of iron "has more mass"—or "is more massive"—than a baseball. You might say that the ball of iron is made up of "more stuff" than the baseball; but again, what is meant by "more stuff"? If you grapple with this question for a time, you may find yourself

Newton's Second Law:

$$\boxed{\text{Acceleration } (a) = \frac{\text{Force } (F)}{\text{Mass } (m)}}$$

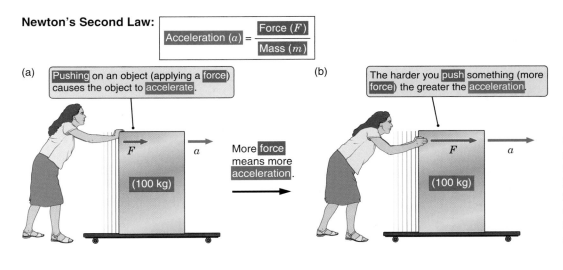

(a) Pushing on an object (applying a force) causes the object to accelerate.

F   a

(100 kg)

More force means more acceleration.

(b) The harder you push something (more force) the greater the acceleration.

F   a

(100 kg)

FIGURE 3.8 Newton's second law of motion says that the acceleration experienced by an object is determined by the force acting on the object divided by the object's mass. (Throughout the text force arrows will be shown as blue.)

More mass means less acceleration.

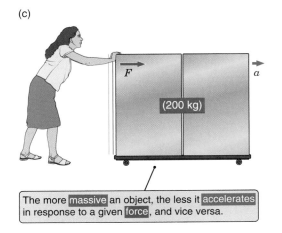

(c)

F   a

(200 kg)

The more massive an object, the less it accelerates in response to a given force, and vice versa.

Instead of spelling this out in words every time, we can introduce a convenient bit of shorthand—*a* for acceleration, *F* for force, and *m* for mass—so we get

$$a = \frac{F}{m}.$$

This is the succinct mathematical statement of Newton's second law of motion.[4] If you are comfortable with mathematics, this elegant expression may speak to you clearly and directly. If not, when you see this equation, remind yourself that Newton's second law is nothing more than the embodi-

---

Acceleration is force divided by mass.

---

ment of three commonsense ideas: (1) When you push on an object, that object accelerates in the direction in which you are pushing; (2) the harder you push on an object, the more the object accelerates; and (3) the more massive the object is, the harder it is to change its state of motion.

chasing the question around in circles. When it comes right down to it, *the property of matter that we refer to as "mass" is nothing more and nothing less than the degree to which an object resists changes in its motion.* (Mass is measured in units of kilograms. An object with a mass of 2 kg is twice as hard to accelerate as an object with a mass of 1 kg. An object with a mass of 9 kg is three times as hard to accelerate as an object with a mass of 3 kg.)

So if we want to know how an object's motion is changing, we need to know two things: What unbalanced force is acting on the object, and what is the resistance of the object to that force? We can put this into equation form as follows:

$$\begin{array}{c}\text{The}\\\text{acceleration}\\\text{experienced}\\\text{by an object}\end{array} = \dfrac{\begin{array}{c}\text{The force acting to change}\\\text{the object's motion}\end{array}}{\begin{array}{c}\text{The object's resistance}\\\text{to that change}\end{array}} = \dfrac{\text{Force}}{\text{Mass}}$$

Newton's Third Law: Whatever Is Pushed, Pushes Back

Imagine you are a child again, sitting in a wagon or standing on a skateboard and pushing yourself along with your foot. Each shove of your foot against the ground sends you faster along your way. But why does this happen? Your muscles flex and your foot exerts a force on the ground. (Earth does not respond much to that force because its great mass gives it great inertia.) Yet this does not explain why *you* experience an acceleration. The fact that you accelerate at

---

[4] Newton's second law is often written as $F = ma$, giving force as units of mass times units of acceleration, or kg m/s^2. These units are aptly named "newtons," abbreviated N.

FOUNDATIONS 3.1

Proportionality

Often in this text we will say that one quantity is *proportional* to another. Proportionality is a way of getting the gist of how something works—understanding the relationships between things—without having to actually calculate the details of one case after another.

PROPORTIONALITY

If two quantities are **proportional** to each other, then making one of them larger means making the other quantity larger by the same factor. In other words, the ratio between the two remains constant. For example, think about the weight of a bag of apples and how much the bag costs. Double the weight of the bag of apples and you double the cost. Increase the weight of the bag of apples by a factor of 5, and the cost goes up by a factor of 5 as well. *The cost of a bag of apples is proportional to the weight of the bag of apples.* We write this relationship as

$$\text{Cost} \propto \text{Weight},$$

where the symbol $\propto$ means "is proportional to." This expression captures the essence of the relationship between the cost and the weight of apples. It tells us that the more apples we buy, the more they will cost us.

CONSTANTS OF PROPORTIONALITY

Sometimes it is enough to know that two quantities are proportional to each other, but sometimes it is not. What if you need to know how much one of those bags of apples will actually set you back? We know that the full relationship between the cost and weight of a bag of apples is that the cost is equal to the price per pound of apples times the weight of the bag. We write

$$\text{Cost} = \text{Price per pound} \times \text{Weight}.$$

Compare this expression with the previous one. When we say that two quantities are proportional to each other, what we mean is that one quantity equals some number *times* the other quantity. The number by which one quantity is multiplied to get the other number is called the **constant of proportionality**. In our example, the constant of proportionality is just the price per pound of apples.

Look at the difference between the two expressions. The fact that the cost of a bag of apples is proportional to the weight of the apples is a statement about the *relationship* between things. It is a statement about how things work. More apples do not cost *less* than fewer apples. More apples cost *more* than fewer apples. The constant of proportionality—here the price per pound of apples—means something very different. Hidden within the price per pound of apples is a great deal of information, such as the cost of growing apples, the cost of transporting them from the orchard, and the profit margin the grocer needs to stay in business. The constant of proportionality carries information about this aspect of the world.

Very often physical laws work in this same fashion. Proportionalities tell us about *relationships*—how two things vary with one another. They let us get a feeling for the "how" in how something works. In this chapter, for example, we find that gravitational force is proportional to an object's mass. Constants of proportionality more precisely tell us about the way the universe is. The universal gravitational constant G is a constant of proportionality that tells us about the intrinsic strength of gravitational interactions and allows calculation of the numerical value of this force. Constants of proportionality are needed if we are to turn an understanding of relationships into hard numbers.

the same time means that as you push on the ground, *the ground must be pushing back on you.*

Part of Newton's genius was his ability to see sublime patterns in such mundane events. Newton realized that *every* time one object exerts a force on another, a matching force is exerted by the second object on the first. That second force is exactly as strong as the first force but is in

exactly the *opposite* direction. The child pushes back on Earth, and Earth pushes the child forward. A canoeist's paddle pushes backward through the water, and the water pushes forward on the paddle, sending the canoe along its way. A rocket engine pushes hot gases out of its nozzle, and those hot gases push back on the rocket, propelling it into space.

For every force there is an equal and opposite force.

All of these are examples of **Newton's third law of motion**, which says that *forces always come in pairs, and those pairs are always equal in magnitude but opposite in direction.* The forces in these action–reaction pairs always act on two different objects. Your weight pushes down on the floor, and the floor pushes back on you with the same amount of force. For every force there is *always* an equal and opposite force. This is one of the few times when we can say "always" and really mean it. **Figure 3.9** gives a few examples. There is a great game hiding in Newton's third law. It is called "find the force." Look around you at all the forces at work in the world, and for each force find its mate. It will *always* be there!

To see how Newton's three laws of motion work together, think about the situation shown in **Figure 3.10**. An astronaut is adrift in space, motionless with respect to the nearby space shuttle. With no tether to pull on, how can the astronaut get back to the ship? The answer? Throw something. Suppose the 100 kg astronaut throws a 1 kg wrench directly away from the shuttle at a speed of 10 m/s. Newton's second law says that in order to cause the motion of the wrench to change, the astronaut has to apply a force to it in the direction away from the shuttle. Newton's third law says that the wrench must therefore push back on the astronaut with as much force but in the opposite direction. The force of the wrench on the astronaut causes the astronaut to begin drifting toward the shuttle. How fast will the astronaut move? Turn to Newton's second law again. A force that causes the 1 kg wrench to accelerate to 10 m/s will not have much effect on the 100 kg astronaut. Because acceleration equals force divided by mass, the 100 kg astronaut will experience only 1/100 as much acceleration as the 1 kg wrench. The astronaut will drift toward the shuttle at the leisurely rate of $1/100 \times 10$ m/s, or 0.1 m/s.

3.4 Gravity Is a Force between Any Two Objects Due to Their Masses

Drop a ball and the ball falls toward the ground, picking up speed as it falls. It accelerates toward Earth. Newton's second law says that where there is acceleration, there is

FIGURE 3.9 Newton's third law states that for every force there is always an equal and opposite force. These opposing forces always act on the two different objects in the same pair.

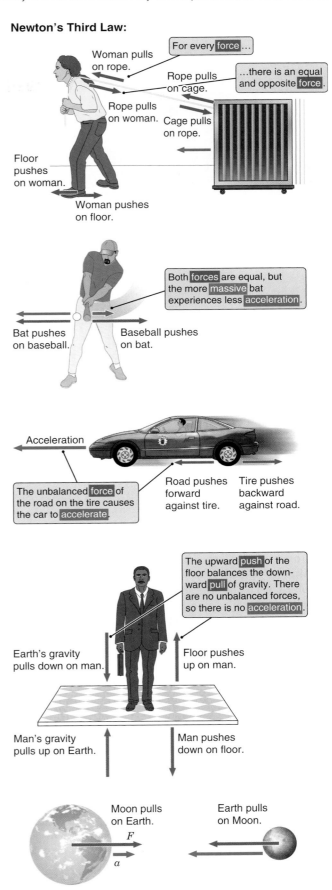

Newton's Third Law:

Woman pulls on rope.

For every force...

Rope pulls on cage.

...there is an equal and opposite force.

Rope pulls on woman.

Cage pulls on rope.

Floor pushes on woman.

Woman pushes on floor.

Both forces are equal, but the more massive bat experiences less acceleration.

Bat pushes on baseball.

Baseball pushes on bat.

Acceleration

The unbalanced force of the road on the tire causes the car to accelerate.

Road pushes forward against tire.

Tire pushes backward against road.

The upward push of the floor balances the downward pull of gravity. There are no unbalanced forces, so there is no acceleration.

Earth's gravity pulls down on man.

Floor pushes up on man.

Man's gravity pulls up on Earth.

Man pushes down on floor.

Moon pulls on Earth.

Earth pulls on Moon.

F

a

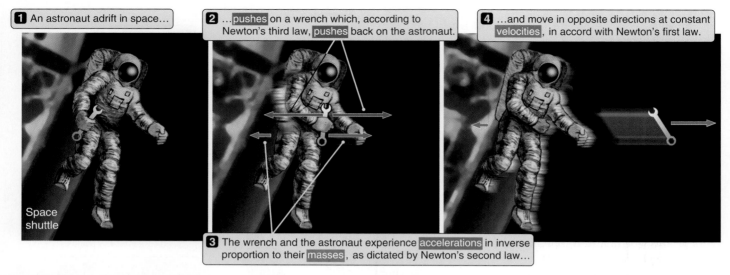

1 An astronaut adrift in space...

2 ...pushes on a wrench which, according to Newton's third law, pushes back on the astronaut.

4 ...and move in opposite directions at constant velocities, in accord with Newton's first law.

Space shuttle

3 The wrench and the astronaut experience accelerations in inverse proportion to their masses, as dictated by Newton's second law...

FIGURE 3.10 According to Newton's laws, if an astronaut throws a wrench the two will move in opposite directions at speeds that are inversely proportional to their masses. (Acceleration and velocity arrows are not drawn to scale.)

force. But where is the force that causes the ball to accelerate? Many forces that we see in everyday life involve "direct contact" between objects.[5] The cue ball slams into the eight ball, knocking it into the pocket. The shoe of the child in the wagon shoves directly onto the surface of the pavement. In cases where there is physical contact between two objects, the source of the forces between them is easy

Gravity is "force at a distance."

to see. But the ball falling toward Earth is an example of a different kind of force, one that acts at a distance across the intervening void of space. The ball falling toward Earth is accelerating in response to the force of **gravity**. We began this chapter with a qualitative discussion of the fundamental role that gravity plays in the universe. Having explored both Kepler's empirical description of the motions of planets about the Sun and Newton's laws of motion, it is time to return to gravity, for it is gravity that unites these two pillars of empirical and theoretical science.

You probably will not be surprised to learn that once again it is Newton we turn to for a **law of gravitation**. At this point in an introductory textbook it is customary to simply present Newton's law of gravitation as a "done deal" and go straight to its application, but such a leap misses one of the most interesting aspects of this stretch of our journey. A common misconception about how science works is the no-

tion that new theories just spring fully formed into the mind of a scientist as if by magic. This idea is certainly supported by the grade school story of the apple falling on Newton's head, literally knocking the idea of gravity into his brain. One could almost get the idea that scientific theories are arbitrary—that Newton could have invented some *other* law of gravity that would have worked just as well. (The idea that scientific knowledge is a cultural construct is discussed in **Excursions 3.2**.) Although it might seem at first glance that his work was arbitrary, nothing could be further from the truth. Where did Newton get his ideas about gravity? What guided him in his development of those ideas, and how did he turn them into a theory with testable predictions? How did he confront that theory in the crucible of experiment and observation? By answering these questions, rather than simply stating Newton's law of gravitation, we will gain some insight into what science is.

Where Do Theories Come From? Newton Reasons His Way to a Law of Gravity

As with inertia, the story of gravity begins with the insight and observation of Galileo. Galileo discovered that all freely falling objects accelerate toward Earth at the same rate, regardless of their mass. Drop a marble and a cannonball, at the same time and from the same height, and they will hit the ground together. If proof were needed, it was provided by astronaut David Scott on the lunar surface (**Figure 3.11**). The gravitational acceleration near the surface of Earth is

---

[5] Actually the "direct contact" between billiard balls or other "solid" objects is also force at a distance—electric force acting at a distance between the electrons and protons in the atoms of which the objects are made.

EXCURSIONS 3.2

Science and Culture

In recent years some critics of science have drawn attention to how science is influenced by culture. It is hard to avoid the conclusion that political and cultural considerations strongly influence which scientific research projects are funded. This choice of funding channels the directions in which scientific knowledge advances and can lead to serious ethical issues. For example, moral judgments about nontraditional lifestyles greatly restricted the funding available for AIDS research during the decade or so after its discovery.

Some critics even carry this view a step further, arguing that scientific knowledge itself is an arbitrary cultural construct. Yet, as illustrated by the discussion of Newton's law of gravity, successful scientific theories are *never* arbitrary. Scientific theories must be consistent with all that we know of how nature works, and turning a clever idea into a real theory with testable predictions is a matter of careful thought and effort. One of the most remarkable aspects of scientific knowledge is its *independence* from culture. Scientists are people, and politics and culture enter into the day-to-day practice of science. But in the end, *scientific theories are judged not by cultural norms but by whether their predictions are borne out by observation and experiment.* As long as the results of experiments are repeatable and do not depend on the culture of the experimenter—that is, as long as there is such a thing as objective physical reality—scientific knowledge cannot be called a cultural construct.

Nor does it seem that the path to knowledge embodied in science is any more arbitrary than the logic it is built on. It is significant that no philosopher critical of science has ever offered a viable alternative for obtaining reliable knowledge of the workings of nature. Had science not arisen when and where it did, something much like it would have arisen at some time and in some location. Furthermore, no other category of human knowledge is subject to standards as rigorous and unforgiving as those of science. For this reason, scientific knowledge is reliable in a way that no other form of knowledge can claim. Whether you want to design a building that will not fall over, choose the best treatment for a disease, or calculate the orbit of a spacecraft on its way to the Moon, you had better consult a scientist rather than a psychic —regardless of your cultural heritage.

Finally, we must admit that some disreputable scientists purposefully try to influence results by inventing or ignoring data, often when claiming "a major breakthrough" or challenging a well-established scientific principle. Fortunately, attempts by others to repeat the experiment will eventually expose such scientific misconduct.

---

usually written as g and has a value of 9.8 m/s^2. Whether you drop a marble or a cannonball, after 1 second it will be falling at a speed of 9.8 m/s, after 2 seconds at 19.6 m/s, and

> **All objects on Earth fall with the same acceleration, g.**

after 3 seconds at 29.4 m/s. (These numbers assume that we can neglect air resistance, which is reasonably negligible for relatively dense objects.)

Having worked out the laws governing the motion of objects, Newton saw something deeper in Galileo's findings. Newton realized that if all objects fall with the same acceleration, then the gravitational *force* on an object must be determined by the object's *mass*. To see why, look back at Newton's second law (acceleration equals force divided by mass). The only way gravitational acceleration can be the same for all objects is if the value of the force divided by the mass is the same for all objects. A greater mass *must*, therefore, be accompanied by a stronger gravitational force. In other words, the gravitational force on an object on Earth is, according to Newton's second law, the object's mass times the acceleration due to gravity, or $F_{grav} = mg$. Make an object twice as massive, and you double the gravitational force acting on it. Make an object three times as massive, and you triple the gravitational force acting on it.

The gravitational force acting on an object is commonly referred to as the object's **weight**. It is easy to see why people often confuse mass and weight. On the surface of Earth, weight is just mass times the constant g. The situation is not helped by the sloppy way we use language. We often say that an object with a mass of 2 kg "weighs 2 kg," but

FIGURE 3.11 Astronaut Alan Bean's portrait of fellow astronaut David Scott standing on the Moon and dropping a hammer and falcon feather together. Both reached the lunar surface simultaneously. (Their lunar module was nicknamed "Falcon.")

it is more correct to express a weight in terms of **newtons** (N): Thus an object with a *mass* of 2 kg has a *weight* of 2 kg × 9.8 m/s² or 19.6 N.

Newton's next great insight came from applying his third law of motion to gravity. For every force there is an equal and opposite force. If Earth exerts a force of 19.6 newtons on a 2 kg mass sitting on its surface, then that 2 kg mass must exert a force of 19.6 newtons on Earth as well. Drop a 20 kg cannonball, and it falls toward Earth, but at the same time Earth falls toward the 20 kg cannonball! The reason we do not notice the motion of Earth is because Earth is very massive. It has a lot of resistance to a change in its motion. In the time it takes a 20 kg cannonball to fall to the ground from a height of 1 km, Earth has "fallen" toward the cannonball by about 3.4×10^{-21} meters, which is only about 1/300 of the diameter of a hydrogen atom!

Newton reasoned that this should work both ways. If doubling the mass of an object doubles the gravitational force between the object and Earth, then doubling the mass of Earth ought to do the same. In short, the gravitational force between Earth and an object must be equal to the product of the two masses times something:

$$\text{Gravitational force} = \text{Something} \times \text{Mass of Earth} \times \text{Mass of object.}$$

If the mass of the object is three times greater, then the force of gravity will be three times greater. Likewise, if the mass of Earth were three times what it is, the force of gravity would have to be three times greater as well. If *both* the mass of Earth *and* the mass of the object were three times greater, the gravitational force would increase by a factor of 3×3, or 9 times. Because objects fall toward the center of Earth, we know that this force is an attractive force acting along a line between the two masses.

"And by the way," reasoned Newton, "why are we restricting our attention to Earth's gravity?" If gravity is a force that depends on mass, then there should be a gravitational

---

The force of gravity is proportional to the product of two masses.

---

force between *any* two masses. Say we have two masses—call them mass 1 and mass 2, or m_1 and m_2 for short. The gravitational force between them is something times the product of the masses:

$$\text{Gravitational force between two objects} = \text{Something} \times m_1 \times m_2.$$

Realize that we have gotten this far just by combining Galileo's observations of falling objects with (1) Newton's laws of motion and (2) Newton's belief that Earth is a mass just like any other mass. There has been no wiggle room —there is nothing arbitrary in what we have done. But what about that "something" in the previous expression? Today we have sensitive enough instruments to allow us to put two masses close to each other in a laboratory, measure the force between them, and determine that something directly. Yet Newton had no such instruments. He had to look elsewhere to go further with his exploration of gravity.

It turns out that Kepler had already thought about this question. He reasoned that because the Sun is the focal point for planetary orbits, the Sun must be responsible for exerting an influence over the motions of the planets. Kepler speculated that whatever this influence is, it must grow weaker with distance from the Sun. (After all, it must surely require a stronger influence to keep Mercury whipping around in its tight, fast orbit than it does to keep the outer planets lumbering along their paths around the Sun.) Kepler's speculation went even further. Although he did not know about forces or inertia or gravity, he did know quite a lot about geometry, and geometry alone suggested how this solar "influence" might change for planets progressively farther from the Sun.

Imagine you have a certain amount of plaster to spread over the surface of a sphere. If the sphere is small, then when you spread the plaster over the sphere you get a thick coat. But if the sphere is larger, the plaster has to spread farther, and so you get a thinner coat. The surface area of a sphere depends on the square of the sphere's radius. Double the radius of a sphere, and the sphere's surface becomes four times what it was. If you plaster this new, larger sphere, the plaster must cover four times as much area, and so the thickness of the plaster will only be a fourth of what it was on the smaller sphere. Triple the radius of the sphere, and the sphere's surface is nine times as large, and the thickness of the coat of plaster will be only a ninth as thick.

Kepler thought that the influence that the Sun exerts over the planets might be like the plaster in this example. As the influence of the Sun extends farther and farther into space, it would have to spread out to cover the surface of a larger and larger imaginary sphere centered on the Sun. (We will learn later that light works in exactly this way.) If so, then, like the thickness of the plaster, the influence of

Gravity is an inverse square law.

the Sun should be proportional to 1 divided by the square of the distance between the Sun and a planet. Double the distance between the Sun and a planet, and this influence declines by a factor of $2 \times 2 = 4$, to ¼ of its original strength. Triple the distance, and this influence declines by a factor of $3^2 = 9$, becoming ⅑ of its initial strength. When something (like the thickness of the plaster or the strength of Kepler's solar "influence") changes in proportion to 1 divided by the square of the distance, we refer to this relationship as an **inverse square law** (see **Connections 3.1**).

Kepler had an interesting idea, but not a scientific theory with testable predictions. What he lacked was a good idea of the true source of this influence and the mathematical tools to calculate how an object would move under such an influence. Newton had both. If gravity is a force between *any* two objects, then there should be a gravitational force between the Sun and each of the planets. Might this gravitational force be the same as Kepler's "influence"? If so, then the something in Newton's expression for gravity might be a term that diminishes according to the square of the distance between two objects. Gravity might behave according to an inverse square law. Newton's expression for gravity now came to look like this:

$$\text{Gravitational force between two objects} =$$
$$\text{Something} \times \frac{m_1 \times m_2}{(\text{Distance between objects})^2}.$$

There is still a "something" left in this expression. Newton guessed that it was a measure of the intrinsic strength of gravitation interactions and that it would turn out to be the same for all objects. He named this something the **universal gravitational constant**. This quantity is now written as G.

Putting the Pieces Together: A Universal Law for Gravitation

Newton had good reasons every step of the way in his thinking about gravity—reasons directly tied to observations of how things in the world behave. Newton's chain of logic and reason brought him to what has come to be known as

---

CONNECTIONS 3.1

Inverse Square Laws

In this chapter we discover that gravity obeys what is called an *inverse square law*. This means that the force of gravity is proportional to 1 divided by the square of the distance between two objects, or

$$F_{\text{grav}} \propto \frac{1}{r^2}.$$

If two objects are moved so that they are twice as far apart as they were originally, the force of gravity between them becomes only ¼ of what it was. If two objects are

moved three times as far apart, the force of gravity drops to ⅑ of its original value.

Gravity is only one of several inverse square laws found in nature. The other important one that we will deal with in this book involves radiation. The intensity of radiation from an object is also proportional to 1 over the square of the distance between the objects. Our discussion of radiation in Chapter 4 will present a clear picture of why an inverse square law applies in that case.

Newton's Universal Law of Gravitation:

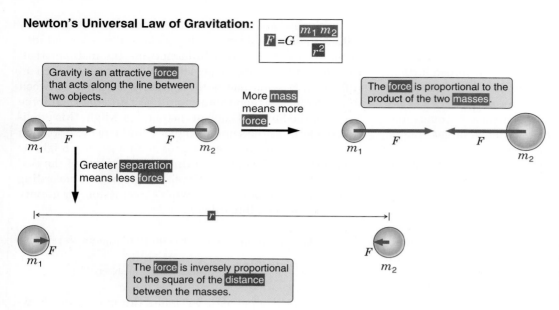

FIGURE 3.12 Gravity is an attractive force between two objects. The force of gravity depends on the masses of the objects and the distance between their centers of mass.

Newton's **universal law of gravitation**. This law, illustrated in **Figure 3.12**, states that gravity is a force between any two objects and has these properties:

1. It is an attractive force acting along a straight line between the two objects.

2. It is proportional to the mass of one object times the mass of the other object:

$$F_{grav} \propto m_1 \, m_2.$$

3. It decreases in proportion to 1 divided by the square of the distance between the two objects:

$$F_{grav} \propto \frac{1}{r^2}.$$

Written as a mathematical formula, the universal law of gravitation states that

$$F_{grav} = G \times \frac{m_1 \times m_2}{r^2},$$

where F is the force of gravity between two objects, m_1 and m_2 are the masses of objects 1 and 2, r is the distance between the centers of mass of the two objects, and G is the universal gravitational constant.

Any time you run across a statement like this, get into the habit of pulling it apart to be sure that it makes sense. First, gravity is an attractive force between two masses that acts along the straight line between the two masses.[6] Regardless

of where you stand on the surface of Earth, Earth's gravity pulls you *toward the center of Earth*. Second, the force of gravity depends on the product of the two masses. If you make m_1 twice as large, then the gravitational force between m_1 and m_2 becomes twice as large. Again, this should make sense. Doubling the mass of an object also doubles its weight. A subtle—but in retrospect very important—point lurks in this statement. The mass that appears in the universal law of gravitation is the *same* mass that appears in Newton's laws of motion. *The same property of an object that gives it inertia is the property of the object that makes it interact gravitationally.* (This equivalence between the effect of gravitation and the effect of inertia later became the basis for Einstein's general theory of relativity, in which mass literally warps space and time. We will return to this idea later in our journey when we discuss *spacetime* in Chapter 17.)

The third part of Newton's universal law of gravitation tells us that the force of gravity is inversely proportional to the *square* of the distance between two objects. Doubling the distance between two objects reduces the strength of gravity to ½², or ¼ of its original value. Tripling the distance between two objects reduces the strength of gravity to ⅓², or ⅑ of its original value (see **Figure 3.13**). Gravity is only one of several laws that we will see in which the strength of some effect diminishes in proportion to the square of the distance.

Playing with Newton's Laws of Motion and Gravitation

If you have ever watched a child play with blocks, you have probably noticed how the child will put the blocks

---

[6] This is not obvious from the equation alone. More properly, this equation should be written using *vector notation,* which would include information about the direction of the force.

together in different ways, seeing what she can build. Perhaps if you are an artist, you play with colors and patterns of light and dark as you create new works. If you are a musician, you might play with tones and rhythms as you compose. Writers play with combinations of words. All of these uses of the term *play* mean the same thing. Play is very serious business because it is by playing that we explore the world. And in exactly the same sense, scientists often play with the equations describing natural laws as they seek new insights into the world around them. We can play a bit with Newton's laws of gravitation and motion and see what interesting things turn up. (If you have difficulty following the math here, do not worry too much. Pay attention instead to the sense of what we are doing.)

The universal gravitational constant G has a value of 6.67×10^{-11} Nm^2/kg^2. This makes gravity a *very* weak force. The gravitational force between two 16-pound (7.26 kg) bowling balls sitting 0.3 m (about a foot) apart is only

$$F_{grav} = 6.67 \times 10^{-11}\,\frac{Nm^2}{kg^2} \times \frac{7.26\ kg \times 7.26\ kg}{(0.3\ m^2)}$$

$$= 3.9 \times 10^{-8}\ N,$$

or 0.000000039 newtons. This is about equal to the weight on Earth of a single bacterium! Gravity is such an important force in our everyday lives only because Earth is so very massive.

There are two different ways to think about the gravitational force that Earth exerts on an object with mass m. The first is to look at gravitational force from the perspective of Newton's second law of motion: Gravitational force equals mass times gravitational acceleration, or

$$F_{grav} = mg.$$

The other way to think about the force is from the perspective of the universal law of gravitation, which says that

$$F_{grav} = G\,\frac{M_\oplus m}{(R_\oplus)^2}.$$

(Here $M_\oplus$ is the mass of Earth and $R_\oplus$ is the radius of Earth.) Because the force of gravity on an object is what it is, the two expressions describing this force must be equal to each other. $F_{grav} = F_{grav}$, so

$$mg = G\,\frac{M_\oplus m}{(R_\oplus)^2}.$$

The mass m is there on both sides of the equation, so we can divide it out. The equation then becomes

$$g = G\,\frac{M_\oplus}{(R_\oplus)^2}.$$

This is interesting. We started out to calculate the gravitational acceleration experienced by an object of mass m on the surface of Earth. The expression that we arrived at says

FIGURE 3.13 The gravitational force between two objects falls off as 1 divided by the square of the distance between them.

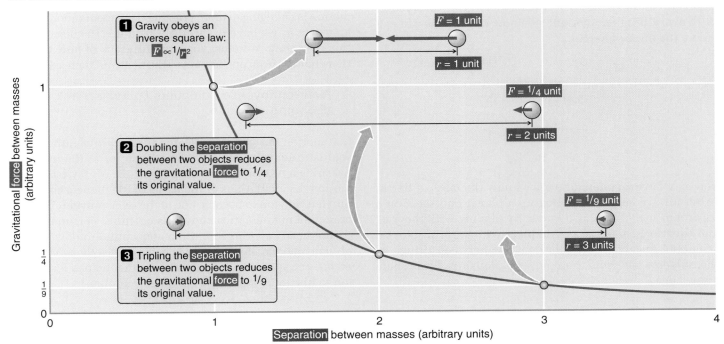

that this acceleration (g) is determined by the mass of Earth ($M_\oplus$) and by the radius of Earth ($R_\oplus$). But the mass of the object itself (m) appears nowhere in this expression. So according to this equation, changing m has no effect on the gravitational acceleration experienced by an object on Earth.

Galileo's discoveries are contained within Newton's laws.

In other words, our play with Newton's laws has shown us that all objects experience the same gravitational acceleration, regardless of their mass. *This is just what Galileo found in his experiments with falling objects!* We already saw that Galileo's work *shaped* Newton's thinking about gravity. Here we find that Galileo's discoveries about gravity are *contained within* Newton's laws of motion and gravitation.

What else can we discover? If we rearrange that last equation a bit so that the mass of Earth is on the left and everything else is on the right, we get

$$M_\oplus = \frac{g(R_\oplus)^2}{G}.$$

Everything on the right side of this equation is known. Galileo measured a value for g, the acceleration due to gravity on the surface of Earth, almost 400 years ago, and in about 235 B.C. Erastosthenes measured the radius of Earth

Newton's laws provide us with the tools to calculate the mass of Earth.

in the manner described in Chapter 2. The universal gravitational constant G is a bit tougher, but it too can be measured in the laboratory—for example, by measuring the slight gravitational forces between two large metal spheres. With everything on the right side now known, we can calculate the mass of Earth:

$$M_\oplus = \frac{gR_\oplus^2}{G}$$

$$= \frac{\left(9.80\,\frac{\text{m}}{\text{s}^2}\right) \times (6.38 \times 10^6\,\text{m})^2}{6.67 \times 10^{-11}\,\frac{\text{m}^3}{\text{kg s}^2}}$$

$$= 5.98 \times 10^{24}\,\text{kg}$$

You may have wondered how we know the mass of Earth. (After all, we cannot just pick up a planet and set it on a bathroom scale.) Now you know. By playing with theories and equations, much as a child plays with building blocks, scientists discover new relationships between things in the universe and from those new relationships comes new knowledge.[7]

3.5 Orbits Are One Body "Falling Around" Another

If you have been following our discussion closely, you may be about ready to take us to task. Kepler may have speculated about the dependence of the solar "influence" that holds the planets in their orbits, and Newton may have speculated that this influence is gravity, but physical law is *not* a matter of speculation! Newton could not measure the gravitational force between two objects in the laboratory directly, so how did he test his universal law of gravitation? Again it was Kepler who provided what Newton lacked. Newton used his laws of motion and his proposed law of gravity to *calculate* the paths that planets should follow as they move around the Sun. When he did so, his calculations predicted that planetary orbits should be ellipses with the Sun at one focus, that equal areas should be swept out during equal times, and that the square of the period of a planet's orbit should vary as the cube of the semimajor axis of that ellipse. In short, Newton's universal law of gravitation *predicted* that planets should orbit the Sun in just the way that Kepler's empirical laws described. This was the moment when it all came together. By *explaining* Kepler's laws, Newton found important corroboration for his law of gravitation. And in the process he moved the cosmological principle out of the realm of interesting ideas and into the realm of testable scientific theories. To see how this happened, we need to look below the surface of how scientists go about connecting their theoretical ideas with events in the real world.

Newton's laws tell us how an object's motion changes in response to forces and how objects interact with each other through gravity. To go from statements about how an object's motion is *changing* to more practical statements about where an object *is*, we have to carefully "add up" the object's motion over time. We must keep careful track of how the motion changes from one instant to the next and then ask where

Newton invented calculus to keep track of changing motions.

that motion has gotten us. Doing this requires rather more sophisticated play than we have time for at the moment. Learning how to make the jump from laws of gravitation and motion to calculations of the paths of the planets about the Sun led Newton to become one of the two coinventors of the branch of mathematics known as *calculus*. Fortunately we do not need calculus to build a conceptual understanding of such motions. Instead we can begin with a *thought experiment*—the same thought experiment that helped lead Newton to his understanding of planetary motions.

---

[7] Newton actually turned this around. He *guessed* at the mass of Earth by assuming it had about the same density as typical rocks. Then he used this mass and the previous equation to get a rough idea of the value of G.

Newton Fires a Shot around the World

Drop a cannonball and it falls directly to the ground, just as any mass does. However, if instead we fire the cannonball out of a cannon that is level with the ground, as shown in **Figure 3.14(a)**, it behaves differently. The ball still falls to the ground in the same time as before, but while it is falling, it is also traveling *over* the ground, following a curved path that carries it some horizontal distance before it finally lands. The faster the cannonball is fired from the cannon (**Figure 3.14(b)**), the farther it will go before hitting the ground.

In the real world this experiment reaches a natural limit. To travel through air the cannonball must push the air out of its way—an effect we normally refer to as *air resistance*—which slows it down. *Air resistance* increases very rapidly with increasing speed. (For example, doubling the speed increases air resistance by much more than two times.) But because this is only a thought experiment, we can ignore such real-world complications. Instead imagine that, having inertia, the cannonball continues along its course until it runs into something. As the cannonball is fired faster and faster, it goes farther and farther before hitting the ground. If the cannonball flies far enough, the curvature of Earth starts to matter. As the cannonball falls toward Earth, Earth's surface "curves out from under it" (**Figure 3.14(c)**). Eventually we reach a point where the cannonball is flying so fast that the surface of Earth curves away from the cannonball at exactly the same rate at which the cannonball is falling toward Earth. This is the case shown in **Figure 3.14(d)**. At this point the cannonball, which always falls *toward the center of Earth,* is literally "falling around the world."

In 1957 the Soviet Union used a rocket to lift an object about the size of a basketball high enough above Earth's blanket of air that wind resistance ceased to be a concern, and Newton's thought experiment became a matter of great practical importance.[8] This object, called *Sputnik I,* was moving so fast that it fell around Earth, just as the cannonball did in Newton's mind. *Sputnik I* was the first human-made object to **orbit** Earth. This is what an orbit is: one object freely falling around another.

In Section 3.1 we asked why astronauts float freely about the cabin of a spacecraft. We now see that it is *not* because they have escaped Earth's gravity. It is Earth's gravity that holds them in their orbit. Instead the answer lies in Galileo's early observation that any object falls in just the same way, regardless of its mass. The astronauts and the spacecraft are both moving in the same direction, at the same speed, and are experiencing the same gravitational acceleration,

---

[8]Actually, wind resistance did not totally cease to be a concern. Objects in orbit within a few hundred kilometers of Earth are moving through the thin outer part of Earth's atmosphere. Friction caused by this thin atmosphere will oppose the object's motion and cause its orbit to *decay*.

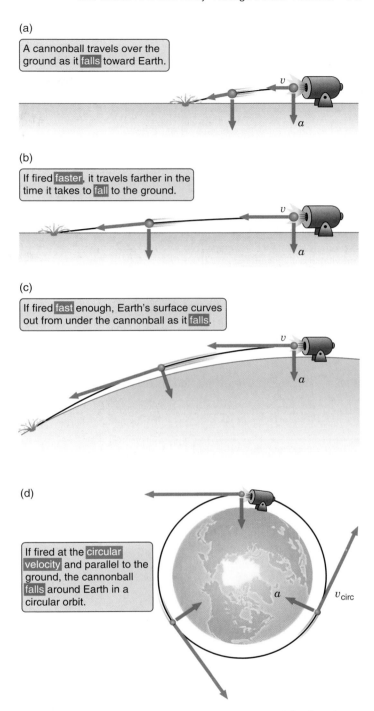

(a)

A cannonball travels over the ground as it falls toward Earth.

(b)

If fired faster, it travels farther in the time it takes to fall to the ground.

(c)

If fired fast enough, Earth's surface curves out from under the cannonball as it falls.

(d)

If fired at the circular velocity and parallel to the ground, the cannonball falls around Earth in a circular orbit.

FIGURE 3.14 Newton realized that a cannonball fired at the right speed would fall around Earth in a circle.

so they fall around Earth together. **Figure 3.15** demonstrates the point. The astronaut is orbiting Earth just as the spacecraft is orbiting Earth. On the surface of Earth our bodies try to fall toward the center of Earth, but the ground gets in the way. We experience our weight when we are standing on Earth because the ground pushes on us hard enough to

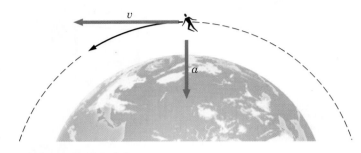

Like Newton's cannonball, an astronaut falls freely around Earth.

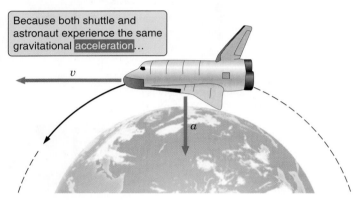

Because both shuttle and astronaut experience the same gravitational acceleration…

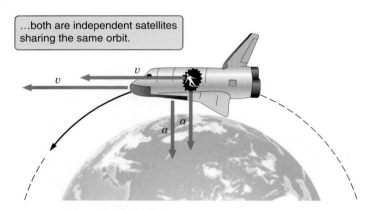

…both are independent satellites sharing the same orbit.

FIGURE 3.15 A "weightless" astronaut has not escaped Earth's gravity. Rather, an astronaut and a spacecraft share the same orbit as they fall around Earth together.

counteract the force of gravity, which is trying to pull us down. In the spacecraft, however, nothing interrupts the astronaut's fall because the spacecraft is falling around

Orbiting means falling around the world.

Earth in just the same orbit. The astronaut is not truly weightless. Instead the astronaut is in **free fall**.

When one object is falling around another, much more massive object, we say that the less massive object is a **satellite** of the more massive object. Planets are satellites of the

Sun, and **moons** are natural satellites of planets. Newton's imaginary cannonball is a satellite. *Sputnik I,* the first artificial satellite (*sputnik* means "satellite" in Russian), was the early forerunner to the spacecraft and the astronauts, which are independent satellites of Earth that conveniently happen to share the same orbit.

How Fast Must Newton's Cannonball Fly?

If fired fast enough, Newton's cannonball falls around the world; but just how fast is "fast enough"? Newton's orbiting cannonball moves along a circular path at constant speed. This type of motion, referred to as **uniform circular motion**, is discussed in more depth in Appendix 7. You are probably familiar with other examples of uniform circular motion. For example, think about a ball whirling around your head on a string, as shown in **Figure 3.16(a)**. If you were to let go of the string, the ball would fly off in a straight line in what-

Centripetal forces maintain circular motion.

ever direction it was traveling at the time, just as Newton's first law says. It is the string that keeps this from happening. The string exerts a steady force on the ball, causing it constantly to change the direction of its motion, always bending its flight toward the center of the circle. This central force is called a **centripetal force**. Using a more massive ball, speeding up its motion, or making the circle smaller so the turn is tighter all increase the force needed to keep the ball from being carried off in a straight line by its inertia.

In the case of Newton's cannonball, there is no string to hold the ball in its circular motion. Instead the centripetal force is provided by gravity, as illustrated in **Figure 3.16(b)**.

Gravity provides the centripetal force that holds a satellite in its orbit.

For Newton's thought experiment to work, the force of gravity must be just right to keep the cannonball moving on its circular path. In this case we can say that

$$\frac{\text{The force needed for}}{\text{uniform circular motion}} = \frac{\text{Force provided}}{\text{by gravity}}.$$

In Appendix 7 we derive an expression for the centripetal force needed to keep an object moving in a circle at a steady speed. If we put that expression on the left side of this equation and the universal law of gravitation on the right side (and then do some algebra), we arrive at

$$v_{\text{circ}} = \sqrt{\frac{GM}{r}},$$

(a)

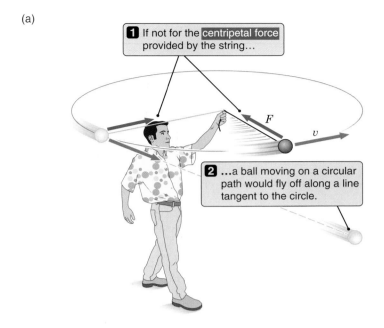

(b)

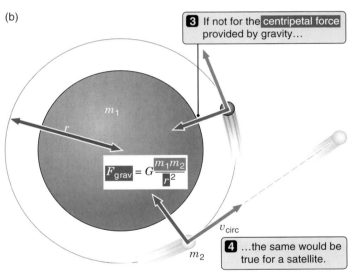

FIGURE 3.16 (a) A string provides the centripetal force that keeps a ball moving in a circle. (We have ignored the smaller force of gravity that also acts on the ball.) (b) Similarly, gravity provides the centripetal force that holds a satellite in a circular orbit.

where M is the mass of the orbiting object and r is the radius of the circular orbit. This value is called the **circular velocity**.

Here is the result we were looking for. If a satellite is in a stable circular orbit, then it *must* be moving at a velocity v_{circ}, where v_{circ} is given by this expression. *If the satellite were moving at any other velocity, it would not be moving in a circular orbit.* Remember the cannonball. If the cannonball were moving too slowly, it would drop below the circular path and hit the ground. Similarly, if the cannonball

were moving too fast, its motion would carry it above the circular orbit. Only a cannonball moving at just the right velocity—the circular velocity— will fall around Earth on a circular path. The circular velocity at Earth's surface is about 8 km/s (see **Foundations 3.2**).

Planets Are Just Like Newton's Cannonball

We can apply this same idea to the motion of Earth around the Sun. In the case of Earth's orbit, we already know from our discussion of the aberration of starlight that Earth travels at a speed of 2.98×10^4 m/s or (29.8 km/s) on its orbit

---

We use Earth's orbit to calculate the Sun's mass.

---

about the Sun. We also know that the radius of Earth's orbit is 1.50×10^{11} m. So we know everything about the circular orbit except for the mass of the Sun. If we can skip a step or two, a little algebra applied to the equation for v_{circ} gives the mass of the Sun ($M_\odot$) as

$$M_\odot = \frac{(v_{circ})^2 \times r}{G}$$

$$= \frac{\left(2.98 \times 10^4 \, \frac{m}{s}\right)^2 \times (1.50 \times 10^{11} \, m)}{6.67 \times 10^{-11} \, \frac{m^3}{kg \, s^2}}$$

$$= 1.99 \times 10^{30} \, kg.$$

Once again a bit of play has taken us places we might not have imagined. We began with Newton's thought experiment about a cannonball fired around the world, and ended up knowing the mass of the star our planet orbits.

As long as we are at it, we can carry our game one step further. Kepler's third law talks about the time for a planet to complete one orbit about the Sun, known as the period P of a planet's orbit. The time it takes an object to make one trip around a circle is just the circumference of the circle ($2\pi r$) divided by the object's speed. (Time equals distance divided by speed.) If the object is a planet in a circular orbit about the Sun, then its speed must be equal to the circular velocity that we calculated. Bringing this together, we get

$$\text{Period } P = \frac{\text{Circumference of orbit}}{\text{Circular velocity}} = \frac{2\pi r}{\sqrt{\frac{GM_\odot}{r}}}.$$

Some algebra gives us

$$P^2 = \frac{4\pi^2}{GM_\odot} \times r^3.$$

Once again our play has really gotten us somewhere interesting. The square of the period of an orbit is equal to a

FOUNDATIONS 3.2

Circular Velocity

It is interesting to put some values into the equation for circular velocity to see how fast Newton's cannonball would really have to travel. The radius of Earth is 6.38 × 10^6 m, the mass of Earth is 5.98 × 10^{24} kg, and the gravitational constant is 6.67 × 10^{11} m³/kg s². (We have seen how each of these values is measured.) Putting these values into the expression for v_{circ} gives

$$v_{circ} = \sqrt{\frac{\left(6.67 \times 10^{-11} \frac{m^3}{kg\,s^2}\right) \times (5.98 \times 10^{24}\,kg)}{6.38 \times 10^6\,m}}$$

$$= 7.9 \times 10^3 \text{ m/s.}$$

Newton's cannonball would have to be traveling about 8 km/s—over 28,000 km/h—to stay in its circular orbit. That's well beyond the reach of a typical cannon but just what we routinely accomplish with rockets.

constant ($4\pi^2/GM_\odot$) times the cube of the radius of the orbit. *This is just Kepler's third law applied to circular orbits.* This is how Newton showed, at least in the special case of a circular orbit, that Kepler's third law—his beautiful "harmony of the worlds"—is a direct consequence of the way objects move under the force of gravity. A more complete treatment of the problem—the problem for which Newton invented calculus—shows that Newton's laws of motion and gravitation predict *all* of Kepler's empirical laws of planetary motion—for elliptical as well as circular orbits.

You may well be wondering why we are taking such a long and sometimes strenuous excursion through the work of Galileo, Kepler, and Newton. Sometimes when we are hiking in the mountains it is hard to see the summit from the perspective of the trail. Now that we have arrived at the top of this particular pass, we can look around and appreciate what we have gained. When Newton carried out his calculations, he found that his laws of motion and

Kepler's laws provided an empirical test of Newton's laws.

gravitation predicted elliptical orbits that agree exactly with Kepler's empirical laws. *This is how Newton tested his theory that the planets obey the same laws of motion as cannonballs and how he confirmed that his law of gravitation is correct.* Had Newton used *any* other rule for gravity, he would have predicted something different for the way planets orbit the Sun, and these predictions would have failed Kepler's observational test. If Newton's laws

had failed to predict motions that agreed with Kepler—if Newton's predictions had not been borne out by observation—then Newton would have had to throw them out and go back to the drawing board! All of his work would have fallen onto that large heap of beautiful ideas that do not pass nature's test.

We promised we would show you *how* science works. Well, this is how science works. Kepler's empirical rules for planetary motion pointed the way for Newton and provided the crucial observational test for Newton's laws of motion and gravitation. At the same time, Newton's laws of motion and gravitation provided a powerful new understanding of why planets and satellites move as they do. Theory and empirical observation work together hand in hand, and our understanding of the universe strides forward.

Rather than hiding from challenges that might prove their theories wrong, scientists actively seek and confront nature's judgment. And even if one scientist fails to find and probe the potential weaknesses of her favorite theory, she can rest assured that another scientist will. As discussed in **Excursions 3.3**, a well-tested scientific theory is a far cry from the "theory" that Elvis was abducted by aliens. A well-tested scientific theory is about as close to certain knowledge as we humans can come. This is the main difference between science and the pseudosciences—astrology, creationism, intelligent design, quack medicine, homeopathy, parapsychology, numerology, and a host of others—which hide slipshod thinking and untested, untestable, or even disproven speculation behind scientific-sounding jargon, and which actively ignore the wealth of evidence against them. Here is the key to the difference between science and what *pretends* to be science. In our discussion of Newton we have focused on the basis of scientific knowledge and on how science is done. If you run across activities that are *not* done this way—if they do not make testable predictions and do not search for and embrace every observation or experiment that might in principle prove them wrong—then they are *not* science, no matter how many "authorities" would have you believe otherwise.

Real-World Orbits Are Not Circles

So far we have concentrated on circular orbits and simply asserted that everything works out for elliptical orbits as well. As we pointed out before, it is much more difficult to carry out these calculations for ellipses than for circles, so we will leave that particular peak unscaled on this journey. Even so, from our current vantage point we can get an idea of what that peak looks like—of how elliptical orbits differ

---

EXCURSIONS 3.3

"After All, It's Only a Theory"

Science is sometimes misunderstood because of the special ways that scientists use everyday words. An example is the word *theory*. In everyday language a theory may mean something that is little more than a conjecture or a guess: "Have you any theory about who might have done it?" "My theory is that a third party could win the next election." In everyday parlance a theory is something worthy of little serious regard. "After all," we say, "it is only a theory."

In stark contrast, a *scientific* theory is a carefully constructed proposition that takes into account all the relevant data and all our understanding of how the world works, and makes testable predictions about the outcome of future observations and experiments. A theory is a well-developed idea that is ready to be confronted by nature. A well-corroborated theory is a theory that has survived many such tests. Rather than being simple speculation, scientific theories represent and summarize bodies of knowledge and understanding that provide our fundamental insights into the world around us. A successful and well-corroborated theory is the pinnacle of human knowledge about the world.

Theories fill a place in a loosely defined hierarchy of scientific knowledge. In science the word *idea* has its everyday use. An idea is just a notion about how something might be. A **hypothesis** is an idea that leads to testable predictions. A hypothesis may be the forerunner of a scientific theory, or it may be based on an existing theory, or both. When an idea has been thought about carefully enough, has been tied solidly to existing theoretical and experimental knowledge, and makes testable predictions, then that idea has become a **theory**. Scientists build **theoretical models** that are used to connect theories with the behavior of complex systems. Competing theories are ultimately decided among on the basis of the success of their predictions. Some theories become so well-tested and are of such fundamental importance that we come to refer to them as **physical laws**. A scientific **principle** is a general idea or sense about how the universe is that guides our construction of new theories. **Occam's razor**, for example, is a guiding principle in science that says that when faced with two hypotheses that explain some phenomenon equally well, we should adopt the simpler of the two.

Unfortunately, there is room for confusion here because our terminology does not always follow these categories. For example, Newton's laws are physical laws, but Kepler's laws are really empirical rules (based on data and observation, but not on more fundamental principles or laws). Furthermore, science is a living enterprise, and the status of ideas changes in time. For example, principles can themselves become theories. The cosmological principle has been instrumental in shaping countless theories about the universe. In turn, the success of those theories has effectively become the experimental test of the cosmological principle itself.

Newton's theory of motion illustrates the point of our digression. This theory is the basis of our worldwide technological civilization and comes about as close to certain knowledge as humankind can hope for. Yet to scientists this knowledge remains a theory, subject to observational and experimental tests via the formal and rigorous application of the scientific method.

So think twice the next time you hear someone casually dismiss some body of scientific knowledge as being "only a theory."

from circular orbits, and why Newton's calculations work out the way they do.

Begin again with a satellite in a circular orbit about Earth. The satellite is traveling at the circular velocity, so it remains the same distance from Earth at all times, neither speeding up nor slowing down in its orbit. But now change the rules a bit. What if the satellite were in the same place in its orbit and moving in the same direction, but traveling *faster* than the circular velocity? The pull of Earth is as strong as ever, but because the satellite has a greater speed, its path is not bent by Earth's gravity sharply enough to hold it in a circle. So the satellite begins to climb above a circular orbit.

As the distance between Earth and the satellite begins to increase, an interesting thing starts to happen. Think about a ball thrown into the air, as shown in **Figure 3.17(a)**. As the ball climbs higher, the pull of Earth's gravity opposes its motion, slowing the ball down. The ball climbs more and more slowly until its vertical motion stops for an instant, then is reversed; the ball begins to fall back toward Earth, picking up speed along the way. Our satellite does exactly the same thing as the ball. As the satellite climbs above a circular orbit and begins to move away from Earth, Earth's gravity opposes the satellite's outward motion, slowing the satellite down. The farther the satellite pulls away from Earth, the more slowly the satellite moves—just as happened with the ball thrown into the air. And just like the ball, the satellite reaches a maximum height on its curving path, then begins falling back toward Earth. Now as the satellite falls back in toward Earth, Earth's gravity is pull-

ing it along, causing it to pick up more and more speed as it gets closer and closer to Earth.

What is true for a satellite orbiting Earth in an elliptical orbit is also true for any object in an elliptical orbit, including a planet orbiting the Sun. As we saw earlier, Kepler's Law of Equal Areas says that a planet moves fastest when it is closest to the Sun and slowest when it is farthest from the Sun. Now we know why. As shown in **Figure 3.17(b)**, planets lose speed as they pull away from the Sun, then gain that speed back as they fall inward toward the Sun.

Newton's laws do more than explain Kepler's laws. Newton's laws also predict different types of orbits that are beyond Kepler's empirical experience. **Figure 3.18** shows a whole series of satellites, each with the same point of closest approach to Earth but with different velocities at that point. A look at the figure shows that the greater the speed a satellite has at its closest approach to Earth, the farther the satellite is able to pull away from Earth, and the more eccentric its orbit becomes. Yet no matter how eccentric it becomes, as long as it remains elliptical, an orbit will eventually bring a satellite back to the planet that it orbits or bring a planet back to the Sun.

You might imagine, though, that somewhere in this sequence of faster and faster satellites there comes a point of no return—a point when the satellite is moving so fast that gravity is unable to reverse its outward motion, so the satellite coasts away from Earth, never to return. This indeed is possible. The lowest speed at which this happens is called the **escape velocity**. If we were to work through the calculation, we would find that the escape velocity is a factor

FIGURE 3.17 (a) A ball thrown into the air slows as it climbs away from Earth, then speeds up as it heads back toward Earth. (b) A planet on an elliptical orbit around the Sun does the same thing. (Although no planet has an orbit as eccentric as that shown, the orbits of comets can be far more eccentric.)

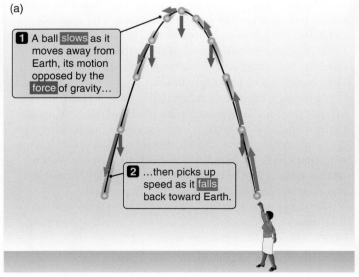

(a)

1 A ball slows as it moves away from Earth, its motion opposed by the force of gravity…

2 …then picks up speed as it falls back toward Earth.

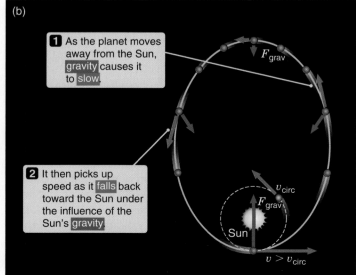

(b)

1 As the planet moves away from the Sun, gravity causes it to slow.

2 It then picks up speed as it falls back toward the Sun under the influence of the Sun's gravity.

F_{grav}

v_{circ}

F_{grav}

Sun

$v > v_{circ}$

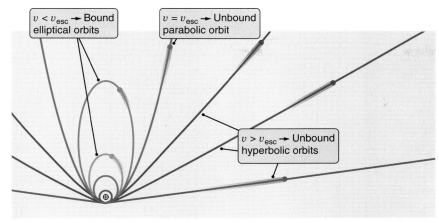

FIGURE 3.18 (a) A range of different orbits that share the same perigee but differ in velocity at that point. (b) Perigee velocities for the orbits in (a). An object's velocity determines the orbit shape and whether the orbit is bound.

A satellite moving fast enough will escape a planet's gravity.

of $\sqrt{2}$ or $1.414\ldots$, larger than the circular velocity. This can be expressed as

$$v_{\text{esc}} = \sqrt{\frac{2GM}{R}} = \sqrt{2}\, v_{\text{circ}}.$$

Look at this equation for a minute to be sure it makes sense to you. The larger the mass (M) of a planet, the stronger its gravity, so it stands to reason that a more massive planet would be harder to escape from than a less massive planet. Indeed, the equation says that the more massive the planet, the greater the required escape velocity. Also, it stands to reason that the closer we are to the planet, the harder it will be to escape from its gravitational attraction. Again, the equation confirms our intuition. As the distance R becomes larger (that is, as we get farther from the planet), v_{esc} becomes smaller (in other words, it is easier to escape from the planet's gravitational pull). For objects at the surface of Earth, the escape velocity is 11.2 km/s (about 40,000 km/h). Once an object reaches escape velocity, the shape of its orbit is no longer an ellipse. But when Newton solved his equations of motion, he found that ellipses are not the only possible shape an orbit can have.

Unbound orbits are hyperbolas or parabolas. If a satellite's velocity is *less* than the escape velocity (v_{esc}), its orbit

Bound orbits are ellipses.

will have an elliptical shape. Elliptical orbits close on themselves. Thus an object traveling in an elliptical orbit is des-

tined to follow the same path over and over again.[9] For this reason *elliptical orbits* are also called **bound orbits**: The satellite is bound to the object it is orbiting about. If a satellite has a velocity *greater* than the escape velocity, it is not bound to the object it is orbiting. Such orbits are called **unbound orbits** and have a hyperbolic shape. A hyperbola does not close like an ellipse but instead keeps opening up forever. A space probe traveling on a *hyperbolic orbit* makes only a single pass around a planet, then is back off into deep

Unbound orbits are hyperbolas or parabolas.

space, never to return. The third type of orbit is the borderline case where the orbiting object moves at exactly the escape velocity. If it had any less velocity it would be traveling in a bound elliptical orbit; any more and it would be moving on an unbound hyperbolic orbit. A body moving with a velocity *equal* to the escape velocity follows a *parabolic orbit*. Like the hyperbolic orbit, the parabolic orbit involves only a single pass by the planet. As an object traveling in a parabolic orbit moves away from a planet, its velocity relative to the planet gets closer and closer to zero. An object traveling in a hyperbolic orbit always has excess velocity relative to the planet, even when it has moved infinitely far away.

[9] Strictly speaking, this is true only for a single body orbiting a second body. The presence of other bodies can cause the ellipse to not quite close, causing the orientation of the ellipse to slowly swing around in the orbital plane. This is referred to as *orbital precession*. Relativistic effects can also cause orbital precession, as we will see in the case of Mercury's orbit in Chapter 17.

Newton's Theory Is a Powerful Tool for Measuring Mass

As mentioned earlier, Kepler's empirical laws describe the motion of the planets but do not explain them. On the basis of Kepler's laws alone we might imagine that angels carry the planets around in their orbits, just as many people believed during the 16th century! Newton's derivation of Kepler's laws changed all of that. Newton showed that the *same* physical laws that describe the flight of a cannonball on Earth—or the fall of the apocryphal apple on his head —also describe the motions of the planets through the heavens. In this way Newton shattered the prevailing concept of the heavens and Earth, and at the same time opened up an entirely new way of investigating the universe. Copernicus may have dislodged Earth from the center of the universe and started us on the way toward the cosmological principle, but it was Newton who moved the cosmological principle out of the realm of philosophy and into the realm of testable scientific theory. And it was through Newton's work that **astrophysics** was born.

Not only is Newton's method more philosophically satisfying than simple empiricism—it is far more powerful as well. We have already seen, for example, how Newton's laws can be used to measure the mass of the Sun and Earth. This could never be done with Kepler's empirical rules. This is especially important when we remember that Newton's laws apply to *all* objects, not just the Sun and Earth. This fact will prove handy as we continue our journey.

Astronomers often rearrange Newton's form of Kepler's third law to read

$$M = \frac{4\pi^2}{G} \times \frac{A^3}{P^2}.$$

Everything on the right side of this equation is either a constant (such as 4, π, and G) or a quantity we can measure (like the semimajor axis A and period P of an orbit). The left side of the equation is the mass of the object at the focus of the ellipse.

It is important to note that we cut a couple of corners to get to this point. For one thing, we arrived at this relationship by thinking about circular orbits, then simply asserting that it holds for elliptical orbits as well. Another corner we cut was assuming that a low-mass object such as a cannonball is orbiting a more massive object such as Earth. Earth's gravity has a strong influence on the cannonball; but as we have seen, the cannonball's gravity has little effect on Earth. For this reason we can imagine that Earth remains motionless while the cannonball follows its elliptical orbit. In the same way it is a good approximation to say that the Sun remains motionless as the planets orbit about it.

This picture changes when two objects are closer to having the same mass. In this case *both* objects experience significant accelerations in response to their mutual gravitational attraction. We now must think of the two objects as falling around *each other,* with each mass moving on its own elliptical orbit around a point (the **center of mass**) located between the two. Yet even in this most general case the equation remains valid. The mass M now refers to the *sum* of the masses of the two objects. So if we can measure the size and period of an orbit—*any* orbit—then we can use this equation to calculate the mass of the orbiting objects. This is true not only for the masses of Earth and the Sun, but also for the masses of other planets, distant stars, our galaxy and distant galaxies, and vast clusters of galaxies. In fact, it turns out that *almost all of our knowledge about the masses of astronomical objects comes directly from the application of this one equation.* In a sense, this single equation even allows us to tackle the question "What is the mass of the universe itself?"

We have come a long way in this chapter. We have gone from Copernicus's simple picture of planets moving on circles around the Sun to the pinnacle of Newton's comprehensive laws of motion and gravitation. The work of Copernicus, Galileo, Kepler, and Newton was not just *a* scientific revolution, it was *the* scientific revolution, which changed forever not only our view of the universe but also our very notion of what it means "to know."

Before getting too carried away, however, we need to put our accomplishment in perspective. Newton's grand theoretical edifice leaves us with the feeling that we understand the "how come" of Kepler's laws and a great deal more—as if we have lifted up the hood of the universe and peeked underneath. And indeed we have. But remember that we have not talked about *why* acceleration equals force divided by mass, or *why* objects have inertia, or *why* one inertial reference frame is as good as any other. These are good, fundamental laws—so fundamental that they form the basis of our understanding of the world we live in and the foundation of our entire technological civilization. Yet the law equating acceleration with force divided by mass, the inverse square law of gravity, and the law of inertia are still basically empirical facts about how objects are observed to move. These are not articles of faith, but rather scientific hypotheses, the validity of which remains to this day subject to test through experimentation and observation.

Summary

- Proportionality describes patterns and relationships in nature.

- The scientific method and nature provide the distinction between science and pseudoscience.

- Newton's three laws govern the motion of all objects.

- Unbalanced forces cause changes in motion.

- Mass is the property of matter that gives it resistance to changes in motion.

- Gravity is a force between any two objects due to their masses.

- *All* objects near Earth's surface fall with the same acceleration, *g*.

- Planets orbit the Sun in elliptical orbits.

- Unbound orbits are parabolas or hyperbolas.

Seeing the Forest through the Trees

The story of planetary motions is also the story of how we as a species learned to do science. No one taught Copernicus or Galileo or Kepler or Newton about the differences between empiricism and theory or about how the scientific method could be used to test their ideas. They had to figure these lessons out on their own as they went along. Yet the accomplishments they made remain near the top of the all-time intellectual feats of humankind, and the trail they blazed pointed the way for all that was to come. Galileo offered powerful insights into the nature of matter and gravity. Kepler built on Copernicus's revolutionary ideas about planetary motions and uncovered three empirical rules that showed the orbits of all the planets to be reflections of the same underlying patterns. Newton combined the two sets of ideas: He built Galileo's insights into a powerful theoretical edifice describing the motions of all objects and then tested this edifice against Kepler's empirical reality. The culmination of this intellectual campaign was far greater than the sum of its parts. The walls separating the heavens and Earth came down once and for all, and the science of astrophysics was born.

The model for science that emerged during this era remains the template for how science is done to this day. Careful empiricism uncovers patterns in nature in need of explanation. Scientific theory seeks to discover the fundamental truths underlying all things. And in the meeting of the two—the test of theory against the unforgiving and stalwart challenge of empirical fact—new knowledge and understanding emerge.

Newton's work became the cornerstone of what is often referred to as classical mechanics. All objects have inertia. An object will continue to move in a straight line at a constant speed unless an unbalanced force acts to change its motion. Mass is the property of matter that resists changes in motion. Every force is matched by another force that is equal in magnitude but opposite in direction. Gravity is a force between any two masses, proportional to the product of the two masses and inversely proportional to the square of the distance between them. Putting all of this together, we find that objects "fall around" the Sun and Earth on elliptical, parabolic, or hyperbolic paths. Orbits are ultimately given their shape by the gravitational attraction of the objects involved, which in turn is a reflection of the mass of these objects. We now understand that it is gravity that holds the universe together, giving planets, stars, and galaxies their very shapes as well as controlling their motions through space. Using Newton's theoretical insight we look backward along this chain, turning observations of the motions of objects throughout the universe into measurements of the masses of objects that no human has ever, or in most cases, will ever visit.

This brings us to the next stage of our journey. If the ancients could have stepped off Earth and touched the planets, they would never have fallen into the conceptual errors that muddled our thinking for millennia—but they had no such luxury. Today we have sent robotic surrogates to all of the planets in the Solar System except one, and humans have walked on the surface of the Moon. Even so, we have made only the most cursory visits to our immediate neighborhood. Even the nearest stars remain thousands of times more distant than the most far-flung of our robotic planetary explorers. For the most part we, like the ancients, are left with nothing on which to base our knowledge of the universe but the signals reaching us from across space. Far and away, the most common of these signals is electromagnetic radiation, which includes light (such as the light by which you are reading this book). Our ability to interpret these signals depends on what we know about light. What is it? How does it originate? How does it interact with matter? What changes does it experience during its journey? In Chapter 4 we turn to these questions.

Key Terms

empirical science, p. 61
Kepler's laws, p. 61

Student Questions

THINKING ABOUT THE CONCEPTS

1. The orbits of the planets around the Sun and the orbits of satellites around these planets are always ellipses rather than perfect circles. Why?

2. When riding in a car, we can sense changes in speed or direction through the forces the car applies on us. Do we wear seat belts in cars and airplanes to protect us from speed or from acceleration? Explain your answer.

3. Weight on Earth is proportional to mass. On the Moon weight is also proportional to mass, but the constant of proportionality is different on the Moon than it is on Earth. Why?

4. Erich von Daniken proposed the theory that Earth was visited by extraterrestrials in the remote past. Would you regard this as scientific or pseudoscientific theory? Is the theory falsifiable? Can you think of any tests that could support or refute the theory?

5. Picture a swinging pendulum. During a single swing, the bob of the pendulum first falls toward Earth and then moves away until the gravitational force between it and Earth finally stops its motion. Would a pendulum swing if it were in orbit? Explain your reasoning.

6. Had Kepler lived on one of a group of planets orbiting a star three times as massive as our Sun, would he have deduced the same empirical laws? Explain your answer.

7. Kepler's and Newton's laws all tell us something about the motion of the planets, but there is a fundamental difference between them. What is the difference?

8. In 1920 a *New York Times* editor refused to publish an article based on rocket pioneer Robert Goddard's paper that predicted space flight, saying that "rockets could not work in outer space because they have nothing to push against" (a statement the *Times* did not retract until July 20, 1969, the date of the *Apollo 11* Moon landing). You, of course, know better. What was wrong with the editor's logic?

9. Aristotle taught that the natural state of all objects is to be at rest. Even though this seems consistent with what we observe around us, explain why Aristotle's conclusion was wrong.

10. Imagine a planet moving in a perfectly circular orbit around the Sun. Because the orbit is circular, the planet is moving at a constant speed. Is this planet experiencing acceleration? Explain your answer.

11. An astronaut standing on Earth could easily lift a wrench having a mass of 1 kg, but not a scientific instrument with a mass of 100 kg. In the International Space Station she is quite capable of manipulating both, although the scientific instrument moves more slowly than the wrench. Explain why.

12. Two comets are leaving the vicinity of the Sun, one traveling in an elliptical orbit and the other in a hyperbolic orbit. What can you say about the future of these two comets? Would you expect either of them to eventually return?

APPLYING THE CONCEPTS

13. During the latter half of the 19th century a few astronomers thought there might be a planet circling the Sun inside Mercury's orbit. They even gave it a name, Vulcan. We now know that Vulcan does not exist. If there were such a planet with an orbit a fourth the size of Mercury's, what would be its orbital period relative to that of Mercury?

14. Earth speeds along at 29.8 km/s in its orbit. Neptune's nearly circular orbit has a radius of 4.5×10^9 km, and the planet takes 164.8 years to make one trip around the Sun. Calculate the rate at which Neptune plods along in its orbit.

15. Venus's circular velocity is 35.03 km/s, and its orbital radius is 1.082×10^8 km. Calculate the mass of the Sun.

16. At the surface of Earth, the escape velocity is 11.2 km/s. What would be the escape velocity at the surface of a very small asteroid having a radius 10^{-4} of Earth's and a mass 10^{-12} of Earth's? If you were standing on the asteroid and threw a baseball with a strong pitch, what would happen to it?

17. How long does it take Newton's mythical cannonball, moving at 7.9 km/s just above Earth's surface, to complete one orbit around Earth?

18. Using values given in the appendixes for G and for Earth's radius and mass, show that the acceleration of gravity at the surface of Earth is 9.80 m/s².

19. What does an acceleration of 9.80 m/s² mean? If you jumped from a stationary balloon, after 1 second you would be falling at a speed of 9.80 m/s. After 2 seconds your speed would be $2 \times 9.80 = 19.6$ m/s or slightly more than 70 km/h! In the absence of any air resistance, how fast would you be falling after 20 seconds?

20. Weight refers to the force of gravity acting on a mass. We often calculate the weight of an object by multiplying its mass by the local acceleration due to gravity. The value of gravitational acceleration on the surface of Mars is 0.39 times that on Earth. Assume your mass is 85 kg. Then your weight on Earth is 833 N ($833\ N = m \times g = 85\ kg \times 9.8\ m/s^2$). What would be your mass and weight on Mars?

StudySpace
wwnorton.com/astro21
provides a Study Plan for each chapter that includes a reading outline, animations, keyword flash cards, and gradebook-enabled multiple-choice quizzes. From StudySpace you can also access premium content in the ebook and SmartWork.

Then God said, "Let there be light,"
and there was light.
And God saw that the light was good;
and God separated the light from
the darkness.
And God called the light day, and the
darkness He called night.
And there was evening and there was
morning, one day.

GENESIS 1:3–5

The setting sun colors clouds over the Australian bush.

CHAPTER 4

Light

4.1 "Let There Be Light"

Light is a fundamental part of our experience of the world because it is through light that we most directly perceive the world beyond our physical grasp. The symbolic nature of light is everywhere in our language. A close companion may be the "light of our lives." To understand something is to "see it." A "bright idea" is symbolized by a lightbulb going off over our heads. As we leave our ignorance behind, we become "enlightened." A symbol of hope is a "light at the end of the tunnel." Throughout our language and culture, light is a metaphor for knowledge and information. Those who have lost their sight face a challenge greater than most of us can imagine. At the same time, light plays another, even more important role in our lives. Light from the Sun warms Earth, drives the wind and rain, and powers photosynthesis in plants, which lie at the bottom of the terrestrial food chain. The energy you expend as you move through your day arrived on the planet in the form of light.

The role of light in astronomy closely parallels the role of light in our everyday existence. Your mental picture of an astronomer is probably of a denizen of the night with head bent to the eyepiece of a telescope, peering at the distant universe. Although this picture is a bit quaint in this modern era of electronic cameras and telescopes orbiting Earth, in many ways it remains on the mark. Whether the telescope is on a mountaintop, in orbit about Earth, or part of a spacecraft hurtling toward a comet makes little difference. Our knowledge of the universe beyond Earth comes overwhelmingly from light given off or reflected by astronomical objects. Fortunately, light is a *very* informative messenger. It carries with it information about the temperatures

KEY CONCEPTS

Unlike the physicist or the chemist, who has control over the conditions in a laboratory, the astronomer must try to glean the secrets of the universe from the light and other particles such as *neutrinos* that reach us from distant objects. On this leg of our journey we turn our attention to light, a most informative messenger, and find that

- Light is an electromagnetic wave with a spectrum extending far beyond the colors of the rainbow.

- Light is *also* a stream of particles called photons.

- Reconciling the wave and particle nature of light and matter points beyond Newton's physics and challenges our everyday ideas about what is "real."

- Measurements of the speed of light also require that we think beyond classical physics and reassess our understanding of time and space.

- The wave/particle nature of light and matter gives different types of atoms unique spectral "fingerprints" that we can use to measure the composition and properties of distant objects.

- Temperature measures the thermal energy of an object and determines the amount and spectrum of light that a dense object emits.

- Light is not only a messenger but is also a way in which energy is carried throughout the universe.

of objects, what they are made of, their speed, and even the nature of the material that the light passed through on its way to Earth.

Yet light plays a far larger role in astronomy than just being a messenger. Light is one of the main ways in which energy is transported throughout the universe. Light carries energy generated in the heart of a star outward through the star and off into space. From stars to planets to vast clouds of gas and dust filling interstellar space—absorption of light heats objects up while emission of light cools them off.

Although light allows us to see the world, we cannot actually "see" light. That is, we do not see light in the same way that we see a tree or we see the stars at night. Light is the messenger but is itself invisible. It is hard to study something that cannot be seen, so an understanding of light was a long time coming. The property of light that is easiest to *try* to measure is the speed at which it travels. This might seem straightforward, yet the answer led 19th- and 20th-century physicists to change how we think about the very fabric of space and time. As we continue our journey we will find that light sets the standard for what we mean by "when," "where," or "how fast" because nothing can travel faster than light.

4.2 Our Picture of Light Evolved with Time

Suppose you are a scientist living in the 16th or 17th century. How might you go about measuring the speed of light? One obvious way would be to have a friend stand on a hilltop far away. You uncover a lantern, and the instant your friend sees the light from your lantern, your friend uncovers her own lantern. The time it takes from when you uncover your lantern to when you see your friend's light will be the light's round-trip travel time—plus, of course, your friend's reaction time. Galileo measured the speed of sound in just this way, but when he tried this method on light, he failed. He could not measure any delay. Galileo concluded that the speed of light must be very great indeed, possibly even infinite.

If light travels so rapidly, then to measure its speed we will need either very large distances over which to measure its flight or very good clocks. Galileo had neither at his disposal, but by the end of the 18th century astronomers had both. The great distances were the distances between the planets, whereas the good clock was provided courtesy of Kepler and Newton. According to Newton's derivation of Kepler's laws, orbital periods should be completely constant, with each orbit taking exactly as much time as the orbit before. This applies to moons orbiting planets just as it applies to planets orbiting the Sun.

In the 1670s **Ole Rømer** (1644–1710) was studying the moons of Jupiter, taking measurements of the times when each moon disappeared behind the planet. Much to his amazement Rømer found that rather than maintaining a regular schedule, the observed times of these events would slowly drift in comparison with predictions. Sometimes the moons disappeared behind Jupiter too soon, and at other times they went behind Jupiter later than expected. Rømer realized that the difference depended on where Earth was in its orbit. If he began tracking the moons when Earth was closest to Jupiter, then by the time Earth was farthest from Jupiter, the moons were a bit over 16½ minutes "late." But if he waited until Earth was once again closest to Jupiter, the moons "made up" the lost time and once again passed behind Jupiter at the predicted times.

It is often the case in science that a difference between theoretical predictions and experimental results points the way to new knowledge, and Rømer's work was no exception. Rømer correctly surmised that rather than a failure of Kepler's laws, he was seeing the first clear evidence that light travels at a finite speed. As shown in **Figure 4.1**, the moons appeared "late" when Earth was farther from Jupiter because of the time needed for light to travel the extra distance between the two planets. Over the course of Earth's yearly trip around the Sun, the distance between Earth and

---

Rømer used Jupiter's moons to measure the speed of light.

Jupiter changes by 2 AU (astronomical units), which is about 3×10^{11} m. The speed of light equals this distance divided by Rømer's 16.7-minute delay, or about 3×10^8 m/s. The value Rømer actually announced in 1676 was a bit on the low side—2.25×10^8 m/s—because the length of 1 astronomical unit was not well known. But Rømer's result was more than adequate to make the point. The speed of light is very great indeed! The orbiting space shuttle moves around Earth at a dazzling speed of about 28,000 km/h (almost 8,000 m/s). Light travels almost 40,000 times faster than this. It could circle Earth in only ⅐ of a second. No wonder Galileo's attempts to measure the speed of light failed!

A good deal of work has been done to improve on Rømer's original result. Modern measurements of the speed of light made with the benefit of high-speed electronics give a value of 2.99792458×10^8 m/s in a vacuum. As of October 1983 the length of a meter is now *defined* as the distance traveled by light in a vacuum in 1/299,792,458 of a second.

The speed of light is 300,000 km/s in a vacuum.

The speed of light in a vacuum, about 300,000 km/s, is one of nature's fundamental constants, usually written as *c*. Keep in mind, however, that this is true *only* in a vacuum.

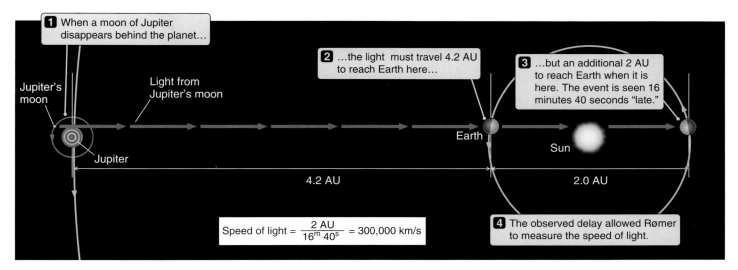

1 When a moon of Jupiter disappears behind the planet...

2 ...the light must travel 4.2 AU to reach Earth here...

3 ...but an additional 2 AU to reach Earth when it is here. The event is seen 16 minutes 40 seconds "late."

Jupiter's moon

Light from Jupiter's moon

Earth

Sun

Jupiter

4.2 AU

2.0 AU

$$\text{Speed of light} = \frac{2 \text{ AU}}{16^m 40^s} = 300{,}000 \text{ km/s}$$

4 The observed delay allowed Rømer to measure the speed of light.

FIGURE 4.1 Ole Rømer measured the speed of light by noting that apparent delays in the orbital motions of Jupiter's moons depend on the distance between Earth and Jupiter.

The speed of light through any medium, such as air or glass, is *always* less than c. We refer to the ratio of light's speed in a vacuum to its speed, v, in a medium as the medium's **index of refraction**:

$$n = \frac{c}{v}.$$

For typical glass n is approximately 1.5. Rearranging this equation we find the speed of light in glass is

$$v = \frac{c}{n} = \frac{300{,}000 \text{ km/s}}{1.5} = 200{,}000 \text{ km/s}.$$

We will come back to the index of refraction in Chapter 5 when we discuss refraction and refracting telescopes.

Recall that in the opening chapter we spoke of distances expressed not in kilometers or miles, but in units of time. For example, the Moon's distance is such that it takes light $1\frac{1}{4}$ seconds to travel between Earth and Moon. In other words, we can say the Moon is $1\frac{1}{4}$ light-*seconds* from Earth.

> A light year is the distance light travels in one year.

The Sun is $8\frac{1}{3}$ light-*minutes* away, and the next nearest star is $4\frac{1}{3}$ light-*years* distant. Light travel time is a convenient way of expressing cosmic distances, and the basic unit is the **light-year**.[1] A light-year is defined as the distance traveled by light in one year, or about 9.5 trillion kilometers. Remember, *a light-year is a measure of distance. It is **not** a*

measure of time. ("My goodness, it's been simply light-years since we last saw them.")

Light Is an Electromagnetic Wave

Since the earliest investigations of light, there has been a good deal of controversy over the question of whether light is composed of particles, as Newton believed, or is instead a wave. (A **wave** is a disturbance that travels from one point to another.) This controversy was seemingly put to rest once and for all in 1873 by the Scottish physicist **James Clerk Maxwell** (1831–1879). One of Maxwell's many accomplishments was the discovery of the fundamental laws that describe electricity and magnetism. The electric force and the magnetic force are actually two aspects of the same electromagnetic phenomenon. The **electric force** is the push and pull between electrically charged particles. Opposite charges attract and like charges repel. The **magnetic force**, on the other hand, is a force between electrically charged particles arising from their motion.

To describe the electric and magnetic forces Maxwell introduced the concepts of the **electric field** and the **magnetic field**. A charged particle creates an electric field that points away from it if the charge is positive, as shown in **Figure 4.2(a)**, or toward it if the charge is negative. To find out how much force is exerted on a charged particle, we multiply the charge of the particle by the strength of the electric field at its location. Because the electric field points directly away from a positively charged particle (or directly toward a negatively charged particle), the force that a second charge feels is either directly toward or directly away from the first charged particle.

---

[1] As we will learn in Chapter 13, professional astronomers frequently use another yardstick to describe stellar and galactic distances, the *parsec*. One parsec is equal to 3.26 light-years.

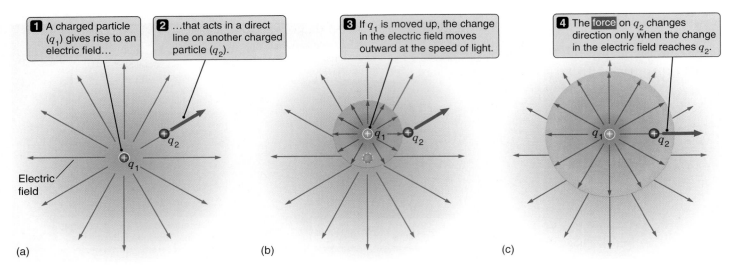

1 A charged particle (q_1) gives rise to an electric field…

2 …that acts in a direct line on another charged particle (q_2).

3 If q_1 is moved up, the change in the electric field moves outward at the speed of light.

4 The force on q_2 changes direction only when the change in the electric field reaches q_2.

Electric field

(a)　　　　　(b)　　　　　(c)

FIGURE 4.2 When a charged particle accelerates, changes in the electric field move outward at the speed of light. (In (b) the charge is shown moving instantly from one place to another for clarity. In reality this could not happen.)

The picture gets a bit more interesting if we quickly move the first charged particle (q_1) by some amount, as shown in **Figure 4.2(b)**. We might expect the force on the second particle (q_2) to change immediately so that it points away from the new position of the first particle. Yet experiments show that it does not. Immediately after the first charge moves, there is *no* change in the force felt by the second charge. Only later does the second particle feel the change in location of the first (**Figure 4.2(c)**). The situation is something like what happens if you are holding onto one end of a long piece of rope and a friend is holding the other end. When you yank your end of the rope up and down, your friend does not feel the result immediately. Instead your yank starts a pulse—a wave—that travels down the rope. Your friend notices the yank only when this wave arrives at his end. Similarly, when you move a charged particle, information about the change travels outward through space as a wave in the electric field. Other charged particles do not know that the first particle has moved until the wave reaches them.

Maxwell summarized the behavior of electric and magnetic fields in four elegant equations. Among other things, these equations say that a changing electric field causes a magnetic field, and that a changing magnetic field causes

Changing electric and magnetic fields lead to a self-sustaining electromagnetic wave.

an electric field. These changes "feed" on themselves. A change in the motion of a charged particle causes a changing electric field, which causes a changing magnetic field, which causes a changing electric field…. Once the process starts, a self-sustaining procession of oscillating electric

and magnetic fields moves out in all directions through space. Instead of a purely electric wave, an accelerating charged particle gives rise to an **electromagnetic wave**.

In addition to predicting that electromagnetic waves should exist, Maxwell's equations also predict how rapidly the disturbance in the electric and magnetic fields should move. In short, Maxwell's equations *predict* the speed at which an electromagnetic wave should travel. When Maxwell carried out this calculation, he discovered that electromagnetic waves should travel at 3×10^8 m/s—which is the speed of light! This agreement could not be simple coincidence. Maxwell had shown that light is an electromagnetic wave.

Maxwell's wave description of light also gives us an idea of how light originates and how it interacts with matter. Imagine a cork floating on a lake on a perfectly calm day. The surface of the lake is as smooth and flat as a mirror until a fish tugs on the hook and line dangling beneath the cork. The motion of the cork causes a disturbance that moves

Accelerating charges cause electromagnetic waves.

outward as a ripple on the surface of the lake (see **Figure 4.3(a)**). In much the same way, an accelerating electric charge (**Figure 4.3(b)**) causes a disturbance that moves outward through space as an electromagnetic wave. (An electric charge that is moving at a constant velocity is stationary in its inertial frame of reference and so does not radiate.) According to Maxwell's equations, *any* time an electrically charged particle is accelerated, the result is an electromagnetic wave. *Accelerating charges are the sources of electromagnetic radiation.*

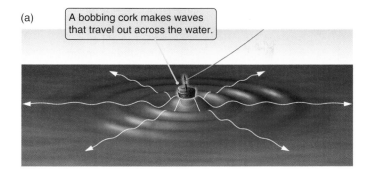

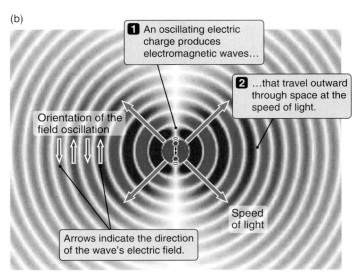

FIGURE 4.3 (a) A fish pulling downward on a cork generates waves that move outward across the water's surface. (b) In similar fashion, an accelerated electric charge generates electromagnetic waves that move away at the speed of light.

Waves Are Characterized by Wavelength, Frequency, Speed, and Amplitude

Along our journey we will encounter waves of different kinds, ranging from electromagnetic waves crossing the vast expanse of the universe to seismic waves traveling through Earth. In the most general sense a wave is a disturbance that travels away from its source. If you drop a pebble in a pond, ripples spread out over the surface of the water. This kind of wave is called a **transverse wave** because the wave's displacement is perpendicular, or "transverse," to its direction of travel (see **Figure 4.4 (a)**). The waves on a plucked guitar string are also transverse waves. Such waves travel because when the material is disturbed, forces try to even out that disturbance. A portion of the material moves back toward its undisturbed position. But because it is still moving and has inertia, when it reaches that position, it overshoots, distorting the material in the opposite way from

A wave is a disturbance that travels away from a source.

how it was originally distorted. Forces now try to push the material back in the other direction, but again there is an overshoot. The material oscillates back and forth from one side of its undisturbed position to the other, and the wave moves along. In the guitar string, the force responsible for the wave is the tension that tries to keep the string straight. In the case of water, the forces responsible for creating the wave are gravity, which tries to pull down the crest of the wave, and water pressure (also caused by gravity), which tries to push the trough of the wave up.

Another kind of wave is called a **longitudinal wave**. Sound waves are an example of longitudinal waves, and so are the waves in a spring (**Figure 4.4(b)**). In a spring, compressed regions try to push into the stretched-out regions to even the spring out. But when a portion of the spring reaches its undisturbed position, it is still moving, and it overshoots. Parts of the spring that were originally compressed are now stretched, and the parts that were originally stretched are now compressed. Regions in which the spring is alternately compressed and stretched move along the length of the spring as the cycle at each point repeats itself. In the case of sound waves, air pressure provides the forces that keep the wave moving.

In these examples the waves result from *mechanical* distortions of the medium the wave travels through (for example, distortion of the surface of the pond or the coils in the spring). Mechanical waves involve distortions measured as distances, and media that have mass. Maxwell showed that light waves are a fundamentally different type of wave. Light waves involve no mechanical distortion of a medium, but instead involve periodic changes in the strength of the electric and magnetic fields. Even so, light waves are generally thought of as transverse waves because the directions of the electric and magnetic fields are perpendicular to the direction in which the wave travels (see **Figure 4.5**).

Waves are generally characterized by four quantities, which are shown in **Figure 4.6**. The **amplitude** of a wave is the maximum excursion from its undisturbed or relaxed position. The wave travels at some speed, which is usually written as v (except in the case of light, where the speed of

Waves are characterized by their wavelength, frequency, speed, and amplitude.

light is written as c). The number of wave crests passing a point in space each second is called the wave's **frequency**, denoted by f. The unit of frequency is cycles per second, which is generally referred to as **hertz** (abbreviated Hz) after the 19th-century physicist **Heinrich Hertz** (1857–1894), who

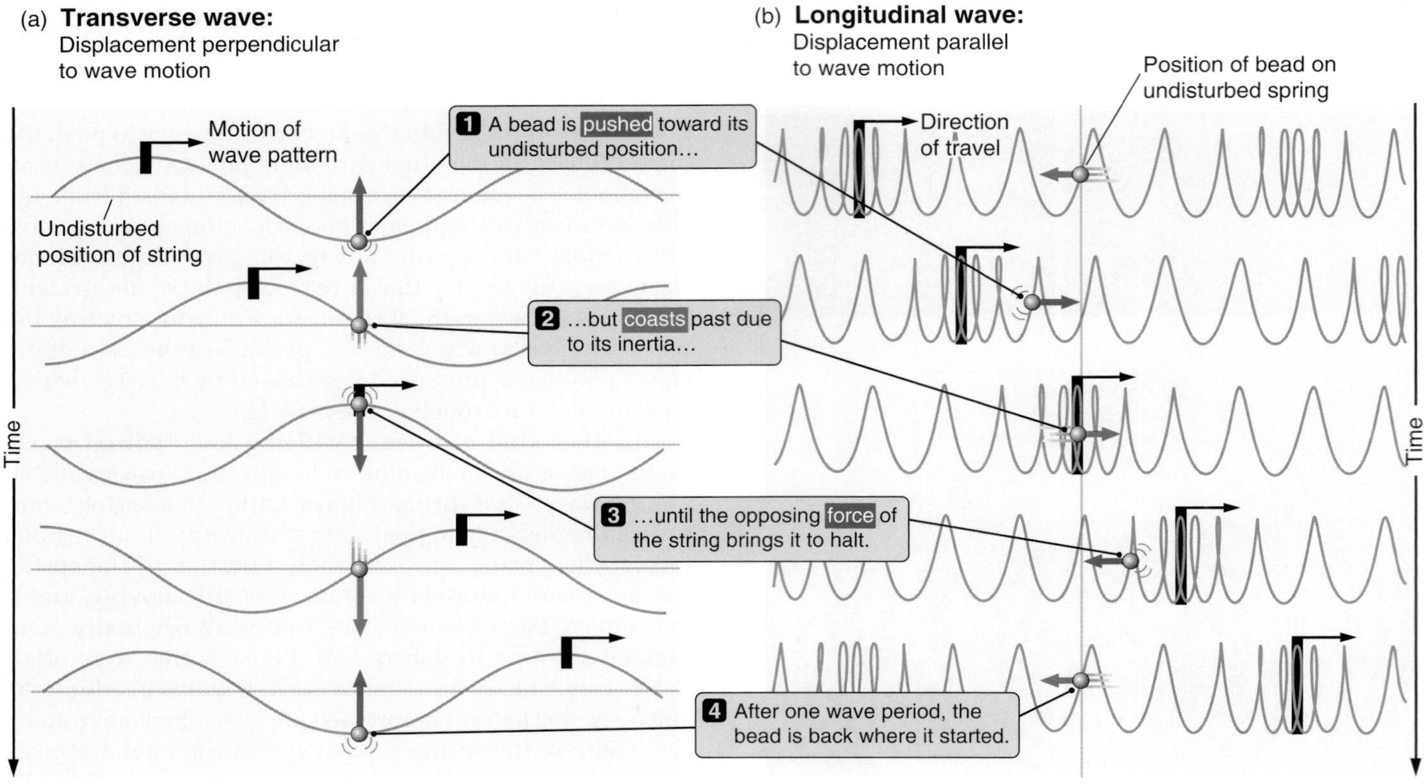

(a) Transverse wave:
Displacement perpendicular
to wave motion

(b) Longitudinal wave:
Displacement parallel
to wave motion

Motion of
wave pattern

Position of bead on
undisturbed spring

1 A bead is pushed toward its
undisturbed position…

Direction
of travel

Undisturbed
position of string

2 …but coasts past due
to its inertia…

Time

3 …until the opposing force of
the string brings it to halt.

Time

4 After one wave period, the
bead is back where it started.

FIGURE 4.4 Mechanical waves result from forces that try to even out disturbances. (a) A transverse wave involves oscillations that are perpendicular to the direction in which the wave travels. (b) A longitudinal wave involves oscillations along the direction of travel of the wave.

was the first to experimentally confirm Maxwell's predictions about **electromagnetic radiation**. The time taken for one complete cycle is called the *period, P,* which is measured in seconds.

The distance a wave travels during one complete oscillation is called the **wavelength**. This is just the distance from one wave crest to the next, or the distance from one wave trough to the next. The wavelength is usually denoted by the Greek letter λ (pronounced "lambda"). There is a clear relationship between the frequency of a wave and its wavelength. If the period of a wave is ½ second—that is, if it takes ½ second for one wave to pass by, crest to crest—then two waves will go by in 1 second. So a wave with a period of ½ second per cycle has a frequency of 2 cycles per second. Similarly, if a wave has a period of 1/100 second per cycle, then 100 waves will pass by each second. This wave has a frequency of 100 Hz. More generally, the frequency of a wave is just 1 divided by its period:

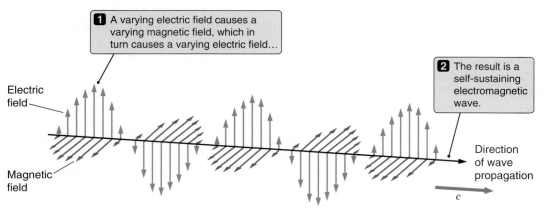

1 A varying electric field causes a
varying magnetic field, which in
turn causes a varying electric field…

2 The result is a
self-sustaining
electromagnetic
wave.

Electric
field

Magnetic
field

Direction
of wave
propagation

c

FIGURE 4.5 Far from its source, an electromagnetic wave consists of oscillating electric and magnetic fields that are perpendicular both to each other and to the direction in which the wave travels.

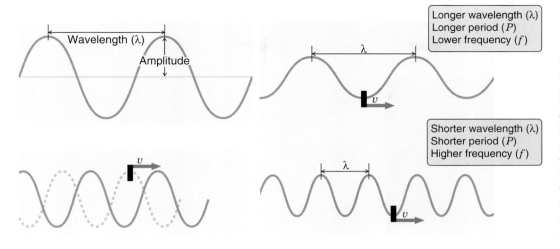

FIGURE 4.6 A wave is characterized by the distance over which the wave repeats itself (called the wavelength, λ), the maximum excursion from its undisturbed state (called the amplitude), and the speed (v) at which the wave pattern travels. In an electromagnetic wave, the amplitude is the maximum strength of the electric field, and the speed of light is written as c.

$$\text{Frequency} = \frac{1}{\text{period}} \quad \text{or} \quad f = \frac{1}{P}$$

There is also a relationship between the period of a wave and its wavelength. The period of a wave is the time between the arrival of one wave crest and the next. During this time the wave travels a distance equal to the separation between the two wave crests, or one wavelength. So far, so good. Now add to the picture the fact that distance traveled equals speed times time taken. Change "distance traveled" to one wavelength and change "time taken" to one period, and we find that the wavelength of a wave equals the speed at which the wave is traveling times the period of the wave:

$$\text{Wavelength} = \text{Speed} \times \text{Period}$$

Physicists use the letter c to represent the speed of light, so we can say that

$$\lambda = c \times P.$$

Using the relationship between period and frequency just given, we can also write

$$\text{Wavelength} = \frac{\text{Speed}}{\text{Frequency}}, \quad \text{or} \quad \lambda = \frac{c}{f}.$$

So if we know the speed of a wave, then knowing one of the three properties—its wavelength, period, or frequency—tells us the other two.

Look at this relationship more closely. The longer the length of a wave, the longer you have to wait between wave crests, so the frequency of the wave will be lower. A shorter wavelength means less distance between wave crests, which means a shorter wait until the next wave comes along. Therefore, a shorter wavelength means a higher frequency. A tremendous amount of information can be carried by waves—intelligible speech, for example, or complex and beautiful music. As we continue our study of the universe,

A long wavelength means low frequency and short wavelengths mean high frequency.

time and again we will find that the information we receive, whether about the interior of Earth or a distant star or galaxy, rides in on a wave.

Maxwell's equations also describe how electromagnetic waves interact with the matter they encounter. Returning to the analogy of the lake in Figure 4.3, imagine now that a second cork is afloat on the lake some distance from the first, as in **Figure 4.7(a)**. The second cork remains stationary until the ripple from the first cork reaches it. As the ripple passes by, the rising and falling of the water causes the cork to rise and fall as well. Similarly, the oscillating electric field of an electromagnetic wave causes an oscillating force on any charged particle that the wave encounters, and this force causes the particle to move about as well (**Figure 4.7(b)**). It takes energy to produce an electromagnetic wave, and that energy is carried through space by the wave. Matter far from the source of the wave can absorb this energy. In this way some of the energy lost by the particles generating the electromagnetic wave is transferred to other charged particles. The emission and absorption of light by matter are the result of the interaction of electric and magnetic fields with electrically charged particles.

Electromagnetic Waves of Different Wavelengths Make Up the Electromagnetic Spectrum

You have almost certainly seen a rainbow like the one in **Figure 4.8** spread out across the sky, or sunlight split into many different colors by a prism. This sorting of light by colors is really a sorting by wavelength. When we talk about light spread out according to wavelength, we refer to the

(a)

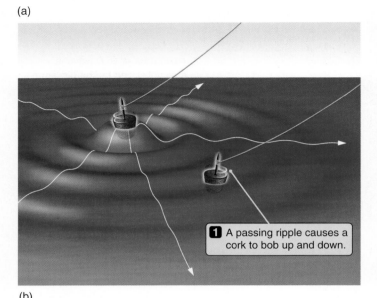

1 A passing ripple causes a cork to bob up and down.

(b)

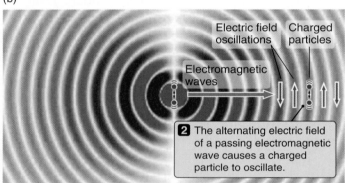

Electric field oscillations

Charged particles

Electromagnetic waves

2 The alternating electric field of a passing electromagnetic wave causes a charged particle to oscillate.

FIGURE 4.7 (a) When waves moving across the surface of water reach a cork, they cause the cork to bob up and down. (b) Similarly, a passing electromagnetic wave causes an electric charge to wiggle in response to the wave.

FIGURE 4.8 The visible part of the electromagnetic spectrum is laid out in all its glory in the colors of this rainbow.

spectrum of the light. On the long-wavelength (and therefore low-frequency) end of the visible spectrum is red light. A **micrometer**, or **micron**, is the unit often used for measuring the wavelength of visible light. A micrometer is a millionth $(1/10^6)$ of a meter. The abbreviation used for a micron

The spectrum of visible light is seen as the colors of the rainbow.

is μm (where μ is the Greek letter "mu"). Another commonly used unit is the nanometer, abbreviated nm. A nanometer is one billionth $(1/10^9)$ of a meter.[2] The wavelengths of the light we perceive as red fall between about 600 and 700 nm.

[2] Astronomers conventionally use nanometers (nm) when referring to wavelengths at visible and shorter wavelengths, micrometers or "microns" (μm) in the infrared, and centimeters (cm) and meters (m) in the microwave and radio regions of the electromagnetic spectrum.

At the other end of the visible spectrum is violet light, which is the bluest of blue light. The shortest-wavelength violet light that our eyes can see has a wavelength of around 350 nm. Stretched out between the two, literally in a rainbow, is the rest of the visible spectrum. The colors in the visible spectrum in order of decreasing wavelength can be remembered as a name: "Roy G. Biv," which stands for

Red Orange Yellow Green Blue Indigo Violet.

The human eye is most sensitive to light in the green to yellow part of the spectrum. This light has a wavelength of around 500 to 550 nm. Green light with a wavelength of 520 nm has a frequency of

$$f = \frac{c}{\lambda} = \frac{3.00 \times 10^8 \frac{\text{m}}{\text{s}}}{520 \times 10^{-9}\,\text{m}} = \frac{5.8 \times 10^{14}}{\text{s}}$$

$$\text{or} \quad 5.8 \times 10^{14}\,\text{Hz}$$

That frequency corresponds to 580 *trillion* wave crests passing by each second!

When we say "visible light," what we mean is "the light that the light-sensitive cells in our eyes respond to." But this is not the whole range of possible wavelengths for elec-

Visible light is only one small segment of the electromagnetic spectrum.

tromagnetic radiation. Radiation can have wavelengths that are much shorter or much longer than our eyes can perceive. The whole range of different wavelengths of light is collectively referred to as the **electromagnetic spectrum**.

Follow along in **Figure 4.9** as we take a tour of the electromagnetic spectrum, beginning with visible light and working our way to shorter and longer wavelengths. We start at the blue end of the visible spectrum. Beyond this short-wavelength, high-frequency side of the visible spectrum, there is light that is "bluer than blue" or actually "more violet than violet." This light, with wavelengths between 40 and 350 nm, is called **ultraviolet (UV) radiation**. You can remember what ultraviolet light is just by looking at the name. The prefix *ultra-* means "extreme," so *ultra*violet light is light that is more "extremely" violet than violet. It is important to remember that ultraviolet light is fundamentally no different from visible light, any more than high C on a piano is fundamentally different from middle C.

As we go to shorter wavelengths of light (and so to higher frequencies), we pass through the ultraviolet part of the spectrum. At a wavelength shorter than 40 nm, or 4×10^{-8} m, we stop calling radiation ultraviolet light and instead start calling it **X-rays**. This distinction comes for historical reasons. When X-rays were discovered in the last part of the 19th century, they were given the name "X" by their discoverer, **Wilhelm Conrad Roentgen** (1845–1923), to indicate they were "a new kind of **ray**." As we continue to even shorter wavelengths, we come to another somewhat arbitrary break. Electromagnetic radiation with the very shortest wavelengths (less than about 10^{-10} m) is referred to as **gamma rays**. Again the reasons are historical. Gamma rays (or γ-rays) were first discovered as a type of radiation given off by radioactive material. It was only later that their true nature became known.

So far we have been considering ever shorter wavelengths and higher frequencies. In principle, there is no limit to this process. We can conceive of gamma rays of arbitrarily short wavelengths and arbitrarily high frequencies (even though practical considerations eventually come into play). We can also go in the other direction. Just as there is light that is more violet than violet, there is also light that is "redder than red." Such light, covering wavelengths longer than about 700 nm and shorter than 500 μm (5×10^{-4} m), is referred to as **infrared (IR) radiation**. Again the key to remembering what infrared light is comes from looking at the word itself. *Infra-* is a prefix that means "below." *Infra*red light is light that has a frequency that is lower than (below) that of red light. When the wavelength of light gets longer than this, we start calling it **microwave radiation**. The longest-wavelength (and therefore lowest-frequency) electromagnetic

FIGURE 4.9 By convention the electromagnetic spectrum is broken into loosely defined regions ranging from gamma rays to radio waves.

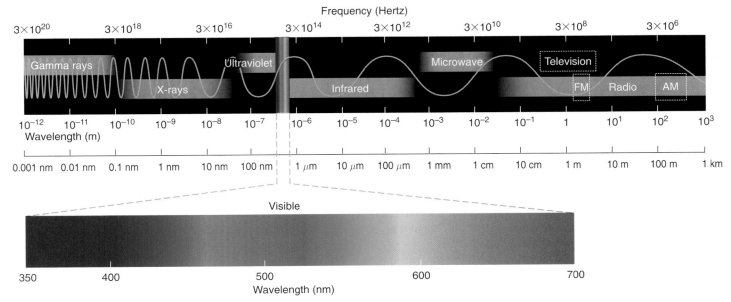

radiation, with wavelengths longer than a few centimeters and ranging up to arbitrarily long wavelengths, is called **radio waves**. Chapter 5 discusses the various kinds of telescopes used by astronomers to capture and analyze the wide range of electromagnetic radiation.

4.3 The Speed of Light Is a Very Special Value

Maxwell's description of light as a wave was a great success, but without realizing it he had also found a flaw in Newtonian physics. To understand this flaw, we need to think back to Chapter 2, where we used a moving car as an example of a moving frame of reference. Imagine that you are sitting in a moving car and there is a ball sitting on the seat beside you. In your frame of reference the ball is at rest. But if the car is moving at 50 mph down the highway, someone standing by the road will say that the ball is also moving at 50 mph. To someone in oncoming traffic moving at 50 mph, the relative speed of both your car and the ball would be 100 mph. There really is no difference between these three perspectives. The laws of physics are the same in *any* inertial frame of reference.

As a variant of the "ball in the car" experiment, imagine that as your car moves down the highway at 50 mph you pitch a fastball forward at 100 mph (see **Figure 4.10(a)**). In your frame of reference the ball is moving at 100 mph, but to an observer standing by the road the ball is moving at 150 mph. (The ball has the original 50 mph speed of the car plus the additional 100 mph that you gave it with your throw.) In the frame of reference of a car in oncoming traffic traveling at 50 mph, the ball is moving at 200 mph. (This is the 150 mph that the ball is moving relative to the ground plus the 50 mph motion of the oncoming car.) In our everyday experience velocities simply add. This is also how Newton's laws say the universe should behave.

Now do exactly the same thought experiment, but with two changes. Instead of a car traveling at 50 mph, imagine you are in a spaceship traveling at half the speed of light, or 0.5*c*, as shown in **Figure 4.10(b)**. Instead of throwing a baseball, you shine a beam of light forward. To you the light is moving at the speed of light, *c*. If we replace "100 mph" with "*c*" in the previous paragraph, we think we know what to expect for other observers. To an observer on a nearby planet, the light should travel by at a speed equal to the speed of your spacecraft plus the speed of light, or 1.5*c*. Similarly, to an observer in an oncoming spacecraft traveling at 0.5*c*, the light should appear to travel at a speed of 2*c*.

We do not have the luxury of performing this experiment while traveling through space at half the speed of light, but physicists are ingenious folk. During the closing years of the 19th century and the early years of the 20th century, physicists were conducting laboratory experiments that were the functional equivalent of our thought experiment. What they found puzzled them greatly. Rather than the speed of the beam of light differing from one observer

---

Surprisingly, experiments showed that the speed of light is the same for all observers.

---

to the next, as expected on the basis of Newton's physics and "common sense," they found instead that *all observers measure exactly the same value for the speed of the beam of light, regardless of their motion!*

As you ride in your spaceship you measure the speed of the beam of light to be *c*, or 3×10^8 m/s. That is as expected because you are holding the source of the light. But the observer on the planet *also* measures the speed of the passing beam of light to be 3×10^8 m/s. Even the passenger in the oncoming spacecraft finds that the beam from your light is traveling at exactly *c* in her own frame of reference. In fact, it turns out that *every observer always finds that light in a vacuum travels at exactly the same speed* c, *regardless of his or her own motion or the motion of the source of the light.*

If at this point you are feeling uneasy and saying to yourself, "This is very bizarre," then you probably have followed what the last few paragraphs have stated. And if this discussion bothers you, imagine the reaction of those physicists! Newton's laws of motion had been the bedrock of science for 200 years, facing every experimental challenge that came their way. Now suddenly that bedrock seemed to turn to sand. How could light have the same speed for *all* observers, regardless of their own velocity? Preposterous! And yet that was the inescapable experimental result. Despite all its spectacular successes, Newtonian physics seemed to be in serious trouble.

Enter a young German-born Swiss patent clerk named **Albert Einstein** (1879–1955). As a 16-year-old schoolboy Einstein had already realized that there was trouble afoot. Light travels in a straight line at a constant speed. Einstein reasoned that according to Newton's laws of motion, there should be a perfectly good inertial frame of reference that moves along with the light and in which the light is stationary. That is, you should be able to "keep up" with light so that you are moving right along with it. But if you could do that, the light would be an oscillating electric and magnetic wave *that does not move*. This was impossible according to Maxwell's equations for electromagnetic waves. There was a contradiction here. Either Maxwell was wrong in his understanding of electricity and magnetism, or Newtonian physics did not apply at very large velocities. As the experimental results rolled in on measurements of the speed of light, it became clear that it was Newtonian physics that needed revision.

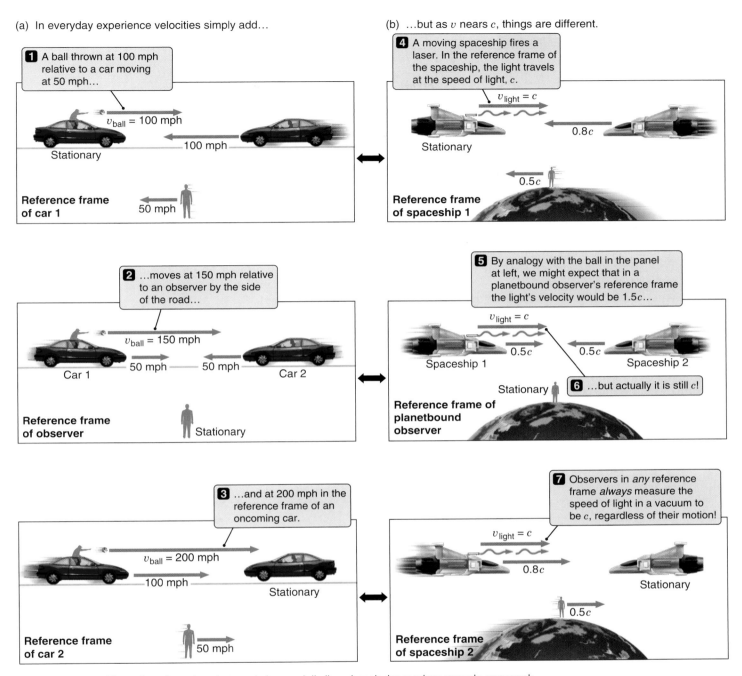

FIGURE 4.10 The rules of motion that apply in our daily lives break down when speeds approach the speed of light. The fact that light itself always travels at the same speed for any observer is the basis of special relativity. (Note that relativity also affects the relative speeds of the two spacecraft.)

Time Is a Relative Thing

Einstein resolved the contradiction between Maxwell and Newton and ushered in a scientific revolution (see **Connections 4.1**) with his theory of **special relativity**, which was published in 1905. Special relativity was Einstein's answer to the question, "What must the universe be like if every observer always measures the same value for the speed of light in a vacuum?" Einstein focused his thinking on pairs of *events*. In relativity, an **event** is something that happens at a particular location in space at a particular time. When you snap your fingers, that is an event. From everyday experience we know that the distance between any two events depends on the frame of reference of the person observing

A Scientific Revolution

Throughout the first three chapters of this book we interlaced our story of the motions of the sky and the discovery of Newton's laws of motion and gravitation with a discussion of the nature of scientific knowledge and the way that science progresses. We stressed that scientific knowledge differs from all other forms of knowledge in that even our most cherished and fundamental knowledge is open to challenge by new observations and experiments. In this chapter we will see this drama play itself out several times over.

By the middle of the 19th century many physicists felt that our fundamental understanding of physical law was more or less complete. For over a century Newtonian physics had withstood the scrutiny of scientists the world over. It seemed that little remained but cleanup work —filling in the details. Some even went so far as to pronounce this period the "end of science." Yet during the late 19th and early 20th centuries, physics was rocked by a series of scientific revolutions that shook the very foundations of our understanding of the nature of reality. In this chapter we have come across several of these revolutions. Einstein's theory of relativity erased the classical distinction between space and time and united our concepts of matter and energy. Quantum mechanics forced us to abandon our everyday understanding of "substance" and even to part with the notion that we live in a universe in which effect follows cause in lockstep. Together these revolutions led to the birth of what has come to be known as **modern physics**. Although modern physics *contains* Newtonian physics, the understanding of the universe offered by modern physics is far more sublime and powerful than the earlier understanding that it subsumed.

As we continue on our journey, we will encounter many other discoveries and successful ideas that forced scientists either to abandon their treasured notions or be left behind, hopelessly locked into a worldview that had ultimately failed the test of observation and experiment. The point is this: In physical science, we are not just paying lip service to a hollow ideal when we say that the rigorous standards of scientific knowledge respect no authorities. No theory, no matter how central or how strongly held, is immune from the rules.

---

them. Suppose you are sitting in a car that is traveling down the highway in a straight line at a constant 60 mph. You snap your fingers (event 1), and a minute later you snap your fingers again (event 2). In your frame of reference *you* are stationary and the two events happened at exactly the same place. They are separated by a minute in *time,* but there is no separation between the two events in *space*. This is very

---

Special relativity concerns the relationship between events in space and time.

---

different from what happens in the frame of reference of an observer sitting by the road. This observer agrees that the second snap of your fingers (event 2) occurred a minute after the first snap of your fingers (event 1), but to this observer the two events were separated from each other in space by a mile. In this everyday, "Newtonian" view, the *distance* between two events depends on the motion of the observer, but the *time* between the two events does not.

Einstein questioned why there was such a distinction between the way Newton treated space and the way New-ton treated time. Einstein realized that the *only* way the speed of light can be the same for all observers is if *the passage of time is different from one observer to the next!* This is a *very* counterintuitive idea, but it is so central to our modern understanding of the universe that it is worth wrestling with a bit. Hang onto your hat while we reconstruct some of the reasoning that led Einstein to this remarkable conclusion.

To measure time, the first thing we need is a clock. The best way to build a clock is to base it on a value that everyone can agree on—such as the speed of light. **Figure 4.11(a)** shows just such a clock as seen by observer 1, who is stationary with respect to the clock. At time t_1 a flashlamp gives off a pulse of light. Call this event 1. The light bounces off a mirror a distance l meters away, then heads back toward its source. At time t_2 the light arrives and is recorded by a photodetector. Call this event 2. The time between events 1 and 2 is just the distance the light travels ($2l$ meters), divided by the speed of light, or $t_2 - t_1 = 2l/c$.

So far so good, but now look at the clock from the perspective of observer 2 in a frame of reference that is moving relative to the clock. In *this* observer's frame of reference he is stationary, and it is the *clock* that is moving at speed *v,* as shown in **Figure 4.11(b)**. (Recall that because any inertial frame of reference is as good as any other, this observer's perspective is as valid as the first observer's perspective.) We see the same two events as before: event 1 when the light leaves the flashlamp and event 2 when the light arrives at the detector. There is a difference, however. In this frame of reference the clock *moves* between the two events, so the light has *farther to go.* (If you do not see this right away, use a ruler to measure the total length of the light path in Figure 4.11(b) and compare it with the total length of the light path in Figure 4.11(a).) The time between the two events is still the distance traveled divided by the speed of light, but now that distance is *longer* than 2*l* meters. Because the speed of light is the same for all observers, the time between the two events must be longer as well!

Go over that again. The two events are the *same two events,* regardless of the frame of reference from which they are observed. The question is, how much time passed between the two events? Because the speed of light is the same for all observers, there *must* be more time between the two events when they are viewed from a frame of reference in which the clock is moving. It takes a moving clock more time than a stationary clock to complete one "tick." Moving clocks *must* run slow, and the passage of time *must* depend on an observer's frame of reference.

To Newton, and to us in our everyday lives, the march of time seems immutable and constant. But in reality the only thing that is truly constant is the speed of light, and even time itself flows differently for different observers.

Here, in a nutshell, is the heart of Einstein's theory of special relativity. In our everyday Newtonian view of the world, we live in a three-dimensional space through which time marches steadily onward. Events occur in space at a certain

---

Space and time together form a four-dimensional "spacetime."

---

time. By the time Einstein finished working out the implications of his insight, he had reshaped this three-dimensional universe into a four-dimensional **spacetime**. Events occur at specific locations within this four-dimensional spacetime, but how this spacetime translates into what we perceive as "space" and what we perceive as "time" depends on our frame of reference.

It is very important to state that Einstein did not throw out Newtonian physics. We were not wasting our time in Chapter 3 when we studied Newton's laws of motion. Instead Einstein found that Newtonian physics is *contained within* special relativity. In our everyday experience we never encounter speeds that approach that of light. Even the breakneck speed of the space shuttle is only about 0.000025*c.* When Einstein's special relativity is restricted

FIGURE 4.11 The "tick" of a light clock as seen in two different reference frames. As Einstein's thought experiment demonstrates, if the speed of light is the same for every observer, then moving clocks *must* run slow.

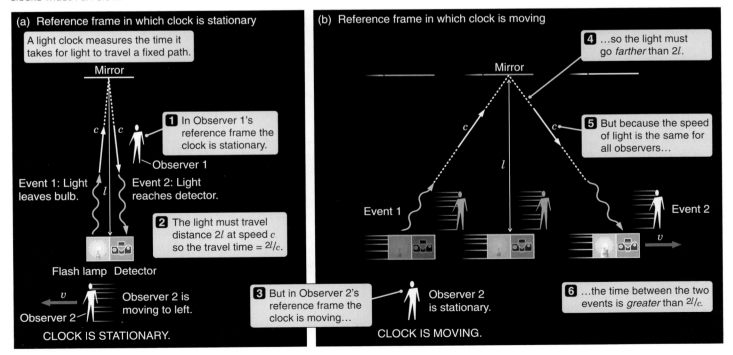

to cases where velocities are much less than the speed of light, then Einstein's equations become the very equations that describe Newtonian physics! In our everyday lives we experience a Newtonian world. Only when relative velocities approach that of light do things begin to depart from the predictions of Newtonian physics. When great velocities cause something to turn out differently than we would expect based on Newtonian physics, this is referred to as a **relativistic** effect.

The Implications of Relativity Are Far-Ranging

The story of special relativity is another case study of how science works. Newton's laws had proven for a long time to be an extraordinarily powerful way of viewing the world. But as science turned its attention to a different phenomenon —the phenomenon of light—difficulties arose. Newton's theory of motion, Maxwell's theory of electromagnetic radiation, and empirical measurements of the speed of light met head on. Such conflicts are what scientists live for: They point the way to new knowledge and new understanding.

Einstein was able to step in and reconcile this conflict, and in the process he changed the way we think about the universe. Einstein's ideas remained controversial well into the 20th century. However, as one experiment after another confirmed the strange and counterintuitive predictions of relativity, scientists came to accept its validity. Today special relativity is an integral and indispensable part of all of physics, shaping our thinking about the motions of the tiniest subatomic particles as well as the motions of the most distant galaxies.

It would be great fun to linger here for a time and explore. But our journey has hardly begun, and there is so much more to see. We hope you will find the time at some point to come back and explore the wonders of the relativistic world in which we live. Puzzling out relativity is time well spent. In the meantime, here are a few of the interesting insights that come from Einstein's work:

1. **What we think of as "mass" and what we think of as "energy" are actually two manifestations of the same thing.** Usually we think of the energy of an object as depending on its speed. The faster it moves, the more energy it has. But Einstein's famous equation $E = mc^2$

says that even a *stationary* object has an intrinsic "rest" energy that equals the mass m of the object times the speed of light, c, squared. The speed of light is a very large number. This relationship between mass and energy says that a single tablespoon of water has a rest energy equal to the energy released in the explosion of over 300,000 tons of TNT! All reactions that produce energy do so by converting some of the mass of the reactants into other forms of energy. But even the most efficient chemical or nuclear reactions release only tiny fractions of the total energy available. Exploding TNT, for example, converts less than a trillionth of its mass into energy. Even the explosion of a hydrogen bomb releases far less than 1 percent of the energy contained in the mass of the bomb.

The equivalence between mass and energy points both ways. In Chapter 3 we defined *mass* as the property of matter that resists changes in motion. Does the energy of an object really increase its resistance to changes in motion? Yes. Even adding to the energy of motion of an object increases its inertia. For example, a proton in a high-energy particle accelerator may approach the speed of light so closely that its total energy is 1,000 times greater than its rest energy. Such an energetic proton is, indeed, harder to "push around" (in other words, it has more inertia) than a proton at rest.

2. **The speed of light is the ultimate speed limit.** There are several ways to think about this. We already discussed the insight that led Einstein to relativity in the first place. Were it possible to travel at the speed of light, then in that frame of reference light would cease to be a traveling wave, and all of the laws of physics would come tumbling down around our ears. We can also think about this limit in terms of the equivalence of mass and energy just discussed. As the speed of an object gets closer and closer to the speed of light, its energy, and therefore its mass, become greater and greater, so it becomes increasingly resistant to further changes in its motion. We can continue to push on it all we like, making it go faster, but we face diminishing returns. The situation is like trying to get from 0 to 1 by halving the remainder again and again. The resulting sequence— $0, \frac{1}{2}, \frac{3}{4}, \frac{7}{8}, \frac{15}{16}, \frac{31}{32}, \frac{63}{64}, \ldots$ —gets arbitrarily close to 1 but never actually reaches it. In the same way, a continuous force applied to an object will cause its velocity to get closer and closer to the speed of light, but it will never actually reach the speed of light. You just cannot get there. It would take an *infinite* amount of energy to accelerate an object with a nonzero rest mass to the speed of light. In short, all the energy in the entire universe is inadequate to accelerate a single electron to the speed of light. We can get the electron arbitrarily close to that number—0.9999999999999999999999... × c is no problem, at least in principle—but there is no getting over

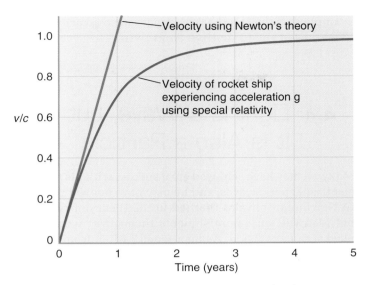

FIGURE 4.12 The speed of a rocket ship experiencing an acceleration equal to Earth's acceleration of gravity. The rocket ship approaches the speed of light but never gets there.

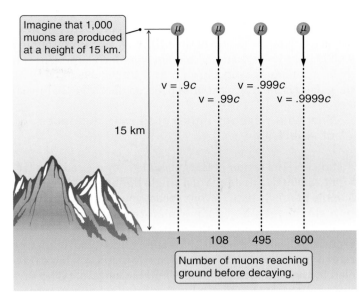

FIGURE 4.13 Plot of the muon lifetime (the time during which half of all muons decay) versus energy, or the velocity at which the muons travel. Also shown is the distance these muons can travel compared to the 15-km height at which they originate.

the hump. In **Figure 4.12** we show how a rocket ship, which experiences a constant acceleration equal to that of gravity on Earth (so its occupants will feel at home), moves faster and faster but never reaches the speed of light. Faster-than-light travel may be a mainstay of science fiction, but "Sorry, Jim, I just cannot make her go faster than *c*!" (We leave the Scottish accent up to your imagination.) Here is one of those cases where wishing that something is physically possible does not necessarily "make it so."

3. **Time passes more slowly in a moving reference frame.** This phenomenon is referred to as **time dilation** because time is "spread out" in the moving reference frame. Were you to compare clocks with an observer moving at $9/10$ the speed of light (0.9*c*), you would find that the other observer's clock was running less than half as fast as your clock (about 0.44 times as fast).[3] You might guess that to the other observer, your clock would be fast, but actually the other observer would find instead that it is *your* clock that was running slow! A bit of thought shows why it must be this way. To you, the other observer may be moving at 0.9*c*, but to the other observer, *you* are moving. Either frame of reference is equally valid, so it stands to reason that if a clock in a moving reference frame runs slow, then you would each find the other's clock to be slow. An everyday scientific experience illustrates this effect. Fast particles

called cosmic ray muons provide an example of time dilation, as illustrated in **Figure 4.13**. Cosmic ray muons are produced at about the 15-km level in Earth's atmosphere when high-energy primary cosmic rays strike atmospheric atoms or molecules. Muons at rest decay very rapidly into other particles. Within 2.2 microseconds, half of all muons will have changed their identity. This happens so quickly that, even traveling at the speed of light, virtually all muons would have decayed long before traveling the 15 km to reach Earth's surface. However, time dilation causes the muon clock to run slower, so they live longer and can travel farther. That is why we are able to detect cosmic ray muons on the ground.

4. **"At the same time" is a relative concept.** Two events that occur at the *same* time for one observer may occur at *different* times for a different observer. Hold out your arms and snap the fingers on both hands at the same time. For you, the two snaps were simultaneous. But to an observer moving by you from right to left at nearly the speed of light, you snapped the fingers of your left hand first and the fingers of your right hand later.

5. **An object in motion is shorter than it is at rest.** More specifically, moving objects are compressed in the direction of their motion. A meter stick moving at 0.9*c* is only 43.6 cm (centimeters) long.[4]

---

[3] The factor by which time is dilated and space is contracted is given by $\frac{1}{\sqrt{1 - \frac{v^2}{c^2}}}$. This factor is often referred to as γ.

[4] See footnote 3.

These different consequences of relativity can be combined in what is often called the *twin paradox.* You head off on a trip to the center of the Milky Way Galaxy, roughly 25,000 light-years distant. Your spectacularly powerful star drive accelerates your ship up to 0.9999999992c. To you, the galaxy is moving by at this speed, so the 25,000 light-year

> **The twin paradox illustrates many aspects of relativity.**

distance to the center of the galaxy is compressed by a factor of 25,000 to a distance of a single light-year (see number 5 in the previous list). At your speed, you cross this distance in a single year. You snap a picture of the gas swirling around the black hole at the center of the galaxy, then turn around to head home and show it to your twin. Again, the return trip takes only a year. So in two years you have traveled to the center of the galaxy and back again. (Who says interstellar travel is such a big deal?) But when you return, you find that your twin died 50,000 years ago. In the reference frame of Earth, your spacecraft crossed the 25,000 light-year distance to the center of the galaxy moving at just under the speed of light. The only reason you survived the journey, according to an Earth-bound observer, is because in your moving frames of reference time ran extremely slow (number 3 in our list). Each leg of the two-way journey took 25,000 years to observers on Earth, and your twin just could not wait that long.

You might puzzle over the twin paradox a bit. Both on the way out and on the way back, in your reference frame it is the clocks on Earth that are running slowly, so you are aging *faster* than your twin. Yet when you return, more time has passed for your twin than for you. How can this be? The answer is that, unlike your twin, you *changed reference frames* during your trip. Event 1 is when you left Earth, and event 2 is when you returned to Earth. Your twin went from one event to the other, riding along in Earth's frame of reference. You, on the other hand, changed reference frames when you left Earth, changed again when you stopped at the center of the galaxy, changed a third time when you left the galactic center to return home, and changed reference frames one final time when you arrived back at Earth. It happens that the path through spacetime that you followed between the two events involved the passage of only two years of what you experienced as time, while your twin's path involved 50,000 years of what your twin experienced as time.

Another way to view this is that the key difference between you and your twin is that you experienced acceleration during your trip, while your twin did not. When two observers are in *uniform motion* relative to one another, *neither* of them can lay claim to being in a unique frame of reference. However, *acceleration is a real phenomenon.* You *feel* acceleration when you are riding in a car, and you would surely *feel* the acceleration of the spaceship in this example. It is the fact that you experienced an acceleration that allowed you to "outlive" your twin.

4.4 Light Is a Wave, but It Is Also a Particle

Maxwell may have achieved great success with his theory of electromagnetic waves, and he may have opened the crack in Newtonian physics that led to the theory of relativity; but Maxwell's accomplishments themselves would soon need serious revision. As mentioned earlier, from early on scientists disagreed over whether light consists of waves or particles. Maxwell's work seemed to put the issue to rest by showing that light is an electromagnetic wave; but before too many years had gone by, the particle description raised its head again. Although the electromagnetic wave theory of light has had many successes in describing phenomena, there are also many phenomena that it does not describe well. These range from the presence of sharp bright and dark "lines" at specific wavelengths in the light from some objects, to the shape of the continuous spectrum of light emitted by a lightbulb. Many of these difficulties with the wave model of light have to do with the way in which light interacts with atoms and molecules.

Scientists working in the late 19th and early 20th centuries discovered that many of the puzzling aspects of light could be better understood if light energy came in discrete packages. In 1905 Einstein published a paper in which he argued that light consists of particles. He based his argument on the **photoelectric effect**, the emission of electrons from surfaces illuminated by electromagnetic radiation above a certain frequency. Einstein showed that the *rate* at which electrons are ejected depends only on the *intensity* of the incident radiation, and that the electron *velocity* depends only on the *frequency* of the incident radiation.[5] Effectively, scientists were reintroducing the particle picture of light. In some ways the particle description of light is easier to think about than its wave description. In this model we think about light as being made up of particles called **photons** (*phot-* means "light," as in *photograph*, and *-on* signifies a particle, as in *electron, neutron,* and *proton*). Photons always travel at the speed of light, and they carry energy. (After our discussion of relativity in Section 4.3, you may wonder how a particle can travel at the speed of light. The answer is that a photon has no mass. A massless particle can travel *only* at the speed of light.)

---

[5] It is interesting to note that it was for his work on the photoelectric effect, not special or general relativity, that Einstein received the Nobel Prize in 1921.

The particle description of light is tied to the wave description of light by a relationship between the energy of a

The energy of a photon is proportional to its frequency.

photon and the frequency or wavelength of the wave. The higher the frequency of the electromagnetic wave, the greater the energy carried by each photon. Specifically we write

$$E = hf \quad \text{or} \quad E = \frac{hc}{\lambda}.$$

The h in this equation is called **Planck's constant** and has the value $h = 6.63 \times 10^{-34}$ joule-second. (Planck's constant is named after the German physicist **Max Planck**, 1858–1947.) According to the particle description of light, the electromagnetic spectrum is a spectrum of photon energies. Photons of shorter wavelength (higher frequency) carry more energy than photons of longer wavelength (lower frequency). For example, photons of blue light carry more energy than photons of longer-wavelength red light. Ultraviolet photons carry more energy than photons of visible light, and X-ray photons carry more energy than ultraviolet photons. The lowest-energy photons are radio wave photons.

The **intensity** of light measures the *total* amount of energy that a beam of the light carries. A beam of red light can be just as intense as a beam of blue light—that is, it can carry just as much energy—but because the energy of a red photon is less than the energy of a blue photon, it will take

Photons are the *quantum mechanical* description of light.

more red photons to reach that intensity than it would take blue photons. This relationship is a lot like money. A hundred dollars is a hundred dollars, but it takes a lot more pennies (low-energy photons) to make up a hundred dollars than it takes 50-cent pieces (high-energy photons).

When physicists speak of the energy of light as broken into discrete packets called photons, they say that the light energy is **quantized**. The word *quantized,* which has the same root as the word *quantity,* means that something is subdivided into discrete units. A photon is referred to as a **quantum of light**. The branch of physics that deals with the quantization of energy and of other properties of matter is called **quantum mechanics**.

Quantum mechanics, like special relativity, is counterintuitive for humans, but its predictions have been confirmed over and over again by experiment. The conflict between everyday, commonsense ideas about the world and the world as revealed through modern science is discussed in **Connections 4.2**.

Atoms Can Occupy Only Certain Discrete Energy States

If we want to understand better how light interacts with matter, we need to start by pinning down exactly what we mean by *matter* in the first place. To a physicist, matter is anything that occupies space and has mass. Virtually *all* of the matter we have direct experience with is composed

Virtually all matter we encounter is composed of atoms.

of **atoms**. The computer keyboard this book is being typed on is made of atoms, and the neurons in your brain that are changing their structure as you read are made of atoms. Atoms are incredibly tiny—so tiny that a single teaspoon of water contains about 10^{23} atoms. (There are more atoms in a single teaspoon of water than there are stars in the observable universe.) When we talk about the interaction of light with matter, what we are really talking about is the interaction of light with atoms, and the things atoms themselves are composed of. So the next question is, "What are atoms?"

Atoms are built from three types of **elementary particles** as illustrated in **Figure 4.14(a)**. Sitting in the center of the atom is the **nucleus**, which is composed of positively charged **protons** and electrically neutral **neutrons**. An atom may have many protons and neutrons in its nucleus. Surrounding the nucleus of the atom are negatively charged **electrons**. For an atom to be electrically neutral, it must have the same number of electrons as protons. Electrons have much less mass than protons or neutrons, so almost all the mass of an atom is found in its nucleus. This naturally leads to a mental picture of an atom as a "tiny solar system," with the massive nucleus sitting in the center and the smaller electrons orbiting about much as planets orbit about the Sun (**Figure 4.14(b)**). We refer to this as the **Bohr model** after the Danish physicist **Niels Bohr** (1885–1962), who proposed it in 1913.

Unless you have thought about atoms a great deal, this is probably your concept of the structure of an atom. It is much the same picture that scientists in the early 20th century held as well. But it has a fatal problem. In this view, an electron whizzing about in an atom is constantly undergoing an acceleration—the direction of its motion is constantly changing. The wave description of electromagnetic radiation says that *any* electrically charged particle that is accelerating must also be giving off electromagnetic radiation. This electromagnetic radiation should be carrying away the orbital energy of the electron. (Imagine that electron as the wiggling electric charge in Figure 4.3(b).) If you calculate how much energy should be carried off by radiation from the electron, you find that only a tiny fraction of a second

CONNECTIONS 4.2

Thinking Outside the Box

As you read about the combined wave and particle description of light, you will likely find yourself scratching your head in confusion over exactly what light really is. If you do, consider yourself in good company. The scientists who invented the seemingly bizarre quantum description of nature had a great deal of trouble thinking about light as well. The wave model of light is clearly the correct description to use in many instances, just as Maxwell has shown. At the same time the particle description of light is also clearly the correct description to use in other cases, as scientists like Planck and Einstein demonstrated. But how can the same thing—light—be both a wave *and* a particle? It is hard for us to imagine a single thing sharing the properties of a wave on the ocean *and* a beach ball, yet light does just that.

Our trouble with thinking of light as both a wave and a particle only hints at the puzzling and philosophically troublesome world of quantum mechanics. As we go further, things only get worse. Light is not the only thing that shares wave and particle properties. In fact, *all* matter shares wave and particle properties. Sometimes a "particle" such as an electron behaves as if it were a wave, while at other times a "wave" of light clearly exhibits the properties of a discrete particle. Early quantum physicists would sometimes joke that on Monday, Wednesday, and Friday, light and matter were particles, whereas on Tuesday, Thursday, and Saturday, light and matter were waves. (And on Sunday it was best just not to think about them at all!)

Light is what light is, and an electron is what an electron is. The trouble with quantum mechanics lies not with the nature of reality, but with what our brains can easily think about. This chapter earlier provided another example of the limitations of our genetic programming. Our brains deal well with objects that are sitting still or even moving as fast as a hard-hit fly ball. Basically our brains cope best with things moving at the speeds of things in nature that we might want to eat or that might want to eat us. (Animals whose brains could not deal with such speeds tended not to survive long enough to pass their genes for those brains on to future generations.) But the brains of our ancestors did not have to deal with things moving at nearly the speed of light, so we should not be surprised that special relativity seems to defy our intuition. Likewise, there was no evolutionary pressure for our ancestors to be able to think easily about the wave/particle duality of light and matter.

Quantum mechanics and special relativity are not the only places where our ease in thinking about nature breaks down. Quantum mechanics deals with the very smallest scales in nature. At the other extreme, our brains did not evolve to think about things as large or as massive as stars and galaxies and the universe. When we move on to these larger scales later in the book, we will find our ideas about the nature of space and time themselves further challenged as we seek ways of visualizing curved spacetime or understanding why the question "What came before the beginning of the universe?" is in some ways much like asking, "What was to the left of last Thursday?"

Our brains exist in a box that is defined by the experiences and circumstances that we and our ancestors had to cope with. One exciting thing about modern physics and astronomy is that they force us to break down the walls of that conceptual box and find tools for understanding what lies beyond its boundaries.

should be needed for the electrons in an atom to lose all their energy and fall into the atom's nucleus! Fortunately for us this does not happen. Atoms exist for very long periods of time, and electrons never "fall into" the nuclei of atoms. So something must be wrong with this concept of an atom. A way out of this difficulty came when scientists realized that, just as waves of light have particlelike properties, so too do particles of matter have wavelike properties. With this realization, the *miniature solar system* model of the atom was modified so that a positively charged nucleus is surrounded *not* by planetlike electrons moving in their orbits, but by electron "clouds" or electron "waves" as illustrated in **Figure 4.14(c)** and discussed further in **Foundations 4.1**.

The strings on a guitar can vibrate only at certain discrete frequencies, giving rise to the discrete notes we hear. In much the same way, the electron waves in an atom can assume only certain specific forms. So instead of being able to take on *any* arbitrary energy, atoms can absorb or emit

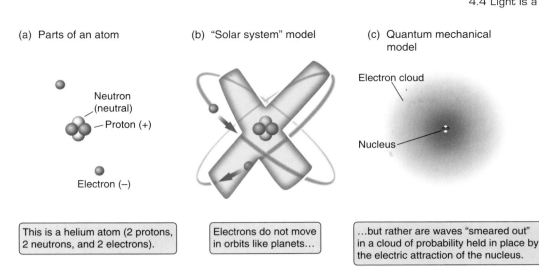

(a) Parts of an atom

Neutron (neutral)

Proton (+)

Electron (−)

This is a helium atom (2 protons, 2 neutrons, and 2 electrons).

(b) "Solar system" model

Electrons do not move in orbits like planets…

(c) Quantum mechanical model

Electron cloud

Nucleus

…but rather are waves "smeared out" in a cloud of probability held in place by the electric attraction of the nucleus.

FIGURE 4.14 (a) An atom is made up of a nucleus consisting of positively charged protons and electrically neutral neutrons, surrounded by less massive negatively charged electrons. (b) Atoms are often drawn as miniature "solar systems," but this model is incorrect. (c) Electrons are actually smeared out around the nucleus in quantum mechanical clouds of probability.

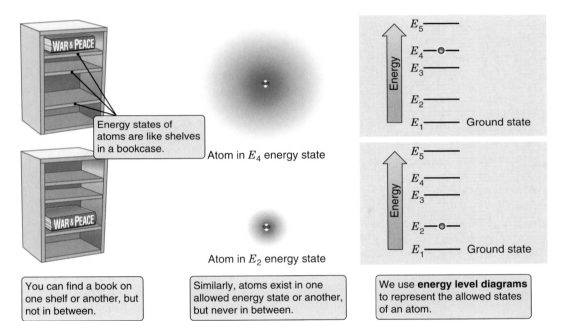

WAR & PEACE

Energy states of atoms are like shelves in a bookcase.

WAR & PEACE

You can find a book on one shelf or another, but not in between.

Atom in E_4 energy state

Atom in E_2 energy state

Similarly, atoms exist in one allowed energy state or another, but never in between.

E_5

E_4

E_3

E_2

E_1 ———— Ground state

Energy

E_5

E_4

E_3

E_2

E_1 ———— Ground state

Energy

We use **energy level diagrams** to represent the allowed states of an atom.

FIGURE 4.15 Atoms can have only certain discrete energies.

only certain specific energies corresponding to the allowed waveforms of their electron clouds. A given atom may have a tremendous number of different energy states available to it, but these states are *discrete*. An atom might have the en-

Atoms can have only certain discrete energies, much as guitar strings can play only certain notes.

ergy of one of these allowed states, or it might have the energy of the next allowed state, *but it cannot have an energy somewhere in between*. We can imagine the energy states of atoms as being a bookcase with a series of shelves as

shown in **Figure 4.15**. The energy of an atom might correspond to the energy of one shelf or to the energy of the next shelf; but the energy of the atom will *never* be found *between* the two shelves.

The lowest possible energy state of an atom—the "floor" —is called the **ground state** of the atom. Allowed states with energies lying above the ground state are called **excited states** of the atom. When the atom is in its ground state, it has nowhere to go. An electron cannot "fall" into the nucleus because there is no allowed state there with less energy for it to occupy. It cannot move up to a higher-energy state without getting some extra energy from somewhere. For this reason an atom will remain in its ground

FOUNDATIONS 4.1

Uncertainty Is Ordinary in the Quantum World

In the world of the very small, nothing seems intuitive. We have learned that all electromagnetic radiation behaves as both waves and particles. Possibly more surprising was the discovery that things like electrons and protons, which you may have visualized as "solid" particles, also have wave characteristics. There is, of course, a nice symmetry here. Waves have particlelike characteristics and particles have wavelike characteristics. This is not just a curious observation. It has huge implications in both science and technology. For instance, the wave–particle property is the principle by which electron microscopes work, as you will see in the next chapter.

Possibly the most significant implication and important outcome of wave–particle duality is the famous **Heisenberg uncertainty principle**, named for the German physicist **Werner Heisenberg** (1901–1976). If particles have wave characteristics, you cannot simultaneously pin down both their exact location and their **momentum**. There will *always* be some uncertainty in one or the other. Momentum (p) is defined as the product of mass and velocity ($p = m \times v$). Keep in mind that *velocity* includes both *speed* and *direction*. We can be more quantitative about this. The product of the uncertainty in a particle's position (Δx) and the uncertainty in its momentum (Δp) is always equal to or greater than a particular constant, which is of the order of Planck's constant, h. We can express this as a simple equation: $\Delta x \times \Delta p \sim h$. In other words, the more you know about *where* something is (Δx approaching zero), the less you can know about *how fast* and *in what direction* it is moving (Δp approaching infinity). Conversely, the better you know the momentum of something, the less you know about its location. This is not a matter of scientists making inferior measurements. You simply *cannot* do better, no matter how precisely you measure!

How does the uncertainly principle work in the real world? You should not be surprised to find that the best examples occur in the realm of subatomic particles. Earlier in this chapter we talked about the Bohr model of the hydrogen atom and imagined electrons as particles sailing around the proton in well-behaved orbits. The Bohr model says the angular momentum of the electron in an orbit has to be given *exactly* by an integer times a constant. There is no room for any uncertainty here. Yet the uncertainty principle must be obeyed. It tells us that if the angular *momentum* has no uncertainty, the angular *position* of the electron must be *completely uncertain*! That is right—you cannot know where the electron is in its orbit. This is why we use a featureless cloud to represent electrons in orbit around an atomic nucleus, as shown in Figure 4.14(c).

Another example involves a different form of the Heisenberg uncertainty principle: $\Delta E \times \Delta t \sim h$, where ΔE is the uncertainty of energy and Δt is the time over which the energy is measured. In Section 4.4 we make the point that atoms can occupy only certain discrete energy states, implying that their electrons are restricted to specific well-defined excited states. But can there really be *no* uncertainty whatsoever in these energy states? The answer becomes apparent if we rewrite the previous equation as $\Delta t \sim h/\Delta E$. If we say that $\Delta E = 0$ (no uncertainty in the energy state), then Δt becomes infinite.[6] If an electron were forced to stay at some excited state forever, it could never drop to a lower level, and we would never see narrow spectral emission lines. We do, of course, see emission lines, so there *must* be a certain amount of uncertainty in the electron's energy. Remember that wavelength is related to energy. A narrow range in energy therefore represents a narrow range in wavelength. So the longer an electron resides in an elevated energy state (Δt is large), the narrower will be the spectral emission line when it finally drops to a lower level (ΔE, and therefore $\Delta \lambda$, is small).

The bottom line here is that we can be absolutely *certain* that there is *uncertainty* at the root of *everything* physical. If you are bothered by this, you are in good company. It *really* bothered Einstein too.

---

[6]If you divide any number by zero, the result will be infinite.

state forever unless something happens to knock it into an excited state. A book sitting on the floor has nowhere left to fall, and it cannot jump to one of the higher shelves of its own accord.

An atom in an excited state is a very different matter, however. Just as a book on an upper shelf might fall to a lower shelf, an atom in an excited state might **decay** down to a lower state by getting rid of some of its extra energy. An important difference between the atom and the book on the shelf, however, is that whereas a snapshot might catch the book between the two shelves, the atom will never be caught between two energy states. When the transition from one state to another occurs, the energy difference between the two states must be carried off all at once. A common way for an atom to do this is to give off a photon. But not just any photon will do. The photon emitted by the atom must carry away exactly the amount of energy lost by that atom as it goes from the higher-energy state to the lower-energy state.

The Energy Levels of an Atom Determine the Wavelengths of Light It Can Emit and Absorb

To better understand the relationship between the energy levels of an atom and the radiation it can emit or absorb, imagine a hypothetical atom that has only two available energy states. Call the energy of the lower-energy state (the ground state) E_1 and the energy of the higher-energy state (the excited state) E_2. The energy levels of this atom can be represented in an energy level diagram like those in Figure 4.15, but with only two levels (see **Figure 4.16(a)**).

To understand the process of *emission,* imagine that the atom begins in the upper state (E_2) and then spontaneously drops down to the lower-energy state (E_1). This is shown in **Figure 4.16(b)**, where the downward arrow indicates that the atom went from the upper state to the lower state. The atom just lost an amount of energy equal to the difference between the two states, or $E_2 - E_1$. However, energy is never truly lost or created, so the energy lost by the atom has to show up somewhere. In this case, the energy

> When an atom drops to a lower energy state, the lost energy is carried away as a photon.

shows up in the form of a photon that is emitted by the atom. The energy of the photon emitted must just match the energy lost by the atom, so the energy of the photon must be $E_{photon} = E_2 - E_1$.

We have already seen the relationship between the energy of a photon and the frequency or wavelength of electromagnetic radiation. Using this relationship we can say that the frequency of the photon emitted by a transition from E_2 to E_1, which we will denote as $f_{2\rightarrow 1}$, is just the energy difference divided by Planck's constant (h):

$$f_{2\rightarrow 1} = \frac{E_{photon}}{h} = \frac{E_2 - E_1}{h}.$$

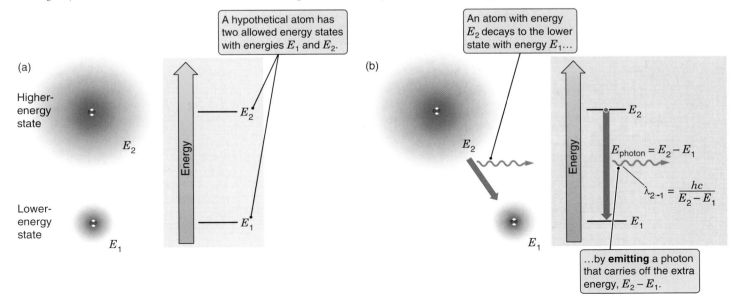

FIGURE 4.16 (a) The energy levels of a hypothetical two-level atom. (b) A photon with energy $hf = E_2 - E_1$ is emitted when an atom in the more energetic state decays to the lower-energy state.

(a)

Higher-energy state

Lower-energy state

A hypothetical atom has two allowed energy states with energies E_1 and E_2.

E_2

E_1

Energy

(b)

An atom with energy E_2 decays to the lower state with energy E_1...

E_2

E_1

Energy

$E_{photon} = E_2 - E_1$

$\lambda_{2\rightarrow 1} = \frac{hc}{E_2 - E_1}$

...by **emitting** a photon that carries off the extra energy, $E_2 - E_1$.

Similarly, the wavelength of the photon is just $\lambda = c/f$, or

$$\lambda_{2\to1} = \frac{c}{f_{2\to1}} = \frac{hc}{E_2 - E_1}.$$

This shows that the wavelengths of photons emitted by an atom—the color of the light that the atom gives off—are determined by the energy level structure of the atom. An atom can emit photons with energies corresponding only to the difference between two of its allowed energy states.

Imagine what the light coming from a cloud of gas consisting of our hypothetical two-state atoms would be like. This case is illustrated in **Figure 4.17**, which shows a collection of our two-state atoms. Any atom that finds itself in the upper energy state (E_2) will quickly decay and emit a photon in some random direction. A cubic meter of the air around you contains about 10^{25} atoms. Even if only a tiny fraction of these atoms emit a photon each second, an enormous number of photons would still come pouring out of the cloud of gas. But instead of containing photons of all different energies (that is, light of all different colors), like sunlight, this light would instead contain only photons with the specific energy $E_2 - E_1$ and wavelength $\lambda_{1\to2}$. In other words, all of the light coming from the cloud would be the same color.

We have all seen what happens to sunlight when it passes through a prism. Sunlight contains photons of all different colors, so when sunlight passes through a prism, it spreads out into all colors of a rainbow. But if we were to pass the light from our cloud of gas through a slit and a prism, as in Figure 4.17, the results would be very different. This time there would be no rainbow. Instead all of the light from the

cloud of gas would show up on the screen as a single bright line. The process we have just described—the production of a photon when an atom decays to a lower-energy state—is

---

The spectrum of a cloud of glowing gas contains emission lines.

---

referred to as **emission**. The bright, single-colored feature in the spectrum of the cloud of gas is referred to as an **emission line**.

So far in this discussion we have ignored an important question: "How did the atom get to be in the excited state E_2 in the first place?" An atom sitting in its ground state will remain in the ground state unless it is somehow given just the right amount of energy to kick it up to an excited state. Most of the time this extra energy comes in one of two forms: (1) The atom absorbs the energy of a photon (we will talk about this possibility shortly); or (2) the atom collides with another atom, or perhaps an unattached electron, and the collision knocks the atom into an excited state. This is how a neon sign works. When a neon sign is turned on, an alternating electric field is set up inside the glass tube that pushes electrons in the gas back and forth through the neon gas inside the tube. Some of these electrons crash into atoms of the gas, knocking them into excited states. The atoms then drop back down to their ground states by emitting photons, causing the gas inside the tube to glow. (In like fashion, an electron beam in a television picture tube collides with atoms in the screen, knocking them into excited states. When those atoms decay they emit the photons, producing what we perceive as the picture on the television.)

FIGURE 4.17 A cloud of gas containing atoms with two energy states, E_1 and E_2, emits photons with an energy $E = hf = E_2 - E_1$, which appear in the spectrogram (right) as an emission line.

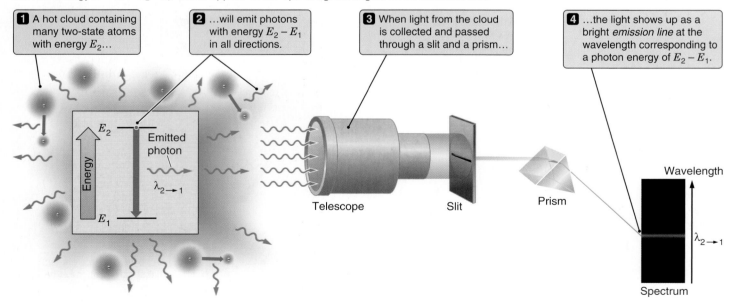

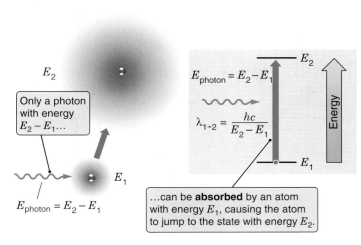

FIGURE 4.18 A photon of energy $hf = E_2 - E_1$ may be absorbed by an atom in the lower-energy state, leaving the atom in the higher-energy state.

So far we have focused on the emission of photons by atoms in an excited state, but what about the opposite process? An atom in a low-energy state can absorb the energy of a passing photon and jump up to a higher-energy state, as shown in **Figure 4.18**, but not just any photon can be absorbed by the atom. As before, the energy that it takes to get from E_1 to E_2 is the difference in energy between the two states, or $E_2 - E_1$. For a photon to cause an atom to jump from E_1 to E_2, it must provide just this much energy. Using the relationship that $E_{photon} = hf$ or $f = E_{photon}/h$, we find that the *only* photons capable of exciting atoms from E_1 to E_2 are photons whose frequency and wavelength are, respectively,

$$f_{1 \rightarrow 2} = \frac{E_{photon}}{h} = \frac{E_2 - E_1}{h}.$$

and

$$\lambda_{1 \rightarrow 2} = \frac{c}{f_{1 \rightarrow 2}} = \frac{hc}{E_2 - E_1}.$$

This is exactly the same energy photon—the same color of light—that is emitted by the atoms when they decay from E_2 to E_1. This is not a coincidence. The energy difference

Atoms can absorb photons with the same
energies as the photons they emit.

between the two levels is the same whether the atom is emitting a photon or absorbing one, so the energy of the photon involved will be the same in either case.

What might the spectrum of light look like when viewed through a cloud composed of our hypothetical gas of two-state atoms? If we shine photons of all different wavelengths (that is, light of all different colors) through the gas from one side, almost all of these photons will pass through the cloud of gas unscathed. So counting how many photons come *out* the other side of the cloud of gas should give us the number of photons that we shined *into* the gas. There is only one exception. Rather than passing through the gas, some of the photons with just the right energy ($E_2 - E_1$) might instead be absorbed by atoms.

If we shine light from a lightbulb directly though a glass prism, we will see that a rainbow of colors comes out as shown in **Figure 4.19(a)**. If we instead shine the light through a cloud of our two-state atoms before putting the light through a prism, the rainbow of colors will be unchanged

When viewed through a cloud of gas,
the spectrum of a lightbulb contains
absorption lines.

except for one detail. Because some of the photons with energies equal to $E_2 - E_1$ will have been absorbed by the gas, these photons will be missing in the light passing through the prism. If we look at the screen, we will see a sharp, dark line at the color corresponding to these photons (**Figure 4.19(b)**). The process of atoms capturing the energy of passing photons is referred to as **absorption**, and the dark feature seen in the spectrum is called an **absorption line**.

There is one final point worth making before leaving the subject of emission and absorption of radiation. When an atom absorbs a photon and jumps up to an excited energy state, there is a good chance that the atom will quickly decay back down to the lower-energy state by emitting a photon with the same energy as the photon it just absorbed. If the atom reemits a photon just like the one it absorbed, you might reasonably ask why the absorption really matters. After all, the photon that was taken out of the passing light was replaced, was it not? The answer is yes and no. The photon was replaced, true enough, but while all of the photons that were absorbed were originally traveling in the *same direction,* the photons that are reemitted travel off in *random directions.* In other words, some of the photons with energies equal to $E_2 - E_1$ are in effect diverted from their original paths by their interaction with atoms. If you look at a lightbulb *through* the cloud, you will notice an absorption line at a wavelength of $\lambda_{1 \rightarrow 2}$; but if you look at the cloud from the side (looking perpendicular to the original beam), you will see it as a glowing light with an emission line at this wavelength.

Emission and Absorption Lines Are the Spectral Fingerprints of Types of Atoms

In the previous section we used a hypothetical atom with only two allowed energy states to help us think about emission and absorption of photons. Real atoms have many more than just two possible energy states that they might occupy,

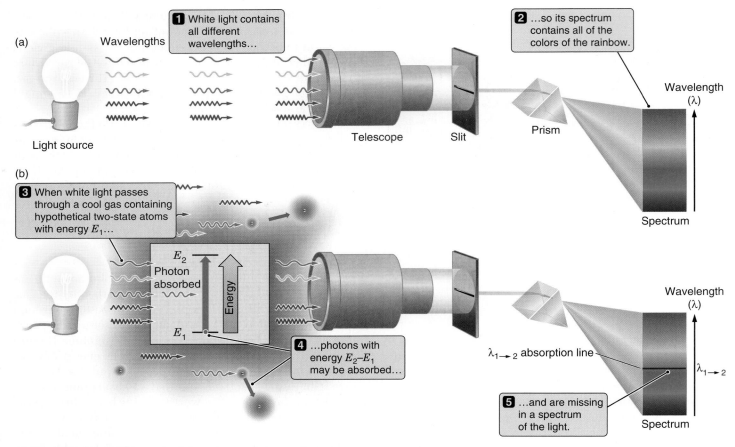

FIGURE 4.19 (a) When passed through a prism, white light produces a spectrum containing all colors. (b) When light of all colors passes through a cloud of hypothetical two-state atoms, photons with energy $hf = E_2 - E_1$ may be absorbed, leading to the dark absorption line in the spectrogram.

so a given type of atom will be capable of emitting and absorbing photons at many different wavelengths. An atom with three energy states, for example, might jump from state 3 to state 2, or from state 3 to state 1, or from state 2 to state 1. The emission lines from such a gas might have wavelengths of $hc/(E_3 - E_2)$, $hc/(E_3 - E_1)$, and $hc/(E_2 - E_1)$.

The allowed energy states of an atom are determined by the complex quantum mechanical interactions among the electrons and the nucleus that comprise the atom. Every hydrogen atom consists of a nucleus containing one proton, plus a single electron in a cloud surrounding the nucleus. As a result, every hydrogen atom has the same energy states available to it. It follows that all hydrogen atoms are capable of emitting and absorbing photons with the same wavelengths. **Figure 4.20(a)** shows the energy level diagram of hydrogen, along with the spectrum of emission lines for hydrogen in the visible part of the spectrum (**Figures 4.20(b)** and **(c)**).

Every hydrogen atom has the *same* energy states available to it, so all hydrogen atoms are in principle capable of producing the same spectral lines. But the energy states of a hydrogen atom are *different* from the energy states avail-

able to a helium atom, a lithium atom, or a boron atom, just as the energy states of these kinds of atoms differ from each other. Each different type of atom has a unique set of available energy states and therefore a unique set of wavelengths

---

The wavelengths at which atoms emit and absorb radiation form unique spectral fingerprints for each type of atom.

---

at which it can emit or absorb radiation. **Figure 4.20(d)** shows the set of emission lines that are given off by discharge tubes (like those in a neon sign) containing different kinds of atoms. These unique sets of wavelengths serve as unmistakable spectral fingerprints for each type of atom.

Spectral fingerprints are of crucial importance to astronomers. They let us figure out what types of atoms (or molecules) are present in distant objects by doing nothing more than looking at the spectrum of light from those objects. If we see the spectral lines of hydrogen, or helium, or carbon, or oxygen, or any other element in the light from a distant object, then we know that some of that element is present in that object. The strength of a line is determined in part by how many atoms of that type are present in the source. By measuring the strength of the lines from different types of atoms in the spectrum of a distant object, astronomers can often infer the relative amounts of different types of atoms of which the object is composed. But it gets even better. The fraction of atoms of a given kind that are in some particular energy state (as opposed to some other energy state) is often determined by factors such as the temperature or the **density** of the gas. By looking at the relative strength of different lines from the same kind of atom, it is often possible to determine the temperature, density, and pressure of the material as well.

How Are Atoms Excited, and Why Do They Decay?

In the last section we sidestepped an aspect of the emission process that has troubled physicists and philosophers alike since the earliest days of quantum mechanics. To appreciate this question, return to the analogy between emission of a photon and a book falling off a shelf. If we place a book on a level shelf and do not disturb it, the book will sit there forever. Once the book is resting on the upper shelf, something must *cause* the book to fall off the shelf. So what about the atom? Once an atom is in an excited state, what causes it to jump down to a lower-energy state and emit a photon? What triggers the event? Sometimes an atom in an upper-energy state can be "tickled" into emitting a photon —a process called *stimulated emission*—but under most circumstances the answer is that *nothing causes the atom to jump to the lower-energy state.* There is no trigger. Instead the atom decays *spontaneously.* And while we can say *about* how long the atom is *likely* to remain in the excited state, the rules of quantum mechanics say (and experiment shows) that we cannot know exactly when a given atom will decay until *after* the decay has happened. The atom decays at some *random* time that is not influenced by anything in the universe and cannot be known ahead of time.

FIGURE 4.20 (a) The energy states of a hydrogen atom are shown. Decays to level E_2 emit photons in the visible part of the spectrum. (b) This is what you might see if you looked at the light from a hydrogen lamp projected through a prism onto a screen. (c) This graph of the brightness of lines versus their wavelength is an example of how spectra are traditionally plotted. (d) Emission lines from several other types of gases: helium, argon, neon, and sodium.

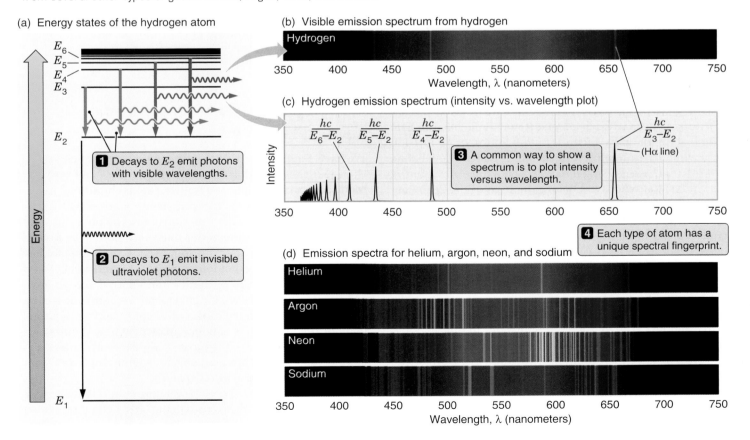

You have seen many examples of this rather amazing phenomenon. For example, you have probably owned a "glow in the dark" toy or a watch with a "glow in the dark" face. Photons in sunlight or from a lightbulb are absorbed by the atoms in the watch face, knocking those atoms into excited energy states. Unlike many excited energy states of atoms that tend to decay in a small fraction of a second, the excited states of the atoms in the watch face instead tend to live for many seconds before they decay. Suppose, for example, that on average these atoms tend to remain in their excited state for 1 minute before decaying and emitting a photon. In other words, suppose that if we wait for 60 seconds, there is a 50–50 chance that any particular atom will have decayed and a 50–50 chance that the atom will remain in its excited state. There are trillions upon trillions of such atoms in the watch face. Although it is impossible to say exactly which atoms in the watch face will decay after a minute, we can say with certainty that *about* half of them *will* decay within 60 seconds. From the standpoint of the glow that we see from the watch, it makes little practical difference which half of the atoms decay and which half do not. All we need to know is that if we wait 1 minute, half of the atoms will have decayed, and the brightness of the glow from the watch will have dropped to half of what it was. If we wait another minute, half of the remaining excited atoms will decay, and the brightness of the glow will be cut in half again. Each 60 seconds, half of the remaining excited atoms decay, and the glow from the watch drops to half of what it was 60 seconds earlier. The glow from the watch slowly fades away.

We have now come upon one of the most philosophically troubling aspects of quantum mechanics. In deep space, where atoms can remain undisturbed for long periods of time, there are certain excited states of atoms that, on average, live for tens of million years or even longer. Envision an atom in such a state. It may have been in that excited state for a few seconds, a few hours, or 50 million years when in an instant it decays to the lower-level state *without anything causing it to do so.* Newton and virtually every physicist who lived before the turn of the 20th century envisioned a clockwork universe in which every effect had a cause. They imagined that if we knew the exact properties of every bit of the universe today, it was just a matter

> Quantum mechanics undermines the orderly, causal universe of Newtonian physics.

of turning the crank on the laws of physics to predict what the state of the universe would be tomorrow. Then quantum mechanics came along and turned this view on its head. Instead of dealing with strict cause-and-effect relationships, physicists found themselves calculating the *probabilities* of certain events taking place and facing fundamental limitations on what can ever be known about the state of the universe. We have mentioned that while Einstein helped start the scientific revolution of quantum mechanics, in the end that revolution left him behind. He could never shake his firm belief in Newton's clockwork, causal universe. "God does not play dice with the universe!" he insisted emphatically. As more of the predictions of quantum mechanics were borne out by experiment, most physicists came to accept the implications of the strange new theory. Einstein, on the other hand, went to his grave looking unsuccessfully for a way to save his notion of order in the universe.

It is interesting to note that although Einstein refused to accept quantum mechanics, our understanding of the quantum mechanical nature of reality owes him a great debt. His was one of the greatest minds of all time, and as he searched tirelessly for flaws in quantum mechanics, he presented challenge after challenge to those who were trying to work out the details of the new theory. It was in responding to Einstein's objections that physicists were forced to confront the full implications of their own work. As an epilogue to this story, this struggle continues to this day. At the dawn of the 21st century a few theoretical physicists are still pursuing Einstein's dream, trying to recast quantum mechanics in a way that recaptures the strict causality that seemed irretrievably lost shortly after the beginning of the 20th century. So far they have had little success, and most physicists doubt that they ever will. However, like Einstein before them, their healthy skepticism has led to ever deeper understanding of the implications and limitations of the theory.

The Doppler Effect—Is It Moving Toward Us or Away from Us?

We have begun to see that to an astronomer light is far more than just the stuff that bounces off the page and lets you read these words. Light is a tightly packed bundle of information that when spread into its component wavelengths can reveal a wealth of information about the physical state of material located tremendous distances away. The nature of light shapes how we think about space and time and has forced physicists to abandon many of their most cherished ideas about the nature of matter and energy. Yet we have only begun to explore what light can tell us. It is time to step back from the precipice of the philosophical implications of quantum mechanics and look instead at how light can be used to measure one of the most straightforward questions about a distant astronomical object: Is it moving away from us or toward us, and at what speed?

Have you ever stood on a street corner and listened as a fire truck sped by with sirens blaring? If so, you might have

much bluer. The detailed answers to these questions were worked out around 1900 by Max Planck. Planck was thinking about a special situation—a hollow, totally enclosed cavity of material at a specific temperature, *T*. The crucial point here is that inside the cavity all the radiation emitted

A blackbody emits thermal radiation that has a Planck spectrum.

by the cavity walls is also absorbed by the walls of the cavity. In this situation a balance is set up, with each bit of the wall emitting just as much thermal radiation as it is absorbing from its surroundings. Physicists refer to such a special situation as a **blackbody**. Planck used this balance to calculate the spectrum of the light inside such a cavity. The result of his calculation, which beautifully matches the results of experiments, is called a **Planck spectrum** or a **blackbody spectrum**.

You might reasonably ask what this hypothetical cavity has to do with the light from the filament of a lightbulb. Surely the filament of a lightbulb is not a cavity of this sort! But in a certain sense it is. The light emitted by charged particles within the filament is mostly absorbed by other charged particles within the filament. This is exactly the assumption that Planck made when calculating the shape of the spectrum of a blackbody. As a result, we expect that the radiation existing *inside the filament of the lightbulb* will have a Planck spectrum. This radiation "leaks out" of the filament, just as light might leak out of a small hole in the side of Planck's cavity. As a result, the radiation from the filament of a lightbulb is very close to a Planck spectrum. The light from stars such as the Sun and the thermal radiation from a planet also often come close to having a blackbody spectrum.

Stefan's Law Says That Hotter Means Much More Luminous

Figure 4.25 shows plots of the Planck spectra for objects at several different temperatures. We now ask the question that scientists must always ask: Does the theoretical prediction agree with observation and experiment? Do these spectra agree with our intuitive ideas and with our experiment with the lightbulb and the dimmer? Begin with luminosity. As the temperature of an object increases,

The luminosity of a blackbody is proportional to T^4.

Planck's theory says that the object gives off more radiation at every wavelength, so the luminosity of the object should increase. In fact, it increases in a hurry. Adding up all of the energy in a Planck spectrum shows that the

increase in luminosity is proportional to the *fourth power* of the temperature: Luminosity $\propto T^4$. This result is known as **Stefan's Law** because it was discovered in the laboratory by **Josef Stefan** (1835–1893) before Planck's theory came along to explain it.

What Stefan's Law actually says is that the amount of energy radiated *by each square meter* of the surface of an object is given by the equation

$$\mathcal{F} = \sigma T^4.$$

In this equation $\mathcal{F}$ is called the **flux**. It is a measurement of the total amount of energy coming through each square meter of the surface each second. The constant σ (pronounced "sigma") is called the **Stefan-Boltzmann constant**, named after the discoverer of this relationship. The value of σ is the same for all cases and is given by 5.67×10^{-8} W/(m^2 K^4) (where W stands for watts). To find the total amount of energy emitted by an object in the form of electromagnetic radiation, multiply $\mathcal{F}$ by the surface area of the object.

Returning to our lightbulb example, we can actually use Stefan's Law to figure out what the surface area of the filament in a lightbulb must be. Suppose the filament in an incandescent bulb operates at a temperature of about 2,500 K. The amount of energy radiated by the bulb is stamped right

FIGURE 4.25 Planck spectra emitted by sources with temperatures of 2,000 K, 3,000 K, 4,000 K, 5,000 K, and 6,000 K. At higher temperatures the peak of the spectrum shifts toward shorter wavelengths, and the amount of energy radiated per second from each square meter of the source increases.

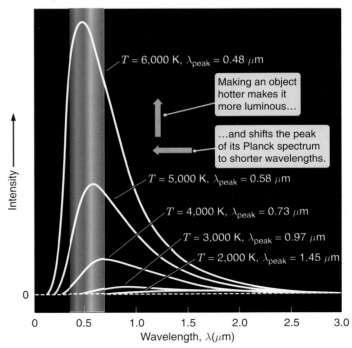

on the face of the bulb—a 100 W bulb has a luminosity of 100 W, which means that it radiates away 100 joules each second. (A **joule** is a unit of energy, abbreviated J.) If the total amount of light from the filament is equal to the flux ($\mathcal{F}$) times the surface area of the filament (A), we can write

$$100 \text{ W} = A \times \sigma T^4.$$

Solving this equation for the area, we get

$$A = \frac{100 \text{ W}}{\sigma T^4} = \frac{100 \text{ W}}{\left(5.67 \times 10^{-8} \frac{\text{W}}{\text{m}^2\text{K}^4}\right) \times (2{,}500 \text{ K})^4}$$

$$= 4.5 \times 10^{-5} \text{ m}^2.$$

Note that not much filament surface area is needed to provide the light that turns night into day in our homes and cities.

Stefan's Law says that an object rapidly becomes more luminous as the temperature increases. If the temperature of an object goes up by a factor of 2, the amount of energy being radiated each second increases by a factor of 2^4 or 16. If the temperature of an object goes up by a factor of 3, then

---

Slight changes in temperature mean large changes in brightness.

---

the energy being radiated by the object each second goes up by a factor of $3^4 = 81$! A lightbulb with a filament temperature of, say, 3,000 K radiates 16 times as much light as it would if the filament temperature were 1,500 K. Even modest changes in temperature can result in large changes in the amount of power radiated by an object.

Wien's Law Says That Hotter Means Bluer

Look again at Figure 4.25, but this time instead of paying attention to how high each curve is, notice where the peak of each curve falls along the horizontal axis. As the temperature increases, the *peak* of the Planck spectrum shifts toward shorter wavelengths, which means the average energy of the photons becomes greater. Just as we surmised, increasing the temperature causes the light from the object to get bluer. The shift in the location of the peak of the Planck spectrum with increasing temperature is given by the equation

$$\lambda_{\text{peak}} = \frac{2{,}900 \ \mu\text{m K}}{T}.$$

This result is referred to as **Wien's Law.** In this equation λ_{peak} (pronounced "**lambda peak**") is the wavelength where the Planck spectrum is at its peak. It is the wavelength where

the electromagnetic radiation from an object is greatest. Wien's Law says that the location of the peak in the spectrum is inversely proportional to the temperature of the

---

The peak wavelength of a blackbody is inversely proportional to its temperature.

---

object. If you increase the temperature by a factor of 2, the peak wavelength becomes half of what it was. If you increase the temperature by a factor of 3, the peak wavelength becomes a third of what it was.

It is useful to put a few numbers into Wien's Law. The surface of the Sun, for example, has a temperature of about 5,800 K. Wien's Law says that the peak in the light from the Sun occurs at a wavelength of

$$\lambda_{\text{peak},\odot} = 0.5 \ \mu\text{m}.$$

The light given off by the Sun is concentrated at a wavelength of about 0.5 μm or 500 nm, which is in the middle of what we refer to as the visible part of the spectrum.

Wien's Law will prove handy as we continue our study of the universe. If we can measure the spectrum of an object emitting thermal radiation and find where the peak in the spectrum is, we can use Wien's Law to calculate the temperature of the object. Turning our previous example around, we have no way of dropping a thermometer into the Sun and directly measuring its temperature, but we *can* observe the spectrum of the light coming from the Sun. When we do so, we find that the peak in the spectrum occurs at a wavelength of about 0.5 μm. Wien's Law can be rewritten as

$$T = \frac{2{,}900 \ \mu\text{m K}}{\lambda_{\text{peak}}}.$$

If we plug the observed peak of the spectrum of the Sun ($\lambda_{\text{peak}} \approx 0.5 \ \mu$m) into this equation, we get

$$T = \frac{2{,}900 \ \mu\text{m K}}{0.5 \ \mu\text{m}} = 5{,}800 \text{ K}.$$

This is how we know the temperature of the Sun.

4.6 Twice as Far Means One-Fourth as Bright

You might have noticed that we have consistently spoken of the "luminosity" of objects, where in everyday language we probably would have just said that one object is "brighter" than another. This is a case where everyday language is too sloppy for science. To a physicist or an astronomer, the **brightness** of electromagnetic radiation refers to the amount of light that is *arriving* at some location, such as the page of the book you are reading or the pupil of your eye. Luminosity refers to the amount of light *leaving* a source. The concept of brightness is certainly related to the concept of luminosity. For example, replacing a lightbulb with a luminosity of 50 W with a 100 W bulb succeeds in making a room twice as bright because it doubles the light reaching any point in the room. But brightness also depends on the distance from a source of electromagnetic radiation. If you needed more light to read this book by, you could replace the bulb in your lamp with a more luminous one, but it would probably be easier to just move the book closer to the light. Conversely, if a light were too bright for you, you would move away from it. Our everyday experience says that as we move away from a light, its brightness decreases.

The particle description of light provides a convenient way to think about the brightness of radiation and how brightness depends on distance. Suppose you had a piece

---

Brightness measures how much light falls per square meter per second.

---

of cardboard that was 1 meter on a side. Intuitively you might imagine that making the light that falls on the cardboard twice as bright would mean doubling the number of photons that hit the cardboard each second. Tripling the brightness of the light would mean increasing the number of photons hitting the cardboard each second by a factor of 3, and so on. Here is a beginning point for understanding brightness. Brightness depends on the number of photons falling on each square meter of a surface each second.

Working with this idea of brightness, now imagine a lightbulb sitting at the center of a spherical shell, as shown in **Figure 4.26**. Photons from the bulb travel in all directions and land on the inside of the shell. To find the number of photons landing on each square meter of the shell during each second (that is, to find the brightness of the light), take the *total* number of photons given off by the lightbulb each second and divide by the number of square meters those photons have to be spread over. The surface area of a sphere is given by the formula $A = 4\pi r^2$, where r is the distance between the bulb and the surface of the sphere (thus r = the radius of the sphere). When this is written as a formula, we find that

$$\begin{array}{l} \text{Number of photons} \\ \text{striking one square} \\ \text{meter each second} \end{array} = \frac{\begin{array}{c} \text{Total number of photons} \\ \text{emitted per second} \end{array}}{\begin{array}{c} \text{Number of square meters} \\ \text{the photons are spread over} \end{array}}$$

$$= \frac{\begin{array}{c} \text{Total number of photons} \\ \text{emitted each second} \end{array}}{4\pi r^2}.$$

The next step in building an understanding of brightness is to change the size of the spherical shell while keeping the total number of photons given off by the lightbulb

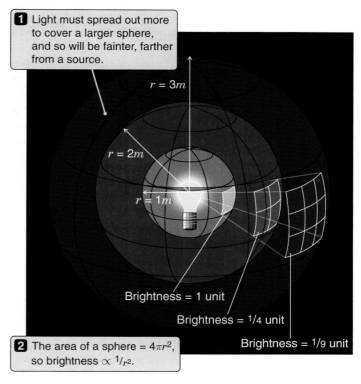

1 Light must spread out more to cover a larger sphere, and so will be fainter, farther from a source.

$r = 3m$

$r = 2m$

$r = 1m$

Brightness = 1 unit

Brightness = 1/4 unit

2 The area of a sphere = $4\pi r^2$, so brightness $\propto 1/r^2$.

Brightness = 1/9 unit

FIGURE 4.26 Light obeys an inverse square law as it spreads out away from a source. Twice as far means one-fourth as bright.

the same. As the shell becomes larger, the photons from the lightbulb must spread out to cover a larger surface area. Each square meter of the shell receives fewer photons each second, so the brightness of the light falls. If the shell's surface is moved twice as far from the light, the area over which the light must spread increases by a factor of $2^2 = 2 \times 2 = 4$.

Like gravity, light obeys an inverse square law.

The photons from the bulb spread out over four times as much area, so the number of photons falling on each square meter each second becomes ¼ of what it was. If the surface of the sphere is three times as far from the light as illustrated in Figure 4.26, the area over which the light must spread increases by a factor of $3 \times 3 = 3^2 = 9$, and the number of photons per second falling on each square meter becomes ⅑ of what it was originally. We encountered just this kind of relationship earlier when we talked about gravity (Chapter 3). Just like gravity, light obeys an inverse square law. The brightness of the light from an object is inversely proportional to the square of the distance from the object. *Twice as far means one-fourth as bright.*

It is nice to think of brightness in terms of photons streaming onto a surface from a light because this gives us a nice mental picture of the physical nature of brightness and why brightness follows an inverse square law. In practice, however, it is usually more convenient to speak of the *energy* coming to a surface each second, rather than the number of photons received.

The luminosity of an object is the total number of photons given off by the object times the energy of each photon. Instead of talking about how the number of photons must spread out to cover the surface of a sphere (brightness), we now talk about how the *energy* carried by the photons must spread out to cover the surface of a sphere. When speaking of brightness in this way, we mean the amount of energy falling on a square meter in a second. If L is the luminosity of the bulb, then the brightness of the light at a distance r from the bulb is given by

$$\text{Brightness} = \frac{\text{Energy radiated per second}}{\text{Area over which energy is spread}}$$

$$= \frac{L}{4\pi r^2}.$$

Before moving on we offer the following aside. Usually the only information that astronomers have to work with is the light from a distant object. For this reason we will use our understanding of radiation over and over again throughout our journey. Time spent now thinking carefully about the electromagnetic spectrum, emission and absorption of photons, Planck radiation, and the inverse square law for brightness will be a *very* good investment for what is to come.

4.7 Radiation Laws Allow Us to Calculate the Equilibrium Temperatures of the Planets

We began our discussion of thermal radiation by asking a straightforward question: "Why does a planet have the temperature that it does?" In a qualitative way we said that the temperature of a planet is determined by a balance between the amount of sunlight being absorbed and the amount of energy being radiated back into space. We now have the tools we need to turn this qualitative idea into a real prediction of the temperatures of the planets.

Begin with the amount of sunlight being absorbed. The amount of energy absorbed by a planet is just the area of the planet that is absorbing the energy times the brightness of sunlight at the planet's distance from the Sun. When we look at a planet, we see a circular disk with a radius equal to the radius of the planet. The area of this circular disk is

πR^2, where R is the radius of the planet. We found in our discussion in Section 4.6 that the brightness of sunlight at a distance d from the Sun is equal to the luminosity of the Sun ($L_\odot$ in watts) divided by $4\pi d^2$. (This d is the same as the r in the previous section. We use d here to avoid confusion with the planet's radius, R.) We must consider one additional factor. Not all of the sunlight falling on a planet is absorbed by the planet. The fraction of the sunlight that is reflected from a planet is called the **albedo**, a, of the planet. The corresponding fraction of the sunlight that is absorbed by the planet is 1 minus the albedo. A planet with an albedo of 1 reflects all the light falling on it. A planet that absorbs 100 percent of the sunlight falling on it has an albedo of 0.

Writing this as an equation, we say that

$$\begin{pmatrix} \text{Energy absorbed} \\ \text{by the planet} \\ \text{each second} \end{pmatrix} = \begin{pmatrix} \text{Absorbing} \\ \text{area of} \\ \text{planet} \end{pmatrix} \times \begin{pmatrix} b = \text{Brightness} \\ \text{of sunlight} \end{pmatrix} \times \begin{pmatrix} \text{Fraction} \\ \text{of sunlight} \\ \text{absorbed} \end{pmatrix}$$

$$= \pi R^2 \times \frac{L_\odot}{4\pi d^2} \times (1-a)$$

where a is the albedo of the planet.

Moving to the other piece of the equilibrium, the amount of energy that the planet radiates away into space each second is just the number of square meters of surface area that the planet has times the power radiated by each square meter. The surface area for the planet is given by $4\pi R^2$. Stefan's Law tells us that the power radiated by each square meter is given by σT^4. So we can say that

$$\begin{pmatrix} \text{Energy radiated by} \\ \text{planet per second} \end{pmatrix} = \begin{pmatrix} \text{Surface area} \\ \text{of planet} \end{pmatrix} \times \begin{pmatrix} \text{Energy radiated by} \\ \text{each m}^2 \text{ each second} \end{pmatrix}$$

$$= 4\pi R^2 \times \sigma T^4.$$

If the planet's temperature is to remain stable—if it is to keep from heating up or cooling off—then it must be radiating away just as much energy into space as it is absorbing in the form of sunlight, as indicated in **Figure 4.27**. That means that we can equate these two expressions. We can set the quantity "Energy radiated by planet" equal to the quantity "Energy absorbed by planet." When we do this, we arrive at the expression

$$\begin{array}{c} \text{Energy radiated by} \\ \text{the planet each second} \end{array} = \begin{array}{c} \text{Energy absorbed by} \\ \text{the planet each second} \end{array}$$

or

$$4\pi R^2 \sigma T^4 = \pi R^2 \frac{L_\odot}{4\pi d^2}(1-a).$$

Look at this equation for a moment. It may seem rather complex, but when broken into pieces it becomes more digestible. On the left side of the equation, $4\pi R^2$ tells how many square meters of the planet's surface are radiating energy back into space, while σT^4 tells how much energy

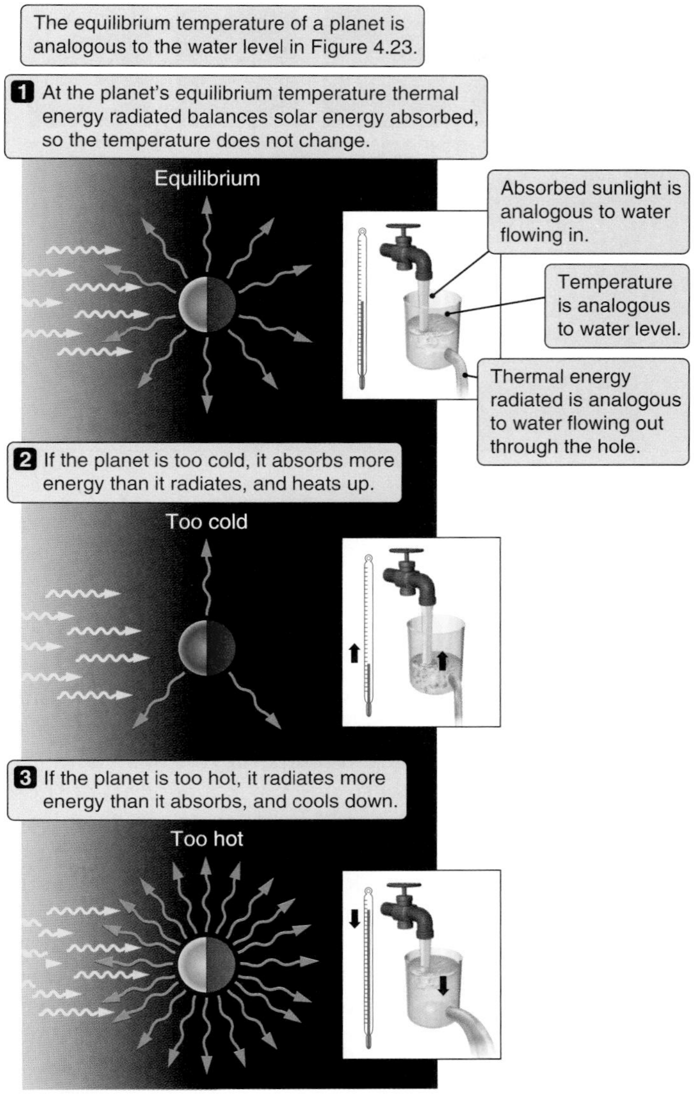

The equilibrium temperature of a planet is analogous to the water level in Figure 4.23.

1 At the planet's equilibrium temperature thermal energy radiated balances solar energy absorbed, so the temperature does not change.

Equilibrium

Absorbed sunlight is analogous to water flowing in.

Temperature is analogous to water level.

Thermal energy radiated is analogous to water flowing out through the hole.

2 If the planet is too cold, it absorbs more energy than it radiates, and heats up.

Too cold

3 If the planet is too hot, it radiates more energy than it absorbs, and cools down.

Too hot

FIGURE 4.27 Planets are heated by sunlight and cooled by emitting thermal radiation into space. If there are no other sources of heating or means of cooling, then the equilibrium between these two processes determines the temperature of the planet.

each one of those square meters radiates each second. Put them together, and you get the total amount of energy radiated away by the planet each second. On the right side of the equation, πR^2 is the area of the planet as seen from the Sun. That amount times the brightness of the sunlight reaching the planet, $L_\odot/4\pi d^2$, tells how much energy is falling on the planet each second. The final $1 - a$ tells how much of that energy the planet actually absorbs. Put everything on the right side of the equation together, and you get the amount of energy absorbed by the planet each second. The equal sign says that the energy radiated away needs to balance the sunlight absorbed. There is no magic here. In

fact, when broken down, this formidable equation embodies little more than a few straightforward ideas such as "hotter means more luminous," "twice as far means one-fourth as bright," and "heating and cooling must balance each other." The math just gives us a convenient way to work with these concepts.

We started down this path hoping to find a way to predict the temperatures of the planets, and a bit of algebra gets us the rest of the way there. Rearranging the previous equation to put T on one side and everything else on the other gives

$$T^4 = \frac{L_\odot(1-a)}{16\sigma\pi d^2}.$$

If we take the fourth root of each side, we wind up with

$$T = \left[\frac{L_\odot(1-a)}{16\sigma\pi}\right]^{\frac{1}{4}} \times \frac{1}{\sqrt{d}}.$$

We have now produced a full-fledged physical model for why the temperatures of planets are what they are. Restating the meaning in words, T is the temperature at which the energy radiated by a planet exactly balances the energy absorbed by the planet. If the planet were hotter than this equilibrium temperature, it would radiate energy away faster than the planet absorbed sunlight, and the temperature would fall. If the planet were cooler than this temperature, it would radi-

Balancing cooling and heating sets an equilibrium temperature.

ate away less energy than was falling on it in the form of sunlight, and the temperature of the planet would rise. Only at this equilibrium temperature do the two balance.

The equation tells us that as the distance from the Sun increases—in other words, as d gets bigger—the temperature of the planet decreases. No surprise there. But now we know how much the temperature should decrease. It should be inversely proportional to the square root of the distance. Using this formula can turn our intuition about why planets that are close to the Sun are hot into a prediction of just how hot they should be.

Figure 4.28 shows a graph of the predicted temperatures of the planets. The vertical bars show the range of temperatures found on the surfaces of each planet (or, in the case of the giant planets, at the tops of their clouds). The black dots show our predictions using the equation above. From the figure, you can see that overall we are not too far off. That should give us a sense of accomplishment: It says that our basic understanding of *why* planets have the temperatures that they do is probably not too far off. Mercury, Mars, and Pluto agree particularly well. (The agreement for Mercury would improve if we took into account the huge difference in temperature between the daytime and nighttime sides of the planet and recomputed our equilibrium accordingly.)

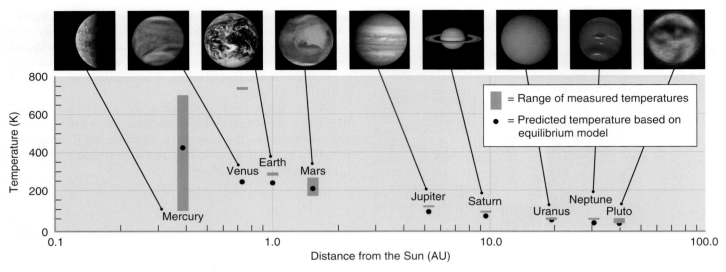

FIGURE 4.28 Predicted temperatures for the classical planets and Pluto, based on the equilibrium between absorbed sunlight and thermal radiation into space, are compared with ranges of observed surface temperatures. Some predictions are correct. Interestingly, others are not.

In other cases, however, our predictions are wrong. For Earth and the giant planets the actual temperatures are a bit higher than the predicted temperatures. In the case of Venus the actual surface temperature is wildly higher than our prediction. Rather than cause for despair, these discrepancies between theory and observation are cause for excitement. As we built our physical model for the equilibrium temperatures of planets, we made a number of assumptions. For example, we assumed that the temperature of the planet was the same everywhere. This is clearly not true: We might expect planets to be hotter on the day side than on the night side. We also assumed that a planet's only source of energy is the sunlight falling on it. Finally,

we assumed that a planet is able to radiate energy into space freely as a blackbody. The discrepancies between our theory and the measured temperatures of some of the planets tell us that for these planets, some or all of these assumptions must be incorrect. In other words, the places where the predictions of our theory are not confirmed by observation point to areas where there is something still to be discovered and understood. The question of *why* these planets are hotter than the prediction will lead us to a number of new and interesting insights into how these planets work. Scientific theories sometimes succeed and sometimes fail, but even when they fail they can teach us a lot about the universe.

Summary

- From gamma rays to visible light to radio waves, all radiation is an electromagnetic wave.

- Light is also a stream of particles called photons.

- The speed of light in a vacuum is 300,000 km/s, and nothing can travel faster.

- Like gravity, light obeys the inverse square law.

- Light from receding objects is redshifted. Light from approaching objects is blueshifted.

- Special relativity concerns the relationship between events in space and time.

- Space and time together form a four-dimensional spacetime.

- Nearly all matter is composed of atoms.

- Atoms absorb and emit radiation at unique wavelengths like spectral fingerprints.

- Temperature is a measure of the thermal energy of an object.

Seeing the Forest through the Trees

In our daily lives, light is the ideal messenger, faithfully telling us about the world around us. As with any good courier, we usually concentrate on the information that light carries, taking the messenger itself for granted. It is easy to forget that our seemingly immediate visual perception of reality is actually a derived experience—the result of a sophisticated interplay between our eyes, our brains, and the flood of electromagnetic radiation that is emitted, absorbed, transmitted, and reflected by objects in the world around us.

The light that our eyes see is only a tiny portion of the full span of the electromagnetic spectrum. Electromagnetic radiation carries with it a wealth of information about the temperature, density, composition, state of motion, and other physical characteristics of the place of its origin and the material it interacts with en route. In place of the carefully controlled laboratory experiments of many other sciences, the astronomer uses a combination of ingenuity and technology to "slice up" the light reaching us and tease out the information it carries about conditions throughout the universe. The role of electromagnetic radiation in astronomy is far more than that of messenger. Radiation is also a participant in the processes that we study. For example, light carries energy from the Sun outward through the Solar System, heating the planets, and light carries energy away from each planet, allowing it to cool. The balance between these two processes establishes the conditions of our existence.

Light may be both an informative messenger and an important player in the ebb and flow of the universe, but the very nature of light itself plays havoc with our commonsense ideas about the world. When we explored the motions of Earth, the Moon, and the planets, we relied heavily on our intuition about the world. The pull or shove of one object on another and the force of gravity that holds us tightly to the surface of Earth are well within the realm of our everyday experience. But when we consider the properties of light, the boundaries of our experience and intuition are shattered. We are forced beyond the confines of the "box" within which our brains evolved. Answering a simple question like "How fast does light travel?" demands that we abandon our most cherished ideas about the nature of space, time, matter, and energy. We ask, "What is light?" and confront abstract concepts like electric and magnetic fields while running headlong into the seemingly impossible question of how something can be both a wave *and* a particle. We delve into the interaction between light and matter and find that at the scale of atoms and photons, Newton's clockwork universe crumbles. In its place we discover a world of random chance and uncertainty, governed not by the strict march of cause and effect but by laws of probability and statistics. We ask about the nature of light—and we collide squarely with the shortcomings of our intuitive ideas about the nature of reality itself.

If electromagnetic radiation is the messenger, how do we hear and interpret the message? In the next chapter we'll learn about the many and varied tools that astronomers use to detect and decipher the meaning of these communications that come to us from the Solar System and beyond.

Key Terms

4. Imagine a future cosmonaut traveling in a spaceship at 0.866 times the speed of light. Special relativity says that the length of his spaceship along the direction of flight is only half of what it was when it was at rest on Earth. He checks this with a meter stick that he brought along with him. Would his measurement confirm the contracted length of his spaceship? Explain your answer.

5. Patterns of emission or absorption lines in spectra can uniquely identify individual atomic elements, just as DNA testing uniquely identifies individual human beings. Explain how positive identification of atomic elements can be used as one way of testing the validity of the cosmological principle.

6. Many physical properties are proportional to the temperature of an object, raised to some power. For example, as we have seen in this chapter, the luminosity of a radiating body is proportional to T^4. Why must the temperature T be expressed in kelvins rather than degrees Celsius or Fahrenheit?

7. Consider two hypothetical planets with no atmospheres. One orbits the Sun at an average distance of 5.0 AU and the other at an average distance of 10.0 AU, yet both have the same average surface temperature. Explain how this could be possible.

8. During a popular art exhibition, the museum staff finds that to protect the artwork they must limit the total number of viewers in the museum at any time. Therefore, new viewers are admitted at the same rate that others leave. Is this an example of static or dynamic equilibrium? Explain.

9. The difference between brightness and luminosity can confuse many people. How would you explain the difference to a family member or a friend who is not taking this class?

Student Questions

THINKING ABOUT THE CONCEPTS

1. Our eyes are not sensitive to electromagnetic radiation with wavelengths much shorter than 400 nm, which lies in the ultraviolet portion of the electromagnetic spectrum. Why is this the case?

2. Einstein's theory of special relativity tells us that no object can travel faster than, or even at, the speed of light. We know that light is an electromagnetic wave, but we know that it is also a particle called a photon. If it acts as a particle, how can a photon travel at the speed of light?

3. The Sun's mass declines by more than 4 million tons of mass each second. What happens to this mass?

APPLYING THE CONCEPTS

10. You are tuned to 790 on AM radio. This station is broadcasting at a frequency of 790 kilohertz (7.9×10^5 Hz). What is the wavelength of the radio signal? You switch to 98.3 on FM radio. This station is broadcasting at a frequency of 98.3 megahertz (9.83×10^7 Hz). What is the wavelength of this radio signal?

11. If a spaceship approaching us at 0.9 times the speed of light shines a laser beam at Earth, how fast will the photons in the beam be moving when they arrive at Earth?

12. We are given a bar of tungsten, a metal with a melting point of 3,640 K and a boiling point of 6,170 K. We heat

the bar until it starts to melt and plot a graph of its emitted Planck energy distribution. We then continue to heat the tungsten until it begins to boil and plot a similar graph.

   a. Sketch and label these graphs on the same set of axes. At what wavelength does the spectrum of each peak?

   b. How many times more luminous is the boiling tungsten than the melting tungsten?

13. Imagine that you have been transported to Neptune, 30 AU from the Sun. How bright would the Sun appear compared to its brightness as seen from Earth?

14. A planet with no atmosphere at 1 AU from the Sun would have an average blackbody surface temperature of 279 K if it absorbed all the Sun's electromagnetic energy falling on it (albedo = 0).

   a. What would be the average temperature on this planet if its albedo were 0.1, typical of a rock-covered surface?

   b. What would be the average temperature if its albedo were 0.9, typical of a snow-covered surface?

15. On a dark night you notice that a distant lightbulb happens to be the same brightness as a firefly that is 5 m away from you. If the lightbulb is a million times more luminous than the firefly, how far away is the lightbulb?

16. Consider a hypothetical planet named Vulcan. If such a planet were in an orbit ¼ the size of Mercury's and had the same albedo as Mercury, what would be the average temperature on Vulcan's surface? Assume that the average temperature on Mercury's surface is 450 K.

17. The average temperature of Earth's surface is approximately 290 K. If the Sun were to become 5 percent more luminous (1.05 times its present luminosity), how much would Earth's temperature increase? (You might consider these to be only modest increases in solar luminosity and Earth's average temperature, but they would nevertheless have a drastic effect on our planet's climate.)

18. Your body, at a temperature of about 37°C (98.6°F), emits radiation in the infrared region of the spectrum.

   a. What is the peak wavelength, in μm, of your emitted radiation?

   b. Assuming an exposed body surface area of 0.25 m², how many watts of power do you radiate to your environment?

   c. In general, the radiating surface of your body will have a temperature that is different from your internal body temperature. Why is this so?

StudySpace
wwnorton.com/astro21
provides a Study Plan for each chapter that includes a reading outline, animations, keyword flash cards, and gradebook-enabled multiple-choice quizzes. From StudySpace you can also access premium content in the ebook and SmartWork.

All truths are easy to understand
once they are discovered. The point
is to discover them.

GALILEO GALILEI (1564–1642)

Robotic rovers *Spirit* and *Opportunity* have roamed the surface of Mars.

The Tools of the Astronomer

5.1 The Optical Telescope—An Extension of Our Eyes

Do you remember the first time you saw the Moon up close? Perhaps your family or a neighbor had a backyard telescope like the one shown in **Figure 5.1**. Or the Moon might have been the featured attraction during a visit to your local planetarium. With that first view came recognition of what the Moon really is—a nearby planetary world covered with craters and vast lava-flooded basins. You might also have been treated to a breathtaking look at the Orion Nebula,

FIGURE 5.1 Amateur astronomers with their telescope. There are several hundred thousand amateur astronomers in the United States alone.

KEY CONCEPTS

In the previous chapter we learned how our understanding of the physical and chemical properties of distant planets, stars, and galaxies comes to us in the form of electromagnetic radiation. But this information must first be collected and processed before it can be analyzed and converted to useful knowledge. Here we will learn about the tools astronomers use to capture and scrutinize that information. We will find that

- Telescopes of various types collect radiation over the entire range of the electromagnetic spectrum—from gamma rays to radio signals.

- Optical telescopes come in two basic types, refractors and reflectors; but all of the larger astronomical telescopes are reflectors.

- Telescope resolution increases with aperture; image size is proportional to a telescope's focal length.

- Earth's atmosphere distorts telescopic images and prevents large parts of the electromagnetic spectrum from reaching the ground. Telescopes in orbit overcome this problem.

- The most effective way to study the planets and moons of our Solar System is to go there.

EXCURSIONS 5.1

A Brief History of the Telescope

As long ago as 1350, craftsmen in Venice were making small disks of glass that could be mounted in frames and worn over the eyes to improve vision. The glass disks were convex on both sides, shaped something like lentils. And so they became known as *lenses*, from *lens,* which is Latin for "lentil." Looking back, it's rather remarkable that more than 250 years would pass before these lenses would be employed for something other than spectacles!

Hans Lippershey (1570–1690) was a German-born spectacle maker living in the Netherlands around the turn of the 17th century. In 1608 legend has it that children, playing with his lenses, put two of them together and saw a distant object magnified. Lippershey looked for himself and mounted the lenses together in a tube to produce a *kijker* (or "looker"), as he called it. As you might imagine, news of his invention spread rapidly, eventually reaching the Italian instrument maker Galileo Galilei. Galileo at once saw the potential of the "looker" for studying the heavens and constructed one of his own, as seen in **Figure 5.2**. By 1610 Galileo became the first to see craters on the Moon, the phases of Venus, and the moons of Jupiter. He was also the first to realize that the Milky Way is made up of countless numbers of individual stars. As the story goes, Galileo was demonstrating

FIGURE 5.3 Newton's reflecting telescope.

his instrument to guests when one of them christened it the "telescope," from the Greek meaning "farseeing," and the name stuck. With its ability to see far beyond the range of the human eye, the refracting telescope quickly revolutionized the science of astronomy.

Unfortunately, all simple-lens telescopes suffer from a serious problem called *chromatic aberration* (see Foundations 5.1). Realizing this, Sir Isaac Newton in 1671 designed a telescope using mirrors instead of lenses. He cast a 2-inch mirror made of speculum (basically copper and tin) and polished it to spherical curvature. He then placed the primary mirror at the bottom of a tube with a secondary flat mirror mounted above it at a 45° angle, which directed the focus to an eyepiece on the outside of the tube (see **Figure 5.3**). As it turned out, others did not share Newton's talent for instrument making, and the reflecting telescope remained a curiosity for decades. The spherical mirror surface used by Newton works only for small mirrors. Larger mirrors require a *parabolic* surface to produce a sharply focused image, and parabolic surfaces are much more difficult to fabricate. It was not until the latter half of the 18th century that large reflecting telescopes came into their own.

Throughout the 19th century both refracting and reflecting telescopes continued to grow in size. By 1897,

FIGURE 5.2 A replica of Galileo's refracting telescope.

(a)

(b)

Yerkes: 1-meter diameter lens
World's largest refracting telescope

Keck: 10-meter diameter mirrors
World's largest reflecting telescopes

FIGURE 5.4 (a) The Yerkes 1-m refractor uses a lens to collect light. (b) The twin Keck 10-m reflectors are more compact and use mirrors to collect light.

though, refracting telescopes had reached their limit with the completion of the Yerkes 1-m (40-inch) refractor (see **Figure 5.4(a)**). Gravitational distortion of a massive lens and a long telescope tube severely limits the size of refractors. The Yerkes telescope was destined to become the world's largest. Reflecting telescopes, on the other hand, seem to have no such size limits. Today the world's largest reflecting telescopes are the 10-m twin Keck telescopes located on 4-km high Mauna Kea in Hawaii (see **Figure 5.4(b)**). Each of the two Keck telescopes has a mirror with a diameter of 10 meters, giving it 4,000,000 times the light-gathering power of the human eye! And even larger reflecting telescopes are in the works. Several organizations are considering telescopes with apertures in the 30-m range, and ESO's 100-m OWL (OverWhelmingly Large) telescope may be in operation by 2020, at an estimated cost of more than €1.0 billion (see **Figure 5.5**). It seems that the only limitation on the size of reflecting telescopes is the cost of fabricating them.

FIGURE 5.5 The proposed OWL 100-m reflecting telescope.

a giant assemblage of gas and dust 1,500 light-years away, or the larger and more remote Andromeda Galaxy. These are but a few of many celestial wonders that come alive in the eyepiece of even a small telescope. As it has for so many before you, the telescope can change the way you view the heavens. From such experiences you might understand why the **telescope** is the astronomer's most important instrument (see **Excursions 5.1**). Yet it is only within the past century and a half that its capabilities have been fully exploited.

Turn a telescope or binoculars on a field of stars and what do you see? As you might expect, the stars appear both closer and brighter. Now look at a distant landscape. The scene seems closer, but its surface is no brighter. What is going on here? It turns out that only *point sources* such as stars appear brighter in a telescope. Like the distant landscape, the Orion Nebula and other extended astronomical

---

The telescope is the astronomer's most important tool.

---

objects look bigger in the eyepiece, but their surfaces are no brighter than they appear to the unaided eye. A telescope gives you a closer view of the Moon, but it does not increase the Moon's surface brightness. For more than two centuries after the invention of the telescope, astronomers struggled with this **surface brightness** problem. No matter how big they built their telescopes, nebulae and galaxies might appear larger, but their faint detail remained elusive. The problem, of course, was not with their telescopes but with the limitations of optics and the human eye. Only with the discovery of photography and the later development of electronic cameras were astronomers finally able to discern the faint but intricate fabric of the cosmos.

Here is a professional secret we can share with you. Today's working astronomers never—well, hardly ever—look through the eyepiece of a telescope. Why ignore such an opportunity? The reason is that they cannot afford to waste valuable observing time. Telescope time can be very expensive: Operating a large telescope for just a single night can cost tens of thousands of dollars! So although it might be exhilarating to glimpse Saturn through the eyepiece of a really big telescope, astronomers can learn much more and make better use of precious observing time by permanently recording the planet's image at a variety of wavelengths or seeing its light spread out into a revealing spectrum. Long after the observing session is over, the wistful peek through the eyepiece would be just a distant memory, but the recorded data remain as a permanent quantitative record for subsequent analysis.

As important as telescopes are, they are not the only instruments in the astronomer's bag of gear. The tools used by modern astronomers are many and remarkably diverse, ranging from physics laboratories to powerful supercom-

puters to robotic probes sent to cruise the Solar System. Astronomy's tools are not only diverse; they are also changing rapidly. It seems that every few months we are likely to see the commissioning of a new mountaintop telescope, the launch of a satellite observatory, or the arrival of a spacecraft at some remote planetary destination. In fact, in the time it takes between the final changes made by the authors of this volume and its appearance on your bookshelf, much of what we might say about the latest astronomy tools will already have become yesterday's news. Rather than give in to instant obsolescence, we will instead make use of the technology of the 21st century to bring you a discussion of (what else?) the technology of the 21st century!

Refractors and Reflectors

When most people think of astronomy, the mental image that comes to mind is a toy store telescope pointed at the night sky. The image is fitting. Some form of telescope forms the heart of almost every tool that astronomers have used to directly observe the heavens. In fact, humanity's use of telescopes dates back to far earlier than the night Galileo first turned his telescope skyward. You were born with two telescopes of your own—your eyes. The human eye is a refracting telescope, which means that it uses a lens to bend the light passing through it, bringing that light to a sharp **focus** on the retina, as shown in **Figure 5.6**. Eyes are such amazingly useful things that biologists believe they have evolved *independently* as many as 60 different times during the history of terrestrial life! Although the human eye is wonderfully evolved to meet our daily needs, it is not so well suited for doing astronomy. For that we need large, powerful telescopes.

Astronomical telescopes come in two basic types, *refracting* and *reflecting*, illustrated here in Figure 5.4. As we noted in our discussion of the eye, a **refracting telescope** forms an image in the **focal plane** when light from a distant

---

There are two types of optical telescopes: refractors and reflectors.

---

object is refracted by the objective lens (see **Figure 5.7**). The diameter of the objective lens defines the telescope's **aperture**, which in turn defines its light-gathering power. (Keep in mind that the light-gathering power of a telescope is proportional to the *area* of its aperture—that is, to the *square* of its diameter.) The distance between the telescope lens and the images formed is referred to as the **focal length** of the telescope. Now we know the two most important parameters of a telescope. The aperture determines a telescope's light-collecting power, and the focal length establishes the size of the image. But herein lies a major

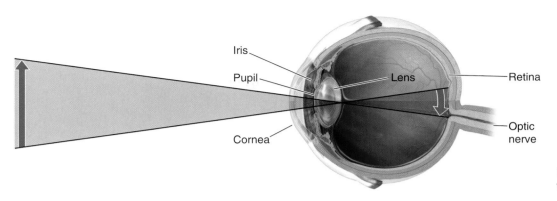

FIGURE 5.6 A schematic view of the human eye.

problem with refractors. To get the most light-gathering power and produce the largest images, a refractor must suspend a massive piece of glass (the objective lens) at the end

A telescope's aperture determines its light-gathering power and resolution.

of a very long tube without sagging unduly under the force of gravity. Take a look at the 1-m Yerkes telescope, shown in Figure 5.4(a). It carries a 450-kg objective lens mounted at the end of a 19.2-m tube. This is as big as refractors get.

The structural limitation of size is not the only problem with refractors. **Refraction** depends on the wavelength of light. (See **Foundations 5.1**.) This means that the focal length of a simple lens is different for red light than it is for blue light, an effect called **chromatic aberration** (see Figure 5.12(a). Refracting telescopes typically use **compound lenses**, such as the one shown in Figure 5.12(b), which partially

correct for chromatic aberration, although some residual effects always remain.

A **reflecting telescope** forms an image in its focal plane when light is reflected from a specially curved mirror, as seen in Figure 5.9. We call this the **primary mirror** because, as we will see, modern reflecting telescopes usually have two or more mirrors. Reflectors have a number of important advantages over refractors. Because the direction of a reflected ray does not depend on the wavelength of light,

All of the world's largest telescopes are reflecting telescopes.

chromatic aberration is no longer a problem. Primary mirrors can be made thinner and therefore less massive than objective lenses. This becomes a distinct advantage when dealing with astronomers' insatiable appetite for ever larger telescopes. Finally, the light path from the primary mirror

FIGURE 5.7 (a) A refracting telescope uses a lens to collect and focus light from two stars, forming images of the stars in its focal plane. (b) Longer focal length telescopes produce larger, more widely separated images.

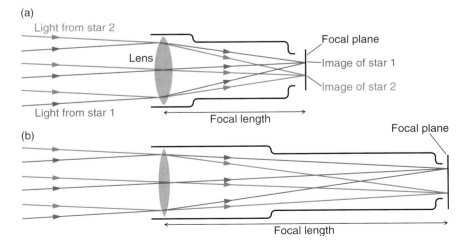

FOUNDATIONS 5.1

When Light Doesn't Go Straight

REFLECTION

Picture a beam of light striking a piece of glass. When light encounters a different medium, in this case going from air to glass, there will always[1] be a certain amount of **reflection** from the surface of the new medium. In other words, some of the light will change its direction of travel. If the medium is smooth and shiny, reflection is easier to understand. The most common example occurs when light encounters an ordinary flat mirror. In **Figure 5.8** an incoming or incident **ray**, *AB*, reflects from the surface, becoming the reflected ray, *BC*. The angle between *AB* and *PB*, the perpendicular to the surface, is called the *angle of incidence* (*i*). The angle between *BC* and *PB*, is called the *angle of reflection* (*r*). In the case of a flat mirror, the angles of incidence and reflection are always equal. What reflects *from* the mirror is a good representation of what falls *on* it, although left and right are interchanged. That's what makes a flat mirror so convenient for admiring our appearance.

Curved mirrors can also be very useful, especially in astronomical telescopes. The same rules of incidence and reflection hold here for each ray, but in this case the reflected rays do not maintain the same angle with respect to each another as they do with a flat mirror. A mirror that is concave toward the incoming light and has a parabolic surface (see **Figure 5.9**) will reflect the rays so that they converge to form an image. If the incoming rays are parallel, as from a distant source—think "star" —the reflected rays cross at a distance from the mirror called the focal length of the mirror. Rays from a distant source on the axis of the mirror will cross on the axis at a point called the *focus*. The surface at which all parallel rays cross is called the *focal plane* (see Figure 5.9).

REFRACTION

Returning to our light beam and the piece of glass, what can we say about the light that is not reflected? When a light wave enters a new medium its speed changes. Remember from Chapter 4 that the speed of light is always

[1] We have to be careful here. In those rare cases in which the index of refraction (see Chapter 4) is exactly the same in both media, there will be no reflection or refraction at the surface.

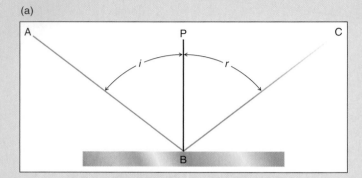

(a)

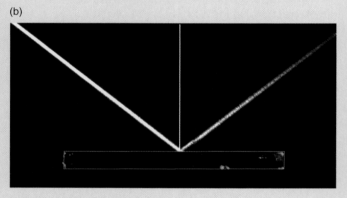

(b)

FIGURE 5.8 (a) Light incident on and reflecting from a flat surface. The angle of incidence (*i*) equals the angle of reflection (*r*). (b) Light from a laser beam is reflected from a flat glass surface.

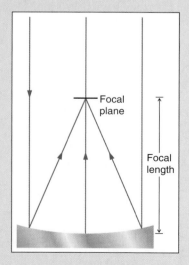

FIGURE 5.9 Parallel rays of light incident on a concave parabolic mirror are brought to a focus in the mirror's focal plane.

Focal plane

Focal length

less in any material medium than it is in a vacuum. If the incident light is perpendicular to the surface, the speed changes but the direction of travel does not. If the incident light encounters the surface at some other angle, the direction of travel also changes. This change in the direction of the incident light is called *refraction*. The amount of refraction depends both on the incidence angle and the relative speeds in the two media, which are defined by their respective indices of refraction (see Chapter 4). In **Figure 5.10(a)** light waves are depicted as coming in from the upper left and hitting the surface of the new medium at a certain angle to the perpendicular. The part of each wave that enters the new medium first is slowed before the other part of the wave enters. This causes the wave to bend, taking up a new direction of travel. If the speed of light in the new medium is less than that in the initial medium (a higher refractive index), the light bends toward the perpendicular. If the speed of light in the new medium is greater than that in the initial medium (a lower refractive index), the light bends away from the perpendicular. **Figure 5.10(b)** shows a green laser beam being refracted by a plastic block.

A convex lens uses its curved surface and refraction to form images by making the rays cross at the focus of the lens. Figure 5.7 illustrates the path of light rays from a distant source passing through a simple convex lens and coming to a focus at the focal plane. Our eyes employ simple convex lenses.

DISPERSION

Shine white light through a glass prism, as shown in **Figure 5.11**, and you'll get a rainbowlike spectrum. This demonstrates that the refraction, and therefore the

(a)

FIGURE 5.10 (a) Light waves are refracted (bent) when entering a medium with a higher index of refraction. They are refracted again as they reenter the medium with a lower index of refraction. (b) Light from a green laser beam is refracted as it enters and exits a plastic block.

(b)

FIGURE 5.11 White light is dispersed into its component colors as it passes through a glass prism.

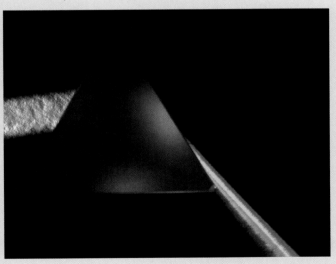

(continued on next page)

FOUNDATIONS 5.1

speed of light in glass, depends on wavelength. Glass, like most transparent materials, has a refractive index (see Chapter 4) that increases with decreasing wavelength. This means that shorter wavelengths (those toward the blue) are refracted more strongly than longer wavelengths (those toward the red). This wavelength-dependent difference in refraction, which spreads the white light out into its spectral colors, is what we call **dispersion**. Although dispersion is helpful in creating prism spectra, it creates a serious problem called *chromatic aberration* in refractive optics, as seem in **Figure 5.12(a)**. Chromatic aberration causes blue light to come to a shorter focus than the longer visible wavelengths. You can see this effect when you look at a bright object such as a distant streetlight through an inexpensive telescope. (Low-priced telescopes usually have only a simple convex objective lens.) The streetlight will appear to be surrounded by a blue halo, which is caused by the blue component of the light being out of focus. Manufacturers of quality cameras and telescopes avoid the use of simple convex lenses in favor of the *compound lens*, similar to that seen **Figure 5.12(b)**. By using two types of glass, a compound lens corrects for chromatic aberration.

FIGURE 5.12 (a) Light of different wavelengths (different colors) comes to different foci along the optical axis of a simple lens, causing chromatic aberration. (b) A compound lens using two types of glass (crown and flint) with different indices of refraction can compensate for much of the chromatic aberration.

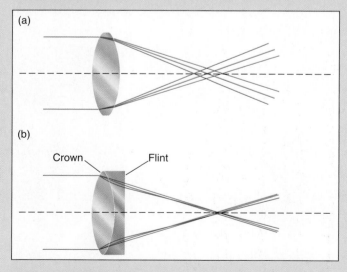

(a)

(b)

Crown   Flint

INTERFERENCE

The intersection of two sets of electromagnetic waves can produce patterns of high and low intensity called **interference**. Say we have a pair of slits and an opaque screen. **Figure 5.13(a)** illustrates monochromatic light (light having a single wavelength or a very narrow range of wavelengths) going through the slits. Each slit now becomes a source of wavefronts. Notice the regular pattern on the screen where the wavefronts from the two slits intersect. If the intersection point occurs where the amplitudes of both waves are at their maximum positive or maximum negative value, the two add and the light will be bright.[2] We call this **constructive interference**. If, on the other hand, one wave is at its maximum *positive* value and the other is at its maximum *negative* value, the sum is zero and the result will be darkness. We call this **destructive interference**. When we replace the two slits with a large number of very narrow, very closely spaced parallel slits, we call it a **grating**. The same effect can be produced by engraving closely spaced lines on a mirror.

If we now substitute a multiwavelength source of light, we get a similar pattern for each and every wavelength, as shown in **Figure 5.13(b)** for a reflective grating. For each wavelength, there will be a different point on the screen where constructive interference takes place. In other words, the grating produces a spectrum. Modern spectrographs use a grating to disperse incoming light into its constituent wavelengths. You can see this effect for yourself. Look at light reflected from a CD or DVD. The closely spaced tracks act as a grating and create a respectable spectrum **(Figure 5.13(c))**.

DIFFRACTION

When light passes through a small opening (or near an opaque edge), the effects of interference come into play. This is called **diffraction. Figure 5.14(a)** shows what happens when monochromatic light from a distant source passes through a lens and is brought to a focus. The image pattern is not a single point as we might expect, but rather is smeared out into a series of bright and dark concentric rings (as seen in **Figure 5.14(b)**) representing constructive and destructive interference from the edge of the aperture. If we make the aperture smaller, the diffracted image grows larger, causing more blurring and limiting how close two

[2]Don't be concerned about the negative value of the wave's amplitude. The intensity of light is actually equal to the *square* of the amplitude, so the negative sign goes away.

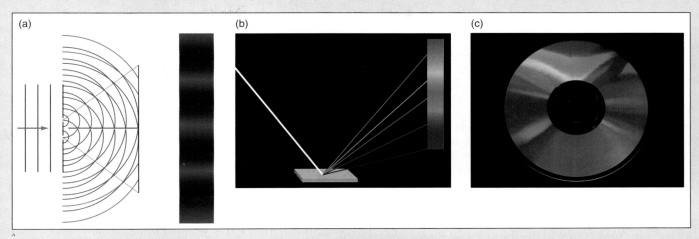

FIGURE 5.13 (a) Constructive and destructive interference patterns are created when monochromatic light passes through a pair of narrow slits. (b) Spectral dispersion is produced by interference when multiwavelength (white) light reflects from a grating. (c) A spectrum is created by the reflection of light from the closely spaced tracks of a CD.

images can be and still be resolved. The angular size of the diffraction pattern depends on the ratio (λ/D) between the wavelength of light (λ) and the aperture of the hole (D). At any given wavelength, larger apertures produce smaller diffraction patterns and therefore higher resolution. And at any given aperture, shorter wavelengths produce smaller diffraction patterns and higher resolution. Electron microscopes take advantage of this property of diffraction to achieve very high resolution by using (what else?) electrons instead of photons to illuminate the target. Recall the dual wave–particle nature of electromagnetic radiation, which we discussed in Chapter 4. Electrons can behave both as particles and as waves, with wavelengths shorter than 0.1 nm. This means that electron microscopes have more than five *thousand* times better resolution than conventional microscopes, which use visible light ($\lambda\sim550$ nm).

FIGURE 5.14 (a) Light waves from a star are diffracted by the edges of a telescope's lens or mirror. (b) This diffraction causes the stellar image to be blurred, limiting a telescope's ability to resolve objects. (c) Diffraction from a circular aperture illuminated by green laser light.

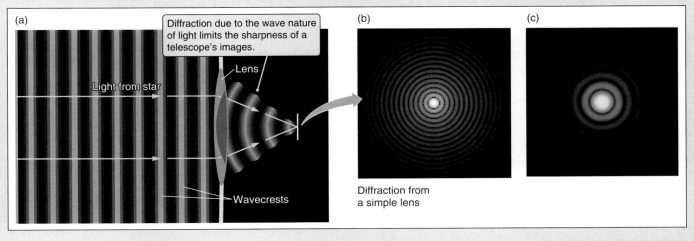

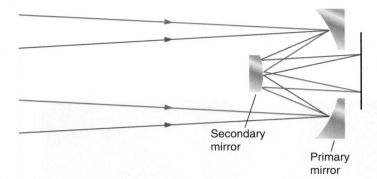

FIGURE 5.15 Reflecting telescopes use mirrors to collect and focus light. Large telescopes typically use a secondary mirror that directs the light back through a hole in the primary mirror to an accessible focal plane behind the primary mirror.

to the focal plane can be folded by introducing a **secondary mirror** (see **Figure 5.15**). This significantly reduces the length and weight of the telescope. For example, the focal length of each of the 10-m Keck telescopes, currently the world's largest reflectors, can be as long as 250 meters, even though the overall length of each telescope is a mere 25 meters. **Table 5.1** lists the world's largest optical telescopes. As you can see, all are reflecting telescopes.

Resolution

One of the eye's major limitations as an astronomical telescope is **resolution.** When we speak of resolution, we are referring to how close two points of light can be to each other before a telescope is no longer able to split the light into two separate images. Unaided, the human eye can resolve objects separated by an angular distance of 1 arcminute,[3] or a 30th the diameter of the full Moon.[4] This may seem small, and in our daily lives it is; yet when we look at the sky, thousands of stars and galaxies may hide within the smallest area the unaided human eye can resolve. Figure 5.7(a) shows the path followed by rays of light from two distant stars as they pass through the lens of a refracting telescope. Comparison with Figure 5.7(b) illustrates that the longer the focal length, the greater the separation between the images. The focal length of a human eye is typically about 20 mm. In comparison, telescopes used by professional astronomers often have focal lengths of tens or even hundreds of meters. Such telescopes make images that are far larger than those formed by your eye, and consequently they contain far more detail.

[3] A description of angular units—radians, degrees, arcminutes, and arcseconds—can be found in Chapter 13, Section 13.2, and in Appendix 1.
[4] Only the sharpest of human eyes achieve 1 arcminute resolution. Typical resolution for many of us would be more like 2 arcminutes.

Focal length explains only one difference between the resolution of telescopes and the unaided eye. The other results from the wave nature of light. As waves of light pass through the lens of a telescope, they spread out from the edges of the lens, as illustrated in Figure 5.14. The distortion of the wavefront as it passes the edge of an opaque object is called **diffraction** (see Foundations 5.1). Diffraction

---

Diffraction, or blurring of an image, depends on the ratio of wavelength to telescope aperture.

"diverts" some of the light from its path, slightly blurring the image made by the telescope. The degree of blurring depends on the wavelength of the light in comparison with the diameter of the telescope lens. The larger the lens relative to the wavelength of the light it is focusing, the less of a problem is posed by diffraction. The ultimate limit on the angular resolution of a telescope, called the **diffraction limit**, is determined by the ratio of the wavelength of light passing through it to the diameter of the lens:

$$\theta = 2.06 \times 10^5 \left(\frac{\lambda}{D}\right),$$

where θ is the diffraction-limited angular resolution in arcseconds,[5] λ is the wavelength of light, and D is the diameter of the telescope. Both λ and D are expressed in the same units, usually meters. As we can see, the smaller the ratio of λ/D, the better will be the resolution of the telescope. We can apply this relationship to the **Hubble Space Telescope (HST)** operating in the visible part of the spectrum. The space telescope's primary mirror has a diameter (D) of 2.4 m. Visible (green) light has a wavelength (λ) of 550 nm or 5.5×10^{-7} m. Substituting into the previous equation, we have

$$\theta = 2.06 \times 10^5 \left(\frac{5.5 \times 10^{-7}}{2.4}\right) = 0.047 \text{ arcseconds}$$

or about 1,000 times better than the resolving power of the human eye.

Atmospheric Distortions—Seeing

The previous equation for the diffraction limit tells us that larger telescopes get better resolution. Theoretically the 10-m Keck telescopes have a diffraction-limited resolution of 0.0113 arcseconds in visible light, which would allow you to read newspaper headlines 60 kilometers away. However, for telescopes with apertures larger than about a meter, Earth's atmosphere stands in the way of better resolution. If you have ever looked out across the desert on a summer day, you have seen the distant horizon shimmer as light from that

[5] See footnote 3.

TABLE 5.1

The World's Largest Optical Telescopes

Mirror Diameter	Telescope	Sponsor	Location	Operational Date
10.4 m	Gran Telescopio Canarias	Spain, Mexico, University of Florida	Canary Islands	2006
10 m	Keck I	Caltech, University of California, NASA	Mauna Kea, Hawaii	1993
10 m	Keck II	Caltech, University of California, NASA	Mauna Kea, Hawaii	1996
10 m	Southern African Large Telescope	11 international partners	Southerland, South Africa	2005
9.2 m	Hobby-Eberly	University of Texas, Penn State, Stanford, Germany	Mount Lock, Texas	1997
2 × 8.4 m	Large Binocular Telescope	University of Arizona, Ohio State, Italy, Germany	Mt. Graham, Arizona	2006
4 × 8.2 m	Very Large Telescope	European Southern Observatory	Cerro Paranal, Chile	2000
8.3 m	Subaru	Japan	Mauna Kea, Hawaii	1999
8 1 m	Gemini North	USA, UK, Canada, Chile, Brazil, Argentina	Mauna Kea, Hawaii	1999
8.1 m	Gemini South	USA, UK, Canada, Chile, Brazil, Argentina	Cerro Panchon, Chile	2000
6.5 m	Magellan I	Carnegie Institute, University of Arizona, Harvard, University of Michigan, MIT	Las Campanas, Chile	2000
6.5 m	Magellan II	Carnegie Institute, University of Arizona, Harvard, University of Michigan, MIT	Las Campanas, Chile	2002

horizon is constantly bent this way and that by turbulent bubbles of warm air rising off the hot desert floor.

The problem is less pronounced when we look overhead, but the twinkling of stars in the night sky tells us the phenomenon is still there. As telescopes magnify the angular diameter of a planet, they also magnify the shimmering effects of the atmosphere. The limit on the resolution of a telescope on the surface of Earth caused by this atmospheric distortion is called **astronomical seeing**. One advantage of launching telescopes such as the Hubble Space Telescope

Earth's atmosphere distorts images.

into orbit around Earth is that from their vantage point above the atmosphere, telescopes get a much clearer view of the universe, unhampered by seeing. Does that mean that groundbased telescopes are becoming obsolete? Not at all. Modern technology has come to their rescue with computer-controlled **adaptive optics**, which compensate for much of the atmosphere's distortion.

To better understand how adaptive optics work, we need to look more closely at how Earth's atmosphere smears out an otherwise perfect stellar image. Look again at Figure 5.14(a). Light from a distant star arrives at the top of Earth's atmosphere as a series of flat, parallel waves called a **wavefront**. If Earth's atmosphere were perfectly homogeneous, the wavefront would remain flat as it reached the objective lens or primary mirror of a groundbased telescope. After making its way through the telescope's optical system, the wavefront would produce a tiny diffraction disk in the focal plane, as shown in Figure 5.14. But Earth's atmosphere is not homogeneous. It is filled with small bubbles of air that have slightly different temperatures than their surroundings. Different temperatures mean different densities, and different densities mean different refractive properties. The air bubbles act as weak lenses, and by the time the wavefront reaches the telescope it is far from flat, as shown in **Figure 5.16**. Instead of a tiny diffraction disk, the image in the telescope's focal plane is distorted and swollen, degrading the resolution. Now suppose we could measure the

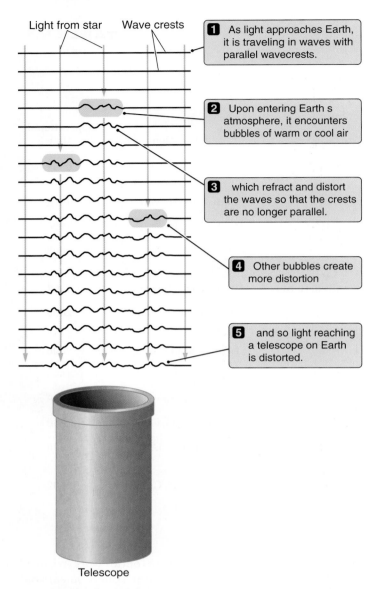

Light from star   Wave crests

1 As light approaches Earth, it is traveling in waves with parallel wavecrests.

2 Upon entering Earth s atmosphere, it encounters bubbles of warm or cool air

3 which refract and distort the waves so that the crests are no longer parallel.

4 Other bubbles create more distortion

5 and so light reaching a telescope on Earth is distorted.

Telescope

FIGURE 5.16 Distortion of the wavefront from a distant object after passing through bubbles of warmer or cooler air in Earth's atmosphere.

amount of distortion in the wavefront and somehow flatten it out. This is how adaptive optics work. First an optical device within the telescope constantly samples the wavefront, measuring its departure from flatness. Then, before

Adaptive optics can correct for atmospheric distortion of telescopic images.

reaching the telescope's focal plane, light is reflected from yet another mirror that has a deformable surface. (Astronomers sometimes call this a "rubber" mirror, although it is actually made of glass.) A computer analyzes the wavefront distortion and sends a signal to mechanisms that bend the

deformable mirror's surface so that it accurately corrects for the distortion of the wavefront. An example of an image corrected by adaptive optics is shown in **Figure 5.17**. The widespread use of adaptive optics has now made the image quality of groundbased telescopes competitive with those of the Hubble Space Telescope. But image distortion is not the only problem caused by Earth's atmosphere. Large regions of the electromagnetic spectrum are partially or completely absorbed by various atmospheric molecules.

Atmospheric Transmission— Windows and Blinds

The final limitation on the human eye is that it is sensitive only to light in the visible part of the electromagnetic spectrum. (That is, after all, why we call it the "visible" part of the spectrum!) Even though visible light is only a small part of the electromagnetic spectrum, it is anything but happenstance that our eyes work in this range of wavelengths. Our atmosphere is transparent in the visible part of the spectrum, but for most of the spectrum outside this restricted wavelength band, trying to see through our atmosphere is like trying to see through a brick wall. Almost all of the X-ray, ultraviolet, and infrared light arriving at Earth is blocked

Earth's atmosphere blocks much of the electromagnetic spectrum.

before it reaches the ground by the layer of atmosphere that surrounds our planet. The visible part of the spectrum, which is not blocked by Earth's atmosphere, is a fairly narrow range of wavelengths, or window, through which we can look at the universe. There are a few other **atmospheric windows** in the spectrum as well, as shown in **Figure 5.18**. Do not make the mistake of thinking that light that fails to reach the surface of Earth is uninteresting. There are many things that we can learn only by observing the universe outside the visible window. Although radio observations are also possible from the ground, we owe a large fraction of what we know about the universe to a host of ultraviolet, X-ray, gamma ray, and infrared telescopes that, beginning in the 1960s, were carried above Earth's atmosphere by rockets. We'll discuss space telescopes in more detail later in the chapter.

5.2 Optical Detectors and Instruments

Detectors are devices placed in a telescope's focal plane to transform images into something that we can see and record. The detector in the human eye is the retina (see Fig-

(a)

(b)

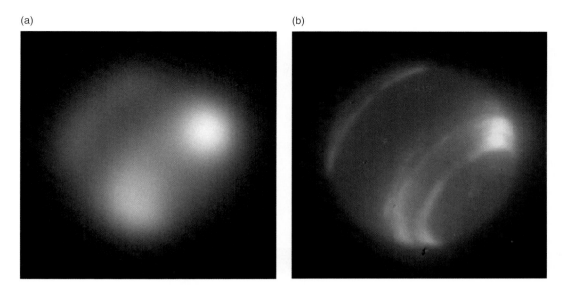

FIGURE 5.17 An image of Neptune taken (a) without and (b) with adaptive optics.

ure 5.6), and the individual receptor cells that respond to light falling on the retina are called *rods* and *cones*. Cones are located near the eye's optical axis at the center of our vision. They provide the highest resolution and allow us to recognize color. The size and spacing of cones determine the 1-arcminute resolution of the human eye, not its 7-mm pupil. Rods, located in our peripheral vision, provide the highest sensitivity to low light levels, but they have poorer resolution and cannot distinguish color. The photons to which the human eye is sensitive have wavelengths ranging from about 400 nm (deep violet) to 700 nm (far red).

So what is it that limits the faintest stars we can see with our unaided eyes, assuming a clear dark night and

good eyesight? This limit is determined in part by two factors that are characteristic of all detectors: *integration time* and *quantum efficiency*. As photons from a star enter the aperture or pupil of our eyes, they fall on and excite cones at the center of our vision. The cones then send a signal to our brains, which interpret this message as "I see a star." We might now ask, "How many photons does it take to send that signal to the brain?" It turns out that your retina is hindered by short-term memory. The eye can add up photons for only a limited interval called the **integration time**. For the human eye, the integration time is about 100 ms. If two images on a television or computer screen appear 30 ms apart, you will see them as a single image because

FIGURE 5.18 Earth's atmosphere blocks most electromagnetic radiation.

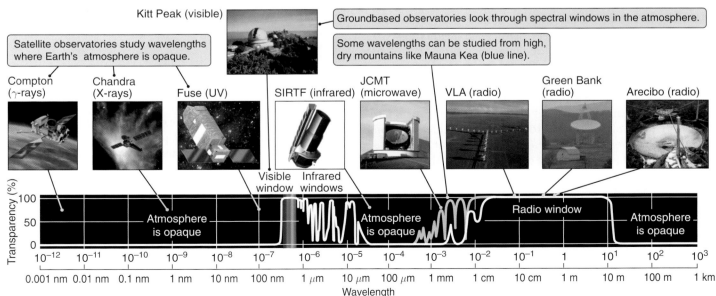

your eyes will sum up whatever they see over an interval of 100 ms. If the images occur, say, 200 ms apart, you will see them as separate images. So the signals the cones send to your brain include only those photons that arrive within an interval of 100 ms. This relatively brief integration time is the biggest factor limiting our nighttime vision. There is another effect called **quantum efficiency** that also restricts our nighttime vision. As the name implies, quantum efficiency is the likelihood that a particular photon landing on the retina will, in fact, produce a response. For the human eye, it takes about 10 photons landing on a cone to activate a single response. In other words, the quantum efficiency of our eyes is about 10 percent. Together integration time and quantum efficiency determine the rate at which photons must arrive on the retina before your brain says, "Aha, I see something." For more than two centuries after the invention of the telescope, the retina of the human eye was the only detector. Permanent records of astronomical observations were limited to what an experienced observer could sketch on paper while working at the eyepiece of a telescope, as illustrated in **Figure 5.19(a)**. Photography would eventually change all that.

Photographic Plates

In 1839 John W. Draper, a New York chemistry professor, created the earliest known astronomical photograph. His subject was the Moon, shown here in **Figure 5.19(b)**. Photography was not quick to catch on among astronomers, though, because this early *daguerreotype* process was slow and very messy. The relatively simple dry emulsion process

---

Photography opened the door to modern astronomy.

---

finally came along in the late 1870s, and with that, astronomical photography took off. Astronomers could now create permanent images of planets, nebulae, and galaxies with ease. Thousands of photographic plates soon filled the "plate vaults" of major observatories. Photography had created its own astronomical revolution.

In the dry emulsion process, a layer of gelatin containing tiny crystals of silver halide is coated onto glass plates or film.[6] During an exposure, photons landing on the emulsion energize the silver halide crystals, creating what is called a *latent* image. "Developing" the emulsion turns these small crystals into black grains of metallic silver, forming a permanent image. The highest density of silver grains occurs where the telescope's image was the brightest. So bright

---

[6]Glass plates, although far more expensive than film, are generally used for imaging and spectroscopy because they have greater geometric stability.

(a)

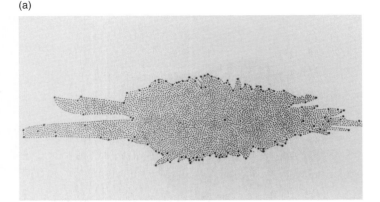

(b)

FIGURE 5.19 (a) Drawing of the Milky Way Galaxy made by William Herschel in the early 19th century. (b) Photograph of the Moon taken by J. W. Draper in 1839.

becomes black, and the photographic image is negative, as illustrated in **Figure 5.20**. The quantum efficiency of most photographic emulsions used in astronomy is very low: typically 1–3 percent, which is even poorer than that of the human eye. But unlike the eye, photographic emulsions can overcome poor quantum efficiency by integrating photons over intervals of many hours. Photography made it possible for astronomers to record and study objects much fainter than the human eye can see.

Photography is not without its own problems. Very faint objects often require long exposures that can take up much of an observing night. (Imagine how you would feel if your 10-hour exposure was spoiled due to some mishap.) Also, the spectral range of photographic emulsions is hardly

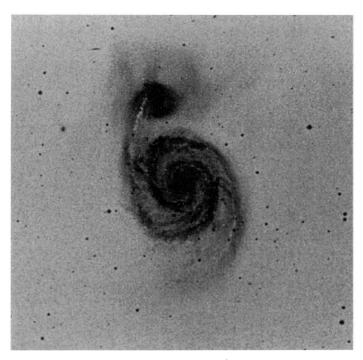

FIGURE 5.20 An image of Galaxy M 51 on a photographic plate.

broader than that of the human eye. In fact, for many years photographic plates were sensitive only to violet and blue light. Another problem is their nonlinear response to light, meaning that the optical density of the processed emulsion is not proportional to the intensity of light falling on it. Finally, there is the nontrivial matter of economics. Each photographic plate can be used only once, and they are expensive. By the middle of the 20th century the search was on for electronic detectors that would overcome many of the deficiencies of photographic plates.

Charge-Coupled Devices (CCD)

Throughout the latter half of 20th century, astronomers employed various electronic detectors to overcome the sensitivity, spectral range, and nonlinearity problems of photography. Some, such as *photoelectric photometers*, are nonimaging devices. They work extremely well for precision stellar photometry because they have excellent linearity, which means that their electronic output is directly proportional to the intensity of light falling on the photometric detector. But they can measure only one stellar image at a time. This resulted in truly labor-intensive observing. Other detectors, such as *vidicons*, are electronic imaging devices with sensitivity far superior to photographic emulsions. Vidicons unfortunately suffer from an electronic instability that causes small geometric distortions of the image. You

may have seen pictures taken by early spacecraft that had little + shaped marks superimposed on them. These *fiducial* marks were engraved on the faceplates of the vidicons to help remove geometric distortions in the image. The search for a better detector continued.

In 1969 scientists at Bell Laboratories were developing "picture phones"—telephones containing a small camera and viewing screen that could display an image of the person at the other end of the conversation. As it turned out, public opinion declared Bell's picture phones an invasion of personal privacy and they were never commercially produced, but the research led to the invention of a remarkable detector called a **charge-coupled device** or **CCD**. Astronomers soon realized that this was the detector they had been looking for. Shortly after it was first applied to astronomical imaging in the mid-1970s, the CCD became the detector of choice in almost all astronomical imaging applications. Gone were the problems associated with photographic emulsions, photoelectric photometers, and vidicon-type imagers. The CCD is a photometrically linear imaging device, able to perform precise photometry over large regions of sky. It responds over a wide spectral range from 200 nm to 1,200 nm and has a high quantum efficiency, typically 80 percent or greater. The output from a CCD is a computer-ready, digital signal that can be sent directly from the telescope to image-processing software or stored on disk for later analysis.

CCDs consist of an ultrathin wafer of silicon, less than the thickness of a human hair, which is divided into a two-dimensional array of picture elements, or **pixels**, as seen in **Figure 5.21(a)**. When a photon strikes a pixel, it creates a small electrical charge within the silicon. As each CCD pixel is "read out," the digital signal that flows to the computer is almost precisely proportional to the accumulated charge. This is what we mean when we say the CCD is a very linear device. Like many electronic detectors, CCDs are subject to thermal noise, but this can be minimized by cooling them down to liquid nitrogen temperatures (~80 K). The first astronomical CCDs were small arrays containing no more than a few hundred thousand pixels. The larger CCDs used in astronomy today may contain as many as 100 million pixels, like the one seen in **Figure 5.21(b)**.

The impact that CCDs have made on astronomy cannot be understated. Conventional photography is now a distant second for imaging at the telescope. As you surf the Internet, nearly every spectacular astronomical image that pops

The CCD is the astronomer's detector of choice.

up on your screen was made with a CCD, whether from groundbased telescopes or from those in space. And professional astronomers are not the only ones making spectacular photos with CCDs. Amateur astronomers are now using commercially available, thermoelectrically cooled CCD

(a)

(b)

FIGURE 5.21 (a) A simplified diagram of a CCD. Photons from a star land on pixels (gray squares) and create electrons within the silicon. The electron charges are electronically moved sequentially to the collecting register at the bottom. Each row is then moved out to the right to an electronic amplifier, which converts the electrical charge of each pixel into a digital signal. (b) A very large charge-coupled device (CCD).

imagers with impressive results. **Figure 5.22(a)** is an image of Saturn taken by an amateur astronomer with a 36-cm telescope. This image shows more detail than the best professional photographs of Saturn taken before CCDs became available to astronomers, an example of which is shown in **Figure 5.22(b)**.

You may never have seen a CCD, but in recent years they have found their way into many devices that we now take for granted—digital cameras, video cameras, and the ubiquitous picture phones, just to name a few.

Spectrographs

Spectrographs, or **spectrometers** as they are often called, are another of the astronomer's essential tools. As we learned in Chapter 4, we can probe the chemistry and physical properties of distant objects by studying their spectra. Early spectrographs used glass prisms to disperse the incoming light into its component wavelengths (see **Figure 5.23**), creating a spectrum like the ones shown in Figure 4.19 and Figure 5.11. A photographic plate recorded the spectrum for precise measurement of the wavelengths of its spectral lines. One disadvantage of glass prisms is that **dispersion** is not uniform with wavelength. Prism spectrographs produce spectra with more dispersion at the shorter-wavelength (violet) end of the spectrum than at the longer-wavelength (red) end. Another drawback is that glass is opaque to ultraviolet and long-wavelength infrared light. Prism spectrographs are more or less limited to the visible part of the spectrum. Most modern spectrographs use a *diffraction grating* to disperse the light (see Figure 5.13(b)) and a CCD to record

FIGURE 5.22 (a) A CCD image of Saturn taken in 2005 by an amateur astronomer with a 36-cm telescope. (b) A pre-CCD photograph of Saturn taken in 1974 with a 1.5-m telescope at the Catalina Observatory of the University of Arizona.

(a)

(b)

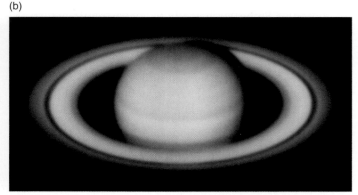

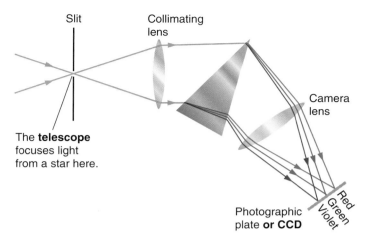

Slit Collimating lens

The **telescope** focuses light from a star here.

Camera lens

Red
Green
Violet

Photographic plate **or CCD**

FIGURE 5.23 Diagram of a prism spectrograph.

the spectrum. Refer to Foundations 5.1 for a more detailed discussion of dispersion and diffraction.

Spectrographs may be designed for either low or high dispersion. Low-dispersion spectrographs are most often used to identify the chemical components of feeble light sources such as nebulae or to measure the reflected spectral energy distribution of faint Solar System objects such as small or distant asteroids. Measurements of temperature and radial velocity, on the other hand, usually require very high dispersion. Today's high-dispersion spectrographs have evolved into substantial scientific instruments, hardly the sort of thing that one would transport from one telescope to another. The high-dispersion spectrographs now associated with most major telescopes tend to be huge and weigh several metric tons—think SUV. For obvious reasons, they are not attached directly to the telescope! A system of mirrors feeds light from the telescope's focal plane into the spectrograph, which is located nearby.

We'll encounter many applications of **spectroscopy** throughout the chapters to come as we continue in our journey through the Solar System and beyond.

5.3 Radio Telescopes

Karl Jansky (1905–1950) was a young physicist working for Bell Telephone Laboratories in the early 1930s when he was assigned the job of identifying sources of static in transatlantic radiotelephone service. He built a pointable antenna and soon identified the major sources of static as nearby and distant thunderstorms; but one source was mysterious. A faint steady hiss rose and fell once every 23 hours and 56 minutes, a characteristic of celestial objects far beyond our Solar System. In 1932 Jansky identified the mysterious source. It was in the Milky Way in the direction of Sagittarius, the galactic center. Excited by his discov-

ery, he submitted a request to build a large dish antenna, a **radio telescope**, to study these signals in more detail. Bell Labs turned down the request. After all, Jansky had already given them the information they needed. Nevertheless, Jansky's discovery marked the birth of radio astronomy. In his honor, the basic unit for the strength of a radio source is called the **jansky**.

In 1937 Grote Reber, a radio engineer and ham radio operator, decided to build his own radio telescope. It consisted of a parabolic sheet of metal, 9 meters in diameter, with a radio receiver mounted at the focus. With this instrument he conducted the first survey of the sky at radio frequencies, and he published the first radio frequency map in 1941. Reber was largely responsible for the rapid advancement in radio astronomy that blossomed in the post–World War II era.

Radio telescopes are yet another of astronomy's indispensable tools. From our Solar System to the most distant galaxies, the penetrating power of radio waves unlocks secrets not possible with shorter-wavelength optical or infrared telescopes. Look back at Figure 5.18 and notice the wide radio window in Earth's atmosphere, covering wavelengths ranging all the way from a centimeter to 10 meters.[7] This ability of radio waves to pass unattenuated through our atmosphere is also the property that allows us to peer through

Radio telescopes allow astronomers to "see" through obscuring gas and dust.

the vast amounts of gas and dust found in many galaxies. Most radio telescopes are large steerable parabolic dishes, typically tens of meters in diameter such as the one shown in **Figure 5.24(a)**. The world's largest radio telescope is the 305-m Arecibo dish built into a natural bowl-shaped depression in Puerto Rico, as seen in **Figure 5.24(b)**. But there can be a price to pay for size. As you might guess from looking at the picture, this huge structure is too big to steer. Instead it must point by moving its radio receiver, suspended in the focal plane above the dish. Arecibo's targets are therefore limited to those celestial sources that pass within 20° of the zenith as Earth's rotation carries them overhead.

As large as radio telescopes are, they have relatively poor angular resolution. Recall our earlier discussion about diffraction. A telescope's angular resolution is determined by the ratio λ/D, where λ is the wavelength of electromagnetic radiation and D is the telescope's aperture. (Keep in mind that a larger ratio means poorer resolution.) Radio telescopes have diameters much larger than the apertures of most optical telescopes, and that helps. But the wavelengths of radio waves are typically several hundred times greater than the

[7] Microwave astronomy is considered a branch of radio astronomy. Microwaves are very high-frequency radio waves with wavelengths ranging from about 1 mm to 10 cm at the short-wavelength end of the radio spectrum. As seen in Figure 5.18, Earth's atmosphere is only partially transparent to microwaves.

(a)

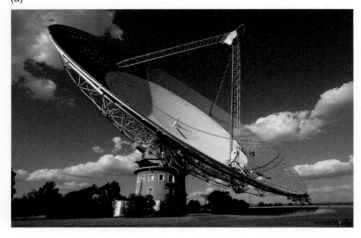

(b)

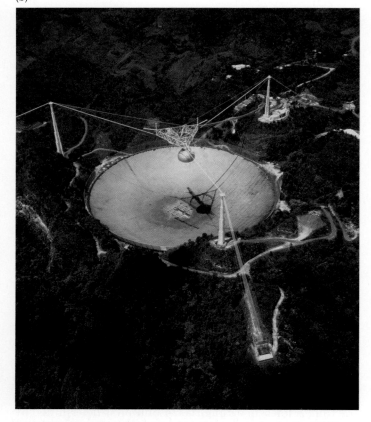

FIGURE 5.24 (a) A large radio telescope in Australia. (b) The Arecibo radio telescope is the world's largest. The steerable receiver suspended above the dish permits limited pointing toward celestial targets as they pass close to the zenith.

wavelengths of visible light, and that hurts. Radio telescopes are thus hampered by the very long wavelengths they are designed to receive. Consider the huge Arecibo dish. Its resolution is typically about 1 arcminute, no better than the unaided human eye! So radio astronomers have had to

develop their own bag of tricks, and one of the cleverest is the interferometer.

Single radio telescopes have relatively poor resolution...

When we combine the signals from two radio telescopes in a certain way, the separation between them—not the diameters of the individual telescopes—determines the angular resolution. For example, if two 10-m telescopes are located 1,000 meters apart, the D in λ/D is 1,000, not 10. Such an arrangement is called an **interferometer** because it makes use of the wavelike properties of electromagnetic radiation, in which signals from the individual telescopes *interfere* with one another (see Foundations 5.1). Usually several telescopes are employed, an arrangement called an **interferometric array**. Through the use of very large arrays, radio astronomers can attain and exceed the angular resolution enjoyed by their optical colleagues. One of the larger radio interferometric arrays is the Very Large Array (VLA) in New Mexico, shown in **Figure 5.25**. The VLA is made up of 27 individual movable dishes spread out in a Y-shaped configuration 30 km across. At a

...but interferometric arrays overcome this problem.

wavelength of 10 cm, this array can achieve resolutions of less than 1 arcsecond. Not satisfied, radio astronomers have sought still larger arrays—and no one can accuse them of having limited imagination. The Very Long Baseline Array (VLBA) employs 10 radio telescopes spread out over more than 8,000 km from the Virgin Islands in the Caribbean to Hawaii in the Pacific. At a wavelength of 10 cm, this array can reach resolutions better than 0.003 arcseconds. It might seem that Earth's diameter would set the ultimate limit on resolution for radio astronomers, but plans are under way to build the Very Long Baseline Interferometer (VLBI) in which one of the radio telescopes is put into near-Earth space. The combined Earth and space-based array would extend over 30,000 km, yielding resolutions far exceeding those of any existing optical telescope.

Before leaving our discussion of interferometers, we should point out that radio astronomers are not the only ones using the interferometer's greater resolving power. Optical telescopes can also be arrayed to yield resolutions greater than those of single telescopes, although for technical reasons the individual units cannot be spread as far apart as radio telescopes. The Very Large Telescope (VLT), operated by the European Southern Observatory (ESO) in Chile, consists of the four VLT 8-m telescopes (see **Figure 5.26**) and four movable 1.8-m auxiliary telescopes. When fully operational it will have a baseline of up to 200 m, yielding angular resolution in the milli-arcsecond range.

FIGURE 5.25 The Very Large Array (VLA) in New Mexico.

5.4 Neutrino and Gravity Wave Detectors

In learning about the Sun in Chapter 14 you will be introduced to the **neutrino**, an elusive particle that plays a major role is the physics of stellar interiors. Of course we can't see beneath the Sun's surface, but observations of neutrinos can give us important insight into what is happening deep within. There's only one problem, and it's a big one: Neutrinos are extremely difficult to detect. To study a neutrino you first have to grab one, and they are nearly impossible to catch. In a sense this is fortunate for us. In less time than it takes you to read this sentence, a thousand trillion (10^{15}) solar neutrinos from the Sun are passing through your body. It doesn't matter a bit if you are reading this at night. Neutrinos are so nonreactive with matter that they can pass right through Earth (and you) as though it (or you) weren't there at all. In fact, half of the neutrinos produced by the Sun would make it through a slab of lead one light-year thick. For us to detect a neutrino, it has to interact with a detector. Several neutrino detectors are in operation today. All are buried deep underground to avoid false detection of cosmic background radiation. Neutrino detectors typically record only one out of every 10^{22} (10 billion trillion) neutrinos passing though them, but that's enough to reveal processes deep within the Sun or witness the violent death of a star 160,000 light-years away.

Another elusive phenomenon is the gravity wave. In fact, **gravity waves** are so elusive we've never actually observed

FIGURE 5.26 The Very Large Telescope (VLT) operated by the European Southern Observatory in Chile. Movable auxiliary telescopes allow the four large telescopes to operate as an optical interferometer.

them. You might even say the several facilities constructed to detect gravity waves have been built on faith. But there is strong, although indirect, observational evidence for their existence, as we will see later in Chapter 17 when we discuss *binary pulsars*. Gravity waves are disturbances in a gravitational field, similar to the waves that spread out from the disturbance you create when you toss a pebble into the quiet surface of a pond. Scientists are eager to detect gravity waves, not so much to confirm their existence but to study the physical phenomena they are likely to reveal. To understand the importance of gravity waves and what they might tell us about disturbances in the fabric of spacetime (refer back to the discussion of special relativity in Chapter 4), we will have to wait until Chapter 17 for a discussion of **general relativity**.

5.5 Getting above Earth's Atmosphere: Airborne and Orbiting Observatories

Imagine strolling through a Hawaiian rainforest while taking in the fragrance of the warm, humid tropical air. In your bliss you might be unaware that 4 km above you on the summit of Mauna Kea astronomers are at war with the same atmospheric water vapor that has so stimulated your senses. Water vapor is the enemy of the infrared astronomer. We have already seen that Earth's atmosphere distorts telescopic images and that certain molecules in Earth's atmosphere, including water, block large parts of the electromagnetic spectrum from getting through to the ground. It shouldn't surprise us then to find that astronomers have put considerable effort into getting their instruments above as much of the atmosphere as possible. Look for an astronomical observatory and you'll be looking at the summit of a tall mountain. Most of the world's larger astronomical telescopes are located 2,000 m and more above sea level. Mauna Kea, a dormant volcano and home of the Mauna Kea Observatory, rises 4,200 m above the Pacific Ocean. At this altitude the MKO telescopes sit above 40 percent of Earth's atmosphere; but more important, 90 percent of Earth's atmospheric water vapor lies below. Still, for the infrared astronomer, the remaining 10 percent is troublesome.

One way to solve the water vapor problem is to make use of high-flying aircraft. NASA's Kuiper Airborne Observatory (KAO), a modified C-141 cargo aircraft, carried a 90-cm telescope and was among the first of these flying observatories. It could cruise at an altitude of 14 km, above 98 percent of Earth's water vapor. NASA retired KAO in 1995 and will replace it with the Stratospheric Observa-

tory for Infrared Astronomy (SOFIA), expected to become operational in 2007. SOFIA will carry a 2.5-m telescope and work in the far infrared region of the spectrum, from 30 μm to 350 μm.

Having full access to the complete electromagnetic spectrum is yet another matter. This means getting completely above Earth's atmosphere.[8] In the late 1940s scientists put ultraviolet and cosmic ray instruments in the nose cones of captured German V-2 rockets and launched them from the White Sands Proving Grounds in New Mexico to altitudes greater than 100 km. Of course, such observations had to be brief because, thanks to gravity, the rockets and their scientific instruments invariably came back down. The next step was to put astronomical instruments into orbit. The first astronomical satellite was the British Ariel 1, launched in 1962 to study solar ultraviolet and X-ray radiation and the energy spectrum of primary cosmic rays. Today we have a multitude of orbiting astronomical telescopes covering the electromagnetic spectrum from gamma rays to microwaves, with many more in the planning stage (see **Table 5.2**).

Optical telescopes, such as the Hubble Space Telescope (HST), can operate successfully at modest altitudes in what is called low earth orbit (LEO), 600 km above Earth's surface. This is also the region where the International Space Station (ISS) and many scientific satellites orbit. For others 600 km is not nearly high enough. Chandra, an X-ray telescope, cannot tolerate even the tiniest traces of atmosphere and so flies

---

Orbiting observatories explore regions of the spectrum inaccessible from the ground.

---

in an orbit that keeps it more than 16,000 km above Earth's surface. And even this is not distant enough for some telescopes. Spitzer, an infrared telescope, is so sensitive it needs to be completely free from Earth's own infrared radiation. The solution was to put it into a *solar* orbit, trailing tens of millions of kilometers behind Earth. Many future space telescopes, including NASA's replacement for HST, will orbit free of Earth, bound only to the Sun.

5.6 Getting Up Close with Planetary Spacecraft

As we have stressed, we live in a remarkable time of discovery, when our newfound technological prowess has allowed us to begin the process of exploring our local corner of space.

---

[8] In a sense, there is no definable upper limit to Earth's atmosphere. As we will learn in Chapter 8, our atmosphere simply blends into outer space at an altitude of about 10,000 kilometers.

TABLE 5.2

Selected Present and Future Space Observatories

Telescope	Space Agency	Description	Launch Year
Hubble Space Telescope	NASA	Optical, infrared, ultraviolet observations	1990
Far Ultraviolet Spectroscopic Explorer	NASA	Ultraviolet spectroscopy	1999
Chandra X-Ray Observatory	NASA	X-ray imaging and spectroscopy	1999
Wilkenson Microwave Anisotropy Probe	NASA	Cosmic background radiation	2001
Spitzer	NASA	Infrared observations	2004
Swift	NASA	Gamma ray bursts	2006
Herschel	ESA	Far-infrared and submillimeter observations	2007
Kepler	NASA	Planet finder	2008
James Webb Telescope	NASA	Replacement for HST	2013

The general strategy for exploring our Solar System begins with a reconnaissance phase, using spacecraft that fly by or orbit a planet or other body. At the opening of the 21st century, we have conducted preliminary reconnaissance of much of the Solar System. We have sent spacecraft flying by all of the classical planets, giving humanity its first ever close-up views of these distant worlds and their moons. We have even seen comets and asteroids at close range. As they sped by, instruments aboard these spacecraft briefly probed the physical and chemical properties of their targets and their environments.

Reconnaissance spacecraft use **remote sensing** instrumentation much like the remote sensing techniques used by Earth-orbiting satellites to study our own planet. These include tools such as cameras capable of taking images in different wavelength ranges, radar for mapping surfaces hidden beneath obscuring layers of clouds, and spectrom-

eters that spread out the target's light into a diagnosable spectrum. Remote sensing allows planetary scientists to map other worlds, measure the heights of mountains, identify geological features, learn about types of rocks present, watch weather patterns develop, measure the composition

---

Planetary spacecraft take our instruments directly to the planets.

---

of atmospheres, and in general get a feeling for the "lay of the land." Still other instruments make *in situ* measurements of the extended atmospheres and space environment through which they travel.

The study of our Solar System from space is a truly international collaboration involving NASA, the European Space Agency (ESA), and the Japanese Space Agency. Other countries, including China and India, may soon join the endeavor.

Flybys and Orbiters

Since the dawn of history no human had ever seen the far side of the Moon. This is because, as we learned in Chapter 2, the orbital and rotational periods of the Moon are equal to one another. This keeps one side of the Moon permanently facing Earth and the other side forever hidden. Hidden, that is, until October 18, 1959. On that date the Soviet **flyby** probe *Luna 3* sent back humanity's first view of the far side of our nearest celestial neighbor (see **Figure 5.27**). No matter how powerful we make our groundbased or Earth orbiting telescopes, *Luna 3* showed us there is nothing quite like going there.

Flyby missions have several distinct advantages in the reconnaissance phase of exploration. First, they are relatively inexpensive and the easiest missions to design and execute. Second, flyby spacecraft such as *Voyager*, shown in **Figure 5.28(a)**, may be able to visit several different worlds during their travels. The downside of flyby missions is that, thanks to the physics of orbits, these spacecraft must move by very swiftly. They are limited to just a few hours or at most a few days in which to conduct close-up studies of their targets. Yet flyby spacecraft give us our first intimate views of our planetary neighbors and provide the details we need to plan follow-up studies.

More detailed reconnaissance work uses spacecraft that orbit around planets. These are intrinsically more difficult missions than flyby missions; but **orbiters** can linger, looking in detail at more of the surface of the object they are orbiting and studying things that change with time, like planetary weather. Spacecraft have orbited the Moon, Venus, Mars, Jupiter, Saturn, and even an asteroid. **Figure 5.28(b)** shows the *Cassini* spacecraft, which, as this book goes to print, is still sending us data from its orbit around Saturn.

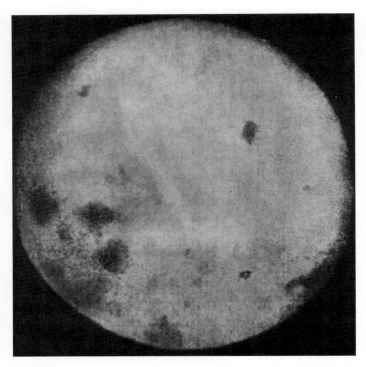

FIGURE 5.27 Humanity's first view of the far side of the Moon seen in this image sent back by the Soviet probe *Lunar-3* in 1959.

landing site. Such remote-controlled vehicles, called **rovers**, were used first by the Soviet Union on the Moon more than a quarter century ago, and more recently by the United States on Mars. The opening photo for this chapter shows an artist's view of one of two rovers still roaming about the Martian landscape.

We have also sent probes into the atmospheres of Venus, Jupiter, and Titan. As they descend, **atmospheric probes** continuously measure and send back physical properties such as temperature, pressure, and wind speed along with other properties, such as chemical composition. Meteorologists take measures of our terrestrial atmosphere from the surface up by sending their instruments aloft in balloons. Planetary scientists must work from the top down by suspending their instruments from parachutes. The end result is much the same. Atmospheric probes have survived all the way to the solid surfaces of Venus and Titan, sending back streams of data during their descent. An atmospheric probe sent into Jupiter's atmosphere never reached that planet's surface because, as we will learn later, Jupiter does not have a solid surface in the same sense that terrestrial planets and moons do. After sending back its data, the Jupiter probe eventually melted and vaporized as it dropped into the hotter layers of the planet's atmosphere.

Landers, Rovers, and Atmospheric Probes

Reconnaissance spacecraft provide a wealth of information about a planet, but there is no better way of obtaining "ground truth" than to put our instruments where they can get right to the heart of it—within a planet's atmosphere or on solid ground. We have landed spacecraft on the Moon, Mars, Venus, Saturn's large moon Titan, and the asteroid Eros. One spacecraft even shot a massive bullet into a comet nucleus to observe the splash. These spacecraft have returned pictures of the surfaces, measured surface chemistry, and conducted experiments to determine the physical properties of the surface rocks and soils.

One disadvantage of using landed spacecraft is that only a few landings in limited areas are practical because of the expense, and the results may apply only to the small area around the landing site. Imagine, for example, what a different picture of Earth we might get from a spacecraft that landed in Antarctica, as opposed to a spacecraft that landed in the caldera (the summit crater) of a volcano or the floor of a dry riverbed. Sites to be explored with landed spacecraft must be very carefully chosen on the basis of reconnaissance data if we are to know what to make of the information they provide. We can mitigate some of the limitations of **landers** by putting their instruments on wheels and sending them from place to place, exploring the vicinity of the

Sample Returns

If you pick up a rock from a road cut, there is a lot you might learn from the rock using the tools that you could easily carry in your pocket. On the other hand, the sophistication of the tools you could carry with you would be limited. It would be much better to pick up a few samples and carry them back to a laboratory equipped with a full range of state-of-the-art instruments capable of measuring chemical compositions, mineral types, radiometric ages (see Foundations 7.1), and other information needed to reconstruct the story of their origin and evolution. So, too, is the case in Solar System exploration. One of the most powerful methods for investigating remote objects is to collect samples of the objects and bring them back to Earth for detailed study. So far, only samples of the Moon, a comet, and the **solar wind** (a stream of charged particles from the Sun) have been collected and returned to Earth.

As we will learn in Chapter 12, we do have meteorites that are considered to be parts of Mars, but there is a problem in putting them in their proper geological perspective.

Sample returns provide "ground truth."

We just talked about picking up a rock from a road cut. This is quite different from simply grabbing any old rock by the side of the road because that rock could have come from anywhere. In geological sampling it is important to have

(a)

(b)

FIGURE 5.28 Explorers of the planets. (a) *Voyager* flew past Jupiter, Saturn, Uranus, and Neptune. (b) *Cassini* orbits Saturn.

samples from a source of known context. Some have claimed that we do not need samples returned from Mars because we already have the Martian meteorites. The problem is that we do not know where on Mars they came from and, therefore, how they fit into the planet's global geology. Plans are currently under way for unmanned "sample and return" missions to Mars.

Of course, we could not collect specimens in a national park without permission and a scientifically valid reason. Similarly, the return of extraterrestrial samples to Earth is governed by international treaties and standards to ensure that contamination of Earth does not occur. For example, before the lunar samples brought back by the Apollo missions could be studied, they (and the astronauts) were placed in quarantine and tested for alien life forms. The same international standards apply to spacecraft landing on planets. The goal of these standards is to avoid *forward contamination,* or transporting life forms from Earth to another planet. If there is life on other planets, then not only is there concern about introducing potential harm, but from a scientific perspective we do not want to "discover" life that we, in fact, have introduced.

With numerous missions under way and others on the horizon, unmanned exploration of the Solar System is an ongoing, dynamic activity. In our journey we will frequently refer to space missions and the information they return, but today's hot results may be tomorrow's old news in light of other, even more exciting discoveries. We hope that you will make use of the *21st Century Astronomy* website as a gateway to the wealth of exciting results that the future holds.

5.7 High-Energy Colliders

Ever since the early years of the 20th century, physicists have been peering into the structure of the atom by observing what happens when small particles collide. In 1906 **Ernest Rutherford** (1871–1937) found that positively charged **alpha particles**[9] are deflected when they pass through a thin sheet of mica, a shiny mineral that readily splits into thin layers. This discovery proved for the first time that atoms must contain heavy, centrally concentrated nuclei. By the 1930s physicists had developed the means to accelerate charged particles such as protons to very high speeds and then observe what happens when they slam into a target. From such experiments (which are continuing even today) physicists have discovered many kinds of **elementary particles**[10] and learned about their physical properties. High-energy particle colliders have proven to be an essential tool for physicists.

Why then do we regard the high-energy particle collider as a tool of the astronomer? How does the astronomer's interest in the largest objects in the universe relate to what happens on the very smallest scales? As we will see in

---

[9] Alpha particles are positively charged He[4] nuclei emitted in certain types of radioactive decay.

[10] In Chapter 4 we learned about three elementary particles: the proton, neutron, and electron. In Section 5.4 we were introduced to another, the neutrino, and in Foundations 14.1 we will meet still another, the positron. However, many other elementary particles are now known to physicists—in fact too many to list here.

FIGURE 5.29 The Fermi National Accelerator Laboratory in Batavia, Illinois.

facility, when it becomes operational in 2007, will produce collisions with nearly 20 times more energy than the Fermi Collider—about 3×10^{-6} joules.

5.8 High-Speed Computers

Imagine your life without computers. From laptops to desktops to larger servers and computers, we depend on computers to surf the Internet, process our data, and organize our daily lives. Tiny computer processors control every modern convenience from automobiles to cameras to washing machines. And as you might assume, computers are essential in the world of science. Data gathering, analysis, and interpretation are entirely dependent on computers—and the more powerful, the better. Consider, for example, analyzing a night's worth of astronomical images recorded by a very large CCD. A single image may contain as many

Chapter 21, to understand the very largest structures we see in the universe—indeed the large-scale universe itself—we need to understand the physics that took place during the earliest moments in the universe, when everything was unbelievably hot and dense. Although we have not yet reached that level of comprehension, the high-energy particle colliders that physicists use today are designed to lead

> **Colliders teach us the physics we need to understand the early universe and the formation of structure.**

us there. As we will see in Chapters 20 and 21, this knowledge may help us to understand such issues as the nature of the dominant matter and energy in the universe, why the universe consists of matter rather than **antimatter**, and whether there really is a beginning or an end to our universe.

Two factors determine the effectiveness of particle accelerators: the energy they can achieve and the number of particles they can accelerate. Whereas the first particle accelerator, the *cyclotron*, attained an energy of about 10^{-13} joules, modern particle colliders now reach much higher energies. For example, the accelerator at the Fermi National Accelerator Laboratory (see **Figure 5.29**) can accelerate protons up to 1.6×10^{-7} joules. This may seem like a small number, but it corresponds to more than a thousand times the rest mass energy (mc^2) of the proton. To put it in perspective, this is the energy of a flying mosquito all concentrated into a single tiny proton. Yet this energy will be dwarfed by the new Large Hadron Collider at CERN, the European Organization for Nuclear Research, shown in **Figure 5.30**. This

FIGURE 5.30 The tunnels and magnets for the Large Hadron Collider at CERN.

as 100 million pixels, with each pixel displaying roughly 10,000 levels of brightness. That adds up to a *trillion* pieces of information in each image! And that's only one image. To analyze their data, astronomers typically do calculations on *every single pixel* of an image in order to remove unwanted contributions from Earth's atmosphere or correct for instrumental effects. From the astronomer's point of view, without high-speed computers, the CCD would be just another electronic curiosity.

High-speed computers also play an essential role in generating and testing theoretical models of astronomical objects. Even when we completely understand the underlying physical laws that govern the behavior of some particular object, it is frequently the case that the object is so complex that it would be impossible to calculate its properties and behavior without the assistance of high-speed computers. For example, as we learned in Chapter 3, we can use Newton's laws to easily compute the orbits of two stars that are gravitationally bound to one another because their orbits take the form of simple ellipses. However, it is not so easy to understand the orbits of the hundred billion (10^{11}) stars that comprise our Milky Way Galaxy, even though *the underlying physical laws remain the same*. If that were not complicated enough, consider the problem involving the collision of *two* such galaxies—and, for good measure, throw in some gas in addition to the stars. We can see the result in **Figure 5.31**. When we apply even the fastest computers available to this problem, the sequence shown in Figure 5.31 is only an

FIGURE 5.31 Numerical simulation of the merging of two galaxies, including the gravitational attraction of all forms of matter. (a) Just before the merge. (b) After passing through one another. (c) As the cores orbit each other and merge. (d) When the central core begins to settle down.

(a)

(b)

(c)

(d)

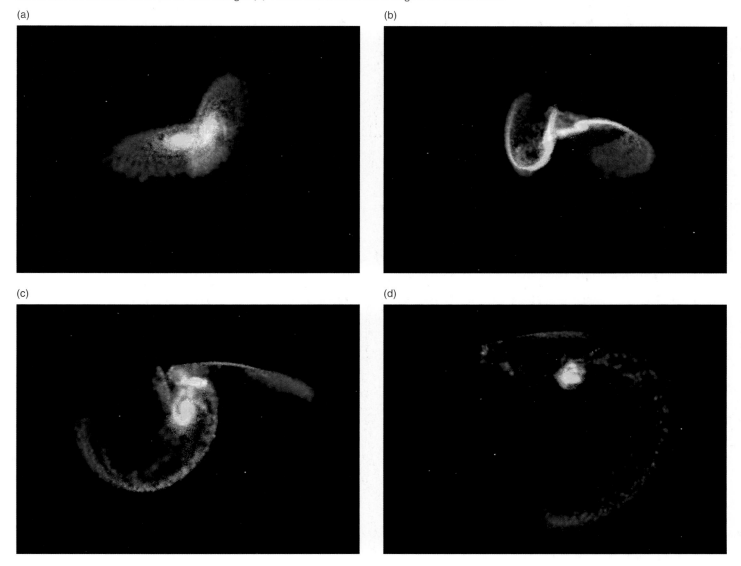

approximation of what we believe really happens. Modern computers have enough speed and memory to handle the behavior of a few million stars at best, so we are forced to assume that a single star in the computer simulation really represents hundreds of thousands of real stars.

Similar modeling procedures have worked well in determining the interior properties of stars and planets, including our own Earth. Although we cannot "see" beneath their surfaces, we have a surprisingly good understanding of their interiors, as we will learn in later chapters. We begin a model by assigning well-understood physical properties to tiny volumes within a planet or star. The computer assembles an enormous number of these individual elements into an overall representation of the complete body. When it is all put together, we have a rather good picture of what the interior of the star or planet is like.

Summary

- Optical telescopes come in two basic types, refractors and reflectors.

- All large astronomical telescopes are reflectors.

- Large telescopes collect more light and have greater resolution.

- The CCD is today's astronomical detector of choice.

- Earth's atmosphere blocks many spectral regions and distorts telescopic images.

- Putting telescopes in space solves problems created by Earth's atmosphere.

- Most of what we know about the planets comes from spacecraft we have sent there.

- Infrared and radio telescopes can see through vast clouds of cosmic gas and dust.

- Radio and optical telescopes can be arrayed to greatly increase angular resolution.

- High-speed computers are essential to the acquisition, analysis, and interpretation of astronomical data.

Seeing the Forest through the Trees

From our perspective in the 21st century, we can only imagine what went through the minds of our distant ancestors as they gazed upward toward the heavens. We might guess that they felt completely comfortable with the daily movements of the Sun and the Moon, yet experienced fear when confronted with an eclipse or the appearance of an occasional comet. They probably took for granted the tiny points of light that filled the nighttime sky, but must have wondered about those mysterious few that seemed to move freely among the others. With nothing more than their eyes to fulfill their curiosity, our early ancestors could only observe and wonder. But wondering alone does not give rise to comprehension. Insight far beyond personal experience was necessary, and it was a long time coming. Somewhat more than two millennia ago enlightenment of a sort came to the classical world. Greek philosophers concluded that these "wandering stars" or planets were unlike the distant canopy of fixed stars—that they were much closer—but stopped short of claiming to know *what* they were. And so these and other heavenly mysteries lingered on until the turn of the 17th century, when two historic events took place.

It all started when an obscure Flemish spectacle maker put a pair of lenses together and saw that distant objects appeared closer. Although Lippershey may have regarded his "looker" as an amusing toy, it was a Tuscan instrument maker who first recognized its potential for studying the heavens. It didn't take Galileo Galilei long to discover that the Moon was a nearby world covered with craters, and that those "mysterious moving points of light" were tiny disks of planets. This was the turning point. The telescope had changed forever our fundamental perception of the heavens. Throughout the four centuries since Galileo's historic discoveries, astronomers have continued to perfect the astronomical telescope. Reaching far outside the visible, telescopes now cover the entire electromagnetic spectrum from gamma rays to radio waves. Some telescopes reside on mountaintops, while others make their home in Earth orbit or nearby space. Still others take the long and arduous journey through interplanetary space to observe Earth's neighbors up close. But telescopes alone cannot provide all the answers. Tools of a different type have recently joined the astronomer's repertoire of gear. Some are buried deep underground quietly observing one of nature's most elusive particles. Others sit on desktops crunching numbers. All are devoted to decoding the cryptic messages sent to us from the cosmos.

Having stopped briefly to examine the many tools used by astronomers, we are now ready to resume our outward journey, armed with a new appreciation of how important it is to proceed carefully. Common sense is not enough. We must instead rely on our growing understanding of physical law and on rigorously tested predictions of carefully constructed theories. These also are useful tools that will allow us to look beyond the surface of spectacular vistas that otherwise would be devoid of sense or meaning or connection. Kepler's laws, Newton's laws, relative motion, Doppler shifts, Wien's Law, Stefan's Law, energy states, spectral lines, and the rest—these are the keys we will use to build an understanding of planets, stars, galaxies, and ultimately the universe itself. The first place in which we will bring these tools to bear will be in our immediate neighborhood as we consider the nature and origin of our Solar System. And to put our own Solar System in perspective, we'll take a look at some of the many other planetary systems that lie far beyond.

Key Terms

surface brightness, p. 136
refracting telescope, p. 136
aperture, p. 136
focal length, p. 137
reflecting telescope, p. 137
primary mirror, p. 137
reflection, p. 138
ray, p. 138
dispersion, p. 140
interference, p. 140
constructive interference, p. 140
destructive interference, p. 140
grating, p. 140
secondary mirror, p. 142
resolution, p. 142
diffraction, p. 142
diffraction limit, p. 142
astronomical seeing, p. 143
adaptive optics, p. 143
wavefront, p. 143
atmospheric window, p. 144
integration time, p. 145
quantum efficiency, p. 146
charge-coupled device (CCD), p. 147
pixel, p. 147
spectrograph, p. 148
radio telescope, p. 149

jansky, p. 149
interferometer, p. 150
interferometric array, p. 150
flyby, p. 153
orbiter, p. 153
lander, p. 154
rover, p. 154
atmospheric probe, p. 154

Student Questions

THINKING ABOUT THE CONCEPTS

1. Optical telescopes reveal much about the nature of astronomical objects. Why do astronomers also need information provided by gamma ray, X-ray, infrared, and radio telescopes?

2. The largest astronomical refractor has an aperture of 1 m. List several reasons why it would be impractical to build a still larger refractor with, say, twice the aperture.

3. Your camera may have a zoom lens, ranging between wide angle (short focal length) and telephoto (long focal length). How would the size of an object in the camera's focal plane differ between wide angle and telephoto?

4. CCDs are now the most commonly used astronomical detector. List three advantages that CCDs have over photographic emulsions.

5. Why do we not have groundbased gamma ray and X-ray telescopes?

6. Some people believe that we put astronomical telescopes in space because it gets them closer to the objects they are observing. As an enlightened student taking this class, you know better. Explain what is wrong with this popular misconception.

7. We have now sent various kinds of spacecraft—including flybys, orbiters, and landers—to all of the classical planets. Explain the advantages and disadvantages of each of these types of spacecraft.

8. Why are the world's largest telescopes located on high mountains?

9. "Twinkle, twinkle, little star. How I wonder what you are." Explain *why* stars twinkle. (In a later chapter we will find out *what* the stars are.)

10. Humans had our first look at the far side of the Moon only as recently as 1959. Why had we not been able to see it earlier—say, when Galileo first observed the Moon with his telescope in 1610?

APPLYING THE CONCEPTS

11. Compare the light-gathering power of a large astronomical telescope (aperture = 10 m) with that of the dark-adapted human eye (aperture = 7 mm).

12. Compare the angular resolution of the Hubble Space Telescope (aperture = 2.4 m) with that of a typical amateur telescope (aperture = 20 cm).

13. Assume that the maximum aperture of the human eye, D, is approximately 7 mm and the average wavelength of visible light, λ, is 5.5×10^{-4} mm.
    a. Calculate the diffraction limit of the human eye in visible light.
    b. How does this compare with its actual resolution of 1 to 2 arcminutes (60 to 120 arcseconds)?
    c. To what do you attribute the difference?

14. The diameter of the full Moon's image in the focal plane of an average amateur's telescope (focal length = 1.5 m) is 13.8 mm. How big would the Moon's image be in the focal plane of a very large astronomical telescope (focal length = 250 m)?

15. One of the earliest astronomical CCDs had 160,000 pixels, each recording 8 bits (256 levels of brightness). Today's CCDs may contain 100 million pixels, each recording 15 bits (32,768 levels of brightness). Compare the number of bits of data that each produces in single image.

16. The VLBA employs an array of radio telescopes ranging across 8,000 km of Earth's surface from the Virgin Islands to Hawaii.
    a. Calculate the angular resolution of the array when radio astronomers are observing interstellar water molecules at a microwave wavelength of 1.35 cm.
    b. How does this compare with the angular resolution of two large optical telescopes separated by 100 m and operating as an interferometer at a visible wavelength of 550 nm?

17. Rovers *Spirit* and *Opportunity* can move across the Martian landscape at speeds of up to 5 cm/s. In contrast, our typical walking speed is about 4 km/h.
    a. How quickly could *Spirit* move the length of a football field (91.44 m)?
    b. Compare this with the time it would take to walk the same distance.

18. *Voyager 1* is now about 100 AU from Earth, continuing to record its environment as it approaches the limits of our Solar System.
    a. What is the distance of *Voyager 1* expressed in kilometers?
    b. How long does it take observational data to get back to us from *Voyager 1*?
    c. How does its distance compare with that of the nearest star (other than the Sun)?

StudySpace
wwnorton.com/astro21
provides a Study Plan for each chapter that includes a reading outline, animations, keyword flash cards, and gradebook-enabled multiple-choice quizzes. From StudySpace you can also access premium content in the ebook and SmartWork.

PART III

Stars and Stellar Evolution

To man, that was in th' evening made,
Stars gave the first delight;
Admiring, in the gloomy shade,
Those little drops of light.

EDMUND WALLER (1606–1687)

Stars on the Hertzprung-Russell (H-R) diagram.

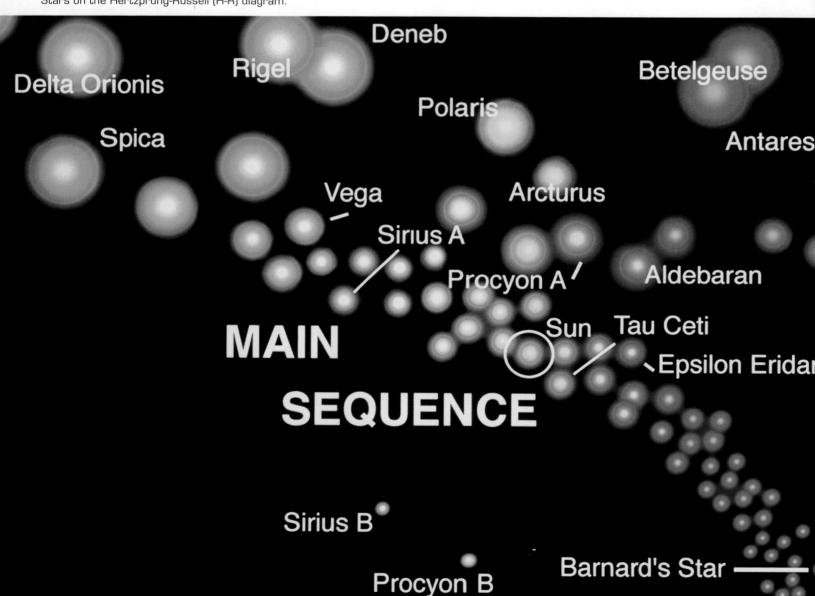

CHAPTER 13

Taking the Measure of Stars

13.1 Twinkle, Twinkle, Little Star, How I Wonder What You Are

As children we look at the sky, see the stars, and wonder. What are those points of light? How far away are they? How bright are they? How hot is a star? How big is a star? How much does a star "weigh"? What are stars made of? How long does a star last? What makes a star shine? These questions have always been a part of the human experience.

For most of human history **stars**, like so much of nature, have seemed mysterious and unknowable. Even so, their passage follows the daily and annual rhythms that shape our lives. It is no wonder that in the absence of any real understanding of the stars, ancient humans turned their imaginations loose and viewed the stars as the province of gods and magic, with power over our lives. Today we understand that the patterns of the constellations have no more mystical influence over the course of events than the random toss of a coin (even though we persist in paying the salaries of astrologers and reading their ramblings).

The story of how we know what we know about the stars is a wonderful tale. It starts with simple but ingenious observations of the sky, borrows from lessons learned about the behavior of matter on Earth and in the Solar System, and ends with clear, straightforward answers to the very questions that a child might ask while gazing at the sky. The path to knowledge of the stars is also wonderful because, unlike many stories of modern science, it is a path that can be traveled and appreciated by anyone willing to take the time to follow it. A bit of geometry, a bit about radiation, a

KEY CONCEPTS

To even the most powerful of telescopes, a star is just a point of light in the night sky. However, by applying our understanding of light, matter, and motion to what we see, we are able to build a remarkably detailed picture of the physical properties of stars. As we take this first step beyond our local Solar System, we will

- Apply a form of "stereoscopic vision," extended by Earth's orbit, to measure distances to nearby stars.

- Use the brightness of stars and their distances from Earth to discover how luminous they are.

- Use our knowledge of thermal radiation to infer the temperatures and sizes of stars from their colors.

- Classify stars, and organize this information on a plot of luminosity versus temperature, called the Hertzsprung-Russell or H-R diagram.

- Measure the composition of stars from spectra.

- Discover that 90 percent of the stars we see lie along a well-defined "main sequence" in the H-R diagram.

- Study the orbits of binary stars and use Kepler's laws to calculate their masses.

- Discover that the mass and composition of a main sequence star determines its luminosity, temperature, and size.

- Explore the range of stellar properties, learning how our Sun compares to other stars.

bit about orbits—all things that we have studied in earlier chapters—and we begin to find ourselves with solid answers to age-old questions.

13.2 The First Step Is Measuring the Distance, Brightness, and Luminosity of Stars

In a sense, one of the most amazing feats that a human can perform is to catch a fly ball. Here comes the ball, traveling at speeds of 150 km/h or more. To put a glove on the ball the fielder must judge not only the direction to the ball, but also its distance. But how does the fielder do it? Like most predators, we have two forward-looking eyes, one on each side of our face. Each of our eyes has a somewhat different view of the world. Exactly how different these perspectives are depends on the distance to the object you are looking at. If you are looking at a house down the street and you blink back and forth between your two eyes, the view that you get changes very little. But if you are looking at a nearby object—perhaps a finger held up at arm's length—the perspectives of your two eyes differ quite a lot. This difference in our eyes' perspectives on objects at different distances is the basis of our **stereoscopic vision**. Put another way, stereoscopic vision is the major key to the way we perceive distances. (**Figure 13.1** shows that the brain can even be fooled into perceiving distance where none exists by showing each eye a different view.)

Our stereoscopic vision allows us to judge the distances of objects as far away as a few hundred meters, but beyond that it is of little use. Our eyes are separated by a distance of only about 6 cm, so the view that your right eye gets of a mountain several kilometers away is indistinguishable from the view seen by your left eye. Comparing the two views tells your brain only that the mountain is too far away for you to judge its distance. (Evolution provided us with only enough depth perception to judge distances to things that we might be trying to eat or that might be trying to eat us!) The distance over which our stereoscopic vision works is limited by the separation between our two eyes. If you wanted to increase the differences between the views your two eyes see, the obvious thing to do would be to somehow move them farther apart. If you could separate your eyes by several meters instead of only a few centimeters, their perspectives would be different enough to allow you to judge the distances to objects that are kilometers away.

Of course, we cannot literally take our eyes out of our heads and hold them apart at arm's length, but we can compare pictures taken from two widely separated locations.

The greatest separation we can get without leaving Earth is to let Earth's orbital motion carry us from one side of the Sun to the other. If we take a picture of the sky tonight, then wait six months and take another picture, our point of view between the two pictures will have changed by the diameter of Earth's orbit, or two astronomical units (AU). With 2 AU

We measure distances to nearby stars by comparing the view from opposite sides of Earth's orbit.

---

separating our two "eyes," we should have very powerful stereoscopic vision indeed. **Figure 13.2** shows how our view of a field of stars changes as our perspective changes during the year. This change in perspective is what allows us to measure the distances to nearby stars.

Distances to Nearby Stars Are Measured Using Parallax

The eye cannot detect the changes in position of a nearby star throughout the year, but telescopes can reveal these small shifts relative to the background stars. **Figure 13.3** shows Earth, the Sun, and three stars. Look first at star 1. When Earth, the Sun, and the star are in this position, they form a long, skinny right triangle. The short leg of the triangle is the distance from Earth to the Sun, or 1 AU. The long side of the triangle is the distance from the Sun to the star. The small angle at the end of the triangle is called the *parallactic angle* of the star, or simply the **parallax** of the star. Over the course of a year, the star's position in the sky appears to shift back and forth, returning to its original position one year later. The amount of this shift—the angle between one extreme in the star's apparent motion and the other—is equal to twice the parallax.

The more distant the star, the longer and skinnier the triangle that it forms, and the smaller the star's parallax. Look again at Figure 13.3. Star 2 is twice as far away as star 1, and its parallax is only half as great. If you were to draw a number of such triangles for different stars, you would find that increasing the distance to the star always reduces the star's parallax. Move a star three times farther away, as with star 3 in Figure 13.3, and you reduce its parallax to one-third of its original value. Move a star 10 times farther away, and you reduce its parallax to $\frac{1}{10}$ of its original value. The parallax of a star (p) is inversely proportional to its distance (d):[1]

$$p \propto \frac{1}{d} \quad \text{or} \quad d \propto \frac{1}{p}.$$

---

[1] Note that parallax is inversely proportional to distance only if the parallax is tiny, as it is for stars. For nearby objects, it is the *tangent* of the parallax that is inversely proportional to the distance.

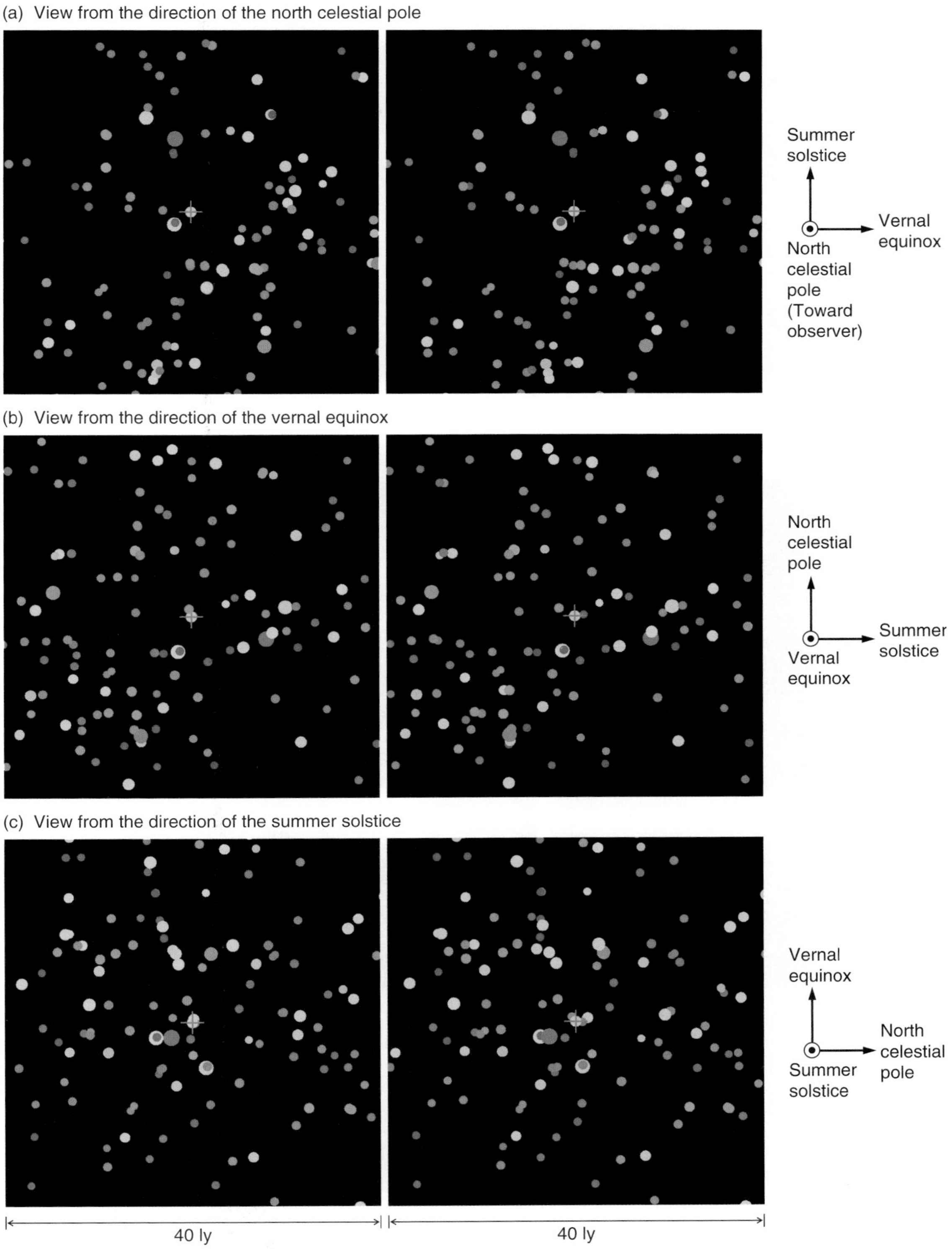

(a) View from the direction of the north celestial pole

Summer
solstice

Vernal
equinox

North
celestial
pole
(Toward
observer)

(b) View from the direction of the vernal equinox

North
celestial
pole

Summer
solstice

Vernal
equinox

(c) View from the direction of the summer solstice

Vernal
equinox

North
celestial
pole

Summer
solstice

40 ly   40 ly

FIGURE 13.1 Your brain uses the slightly different views offered by your two eyes to "see" the distances and three-dimensional character of the world around you. These stereoscopic pairs show the stars in the neighborhood of the Sun as viewed (a) from the direction of the north celestial pole, (b) from the direction of the vernal equinox, and (c) from the direction of the summer solstice. The field shown is 40 light-years on a side. The Sun is at the center, marked with a green cross. The observer is 400 light-years away and has "eyes" separated by about 30 light-years. To view these stereoscopic images, hold a card between the two images and look at them from about a foot and half away. Relax and look straight at the page until the images merge into one.

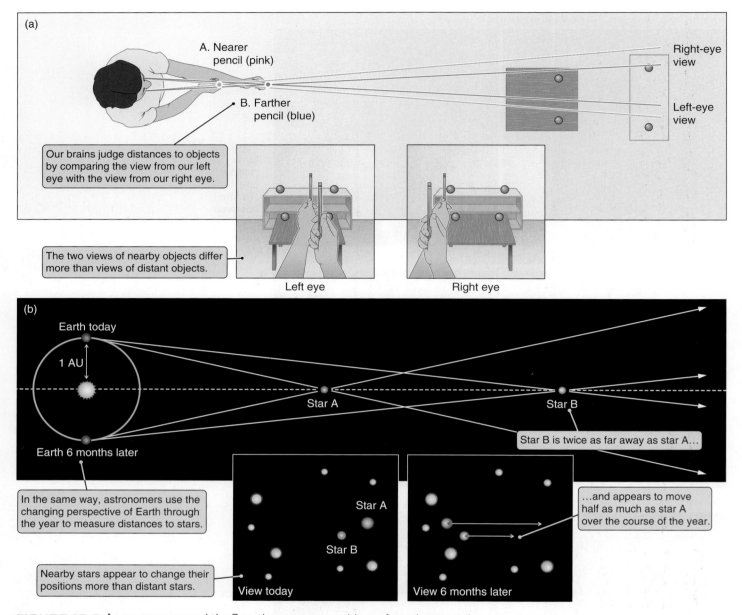

FIGURE 13.2 As we move around the Sun, the apparent positions of nearby stars change more than the apparent positions of more distant stars. This is the starting point for measuring the distances to stars.

The parallaxes of real stars are tiny. Rather than talking about parallaxes of 0.0000028° or 4.8 10⁻⁸ radians (1 **radian =** 57.3°), astronomers normally measure parallaxes in units of seconds of arc. Just as an hour on the clock is divided into minutes and seconds of time, a degree can be divided into minutes and seconds of arc. **A minute of arc (or arcminute)** is 1/60 of a degree, and a **second of arc (or arcsecond)** is 1/60 of a minute of arc. That makes a second of arc 1/3,600 of a degree or 1/1,296,000 of a complete circle. An arcsecond is about equal to the angle formed by the diameter of a golf ball at the distance of 5 miles. (Arcseconds are often denoted by the symbol ″; one second of arc is written as 1″,

and one arcminute is written as 1′. However, we will spell out these units in this text.)

If the angle at the apex of a triangle is 1 arcsecond, and the base of the triangle is 1 AU, then the length of the tri-

---

A parsec is defined as the distance at which the parallax equals 1 arcsecond.

---

angle is 206,264.81 AU. This distance, which corresponds to 3.09 × 10¹⁶ m or 3.26 light-years, is referred to as a **parsec** (abbreviated pc). The relationship between distance measured in parsecs and parallax measured in arcseconds is

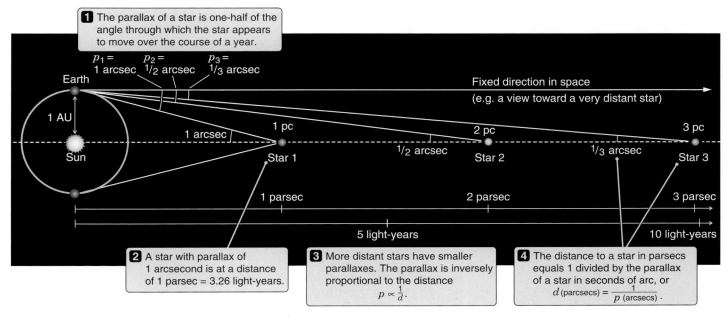

1 The parallax of a star is one-half of the angle through which the star appears to move over the course of a year.

$p_1 =$ 1 arcsec $p_2 =$ 1/2 arcsec $p_3 =$ 1/3 arcsec

Earth

1 AU

Sun

1 arcsec

Fixed direction in space (e.g. a view toward a very distant star)

1 pc   Star 1   1/2 arcsec   Star 2   2 pc   1/3 arcsec   Star 3   3 pc

1 parsec   2 parsec   3 parsec

5 light-years   10 light-years

2 A star with parallax of 1 arcsecond is at a distance of 1 parsec = 3.26 light-years.

3 More distant stars have smaller parallaxes. The parallax is inversely proportional to the distance $p \propto \frac{1}{d}$.

4 The distance to a star in parsecs equals 1 divided by the parallax of a star in seconds of arc, or $d \, (\text{parcsecs}) = \dfrac{1}{p \, (\text{arcsecs})}$.

FIGURE 13.3 The parallax of three stars at different distances. Parallax is inversely proportional to distance.

illustrated in Figure 13.3. Using the fact that a star with a parallax of 1 arcsecond is at a distance of 1 parsec, we can turn the inverse proportionality between distance and parallax into an equation:

$$\left(\begin{array}{c} \text{Distance measured} \\ \text{in parsecs} \end{array} \right) = \frac{1}{\left(\begin{array}{c} \text{Parallax measured} \\ \text{in arcseconds} \end{array} \right)}$$

If you measure the parallax of a star to be 0.5 arcsecond, then you know the star is located at a distance of 1/0.5 = 2 parsecs. A star with a parallax of 0.01 arcsecond is located at a distance of 1/0.01 = 100 parsecs.

In this book we will usually use units of *light-years* to indicate distances to stars and galaxies. One light-year is the distance that light travels in 1 year—about 9 trillion kilometers. We use this unit because it is the unit you are most likely to see in a newspaper article or a popular book about astronomy. But astronomers seldom use light-years when talking among themselves. When astronomers discuss distances to stars and galaxies, the unit they use is the parsec. (You can always convert between the two units. One parsec is 3.26 light-years.)

The star closest to us (other than the Sun) is Proxima Centauri. Proxima Centauri is located at a distance of 4.22 light-years, or 1.3 parsecs, and is a faint member of a system of three stars called Alpha Centauri. This star has a parallax of only about ¾ arcsecond. It is no wonder that ancient astronomers were unable to detect the apparent motions of the stars over the course of a year!

When astronomers began to apply this technique, they discovered that stars are very distant objects indeed. The first successful measurement of the parallax of a star was made by F. W. Bessel, who in 1838 reported a parallax of 0.314 arcsecond for the star 61 Cygni. This implied that 61 Cygni was 3.2 parsecs away, or 660,000 times as far away as the Sun. With this one measurement, Bessel increased the known size of the universe 10,000-fold! Today we know of 55 stars in 38 single-, double-, or triple-star systems within 15 light-years of Earth. A sphere with a radius of 15 light-years has a volume of about 14,000 cubic light-years, so this corresponds to a local density of 37 systems per

Stars are few and far between in our neighborhood.

14,000 cubic light-years. That is about 0.0026 star systems per cubic light-year. Stated another way, in the neighborhood of the Sun, each system of stars has on average about 380 cubic light-years of space (a volume about 4.5 light-years in radius) all to itself.

Knowledge of our stellar neighborhood took a tremendous step forward during the 1990s with the completion of the Hipparcos mission. The Hipparcos satellite measured the positions and parallaxes of 120,000 stars. These measurements, taken from a satellite well above Earth's obscuring atmosphere, are better than the measurements that can typically be made from telescopes located on the surface of Earth. But even this catalog has its limits. The accuracy of any given Hipparcos parallax measurement is about ±0.002 arcsecond. Because of this **observational uncertainty**, our measurements of the distances to stars are not perfect. For example, a star with a parallax measured by Hipparcos of 0.004 arcsecond might really have a parallax of anywhere between 0.002 and 0.006 arcsecond. So instead of knowing that the distance to the star is exactly 250 parsecs (1 divided by 0.004 arcsecond), we know only that the star is probably between about 170 parsecs (1/0.006 arcsecond) and 500 parsecs (1/0.002 arcsecond). With current technology, parallax becomes useless as a way of measuring stellar distances for stars more than a few hundred parsecs away. If you are measuring stellar distances using parallax, and you need to know the distance to an accuracy of 10 percent or better, you are restricted to stars that are less than about 50 parsecs (160 light-years) away. Yet technology keeps marching onward. Plans are under way to launch the *Space Interferometry Mission (SIM)* in 2010—an orbiting spacecraft that will measure parallax with an accuracy of 4 microarcseconds. This will allow astronomers to determine distances out to 25,000 parsecs (80,000 light-years) with an uncertainty of 10 percent.

Once Distance and Brightness Are Known, Luminosity Can Be Calculated

When we talk about the brightness of an object, we are making a statement about how that object appears to us. In Chapter 4 we saw that brightness corresponds to the amount of energy falling on a square meter of area each second in the form of electromagnetic radiation. (When astronomers talk about the brightness of stars, they usually use a system

> Brightness depends on the observer's perspective, whereas luminosity does not.

called **magnitudes**, which is discussed in Appendix 6.) But while the brightness of a star is directly measurable, that does not immediately tell us much about the star itself. As illustrated in **Figure 13.4**, a bright star in the night sky may in fact be a dim one, appearing bright only because it is nearby. Conversely, a faint star may be a powerful beacon, still visible despite its tremendous distance.

To learn about the stars themselves, we need to know the total energy radiated by a star each second—the star's luminosity. In Chapter 4 we studied the relationship between the brightness, luminosity, and distance of objects. Borrowing from that earlier work, we recall that the brightness of an object that has a known luminosity and is located at a distance d is given by

$$\text{Brightness} = \frac{\text{Total light per second}}{\text{Area of a sphere of radius } d}$$

$$= \frac{\text{Luminosity}}{4\pi d^2}.$$

We can rearrange this equation, moving the quantities we know how to measure (distance and brightness) to the right side and the quantity we would like to know (luminosity) to the left, giving us

$$\text{Luminosity} = 4\pi d^2 \times \text{Brightness}.$$

This equation answers the question "How much total light must a star be giving off to be as distant as it is while looking as bright as it does?" In the process we take two measurable quantities that depend on our particular perspective (distance and brightness) and from them obtain a quantity that is a property of the star itself, namely luminosity.

When measuring the luminosity of stars as just described, we find that stars vary tremendously in the amount of light they give off. The Sun provides a convenient yardstick for measuring the properties of stars, including their

> Some stars are 10 billion times more luminous than other stars.

luminosity. The luminosity of the Sun is written as $L_\odot$. The most luminous stars can exceed 1,000,000 $L_\odot$ or 10^6 $L_\odot$ (a million times the luminosity of the Sun). The least luminous stars have luminosities less than 0.0001 $L_\odot$, or 10^{-4} $L_\odot$ (1/10,000 that of the Sun). The most luminous stars are over 10 billion (10^{10}) times more luminous than the least luminous stars. Only a very small fraction of stars are near the upper end of this range of luminosities. The vast majority

> There are many more low-luminosity stars than high-luminosity stars.

of stars are at the faint end of this distribution, less luminous even than our Sun. **Figure 13.5** shows the distribution of luminosities for the known stars within 1,000 parsecs (3,260 light-years) of Earth.

13.3 Radiation Tells Us the Temperature, Size, and Composition of Stars

Two everyday concepts—stereoscopic vision, and the fact that the closer an object is, the brighter it appears—have given us the tools we need to measure the distance and luminosity of stars. In these two steps stars have gone from being mere faint points of light in the night sky, to being extraordinarily powerful beacons located at distances almost impossible for the mind to comprehend. To go further along this journey of discovery, we again turn to the laws of radiation that we studied in Chapter 4.

Stars are gaseous, but they are fairly dense—dense enough that the radiation from a star comes close to obeying the same laws as the radiation from objects like the heating element on an electric stove or the filament in a lightbulb. That means we can use our understanding of Planck radia-

(a)

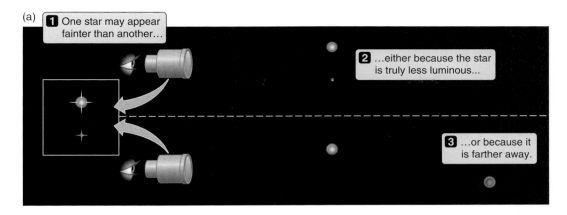

(b)

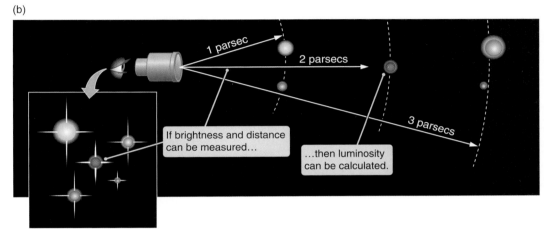

FIGURE 13.4 The brightness of a visible star depends on both its luminosity and its distance. If brightness and distance are measured, luminosity can be calculated.

tion—results such as Stefan's Law (hotter at same size means more luminous) and Wien's Law (hotter means bluer)—to understand the radiation from stars. In particular, Wien's Law and Stefan's Law will allow us to measure the temperatures and sizes of our stellar neighbors.

Wien's Law Revisited: The Color and Surface Temperature of Stars

The hotter the surface of an object, the bluer the light that it emits. As illustrated in **Figure 13.6(a)**, stars with especially hot surfaces are blue, stars with especially cool sur-

---

Measuring the color of a star tells us its surface temperature.

---

faces are red, and our Sun is a middle-of-the-road yellow. More formally, Wien's Law states that

$$T = \frac{2,900 \mu\text{m K}}{\lambda_{\text{peak}}},$$

where λ_{peak} is the wavelength at which the electromagnetic radiation from a star is most intense. Obtain a spectrum of a star, measure the wavelength at which the spectrum peaks, and Wien's Law will tell you the temperature of the star's surface.[2]

In practice, it is usually not necessary to obtain a complete spectrum of a star to determine its temperature. Instead, stellar temperatures are often measured in a way that is similar to the way that your eye sees color. Your eye and brain distinguish color by comparing how bright an object is at one wavelength with how bright it is at another wavelength. Some of the light-sensitive cells in your eye are sensitive primarily to red light, others are sensitive mostly to green light, and still others are sensitive to blue light. If you look at a very red lightbulb, the red-sensitive cells in your eye report a bright light while the blue- and green-sensitive cells see less light. If you look at a yellow light, the green- and red-sensitive cells are strongly stimulated, but the blue-sensitive cells are not. White light stimulates all three kinds of cells. By combining the signals from cells that are sensitive to different wavelengths of light, your brain can distinguish between fine shades of color.

---

[2] It is important to remember that the color of a star tells us only about the temperature at the star's *surface* because the surface is the part of the star giving off the radiation that we see. Star surfaces may be hot, but later in our journey we will discover that star interiors are far hotter still.

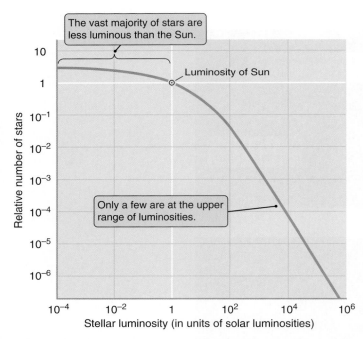

FIGURE 13.5 The distribution of luminosities of known stars within 1,000 pc (3,260 ly) of Earth.

Astronomers often measure the colors of stars in much the same way. The brightness of a star is usually measured through a **filter**—sometimes just a piece of colored glass—that lets through only a certain range of wavelengths. Two of the most common filters used by astronomers are a blue filter that transmits light with wavelengths around 440 nm, and a yellow-green filter that passes light with wavelengths around 550 nm. The first of these filters is (sensibly enough) called a "blue" filter. However, by convention the second of these filters is referred to as a "visual" filter rather than a "yellow-green" filter because it transmits roughly the range of wavelengths to which our eyes are most sensitive.

Figure 13.7 shows Planck spectra with temperatures of 2,500 K through 20,000 K, adjusted so that they are all the same brightness at 550 nm (the wavelength of the center of the range transmitted by the visual filter). A hot star, with a spectrum like the 20,000 K Planck spectrum shown, gives off more light in the blue part of the spectrum than in the visual part of the spectrum. If we were to divide the brightness of the star as seen through the blue filter (written b_B) by the brightness of the star as seen through the visual filter (written b_V), the ratio b_B/b_V would be greater than 1. On the other hand, a cool star with a spectrum more like the 2,500 K Planck spectrum shown is much fainter in the blue part of the spectrum than in the visual part of the spectrum. The ratio of blue light to visual light b_B/b_V is less than 1 for the cool star. This ratio of brightness between the blue and visual filters is referred to as the **b_B/b_V color** of the star. By convention astronomers usually adjust the way they measure the blue and visual brightnesses of stars so that the star Vega (which has a surface temperature of around 10,000 K) has a b_B/b_V color of 1. On this scale, the Sun has a b_B/b_V color of 0.56.

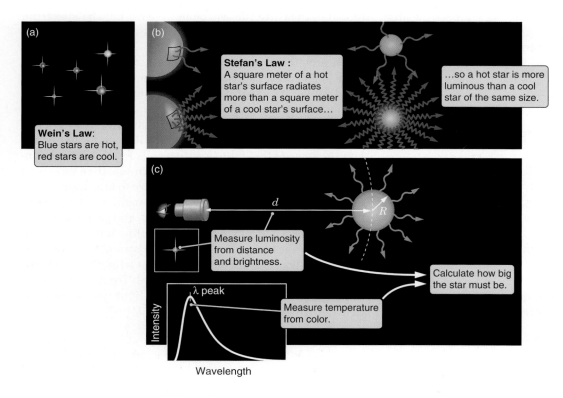

FIGURE 13.6 The temperature and size of stars are measured by applying our understanding of Planck radiation.

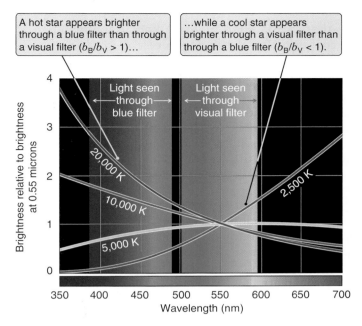

A hot star appears brighter through a blue filter than through a visual filter ($b_B/b_V > 1$)...

...while a cool star appears brighter through a visual filter than through a blue filter ($b_B/b_V < 1$).

FIGURE 13.7 A star's b_B/b_V color depends on its temperature. The Planck spectra shown here are adjusted so they have the same brightness at 0.55 microns.

The fact that the b_B/b_V color of a star depends on the star's temperature is an extremely handy tool for astronomers. It means that a single pair of snapshots of a group of stars, each taken through a different filter, is enough to allow us to measure the surface temperature of every star in our camera's field of view! In this way it is often possible to measure the temperatures of hundreds or even thousands of stars at once. When we do, we find that just as there are many more low-luminosity stars than high-luminosity stars, there are also many more cool stars than hot stars. Most stars have surface temperatures lower than that of the Sun, as you can see in Figure 13.5.

Stefan's Law Revisited: Finding the Sizes of Stars

Once we know the temperature of a star, we also know how much radiation each square meter of the star is giving off each second. As shown in Figure 13.6(b), each square meter of the surface of a hot blue star gives off more radiation per second than a square meter of the surface of a cool red star. A hot star will be more luminous than a cool star of the same size. A small hot star might even be more luminous than a larger cool star. As noted in Figure 13.6(c), we can use the relationship between temperature and luminosity of each square meter of a surface to infer the sizes of stars.

According to Stefan's Law (see Chapter 4), the amount of energy radiated each second by each square meter of the surface of a star is equal to the constant σ times the surface temperature of the star raised to the fourth power. Written as an equation, this says

$$\text{Energy radiated each second by 1 m}^2 \text{ of surface} = \sigma T^4.$$

To find the total amount of light radiated each second by the star, we need to multiply the radiation per second from each square meter times the number of square meters of the star's surface:

$$\begin{array}{ccc} \text{Total energy} & \text{Energy radiated} & \text{Number of square} \\ \text{radiated by the} = \text{by a square meter} \times \text{meters of surface} . \\ \text{star each second} & \text{each second} & \text{of the star} \end{array}$$

The left item in this equation—the total energy emitted by the star per second (in units of joules per second, abbreviated J/s)—is the star's luminosity L. The middle item in this equation—the energy radiated by each square meter of the star in a second [in units of joules per square meter per second, or J/(m²/s)]—can be replaced with the σT^4 from Stefan's Law. The remaining item in the equation—the number of square meters covering the surface of the star—is the surface area of a sphere, $A_{\text{sphere}} = 4\pi R^2$ (in units of m²), where R is the radius of the star. If we replace the words in the equation with the appropriate mathematical expressions for Stefan's Law and the area of a sphere, our equation for the luminosity of a star looks like this:

$$\text{Luminosity} = \left(\frac{\text{J}}{\text{m}^2\,\text{s}}\right) \times \text{m}^2$$

$$= (\sigma T^4) \times (4\pi R^2)$$

Combining the right sides of these two equations, we get

$$L = 4\pi R^2 \sigma T^4 \quad \text{in units of J/s (watts)}.$$

It is worth staring at this equation for a minute to see the sense behind it. Because the constants (4, π, and σ) do not change, the luminosity of a star is proportional only to $R^2 T^4$. The R^2 makes sense: The surface area of a star is proportional to the square of the radius of the star. Make a star three times as large, and its surface area becomes $3^2 = 9$ times as large. There is nine times as much area to radiate,

> **The larger and hotter a star is, the more luminous it will be.**

so there is nine times as much radiation. The effect of T^4 is understandable as well: Make a star twice as hot, and each square meter of the star's surface radiates $2^4 = 16$ times as much energy. (As we have seen before, an apparently complex mathematical expression is really a shorthand way of writing our understanding of what is going on physically

—in this case the commonsense result that larger, hotter stars are more luminous than smaller, cooler stars.)

The previous equation answers the question "How luminous will a star of a given size and a given temperature be?" Now turn this question around and ask, "How large does a star of a given temperature need to be to have a total

If the temperature and luminosity of a star are known, its size can be calculated.

luminosity of L?" Rearrange the last equation, moving the things that we know how to measure (temperature and luminosity) to the right side of the equation, and the thing that we would like to know (the radius of the star) to the left side. After a couple of steps of algebra we find

$$R = \sqrt{\frac{L}{4\pi\sigma}} \times \frac{1}{T^2}.$$

Again, the right side of the equation contains only things that we know or can measure. The constants 4, π, and σ are always the same. L is the luminosity of the star and can be found by combining measurements of the star's brightness and parallax. T is the surface temperature of the star, which can be measured from its color. From these we now know something new: the size of the star. We will refer to this last equation as the **luminosity–temperature–radius relationship** for stars.

The luminosity–temperature–radius relationship has been used to measure the sizes of many thousands of stars. When we talk about the sizes of stars, we again use our Sun as a yardstick. The radius of the Sun, written $R_\odot$, is 696,000 km, or about 700,000 km. When we look at stars around us, we find that the smallest stars that we see, called

There are many more small stars than large stars.

white dwarfs, have radii that are only about 1 percent of the Sun's radius ($R = 0.01\ R_\odot$). The largest stars that we see, called *red supergiants,* can have radii more than a thousand times that of the Sun. There are many more stars toward the small end of this range, smaller than our Sun, than there are giant stars.

Stars Are Classified According to Their Surface Temperature

So far we have concentrated on what we can learn about stars by applying our understanding of thermal radiation, an approach that seems from the previous section not to have led us too far astray. However, the spectra of stars are far from perfect Planck spectra. Instead, when we look at the spectra of stars, we see a wealth of absorption lines and

Absorption lines form as light escapes through the atmosphere of the star.

occasionally emission lines. In Chapter 4 we found that absorption lines occur when light passes through a cloud of gas: The atoms and molecules of the gas absorb light of certain specific wavelengths, characteristic of the kind of atom or molecule that is doing the absorbing. Similarly, the atoms and molecules in diffuse hot gas will emit light of specific wavelengths as well. Both of these processes are at work in stars.

Although the hot surface of a star emits radiation with a spectrum very close to a smooth Planck curve, this light must then escape through the outer layers of the star's atmosphere. The atoms and molecules in the atmosphere of the star leave their absorption line fingerprints in this light, as shown in **Figure 13.8**. The atoms and molecules in the star's atmosphere and any gas that might be found in the vicinity of the star can also produce emission lines in stellar spectra. Although absorption and emission lines complicate how we use the laws of Planck radiation to interpret light from stars, spectral lines more than make up for this trouble by providing a wealth of information about the state of the gas in a star's atmosphere.

The spectra of stars were first classified during the late 1800s, well before stars, atoms, or radiation were well un-

Stars are classified by the appearance of their spectra.

derstood. Stars were classified not on the basis of their physical properties, but on the appearance of the dark bands (now known as absorption lines) seen in their spectra. The original ordering of this classification was arbitrarily based on the prominence of particular absorption lines known to be associated with the element hydrogen. Stars with the strongest hydrogen lines were labeled *A stars,* stars with somewhat weaker hydrogen lines were labeled *B stars,* and so on.

This classification scheme was refined and turned into a real sequence of stellar properties in the early 20th century. The new system, originally proposed in 1901, was the work of **Annie Jump Cannon** (1863–1941), who led an effort at the Harvard Observatory to systematically examine and classify the spectra of hundreds of thousands of stars. She dropped many of the earlier spectral types, keeping only seven, which she reordered into a sequence that was no longer arbitrary, but instead was based on surface temperatures. Spectra of stars of different types are shown in **Figure 13.9**. The hottest stars, with surface temperatures over 30,000 K, are labeled O stars. O stars show relatively featureless spectra, with only weak absorption lines from hydrogen and helium. The coolest stars—M stars—have temperatures as low as about 2,800 K. M stars show myriad lines from many different types

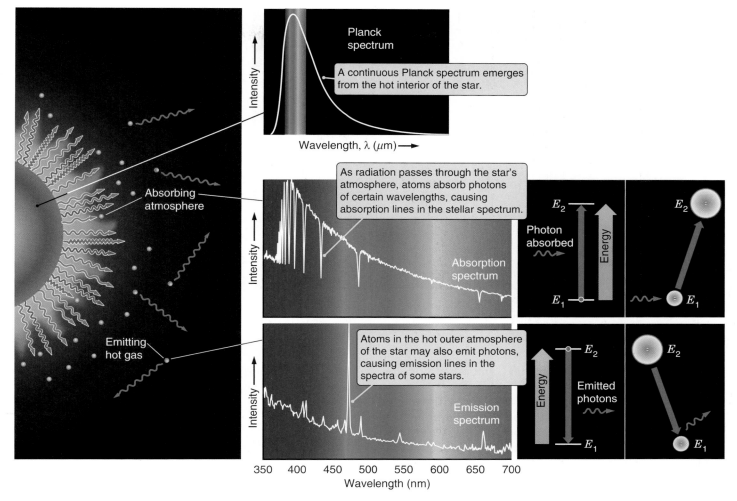

FIGURE 13.8 The formation of absorption and emission lines in the spectra of stars.

of atoms and molecules. The complete sequence of **spectral types** of stars, from hottest to coolest, is O, B, A, F, G, K, M; a mnemonic offered by Henry Norris Russell for remembering this sequence—*Oh Be A Fine Girl, Kiss Me*—has been a tool for students of astronomy for over 70 years. (About half of a typical class may want to substitute the word *Guy* in the appropriate location.) The spectral sequence of OBAFGKM has undergone several modifications over time. Recently two new spectral types, L and T, have been added to classify stars that are even cooler than M stars.

The scheme of spectral classification is discussed in more depth in **Foundations 13.1**. Because different spectral lines are formed at different temperatures, we can use the absorption lines in a star's spectrum to measure the star's temperature directly. The surface temperatures of stars measured in this way agree extremely well with the surface temperatures of stars measured using Wien's Law, again confirming a prediction of the cosmological principle: The same physical laws that apply on Earth apply to stars as well.

Stars Are Composed Mostly of Hydrogen and Helium

The most obvious differences in the lines seen in stellar spectra are due to temperature, but the details of the absorption and emission lines found in starlight carry a wealth of other information as well. By applying our knowledge of the physics of atoms and molecules to the study of stellar

---

Spectral lines are used to measure many properties of stars, including chemical composition.

---

absorption lines, we can determine accurately not only surface temperatures of stars, but also pressures, chemical compositions, magnetic field strengths, and other physical properties of stars. Also, by making use of the Doppler shift of emission and absorption lines, we can measure rotation rates, motions of the atmosphere, expansion and contrac-

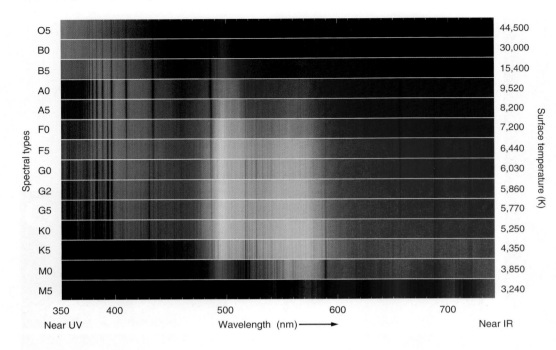

FIGURE 13.9 Spectra of stars with different spectral types, ranging from hot blue O stars to cool red M stars. Hotter stars are brighter at shorter wavelengths. The dark lines are absorption lines.

tion, "winds" driven away from stars, and other dynamic properties of stars.

As we continue to learn about stars, we will come to appreciate that one of the most interesting and important things about a star is its chemical composition. We already saw in Chapter 4 that if we look at a source of thermal radiation through a cloud of gas, the strength of various absorption lines tells us what kinds of atoms are present in the gas and in what abundance. Although we must take great care in interpreting spectra to properly account for the temperature and density of the gas in the atmosphere of a star, in most respects stars are ready-made laboratories for carrying out just such an experiment.

When analyzed in this way, most stars are found to have atmospheres that consist predominantly of the least massive elements. Hydrogen typically makes up over 90 percent of the atoms in the atmosphere of a star, with helium accounting for most of what remains. All of the other chemical elements,[3] which are collectively referred to as **heavy elements** or (more properly) as *massive elements,* are present in only trace amounts. (This would be a good time to go back and review Figure 9.5, the "astronomer's periodic table.") **Table 13.1** shows the chemical composition of the atmosphere of the Sun, which is fairly typical for stars in our vicinity. On the other hand, chemical composition can vary tremendously from star to star. In particular, some stars show even smaller amounts of elements other than hydrogen and helium in their spectra. The existence of such stars, all but devoid of more massive elements, provides important

clues about the origin of chemical elements and the chemical evolution of the universe.

13.4 Stellar Masses Are Measured by Analyzing Binary Star Orbits

The most important property of stars that we have yet to investigate is their mass. Determining the mass of an object can be a tricky business. We certainly cannot rely on the amount of light from an object or the object's size as a measure of its mass. Massive objects can be large or small, faint

> To measure mass, astronomers look for the effects of gravity.

or luminous. The only thing that *always* goes with mass is gravity. Mass is responsible for gravity, and gravity in turn affects the way masses move. When astronomers are trying to determine the masses of astronomical objects, they almost always wind up looking for the effects of gravity.

When discussing the orbits of the planets in Chapter 3, we found that Kepler's laws of planetary motion are the result of gravity. We even went so far as to show that the properties of the orbit of a planet about the Sun can be used to measure the mass of the Sun. If we could find a planet orbiting a distant star, we could apply the same technique to determine the mass of that star. Unfortunately, today's

---

[3] Astronomers often (although incorrectly) refer to all elements more massive than hydrogen and helium as *metals.*

FOUNDATIONS 13.1

Spectral Type

Figure 13.9 shows representative spectra of stars of different spectral types. You can see in this figure that not only are hot stars bluer than cool stars, but the absorption lines in their spectra are quite different as well. This is because differences in the temperature of the gas in the atmosphere of a star affect the state of the atoms in that gas, which in turn affects the energy level transitions available to absorb radiation.

At the hot end of the spectral classification scheme, the temperature in the atmospheres of O stars is so high that most atoms have had one or more electrons stripped from them by energetic collisions within the gas. There are few transitions available in these ionized atoms that cause absorption lines in the visible part of the electromagnetic spectrum, so the visible spectrum of an O star is relatively featureless. At lower temperatures there are more atoms in energy states that can absorb light in the visible part of the spectrum, so the visible spectra of cooler stars are far more complex than the spectra of O stars. Most absorption lines have an optimum temperature at which they are formed most strongly. For exam-

ple, absorption lines from hydrogen are most prominent at surface temperatures of about 10,000 K, which is the surface temperature of an A star. (This should be no surprise. A stars were so named because they are the stars with the strongest lines of hydrogen in their spectra.) At the very lowest stellar temperatures, atoms in the atmosphere of the star begin to react with each other, forming molecules. Molecules such as titanium oxide (TiO) are responsible for much of the absorption in the atmospheres of cool M stars.

It should be stressed that the sequence of spectral classification is really a sequence of surface temperatures, and that the boundaries between types are completely artificial. A hotter-than-average G star is very similar to a cooler-than-average F star. Astronomers break the main spectral types down into a finer sequence of subclasses by adding numbers to the letter designations. For example, the hottest B stars are called B0 stars, slightly cooler B stars are called B1 stars, and so on. The coolest B stars are B9 stars, which are only slightly hotter than A0 stars. The Sun is a G2 star.

---

telescopes are not yet powerful enough to directly see planets orbiting other stars, but we can watch as two *stars* orbit about each other. About half of the higher-mass stars in the sky are actually multiple systems consisting of several stars moving about under the influence of their mutual gravity. Most of these are **binary stars** in which two stars orbit each other in elliptical orbits as predicted by Kepler's laws. However, most low-mass stars are single, and low-mass stars far outnumber their higher-mass brethren. This means that *most stars are single*.

Binary Stars Orbit a Common Center of Mass

When discussing the motions of the planets, we usually say that the planets orbit around the Sun without worrying about the effect of their gravity on the Sun's motion. On the other hand, if two objects—say, two stars—are closer to the same mass, we have to abandon this simple mental

picture. Each object's motion will be noticeably affected by the gravitational force from the other, and we must worry about what happens when the two objects *orbit each other*. (In fact, as we learned in Chapter 6, it is primarily from the wobbles that they cause in the motions of stars that we know about planets beyond the Solar System.)

Think about what would happen if we were to set two unequal masses—m_1 and m_2—adrift in space with no motion of one mass relative to the other. As soon as we release the two masses, gravity will begin to pull them together. The force of mass 1 on mass 2 equals the force of mass 2 on mass 1, and each mass will begin to fall toward the other. But even though the forces on each mass are equal, the accelerations experienced by the two masses are not. Acceleration equals force divided by mass, so the *less massive object will experience the greater acceleration*.

Suppose, for the moment, that m_1 is three times as massive as m_2. That means that the acceleration of m_2 will be three times as great as the acceleration of m_1. At any given point in time, m_2 will be moving toward m_1 three times as fast as m_1 is moving toward m_2. When the two objects

TABLE 13.1

The Chemical Composition of the Sun*

Element	Percentage by Number	Percentage by Mass
Hydrogen	92.5%	74.5%
Helium	7.4%	23.7%
Oxygen	0.064%	0.82%
Carbon	0.039%	0.37%
Neon	0.012%	0.19%
Nitrogen	0.008%	0.09%
Silicon	0.004%	0.09%
Magnesium	0.003%	0.06%
Iron	0.003%	0.13%
Sulfur	0.001%	0.04%
Total of others	0.001%	0.03%

*The relative amounts of different chemical elements in the atmosphere of the Sun. The second column tells what fraction of the Sun's atoms are the listed element. The third column tells what fraction of the Sun's mass consists of a listed chemical element.

collide, m_2 will have fallen three times as far as m_1. The point where the two objects meet, called the *center of mass* of the two objects, will be three times as far from the original position of the less massive m_2 as than from the original position of the more massive m_1. (If the two objects were sitting on a seesaw in a gravitational field, the support of the seesaw would have to be directly under the center of mass for the objects to balance, as shown in **Figure 13.10**.)

If we take our two masses and give them some motion perpendicular to the line between them, then instead of simply falling into each other, they will instead orbit

---

Each star in a binary system follows an elliptical orbit around the system's center of mass.

---

around each other. When Newton applied his laws of motion to the problem of orbits, he found that two objects move in elliptical orbits with their common center of mass at one focus of both of the ellipses, as shown in **Figure 13.11**. The center of mass, which lies along the line between the two

objects, remains stationary. The two objects will always be found on exactly opposite sides of the center of mass, and the elliptical orbit of the more massive object is just a smaller version of the elliptical orbit of the less massive object.

Because the perimeter of the orbit of the less massive star is larger than that of the more massive star's orbit, the less massive star must be moving *faster* than the more mas-

---

The less massive star has a larger orbit and so must move faster.

---

sive star. The less massive star has farther to go than the more massive star, but it must cover that distance in the same amount of time. As a result, the velocity v of a star in a binary system is inversely proportional to its mass:

$$\frac{v_1}{v_2} = \frac{m_2}{m_1}.$$

FIGURE 13.10 The center of mass of two objects is the "balance" point on a line joining the centers of two masses.

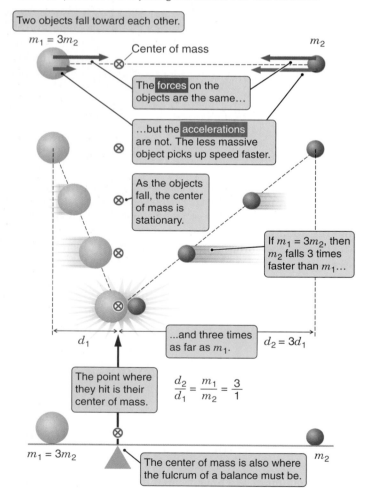

Two objects fall toward each other.

$m_1 = 3m_2$ Center of mass m_2

The forces on the objects are the same...

...but the accelerations are not. The less massive object picks up speed faster.

As the objects fall, the center of mass is stationary.

If $m_1 = 3m_2$, then m_2 falls 3 times faster than m_1...

...and three times as far as m_1.

d_1 $d_2 = 3d_1$

The point where they hit is their center of mass.

$$\frac{d_2}{d_1} = \frac{m_1}{m_2} = \frac{3}{1}$$

$m_1 = 3m_2$ m_2

The center of mass is also where the fulcrum of a balance must be.

from star 2 would be redshifted and lines from star 1 would be Doppler shifted to the blue. Because the two stars are exactly on opposite sides of their orbits, the two stars are always moving in opposite directions. The ratio of the masses of the two stars can be measured by comparing the size of the Doppler shift in the spectrum of star 1 with the size of the Doppler shift in the spectrum of star 2.

Kepler's Third Law Gives the Total Mass of a Binary System

In Chapter 3 we conveniently ignored all of this complexity having to do with the motion of two objects about their common center of mass. Now, however, it is this very complexity that allows us to measure the masses of the two stars in a binary system. In his derivation of Kepler's laws, Newton showed that if two objects with masses m_1 and m_2 are in orbit about each other, then the period of the orbit, P, is related to the average distance between the two masses, A, by the equation

$$P^2 = \frac{4\pi^2 A^3}{G(m_1 + m_2)}.$$

Rearranging this a bit turns it into an expression for the sum of the masses of the two objects:

$$m_1 + m_2 = \frac{4\pi^2}{G} \times \frac{A^3}{P^2}.$$

We can cleverly ignore the value of $4\pi^2/G$ by using what we know about Earth's orbit around the Sun. If $A = 1$ AU and $P = 1$ year, then we know that $m_1 + m_2 = M_\odot + M_\oplus \approx M_\odot$ (because $M_\odot$ is much larger than $M_\oplus$). So if we express the masses, A, and P in that equation in terms of Solar System units [such as $m_1 (M_\odot) = m_1/(1\ M_\odot)$, $A_{AU} = A/(1\ \text{AU})$, and $P_{years} = P/(1\ \text{year})$], then this equation simplifies to

$$m_1(M_\odot) + m_2(M_\odot) = \frac{(A_{AU})^3}{(P_{years})^2}.$$

We now know all that we need to know to measure the masses of two stars in a binary system. If we can measure the period of the binary system and the average separation between the two stars, then Kepler's third law gives us the total mass. Referring back to the previous subsection, if we can measure the sizes of the orbits of the two stars independently, or independently measure their velocities, then we

Kepler's third law gives the total mass of a binary system.

can determine the ratios of the masses of the two stars. If we know the total mass of the system as well as the ratio of the two masses, we have all we need to determine the mass of each star separately. To express this in symbols, if we

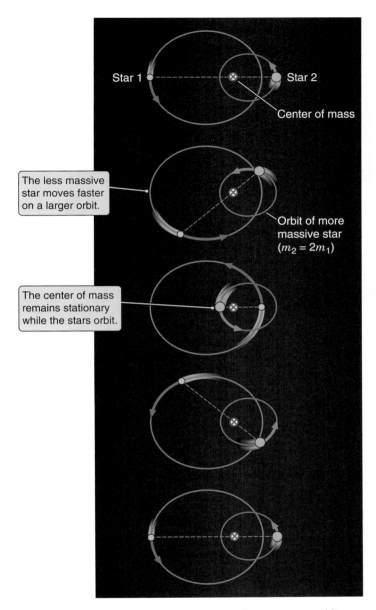

FIGURE 13.11 In a binary star system, the two stars orbit on elliptical paths about their common center of mass. In this case star 2 has twice the mass of star 1. The eccentricity of the orbits is 0.5. There are equal time steps between the frames.

This relationship between the velocities and the masses of the stars in a binary system is a crucial part of how binary stars are used to measure stellar masses. Imagine that you are watching a binary star edge-on, as shown in **Figure 13.12**. As one star is moving toward you, the other star will be moving away from you, and vice versa. If you were to look at the spectra of the two stars, you might see that the absorption lines from star 1 are redshifted relative to the overall center-of-mass velocity, while the absorption lines from star 2 are blueshifted. If you were to look again half an orbital period later, the situation would be reversed. Lines

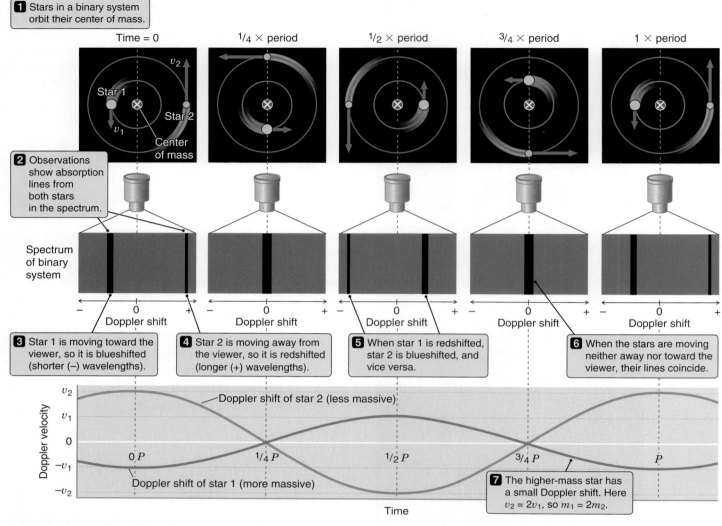

FIGURE 13.12 The orbits of two stars in a binary system are shown, along with the Doppler shifts of the absorption lines in the spectrum of each star. Star 1 is twice as massive as star 2 and so has half the Doppler shift.

know both m_1/m_2 and $m_1 + m_2$, a bit of algebra yields separate values for m_1 and m_2.

There are two ways to go about measuring the orbital properties that we need to determine the masses of stars. If a binary system is a **visual binary**—that is, if we can take pictures that show the two stars separately—then we can watch over time as the stars orbit each other. From these observations we can measure the shapes and period of the orbits of the two stars. When combined with Doppler measurements of the line-of-sight velocities of the stars, this is all the information we need. But in many binary systems, the two stars are so close together and so far away from us that we cannot actually see the stars separately. We know these stars belong to binary systems only because we see the spectral lines of the two stars as they are Doppler shifted away from each other first in one direction, and then in the

other direction. Yet even those binary stars that are known only by the Doppler shifts in their spectra can sometimes be used to measure the masses of the stars. If a binary star is an **eclipsing binary**, in which we see a dip in brightness as one star passes in front of the other (see **Figure 13.13**), then we know we are looking at the system edge-on. In this case, the peak Doppler shift of each star gives its total orbital velocity. The period of the orbit is given by the time it takes for a set of spectral lines to go from approaching to receding and back again. If we know the orbital velocities of the stars, and we know the period of the orbit, we also know the size of the orbit because distance equals velocity times time.

Historically most stellar masses were measured for stars in eclipsing binary systems because all that we need to determine the mass is a series of spectra of the system. Re-

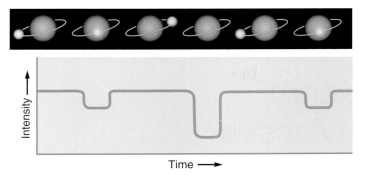

FIGURE 13.13 The shape of the light curve of an eclipsing binary can reveal information about the relative size and surface brightness of the two stars.

cently, however, new observational capabilities have greatly improved our ability to see the stars in a binary directly. As of this writing, accurate measurements of masses have been obtained for about 250 binary stars, about half of which are eclipsing binaries. (**Foundations 13.2** gives a concrete example of how the masses of two stars in an eclipsing binary are determined.) These measurements show that some stars are less massive than the Sun, whereas others are more massive. However, the range of stellar masses is not nearly as great as the range of stellar luminosities. The least massive stars have masses of about 0.08 $M_\odot$, while the most massive

stars probably have masses somewhere between 130 $M_\odot$ and 150 $M_\odot$. Thus while stellar luminosities change by a factor of 10^{10}, or 10 billion, from most luminous to least luminous, the most massive stars are only about 10^3, or 1,000, times more massive than the least massive stars.

You might wonder why the mass of a star should have any limits, but these lower and upper limits are determined solely by the physical processes that go on deep within the interior of the star. We will learn in the chapters ahead that a minimum stellar mass equal to 0.08 times that of the Sun is necessary to ignite the nuclear furnace that keeps a star shining, but the furnace can run out of control if the stellar mass is much greater than 100 times that of the Sun.

13.5 The H-R Diagram Is the Key to Understanding Stars

We have come a long way in our effort to measure the physical properties of stars. **Table 13.2** summarizes the techniques we have used. On the other hand, just knowing some of the basic properties of stars does not mean that we under-

TABLE 13.2

Taking the Measure of Stars*

Property	Method
Distance	Measure the parallax of the star over the course of the year.
Luminosity	For a star with a known distance, measure the brightness, then apply the inverse square law of radiation: $$\text{Luminosity} = 4\pi \times \text{Distance}^2 \times \text{Brightness}$$
Temperature	Measure the color of the star using b_B/b_V or from the star's spectrum. Use Wien's Law to relate the color to a temperature.
Size	For a star with known luminosity and temperature, use Stefan's Law to compute how large the star must be (the luminosity–temperature–radius relationship).
Mass	Measure the motions of the stars in a binary system, use these to determine the orbits of the stars, then apply Newton's form of Kepler's laws.
Composition	Knowing the temperature of a star, analyze the lines in its spectrum to measure chemical composition.

*A brief summary of the methods used to measure basic properties of stars.

FOUNDATIONS 13.2

Using a Celestial Bathroom Scale to Measure the Masses of an Eclipsing Binary Pair

It is worth looking at an example to see how the relationships discussed in the text are actually put to work. Suppose you are an astronomer studying a binary star system. After observing the star for several years, you build up the following information about the binary system:

1. The star is an eclipsing binary.

2. The stars are in circular orbits.

3. The period of the orbit is 2.63 years.

4. Star 1 has a Doppler velocity that varies between +20.4 km/s and –20.4 km/s.

5. Star 2 has a Doppler velocity that varies between +6.8 km/s and –6.8 km/s.

These data are summarized in **Figure 13.14**. You begin your analysis by noting that the star is an eclipsing binary, which tells you that the orbit of the star is edge-on to your line of sight. The Doppler velocities tell you the total orbital velocity of each star, and you determine the size of the orbits using the relationship Distance = Speed × Time. In one orbital period, star 1 travels a distance of

$$\left(20.4 \tfrac{\text{km}}{\text{s}}\right) \times \left(2.63 \text{ years}\right) \times \left(3.156 \times 10^7 \tfrac{\text{s}}{\text{year}}\right)$$

$$= 1.69 \times 10^9 \text{ km}.$$

This distance is 2π times the radius of the star's orbit, so star 1 is following an orbit with a radius of 2.7×10^8 km, or 1.8 AU. A similar analysis of star 2 shows that its orbit has a radius of 0.6 AU.

Next you apply Kepler's third law, which says that

$$m_1(M_\odot) + m_2(M_\odot) = \frac{(A_{\text{AU}})^3}{(P_{\text{years}})^2},$$

where A is the average distance between the two stars. Here you have A = 1.8 AU + 0.6 AU = 2.4 AU. Because you know A and the period P, you can calculate the total mass of the two stars:

$$m_1(M_\odot) + m_2(M_\odot) = \frac{(A_{\text{AU}})^3}{(P_{\text{years}})^2} = \frac{2.4^3}{2.63^2} = 2.0 \, M_\odot.$$

To sort out the individual masses of the stars, you use the fact that the more massive the star, the less it moves in response to the gravitational attraction of its companion:

$$\frac{m_2}{m_1} = \frac{v_1}{v_2} = \frac{20.4 \tfrac{\text{km}}{\text{s}}}{6.8 \tfrac{\text{km}}{\text{s}}} = 3.0,$$

stand stars. The next step in our journey involves looking for patterns in the properties we have determined. The first astronomers to take this step were a Dane by the name of **Einar Hertzsprung** (1873–1967) and the American astronomer mentioned earlier, **Henry Norris Russell** (1877–1957). In the early part of the 20th century (from 1906 to 1913, to be precise), Hertzsprung and Russell were independently studying the properties of stars. Each scientist plotted the luminosities of stars versus some measure of their surface temperatures (such as color or spectral type). The resulting plot is referred to as the **Hertzsprung-Russell diagram** or

simply the H-R diagram. The **H-R diagram** proved to be the Rosetta stone for understanding stars.

Take Time to Understand the H-R Diagram

Before going further in the text, it would be very much worth your while to study the H-R diagram carefully until you understand what it can tell you. *The H-R diagram is the*

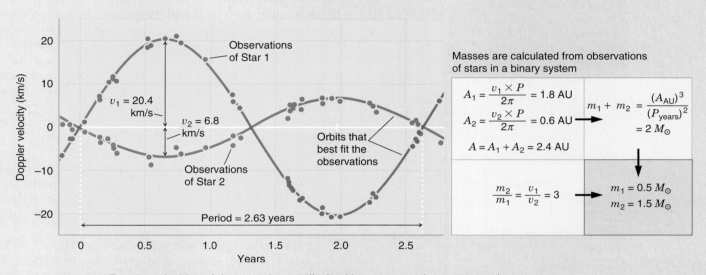

FIGURE 13.14 Doppler velocities of the stars in an eclipsing binary are used to measure the masses of the stars.

so star 2 is three times as massive than star 1. Algebra gets you the rest of the way. Substituting $m_2 = 3 \times m_1$ into the equation $m_1 + m_2 = 2.0\ M_\odot$ gives you

$$m_1 + (3 \times m_1) = 4 \times m_1 = 2.0\ M_\odot,$$

or $m_1 = 0.5\ M_\odot$. Substituting $m_1 = 0.5\ M_\odot$ into the equation $m_2 = 3 \times m_1$ gives you $m_2 = 1.5\ M_\odot$.

Star 1 has a mass of 0.5 $M_\odot$, and star 2 a mass of 1.5 $M_\odot$. You can tell your stars to step off the scale now.

The H-R diagram is the single most used and useful diagram in astronomy.

single most used and useful diagram in astronomy. Everything about stars, from their formation to their old age and eventual death, is discussed using the H-R diagram.

Begin with the layout of the H-R diagram itself, shown in **Figure 13.15**. Along the horizontal axis (the *x*-axis, if you will) we plot the surface temperature of stars, but it is plotted backward: Temperature starts out hot on the left side of the

diagram and *decreases* going to the right. Remember that. Hot blue stars are on the left side of the H-R diagram, while cool red stars are on the right side of the H-R diagram. This sequence is plotted *logarithmically*, which is to say that a step along the axis from a point representing a star with a surface temperature of 30,000 K to one with a surface temperature of 10,000 K (that is, a temperature change by a factor of 3) is the same as a step between points representing a star with a temperature of 9,000 K and a star with a temperature of 3,000 K (also a factor-of-3 change). The temperature axis can also be labeled with the spectral types or colors

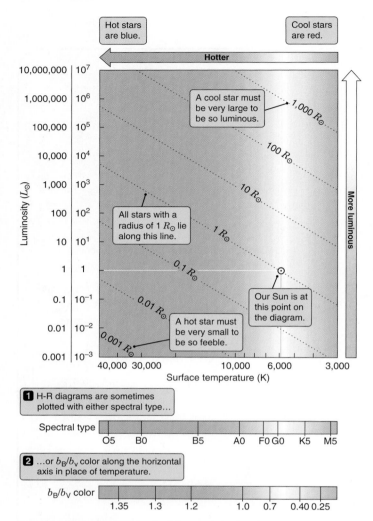

FIGURE 13.15 This is the layout of the Hertzsprung-Russell, or H-R, diagram used to plot the properties of stars. More luminous stars are at the top of the diagram. Hotter stars are on the left of the diagram. Stars of the same radius (R) lie along the dotted lines moving from upper left to lower right.

It turns out that the H-R diagram tells us about more than just the surface temperatures and luminosities of stars. Earlier in the chapter we discovered that if the temperature and the luminosity of a star are known, then we can calculate its radius. Because each point in the H-R diagram is specified by a surface temperature and a luminosity, we can use the luminosity–temperature–radius relationship to find the size of a star at that point as well. We can think through how this works using what we know about radiation. A star in the upper right corner of the H-R diagram is very cool, which according to Stefan's Law means that each square meter of its surface is radiating only a small amount of energy. But at the same time this star is extremely luminous. Such a star must be huge to account for its high overall luminosity despite the feeble radiation coming from each square meter of its surface. Conversely, a star in the lower left corner of the H-R diagram is very hot, which means that a large amount of energy is coming from each square meter of its surface. However, this star has a very low overall luminosity. The conclusion is that its surface area cannot be very large. Stars in the lower left corner of the H-R diagram are small.

This result persists everywhere in the H-R diagram. Moving up and to the right takes you to larger and larger stars. Moving down and to the left takes you to smaller and smaller stars. All stars of the same radius lie along lines running across the H-R diagram from the upper left to the lower right, as shown in Figure 13.15.

Like an experienced surveyor who "sees" the lay of the land when looking at a topographical map, or a classical musician who "hears" the rhythms and harmonies when looking at a piece of sheet music, you should get to the point where you automatically "see" the properties of a star—its temperature, color, size, and luminosity—from a glance at its position on the H-R diagram.

The Main Sequence Is a Grand Pattern in Stellar Properties

Figure 13.16(a) shows the H-R diagram for 16,600 nearby stars based on observations obtained by the Hipparcos satellite. A quick look at this diagram immediately reveals a remarkable fact. Instead of being strewn all about the diagram, we find instead that about 90 percent of the stars in

> Most stars lie along the main sequence of the H-R diagram.

the sky lie along a well-defined sequence running across the H-R diagram from lower right to upper left. This sequence of stars is called the **main sequence**. On the left end of the main sequence are the O stars: hotter, larger, and more luminous than the Sun. On the right end of the main sequence are the M stars: cooler, smaller, and fainter than the

of stars. Hot blue O stars with large b_B/b_V colors are on the left side of the diagram, while cool red M stars with small b_B/b_V colors are on the right side of the diagram.

Along the vertical axis (the y-axis) we plot the luminosity of stars—the total amount of energy a star radiates each second. This time the plot is the "right way" around: More luminous stars are toward the top of the diagram, while less luminous stars are toward the bottom. As with the temperature axis, luminosities are plotted logarithmically, in this case with each step along the axis corresponding to a multiplicative factor of 10 in the luminosity. To understand why the plotting is done this way, recall that the most luminous stars are 10 billion times more luminous than the least luminous stars, yet all of these stars must fit on the same plot.

Sun. If you know where a star lies on the main sequence, then you know its approximate luminosity, surface temperature, and size.

From a practical standpoint, the main sequence is extremely handy. For example, you may have wondered in our discussion of stellar parallax how we measure the distances to stars that are farther away than a few hundred light-years. The main sequence comes to the rescue. We can determine whether a star is a main sequence star by looking at the absorption lines in its spectrum. If a star is on the main sequence, then by virtue of its location on the H-R diagram you know its luminosity. If you know its luminosity and you measure its brightness, you can use the inverse square law of radiation to find

its distance. (How far away must a star of that luminosity be to have the brightness we measure?) This method of determining distances to stars is called **spectroscopic parallax**. Parallax and spectroscopic parallax are the first two steps along a chain of reasoning that will let us build our knowledge of distances all the way to the edge of the observable universe.

Figure 13.16(b) makes another important point for astronomy, as well as for all sciences. The red symbols are 46 of the nearest stars to Earth, whereas the blue symbols show the 97 brightest stars as seen in our sky. There are many more cool, low-luminosity stars in the sky than there are hot, high-luminosity stars. If we restrict our attention to stars that are near the Sun, these are all that we see. Yet if

FIGURE 13.16 (a) An H-R diagram for 16,600 stars obtained by the Hipparcos satellite. Most of the stars lie in a band running from the upper left of the diagram toward the lower right called the main sequence. (Dot color represents number of stars, with blue indicating the fewest and red the most.) (b) An H-R diagram for two different samples of stars. The red symbols show the H-R diagram for 46 stars that are especially close to the Sun. The blue symbols show the 97 brightest stars in the sky. Note that because these are observational H-R diagrams they are plotted against observed quantities, b_B/b_V color in (a) and spectral type in (b).

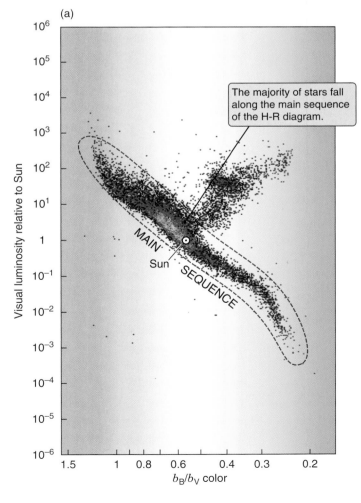

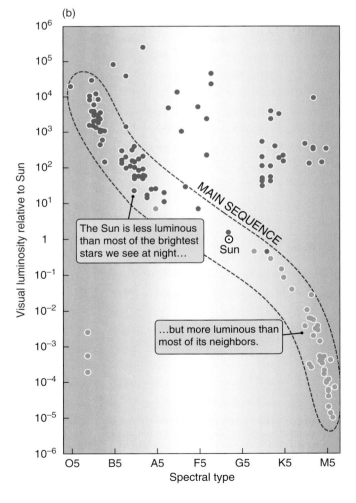

we look at the brightest stars in the sky, we get a very different picture. Even though these stars are farther away, they are so luminous that they dominate what we see. Think about how different your impression of the properties of stars would be if all you knew about were the nearest stars, or if all you knew about were the brightest stars. Neither of these groups alone accurately represents what stars as a whole are like. Whether you are an astronomer studying the properties of stars or a political pollster measuring the sense of the people, the validity of your results depends on the care with which you choose the sample (of stars or people) that you study.

When we add mass to the picture, the main sequence of the H-R diagram gets even more interesting. Stellar mass increases smoothly as we go from the lower right to the upper left along the main sequence. The faint, cool stars on the right side of the main sequence have low masses, whereas the luminous, hot stars on the left side of the main sequence are high-mass stars. Stated another way, if we know the mass of a main sequence star, then we know where on the main sequence the star is, which means we know the star's approximate temperature, size, and luminosity. If a main

---

The mass of a main sequence star determines what its fate will be.

---

sequence star is less massive than the Sun, it will be smaller, cooler, redder, and less luminous than the Sun. It will be located to the lower right of the Sun on the main sequence. On the other hand, if a star is more massive than the Sun, it will be larger, hotter, bluer, and more luminous than the Sun. It will be located to the upper left of the Sun on the

TABLE 13.3

The Main Sequence of Stars*

Spectral Type	b_B/b_V Color[+]	Temperature (K)	Mass ($M_\odot$)	Radius ($R_\odot$)	Luminosity ($L_\odot$)
O5	1.36	44,500	60	17.8	794,000
B0	1.32	30,000	18	9.3	52,500
B5	1.17	15,400	5.9	3.8	832
A0	1.02	9,520	2.9	2.5	54
A5	0.87	8,200	2.0	1.74	14
F0	0.76	7,200	1.6	1.35	6.5
F5	0.67	6,440	1.3	1.2	3.2
G0	0.59	6,030	1.05	1.05	1.5
G2 (Sun)	0.56	5,860	1.00	1.00	1.0
G5	0.53	5,770	0.92	0.93	0.8
K0	0.47	5,250	0.79	0.85	0.4
K5	0.35	4,350	0.67	0.74	0.15
M0	0.27	3,850	0.51	0.63	0.08
M5	0.22	3,240	0.21	0.32	0.011
M8	0.19	2,640	0.06	0.13	0.0012

*Approximately 90 percent of the stars in the sky fall on the main sequence of the H-R diagram. This table gives the properties of stars along the main sequence.
[+]Normally astronomers refer to the "B-V color" of a star as the difference between the B and V magnitudes of the star. To avoid the complication of magnitudes, in this text we instead refer to the brightness ratio b_B/b_V as the "b_B/b_V color" of the star.

Low-mass main sequence stars are faint and cool. High-mass main sequence stars are hot and luminous.

main sequence. *The mass of a star determines where on the main sequence the star will lie.*

To see this directly, look at **Table 13.3** and **Figure 13.17**, which show the properties of stars along the main sequence. If, for example, a main sequence star has a mass of 18 $M_\odot$, it will be a B0 star. It will have a surface temperature of 30,000 K, a radius of over 9 $R_\odot$, and a luminosity around 50,000 times that of the Sun. If a main sequence star instead has a mass of 0.21 $M_\odot$, it will be an M5 star. It will have a surface temperature of 3,240 K, a radius of about 0.32 $R_\odot$, and a luminosity of about 0.01 $L_\odot$. A main sequence star with a mass of 1 $M_\odot$ will be a G2 star like the Sun and will have the same surface temperature, size, and luminosity as the Sun.

It is difficult to overemphasize the importance of this result, so we state it again for clarity. For stars of similar chemical composition, *the mass of a main sequence star alone determines all of its other characteristics.* **Figure 13.18** illustrates this result. Knowing the mass (and chemical composition) of a main sequence star tells us how large it is, what its surface temperature is, how bright it is, what its internal structure is, how long it will live, how it will evolve, and what its final fate will be!

Upon reflection, this result is both sensible and possibly the most important and fundamental result in all of astrophysics. If you have a certain amount and type of material to make a star, there is only one kind of star you can make. As we go on to discuss what physical processes give a star its structure, this will make even more sense. We will find that a star is a "battle" between gravity (which is trying to pull the star together) and the energy released by nuclear reactions in the interior of the star (which are trying to blow

it apart). The mass of the star determines the strength of its gravity, which in turn determines how much energy must be generated in its interior to prevent it from collapsing under its own weight. The mass of a star determines where the balance is struck.

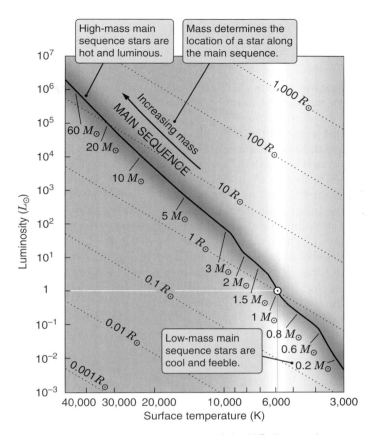

FIGURE 13.17 The main sequence of the H-R diagram is a sequence of masses.

FIGURE 13.18 Plots of luminosity, radius, and temperature versus mass for stars along the main sequence. The mass (and chemical composition) of a main sequence star determines all of its other properties.

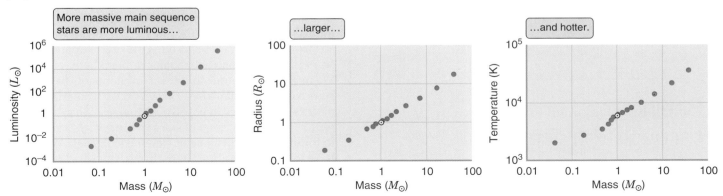

Not All Stars Are Main Sequence Stars

Before leaving our discussion of the observed properties of stars, we need to point out that although most stars in the sky are main sequence stars, some are not. Some stars are found in the upper right portion of the H-R diagram, well above the main sequence. From their position we know that they must be bloated, luminous, cool giants, with radii hundreds or thousands of times the radius of the Sun. If the Sun were such a star, its atmosphere would swallow the orbits of the inner planets, including Earth. At the other extreme are stars found in the extreme lower left corner of the H-R diagram. These stars must be tiny, with sizes comparable to the size of Earth. Their small surface areas explain why they have such low luminosities despite having temperatures that can rival or even exceed the surface temperature of the hottest main sequence O stars.

The existence of the main sequence, together with the fact that the mass of a main sequence star controls where on the main sequence it will lie, are grand patterns that point to the possibility of some deep understanding of what stars are and what makes them tick. By the same token, the existence of stars that do *not* follow these grand patterns raises yet more questions. What is it about a star that determines whether or not it is part of the main sequence? In the decades that followed the discovery of the main sequence, few problems in astronomy attracted more attention than these questions. Their answers turned out to be as fundamental as stellar pioneers Annie Jump Cannon, Einar Hertzsprung, and Henry Norris Russell could ever have hoped. The existence of the main sequence holds the essential clue to what stars are and how they work. The properties of stars that are not on the main sequence point to an understanding of how stars form, how they evolve, and how they die. Much of the rest of the text discussing stars will be spent on the lessons learned from patterns in the H-R diagram.

Summary

- The distances to nearby stars are measured stereoscopically by their parallax.

- The nearest star (other than the Sun) is Proxima Centauri at a distance of 4.22 light-years.

- When parallax measurements fail, distances to stars must be inferred from physical characteristics.

- Radiation tells us the temperature, size, and composition of stars.

- There are many more small, cool stars than large, hot stars.

- The H-R diagram is the single most useful diagram in astronomy.

- The mass and composition of a main sequence star determine its luminosity, temperature, and size.

- The mass of any star determines its ultimate fate.

Seeing the Forest through the Trees

When you walk down a path, each step you take covers only a short distance. However, by persistently putting one foot in front of the other, you find after a time that you have gotten somewhere. The same is true for the road we have traveled in this chapter. Each step along the way has been relatively small and understandable—involving the application of a physical principle that we see at work in the world around us, or the use of a tool from algebra or geometry. However, when we compare our understanding of stars reached by the end of the chapter with the understanding we started out with, it is amazing how far we have come.

Our path in this chapter followed the course of the triumph of our physical understanding of the universe. The properties of electromagnetic radiation, the structure of atoms and molecules, the gravitational attraction between masses—all of these and more came into play as we built up our understanding of the physical nature of stars one piece at a time. We saw many occasions in this chapter when the cosmological principle might have collapsed. We might have seen emission and absorption lines in the spectra in stars that were different from those measured in terrestrial laboratories. We might have found that the temperatures of stars inferred from their absorption spectra disagreed with the temperatures of stars inferred from their peaks in blackbody emission. We might have found that measurements of the sizes of stars based on Stefan's Law disagreed with measurements of the sizes of stars in eclipsing binaries. We might have found that the motions of stars in binary systems failed to follow the predictions of Newton's physics. We might have found any or all of these things, *but we did not!* The successes of this chapter give us strong reason to believe that the same physical laws at work right here on Earth and in our Solar System also describe the fundamental character and behavior of matter and energy throughout the rest of the universe.

With an understanding of the basic physical properties of stars in place, we are now ready to ask much more fundamental questions about stars. We are ready, figuratively speaking, to "lift the hood" and see what lies within. How do stars work? How do they form? How do they evolve? How do they die? We will begin to address these questions by investigating the star that serves as the standard by which we measure other stars: the star we know best, our Sun.

Key Terms

Student Questions

THINKING ABOUT THE CONCEPTS

1. Distance to stars can be measured in inches, miles, kilometers (km), astronomical units (AU), light-years (ly) and parsecs (pc). Why do most astronomers prefer to use parsecs?

2. The distances of nearby stars are determined by their parallaxes. Why is the uncertainty in our knowledge of a star's distance greater for stars that are farther from Earth?

3. To know certain properties of a star, you first must determine the star's distance. For other properties, knowledge of distance is not necessary. Into which category would you place each of the following properties: size, mass, temperature, color, spectral type, and chemical composition? In each case state your reason(s).

4. Albiero, a star in the constellation of Cygnus, is a binary system whose two components can be seen easily with even a small amateur telescope. Viewers describe the brighter star as "golden" and the fainter one as "sapphire blue."
   a. What does this tell you about the relative temperatures of the two stars?
   b. What does it tell you about their respective sizes?

5. Explain why the stellar spectral types (O, B, A, F, G, K, M) are not in alphabetical order. Also explain the sequence of temperatures defined by these spectral types.

6. Logarithmic (log) plots show major steps along an axis scaled to represent equal factors, most often factors of 10. Why, in astronomy, do we sometimes use a log plot instead of the more conventional linear plot?

7. How do we estimate the mass of stars that are not in eclipsing or visual binary systems?

8. Other than the Sun, the only stars whose mass we can measure directly are those in eclipsing or visual binary systems. Explain why.

9. Although we tend to think of our Sun as an "average" main sequence star, it is actually hotter and more luminous than average. Explain.

10. Star masses range from 0.08 $M_\odot$ to about 100 $M_\odot$.
    a. Why are there no stars with masses less than 0.08 $M_\odot$?
    b. Why are there no stars with masses much greater than 100 $M_\odot$?

APPLYING THE CONCEPTS

11. Sketch a circle and mark an arc along the circumference equal to the radius of the circle. The size of the angle subtended by the arc, measured in *radians,* is given by the length of the arc divided by the radius (r) of the circle. Because the circumference of a circle is $2\pi r$, there must be $2\pi r/r$ or 2π radians in a circle.
    a. How many degrees are there in one radian?
    b. How many arcseconds are there in one radian? How does this compare to the number of AU in one parsec (206,264.81) that we cite in the text?
    c. What angle would a round object that has a diameter of 1 parsec make in our sky if we see it at a distance of 1 parsec? A distance of 10 parsecs?

d. What do your answers to parts (b) and (c) of this question tell you about how the actual size of an object is related to its angular size measured in units of radians?

12. Our eyes are typically 6 cm apart. Suppose your eye separation is average, and you see an object jump from side to side by one-half degree as you blink back and forth between your eyes. How far away is that object?

13. Sirius, the brightest star in the sky, has a parallax of 0.379 arcseconds. What is its distance in parsecs? In light-years?

14. Sirius is 22 times more luminous than the Sun, and Polaris (the "North Pole Star") is 2,350 times more luminous than the Sun. Sirius appears 23 times brighter than Polaris. How much farther away from us is Polaris than is Sirius? What is the distance of Polaris in light-years?

15. Proxima Centauri, the star nearest to Earth other than the Sun, has a parallax of 0.772 arcseconds. How long does it take light to reach us from Proxima Centauri?

16. The Sun is about 16 trillion (1.6×10^{13}) times brighter than the faintest stars visible to the naked eye.
    a. How far away (in AU) would an identical solar-type star be if it were just barely visible to the naked eye?
    b. What would be its distance in light-years?

17. Sirius and its companion orbit around a common center of mass with a period of 50 years. The mass of Sirius is 2.35 times the mass of the Sun.
    a. If the orbital velocity of the companion is 2.35 times greater than that of Sirius, what is the mass of the companion?
    b. What is the semimajor axis of the orbit?

18. Assume that the main sequence for stars is determined by the relation $L = \text{Constant} \times T^7$. How would you then express the main sequence as a relation between radius and temperature?

It is stern work, it is perilous work to
    thrust your hand in the sun
And pull out a spark of immortal flame
    to warm the hearts of men.

JOYCE KILMER (1886–1918)

A sunset in New York City.

CHAPTER 14

A Run-of-the-Mill G Dwarf: Our Sun

14.1 The Sun Is More Than Just a Light in the Sky

How wonderful, after a long cold night, to see the light and feel the warmth of the rays of the Sun. Energy from the Sun is responsible for daylight, for our weather and seasons, and for terrestrial life itself. No object in nature has been more revered or more worshipped than the Sun. In fact, the modern symbol for the Sun, ☉, is nothing other than the ancient Egyptian hieroglyph for the Sun god Ra. Yet although the Sun may have dominated human consciousness since the dawn of our species, the discussion in Chapter 13 offers a very different perspective. To an astronomer at the opening of the 21st century, the Sun is the prototype for main sequence stars. With a middle-of-the-road spectral type of G2, the Sun is all but indistinguishable from billions of other stars in our galaxy. It is also the star against which all other stars are measured. The mass of the Sun, the size of the Sun, the luminosity of the Sun—these basic properties of the Sun provide the yardsticks of modern astronomy.

The Sun may be run-of-the-mill as far as stars go, but that makes it no less awesome an object on a human scale. The mass of the Sun, 1.99×10^{30} kilograms, is over 300,000 times that of Earth. The Sun's radius of 696,000 km is over 100 times that of Earth. At a luminosity of 3.85×10^{36} watts, the Sun produces more energy in a second than all of the electrical power plants on Earth could generate in 10 million years. The Sun is also the only star we can study at close range. Much of the detailed information that we know about stars has come only by studying our local star. Even though the Sun has been intensively studied for quite some time,

KEY CONCEPTS

To most humans the Sun is the most important object in the heavens. It lights our days, warms our planet, and provides the energy for life. But to astronomers the Sun is a typical main sequence star, located conveniently nearby for detailed study. In this chapter, as we take a closer look at our local star, we will learn about

- The balances between pressure and gravity and between energy generation and loss that determine the structure of the Sun.

- Fusion of hydrogen to helium, and how mass is efficiently converted into energy in the Sun's core.

- The different ways that energy moves outward from the Sun's core toward its surface.

- Physical models of the Sun's interior and how they are tested using observations of solar neutrinos and seismic vibrations on the surface of the Sun.

- The structure of the Sun's atmosphere, from its 5,770 K photosphere to its 1 million K corona.

- Sunspots, flares, coronal mass ejections, and other consequences of magnetic activity on the Sun.

- Eleven- and 22-year cycles in solar activity.

- The solar wind streaming away from the Sun.

- How solar activity affects Earth.

the wealth of phenomena displayed by the Sun will keep solar astronomers busy for many years to come.

In the previous leg of our journey, we looked at the gross physical properties of distant stars, including their mass, luminosity, size, temperature, and chemical composition. Now, as we turn our attention toward our own local star, we face more fundamental questions. How does the Sun work? Where does it get its energy? Why does it have the size, temperature, and luminosity that it does? How has it been able to remain so constant over the billions of years since the Solar System formed? In short, we now confront the question "What is a star?"

14.2 The Structure of the Sun Is a Matter of Balance

In Chapter 7 we tackled the question of how we know about the interior of Earth, despite the fact that no machine has ever done more than scratch the surface of our planet. The answer was a combination of physical understanding, detailed computer models, and clever experiments that test the predictions of those models. The task of exploring the interior of the Sun is much the same. As with Earth, the structure of the Sun is governed by a number of physical processes and relationships. Using our understanding of physics, chemistry, and the properties of matter and radiation, we express these processes and relationships as mathematical equations. High-speed computers are then used to simultaneously solve these equations and arrive at a model of the Sun. One of the great successes of 20th century astronomy was the successful construction of a physical model of the Sun that agrees with our observations of the mass, composition, size, temperature, and luminosity of the real thing.

Our current model of the interior of the Sun is the culmination of decades of work by thousands of physicists and astronomers. Understanding the details that lie within this model is the work of lifetimes. Even so, the essential ideas underlying our understanding of the structure of the Sun are found in a few key insights. In turn, these insights can be summed up in a single statement: *The structure of the Sun is a matter of balance.*

The first key balance within the Sun is the balance between pressure and gravity illustrated in **Figure 14.1**. The Sun is a huge ball of hot gas. If gravity were stronger than pressure within the Sun, the Sun would collapse. Likewise, if pressure were stronger than gravity, the Sun would blow itself apart. We have seen this balance before and have given it a name—*hydrostatic equilibrium* (see Foundations 7.2). Hydrostatic equilibrium sets the pressure at each point within a planet and determines the atmospheric pressure

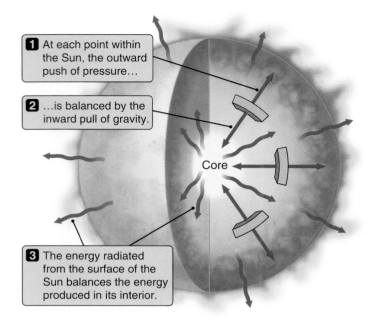

1 At each point within the Sun, the outward push of pressure…

2 …is balanced by the inward pull of gravity.

Core

3 The energy radiated from the surface of the Sun balances the energy produced in its interior.

FIGURE 14.1 The structure of the Sun is determined by balances between forces and in the outward flow of energy.

at Earth's surface. Hydrostatic equilibrium says that the pressure at any point within the Sun's interior must be just enough to hold up the weight of all the layers above that point. If the Sun were not in hydrostatic equilibrium, then

The Sun is in hydrostatic equilibrium.

forces within the Sun would not be in balance, so the surface of the Sun would *move*. The Sun today is the same as it was yesterday and the day before. That is all the observation we need to infer that the interior of the Sun is in hydrostatic equilibrium.

Hydrostatic equilibrium becomes an even more powerful concept when combined with what we know about the way gases behave. As we move deeper into the interior of the Sun, the weight of the material above us becomes greater, and hence the pressure must increase. As we have learned before (see Foundations 8.2), in a gas, higher pressure means higher density and/or higher temperature. **Figure 14.2(a)** shows how conditions vary as distance from the center of the Sun changes. As we go deeper into the Sun, the pressure climbs; as it does, the density and temperature of the gas climb as well.

A second fundamental balance within the Sun is a balance of energy. Stars like the Sun are remarkably stable objects. Geological records show that the luminosity of the Sun has remained nearly constant for billions of years. In fact, the very existence of the main sequence says that stars do not change much over the main part of their lives. To remain in balance, the Sun must produce just enough en-

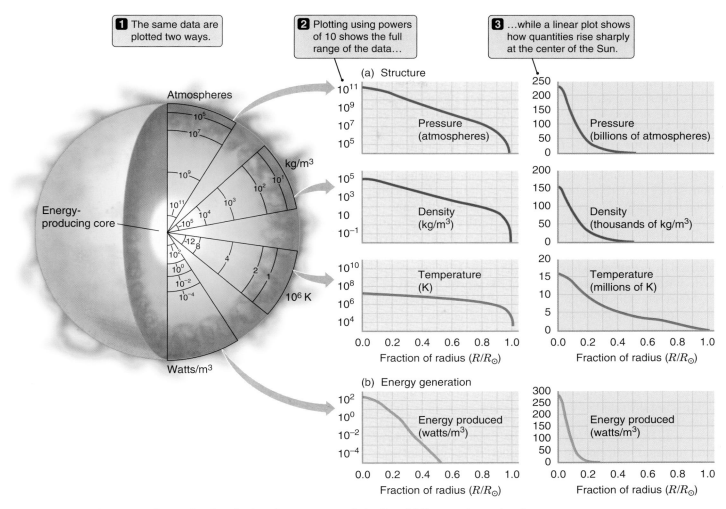

1 The same data are plotted two ways.

2 Plotting using powers of 10 shows the full range of the data...

3 ...while a linear plot shows how quantities rise sharply at the center of the Sun.

(a) Structure

(b) Energy generation

FIGURE 14.2 A cutaway figure showing the interior structure of the Sun. (a) Temperature, density, and pressure increase toward the center of the Sun. (b) Energy is generated in the Sun's core.

Solar energy production must balance what is radiated away.

ergy in its interior each second to replace the energy that is radiated away by its surface each second. This is a new type of balance, one we have not dealt with before. Understanding the balance of energy within the Sun requires thinking about how energy is generated in the interior of the Sun, and how that energy finds its way from the interior to the Sun's surface, where it is radiated away.

The Sun Is Powered by Nuclear Fusion

One of the most basic questions facing the pioneers of stellar astrophysics was where the Sun and stars get their energy. The answer to this question came not from astronomers' telescopes, but from theoretical work and the laboratories of nuclear physicists. At the heart of the Sun lies a nuclear furnace capable of powering the star for billions of years.

The nucleus of most hydrogen atoms consists of a single proton. Nuclei of all other atoms are built from a mixture of protons and neutrons. Most helium nuclei, for example, consist of two protons and two neutrons. Most carbon nuclei consist of six protons and six neutrons. Protons have a

Atomic nuclei are held together by the strong nuclear force.

positive electric charge, and neutrons have no net electric charge. Like charges repel, so all of the protons in an atomic nucleus must be pushing away from each other with a tremendous force because they are so close to each other. If electric forces were all there was to it, the nuclei of atoms would rapidly fly apart—yet atomic nuclei *do* hold together. We conclude that there must be some other force in nature, even stronger than the electric force that "glues" the protons and neutrons in a nucleus together. That force, which acts only over short distances, is called the **strong nuclear force**.

The strong nuclear force is indeed a very powerful force. It would take an enormous amount of energy to pull apart the nucleus of an atom such as helium into its constituent parts. The reverse of this process says that when you assemble an atomic nucleus from its component parts, this same enormous amount of energy is released. The process of combining two less massive atomic nuclei into a single more massive atomic nucleus is referred to as **nuclear fusion**. Many nuclear processes are possible, and as we continue our study of stars, we will find that a wide range of nuclear reactions can occur in stars. In the Sun, as in all main sequence stars, the only significant process going on is the fusion of hydrogen to form helium—a process often referred to as **hydrogen burning** (even though it has nothing to do with fire in the usual sense of the word).

Main sequence stars get their energy by fusing hydrogen atoms together to make helium. To judge the effectiveness of this reaction, we can make use of one of the key results of special relativity (Chapter 4): the equivalence between mass and energy. Mass can be converted to energy and energy can be converted to mass, with Einstein's famous equation $E = mc^2$ providing the exchange rate between the two. Comparing the mass of the *products* of a reaction with the mass of the *reactants* tells us what fraction of the original mass was turned into energy in the process. The mass of four separate hydrogen atoms is 1.007 times greater than the mass of a single helium atom; so when hydrogen fuses to make helium, 0.7 percent of the mass of the hydrogen is converted to energy.

Conversion of 0.7 percent of the mass of the hydrogen into energy might not seem very efficient—until we compare it with other sources of energy and discover that it is millions of times more efficient than even the most efficient chemical reactions. Fusing a *single gram* of hydrogen into helium releases about 6×10^{11} joules of energy, which is enough to boil all of the water in about 10 average backyard swimming pools. Scale this up to converting roughly 600

Nuclear fusion is a very efficient source of energy.

---

million metric tons of hydrogen into helium every second (with 4 million metric *tons* of matter converted to energy in the process), and you have our Sun. The sunlight falling on Earth may be responsible for powering almost everything that happens on our planet, but it amounts to only about a hundred-billionth of the energy radiated by our local star. The Sun has been burning hydrogen at this prodigious rate for 4.6 billion years, during which time it has converted about half of the hydrogen at its center into helium. A favorite theme of science fiction is the fate that awaits Earth when the Sun dies, but we need not worry anytime soon.

The Sun is only about halfway through its 10-billion-year lifetime as a main sequence star.

Whether a ball rolling down hill, a battery discharging itself through a lightbulb, or an atom falling to a lower state by emitting a photon, most systems in nature tend to seek the lowest-energy state available to them. Going from hydrogen to helium is a big ride downhill in energy, so we might imagine that hydrogen nuclei would naturally tend to fuse together to make helium. Fortunately for us, however, a major roadblock stands in the way of nuclear fusion. The strong nuclear force responsible for binding atomic nuclei together can act only over very short distances—10^{-15} meter or so, or about a hundred-thousandth the size of an atom. To get atomic nuclei to fuse, they must be brought close enough to each other for the strong nuclear force to assert itself, but this is hard to do. All atomic nuclei have positive electric charges, which means that any two nuclei will repel each other. This electric repulsion, illustrated in **Figure 14.3**, serves as a barrier

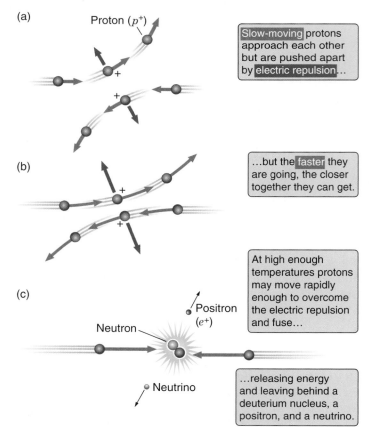

FIGURE 14.3 Atomic nuclei are positively charged and so repel each other. If two nuclei are moving toward each other, the faster they are going, the closer they will get before veering away, as shown in (a) and (b). (c) At the temperatures and densities found in the centers of stars, thermal motions of nuclei are so energetic that nuclei can overcome this electric repulsion, so fusion takes place.

(a) Proton (p^+)

Slow-moving protons approach each other but are pushed apart by electric repulsion...

(b) ...but the faster they are going, the closer together they can get.

(c) Positron (e^+)
Neutron
Neutrino

At high enough temperatures protons may move rapidly enough to overcome the electric repulsion and fuse...

...releasing energy and leaving behind a deuterium nucleus, a positron, and a neutrino.

against nuclear fusion. Fusion cannot take place unless this electric barrier is somehow overcome.

As shown in Figure 14.2(b), energy in the Sun is produced in its innermost region, called the Sun's *core*. Conditions in the Sun's core are extreme. Matter at the center of the Sun has a density about 150 times the density of water (which is 1,000 kg/m³), and the temperature at the center of the Sun is about 15 million K. The thermal motions of atomic nuclei in the Sun's core are tens of thousands times more energetic than the thermal motions of atoms at room temperature. As illustrated in Figure 14.3(c), under these conditions atomic nuclei slam into each other hard enough to overcome the electric repulsion between them and allow short-range nuclear forces to act. The hotter and denser a

Hydrogen fuses to helium in the core of the Sun.

gas, the more of these energetic collisions will take place each second. For this reason, the rate at which nuclear fusion reactions occur is extremely sensitive to the temperature and the density of the gas. Half of the energy produced by the Sun is generated within the inner 9 percent of the Sun's radius, or less than 0.1 percent of the volume of the Sun (Figure 14.2(b)).

There are several reasons that hydrogen burning is the most important source of energy in main sequence stars. Hydrogen is the most abundant element in the universe, so it offers the most abundant source of nuclear fuel at the beginning of a star's lifetime. Hydrogen burning is also the most efficient form of nuclear fusion, converting a larger fraction of mass into energy than any other type of reaction. But the most important reason why hydrogen burning is the dominant process in main sequence stars

is that hydrogen is also the easiest type of atom to fuse. Hydrogen nuclei—protons—have an electric charge of +1. The electric barrier that must be overcome to fuse protons is the repulsion of a single proton against another. Compare this to the force required, for example, to get two

Hydrogen burns mostly via the proton–proton chain.

carbon nuclei close enough to fuse. To fuse carbon we must overcome the repulsion of six protons in one carbon nucleus pushing against the six protons in another carbon nucleus. The resulting force is proportional to the product of the charges of the two atomic nuclei, making the repulsion between two carbon nuclei 36 times stronger than that between two protons. For this reason, hydrogen fusion occurs at a much lower temperature than any other type of nuclear fusion. In the cores of low-mass stars such as the Sun, hydrogen burns primarily through a process called the **proton–proton chain**. The dominant branch of the proton–proton chain is illustrated in **Figure 14.4** and discussed in **Foundations 14.1**.

Energy Produced in the Sun's Core Must Find Its Way to the Surface

Some of the energy released by hydrogen burning in the core of the Sun escapes directly into space in the form of neutrinos (see Foundations 14.1), but most of the energy goes instead into heating the solar interior. To understand the structure of the Sun we must understand how thermal energy is able to move outward through the star. The nature

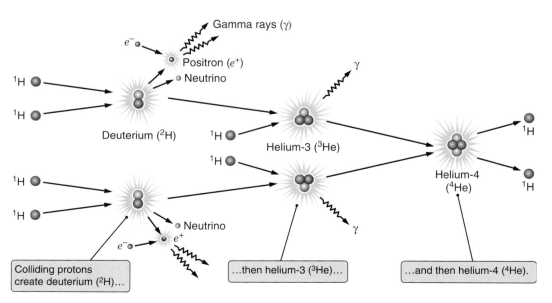

FIGURE 14.4 The Sun and all main sequence stars get their energy by fusing the nuclei of four hydrogen atoms together to make a single helium atom. In the Sun, about 85 percent of the energy produced comes from the branch of the proton–proton chain shown here.

of **energy transport** within a star is one of the key factors determining the star's structure.

Thermal energy can be transported from one place to another by a number of methods. Pick up a bucket of hot coals and carry it from one side of the room to the other, and you have transported thermal energy. A common way in which energy is transported in our everyday lives is by **thermal conduction**. For example, hold one end of a metal rod while you put the other end into a fire. Soon the end of the rod that is in your hand becomes too hot to hold. Thermal conduction occurs as the energetic thermal vibrations of atoms and molecules in the hot end of the rod cause their cooler neighbors to vibrate more rapidly as well. However, while thermal conduction is the most important way energy is transported in *solid* matter, it is typically ineffective in a *gas*. Thermal conduction is unimportant in the transport of energy from the core of the Sun to its surface. Thermal energy is instead carried outward from the center of the Sun by two other mechanisms: radiation and convection.

The transport of energy from one place to another by **radiative transfer** involves photons moving from hotter regions to cooler regions, carrying energy with them. Imagine a hotter region of a star sitting next to a cooler region, as shown in **Figure 14.5**. Recall from our study of radiation in

FOUNDATIONS 14.1

The Proton–Proton Chain

Hydrogen burning in the Sun and other low-mass stars takes place through a series of nuclear reactions called the proton–proton chain. There are three different "branches" to the proton–proton chain. The most important of these (Figure 14.4), responsible for about 85 percent of the energy generated in the Sun, consists of three steps. In the first step, two hydrogen nuclei fuse. In the process, one of the protons is transformed into a neutron by emitting a positively charged particle called a **positron** and another type of elementary particle called a **neutrino**. The conversion of a proton into a neutron by the emission of a positron and a neutrino is one variety of a process referred to as **beta decay**.

The positron is expelled at a great velocity, carrying away some of the energy released in the reaction. Electrons and positrons have opposite electric charges, so they attract each other. As a result, our expelled positron will soon collide with one of the many electrons moving freely about in the center of the Sun. But the positron is the **antiparticle** of the electron, and as we'll learn later in Chapter 21, when particle and antiparticle collide they annihilate each other, with their total mass being converted into energy. Thus the annihilation of electrons and positrons in the Sun's core produces energy in the form of gamma ray photons. These photons carry part of the energy released when the two protons fused, thereby helping to heat the surrounding gas. The neutrino, on the other hand, is a very elusive particle. Its interactions with matter are so feeble that its most likely fate is to escape from the Sun without further interactions with matter.

The new atomic nucleus formed by the first step in the proton–proton chain consists of a proton and a neutron. This is the nucleus of a heavy isotope of hydrogen called *deuterium*, or ^{2}H. The second step of the proton–proton chain occurs when another proton slams into the deuterium nucleus, fusing with it to form the nucleus of a light isotope of helium, ^{3}He, consisting of two protons and a neutron. The energy released in this step is carried away as a gamma ray photon. The third and final step in the proton–proton chain occurs when two ^{3}He nuclei collide and fuse together, producing an ordinary ^{4}He nucleus and ejecting two protons in the process. The energy released in this step shows up as kinetic energy of the helium nucleus and two ejected protons.

This dominant branch of the proton–proton chain can be written symbolically as

$$^1H + {}^1H \rightarrow {}^2H + e^+ + \nu,$$

followed by

$$e^+ + e^- \rightarrow \gamma + \gamma$$

$$^2H + {}^1H \rightarrow {}^3He + \gamma$$

$$^3He + {}^3He \rightarrow {}^4He + {}^1H + {}^1H$$

Here the symbols are e^- for an electron, e^+ for a positron, ν (the Greek letter "nu") for a neutrino, and γ ("gamma") for a gamma ray photon.

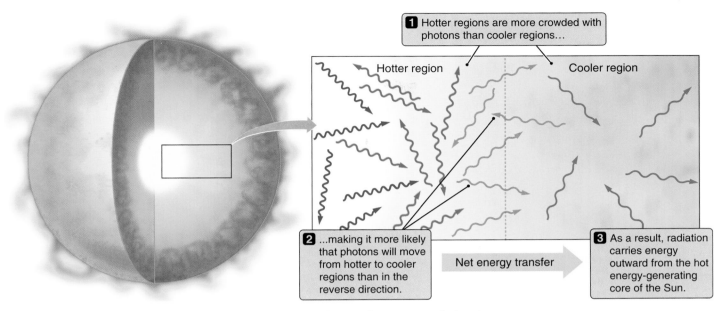

FIGURE 14.5 Higher-temperature regions deep within the Sun produce more radiation than lower-temperature regions farther out. Although radiation flows in both directions, more radiation flows from the hot regions to the cooler regions than from the cooler regions to the hot regions. In this way radiation carries energy outward from the inner parts of the Sun.

Chapter 4 that the hotter region will contain more (and more energetic) photons than the cooler region. (There will be a Planck spectrum of photons in both regions, so the total energy carried by photons will be proportional to the fourth power of the temperature in each region.) More photons will

Radiation carries energy from hotter regions to cooler regions.

move by chance from the hotter ("more crowded") region to the cooler ("less crowded") region than in the reverse direction. Thus there is a net transfer of photons and photon energy from the hotter region to the cooler region; and in this way radiative transfer carries energy from hotter regions to cooler regions.

If the temperature differs by a large amount over a short distance, then the concentration of photons will differ sharply as well, which favors rapid radiative energy transfer. The transfer of energy from one point to another by radiation also depends on how freely radiation can move from one point to

Opacity impedes the outward flow of radiation.

another within a star. The degree to which matter impedes the flow of photons through it is referred to as **opacity**. The opacity of material depends on many things including the density of material, its composition, its temperature, and the wavelength of the photons moving through it.

Radiative transfer is most efficient in regions where opacity is low. In the inner part of the Sun, where temperatures are high and atoms are ionized, opacity comes mostly from the interaction between photons and free electrons (electrons not attached to any atom). Here opacity is relatively low, and radiative transfer is capable of carrying the energy produced in the core outward through the star. The region in which radiative transfer is responsible for energy transport extends 71 percent of the way out toward the surface of the Sun. This region (see **Figure 14.6**) is referred to as the Sun's **radiative zone**. Even though the opacity of the radiative zone is relatively low, photons are still able to travel only a short distance before interacting with matter. The path that a photon follows is so convoluted that on average it takes the energy of a gamma ray photon produced in the interior of the Sun about 100,000 years to find its way to the outer layers of the Sun. Opacity serves as a blanket, holding energy in the interior of the Sun and letting it seep away only slowly.

From a peak of 15 million K at the center of the Sun, the temperature falls to about 100,000 K at the outer margin of the radiative zone. At this temperature atoms are no longer completely ionized, which increases the opacity. As the opacity becomes greater, radiation becomes less efficient in carrying energy from one place to another. The energy that is flowing outward through the Sun "piles up." The physical sign that energy is piling up is that the *temperature gradient*—that is, how rapidly temperature drops with increasing distance from the center of the Sun—becomes

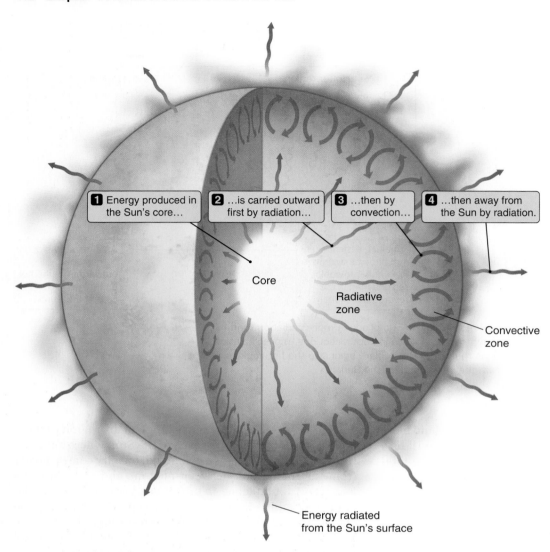

1 Energy produced in the Sun's core...

2 ...is carried outward first by radiation...

3 ...then by convection...

4 ...then away from the Sun by radiation.

Core

Radiative zone

Convective zone

Energy radiated from the Sun's surface

FIGURE 14.6 The interior structure of the Sun is divided into zones on the basis of where energy is produced and how it is transported outward.

very steep. (Radiative transfer carries energy from hotter regions to cooler regions, smoothing out temperature differences between them. As the opacity increases, radiation becomes less effective in smoothing out temperature differences, so temperature differences between one region and another become greater.)

As we move farther toward the surface of the Sun, radiative transfer becomes so inefficient (and the temperature

> **In the outer part of the Sun, energy is carried by convection.**

gradient so steep) that a different way of transporting energy takes over. Like a hot-air balloon, cells (or packets) of hot gas become buoyant and rise up through the lower-temperature gas above them, carrying energy with them. Thus convection begins. Just as convection carries energy from the interior of planets to their surfaces, or from the Sun-heated surface of Earth upward through Earth's atmosphere, convection also plays an important role in the trans-

port of energy outward from the interiors of many stars, including the Sun. The solar **convective zone** extends from the outer boundary of the radiative zone out to just below the visible surface of the Sun.

In the outermost layers of stars, radiation again takes over as the primary way that energy is transported. (This must be the case. After all, it is radiation that transports energy from the outermost layers of a star off into space.) Even so, the effects of convection can be seen as a perpetual roiling of the visible surface of the Sun.

What If the Sun Were Different?

At this point we have seen the pieces that go into calculating a model of the interior of the Sun. To put these pieces together, we again stress that the key is *balance*. The temperature and density in the core of the model Sun must be just right to produce the same amount of energy as would be radiated away by a star of the same size and surface temperature as the Sun. The temperature and density at each point within the model Sun must be just right so that transport of energy away from the core by radiation and convection just balances the amount of energy produced by fusion in the core. The density, temperature, and pressure of the model Sun must vary from point to point in such a way that the outward push of pressure is everywhere balanced by the inward pull of gravity. Finally, the whole model must depend on only two things: the total mass of gas from which the star is made, and the chemical composition of that gas.

Let us restate that last idea to drive the point home. *In order to be successful, our model of the Sun must start with no more information than the known mass and chemical composition of the real Sun, and from these it must correctly predict all the other observed properties of the real Sun.* The amazing thing is that our model does exactly that. Computer modeling shows that a ball of gas containing one solar mass with the same composition as our Sun can have only *one* structure and still satisfy all the different balances just listed *at the same time.* Our model predicts what the size, temperature, and luminosity of the resulting star should be—predictions that agree remarkably well with the observed properties of the real Sun.

To better understand *why* there is only one possible structure that a 1 $M_\odot$ star can have, we can "what-if" the Sun. For example, we might ask, "What if a star had the same mass, surface temperature, and composition as the Sun, but was somehow larger than the Sun? What properties would such a hypothetical star have? Specifically, what would happen to the balance between the amount of energy generated within such a hypothetical star and the amount of energy that it radiates away into space?" Follow along in **Figure 14.7** as we consider what would happen if the Sun were "too large."

We begin with the second part of this balance. Because our hypothetical star would have more surface area than the Sun, it would be able to more effectively radiate its energy into space. To keep a one-solar-mass star inflated to a size larger than the Sun would require that the star be more luminous than the Sun.

But now let us consider what is going on in the interior of our hypothetical star. Because the star is larger than the Sun but contains the same amount of mass as the Sun, the force of gravity at any point within our hypothetical star would be less than the force of gravity at the corresponding location within the Sun. (This is a result of the inverse square law of gravitation. If the radius R is larger in our hypothetical star, then $1/R^2$ must be smaller.) Weaker gravity means that the weight of matter pushing down on the interior of our hypothetical star would be less than in the Sun. Because according to hydrostatic equilibrium the pressure at any point within a star is equal to the weight of overlying matter, the

FIGURE 14.7 A star like the Sun can have only the structure that it has. Here we imagine the fate of a Sun with too large a radius.

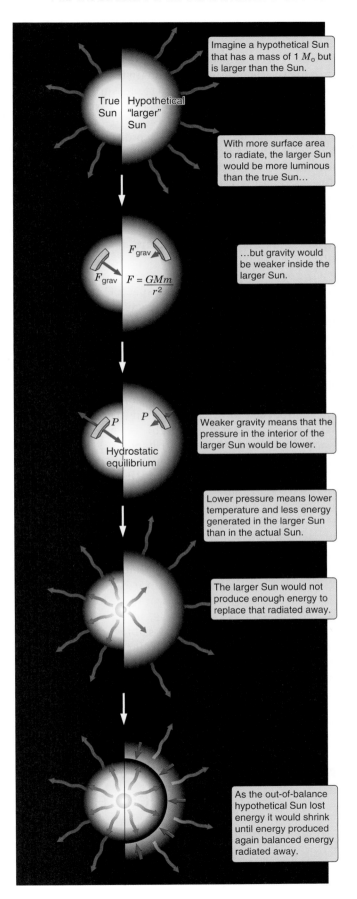

pressure at any point in the interior of our hypothetical star would be less than the pressure at the corresponding point in the Sun. This would affect the amount of energy that our star produces. The proton–proton chain runs faster at higher temperature and density, so the lower pressure in the interior of hypothetical star means that less energy would be generated there than in the core of the Sun.

But wait a minute—there is a contradiction here. We found that our hypothetical star would have to be more luminous than the Sun, but at the same time it would be producing less energy in its interior than the Sun does. This violates the balance that must exist in any stable star between the amount of energy generated within the star and

---

If the Sun were any different, it would not be in balance.

---

the amount of energy radiated into space. Our hypothetical star cannot exist! Stated another way, even if we could somehow magically pump up the Sun to a size larger than it actually is, it would not remain that way. Less energy would be generated in its core, while more energy would be radiated away at its surface. The Sun would be out of balance. As a result, the Sun would lose energy, the pressure in the interior of the Sun would decline, and the Sun would shrink back toward its original (true) size.

We could do the same thought experiment the other way around, asking what would happen if the Sun were smaller than it actually is. With less surface area it would radiate less energy. At the same time the Sun's mass would be compacted into a smaller volume, driving up the strength of gravity and therefore the pressure in its interior. Higher pressure implies higher density and temperature, which in turn would cause the proton–proton chain to run faster, increasing the rate of energy generation. Again there is a contradiction—an imbalance. This time, with more energy being generated in the interior than is radiated away from the surface, pressure in the Sun would build up, causing it to expand toward its original (true) size.

In the end, there is only one possible Sun. If a star contains one solar mass of material of solar composition, there is only one structure of that star that maintains all of the balances that must be maintained. Our stable, reliable Sun is the result.

14.3 The Standard Model of the Sun Is Well Tested

The standard model of the Sun correctly predicts such global properties of the Sun as its size, temperature, and luminosity. This is a remarkable feat, but the model predicts much more than these properties. In particular, the standard model of the Sun predicts exactly what nuclear reactions should be occurring in the core of the Sun, and at what

---

Neutrinos escape freely from the core of the Sun.

---

rate. The nuclear reactions that make up the proton–proton chain produce copious quantities of neutrinos. As we have noted, neutrinos are extremely elusive beasts—so elusive that almost all of the neutrinos produced in the heart of the Sun travel freely through the outer parts of the Sun and on into space as if the Sun were not there. The core of the Sun lies buried beneath 700,000 km of dense, hot matter, seemingly buried forever away from our view. Yet the Sun is *transparent* to neutrinos.

It may take thermal energy produced in the heart of the Sun 100,000 years to find its way to the Sun's surface, but the solar neutrinos streaming through you as you read these words were produced by nuclear reactions in the very heart of the Sun only $8\frac{1}{3}$ minutes ago. If we could find a way to capture and analyze these neutrinos, think of what we might learn! In principle, neutrinos offer a direct window into the very heart of the Sun's nuclear furnace.

Using Neutrinos to Observe the Heart of the Sun

Turning the promise of neutrino astronomy into reality turns out to be a formidable technical challenge. The same property of neutrinos that makes them so exciting to astronomers—the fact that their interaction with matter is so feeble that they can escape unscathed from the interior of the Sun—also makes them notoriously difficult to observe. Suppose we wanted to build a neutrino detector capable of stopping half of the neutrinos falling on it. Our hypothetical detector would need the stopping power of a piece of lead a light-year thick! Yet despite the difficulties, neutrinos offer such a unique and powerful window into the Sun that they are worth going to great lengths to try to detect.

Fortunately, the Sun produces a truly enormous number of neutrinos. As you lie in bed at night, about 400 trillion solar neutrinos pass through your body each second, having already passed through Earth. With this many neutrinos about, a neutrino detector does not have to be very efficient to be useful. Several methods have been devised to measure neutrinos from the Sun and from other astronomical sources such as supernova explosions, and a number of such experiments are under way. These experiments have successfully detected neutrinos from the Sun, and in so doing they have provided crucial confirmation that nuclear fusion reactions indeed are responsible for powering the Sun.

However, as with many good experiments, measurements of solar neutrinos raised new questions while answering others. After their initial joy at confirming that the Sun really is a nuclear furnace, astronomers became troubled that there seemed to be only about a third to a half as many solar neutrinos as predicted by solar models. The difference between the predicted and measured flux of solar neutrinos was referred to as the **solar neutrino problem**.

Understanding the solar neutrino problem was an area of very active research. One possible explanation was that our understanding of the structure of the Sun was somehow wrong. This seemed unlikely, however, because of the many other successes of our solar model. A second possibility was that our understanding of the neutrino itself was incomplete. The neutrino was long thought to have zero mass (like photons) and to travel at the speed of light. However, if neutrinos actually do have a tiny amount of mass, then theories from particle physics predict that solar neutrinos should *oscillate* (alternate back and forth) among three different kinds or *flavors*—the *electron, muon,* and *tau neutrinos,* as shown in **Figure 14.8**. Because only one of these, the electron neutrino, could interact with the atoms in the earlier neutrino detectors (described in **Tools 14.1**), neutrino oscillations provided a convenient explanation for why we saw only about a third of the expected number of neutrinos. And, as seen in Figure 14.8, electron neutrinos should also change flavor as they interact with solar material during their escape from the Sun.

After several decades of work on the solar neutrino problem, this last idea has won out. Work currently under way at high-energy physics labs, nuclear reactors, and neutrino telescopes around the world is showing that neutrinos *do* have a nonzero mass, and this work has uncovered evidence of neutrino oscillations.

FIGURE 14.8 If neutrinos have mass, they should oscillate among the three types of neutrinos (electron, muon, and tau). In sequences (1) through (5) a neutrino oscillates between electron and muon types. Changes from one type to another can also take place in the presence of matter at a certain density found within the Sun. In (6) an electron neutrino created in the Sun's core converts (7) to a muon neutrino, which then arrives at Earth. Early neutrino detectors would not have recognized this muon neutrino.

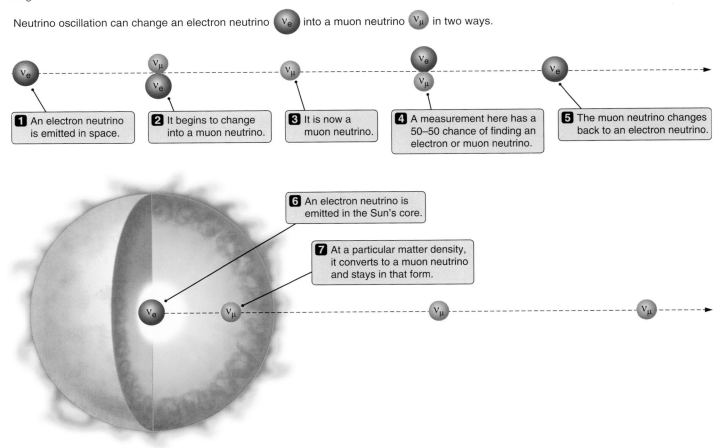

Neutrino oscillation can change an electron neutrino v_e into a muon neutrino v_μ in two ways.

1 An electron neutrino is emitted in space.

2 It begins to change into a muon neutrino.

3 It is now a muon neutrino.

4 A measurement here has a 50–50 chance of finding an electron or muon neutrino.

5 The muon neutrino changes back to an electron neutrino.

6 An electron neutrino is emitted in the Sun's core.

7 At a particular matter density, it converts to a muon neutrino and stays in that form.

TOOLS 14.1

Neutrino Astronomy

A neutrino telescope hardly fits anyone's expectation of what a telescope should look like. The first experiment designed to detect solar neutrinos consisted of a cylindrical tank filled with 100,000 gallons of dry cleaning fluid—C_2Cl_4, or perchloroethylene—buried 1,500 meters deep within the Homestake gold mine in Lead, South Dakota. A tiny fraction of neutrinos passing through this fluid interact with chlorine atoms, causing the reaction

$$^{37}Cl + \nu \rightarrow {}^{37}Ar + e^-$$

to take place. The ^{37}Ar formed in the reaction is a radioactive isotope of argon. The tank must be buried deep within Earth to shield the detector from the many other types of radiation capable of producing argon atoms. The argon is flushed out of the tank every few weeks and measured.

The Homestake detector (**Figure 14.9(a)**) began operations in 1965. Over the course of two days, roughly 10^{22} (10 billion trillion) solar neutrinos pass through the Homestake detector. Of these, on average only *one* neutrino interacts with a chlorine atom to form an atom of argon. Even so, this is enough of a signal to measure, and the effects of solar neutrinos are seen. Since then over a dozen and a half neutrino detectors have been built, each using different reactions to detect neutrinos of different energies. For example, the Soviet–American Gallium Experiment (SAGE) and the European GALLEX experiment use reactions involving conversion of gallium atoms into germanium atoms ($^{71}Ga + \nu \rightarrow {}^{71}Ge + e^-$) to detect solar neutrinos. (SAGE uses 60 tons of metallic gallium. GALLEX uses 30 tons of gallium in the form of gallium chloride. These two experiments account for most of the world's current supply of gallium.)

Among the more ambitious detectors capable of detecting all three flavors of neutrinos are the Super Kamiokande detector and the Sudbury Neutrino Observatory (SNO). Super Kamiokande is located in an active zinc mine 2,700 m under Mount Ikena, near Kamioka, Japan. It is a 50,000-ton tank of ultrapure water, surrounded by 13,000 photomultiplier tubes capable of registering extremely faint flashes of light. When a neutrino interacts with an atom in the tank, a faint conical flash of blue light is produced. This flash is seen by some of the photomultipliers. **Figure 14.9(b)** shows a picture of Super Kamiokande, while **Figure 14.9(c)** shows a map of the flash of light from a single neutrino. SNO uses 1,000 tons of heavy water (D_2O) contained in a 12-m sphere surrounded by light detectors (**Figure 14.9(d)**) and buried 2,000 m deep in an active nickel mine near Sudbury, Ontario.

Neutrino telescopes observe neutrinos produced in the heart of the Sun, allowing us to directly observe the results of the nuclear reactions going on there. While these observations have provided crucial confirmation that stars are powered by nuclear reactions, they have also challenged our models of the solar interior and led us to change our ideas about the nature of the neutrino itself (as is evident from the discussion of the solar neutrino problem in the text). In addition to solar neutrinos, a number of experiments detected neutrinos from supernova 1987A. As we will see in Chapter 17, this was an explosion marking the end of the life of a massive star located 160,000 light-years away in a small galaxy called the Large Magellanic Cloud. Neutrino astronomy was one of the great innovations of 20th century astronomy and is certain to have many applications in the 21st century.

Helioseismology Observes Echoes from the Sun's Interior

In Chapter 7 we found that models of Earth's interior predict how density and temperature change from place to place within our planet. These differences affect the way pressure waves travel through Earth, bending the paths of these waves. Models of Earth's interior are tested by comparing measurements of seismic waves from earthquakes with model predictions of how seismic waves should travel through the planet.

The same basic idea has now been applied to the Sun. Detailed observations of motions of material from place to place across the surface of the Sun show that the Sun vibrates or "rings," something like a struck bell. Compared

FIGURE 14.9 Neutrino "telescopes" do not look much like visible-light telescopes. (a) The Homestake neutrino detector is a 100,000-gallon tank of dry cleaning fluid located deep in a mine in South Dakota. (b) The Super Kamiokande detector (shown while being filled) is a tank containing 50,000 tons of pure water surrounded by about 13,000 photomultiplier tubes that record flashes of light from reactions within the tank. (c) A map of the flash of light from a single neutrino detected by the Super Kamiokande detector. (d) The Sudbury neutrino detector, buried 2 km deep in a Canadian nickel mine.

to a well-tuned bell, however, the vibrations of the Sun are very complex, with many different frequencies of vibrations occurring simultaneously. These motions are echoes of what lies below. Just as geologists use seismic waves from earthquakes to probe the interior of Earth, solar physicists use the surface oscillations of the Sun to test our understanding of the solar interior. This science, new to the later years of the 20th century, is called **helioseismology**. Like neutrino astronomy, helioseismology has created quite a stir among astronomers by letting us "see" into the invisible heart of the Sun (**Figure 14.10**).

Observing the "ringing" motion of the surface of the Sun is a difficult business. To detect the disturbances of helioseismic waves on the surface of the Sun, astronomers must use instruments capable of measuring Doppler shifts of less than 0.1 m/s while detecting changes in brightness

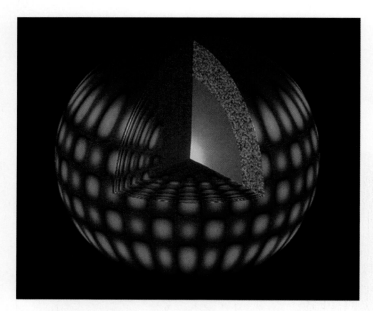

FIGURE 14.10 The interior of the Sun rings like a bell as helioseismic waves move through it. This figure shows one particular "mode" of the Sun's vibration. Red shows regions where gas is traveling inward, blue where gas is traveling outward. We can observe these motions by using Doppler shifts.

of only a few parts per million at any given location on the Sun. In addition, there are tens of millions of different wave motions possible within the Sun. Some waves travel around the circumference of the Sun, providing information about the density of the upper convection zone. Other waves travel through the interior of the Sun, revealing the density structure of the Sun close to its core. Still others travel inward toward the center of the Sun until they are bent by the changing solar density and return to the surface. All of these wave motions are going on at the same time. Sorting out this jumble requires computer analysis of long, unbroken strings of solar observations.

Helioseismology studies of the Sun got a huge boost in the closing years of the 20th century with two very successful projects. The Global Oscillation Network Group, or *GONG,* is a network of six solar observation stations spread around the world. With this network, solar astronomers are able to observe the surface of the Sun approximately 90 percent of the time. The other project is the Solar and Heliospheric Observatory, or *SOHO* spacecraft, which is a joint mission between NASA and the European Space Agency. By orbiting at the L_1 Lagrangian point of the Sun–Earth system (see Chapter 10), *SOHO* orbits in lockstep with Earth at a location approximately 1,500,000 km (0.01 AU) from Earth on a line directly between Earth and the Sun. *SOHO* carries a complement of 12 scientific instruments designed to monitor the Sun and measure the solar wind upstream of Earth. *SOHO* observations have dramatically improved our detailed knowledge of the Sun.

To interpret helioseismology data, scientists compare the strength, frequency, and wavelengths of observed vibrations with predicted vibrations calculated from models of the solar interior. This technique provides a powerful test of our understanding of the solar interior, and it has led both to some surprises and to improvements in our models.

Helioseismology confirms the predictions of solar models.

For example, some scientists had proposed that the solar neutrino problem might be solved if there were less helium in the Sun than generally imagined—an explanation that was ruled out by analyzing the waves that penetrate to the core of the Sun. Helioseismology also showed that the value for opacity used in early solar models was too low. This led astronomers to recalculate the location of the bottom of the convective zone. Both theory and observation now put the base of the convective zone at 71.3 percent of the way out from the center of the Sun, with an uncertainty in this number of less than half a percent! The amazing agreement between the predictions of computer models of the Sun and the temperature and density structure within the Sun measured by helioseismology is truly remarkable.

14.4 The Sun Can Be Studied Up Close and Personal

The Sun is a large ball of gas, and so it has no solid surface in the sense that Earth does. Instead it has the kind of surface that a cloud on Earth does, or the "surface" of one of the Jovian planets in our Solar System. To help you understand such a surface, imagine a fog bank. The surface of the fog bank is a gradual thing—an illusion really. Imagine watching some people walking into a fog bank. When they disappear from view, you would say that they were definitely inside the fog bank, even though they never passed through a noticeable boundary. The apparent surface of the Sun is defined by the same effect. Light from the surface of

The apparent surface of the Sun is called the photosphere.

the Sun can escape into space, and so we can see it. Light from below the surface of the Sun cannot escape directly into space, and so we cannot see it. The Sun's surface is referred to as the solar **photosphere**. (*Photo* means light; the photosphere is the place light comes from.) There is no instant when you can say that you have suddenly crossed the surface of a fog bank, and by the same token there is no instant when we suddenly cross the photosphere of the Sun.

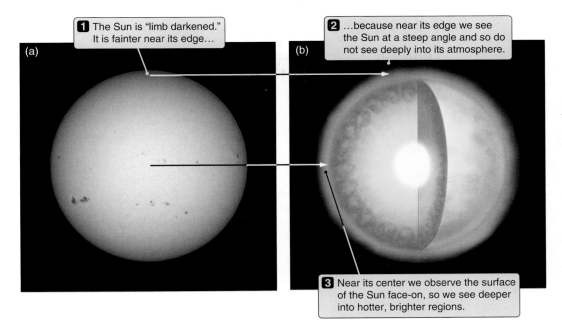

1 The Sun is "limb darkened." It is fainter near its edge…

(a)

2 …because near its edge we see the Sun at a steep angle and so do not see deeply into its atmosphere.

(b)

3 Near its center we observe the surface of the Sun face-on, so we see deeper into hotter, brighter regions.

FIGURE 14.11 (a) When viewed in visible light, the Sun appears to have a sharp outline, even though it has no true surface. The center of the Sun appears brighter while the limb of the Sun is darker, an effect known as limb darkening. (b) Looking at the middle of the Sun allows us to see deeper into the Sun's interior than looking at the edge of the Sun does. Because higher temperature means more luminous radiation, the middle of the Sun appears brighter than its limb.

The surface of the Sun—the photosphere—has an **effective temperature**[1] of 5,770 K and ranges from 6,600 K at the bottom to 4,400 K at the top. It is a zone about 500 km thick, across which the density and opacity of the Sun increase sharply. The reason the Sun appears to have a well-defined surface and a sharp outline when viewed from Earth (*never look at the Sun directly!*) is because this zone is relatively shallow; 500 km does not look very thick when viewed from a distance of 150 million km.

Look at a photograph of the Sun such as **Figure 14.11(a)**, and notice that the Sun appears to be fainter near its edges than near its center. This effect, called **limb darkening**, is an artifact of the structure of the Sun's photosphere. (The *limb* of a celestial body is the outer border of its visible disk.) The cause of limb darkening is illustrated in **Figure 14.11(b)**. Near the edge of the Sun you are looking through the photosphere at a steep angle. As a result, you do not see as deeply into the interior of the Sun as when looking directly down through the photosphere near the center of the Sun's disk. The light you see coming from near the limb of the Sun is from a layer in the Sun that is shallower and hence cooler and fainter.

The Solar Spectrum Is Complex

In one sense the surface of the Sun may be an illusion, but in another sense it is not. The transition between "inside" the Sun and "outside" the Sun is quite abrupt. In the outer-most part of the Sun, the density of the gas drops very rapidly with increasing altitude. This is the region we refer to as the Sun's **atmosphere**. **Figure 14.12** shows how the pressure and temperature change across the atmosphere of the Sun. The Sun's atmosphere is where all visible solar phenomena take place.

Very nearly all the radiation from below the Sun's photosphere is absorbed by matter and cannot escape—it is trapped. This is exactly our definition from Chapter 4 for the conditions under which blackbody radiation is formed. We should not then be surprised that the radiation able to leak out of the Sun's interior has a spectrum very close to being a Planck (blackbody) spectrum. This is why in Chapter 13 we were able to understand much about the physical properties of stars by applying our understanding of blackbody radiation.

As we look in more detail at the structure of the Sun, however, this simple description of the spectra of stars begins to go wrong. The fact that light from the solar photosphere must escape through the upper layers of the Sun's atmosphere affects the spectrum that we see. In Chapter 13 we discussed the presence of absorption lines in the spectra of stars. Now we can take a closer look at how these absorption lines form. As photospheric light travels upward through the solar atmosphere, atoms in the solar atmosphere absorb light at discrete wavelengths. Because from our perspective the Sun is so much brighter than any other star, its spectrum can be studied in far more detail. Today specially designed telescopes and high-resolution spectrographs have been built specifically to study the light from the Sun. The amazing structure present in the solar spectrum can be seen in **Figure 14.13**. Absorption lines from over 70 elements have been identified. Analysis

---

[1] The effective temperature of the photosphere is the temperature at which the Sun appears to radiate.

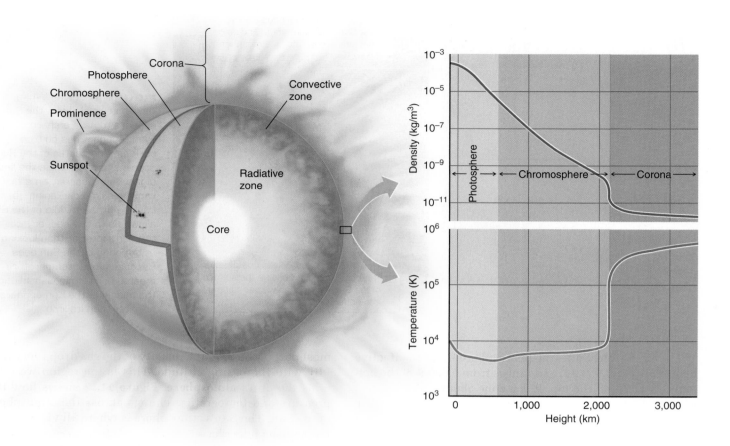

FIGURE 14.12 The components of the Sun's atmosphere, along with plots of how the temperature and density change with height at the base of the Sun's atmosphere.

of these lines forms the basis for much of our knowledge of the solar atmosphere, including the composition of the Sun, and is the starting point for our understanding of the atmospheres and spectra of other stars.

The Sun's Outer Atmosphere: Chromosphere, Corona, and Solar Wind

As we move upward through the Sun's photosphere, the temperature continues to fall, reaching a minimum of about 4,400 K at the top of the photosphere. At this point the temperature trend reverses and slowly begins to climb, rising to about 6,000 K at a height of 1,500 km above the top of the photosphere. This region above the photosphere is called the **chromosphere** (see **Figure 14.14(a)**). The chromosphere was discovered in the 19th century during observations of total solar eclipses (**Figure 14.14(b)**). The chromosphere is seen most strongly as a source of emission lines, especially

The Sun's chromosphere lies above the photosphere.

the Hα line from hydrogen. In fact, it is from the deep red color of the Hα line that the *chromosphere* (the place where color comes from) gets its name. It was also from a spectrum of the chromosphere that the element helium was discovered in 1868. Helium is named after *helios,* the Greek word for "Sun."[2]

At the top of the chromosphere, across a transition region that is only about 100 km thick, the temperature suddenly soars (Figure 14.12). Above this transition lies the outermost region of the Sun's atmosphere, called the **corona**, in which temperatures reach 1 million to 2 million K. The corona is probably heated by magnetic waves and magnetic fields in much the same way the chromosphere is, but why the temperature changes so abruptly at the transition be-

---

[2] It is rather remarkable that helium, the second most common element in the universe, was discovered in the Sun before it was identified on Earth!

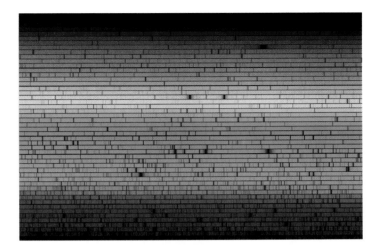

FIGURE 14.13 A high-resolution spectrum of the Sun, stretching from 400 nm (lower left corner) to 700 nm (upper right corner), showing a wealth of absorption lines.

The corona has a temperature of millions of kelvins.

tween the chromosphere and the corona is not at all clear. Since ancient times the Sun's corona has been known—visible during total solar eclipses as an eerie outer glow stretching a distance of several solar radii beyond the Sun's surface (see **Figure 14.14(c)**). Because it is so hot, the solar corona is a strong source of X-rays. Atoms in the corona are also highly ionized. The spectrum of the corona shows

emission lines from ionic species such as Fe^{13+} (iron atoms from which 13 electrons have been stripped) and Ca^{14+} (calcium atoms from which 14 electrons have been stripped).

Solar Activity Is Caused by Magnetic Effects

Virtually all of the structure seen in the atmosphere of the Sun is imposed on the gas by the Sun's magnetic field. High-resolution images of the Sun, such as **Figure 14.15**, show *coronal loops* that make the Sun look as though it were covered with matted, tangled hair. This fibrous or rope-like texture in the chromosphere is the result of magnetic structures called *flux tubes,* much like the flux tube that connects Io with Jupiter (see Chapter 9). Magnetic fields are responsible for much of the structure of the corona as well. Recall from Chapter 8 that "hot" atoms are able to escape from the tops of planetary atmospheres. Shouldn't this happen on the Sun as well? Well, in this case, the answer is yes and no. The corona is far too hot to be held in by the Sun's gravity, but over most of the surface of the Sun coronal gas is confined instead by magnetic loops with both ends firmly anchored deep within the Sun. The magnetic field in the corona acts almost like a network of rubber bands that coronal gas is free to slide along but cannot cross. In contrast, about 20 percent of the surface of the Sun is covered by an ever-shifting pattern of **coronal holes**. These are apparent in X-ray images of the Sun (**Figure 14.16**) as dark regions, indicating that they are cooler and lower in density than

FIGURE 14.14 (a) This image of the Sun, taken in Hα light during a transit of Venus, shows structure in the Sun's chromosphere. The planet Venus is seen in silhouette against the disk of the Sun. (b) The chromosphere seen during a total eclipse. (c) This eclipse image shows the Sun's corona, consisting of million-kelvin gas that extends for millions of kilometers beyond the surface of the Sun.

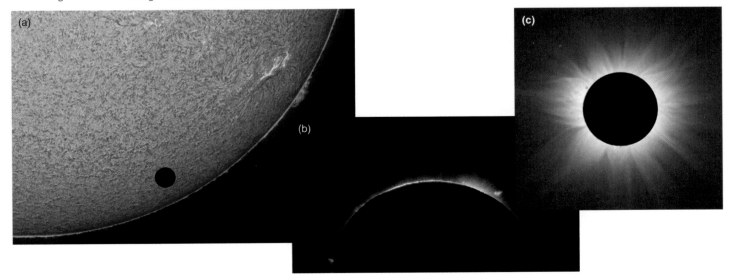

FIGURE 14.15 A close-up image of the Sun showing the tangled structure of coronal loops.

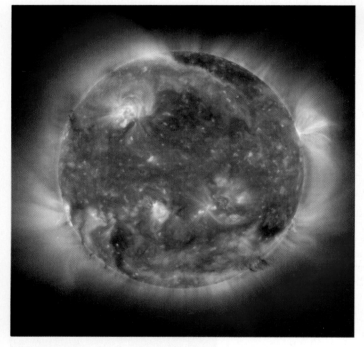

FIGURE 14.16 X-ray images of the Sun show a very different picture of our star than images taken in visible light. The brightest X-ray emission comes from the base of the Sun's corona, where gas is heated to temperatures of millions of kelvins. This heating is most powerful above magnetically active regions of the Sun. Also visible are dark coronal holes.

their surroundings. Coronal holes are large regions where the magnetic field points outward, away from the Sun, and where coronal material is free to stream away into interplanetary space.

We have encountered this flow of coronal material away from the Sun before. This is the same *solar wind* responsible

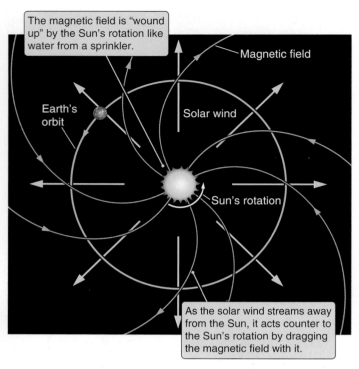

The magnetic field is "wound up" by the Sun's rotation like water from a sprinkler.

Magnetic field

Earth's orbit

Solar wind

Sun's rotation

As the solar wind streams away from the Sun, it acts counter to the Sun's rotation by dragging the magnetic field with it.

FIGURE 14.17 The solar wind streams away from active areas and corona holes on the Sun. As the Sun rotates, the solar wind picks up a spiral structure, much like the spiral of water that streams away from a rotating lawn sprinkler.

for shaping the magnetospheres of planets (Chapters 8 and 9) and for blowing the tails of comets away from the Sun (Chapter 12). The relatively steady part of solar wind consists of lower-speed flows, with velocities of around 350 km/s, and higher-speed flows, with velocities up to about 700 km/s.

A "wind" blows away from the Sun.

The higher-speed flows originate in coronal holes. Depending on their speed, particles in the solar wind take about two to five days to reach Earth. Frequently, two to five days after a coronal hole passes across the center of the face of the Sun, there is an increase in the speed and density of the solar wind reaching Earth. The solar wind drags the Sun's magnetic field along with it. The magnetic field in the solar wind gets "wound up" by the Sun's rotation, as shown in **Figure 14.17**. This gives the solar wind a spiral structure, something like a stream of water from a rotating lawn sprinkler.

The effects of the solar wind are felt throughout the Solar System. As we have seen, the solar wind causes the tails of comets, shapes the magnetospheres of the planets, and provides the energetic particles that power Earth's spectacular auroral displays. Using space probes, we have been able to observe the solar wind extending out to nearly 100 AU from the Sun. But the solar wind does not go on forever. The far-

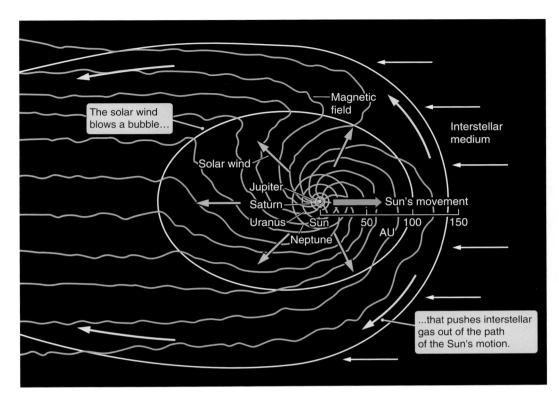

FIGURE 14.18 The solar wind streams away from the Sun for about 100 AU, until it finally piles up against the pressure of the interstellar medium through which the Sun is traveling. At the beginning of the 21st century, the *Voyager 1* spacecraft is approaching the expected location of this boundary.

ther it gets from the Sun, the more it has to spread out. Just like radiation, the density of the solar wind follows an inverse square law. At a distance of around 100 AU from the Sun, the solar wind is assumed no longer to be powerful enough to push the **interstellar medium** (the gas and dust that lie between stars in a galaxy and that surround the Sun) out of the way. There the solar wind will stop abruptly, "piling up" against the pressure of the interstellar medium. **Figure 14.18** shows this region of space over which the wind from the Sun holds sway. Sometime within the next decade the *Voyager 1* spacecraft is expected to cross the outer edge of this boundary and begin sending back our first "on the scene" measurements of true interstellar space.[3]

The best-known features on the surface of the Sun are relatively dark blemishes in the solar photosphere, called **sunspots**, that come and go over time. Early telescopic observations of sunspots made during the 17th century led to the discovery of the Sun's rotation, which has an average period of about 27 days as seen from Earth and 25 days relative to the stars. Observations of sunspots also show that the Sun, like Saturn, rotates more rapidly at its equator than it does at higher latitudes. This effect, referred to as **differential rotation**, is possible only because the Sun is a large ball of gas rather than a solid object. Sunspots appear dark, but only in contrast to the brighter surface of the Sun. Sun-

[3] The truth is, we really don't know just how far from the Sun this boundary lies, but we *will* know it when we see it.

Sunspots are cooler than their surroundings.

spots are about 1,500 K cooler than their surroundings. What does this tell us? Think back to Stefan's Law in Chapter 4. The flux from a blackbody is proportional to the fourth power of the temperature. Let's take round numbers for the temperature of a typical sunspot and the surrounding photosphere as 4,500 K and 6,000 K, respectively. Stefan's Law informs us that the surface brightness of a sunspot is about

$$\left(\frac{4{,}500}{6{,}000}\right)^4 = 0.32$$

or about ⅓ the brightness of the surrounding photosphere.

Figure 14.19 is a photograph of a sunspot group, and **Figure 14.20** shows the remarkable structure of one of these blemishes on the surface of the Sun. Each sunspot consists of an inner dark core called the **umbra**, which is surrounded by a less dark region called the **penumbra**. The penumbra

Magnetic fields cause sunspots.

shows an intricate radial pattern, reminiscent of the petals of a flower. Observations of sunspots show that they are magnetic in origin, with magnetic fields thousands of times greater than the magnetic field at Earth's surface. Sunspots occur in pairs that are connected by loops in the magnetic

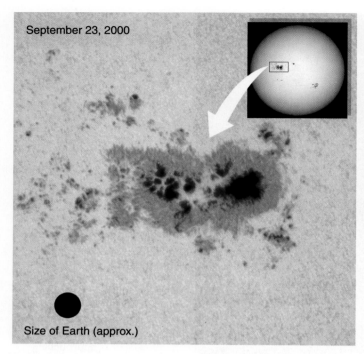

September 23, 2000

Size of Earth (approx.)

FIGURE 14.19 This false-color *SOHO* image shows a group of sunspots. The darkest part of sunspots are called umbrae. The surrounding areas, which look like petals on a flower, are the penumbrae. Sunspots are magnetically active regions that are cooler than the surrounding surface of the Sun.

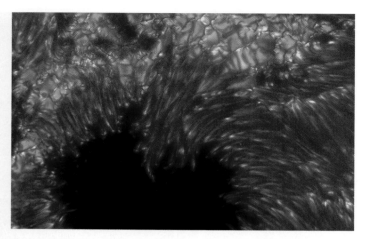

FIGURE 14.20 A very high-resolution view near the central dark region of a large sunspot. Nearly two Earths could fit side by side in this image.

field, much like the shape of a horseshoe. Sunspots range in size from about 1,500 km across up to complex groups that may contain several dozen individual spots and span 50,000 km or more. The largest sunspot groups are so large that they can be seen with the naked eye and have been noted since antiquity. (*But do not look directly at the Sun!* More than one solar observer has paid the price of blindness for a glimpse of a naked-eye sunspot. **Figure 14.21** shows a number of ways to look at the Sun in safety. Direct viewing through welder's glass is also safe.)

Individual sunspots are ephemeral. Although a few sunspots last for 100 days or longer, half of all sunspots come and go in about 2 days, and 90 percent are gone within 11 days. The amount of sunspot activity and the locations on the Sun where sunspots are seen change over time in a pronounced 11-year pattern called the **sunspot cycle (Figure 14.22(a))**. At the beginning of a cycle, sunspots begin to appear at solar latitudes of around 30° to the north and south

Sunspots come and go in an 11-year cycle.

of the solar equator. Over the following years the regions where most sunspots are seen move toward the equator as the number of sunspots increases and then declines. As the last few sunspots near the equator are seen, sunspots again

begin appearing at middle latitudes, and the next cycle begins. A diagram (**Figure 14.22(b)**) showing the number of sunspots at a given latitude plotted against time has the appearance of a series of opposing diagonal bands and is often referred to as the sunspot "butterfly diagram."

Telescopic observations of sunspots date back for almost 400 years, and there were naked-eye reports of sunspots even before that. **Figure 14.23** shows the historical record of sunspot activity. Although astronomers often speak of the 11-year sunspot cycle, the cycle is neither perfectly periodic nor especially reliable. The time between peaks in the number of sunspots actually varies between about 9.7 and 11.8 years. The number of spots seen during a given cycle fluctuates as well. Some cycles are real monsters. There have also been times when sunspot activity has disappeared almost entirely for extended periods. The most recent extended lull in solar activity, called the **Maunder minimum**, lasted from 1645 to 1715. Normally there would be six peaks in solar activity in 70 years, but virtually no sunspots were seen during the Maunder minimum.

In the early 20th century George Ellery Hale was the first to show that the 11-year sunspot cycle was actually half of a 22-year magnetic cycle during which the direction of the Sun's magnetic field reverses from one 11-year sunspot cycle to the next. In one sunspot cycle the leading sunspot in each pair tends to be a north magnetic pole, whereas the trailing sunspot tends to be a south magnetic pole. In the next sunspot cycle this polarity is reversed: The leading spot in each pair is a south magnetic pole. The transition between these two magnetic polarities occurs near the peak of each sunspot cycle (Figure 14.22(c)). The predominant theory for what causes this magnetic cycle involves a *dynamo* in the interior of the Sun, much like the dynamos that generate the magnetic fields of the planets.

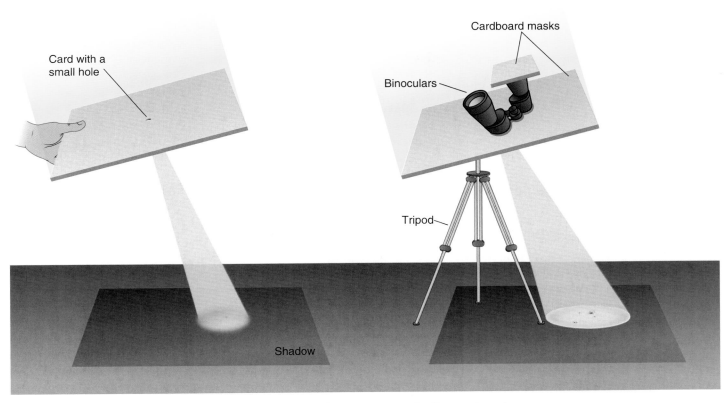

FIGURE 14.21 There are a number of ways of safely viewing sunspots on the Sun or observing a partial solar eclipse. Never look at the Sun directly (except during totality of a total eclipse)!

The effects of magnetic activity on the Sun are felt throughout the Sun's photosphere, chromosphere, and corona. Sunspots are only one of a host of phenomena that follow the Sun's 22-year cycle of magnetic activity. The peaks of the cycle, referred to as **solar maxima**, are times of intense activity, as can be seen in the ultraviolet images of the Sun in Figure 14.22(d). Although sunspots are darker-than-average features, they are often accompanied by a brightening of the solar chromosphere that is seen most clearly in the light of emission lines such as Hα. The magnificent loops seen arching through the solar corona are also anchored in solar active regions. These include solar **prominences** such as those shown in **Figure 14.24**. Prominences are magnetic flux tubes of relatively cool (5,000 to 10,000 K) gas extending through the million-kelvin gas of the corona. Although most prominences are relatively quiescent, others can erupt out through the corona, towering a million kilometers or more over the surface of the Sun and ejecting material into the corona at velocities of 1,000 km/s.

Figure 14.25(a and b) shows a picture of a **solar flare** erupting from a sunspot group. Solar flares are the most energetic form of solar activity. These are violent eruptions in which tremendous amounts of magnetic energy are released over the course of a few minutes to a few hours. Flares can heat gas to temperatures of 20 million K, and they are the source of intense X-ray and gamma ray radiation. Hot plasma is seen to move outward from flares at speeds that can reach 1,500 km/s. These events, called **coronal mass ejections** (**Figure 14.25(c and d)**), send powerful bursts of energetic particles outward through the Solar System. Magnetic effects in flares can accelerate subatomic particles to almost the speed of light.

Solar Activity Affects Earth

The amount of solar radiation at Earth's distance from the Sun is, on average, 1.35 kilowatts per square meter.[4] Satellite measurements of the amount of radiation coming from the Sun (**Figure 14.26**) show that this value can vary by as much as 0.2 percent over periods of a few weeks as

[4] Photons from the Sun provide an important alternative source of energy.

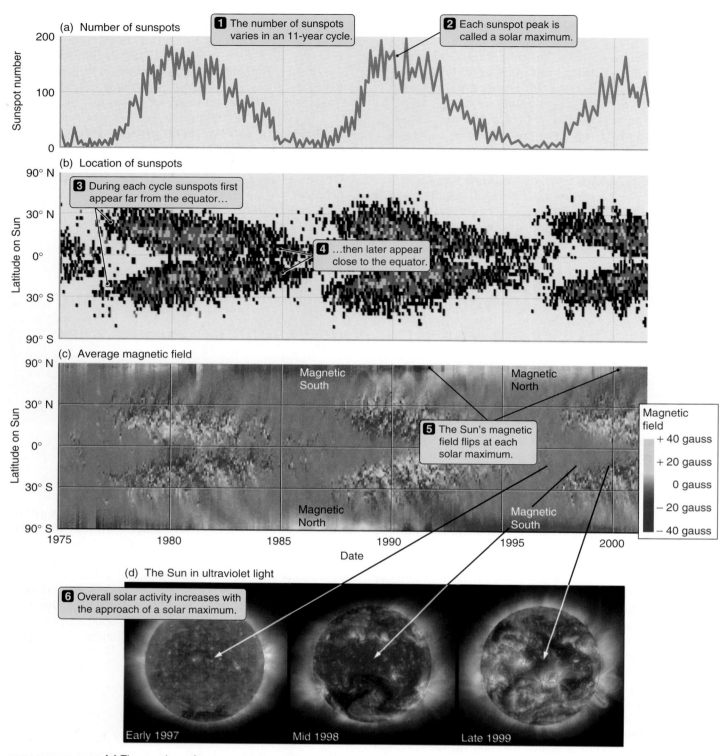

(a) Number of sunspots

1 The number of sunspots varies in an 11-year cycle.

2 Each sunspot peak is called a solar maximum.

(b) Location of sunspots

3 During each cycle sunspots first appear far from the equator…

4 …then later appear close to the equator.

(c) Average magnetic field

Magnetic South

Magnetic North

5 The Sun's magnetic field flips at each solar maximum.

Magnetic North

Magnetic South

Magnetic field

+ 40 gauss

+ 20 gauss

0 gauss

− 20 gauss

− 40 gauss

(d) The Sun in ultraviolet light

6 Overall solar activity increases with the approach of a solar maximum.

Early 1997   Mid 1998   Late 1999

FIGURE 14.22 (a) The number of sunspots versus time for the last few solar cycles. (b) The solar butterfly diagram, showing the fraction of the Sun covered at each latitude. (c) The Sun's magnetic field flips every 11 years. (d) The approach of a solar maximum is apparent in the *SOHO* images taken in ultraviolet light.

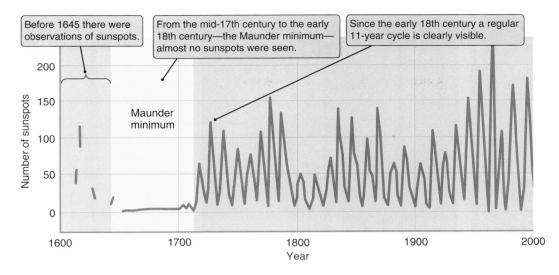

Before 1645 there were observations of sunspots.

From the mid-17th century to the early 18th century—the Maunder minimum—almost no sunspots were seen.

Since the early 18th century a regular 11-year cycle is clearly visible.

Maunder minimum

FIGURE 14.23 Sunspots have been observed for hundreds of years. In this figure the 11-year cycle in the number of sunspots (half of the 22-year solar magnetic cycle) is clearly visible. Sunspot activity varies greatly over time. The period from the middle of the 17th century to the early 18th century, when almost no sunspots were seen, is called the Maunder minimum.

dark sunspots and bright spots in the chromosphere move across the disk. Overall, however, the increased radiation from active regions on the Sun more than makes up for the reduction in radiation from sunspots. On average the Sun seems to be about 0.1 percent brighter during the peak of a solar cycle than it is at its minimum.

Solar activity has many effects on Earth. Solar active regions are the source of most of the Sun's extreme ultraviolet and X-ray emission. This energetic radiation heats Earth's upper atmosphere and during periods of solar activity causes Earth's upper atmosphere to swell. When this happens, it can significantly increase the atmospheric drag on spacecraft orbiting near Earth, causing their orbits to decay. Prior to the launch of the Hubble Space Telescope, many people working on the project were concerned that because it was to be launched into the teeth of a solar maximum, the telescope might not survive for long. One reason for repeated shuttle visits to the HST was to "reboost" it up to its original orbit to make up for the slow decay in its orbit caused by drag in the rarefied outer parts of Earth's

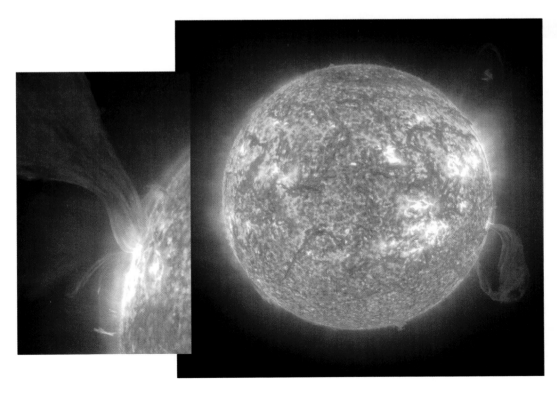

FIGURE 14.24 Solar prominences are magnetically supported arches of hot gas that rise high above active regions on the Sun. The inset shows detail at the base of a large prominence.

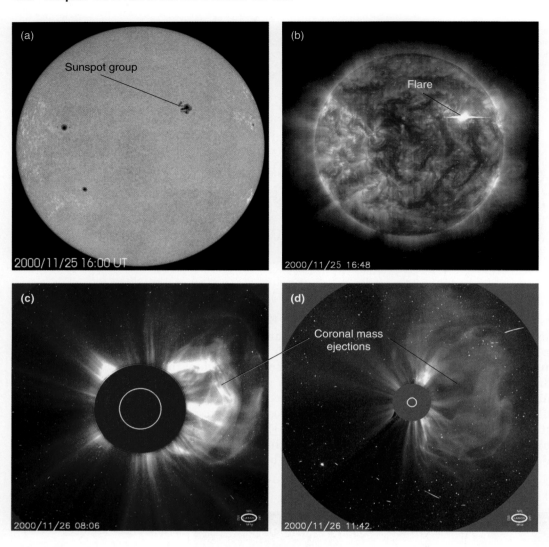

FIGURE 14.25 In this sequence of false-color images taken by the *SOHO* spacecraft a large sunspot group (a) emits a powerful flare, shown (b) in ultraviolet light. The resulting coronal mass ejection is seen streaming away from the Sun in (c) and (d). (The dark circular masks in the last two frames obscure the Sun's disk, allowing *SOHO* to observe the corona. The white circles indicate the size of the Sun.)

atmosphere. As this book goes to press, NASA's plans for a shuttle reboost visit to HST remain in doubt. Left unattended, Earth's atmospheric drag will eventually bring HST back to Earth, although not in one piece.

Earth's magnetosphere is the result of the interaction between Earth's magnetic field and the solar wind. Increases in the solar wind accompanying solar activity, especially coronal mass ejections directed at Earth, can disrupt Earth's

---

Solar storms cause auroras and disrupt electric power grids on Earth.

---

magnetosphere in ways that are obvious even to nonscientists. Spectacular auroras can accompany such events, as can magnetic storms that have been known to disrupt electrical power grids, causing blackouts across large regions. In fact, the first clear evidence that Earth's environment was affected by solar magnetic activity was the observed rela-

tionship between sunspot activity and terrestrial events such as auroras and perturbations in Earth's magnetic field. Solar flares can have an especially dramatic effect on the environment within the Solar System. Energetic particles accelerated in solar flares pose one of the greatest dangers to human exploration of space.

Over the years people have attempted to relate sunspot activity to phenomena ranging from the performance of the stock market to the mating habits of Indian elephants. Few of these supposed relationships have withstood serious scrutiny, however. An especially interesting idea is that variations in Earth's climate might be related to solar activity. We have certainly seen that solar activity affects Earth's upper atmosphere, and it might not be too much of a reach to imagine that it could affect weather patterns as well. It has also been suggested that variations in the amount of radiation from the Sun might be responsible for variations in Earth's climate in the past. Although the causes of global

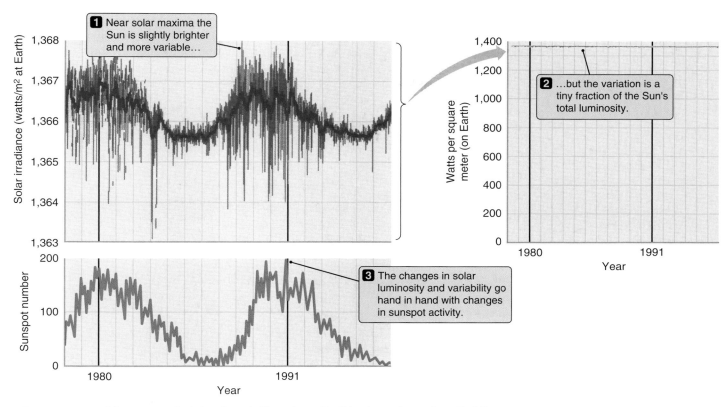

FIGURE 14.26 Measurements taken by satellites above Earth's atmosphere show that the amount of light from the Sun changes slightly over time.

climatic change remain controversial, models suggest that observed variations in the Sun could account for only about 0.1 K differences in Earth's average temperature—less than the effects due to the ongoing buildup of carbon dioxide in Earth's atmosphere. On the other hand, triggering the onset of an ice age may require a drop in global temperatures of only around 0.2 to 0.5 K, so the quest for a definite link between solar variability and changes in Earth's climate persists.

Summary

- Nuclear reactions, converting hydrogen to helium, are the source of the Sun's energy.

- Energy created in the Sun's core moves outward to the surface, first by radiation and then by convection.

- Neutrinos are elusive, almost massless particles that interact only very weakly with other matter.

- The apparent surface of the Sun is called the photosphere.

- Sunspots are photospheric regions that are cooler than their surroundings.

- The Sun's outermost region is called the corona.

- The temperature of the Sun's atmosphere ranges from about 6,000 K near the bottom to a million degrees at the top.

- Sunspots reveal the 11- and 22-year cycles in solar activity.

- Material streaming away from the Sun's corona creates the solar wind.

- Solar storms can produce terrestrial auroras and disrupt power grids.

Seeing the Forest through the Trees

The universe contains more stars than there are grains of sand on all of Earth's beaches. Even "nearby" stars are so distant that at best we see them as faint points of light in the night sky. Only one star is close enough for its complex structure to be studied in great detail. That star is a maelstrom of fierce activity, fueled by nuclear fires burning deep within its interior. Storms rage across its surface. Giant flares burst forth, arching millions of kilometers into space before collapsing back again. That star heats a huge and rarefied corona to millions of kelvins and blows a bubble in interstellar space large enough to swallow our entire Solar System. That star—a main sequence G2 star, different in no way from main sequence G2 stars throughout our galaxy and the universe—is the star we call our Sun.

No machine of human manufacture has ever visited the interior of the Sun, just as no machine has more than scratched the surface of Earth. Even so, our understanding of the interior of the Sun is remarkably complete. Like everything else in the universe, the Sun is governed by the laws of physics. Computer models based on those physical laws show that there is only one possible structure that a 1 $M_\odot$ ball of gas with the chemical composition of the Sun can have while remaining in balance. According to those models, the Sun draws its energy from the fusion of hydrogen to make helium in its core. That energy, carried from the Sun's core to its surface by radiation and convection, arrives at just the rate needed to replace the energy radiated away into space. All of this takes place while maintaining a delicate standoff in the battle between the inward pull of gravity and the outward push of pressure fueled by the nuclear fires in the Sun's interior. Models of the Sun make a host of detailed predictions, ranging from the density and temperature at every point within the Sun's interior to the nuclear reactions that take place in its core. Over the past decades the predictions of models of the Sun have been tested in ever more sophisticated and detailed ways, and those predictions have been confirmed. At the opening of the 21st century, it is fair to say that we *know* what the interior of the Sun is like.

Our knowledge of the Sun also forms the cornerstone of our knowledge of all stars everywhere. Starting with a successful model of the Sun, we can now go on to ask a host of other questions. What happens if we increase the mass of the model star? What happens if we change its chemical abundances? What happens if we run the model forward in time until all of the hydrogen at the star's center is gone? How do we assemble something that looks like the Sun in the first place? Our study of the Sun has set the stage for the next leg of our journey as we turn our attention to how stars form, how they differ from one another, how they change with time, and how they die.

Key Terms

strong nuclear force, p. 406
hydrogen burning, p. 406
proton–proton chain, p. 407
energy transport, p. 408
thermal conduction, p. 408
radiative transfer, p. 408
positron, p. 408
neutrino, p. 408
beta decay, p. 408
antiparticle, p. 408
opacity, p. 409
radiative zone, p. 409
convective zone, p. 410
helioseismology, p. 415
photosphere, p. 416
effective temperature, p. 417
limb darkening, p. 417
atmosphere, p. 417
chromosphere, p. 418
coronal hole, p. 419
interstellar medium, p. 421
sunspot, p. 421
differential rotation, p. 421
umbra, p. 421
penumbra, p. 421
sunspot cycle, p. 422
Maunder minimum, p. 422
solar maximum, p. 423
prominence, p. 423
solar flare, p. 423
coronal mass ejection, p. 423

Student Questions

THINKING ABOUT THE CONCEPTS

1. Engineers and physicists dream of solving the world's energy supply problem by constructing power plants that would convert globally plentiful hydrogen to helium. Our Sun seems to have solved this problem. On

Earth, what is a major obstacle to this environmentally clean solution?

2. Explain the proton–proton chain process in which the Sun generates energy by converting hydrogen to helium.

3. In the proton–proton chain process, the mass of four protons is slightly greater than the mass of a helium nucleus. Explain what this difference in mass means for a star like the Sun.

4. What experiences have you had with energy transmitted to you by radiation alone? By convection alone? By conduction alone? (You are exposed to all three every day.)

5. If neutrinos have mass, as now seems to be the case, they cannot travel at the speed of light. Explain why not.

6. Why are neutrinos so difficult to detect?

7. Discuss the "solar neutrino problem" and how this problem was solved.

8. What techniques do you find in common between how we probe the internal structure of the Sun and the internal structure of Earth?

9. Explain the relationship between the 11-year sunspot cycle and the 22-year magnetic cycle.

10. Solar flares and magnetic storms can adversely affect Earth's power grids, radio communication, and the health of humans in orbit. How might observations of the Sun help us forecast such events?

APPLYING THE CONCEPTS

11. The Sun shines by converting mass into energy according to Einstein's well-known relationship, $E = mc^2$. Show that if the Sun produces 3.78×10^{26} J of energy per second, it must convert 4.2 million metric tons (4.2×10^9 kg) of mass per second into energy.

12. Assume the Sun has been producing energy at a constant rate over its lifetime of 4.5 billion years (1.4×10^{17} s).
    a. How much mass has it lost creating energy over its lifetime?
    b. The present mass of the Sun is 2×10^{30} kg. What fraction of its present mass has been converted into energy over the lifetime of the Sun?

13. Imagine the source of energy in the interior of the Sun were to abruptly change.
    a. How long would it take before a neutrino telescope could detect the event?
    b. When would a visible-light telescope see evidence of the change?

14. On average, how long does it take particles in the solar wind to reach Earth from the Sun if they are traveling at 300 km/s?

15. Sunspots appear relatively dark because their surface temperature typically is 1,500 K cooler than the surrounding photosphere (5,780 K). Compare the surface brightness (power radiated per square meter) of a dark sunspot to that of the surrounding photosphere.

16. The hydrogen bomb represents humankind's efforts to duplicate processes going on in the core of the Sun. The energy released by a 5-megaton hydrogen bomb is 3×10^{14} J.
    a. How much mass did Earth lose each time a 5-megaton hydrogen bomb was exploded?
    b. This textbook, *21st Century Astronomy*, has a mass of about 1 kg. If we converted all of its mass into energy, how many 5-megaton bombs would it take to equal that energy?

StudySpace
wwnorton.com/astro21
provides a Study Plan for each chapter that includes a reading outline, animations, keyword flash cards, and gradebook-enabled multiple-choice quizzes. From StudySpace you can also access premium content in the ebook and SmartWork.

There are no whole truths; all truths are half truths. It is trying to treat them as whole truths that plays the devil.

ALFRED NORTH WHITEHEAD (1861–1947)

A star forming region 3,000 light-years away in the constellation Cepheus.

CHAPTER 15

Star Formation and the Interstellar Medium

15.1 Whence Stars?

To mortal humans, the stars seem fixed and unchanging. Indeed, our local star, the Sun, has remained much the same for long enough to allow life on Earth to arise, evolve, and begin to contemplate its own origin and fate. But our perspective is distorted by our brief existence. All of recorded human history amounts to less than a millionth of the age of the Sun and Earth. In Chapter 6 we told the story of how our Solar System formed some 4.6 billion years ago out of a swirling disk of gas and dust that surrounded the protostellar Sun. We are now ready to return to that story, but with a different focus. First we will step back and look at the interstellar environment that gave birth to the Sun and Solar System. Then we will focus our attention on the protostar at the center of the circumstellar disk and watch as it becomes a star.

The root of the word *astronomy* may mean "star," but if we ask what fraction of space is filled with stars, we get a very different picture of their importance. The volume of our Sun is about 1.3 million times the volume of Earth. Though this volume might seem enormous to us, it is almost incomprehensibly tiny compared with the volume of space. Our Sun is the only star in a region of space with a volume of over 300 cubic light-years. If our Sun were the size of a golf ball sitting on an author's office desk in Tempe, Arizona, its nearest stellar neighbor would be a similar golf ball sitting on another author's desk in Santa Cruz, California, 1,000 km away. The rest of the volume of our galaxy is filled with the *interstellar medium*. It is here that the story of stars begins and ends. Stars are formed from the gas in the interstellar medium; they live their lives there; and when they die, they

KEY CONCEPTS

When you look at the night sky it is stars that you notice, yet stars fill only the tiniest fraction of the volume of space. On this leg of our journey we turn our attention to the tenuous medium between the stars, and we discover that

- There are various phases of the interstellar medium, ranging from cold, relatively dense molecular clouds to hot, tenuous, intercloud gas.

- Different phases of the interstellar medium emit various types of radiation and can be observed at different wavelengths, ranging from radio waves to X-rays.

- Dust in the interstellar medium blocks visible light but glows with infrared radiation.

- Gas in the interstellar medium is heated and ionized by energy from stars and stellar explosions.

- Stars form in clusters from dense cores buried within giant molecular clouds.

- Forming stars are seen by their infrared emission and by the effects they have on their surroundings.

- We can follow the progress of an evolving protostar by its track across the H-R diagram.

- Protostars collapse, radiating away their gravitational energy until fusion starts in their cores, and they settle onto the main sequence.

bequeath some of their chemical and energetic legacy back to the interstellar medium.

Before taking a look at the interstellar medium, however, we need to put it into another context as well. Remember that galaxies are *crowded* in comparison with the vast expanses of intergalactic space. Returning to the golf ball analogy, if we ask the whereabouts of a star in the spiral galaxy nearest to our own, that star would be equivalent to a golf ball somewhere on Jupiter. But that is a story for later in our journey. For now we turn our attention to our "local neighborhood" and the interstellar medium that fills the volume of our own galaxy.

15.2 The Interstellar Medium

The Sun formed from the interstellar medium, so it is not too surprising that the chemical composition of the interstellar medium in our region of the Milky Way is similar to the chemical composition of the Sun (see the "astronomer's periodic table" in Chapter 9). In the interstellar medium about 90 percent of the atomic nuclei are hydrogen, and the remaining 10 percent are almost all helium. More massive elements together account for only 0.1 percent of the atomic

Interstellar gas is extremely tenuous.

nuclei, or about 2 percent of the mass in the interstellar medium. Roughly 99 percent of that interstellar matter is gaseous, consisting of individual atoms or molecules moving freely about, as do the molecules in the air around us. But interstellar gas is far more tenuous than the thick soup that we breathe. Each cubic centimeter of the air around you contains about 2.5×10^{19} molecules. A good terrestrial vacuum pump can reduce this density down to about 10^{10} molecules/cm³—about a billionth as dense. By comparison, the interstellar medium has an average density of less than 1 atom/cm³—one ten-billionth as dense still! To make this more understandable, imagine a square tube 1 cm on a side and a meter long, filled with air and sealed at both ends. Also imagine you can stretch the tube without changing its cross section. As you do so, the air in the tube has to spread out to fill the larger volume and so is less dense. When the tube is stretched to a length of 10 meters, the air inside will be a tenth as dense as the air around you. When the tube is stretched to 100 meters, the air inside will be a hundredth as dense. For the air in the tube to match the average density of interstellar gas—that is, for it to have the same amount of mass per unit volume as the interstellar medium—you would have to stretch it into a tube 25,000 light-years long! Stated another way, there is about as much material in a column of air between your eye and the floor beside you as there is interstellar gas in a column with the same cross

section stretching from the Solar System to the center of the galaxy.

The Interstellar Medium Is Dusty

About 1 percent of the material in the interstellar medium is in the form of solid grains, referred to as **interstellar dust**. "Interstellar soot" might be a better name. Ranging from

About 1 percent of the mass of the interstellar medium is in grains.

little more than large molecules up to particles about 300 nm across, these solid grains more closely resemble the particles of soot from a candle flame than they do the dust that collects on a windowsill. (It would take several hundred "large" interstellar grains to span the thickness of a single human hair.) Interstellar dust gets its start when refractory materials such as iron, silicon, and carbon stick together to form grains in dense, relatively cool environments like the outer atmospheres and "winds" of cool, red giant stars, or in dense material thrown into space by stellar explosions. (More about these topics later.) Once in the interstellar medium, other atoms and molecules may stick to these grains. This process is remarkably efficient: About half of all interstellar matter more massive than helium (1 percent of the total mass of the interstellar medium) is found in interstellar grains.

If there is only as much material between us and the center of the galaxy as there is between your eye and the floor, then you might expect our view of distant objects to be as clear as your view of your own big toe. But nothing

Interstellar dust obscures our view.

could be further from the truth. It turns out that interstellar dust is extremely effective at blocking light. If the air around you contained as much dust as a comparable mass of interstellar material, it would be so dirty that you would be hard pressed to see your hand held up 10 cm in front of your face. Go out on a dark summer night in the Northern Hemisphere (or a dark winter night in the Southern Hemisphere) and look closely at the Milky Way, visible as a faint band of diffuse light running through the constellation Sagittarius. You will see a dark "lane" running roughly down the middle of this bright band, splitting it in two. This dark band is the result of vast expanses of interstellar dust blocking our view of distant stars.

When interstellar dust gets in the way of radiation from distant objects, the effect is called **interstellar extinction**. Not all electromagnetic radiation suffers equally from interstellar extinction, however. **Figure 15.1** shows two images of our galaxy: one taken in visible light, the other

FIGURE 15.1 The top image shows an all-sky picture of the Milky Way taken in visible light. The dark splotches blocking our view are dusty interstellar clouds. The bottom image shows the same field of view as seen in the near infrared. Infrared radiation penetrates interstellar dust, giving us a clearer view of the stars in our galaxy.

taken in the infrared (IR). The dark clouds that block the shorter-wavelength visible light seem to have vanished in the longer-wavelength IR image, allowing us to see through the clouds to the center of our galaxy and beyond.

As a starting point for understanding why short-wavelength radiation is blocked by dust while long-wavelength radiation is not, think about a different kind of wave—waves on the surface of the ocean. Imagine you are on the ocean in a strong swell. If you are floating in a small boat, which is much smaller than the wavelength of ocean waves, the swell causes you to bob gently up and down. But that is about all: There is no other interaction between the waves and the small boat. The story is quite different if you are on a larger boat that is about half as long as the wavelength of the ocean waves. Now the bow of the boat may be on a wave crest while the stern of the boat is in a trough, or vice versa. The boat tips wildly back and forth as the waves go by. If the size of the boat and the wavelength of the waves are a good match, even fairly modest waves will rock the boat. (You might have noticed this if you have ever been in a canoe or rowboat when the wake from a speedboat came by.) Now imagine viewing these two situations from the perspective of the wave. The wave is hardly affected by the small boat, but it is strongly affected by the large boat. The energy to

drive the wild motions of the boat comes from the wave, so the motion of the wave is affected by the interaction.

The interaction of electromagnetic waves with matter is more involved than a boat rocking on the ocean, but the same basic idea often applies. Tiny interstellar dust grains interact most strongly with ultraviolet and blue light, which have wavelengths comparable to the typical size of dust grains. For this reason, ultraviolet and blue light are effectively blocked by interstellar dust. Short wavelengths suffer

Long-wavelength radiation penetrates interstellar dust.

heavily from interstellar extinction. Infrared and radio radiation, on the other hand, have wavelengths that are too long to interact strongly with the tiny interstellar dust grains, so they can travel largely unimpeded across great interstellar distances. To sum up, at visible and ultraviolet wavelengths, most of the galaxy is hidden from our view by dust. In the infrared and radio portions of the spectrum, however, we get a far more complete view.

In the visible part of the spectrum, short-wavelength blue light suffers more from extinction than longer-wavelength red light. As a result, as shown in **Figure 15.2**, an object viewed through dust looks redder (actually less blue) than it really is. This effect is called **reddening**. Correcting for the fact that stars and other objects appear both fainter and redder than they would in the absence of dust can be one of the most difficult parts of interpreting astronomical observations, often adding to our uncertainty in the measurement of an object's properties.

Interstellar extinction may not be much of a concern at infrared wavelengths, but dust still plays an important role in what we see when we look at the sky in the infrared. Grains of dust, like any solid object, glow at wavelengths determined by their temperature. In Chapter 4 we learned

Dust glows in the infrared.

how the equilibrium between absorbed sunlight and emitted thermal radiation determines the temperatures of the terrestrial planets (see Figure 4.27). A similar equilibrium is at work in interstellar space, where dust is often heated by starlight and by the gas in which it is immersed to temperatures of tens to hundreds of kelvins (**Figure 15.3**). At a temperature of 100 K, Wien's Law says that dust will glow most strongly at a wavelength of 29 μm:

$$\lambda_{\max} = \frac{2,900 \ \mu\text{m K}}{T} = \frac{2,900 \ \mu\text{m K}}{100 \text{ K}} = 29 \ \mu\text{m}.$$

Cooler dust—say, at a temperature of 10 K—glows most strongly at a wavelength of 290 μm. Both of these wavelengths are in the far infrared part of the electromagnetic spectrum. Much of what we see when looking at the sky with infrared "eyes" is not really starlight at all, but thermal

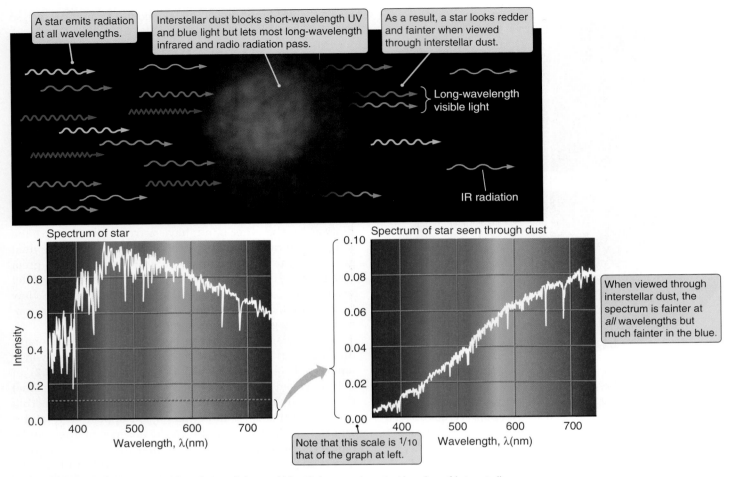

FIGURE 15.2 The wavelengths of ultraviolet and blue light are close to the size of interstellar grains, so the grains effectively block this light. Grains are less effective at blocking longer-wavelength light. As a result, the spectrum of a star (lower left) when seen through an interstellar cloud (lower right) appears fainter and redder.

radiation from dust. When NASA launched the Infrared Astronomical Satellite (IRAS) in 1983, astronomers got their first look at the sky at wavelengths out to 100 μm, and the views (like **Figure 15.4**) were unlike any seen before. IRAS showed the dark clouds that you see when you look at the Milky Way glowing brilliantly in infrared radiation from dust.

Interstellar Gas Has Different Temperatures and Densities

The gas and dust that fill interstellar space within our galaxy are not spread out evenly. About half of all interstellar gas is concentrated into only about 2 percent of the volume of interstellar space. These relatively dense regions are referred to as **interstellar clouds**. (Interstellar clouds are *not* like terrestrial clouds, which are locations where water

Most interstellar material is concentrated in relatively dense clouds.

vapor has condensed to form liquid droplets. Rather, interstellar clouds are places where the interstellar gas is more concentrated than in surrounding regions.) The other half of the interstellar gas is spread out through the remaining 98 percent of the volume of interstellar space. This is called **intercloud gas**, meaning gas that is found between the clouds.

The properties of intercloud gas vary from place to place. Some intercloud gas is extremely hot, with temperatures in the millions of kelvins, close to those found in the centers of stars. Even so, were you to find yourself adrift in an expanse of hot intercloud gas, your first concern would be freezing to death! The gas is hot, which means that the atoms that make it up are moving about very rapidly; so when one of these atoms runs into you, it hits you very hard. But the

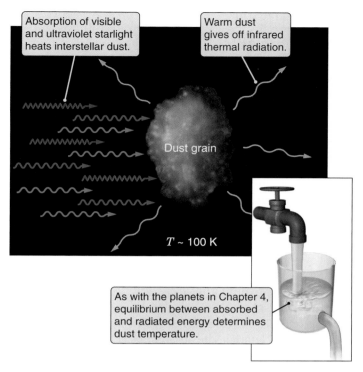

Absorption of visible and ultraviolet starlight heats interstellar dust.

Warm dust gives off infrared thermal radiation.

Dust grain

$T \sim 100$ K

As with the planets in Chapter 4, equilibrium between absorbed and radiated energy determines dust temperature.

FIGURE 15.3 The temperature of interstellar grains is determined by the same type of equilibrium between absorbed and emitted radiation that determines the temperatures of planets.

gas is also extremely tenuous. Typically you would have to search a liter (1,000 cm³) or more of hot intercloud gas to find a single atom. To gather up 1 gram of hot intercloud gas —about the mass found in a liter of air—you would have to collect all of the gas in a cube over 8,000 km on an edge! There are so few atoms in a given volume of hot intercloud gas that atoms would not run into you very frequently, so this million-kelvin gas would do little to keep you warm.

You would radiate energy away and cool off much faster than the gas around you could replace the lost energy.

Extremely hot intercloud gas is thought to occupy about half the volume of interstellar space. This gas is heated mostly by the energy of tremendous stellar explosions called **supernovae**. (We will return to the story of supernovae in Chapters 16 and 17. For now it is enough to know that these are *very* energetic events!) Our Solar System is located inside a rarefied bubble of hot intercloud gas that has a density of about 0.006 hydrogen atoms per cubic centimeter and is

We live in a bubble of million-kelvin intercloud gas.

about 650 light-years across. Some astronomers have suggested that this is the remnant of the hot bubble produced by a supernova explosion about 100,000 years ago. Like the million-kelvin gas in the corona of the Sun, hot intercloud gas glows faintly in the energetic X-ray portion of the electromagnetic spectrum. X-rays cannot penetrate Earth's atmosphere, so it is necessary to get above our atmosphere to observe the radiation coming from hot intercloud gas. For rocket- and satellite-borne X-ray telescopes, however, the entire sky is aglow with faint X-rays coming from the bubble of million-kelvin hot gas in which our Solar System is immersed (see **Figure 15.5**).

Not all intercloud gas is as hot as our local bubble. Most of the remaining intercloud gas has a temperature of about 8,000 K and a density ranging from about 0.01 to 1 atom per cubic centimeter. To gather up a gram of this warm intercloud gas, we would need to collect all of the gas in a cube of interstellar space "only" 800 km on an edge.

Interstellar space is awash with the light from stars. Some of this is ultraviolet light with wavelengths shorter than 91.2 nm. Each photon of this ultraviolet light has enough energy that, if absorbed by a hydrogen atom, it will ionize the atom, kicking its electron free of the nucleus. If

FIGURE 15.4 Far infrared images of the sky show clouds of warm, glowing interstellar dust. (The black bands in this and the following figure are due to missing data.)

FIGURE 15.5 The faint X-ray glow that fills the sky is due largely to emission coming from the bubble of million-kelvin gas that surrounds the Sun. Bright spots are more distant sources, including objects such as bubbles of very hot, high-pressure gas surrounding the sites of recent supernova explosions.

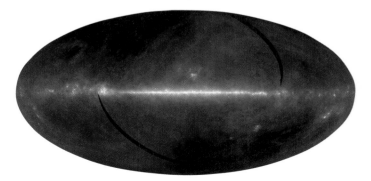

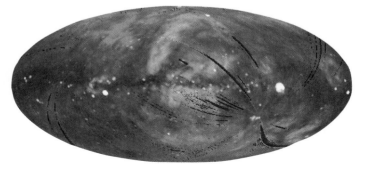

there are enough of these energetic photons around, then atoms in the interstellar gas cannot remain neutral (because any neutral hydrogen atom will soon be ionized by starlight). In this way about half of the volume of warm intercloud gas is kept ionized. But if there is a large enough expanse of warm intercloud gas, then the ionizing photons in the starlight get "used up." About half of the volume of warm intercloud gas is mostly neutral. This gas is usually surrounded by regions of ionized intercloud gas that shield the neutral gas from starlight, much as Earth's ozone layer shields the surface of Earth from harmful ultraviolet radiation from the Sun.

One way to look for interstellar gas, including both warm and hot interstellar gas, is to study the spectra of distant stars. As starlight passes through interstellar gas, absorption lines form. Just as absorption lines formed in the atmosphere of a star tell us much about the star's temperature, density, and chemical composition, interstellar absorption lines provide similar information about the gas the light has passed through. Interstellar gas can also be studied by looking for the radiation that it emits. In regions of warm *ionized* gas, protons and electrons are constantly "finding each other" and recombining to make hydrogen atoms. Because it takes energy to tear a hydrogen atom apart, conservation of energy says that when a proton and electron combine to form a hydrogen atom, this same amount of energy must be freed up. This freed-up energy must go somewhere, and that somewhere is into the form of electromagnetic radiation. Typically when a proton and electron combine,

Interstellar gas can produce both emission and absorption lines.

the resulting hydrogen atom is left in an excited state (see our discussion of energy states of atoms in Chapter 4). The atom then drops down to lower and lower energy states, emitting a photon at each step. Together these photons carry away an amount of energy equal to the ionization energy of hydrogen plus whatever energy of motion the electron and proton had before recombining. The photons emitted in this process have wavelengths corresponding to the energy differences between the allowed energy states of hydrogen. In other words, warm ionized interstellar gas glows in emission lines characteristic of hydrogen (as well as other elements). Usually the strongest emission line given off by warm interstellar gas in the visible part of the spectrum is the Hα (hydrogen alpha) line, which is seen in the red part of the spectrum at a wavelength of 656.3 nm.

Figure 15.6 shows an image of a portion of the sky taken in the light of Hα emission. The faint diffuse emission in the figure comes mostly from warm ionized intercloud gas. The especially bright spots, on the other hand, are quite different. These are regions where intense ultraviolet radiation from massive, hot, luminous O and B stars is able to ionize even relatively dense interstellar clouds. These are

H II regions are ionized by hot, luminous O and B stars.

called **H II regions** (pronounced "H two"), signifying that they are made up of the second, or ionized, form of hydrogen. In the chapters to come we will address the question of how long stars live, and we will discover that O stars live for only a few million years. As a result they usually do not move too far away from where they formed. The glowing clouds seen as H II regions are the very clouds from which the stars were born. Thus H II regions are signposts to regions where active star formation is taking place.

One of the closest H II regions to the Sun is the "Great Nebula" in the constellation Orion (see **Figure 15.7**). In fact, the Orion Nebula is tiny as H II regions go. It is "great" only because it happens to be very close to us. Almost all of the ultraviolet light that powers the **nebula** comes from a single hot star, and all told there are only a few hundred stars forming in the immediate vicinity. In contrast, **Figure 15.8** shows a giant H II region called 30 Doradus (also known as the Tarantula Nebula). The 30 Doradus H II region is located in the Large Magellanic Cloud, a small companion galaxy to the Milky Way, 150,000 light-years distant. This enormous cloud of ionized gas is powered by a dense star cluster containing thousands of hot, luminous stars.

Warm *neutral* hydrogen gas gives off radiation in a different way than does warm ionized hydrogen. Many subatomic particles, including protons and electrons, are magnetized. It is as if each of these particles has a bar magnet, with a north and a south pole, built into it. According to the rules of quantum mechanics, a hydrogen atom can exist in only two different configurations. Either the magnetic "poles" of the proton and electron point in the *same* direction, or they point in *opposite* directions. The "electron's magnet" and the "proton's magnet" push on each other, so the way they line up affects the energy that the

FIGURE 15.6 Warm (about 8,000 K) interstellar gas glows in the Hα line of hydrogen. This image, showing the Hα emission from much of the northern sky, reveals how complex the structure of the interstellar medium is.

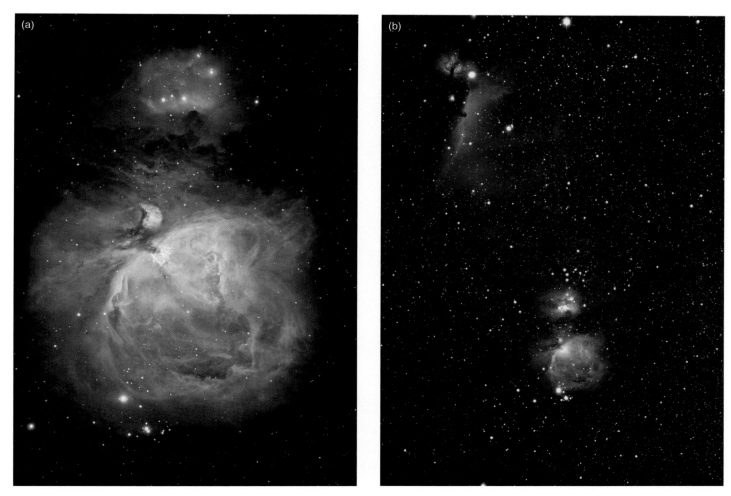

FIGURE 15.7 (a) The Orion Nebula is seen here as a glowing region of interstellar gas surrounding a cluster of young, hot stars. New stars are still forming in the dense clouds surrounding the nebula. (b) The Orion Nebula (at bottom) is only a small part of the larger Orion star-forming region.

atom has. When the two "magnets" point in opposite directions the atom has slightly more energy than when they point in the same direction. It takes energy to change a hydrogen atom from the lower-energy, magnetically aligned state to the higher-energy, magnetically unaligned state. This is no problem because, in warm neutral interstellar gas, interactions between atoms easily supply enough energy to do the trick. But once a hydrogen atom is in its

Neutral hydrogen emits 21-cm radiation.

higher-energy (magnetically unaligned) state, getting back into its lower-energy (magnetically aligned) state requires nothing but time. If left alone long enough, a hydrogen atom in the higher-energy state will spontaneously jump to the lower-energy state, emitting a photon in the process. The energy difference between the two magnetic states of a hydrogen atom is extremely small: less than a two-millionth

of the energy needed to tear the electron away from the proton completely. The radiation given off by these low-energy transitions has a wavelength of 21 cm, which lies in the radio portion of the spectrum.

The tendency for hydrogen atoms to emit 21-cm radiation is extremely weak. On average you would have to wait about 11 million years for a hydrogen atom in the higher-energy state to spontaneously jump to the lower-energy state and give off a photon. Even so, there is a *lot* of hydrogen in the universe. As seen in **Figure 15.9**, the sky is aglow with 21-cm radiation from neutral hydrogen. Because of its long wavelength, 21-cm radiation freely penetrates dust in the interstellar medium, allowing us to see neutral hydrogen throughout our galaxy, while measurements of the Doppler shift of the line tell how fast the source of the radiation is moving. These two attributes make the 21-cm line of neutral hydrogen perhaps the most important kind of radiation there is for understanding the structure of our galaxy.

FIGURE 15.8 This ESO image shows the 30 Doradus Nebula, located in a small companion galaxy to the Milky Way, 150,000 light-years from Earth. Interstellar gas in this region of intense star formation is caused to glow by the ultraviolet radiation coming from the dense cluster of thousands of young, massive stars.

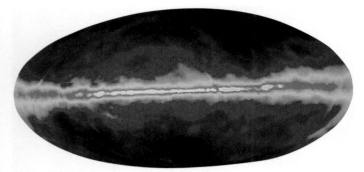

FIGURE 15.9 This radio image of the sky taken at a wavelength of 21 cm shows clouds of neutral hydrogen gas throughout our galaxy. Because radio waves penetrate interstellar dust, 21-cm observations are a crucial probe of the structure of our galaxy.

Regions of Cool, Dense Gas Are Called Clouds

As mentioned earlier, intercloud gas fills 98 percent of the volume of interstellar space but accounts for only half of the mass of interstellar gas. The remaining 50 percent of all interstellar gas is concentrated into much denser interstellar clouds that occupy only 2 percent of the volume of the galaxy. Most interstellar clouds are composed primarily of neutral atomic hydrogen (that is, isolated neutral hydrogen atoms). These clouds are much cooler and denser than the warm intercloud gas. They have temperatures of around 100 K and densities in the range of about 1 to about 100 atoms per cubic centimeter. Even so, at a density of 100 atoms/cm^3 you would still have to gather up all of the gas in a box 170 km on an edge to collect a single gram of this material.

On Earth it is uncommon to find atoms in isolation —most atoms are tied up in molecules. (For example, in Earth's atmosphere only nonreactive gases such as argon are typically found in their atomic form.) Yet so far the interstellar gas that we have encountered consists of isolated atoms rather than molecules. The reason is that throughout most of interstellar space, including most interstellar clouds, molecules do not survive for long. If interstellar gas is too hot, then any molecules that do exist will soon collide with other molecules or atoms with enough energy to break the molecules back into their constituent atoms. The temperature in a neutral hydrogen cloud may be low enough for some molecules to survive, but even here a different process will soon destroy any molecules that might form. Photons of starlight with enough energy to break molecules apart can penetrate into neutral hydrogen clouds. Only in the hearts of the densest of interstellar clouds, where dust effectively blocks even the relatively low-energy photons required to shatter molecules, does interstellar chemistry get its chance. For obvious reasons these dark clouds are referred to as **molecular clouds**.

In images such as **Figure 15.10** molecular clouds are evident from their silhouettes against a background of stars. Inside such clouds it is dark and usually very cold, with a typical temperature of only about 10 K—that is, 10 kelvins above absolute zero. Most of these clouds have densities of

---

Molecular clouds are dense and cold.

---

around 100 to 1,000 molecules/cm^3, but densities as high as 10^{10} molecules/cm^3 have been observed. (Even at 10^{10} molecules/cm^3 this gas is still less than a billionth as dense as the air around you, making it an extremely good vacuum by terrestrial standards.) In this cold, relatively dense environment, atoms combine to form a wide variety of molecules.

By far the most common component of molecular clouds is molecular hydrogen. Molecular hydrogen, which consists of two hydrogen atoms, is the smallest possible molecule. Molecules radiate mostly in the radio and infrared portions of the electromagnetic spectrum. Molecular emission lines are useful to astronomers in much the same way as the spectral lines from atoms: Each type of molecule is unique in its properties, and the energies of its allowed states are unique as well. The wavelengths of emission lines from molecules are an unmistakable fingerprint of the kind of molecule responsible.

In addition to molecular hydrogen, over 140 other molecules have been discovered in interstellar space. These molecules range from very simple structures such as carbon monoxide (CO), one of the most common molecules, to complex organic compounds such as methanol (CH_3OH) and acetone ($(CH_3)_2CO$), to carbon chains as large as $HC_{11}N$. Very large carbon molecules, made of hundreds of individual atoms, bridge the gap between large interstellar molecules and small interstellar grains. As mentioned earlier, visible light cannot escape from the molecular clouds where interstellar molecules are concentrated. Fortunately, however, the radio waves emitted by molecules are unaffected by interstellar dust, so they escape easily from dark molecular clouds. Observations of radio waves from molecules give us a remarkable look at the innermost workings of the densest and most opaque interstellar clouds.

Molecular clouds have masses ranging from a few times the mass of our Sun to 10^7 (10 million) times the mass of our Sun. The smallest molecular clouds may be less than half

Stars form in giant molecular clouds.

a light-year across, whereas the largest molecular clouds may be over a thousand light-years in size. The largest molecular clouds qualify for the title **giant molecular cloud**. These behemoths typically have masses of a few hundred thousand times that of our Sun and on average are about 120 light-years across. There are about 4,000 giant molecular clouds in our galaxy and many more smaller ones. Molecular clouds fill only a tiny fraction of the volume of our galaxy—probably only about 0.1 percent of interstellar space. These clouds may be rare, but they are extremely important to our story because they are the cradles of star formation.

Having looked at the many different phases of the interstellar medium, ranging from vast expanses of tenuous million-kelvin intercloud gas to the cold dense interiors of molecular clouds, an obvious question to ask is "How does it all fit together?" Discussing the structure of the interstellar medium is a lot like discussing the nature of weather on Earth: It is extremely complex, often difficult to understand, and all but impossible to predict precisely.

In the interstellar medium the energy to power the "weather" comes from stars. Warm gas heated by ultraviolet radiation from massive, hot stars pushes outward into its surroundings. Blast waves from supernova explosions sweep out vast, hot "cavities," piling up everything in their path like snow in front of a snowplow. Interstellar clouds are destroyed under the onslaught of these violent events, but they are also formed by these same forces. Swept-up intercloud gas becomes the next generation of clouds. Hot bubbles of high-pressure interstellar gas crush molecular clouds, driving up their densities and perhaps triggering the formation of new generations of stars. The interstellar medium is so well stirred that *on average* interstellar clouds are moving around with random velocities of approximately 20 km/s. The fastest motions of interstellar material are those of very hot gas and are measured in thousands of kilometers per second.

The story of the wide range of phenomena that shape the interstellar medium is fascinating. We could easily stay here and look around for quite some time, but instead we must press on in our journey. The time has come to watch as a molecular cloud begins the slow process of collapsing under its own weight, beginning a chain of events that will culminate with the ignition of nuclear fires within a new generation of stars.

FIGURE 15.10 In visible light, interstellar molecular clouds are seen in silhouette against a background of stars and glowing gas. Interstellar dust in the clouds blocks short-wavelength visible and ultraviolet light.

15.3 Molecular Clouds Are the Cradle of Star Formation

As we have seen, objects such as planets and stars are held together by gravity. The mutual gravitational attraction between each part of a planet or star and every other part of the object results in a net inward force that pulls all parts of the object toward its center. If an object is stable, then this inward force of the object's self-gravity must be balanced by something—often the outward push of pressure within the object. In our discussion of planets and stars we referred to the balance between gravity and pressure in a stable object as *hydrostatic equilibrium.*

All of these same concepts can be applied to clouds in the interstellar medium. As shown in **Figure 15.11**, interstellar clouds also have self-gravity: Each part of the cloud feels a gravitational attraction for every other part of the cloud. However, interstellar clouds are rarely in hydrostatic equilibrium. In most interstellar clouds pressure is much stronger than self-gravity. (Because gravity follows an inverse square law, the more spread out an object's mass is, the weaker is its self-gravity.) Pressure in these clouds would cause them to expand if not for the pressure of the surrounding intercloud medium holding them in. The intercloud medium is much less dense than interstellar clouds, but it is also much hotter, so its pressure is high enough to confine the clouds. Some molecular clouds, on the other hand,

FIGURE 15.11 Parcels of gas within a molecular cloud feel the gravitational attraction of all other parts of the molecular cloud, leading to a net gravitational force toward the cloud's center. This is referred to as the cloud's self-gravity.

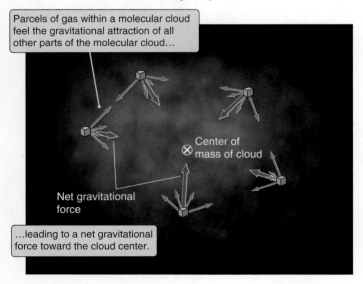

Parcels of gas within a molecular cloud feel the gravitational attraction of all other parts of the molecular cloud...

⊗ Center of mass of cloud

Net gravitational force

...leading to a net gravitational force toward the cloud center.

violate hydrostatic equilibrium in the other direction. These clouds are massive enough and dense enough—that is, there is enough mass packed together in a small enough volume—that their self-gravity becomes important. (More mass along with smaller distances between different parts of the

Some molecular clouds collapse under their own weight.

cloud means that gravity is stronger.) Furthermore, these clouds are cool enough that pressure in the clouds is relatively low despite their high density. In such clouds self-gravity is much greater than pressure, so the clouds collapse under their own weight.

If self-gravity in a molecular cloud is much more significant than pressure, we might expect that gravity should win outright, and the cloud should rapidly collapse toward its center. In practice the process goes very slowly because there are several other effects that stand in the way of the collapse. One such effect that slows the collapse of a cloud is conservation of angular momentum. (We discussed the effects of angular momentum and the flattening of a collapsing cloud in Chapter 6.) Other properties of molecular clouds that prevent them from collapsing rapidly are tur-

Angular momentum, turbulence, and magnetic fields slow the collapse but gravity wins in the end.

bulence and the effects of magnetic fields. Even though these effects may slow the collapse of a molecular cloud, in the end gravity will win. Magnetic fields in the cloud can slowly die away. One part of the cloud can lose angular momentum to another part of the cloud, allowing the part of the cloud with less angular momentum to collapse further. Turbulence dies away. The details of these processes are complex and the subject of much current research. But the crucial point is this: The effects preventing the collapse of a molecular cloud are temporary, but gravity is persistent. As the forces that oppose the cloud's self-gravity gradually fade away, the cloud slowly becomes smaller.

Molecular Clouds Fragment as They Collapse

Crucial to the process of star formation is the fact that as a cloud becomes smaller, the gravitational forces trying to pull it together grow stronger. This is a result of the inverse square law of gravitation discussed in Chapter 3. Suppose a cloud starts out being 4 light-years across. By the time the cloud has collapsed to 2 light-years across, the different parts of the cloud are, on average, only half as far away from

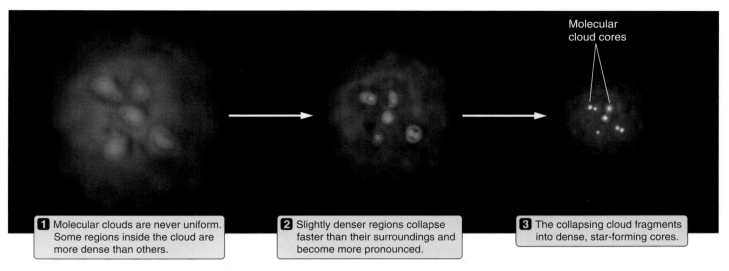

1 Molecular clouds are never uniform. Some regions inside the cloud are more dense than others.

2 Slightly denser regions collapse faster than their surroundings and become more pronounced.

3 The collapsing cloud fragments into dense, star-forming cores.

FIGURE 15.12 As a molecular cloud collapses, denser regions within the cloud collapse more rapidly than less dense regions. As this process continues, the cloud fragments into a number of very dense molecular cloud cores that are embedded within the large cloud. These cloud cores may go on to form stars.

each other as they were when they started. As a result, the gravitational attraction they feel toward each other will be four times stronger than when the cloud was 4 light-years across. When the cloud is ¼ as large as when it started out (or 1 light-year across), the force of gravity will be 16 times as strong. As a collapsing cloud becomes smaller, gravity becomes stronger. As gravity becomes stronger, the collapse picks up speed. And as the collapse picks up speed, gravity increases even more rapidly. The collapse "snowballs."

Molecular clouds are never uniform. Some regions within the cloud are invariably denser than others and so collapse within themselves more rapidly than surrounding regions. As these regions collapse, their self-gravity becomes stronger

Stars form in molecular cloud cores.

because they are more compact, so they collapse even faster. **Figure 15.12** shows the result of this process. What started out as slight variations in the density of the cloud grow to become very dense concentrations of gas. Instead of collapsing into a single object, the molecular cloud fragments into a number of very dense **molecular cloud cores**. A single molecular cloud may form hundreds or thousands of molecular cloud cores. It is from these dense cloud cores, typically a few light-months in size, that stars may form.

As a molecular cloud core becomes smaller, the gravitational forces that are trying to crush the cloud grow stronger. Eventually gravity is able to overwhelm the forces due to pressure, magnetic fields, and turbulence that have been resisting it. This happens first near the center of the cloud because it is there that the material in the cloud is most

strongly concentrated. The inner parts of the cloud core start to rapidly fall inward. This starts a sequence of events much like a chain of dominos in which one domino topples the next, which topples the next. The pressure from the central part of the cloud core had been supporting the weight of the layers above it. When the center of the cloud collapses, this support is suddenly removed. Without this "pressure support," the next outer layer begins to fall freely inward toward the center as well. But what about the layers farther out? Now there is nothing to hold them up, either, so the process continues: Each layer of the cloud core falls inward in turn, thereby removing support from the layers still farther out. As shown in **Figure 15.13**, the cloud core collapses like a house of cards when the bottom layer is knocked out. The whole structure comes crashing down.

We have been here before. It was at this point in the story of star formation that our discussion of the formation of the Solar System in Chapter 6 began. An overview of the process is shown in **Figure 15.14**. Material from the collapsing molecular cloud core falls inward. Because of its angular momentum, this material lands on a flat, rotating accretion disk. Figure 6.1 shows images of several such disks. The

Our Solar System began in a molecular cloud.

dark bands in these images are the disks seen edge-on, while the bright regions in the figure are starlight reflected from the surfaces of the disks. Most of this material eventually finds its way inward onto the growing protostar at the center of the disk, but a small fraction of it remains to become

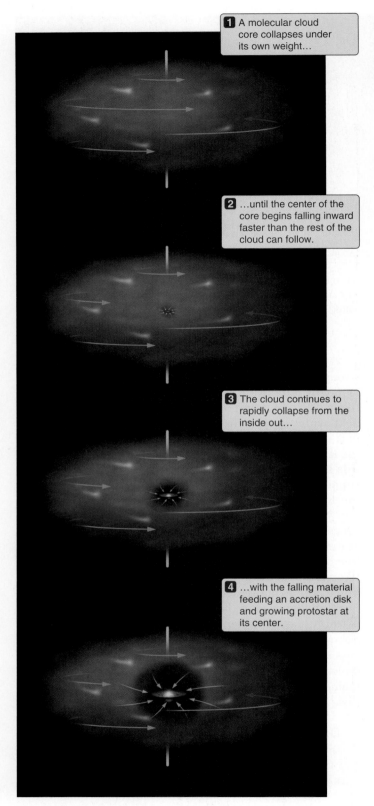

1 A molecular cloud core collapses under its own weight...

2 ...until the center of the core begins falling inward faster than the rest of the cloud can follow.

3 The cloud continues to rapidly collapse from the inside out...

4 ...with the falling material feeding an accretion disk and growing protostar at its center.

FIGURE 15.13 When a molecular cloud core gets very dense, it collapses from the inside out. Conservation of angular momentum causes the infalling material to form an accretion disk that feeds the growing protostar.

the stuff planets are made of. The last time we passed this way, we followed the evolution of the gas and dust left behind in the disk and saw how it leads naturally to a planetary system with the properties of our Solar System. This time through we will instead follow the story of the protostar as it becomes a star.

15.4 The Protostar Becomes a Star

We pick up the story of our protostar at a time when the cloud is still collapsing, and more and more material is falling on the disk. At this point the surface of the protostar has also been heated to a temperature of thousands of kelvins as the gravitational energy of the original molecular cloud is converted into thermal energy. The protostar is also

Protostars are large and luminous.

huge—hundreds of times larger than our Sun—which means that the surface of the protostar is tens of thousands of times larger than the surface of the Sun. Each square meter of that enormous surface is radiating energy away in accordance with Stefan's Law. As a result the protostar is thousands of times more luminous than our Sun, even though the nuclear reactions that will power it through its life have yet to begin.

At this stage in its life a protostar is extremely luminous, yet it probably cannot be seen by astronomers in visible light. There are two reasons for this. First, the protostar is relatively cool as stars go, so most of its radiation is in the infrared part of the spectrum. Even more important, the protostar begins its life buried deep in the heart of a dense and dusty molecular cloud. The dust in the cloud blocks any visible light, so our view of the protostar is obscured. On the other hand, much of the longer-wavelength infrared light from the protostar *is* able to escape from the star's hidden cradle. In addition, the dust in the cloud core is heated by the light from the star, and this dust also glows in the infrared. For this reason protostars are studied mostly in the infrared part of the spectrum.

The study of protostars and other related young stellar objects was revolutionized during the 1980s and 1990s as sensitive new instruments capable of "seeing" in the infrared were trained on regions of known star formation. What we discovered was that the dark clouds, long familiar to us but impenetrable until then, are really entire clusters of dense cloud cores, young stellar objects, and glowing dust when viewed in the infrared. **Figure 15.15** shows infrared pictures of stars forming within columns of gas and dust in the Eagle Nebula.

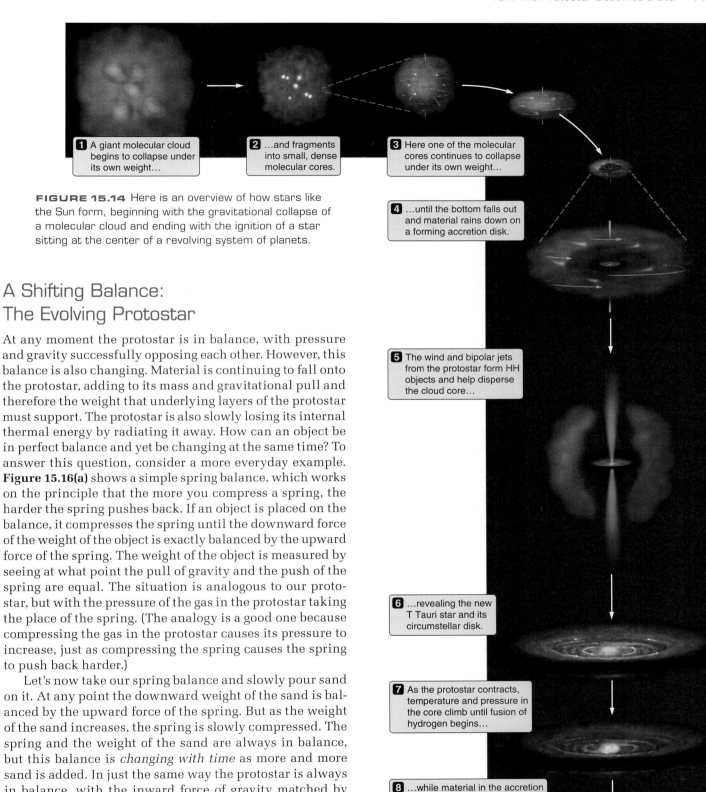

1 A giant molecular cloud begins to collapse under its own weight…

2 …and fragments into small, dense molecular cores.

3 Here one of the molecular cores continues to collapse under its own weight…

FIGURE 15.14 Here is an overview of how stars like the Sun form, beginning with the gravitational collapse of a molecular cloud and ending with the ignition of a star sitting at the center of a revolving system of planets.

4 …until the bottom falls out and material rains down on a forming accretion disk.

5 The wind and bipolar jets from the protostar form HH objects and help disperse the cloud core…

6 …revealing the new T Tauri star and its circumstellar disk.

7 As the protostar contracts, temperature and pressure in the core climb until fusion of hydrogen begins…

8 …while material in the accretion disk may coalesce to form a planetary system.

A Shifting Balance: The Evolving Protostar

At any moment the protostar is in balance, with pressure and gravity successfully opposing each other. However, this balance is also changing. Material is continuing to fall onto the protostar, adding to its mass and gravitational pull and therefore the weight that underlying layers of the protostar must support. The protostar is also slowly losing its internal thermal energy by radiating it away. How can an object be in perfect balance and yet be changing at the same time? To answer this question, consider a more everyday example. **Figure 15.16(a)** shows a simple spring balance, which works on the principle that the more you compress a spring, the harder the spring pushes back. If an object is placed on the balance, it compresses the spring until the downward force of the weight of the object is exactly balanced by the upward force of the spring. The weight of the object is measured by seeing at what point the pull of gravity and the push of the spring are equal. The situation is analogous to our protostar, but with the pressure of the gas in the protostar taking the place of the spring. (The analogy is a good one because compressing the gas in the protostar causes its pressure to increase, just as compressing the spring causes the spring to push back harder.)

Let's now take our spring balance and slowly pour sand on it. At any point the downward weight of the sand is balanced by the upward force of the spring. But as the weight of the sand increases, the spring is slowly compressed. The spring and the weight of the sand are always in balance, but this balance is *changing with time* as more and more sand is added. In just the same way the protostar is always in balance, with the inward force of gravity matched by the outward force of pressure (see **Figure 15.16(b)**). But its balance, too, is changing with time. Just like sand that has fallen onto the spring balance, material that has fallen onto the protostar compresses the protostar more and more, and the protostar evolves.

HST image in visible light shows columns of molecular gas illuminated by nearby stars.

When viewed in infrared radiation we see into the dusty columns where new stars are forming from interstellar gas.

FIGURE 15.15 This HST image of the Eagle Nebula shows dense columns of molecular gas and dust at the edge of an H II region. Infrared images of the same field, also taken with the HST, show young stars forming within these columns.

As material continues to fall onto the protostar, it adds weight to the outer layers of the protostar. This growing weight compresses the material in the interior of the protostar; and as it is compressed, the interior of the protostar grows hotter. (Compressing a gas always causes it to heat up, and letting a gas expand always causes it to cool down. The cooling systems in your air conditioner and refrigerator work by compressing gas to make it hot, then letting this hot gas cool so that when it expands again it gets really cold.) If the interior of the protostar is now denser and hotter, that also means that the pressure is higher—just enough higher to balance the increased weight of the material above it. Balance is always maintained.

Even after material stops falling on the protostar, the protostar *still* goes on evolving in much the same way. Earlier we saw that the protostar is radiating away many times as much energy as the Sun. The source of the energy being radiated away is the thermal energy trapped in the interior of the protostar—the same thermal energy responsible for supporting the protostar against the forces of gravity. So as thermal energy leaks out of the protostar, gravity gains the upper hand, and the protostar slowly contracts. As the protostar shrinks, the forces of gravity become greater (because the parts of the protostar are closer together—remember the inverse square law of gravity). Meanwhile, as the protostar shrinks its interior is compressed, so it grows hotter. As density and temperature climb, so does the pressure—*just enough to continue*

The protostar radiates away thermal energy and shrinks.

to counteract the growing force of gravity. Balance between gravity and pressure is always maintained. This process continues, with the star becoming smaller and smaller and its interior growing hotter and hotter, until the center of the star is finally hot enough for fusion of hydrogen into helium to begin.

Something strange is going on here. When an object radiates energy, it normally gets cooler. When you turn off the electric coil on your stove, the coil cools as it loses the thermal energy within it. Yet we just concluded that as a protostar radiates thermal energy away, it actually grows hotter. How can that be? Once again the answer lies with the concept of conservation of energy. As the protostar contracts, every part of the protostar is slowly falling toward its center, which means that the protostar is *losing gravitational energy*. This gravitational energy has to show up in some other form, and it does so as thermal energy. Some of this thermal energy is radiated away, but not all of it—so the interior of the star grows hotter.

This chain of events is shown in a nutshell in **Figure 15.17**. The protostar radiates away thermal energy, and as it does so, it loses pressure support. Deprived of pressure support, the protostar contracts. As the protostar contracts, gravitational energy is converted to thermal energy, driving up the temperature in the protostar's interior. If the protostar is massive enough, then the temperature in the interior of the protostar will eventually become hot enough for nu-

Fusion begins, and the protostar becomes a star.

clear fusion to begin. Here is the point at which the transition from protostar to star takes place. The distinction between the two is that a *protostar* draws its energy from gravitational collapse, whereas a *star* draws its energy from thermonuclear reactions in its interior.

FIGURE 15.16 (a) A spring balance comes to rest at the point where the downward force of gravity is matched by the upward force of the compressed spring. As sand is added, the location of this balance point shifts. (b) Similarly, the structure of a protostar is determined by a balance between pressure and gravity. Like the spring balance, the structure of the protostar constantly shifts as the protostar radiates energy away and additional material falls onto its surface.

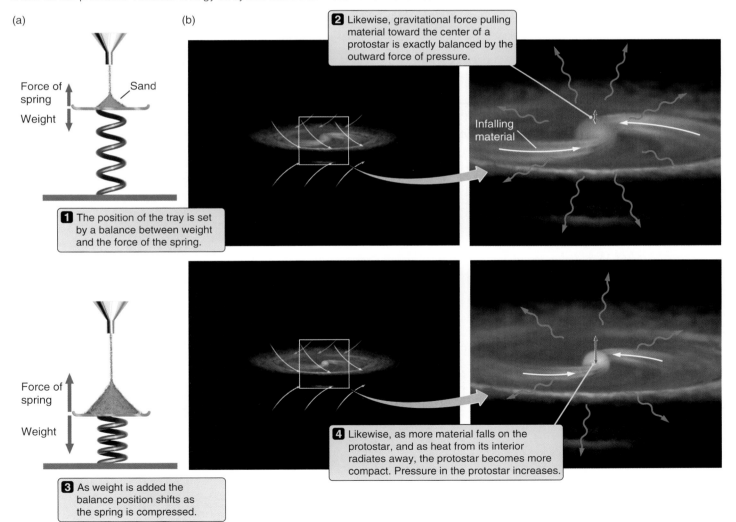

(a)

(b)

Force of spring

Sand

Weight

1 The position of the tray is set by a balance between weight and the force of the spring.

2 Likewise, gravitational force pulling material toward the center of a protostar is exactly balanced by the outward force of pressure.

Infalling material

Force of spring

Weight

3 As weight is added the balance position shifts as the spring is compressed.

4 Likewise, as more material falls on the protostar, and as heat from its interior radiates away, the protostar becomes more compact. Pressure in the protostar increases.

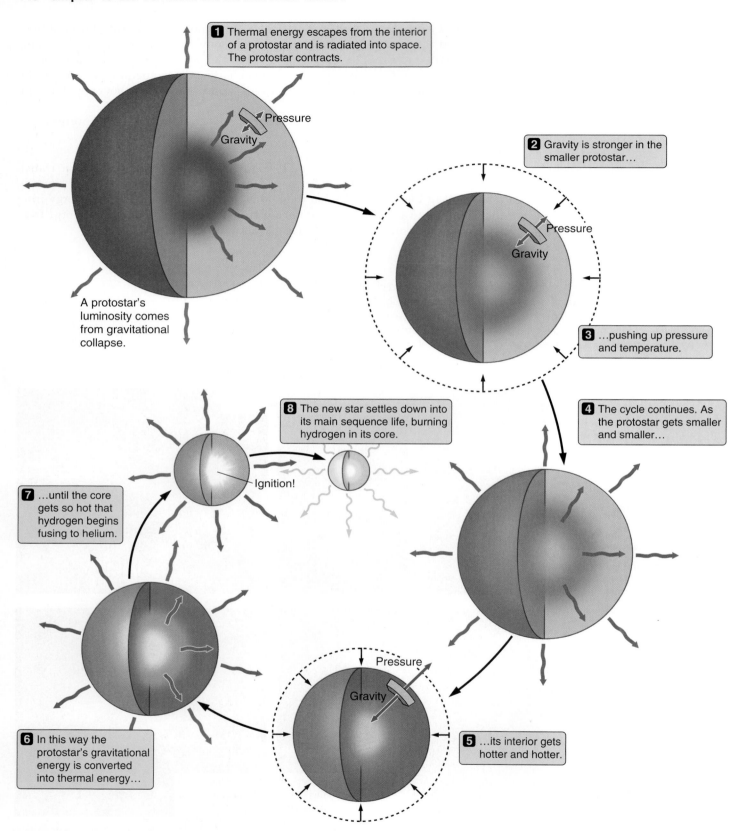

1 Thermal energy escapes from the interior of a protostar and is radiated into space. The protostar contracts.

Pressure

Gravity

A protostar's luminosity comes from gravitational collapse.

2 Gravity is stronger in the smaller protostar...

Pressure

Gravity

3 ...pushing up pressure and temperature.

4 The cycle continues. As the protostar gets smaller and smaller...

5 ...its interior gets hotter and hotter.

Pressure

Gravity

6 In this way the protostar's gravitational energy is converted into thermal energy...

7 ...until the core gets so hot that hydrogen begins fusing to helium.

8 The new star settles down into its main sequence life, burning hydrogen in its core.

Ignition!

FIGURE 15.17 As a protostar radiates away energy, it loses pressure support in its interior and contracts. This contraction drives the interior pressure up. The counterintuitive result is that radiating energy away causes the interior of the protostar to grow hotter and hotter until nuclear reactions begin in its interior.

The protostar's mass determines whether it will ever become a star. As the protostar slowly collapses, the temperature at its center gets higher and higher. If the protostar is more massive than about 0.08 $M_\odot$, the temperature in its

At least 0.08 $M_\odot$ is needed to be a star.

core will eventually reach the 10-million-kelvin mark, and fusion of hydrogen into helium will begin. When this happens the newly born star will once again adjust its structure until it is radiating energy away from its surface at just the rate energy is being liberated in its interior. As it does so, it "settles" onto the main sequence of the H-R diagram, where it will spend the majority of its life. If the protostar has a mass of less than 0.08 $M_\odot$, it will never reach the point where nuclear burning takes place. Such a failed star, the first examples of which were discovered in the closing years of the 20th century, is called a **brown dwarf**. A brown dwarf is neither star nor planet, but something in between. It forms the same way a star forms, yet in many respects it is like a giant Jupiter. Brown dwarf candidates are seen as especially

Brown dwarfs are failed stars.

cool M stars and as members of the L and T spectral types discussed in Chapter 13. A brown dwarf never burns hydrogen in its core but instead glows by continuously cannibalizing its own gravitational energy. As the years pass a brown dwarf gets progressively smaller and fainter.

Evolving Stars and Protostars Follow "Evolutionary Tracks" on the H-R Diagram

Within the evolving protostar it is convection rather than radiation that carries energy outward, keeping the protostar's interior well stirred. Interestingly, while the interior of the protostar grows hotter and hotter as it contracts, its *surface* stays at about the same temperature through most of this phase of its evolution. This distinction is important. The surface temperature of a star or protostar is *not* the same as the temperature deep in its interior. For example, in Chapter 14 we found that the temperature of the surface of the Sun is 5,860 K while the temperature of its interior is millions of kelvins. As the protostar contracts, the temperature deep within the star grows hotter and hotter, but the temperature of the star's *surface* remains nearly constant.

In the 1960s the Japanese theoretician Chushiro Hayashi explained why this is so. Hayashi pointed out that the atmospheres of stars and protostars contain a natural thermostat: the H⁻ ion. (An H⁻ or "H minus" ion is a hydrogen atom that has acquired an extra electron and therefore has a negative charge.) The amount of H⁻ in the atmosphere of

a protostar is highly sensitive to the temperature at the surface of the protostar. The cooler the atmosphere of a star, the more slowly atoms and electrons are moving, and the easier it is for a hydrogen atom to hold onto an extra electron. As a result, the cooler the atmosphere of the star, the more H⁻ there is.

The H⁻ ion, in turn, helps to control how much energy is radiated away by a star or protostar. The more H⁻ there is in the atmosphere of a star or protostar, the more opaque the atmosphere is, and the more effectively the thermal energy of the protostar is trapped in its interior. Imagine that the surface of the protostar is too cool. Too cool means that

H⁻ is a stellar thermostat.

extra H⁻ forms in the atmosphere, making the atmosphere of the protostar more opaque. This traps more of the radiation that is trying to escape, and the trapped energy heats up the star. As the temperature climbs, H⁻ ions are destroyed (that is, changed to neutral H atoms). Now imagine the other possibility—that the protostar is too hot. Then H⁻ in its atmosphere will be destroyed, so the atmosphere will become more transparent, allowing radiation to more freely escape from the interior. Because the protostar cannot hold onto enough of its energy to stay warm, the surface cools. In either case—too cold or too hot—H⁻ is formed or destroyed until the star's atmosphere once again traps just the right amount of escaping radiation.

The H⁻ ion is basically doing the same job that you do with your bed covers at night. If you get too cold, you pile on extra covers to trap your body's thermal energy and keep you warm (more H⁻ ions). If you get too hot, you kick some covers off so you will cool off (fewer H⁻ ions). It is a shame that we have no such "automatic cover" to maintain our body temperature at night as effectively as the H⁻ ion controls the surface temperature of a protostar.

The amount of H⁻ in the atmosphere of a protostar keeps the surface temperature of the protostar somewhere between about 3,000 and 5,000 K, depending on the mass and age of the protostar. Because the temperature of the star is not changing much, the amount of energy per unit time (power) radiated away by each square meter of the surface of the star does not change much either. Recall Stefan's Law from Chapter 4, which says that the amount of power radiated by each square meter of the star's surface is determined by its temperature. But as the star shrinks, the area of its surface shrinks as well. There are fewer square meters of surface to radiate, so the luminosity of the protostar drops. As viewed from the outside, the protostar stays nearly the same temperature and color but gradually gets fainter as it evolves toward its eventual life as a main sequence star.

In Chapter 13 we introduced the H-R diagram and used it to begin to understand how the properties of stars differ. For the next several chapters we will also use the H-R diagram to keep track of how stars change as they evolve through their lifetimes. The path across the H-R diagram that a star follows as it goes through the different stages of its life is called the star's **evolutionary track**. The particular path that an evolving protostar follows as it approaches the main sequence is called its **Hayashi track**. The protostar is brighter than it will be as a true star on the main sequence, so a protostar's Hayashi track is located above the main sequence on the H-R diagram. Because the surface temperature of the protostar stays nearly constant as the protostar contracts, the protostar's Hayashi track prior to the beginning of hydrogen burning is an almost vertical line on the H-R diagram. **Figure 15.18** shows the pre–main sequence evolutionary tracks of stars of several different masses. Astronomers say that an evolving protostar "descends the Hayashi track."

Not All Stars Are Created Equal

In Chapter 13 we found that stars can have a wide range of masses, varying from less than a 10th the mass of the Sun up to perhaps 100 times the mass of the Sun. What determines how massive a star will be? The answer to this question is unclear and is a topic of a great deal of ongoing research. One obvious possibility is that a forming star grows until it uses up all of the gas around it: It becomes no larger simply because it has run out of material. Although this explanation would be easy to understand, it does not match our observations of how star formation actually takes place. At this point in our story the contracting protostar is at the center of a collapsing molecular cloud core, which

FIGURE 15.18 The evolution of pre–main sequence stars can be followed on the H-R diagram. Protostars in the upper right portion of the diagram are large and cool. The roughly vertical, constant-temperature parts of the evolutionary tracks of low-mass protostars are referred to as Hayashi tracks.

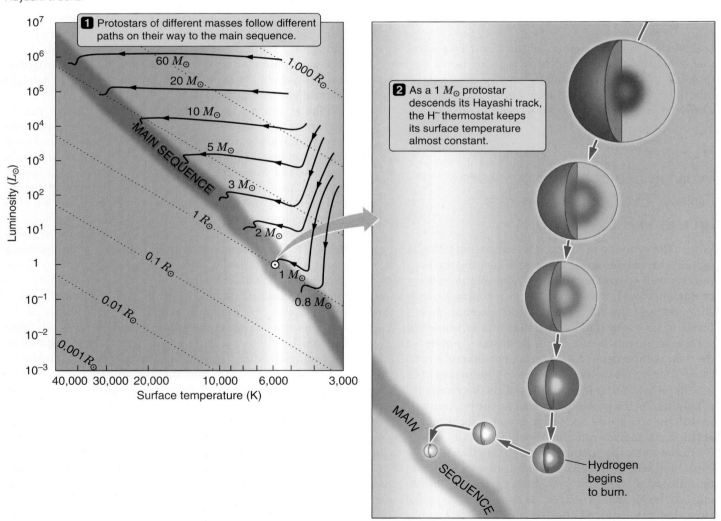

in turn is a denser-than-average region inside a molecular cloud whose total mass may be hundreds of thousands of times greater than the mass of the Sun. Observations indicate that under most circumstances star formation is a very inefficient process. Only a small fraction of the material in a molecular cloud—perhaps a few percent—ends up as part of the stars forming within it. Something must prevent most of the material in a molecular cloud from ever actually falling onto protostars. There are a number of ideas about what this something might be. One intriguing possibility is that forming stars control their own masses.

When we observe young stellar objects, we often find clear signs that material is falling onto the central protostar and accretion disk, as discussed earlier. However, it is also common to observe powerful flows of material moving *away* from forming stars at the same time that material is being accreted *onto* the stars. How can this be? **Figure 15.19** shows how this happens. Material falls onto the accretion disk and moves inward toward the equator of the star while at the same time other material is blown away from the protostar and disk in the two opposite directions away from the plane of the disk. The resulting stream of material away from the protostar is called a **bipolar outflow**.

Some bipolar outflows from young stellar objects are slow and fairly disordered, but others produce remarkable "jets" of material that move away from the central protostar and disk at velocities of hundreds of kilometers per second. The material in these **jets** flows out into the interstellar medium where it heats, compresses, and pushes away surrounding interstellar gas. Knots of glowing gas accelerated

Protostars drive powerful bipolar outflows.

by jets are referred to as **Herbig-Haro objects** (or **HH objects** for short), named after the two astronomers who first identified them and associated them with star formation. **Figure 15.20** shows a Hubble Space Telescope image of the first HH objects discovered, HH 1–2. These two HH objects are the two sides of the bipolar outflow from a single source.

The origin of outflows from protostars is not as well understood as we would like, but current models suggest that they are the result of magnetic interactions between the protostar and the disk. The interior of a protostar on its Hayashi track is convective. Great cells of hot gas are rising from the interior of the star, while other cells of cooler gas are falling toward the center. This convection, coupled with the protostar's rapid rotation, can lead to the formation of a dynamo, similar to the dynamo that drives the Sun's magnetic field. The dynamo in the center of a protostar would be much more powerful than the Sun's dynamo, however. The protostar's resulting strong magnetic field might cause the protostar to begin blowing a powerful wind. It might also act something like the blade in a blender, tearing at the inner edge of the accretion disk and flinging material off into interstellar space.

1 Material moves in toward the protostar in the accretion disk.

2 Bipolar jets and outflows flow away from the young star and disk.

Wind

Jets

3 This structure is seen in images of forming stars.

FIGURE 15.19 Material falls onto an accretion disk around a protostar, then moves inward, eventually falling onto the star. In the process some of this material is driven away in powerful jets that stream perpendicular to the disk.

Powerful protostellar winds, jets, and other outflows could disrupt the cloud core and accretion disk from which the protostar formed, shutting down the flow of material onto the protostar. Up until the time that the protostellar wind begins, the protostar is enshrouded in the dusty molecular cloud core from which it was born. As the wind from the protostar disperses this obscuring envelope, we get our first direct, visible-light view of the protostar: The protostar is "revealed." Once the contracting protostar makes its appearance, it is referred to as a **T Tauri star**. This name comes from the star labeled T in the constellation Taurus. The star T Tauri was the first recognized member of this class of objects.

Astronomers have long known that stars are often found together in closely knit collections called **star clusters**. **Figure 15.21** shows one such star cluster, called the Pleiades or Seven Sisters. If you have normal eyesight (or a pair of binoculars), on a clear winter night in the Northern Hemisphere you can see the brightest of the stars in this cluster as a tight bunch in the constellation Taurus. Star clusters gave astronomers their first evidence that many stars of all different masses can form together at the same place and at about the same time. When we look around the galaxy at large, we see a hodgepodge collection of stars—some that are very old and others very young. If these were the only stars we had to study, it would be extremely difficult to learn much about how stars evolve. Star clusters, on the other hand, are large collections of stars that all formed *at the same time, in the same place, and from the same material.* They provide us with tailor-made samples to study star formation.

Even though the few brightest and most massive stars in a cluster dominate what we see, the vast majority of stars in a cluster are much lower-mass stars like our Sun. In fact, some star-forming regions do not seem to form any especially massive stars at all. There is great interest among astronomers in how and why molecular clouds subdivide themselves into low- and high-mass stars. The details of this division —specifically, what fraction of newly formed stars will be

of what masses—are crucial if we are to use observations of the stars around us today to untangle the history of star formation in our galaxy. Unfortunately we are still far from a detailed understanding of why some cloud cores become 1 $M_\odot$ stars while others become 10 $M_\odot$ stars.

Following a cloud core collapse, the subsequent evolution of a protostar is determined largely by its mass. Calculations suggest that a star with the mass of our Sun probably takes about 10 million years or so to descend the Hayashi

---

Star formation may take millions of years.

---

track and become a star on the main sequence. If we look at the entire history, including the collapse and fragmentation of the molecular cloud itself, the total time is probably more like 30 million years. More massive stars go through this process much faster. A 10 $M_\odot$ star might go from the

FIGURE 15.20 (a) Artist's view of jets from a protostar slamming into surrounding interstellar gas, heating the gas and causing it to glow. (b) This HST image shows the bow shocks formed at the ends of a bipolar jet from a protostar. Enlargements of the bow shock (c) and jet (d) are shown at right. Only one side of the jet itself can be seen because the other side is hidden behind the dark cloud in which the star is embedded.

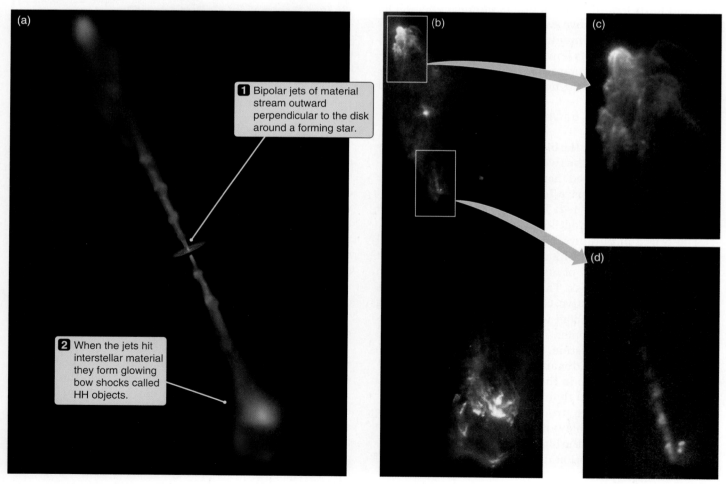

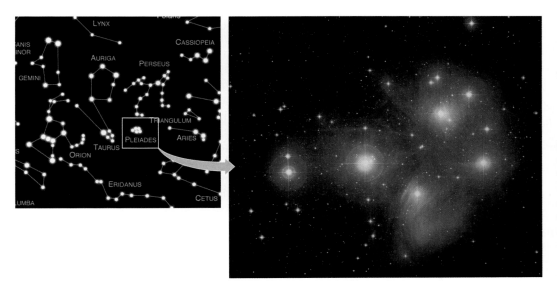

FIGURE 15.21 If you have good eyes (or a pair of binoculars), on a clear winter night you can see the brightest of the stars in this cluster—a tight bunch in the constellation Taurus called the Pleiades or Seven Sisters. The diffuse blue light around the stars is starlight scattered by interstellar dust.

stage of being a molecular cloud core to burning hydrogen in its interior in only 100,000 years. A 100 $M_\odot$ star might make the journey in less than 10,000 years. By comparison, a 0.1 $M_\odot$ star might take 100 million years to finally reach the main sequence.

The 30 million years or so that it took for our Sun to form is a long time, but it is a tiny fraction of the 10 billion years during which the Sun will steadily fuse hydrogen into helium as a main sequence star. It is no wonder that so few of the stars we see in the sky are such young objects. But every star that we see was young *at one time,* including our own Sun.

Not surprisingly, many questions about star formation remain. One question is "How must we modify our story to

Often multiple stars form.

account for the formation of binary stars or other multiple-star systems?" When we observe the sky, we find that about half of the stars we see are part of multiple-star systems. At what point during star formation is it determined that a

collapsing cloud core will form several stars instead of just one? Some models suggest that this split may happen early on in the process, during the fragmentation and collapse of the molecular cloud. The advantage of these ideas is that they provide a natural way of dealing with much of the angular momentum of the cloud core: It goes into the orbital angular momentum of the stars about each other. Other models suggest that additional stars may form from the accretion disk around an initially single protostar.

The picture of star formation presented here is remarkably complete, considering that we have never visited a protostar or watched a star form. Instead astronomers have observed many different stars at different stages in their formation and evolution, then used their knowledge of physical laws to tie these observations together into a coherent, consistent description of how, why, and where stars form. This two-pronged attack—using observations to see what things exist in the universe, and using physics to understand how they work and the relationships between them —is how all astronomy (and in a certain sense all physical science) works.

Summary

- The interstellar medium ranges from cold, relatively dense molecular clouds to hot, tenuous intercloud gas.

- Dust in the interstellar medium blocks visible light but becomes more transparent at longer wavelengths.

- Neutral hydrogen is detected by its 21-cm radio emission.

- Stars form in clusters within collapsing cores inside giant molecular clouds.

- Our Solar System began within a molecular cloud core.

- Protostars glow with thermal energy released from converted gravitation energy as they collapse.

- A protostar lands on the main sequence when nuclear reactions begin in its core.

- A protostar must have at least 0.08 $M_\odot$ to become a true star.

- Brown dwarfs are neither stars nor planets but something in between.

Seeing the Forest through the Trees

We began our discussion of the Solar System in Chapter 6 by describing its formation. Nearly 5 billion years ago a vast interstellar cloud collapsed under its own weight to form a swirling disk of gas and dust. Within that disk small objects stuck together to become parts of larger objects, culminating in the formation of the planets and other solid bodies of our Solar System. We now can see that earlier story—and in some sense our entire discussion of the Sun, Earth, and Solar System—as a "sidebar" to a larger story, the thread of which we picked up again on this leg of our journey. The tenuous expanse of gas and dust that fills the vast reaches of interstellar space sets the stage for the ongoing formation of generations upon generations of stars and planetary systems, of which ours is but one among thousands of billions of billions in the universe.

The interstellar medium is a difficult topic, even for experts in the field. The time that we have for our journey is limited. To go beyond the basic description of the interstellar medium here would be a long and arduous excursion. It would also quickly move onto more speculative ground. For now we suggest that you think of the interstellar medium as you might a complex and subtle ecosystem. It is one thing to catalog the inhabitants of an ecosystem, but it is quite another to understand the interrelationships and dependencies among those organisms well enough to say that you know how the ecosystem works. In like fashion, astronomers at the turn of the 21st century have a fairly complete picture of what makes up the interstellar medium, but we do not fully understand the complex interplay between the phases of the interstellar medium and the stars that are born and die there. Even so, we have learned some important lessons and have developed some important tools. For example, 21-cm radiation from ubiquitous interstellar hydrogen—radiation that easily penetrates the dust that obscures our view of the universe at visible wavelengths—gives us a tool to see out to the far reaches of our galaxy and map out our home in the universe.

Fortunately our understanding of stars is far more complete. In fact, the workings of stars represent such a nicely posed physics problem that we would probably include them on our journey even if this were a pure physics text rather than an astronomy text. Many of the physical principles employed in our discussion of star formation and the evolution of protostars in this chapter should have sounded familiar. The balance between gravity and pressure that gives a protostar its structure is the same balance at work within the Sun. The chain of physical reasoning that we used to understand the collapse of a protostar (gravitational energy that is converted to thermal energy and then radiated away) applies almost unmodified to understanding the ongoing (albeit slow) collapse of Jupiter.

Looking forward in our journey, the physical insight that we have developed will continue to serve us well. Stars are temporal objects. They are born from interstellar gas; they shine by the light of nuclear fires deep within their hearts; and when they exhaust their fuel, they die. The changing balance of the protostar is only the first chapter in a process of evolution that continues throughout the star's life. We found it convenient to follow the changes taking place within an evolving protostar by tracking its progress across the face of the H-R diagram. In doing so we discovered the best way that astronomers have found to draw a road map of a star's life. That is the road map that will be our guide for the next stretch of our journey.

Key Terms

Student Questions

THINKING ABOUT THE CONCEPTS

1. Clouds in our sky consist of water condensed from atmospheric water vapor. How does the process that forms our clouds compare with the process that forms interstellar hydrogen clouds from the interstellar medium?

2. If you placed your hand in boiling water (100°C) for even one second, you would get a very serious burn. If you placed your hand in a hot oven (200°C) for a second or two, you would hardly feel the heat. Explain why this is so and how it relates to million-kelvin regions of the interstellar medium.

3. The interstellar medium is approximately 99 percent gas and 1 percent dust. Why is it the dust and not the gas that blocks our visible-light view of the galactic center?

4. Using your own experience of walking or riding in a dense fog, explain how the existence of large amounts of interstellar dust may have influenced the views held by 19th century astronomers regarding the structure of our galaxy.

5. Explain how the important discovery of 21-cm radio emission has allowed us to detect interstellar clouds of neutral hydrogen (H I), even when large amounts of interstellar dust are in the way.

6. Explain how the development of infrared astronomy has made it possible for astronomers to study the detailed processes of star formation.

7. Molecular clouds are much more massive than the stars they form, and many stars can form from a single large cloud. What does this imply about the environment in which our Sun and Solar System were born?

8. Stellar radiation can convert atomic hydrogen (H I) to ionized hydrogen (H II).
   a. Why does a B8 main sequence star ionize far more interstellar hydrogen in its vicinity than a K0 giant of the same luminosity?
   b. What properties of a star are important in determining whether it can ionize large amounts of nearby interstellar hydrogen?

9. When a hydrogen atom is ionized, a single particle becomes two particles.
   a. Identify the two particles.
   b. If both particles have the same kinetic energy, which moves faster?

10. Molecular hydrogen is very difficult to detect from the ground, but we can easily detect carbon monoxide (CO) by observing its 2.6-cm microwave emission. Describe how observations of CO might help astronomers infer the amounts and distribution of molecular hydrogen within giant molecular clouds.

11. You can think of a brown dwarf as a failed star—that is, one lacking sufficient mass for nuclear reactions to begin. What similarities and differences do you see between a brown dwarf and a giant planet such as Jupiter? Would you classify a brown dwarf as a supergiant planet? Explain your answer.

APPLYING THE CONCEPTS

12. The mass of a proton is 1,850 times the mass of an electron. If a proton and an electron have the same kinetic energy, $KE = \frac{1}{2}\, mv^2$, how many times greater is the velocity of the electron than that of the proton?

13. If a typical hydrogen atom in a collapsing molecular cloud core starts at a distance of 10,000 AU (1.5×10^{12} km) from the core's center and falls inward at an average velocity of 1.5 km/s, how many years does it take to reach the newly forming protostar? Assume that a year is 3×10^7 s.

14. From Table 13.1 we can see that the ratio of hydrogen atoms (H) relative to carbon atoms (C) in the Sun's atmosphere is approximately 2,400 to 1. It would be reasonable to assume that this ratio also applies to molecular clouds. If 2.6-cm radio observations indicate 100 $M_\odot$ of carbon monoxide (CO) in a giant molecular cloud, what is the implied mass of molecular hydrogen (H$_2$) in the cloud? (Carbon represents $\frac{3}{7}$ of the mass of a CO molecule.)

15. Neutral hydrogen emits radiation at a radio wavelength of 21 cm when an atom drops from a higher-energy magnetic state to a lower-energy magnetic state. On average,

each atom remains in the higher-energy state for 11 million years (3.5×10^{14} s).

  a. What is the probability that any given atom will make the transition in 1 second?

  b. If there are 6×10^{59} atoms of neutral hydrogen in a 500 $M_\odot$ cloud, how many photons of 21-cm radiation will the cloud emit each second?

  c. How does this compare with the 1.8×10^{45} photons emitted each second by a solar-type star?

16. The Sun took 30 million years to evolve from a collapsing cloud core to a star, with 10 million of those years spent on the Hayashi track. It will spend a total of 10 billion years on the main sequence. Suppose we were to compress the Sun's main sequence lifetime into just a single year.

  a. How long would the total collapse phase last?

  b. How long would it spend on the Hayashi track?

It is said an Eastern monarch once charged his wise men to invent him a sentence to be ever in view, and which should be true and appropriate in all times and situations. They presented him the words "And this, too, shall pass away."

ABRAHAM LINCOLN (1809–1865),
SEPTEMBER 30, 1859

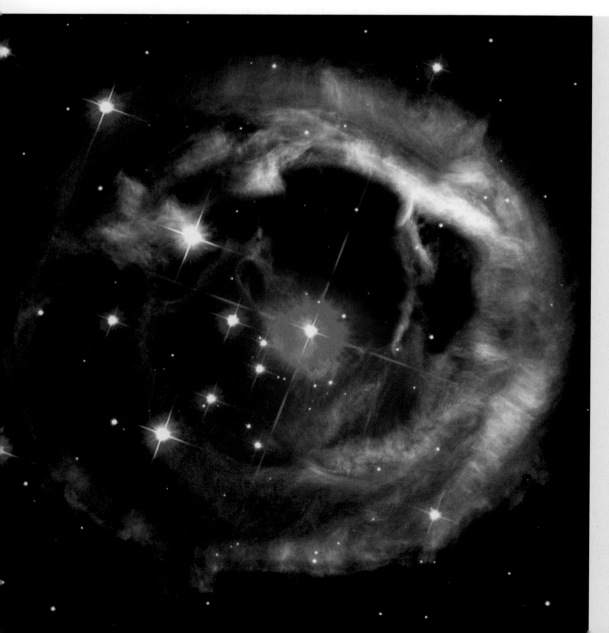

Shells of dust reflect successive outbursts of the supergiant star, V838 Monocerotis.

CHAPTER 16

Stars in the Slow Lane

16.1 This, Too, Shall Pass Away

If you go outside tonight and look up, you will not see the same sky that you would have seen a year ago. The stars themselves probably have not changed, and the Moon might even be in the same place and same phase. Nonetheless, what you see will be radically different because *the way you see it* has changed. A year ago the sky was probably full of points of light, called stars, that were "somewhere off in space." Now you see an expanse of distant suns, each farther away than the mind can easily comprehend, and each shining with a devastating brilliance powered by a nuclear inferno deep in its heart. When you look at the dark splotches that mar the Milky Way's eerie glow, you see vast clouds of interstellar gas and dust. In your mind's eye you look inside these clouds at the new suns and new solar systems being born there, and imagine the birth of our own planet and the star we orbit, 4.6 billion years ago. The sky is like a page from a book. To those who cannot read, written words are so many chicken scratches; but to those who can, thoughts and words and ideas spring off the page to touch our hearts and stimulate our minds and imaginations. The sky is a book worth reading, as is the rest of nature. At this stage in our journey we are well on our way to becoming literate.

So far on our journey, we have seen interstellar clouds collapse under the force of gravity to form immense proto-stars surrounded by swirling disks of gas and dust. We have watched as dust that is left behind in the disk accumulates into planets and moons, asteroids and comets. We have followed along as protostars continue their collapse until the

KEY CONCEPTS

Within its core the Sun fuses over 4 billion kilograms of hydrogen to helium each second; and although the Sun may seem immortal by human standards, eventually it will run out of fuel. When it does, some 5 billion years from now, the Sun's time on the main sequence will come to an end. As we examine what happens when a low-mass star like the Sun nears the end of its life, we will find that

- The more massive a star, the shorter its lifetime.

- We can follow the post–main sequence evolution of stars by tracing their paths on the H-R diagram.

- When the Sun runs out of fuel at its center, it will grow into a larger, more luminous red giant star.

- Red giants, and some other evolved stars, are shaped by dense "degenerate" cores in which atoms have been crushed by gravity.

- Evolving low-mass stars go through a series of stages, eventually burning helium to carbon and building up a core of carbon ash.

- In the end a low-mass star will eject its outer layers, possibly forming a planetary nebula, and leaving behind a tiny degenerate white dwarf.

- Low-mass stars in binary systems may experience more exciting fates as novae or supernovae.

457

nuclear fuel in their cores ignites, and we have recognized a grand pattern—the main sequence—in the stars that those protostars go on to become. We have even looked inside one of these stars, our Sun, and come to appreciate the battle between gravity and pressure that gives our Sun its structure. Our understanding of that battle provides our understanding of all stars along the main sequence. A star's mass sets the strength of its gravity, and the need to balance that gravity in turn sets the pressure in the star's interior. The more massive the star, the higher the pressure needed to hold it up, and the more rapidly the star must burn its nuclear fuel to support its own weight. From luminous O stars on the hot end of the main sequence to faint M stars on the cool end, mass is the fundamental and overriding property that makes a main sequence star what it is.

We have seen stars born and seen how they live. Now the time has come to watch them die. Like all main sequence stars, the Sun gets its energy by converting hydrogen to helium in its core. This is what *defines* the main sequence: Being on the main sequence *means* that a star is burning hydrogen in its core. But a star cannot remain a main sequence star forever. Hydrogen at the core of a star is a consumable resource. Any star eventually exhausts this

Stars eventually exhaust their nuclear fuel.

resource—it "runs out of gas"—and when it does, its structure begins to change dramatically. Just as the balance between pressure and gravity within a protostar constantly changes as it descends the Hayashi track toward the main sequence, new balances must also constantly be found as a star evolves beyond the main sequence, until at last no balance is possible at all.

The mass and composition of a star control the star's life on the main sequence, and they remain at center stage in the closing acts of the star's story as well. The evolutionary course followed by each of the hundred billion or so stars

Mass and composition determine a star's fate.

that make up the Milky Way Galaxy (and each of the stars in every other galaxy throughout the universe) is locked in place when the star forms, determined foremost by the seemingly incidental fact of the amount of mass incorporated into the star at the time of its birth, and secondarily by the chemical composition of the material from which it formed.

Each star is unique. Relatively minor differences in the masses and chemical compositions of two stars can sometimes result in significant, and possibly even dramatic, differences in their fates. The course followed by a star with a mass of 1.1 $M_\odot$ is not identical to the fate of a star with a mass of 0.9 $M_\odot$. Despite that fact, stars can be divided roughly into two broad categories whose members evolve in quali-

tatively different ways. Massive, luminous O and B stars follow a fundamentally different course from cooler, fainter, less massive stars found toward the lower right end of the

Low-mass and high-mass stars evolve differently.

main sequence. These stars, which have masses less than about 3 $M_\odot$, are referred to as **low-mass stars** and are typified by our Sun. In this chapter we begin our discussion of stellar evolution by examining the stages through which low-mass stars progress. What better place to start than by asking what fate awaits our own Sun?

16.2 The Life and Times of a Main Sequence Star

In Chapter 14 we learned that the structure of the Sun is determined by a balance between the inward force of gravity and the outward force of pressure. The pressure within the Sun is in turn maintained by energy released by nuclear fusion in the heart of the star. If you were to add mass to the Sun, the weight of material pushing down on the inner regions of the star would increase. Gravity would gain an advantage. The inner parts of the Sun would be compressed by the added weight, driving up the temperature and density there. This increase in temperature and density would in turn accelerate the pace of the nuclear reactions occurring there.

We have followed this chain of reasoning before, but it is so crucial to what is to come that it is worth reviewing here. Increasing the temperature and pressure at the center of a star means several things. It means that more atomic nuclei are packed together into a smaller volume. You know that you are far more likely to bump into another person while strolling through a crowded shopping mall than through an empty park. Similarly, packing atomic nuclei more tightly together increases the likelihood that they will run into each other and fuse. Higher density means more frequent collisions between atomic nuclei, and more collisions mean faster burning. Higher temperature also drives

Increasing temperature and pressure speed up nuclear burning.

up the rate of nuclear reactions: Higher temperature means that atomic nuclei are moving faster, so they are more likely to bump into each other. More important, higher temperature means that atomic nuclei collide with each other more *violently*, making it more likely they will overcome the electric repulsion that pushes the positively charged nuclei apart. As a result of the combined effects of temperature and density, modest increases in pressure within a star can sometimes lead to dramatic increases in the amount of energy released by nuclear burning.

Here is the key to understanding why the main sequence is primarily a sequence of masses, with low-mass stars on the faint end and high-mass stars on the luminous end. More mass means stronger gravity; stronger gravity means higher temperature and pressure in the star's interior; higher temperature and pressure mean faster nuclear reactions; and faster nuclear reactions mean a more luminous star. If the Sun were more massive it would necessarily have a different balance between gravity and pressure—a balance in which the Sun would burn its nuclear fuel more rapidly and so would be more luminous. In other words, if the Sun were more massive it would be located at a different position on the H-R diagram: farther up and to the left on the main sequence. Mass determines the structure of a star and its place on the main sequence. This, in a nutshell, is the heart of our understanding of stellar structure.

Higher Mass Means a Shorter Lifetime

A main sequence star can live for only so long. But how long is "long"? The question "How long can a main sequence star continue to burn hydrogen in its core?" is much like the question "How long can you drive your car before it runs out of gas?" In the case of the car, the answer depends in part on how much gas your tank holds. The larger the gas tank, the more fuel you have, and so the longer your motor might run. But the answer also depends on the size and efficiency of your motor. An eight-cylinder, four-barrel-carburetor, supercharged 409-cubic-inch gas guzzler drinks fuel a lot faster than a motor scooter. The amount of time your motor will run is determined by a competition between these two effects. The larger motor might run out of gas first, even if it is attached to a much larger tank.

The competition between these two effects—tank size and motor size—is most readily expressed as a ratio. How long your motor runs is given by the amount of gas in the tank, divided by how quickly the motor uses it:

$$\begin{array}{c}\text{Lifetime of}\\\text{tank of gas}\\(\text{hours})\end{array} = \dfrac{\begin{array}{c}\text{Amount of fuel}\\(\text{gallons})\end{array}}{\begin{array}{c}\text{Rate at which fuel is used}\\(\text{gallons/hour})\end{array}}$$

If you have a 15-gallon tank and your motor is burning fuel at a rate of 3 gallons each hour, then your motor will use up all of the gas in just

$$\frac{15 \text{ gallons}}{3 \text{ gallons/hour}} = 5 \text{ hours.}$$

The same principle works for main sequence stars. The "amount of fuel" is determined by the mass of the star. The more massive the star, the more hydrogen there is available to power nuclear burning. The "rate at which fuel is used" is measured by the luminosity of the star. Main sequence

---

How quickly a star runs out of fuel depends on its mass and luminosity.

---

stars are "in balance," so energy is radiated into space from the surface of the star at the same rate at which energy is being generated in its core. (This balance between energy generation and luminosity remains true at almost every stage of a star's evolution.) If one main sequence star has twice the luminosity of another, then it must be burning hydrogen at twice the rate of the other star.

An expression for the lifetime of the star looks very similar to our previous expression for the time it takes for your car to run out of fuel:

$$\begin{array}{c}\text{Lifetime}\\\text{of star}\end{array} = \dfrac{\begin{array}{c}\text{Amount of fuel}\\(\propto \text{mass of star})\end{array}}{\begin{array}{c}\text{Rate fuel is used}\\(\propto \text{luminosity of star})\end{array}}$$

If we use what we know about how much hydrogen must be converted into helium each second to produce a given amount of energy, as well as the fraction of its hydrogen that a star burns, this equation says that the length of time a star spends on the main sequence is given approximately by

$$\tau_{\text{MS}} = (1.0 \times 10^{10}) \times \frac{M\,(M_\odot)}{L\,(L_\odot)} \text{ years.}$$

By definition the Sun has a mass M equal to 1 $M_\odot$ and a luminosity L equal to 1 $L_\odot$. Plugging these numbers into this equation tells us that our Sun will use up the hydrogen in its core after spending 10 billion years on the main sequence. This number, written τ_{MS} in the equation here, is called the **main sequence lifetime** of the star in question, in this case our Sun.

Now compare this number with the lifetime of a more massive star. The relationship between mass and luminosity of stars is very sensitive. Relatively small differences in the masses of stars result in large differences in their main sequence luminosities. From **Table 16.1** you can see that a main sequence O5 star has a mass 60 times the mass of the Sun. But a 60 $M_\odot$ main sequence star is *not* just 60 times as luminous as the Sun—it is 794,000 times as luminous! Putting in $M = 60\ M_\odot$ and $L = 794{,}000\ L_\odot$ means that

$$\tau_{\text{MS}}\,(\text{O5 star}) = (1.0 \times 10^{10})\,\frac{60}{794{,}000}$$

$$= 8 \times 10^5 \text{ years.}$$

So instead of the 10 billion year lifespan of the Sun, an O5 star has a main sequence lifetime of *less than 1 million*

TABLE 16.1

Main Sequence Lifetimes

Spectral Type	Mass ($M_\odot$)	Luminosity ($L_\odot$)	Main Sequence Lifetime (Years)
O5	60	794,000	3.6×10^5
B0	17.5	52,500	1.0×10^7
B5	5.9	832	7.2×10^7
A0	2.9	54	3.9×10^8
A5	2.0	14	1.1×10^9
F0	1.6	6.5	2.1×10^9
F5	1.3	3.2	3.5×10^9
G0	1.05	1.5	8.3×10^9
G2 (our Sun)	1.0	1.0	1.0×10^{10}
G5	0.92	0.8	1.5×10^{10}
K0	0.79	0.4	3.7×10^{10}
K5	0.67	0.15	5.3×10^{10}
M0	0.51	0.08	$7.9 \times 10^{10}*$
M5	0.21	0.011	$2.9 \times 10^{11}*$
M8	0.19	0.0012	$1.1 \times 10^{12}*$

Lifetimes marked with an asterisk (*) are based on the "rule of thumb" discussed in the text. The other lifetimes are based on computer models.

years! Several generations of O stars have lived and died in the time that hominids have walked the surface of Earth. Even though the 60 $M_\odot$ star starts out with 60 times as much

---

Low-mass stars live much longer than high-mass stars.

---

fuel as the Sun, it burns that fuel so much faster that it uses it up in less than a ten-thousandth the time.

The formula just given is a rule of thumb for calculating stellar lifetimes. Better values, such as those in Table 16.1, are based on detailed computer models of stars at each mass. Even so, the values that you get by applying this rule of thumb are reasonably close to the model values. (**Figure 16.1** gives a comparison between the two sets of numbers.) The rule of thumb also lets us see the reasons behind the differences in lifetimes of stars of different masses. While high-mass stars live fast and die young, low-mass stars live their lives in the slow lane.

The lifetimes of stars should be of more than passing interest to us. We don't yet know if life exists on planets orbiting other stars, but we can confidently assume that life would be unlikely on a planet orbiting a massive star with a stable life of only a few million years. Likewise it would be unlikely for life to develop on a planet surrounding a very low-mass star that had yet to initiate the stable nuclear burning phase. (See **Connections 16.1**.)

The Structure of a Star Changes as It Uses Its Fuel

To say that a main sequence star is stable is not to say that it does not continually change. When the Sun formed about 90 percent of its atoms were hydrogen atoms. Since then the Sun has produced its energy by converting hydrogen into helium via the proton–proton chain. As the composition of a star changes, so must its structure. When we discussed the collapse of a protostar toward the main sequence in Chapter 15, we considered the idea of a changing balance between gravity and pressure. The protostar was always in balance; but as it radiated away thermal energy, this balance was constantly changing, shifting toward that of a smaller and denser object. The same concept applies here. As a main sequence star uses the fuel in its core, its structure must continually shift in response to the changing core composition. At any given point in its lifetime, a main sequence star like the Sun is in balance, but the balance in the Sun today is slightly different from the balance in the Sun 4 billion years ago, and slightly different from the balance it will have 4 billion years from now. Between the time the Sun was born and the time it will leave the main sequence, its luminosity will roughly double, with most of this change occurring during the last billion years of its life on the main sequence. Stars evolve even as they "sit" on the main sequence, although this evolution is slow and modest in comparison with the events that follow.

What happens to the helium that is produced in the core of a low-mass main sequence star like our Sun? We might imagine that it begins to burn, with helium atoms fusing to form heavier elements; but such is not the case. For fusion to occur, atomic nuclei must be slammed together with enough energy to overcome the electric repulsion between

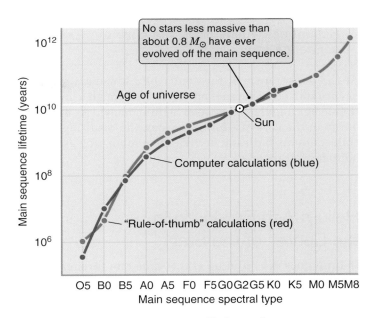

FIGURE 16.1 The main sequence lifetimes of stars.

them. Getting two helium nuclei close enough to fuse involves two protons in one nucleus pushing against two protons in the other nucleus, which is four times the repulsive force of a hydrogen nucleus pushing on a hydrogen nucleus.

CONNECTIONS 16.1

Origins—Choosing the Right Kind of Star

It is no coincidence that we live on a planet orbiting at a comfortable distance from a typical low-mass star. To begin with, we know that liquid water is absolutely essential to life—at least to our kind of life. The human body is a good example: We are about two-thirds water.

Had Earth formed a bit closer to the Sun, it would have experienced the inferno that befell Venus. Had it formed a bit farther out, it would have suffered the deep freeze that fate handed to Mars. Call it the Goldilocks effect. Earth is at just the right place to support life, where it is neither too hot nor too cold. To be sure, stars both less massive and more massive than the Sun have their own comfort zones. Less massive, cooler stars have life-supporting zones that are narrower than the Sun's. In other words, there is not much wiggle room between too hot and too cold. This obviously minimizes the chance

that a habitable planet would just happen to form within that slender zone. Stars that are more massive have wider life zones, but they come with a very different problem. The lifetimes of stars more massive than the Sun are so short that advanced life forms would not have sufficient time to develop. A star with twice the Sun's mass would enjoy relative stability on the main sequence for only about a billion years before the helium flash incinerated everything around it—hardly enough time for anything more than the most primitive life forms to evolve. Although we can't rule out the possibility that life does exist somewhere around a less suitable star, we can tell from our own experience that if you want a planet suitable for sustaining advanced life, you can't do better than to choose one orbiting at a comfortable distance around a solar-type star.

Helium ash builds up in the core of a main sequence star.

At the temperature at the center of a low-mass main sequence star, collisions are not energetic enough to overcome the electric repulsion between helium nuclei.[1] As hydrogen burns in the core of a low-mass star, the resulting helium collects there, building up like the nonburning ash in the bottom of a fireplace.

Helium Ash Builds Up in the Center of the Star

Helium does not build up at the same rate throughout the interior of a star. In Chapter 7 we found that the temperature and pressure within Earth *must* be highest at the center of the planet. No other configuration makes sense. Exactly the same arguments apply equally well to stars. The fact that the temperature and pressure are highest at the center of a main sequence star means that hydrogen burns most rapidly there as well. As a result, nonburning helium "ash" accumulates most rapidly at the center of the star.

If we could cut a star open and watch as it evolves, we would see the chemical composition of the star changing most rapidly at its center and less rapidly as we move outward through the star. **Figure 16.2** shows how the chemical composition inside a star like the Sun changes throughout its main sequence lifetime. When the Sun formed, it had a uniform composition throughout, with hydrogen accounting for about 70 percent of the mass in the Sun and helium accounting for most of the remaining 30 percent. With time, the helium fraction in the center of the Sun climbed. Today, roughly 5 billion years later, only about 35 percent of the mass at the center of the Sun is hydrogen.

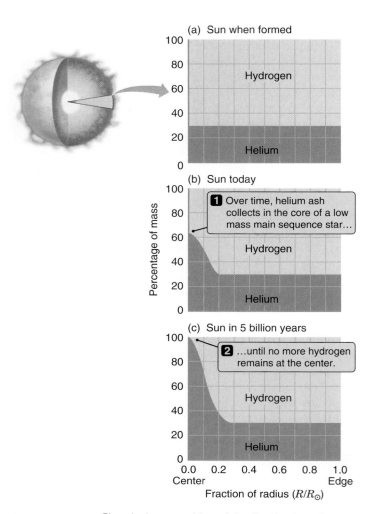

FIGURE 16.2 Chemical composition of the Sun is plotted here as a fraction of mass against distance from the center of the Sun. (a) When the Sun formed 5 billion years ago, about 30 percent of its mass was helium and 70 percent was hydrogen throughout. (b) Today the material at the center of the Sun is about 65 percent helium and 35 percent hydrogen. (c) The Sun's main sequence life will end in about 5 billion years, when all of the hydrogen at the center of the Sun is gone.

16.3 A Star Runs Out of Hydrogen and Leaves the Main Sequence

Hydrogen burning in the core of a star cannot continue forever. Eventually—about 5 billion years from now in the case of the Sun—a star exhausts all of the hydrogen fuel available at its center. At this point the innermost core of the star is composed entirely of helium ash. As thermal energy leaks out of the helium core into the surrounding layers of the star, no more energy is generated within the core to replace it. The balance that has maintained the structure of the star throughout its life is now broken. The star's life on the main sequence has come to an end.

The Helium Core Is Degenerate

Throughout your life you have experienced forms of matter that are mostly *empty space*. Just as the Solar System is mostly empty except for the tiny bit of space occupied by

---

[1] An astute student may be puzzled, remembering from Chapter 14 that one step of the proton–proton chain is the fusion of two ³He nuclei, which have as strong an electric repulsion as two ⁴He nuclei do. The answer is that the strong nuclear force interaction between the two ³He nuclei is much more powerful than the strong force interaction between two ⁴He nuclei, so two ³He nuclei do not have to get as close together as do two ⁴He nuclei in order to fuse.

the Sun and the planets, an atom is mostly empty except for the tiny bit of space occupied by the nucleus and the electrons. The same is true for the matter within the Sun. At the enormous temperatures within the Sun the electrons have almost all been stripped away from their atoms by energetic collisions. (In other words, the gas is completely ionized.) So the gas inside the Sun is a mixture of electrons and atomic nuclei all flying about freely. Even so, the gas that makes up the Sun is still mostly empty space, with the electrons and atomic nuclei filling only a tiny fraction of the volume.

When a low-mass star like the Sun exhausts the hydrogen at its center, the situation changes. As gravity begins to win the shoving match with pressure, the helium core is crushed to an ever smaller size and an ever greater density, but there is a limit to how dense the core can get. The rules of quantum mechanics (the same rules that say that atoms can have only certain discrete amounts of energy and that light comes in packets called photons) limit the number of electrons that can be packed into a given volume of space at a given pressure.[2] As the matter in the core of the star is compressed further and further, it finally bumps up against this limit. The space occupied by the core of the star is no longer mostly empty, but is now effectively "filled"

The crushed helium core is electron degenerate.

with electrons that are smashed tightly together. Matter at the center of the star is now so dense that a single cubic centimeter of this material can have a mass of a metric ton (1,000 kg) or more. Matter that has been compressed to this point is said to be **electron degenerate**.

Hydrogen Burns in a Shell Surrounding a Core of Helium Ash

Once a low-mass star exhausts the hydrogen at its center, nuclear burning may end there, but the story is very different outside the core. The layers surrounding the degenerate core still contain hydrogen, and this hydrogen continues to burn. Astronomers speak of **hydrogen shell burning** because the hydrogen is burning in a shell surrounding a core of helium, like eating the flesh of a fruit around its pit.

The electron degenerate core of a star has a number of fascinating properties. For example, as more and more helium ash piles up on the degenerate core, the size of the core *shrinks*. It does so because the added mass increases the strength of gravity and therefore the weight bearing down

---

[2] If you have taken high school or college chemistry, you may remember the *Pauli exclusion principle*, which limits the number of electrons that can go into a single orbital in an atom. This principle also limits the number of electrons that can be packed into the energy states available in the center of a star.

on the core, which means that the electrons can be smashed together into a smaller volume. The presence of the degenerate core triggers a chain of events that will dominate the evolution of our 1 $M_\odot$ star for the next 50 million years. Follow along as we step through the chain of cause and effect that takes center stage in the post–main sequence evolution of a low-mass star:

1. From our discussion in Chapter 10 (see Figure 10.3) we know that just outside the star's degenerate core the gravitational acceleration g_{core} is given by

$$g_{core} = \frac{GM_{core}}{r_{core}^2},$$

where M_{core} is the mass of the helium core and r_{core} is its radius. As the helium core grows, its larger mass (bigger M_{core}) and its shrinking size (smaller r_{core}) both cause the strength of gravity at the surface of the core to increase. As more helium is added to the core, the strength of gravity at its surface increases dramatically.

2. Increasing the strength of gravity around the core increases the weight of the overlying material pushing down on the hydrogen-burning shell surrounding the core.

3. This increase in weight must be balanced by an increase in pressure in the inner parts of the star. In particular, the pressure in the hydrogen-burning shell must increase.

4. Increasing the pressure in the hydrogen-burning shell drives up the rate of the nuclear reactions occurring in the shell.

5. Faster nuclear reactions release more energy, so the luminosity of the star increases.

This is a very counterintuitive result! We might have imagined that when a star like the Sun uses up the nuclear fuel at its center, it would grow fainter. Yet just the opposite happens. A degenerate core means stronger gravity; stronger

Running out of fuel makes the star grow more luminous.

gravity means higher pressure; higher pressure means faster nuclear burning; and faster nuclear burning means a more luminous star. When the low-mass star "runs out of gas" at its center, it responds by getting more luminous!

Tracking the Evolution of the Star on the H-R Diagram

The changes that occur in the heart of a star with a degenerate helium core are reflected in changes in the overall structure of the star. With time, the mass of the degenerate

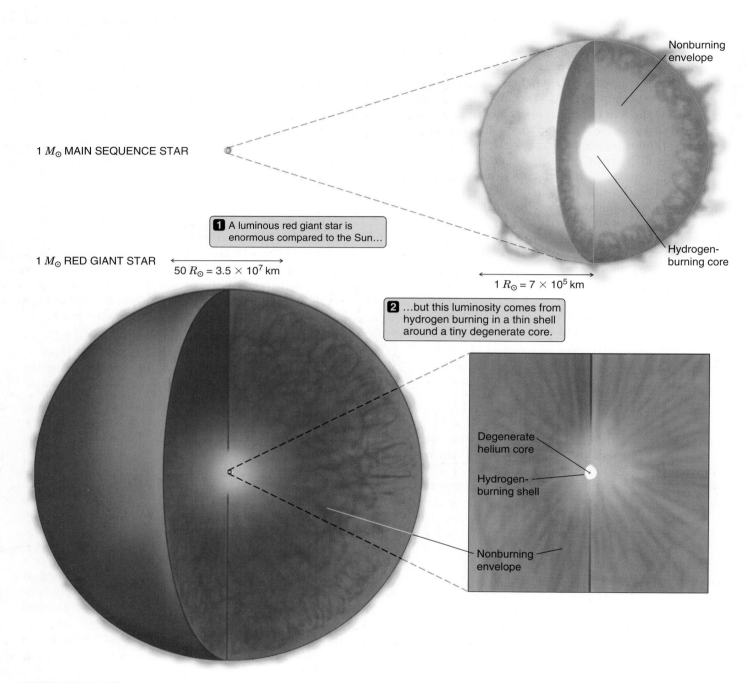

1 $M_\odot$ MAIN SEQUENCE STAR

Nonburning envelope

Hydrogen-burning core

1 $R_\odot$ = 7 × 10⁵ km

1 A luminous red giant star is enormous compared to the Sun…

1 $M_\odot$ RED GIANT STAR

50 $R_\odot$ = 3.5 × 10⁷ km

2 …but this luminosity comes from hydrogen burning in a thin shell around a tiny degenerate core.

Degenerate helium core

Hydrogen-burning shell

Nonburning envelope

FIGURE 16.3 The structure of a star near the top of the red giant branch is compared with the structure of the Sun. Left panels compare the size of the Sun with the size of the red giant. Right panels compare the size and structure of the Sun with the core of the red giant. The panels at right are blown up by about 50 times compared to the panels on the left.

helium core grows as more and more hydrogen is converted into helium ash in the surrounding shell. And as the mass of the degenerate helium core grows, so does the rate of energy generation in the surrounding hydrogen-burning shell. This increase in energy generation heats the overlying layers of the star, causing them to expand. The star becomes a

bloated, luminous giant. As illustrated in **Figure 16.3**, the internal structure of the star is now fundamentally different from when the star was on the main sequence. The giant can grow to have a luminosity hundreds of times the luminosity of the Sun and a radius over 50 $R_\odot$. Yet at the same time the core of the giant star is far more compact than that

A bloated luminous giant surrounds a tiny degenerate core.

of the Sun, with much of the star's mass becoming concentrated into a volume that is only a few times the size of Earth.

From our vantage point outside the star, we are not privy to the changes taking place deep within its interior. All we can see directly is that the star becomes larger and more luminous, and perhaps surprisingly, *cooler and redder* as well. The enormous expanse of the star's surface allows it to cool very efficiently. Even though its *interior* grows hotter and its luminosity higher, the *surface* temperature of the star actually begins to drop.

Just as we used the H-R diagram to follow the changes in a protostar on its way to the main sequence, the H-R diagram is a handy device for keeping track of the changing luminosity and temperature of the star as it evolves away from the main sequence (see **Figure 16.4**). As soon as the star exhausts the hydrogen in its core, it leaves the main sequence and begins to move upward and to the right on the H-R diagram, growing more luminous but cooler. We refer to such a star, which is somewhat brighter and larger than it was on the main sequence, as a **subgiant** star. As the

subgiant continues to evolve, it grows larger and cooler, but after a time its progress to the right on the H-R diagram hits a roadblock: the H^- thermostat. When the temperature of the subgiant star has dropped by about 1,000 K relative to its temperature on the main sequence, H^- ions start to form in great abundance in its atmosphere. We have encountered

An evolving low-mass star moves up and to the right on the H-R diagram.

formation of H^- ions before. In our discussion of protostars in Chapter 15, it was the H^- ion in the protostar's atmosphere that acted as a thermostat, regulating the star's temperature. The H^- ion serves exactly the same role here, regulating how much radiation can escape from the star and preventing it from becoming any cooler.

Because the star can cool no further, it begins to move almost vertically upward on the H-R diagram, growing larger and more luminous but remaining about the same temperature. The star has become a red giant—an obvious name for a star that is now both redder and larger than it was on the main sequence. We can think of the path that a star follows on the H-R diagram as it leaves the main sequence as being a tree "branch" growing out of the "trunk" of the main sequence. Astronomers refer to these tracks as

FIGURE 16.4 The evolution of a red giant star on the H-R diagram. The structure of a red giant star consists of a degenerate core of helium ash surrounded by a hydrogen-burning shell. As the star moves up the red giant branch, it comes close to retracing the Hayashi track that it followed when it was a protostar collapsing toward the main sequence.

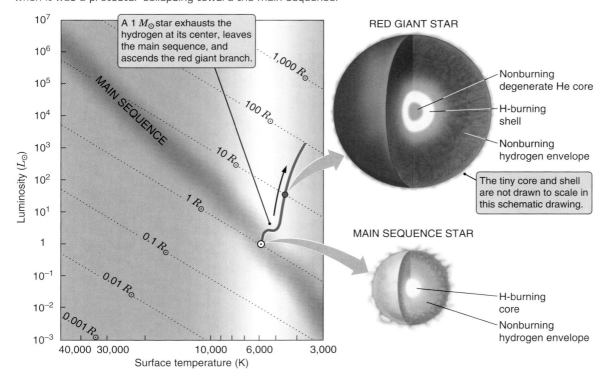

the **subgiant branch** and the **red giant branch** of the H-R diagram. Interestingly, the path that a red giant follows on the H-R diagram closely parallels the path that it followed earlier as a collapsing protostar on its way toward the main sequence—except, of course, in reverse: This time the star is moving up that path rather than coming down it. This similarity is not a coincidence. The same physical processes (such as the H⁻ thermostat) that give rise to the Hayashi track followed by a collapsing protostar also control the relationship between luminosity, size, and surface temperature in an expanding red giant.

As the star leaves the main sequence, the changes in its structure occur sluggishly at first, but then they pick up steam as the star moves up the red giant branch faster and faster. It takes 200 million years or so for a star like the Sun to go from the main sequence to the top of the red giant branch. Roughly the first half of this period is spent on the subgiant branch as the star's luminosity increases to

about 10 $L_\odot$. During the second half of this time the star's luminosity skyrockets from 10 $L_\odot$ to almost 1,000 $L_\odot$. The evolution of the star, illustrated in **Figure 16.5**, is reminiscent of the growth of a snowball rolling downhill. The larger the snowball becomes, the faster it grows, and the faster it grows, the larger it becomes. "Growth" and "size" feed off each other, and what began as a bit of snow at the top of the mountain soon becomes a huge ball that could trigger an avalanche.

The analogy between the evolution of a red giant star and the growth of a snowball is actually a pretty good one. The helium core of the star grows in mass (but not in radius!) as hydrogen is converted to helium in the hydrogen-burning shell. The increasing mass of the ever more compact helium core drives up the force of gravity in the heart of the star. Stronger gravity means higher pressure, and higher pressure accelerates nuclear burning in the shell. But faster nuclear reactions in the shell convert hydrogen into helium more

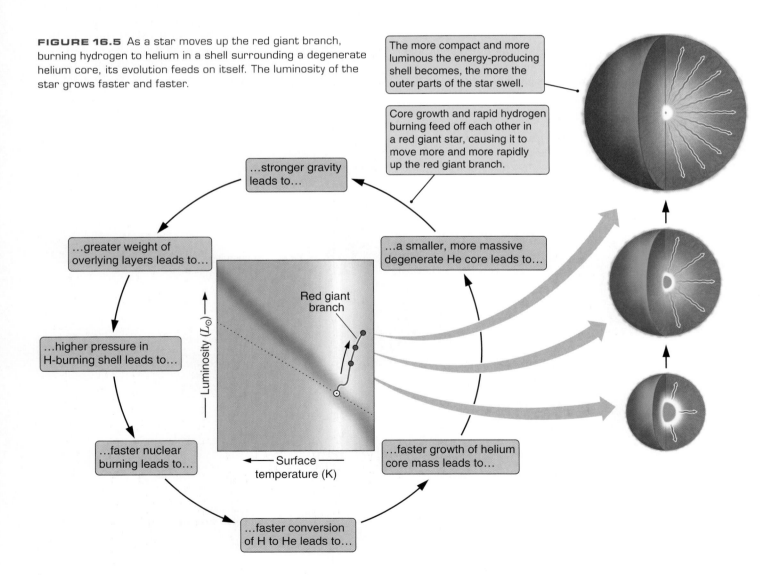

FIGURE 16.5 As a star moves up the red giant branch, burning hydrogen to helium in a shell surrounding a degenerate helium core, its evolution feeds on itself. The luminosity of the star grows faster and faster.

The more compact and more luminous the energy-producing shell becomes, the more the outer parts of the star swell.

Core growth and rapid hydrogen burning feed off each other in a red giant star, causing it to move more and more rapidly up the red giant branch.

...stronger gravity leads to...

...greater weight of overlying layers leads to...

...higher pressure in H-burning shell leads to...

...faster nuclear burning leads to...

Red giant branch

Luminosity ($L_\odot$)

Surface temperature (K)

...a smaller, more massive degenerate He core leads to...

...faster growth of helium core mass leads to...

...faster conversion of H to He leads to...

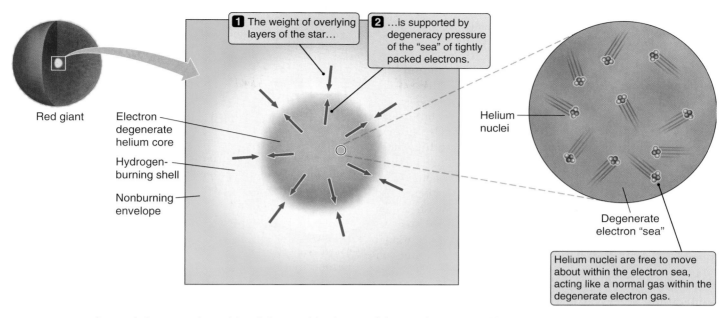

FIGURE 16.6 In a red giant star the weight of the overlying layers of the star is supported by electron degeneracy pressure in the core arising from the fact that electrons are packed together as tightly as allowed by quantum mechanics. Even so, atomic nuclei in the core of the star are able to move freely about within the sea of degenerate electrons, so they behave as a normal gas.

quickly, so the mass of the core grows more rapidly. We have come full circle in a cycle that feeds on itself. Increasing core mass leads to ever faster burning in the shell, and the faster hydrogen is burned in the shell, the faster the core mass grows. As a result, the star's luminosity climbs at an ever faster rate.

16.4 Helium Begins to Burn in the Degenerate Core

The growth of the red giant cannot continue forever, and we find ourselves once again asking a crucial question for understanding the evolution of stars: "What will be the next thing to give?" The answer lies in another unusual property of the degenerate helium core, and we next turn our attention there.

The Atomic Nuclei in the Core Form a "Gas within a Gas"

The core of the red giant star is electron degenerate, which means that as many *electrons* are packed into that space as the rules of quantum mechanics allow at that pressure. However, *atomic nuclei* in the core are still able to move freely

about, as shown in **Figure 16.6**, just as they are throughout the rest of the star.

"Wait a minute!" you say. "That last statement is nonsense for at least two different reasons. First, if the electrons are packed as tightly as possible into the core of the star, then surely the atomic nuclei are packed as tightly as possible into the core as well. After all, we normally think of atomic nuclei as being larger than electrons." But the world behaves in strange ways at the quantum mechanical level. The rules of quantum mechanics say that when packed together tightly, less massive particles like electrons effectively "take up more space" than more massive particles like atomic nuclei. As the core of the star is compressed, electrons become degenerate much sooner than the atomic nuclei do.

"Fine," you say, "but how can the atomic nuclei go moving freely about within a core that is wall-to-wall electrons?" Actually, this is no problem at all: The laws of quantum mechanics place few restrictions on electrons and atomic nuclei occupying the same physical space. As far as the atomic nuclei are concerned, the electron degenerate core of the star is still mostly empty space. The nuclei are free to go flying about through the sea of degenerate electrons almost as if the electrons were not there. The negative charges of the electrons are important because they balance out the positive charges of the nuclei; but apart from that the atomic nuclei in the electron degenerate core of a star are a perfectly normal "gas within a gas," behaving just as the (electrically neutral) atoms and molecules in the air around you do.

Helium Burning and the Triple-Alpha Process

As the star evolves up the red giant branch, its helium core grows not only smaller and more massive, but hotter as well. This is partly because of the gravitational energy released as the core shrinks (just as the protostar's core grew hotter as it collapsed) and partly due to the energy released by the ever faster pace of hydrogen burning in the surrounding shell. The climbing temperature of the core means that the thermal motions of the atomic nuclei in the core become more and more energetic. Eventually, at a temperature of around 10^8 K (a hundred million kelvins), the collisions between helium nuclei in the core become energetic enough to overcome the electric repulsion between them. Helium nuclei are slammed together hard enough for the strong nuclear force to act, and helium burning begins.

Helium burns in a two-stage process, referred to as the **triple-alpha process,** illustrated in **Figure 16.7.** First two helium nuclei (^{4}He) fuse to form a beryllium-8 nucleus (^{8}Be) consisting of four protons and four neutrons. The ^{8}Be nucleus is extremely unstable. Left on its own, it would

---

Helium burns via the triple-alpha process.

---

break apart again after only about a trillionth (10^{-12}) of a second. But if, in that short time, it collides with another ^{4}He nucleus, the two nuclei will fuse into a stable nucleus of carbon-12 (^{12}C) consisting of six protons and six neutrons. The triple-alpha process takes its name from the

FIGURE 16.7 The triple-alpha process: Two ^{4}He nuclei fuse to form an unstable ^{8}Be nucleus. If this nucleus collides with another ^{4}He nucleus before it breaks apart, the two will fuse to form a stable nucleus of carbon-12 (^{12}C). The energy produced is carried off both by the motion of the ^{12}C nucleus and by a high-energy gamma ray emitted in the second step of the process.

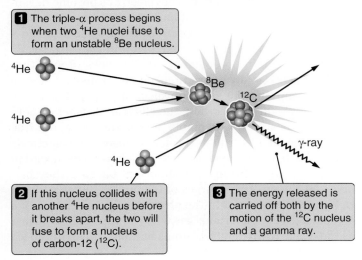

1 The triple-α process begins when two ^{4}He nuclei fuse to form an unstable ^{8}Be nucleus.

^{4}He

^{4}He

^{8}Be

^{12}C

^{4}He

γ-ray

2 If this nucleus collides with another ^{4}He nucleus before it breaks apart, the two will fuse to form a nucleus of carbon-12 (^{12}C).

3 The energy released is carried off both by the motion of the ^{12}C nucleus and a gamma ray.

---

fact that it involves the fusion of three ^{4}He nuclei, which are traditionally referred to as **alpha particles**.

The Helium Core Ignites in a Helium Flash

The next phase of the star's evolution is shown in **Figure 16.8.** Degenerate material is a very good conductor of thermal energy, so any differences in temperature within the core are quickly evened out. As a result, the degenerate core of a red giant star is at almost exactly the same temperature throughout. When helium burning begins at the center of the core, the energy released quickly heats the entire core. Within a few minutes the entire core is burning helium into carbon by the triple-alpha process.

In a normal gas like the air around you, the pressure of the gas comes from the random thermal motions of the atoms. Increasing the temperature of such a gas means that the motions of the atoms become more energetic, so the pressure of the gas increases. If the helium core of a red giant star were a normal gas, the increase in temperature that accompanies the onset of helium burning would lead to an increase in pressure. The core of the star would expand; the temperature, density, and pressure would decrease; nuclear reactions would slow; and the star would settle down into a new balance between gravity and pressure. These are exactly the sorts of changes that are steadily occurring within the cores of main sequence stars like the Sun, as the structure of the stars steadily and smoothly shifts in response to the changing composition in the stars' cores.

However, the degenerate core of a red giant is not a normal gas. In a certain sense the degenerate core is more like a rock than a normal gas. A rock is hard to crush because of how the atoms within it are packed together. Heating a rock does not cause it to swell up like a hot-air balloon, and cooling a rock does not cause it to deflate. Similarly, the pressure that prevents the degenerate core of a white dwarf from collapsing comes from how tightly the electrons in the core are packed together. Heating the degenerate core of a red giant does not change the number of electrons that can be packed into its volume, so heating the core does not change the pressure by much. And if the pressure does not increase, there is nothing to cause the core to expand.

Take another run at that to be sure it clicks. When helium begins to burn in the degenerate core, the temperature of the core goes up, but the pressure does not. So the onset

---

The helium flash is a thermonuclear runaway—an explosion within the star.

---

of helium burning in the degenerate core of a red giant does not cause the core to expand! Yet even though the higher temperature does not change the pressure, it does cause the

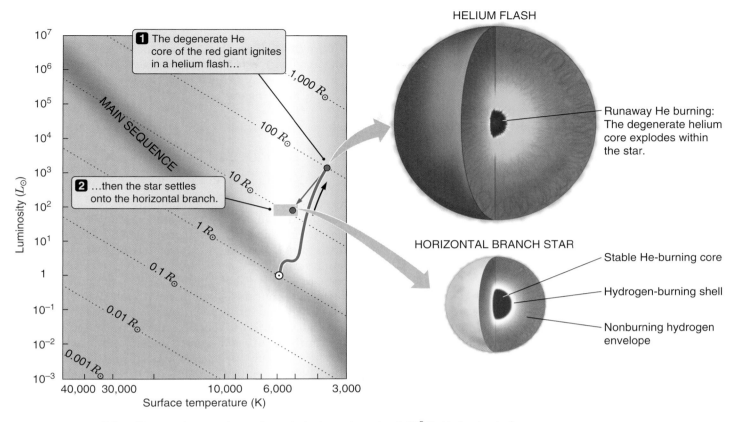

FIGURE 16.8 When the core temperature of a red giant reaches about 10^8 K, He begins to burn explosively in the degenerate core, leading to a helium flash. After a few hours the core of the star begins to inflate, ending the helium flash. Over about the next 100,000 years (a relatively short time) the star settles onto the horizontal branch, where it burns He in its core and H in a surrounding shell.

helium nuclei to be slammed together with more frequency and greater force, so the nuclear reactions become more vigorous. The process begins to snowball again. More vigorous reactions mean higher temperature, and higher temperature means even more vigorous reactions. Thermonuclear burning in the degenerate core runs away with itself, wildly out of control as increasing temperature and increasing reaction rates feed each other. This is the **helium flash**.

It is difficult to imagine the drama of the thermonuclear runaway that takes place during the helium flash. Helium burning begins at a temperature of about 100 million K. By the time the temperature has climbed by a mere 10 percent, to 110 million K, the rate of helium burning has increased to 40 times what it was at 100 million K. By the time the core's temperature reaches 200 million K, the core is burning helium 460 million times faster than it was at 100 million K! As the temperature in the core grows higher and higher, the thermal motions of the electrons and nuclei become more energetic, and the pressure due to these thermal motions becomes greater and greater. Within seconds of ignition the thermal pressure increases to the point that

it is no longer smaller than the degeneracy pressure, at which point the core literally explodes. We do not see the explosion, however, because it is contained within the star. The energy released in this runaway thermonuclear explosion lifts the overlying layers of the star, and as the core expands, the electrons are able to spread out. The drama is over within a few hours. The expanded helium-burning core is no longer degenerate, and the star is on its way toward a new equilibrium.

Following the helium flash, the star once again does something counterintuitive. You might imagine that helium burning in the core would cause the star to grow more luminous, but it does not. The tremendous energy released during the helium flash goes into fighting gravity and puffing up the core. After the helium flash the core (which is no longer degenerate) is much larger, so the acceleration due to gravity within it and the surrounding shell is much smaller. (Again, $g = GM/r^2$, so a larger core radius means smaller values of g.) Weaker gravity means less weight pushing down on the core and the shell, which means lower pressure. Lower pressure in turn slows the nuclear reactions. The net result is that following the helium flash, core helium burning keeps the core of the star puffed up, and the star becomes less luminous than it was as a red giant.

The star spends the next 100,000 years or so settling into a stable structure in which helium burns to carbon in a normal, nondegenerate core while hydrogen burns to helium in a surrounding shell. The star is now about a hundredth as luminous as it was at the time of the helium flash. The lower luminosity means that the outer layers of the star are not as puffed up as they were as a red giant. The star shrinks, and as it does so its surface temperature climbs. (This is just the reverse of the sequence of events that caused the red giant to become larger and redder as it grew more luminous.) At this point in their evolution, low-mass stars with chemical compositions similar to the Sun

---

Horizontal branch stars burn helium in the core and hydrogen in a shell.

---

will bunch up on the H-R diagram just to the left of the red giant branch. Stars that contain much less iron than the Sun tend to distribute themselves away from the red giant branch along a nearly horizontal line on the H-R diagram. This stage of stellar evolution takes its name from this horizontal band. The star is now referred to as a **horizontal branch** star.

16.5 The Low-Mass Star Enters the Last Stages of Its Evolution

The evolution of a solar-type star from the main sequence to the helium flash and horizontal branch is fairly well understood. Just as our understanding of the interior of the Sun comes from computer models of the physical conditions within our local star, our understanding of the evolution of a red giant comes from computer models that look at the changes in structure as the star's degenerate helium core grows. These models show that any star with a mass of about 1 $M_\odot$ will follow the march from main sequence to helium flash, then drop down onto the horizontal branch. However, when we try to use computer models to push our understanding to what happens next, the road that we follow gets a bit trickier. We already noted that differences in chemical composition between stars significantly affect where they fall on the horizontal branch. From this point on, small changes in the properties of a star—in mass, in chemical composition, in the strength of the star's magnetic field, or even in the rate at which the star is rotating—can lead to qualitative differences in how the star evolves. The

farther we go in time beyond the helium flash, the more possible divergent evolutionary paths are open to a low-mass star.

With this caution in mind, we continue our story of the evolution of a 1 $M_\odot$ star with solar composition, following what we currently believe to be the most likely sequence of events awaiting our Sun.

The Star Moves Up the Asymptotic Giant Branch

The structure of a horizontal branch star is much like the structure of a main sequence star in many respects. The biggest difference is that now instead of burning hydrogen into helium in a stable, nondegenerate core, the horizontal branch star is burning helium into carbon in a stable, nondegenerate core. (The other difference, of course, is that hydrogen is continuing to burn in a shell surrounding the core.) The behavior of a star on the horizontal branch is remarkably similar to that of a star on the main sequence. This similarity offers the key to understanding what comes next. If you review our account of what happened as the star evolved off the main sequence, and if you replace "hydrogen" with "helium" and "helium" with "carbon," you will have a pretty good description of how the star evolves as it leaves the horizontal branch.

The star's life on the horizontal branch is, however, much shorter than its life on the main sequence. For one thing, there is now less fuel to burn in its core. In addition, the star is more luminous, so it must be consuming fuel more rapidly. Finally, helium is a much less efficient nuclear fuel than hydrogen. Even so, for 50 million years the horizontal branch star remains stable, burning helium into carbon in its core and hydrogen to helium in a shell.

The temperature at the center of a horizontal branch star is not high enough for carbon to burn,[3] so carbon ash builds up in the heart of the star, just as helium ash accumulates at the center of a main sequence star. When the horizontal branch star has burned all of the helium at its core, gravity once again begins to win. The nonburning carbon ash

---

The star forms a degenerate carbon core as it leaves the horizontal branch.

---

core is crushed by the weight of the layers of the star above it until once again the electrons in the core are packed together as tightly as the laws of quantum mechanics allow at its pressure. The carbon core is now electron degenerate, with physical properties much like those of the degenerate helium core at the center of a red giant.

---

[3] You may recall from Chapter 14 that *burn* as used here refers not to fire but to nuclear fusion.

The small, dense electron degenerate carbon core drives up the strength of gravity in the inner parts of the star, which in turn drives up the pressure, which speeds up the nuclear reactions, which causes the degenerate core to grow more rapidly...we have heard this story before. The internal changes occurring within the star are similar to the changes that took place at the end of the star's main sequence lifetime, and the path the star follows as it leaves the horizontal branch echoes that earlier phase of evolution as well. Just as the star accelerated up the red giant branch as its degenerate helium core grew, the star now leaves the horizontal branch and once again begins to grow larger, redder, and more luminous as its degenerate carbon core grows. The path that the star follows on the H-R diagram (**Figure 16.9**) closely parallels the path it followed as a red giant, getting closer to the red giant branch as the star grows more luminous. That is why this phase of evolution is referred to as the **asymptotic giant branch (AGB)** of the H-R diagram. An AGB star burns helium and hydrogen in nested concentric shells surrounding a degenerate carbon core.

Giant Stars Lose Mass

Building on our analogy between AGB stars and red giants, you might imagine that the next step in the evolution of an AGB star should be a "carbon flash," when carbon burning begins in the star's degenerate core. Yet a carbon flash never happens. Before the temperature in the carbon core becomes high enough for carbon to burn, the star loses its gravitational grip on itself and expels its outer layers into interstellar space.

Red giant and AGB stars are huge objects. The AGB star into which a 1 $M_\odot$ main sequence star evolves can grow to a radius hundreds of times the radius of our Sun. When our Sun becomes an AGB star, its outer layers will swell to the point that they engulf the orbits of the inner planets, possibly including Earth. (Do not lose much sleep over this eventuality, however. It will be 5 billion years before Earth is engulfed by the burgeoning Sun, and well before then Earth will be well toasted by the thousandfold increase in the Sun's luminosity.) When a star expands to such a size, its hold on its outer layers becomes tenuous indeed.

FIGURE 16.9 An asymptotic giant branch star consists of a degenerate carbon core surrounded by helium- and hydrogen-burning shells. As the carbon core grows, the star brightens, accelerating up the asymptotic giant branch just as it earlier accelerated up the red giant branch while its degenerate helium core grew.

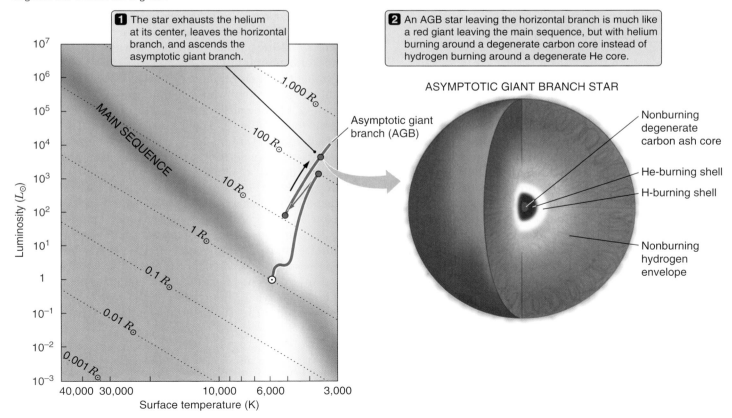

Once again an understanding of what happens to the star rests with Newton's universal law of gravitation. Recalling that $g = GM/r^2$, we know that the acceleration due to gravity (g) at the surface of a star with a mass of 1 $M_\odot$ and a radius of 100 $R_\odot$ is only 1/10,000 as strong as the gravity

Red giants and AGB stars lose mass from their outer layers.

at the surface of the Sun. It takes little in the way of a kick from below to push material near the surface of such a giant star over the edge, driving it completely away from the star. The process of **stellar mass loss** begins when the star is still on the red giant branch; by the time a 1 $M_\odot$ main sequence star reaches the horizontal branch, it may have lost 10 to 20 percent of its total mass. As the star ascends the asymptotic giant branch, it loses another 20 percent or even more of its total mass. By the time it is well up on this branch, a star that began as a 1 $M_\odot$ star probably has a mass less than about 0.7 $M_\odot$, and it may even have lost more than half of its original mass. Mass loss on the asymptotic giant branch can be spurred on by a lack of stability in the star's interior. The extreme sensitivity of the triple-alpha process to temperature can lead to episodes of rapid energy release, which can provide the extra kick needed to expel material from the star's outer layers. It also means that stars that are initially quite similar can behave very differently when they reach this stage in their evolution.

The Post-AGB Star May Cause a Planetary Nebula to Glow

Toward the end of an AGB star's life, mass loss itself becomes a runaway process. When a star loses a bit of its outer layers, it reduces the weight pushing down on the underlying layers of the star. Without this weight holding them down, the outer layers of the star puff up even larger than they were before. The star, which is now both less massive and larger, is even less tightly bound by gravity, so even less energy is needed to push outer layers away from the star. The situation is a bit like taking the lid off a pressure cooker. Mass loss leads to weaker gravity, which leads to faster mass loss, which leads to weaker gravity.... When the end comes, much of the remaining mass of the star is ejected into space, typically at speeds of 20 to 30 km/s.

After ejection of its outer layers, all that is left of the low-mass star itself is a tiny, very hot electron degenerate carbon core, surrounded by a thin envelope in which hydrogen and helium are still burning. This star is now somewhat less luminous than when it was at the top of the asymptotic giant branch, but it is still much more luminous than a horizontal branch star. The remaining hydrogen and helium in the star rapidly burn to carbon, and as more and more of the mass of

the star ends up in the carbon core, the star itself shrinks and becomes hotter and hotter. Over the course of only 30,000 years or so following the beginning of runaway mass loss, the star moves very rapidly from right to left across the top of the H-R diagram as shown in **Figure 16.10**.

The surface temperature of the star may eventually reach 100,000 K or hotter. Wien's Law says that at such temperatures most of the light from the star is in the hard (high-energy) UV part of the spectrum:

$$\lambda_{\text{peak}} = \frac{2,900\ \mu\text{m K}}{100,000\ \text{K}} = 0.029\ \mu\text{m}.$$

The intense UV light from what remains of the star heats and ionizes the expanding shell of gas that was recently ejected by the star. The ultraviolet light causes the shell of gas to glow in the same way that UV light from an O star causes an H II region to glow.

If conditions are right,[4] the mass ejected by the AGB star will pile up in a dense, expanding shell. If you were to look at such a shell through a telescope, you would see it as a round or perhaps oblong patch of light surrounding the remains of the AGB star. At first glance you might almost

A planetary nebula may form around dying low-mass stars.

confuse the glowing shell with the appearance of a planet (**Figure 16.11(a)**), which is why such objects are referred to as **planetary nebulae**. But there is nothing truly "planetary" about a planetary nebula, as is apparent in **Figure 16.11(b)**. Instead it is the remaining outer layers of a star, ejected into space as a dying gasp at the end of the star's ascent of the asymptotic giant branch.

Rather than being simple spherical shells of the sort we might naively predict would surround a nice spherical star, planetary nebulae show a dazzling range of appearances, earning them names like Owl Nebula, Clown Nebula, Cat's Eye Nebula, and Dumbbell. This extraordinary menagerie, illustrated in **Figure 16.12**, serves to rub astronomers' noses in the complexity of the late stages of a star's evolution, telling of eras when mass loss from the star was slower or faster, and of times when mass was ejected primarily from the star's equator or its poles. The gas in the planetary nebula also contains chemical elements that were produced by nuclear burning in the star, offering us our first look at the processes responsible for the chemical evolution of the universe (see **Connections 16.2**). The planetary nebula is visible for 50,000 years or so before the gas ejected by the star disperses so far that the nebula is too faint to be seen.

---

[4] Not all stars form planetary nebulae. Massive stars go through the post-AGB stage too quickly. Stars with insufficient mass take too long, so their envelope evaporates before they can illuminate it. Some astronomers believe that our own Sun will not retain enough mass during its post-AGB phase to form a planetary nebula.

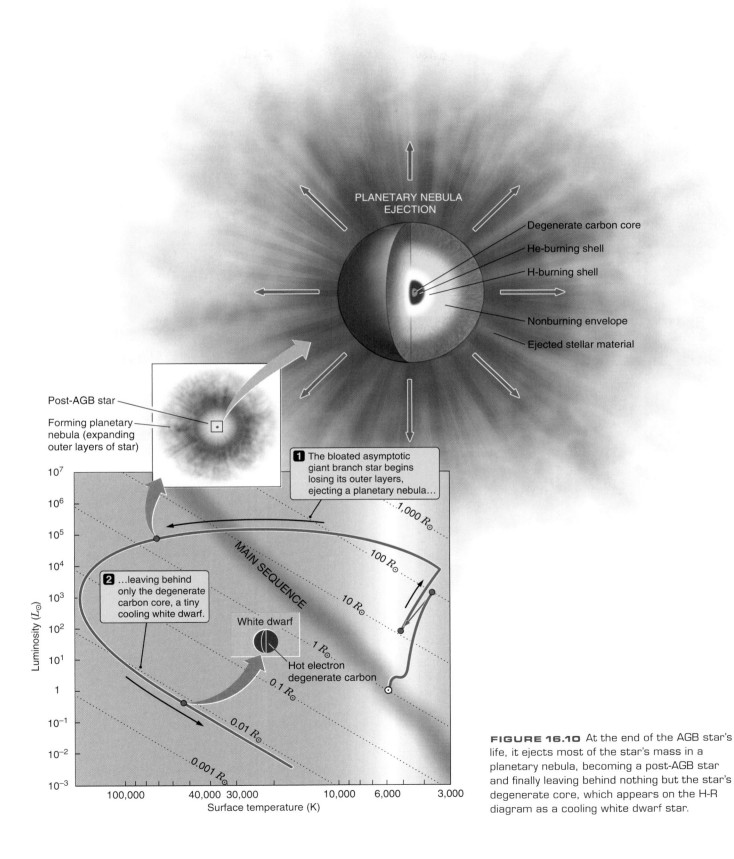

PLANETARY NEBULA
EJECTION

Degenerate carbon core

He-burning shell

H-burning shell

Nonburning envelope

Ejected stellar material

Post-AGB star

Forming planetary
nebula (expanding
outer layers of star)

1 The bloated asymptotic
giant branch star begins
losing its outer layers,
ejecting a planetary nebula…

MAIN SEQUENCE

1,000 $R_\odot$

100 $R_\odot$

10 $R_\odot$

2 …leaving behind
only the degenerate
carbon core, a tiny
cooling white dwarf.

White dwarf

Hot electron
degenerate carbon

1 $R_\odot$

0.1 $R_\odot$

0.01 $R_\odot$

0.001 $R_\odot$

Luminosity ($L_\odot$)

10^7
10^6
10^5
10^4
10^3
10^2
10^1
1
10^{-1}
10^{-2}
10^{-3}

100,000   40,000 30,000   10,000   6,000   3,000
Surface temperature (K)

FIGURE 16.10 At the end of the AGB star's
life, it ejects most of the star's mass in a
planetary nebula, becoming a post-AGB star
and finally leaving behind nothing but the star's
degenerate core, which appears on the H-R
diagram as a cooling white dwarf.

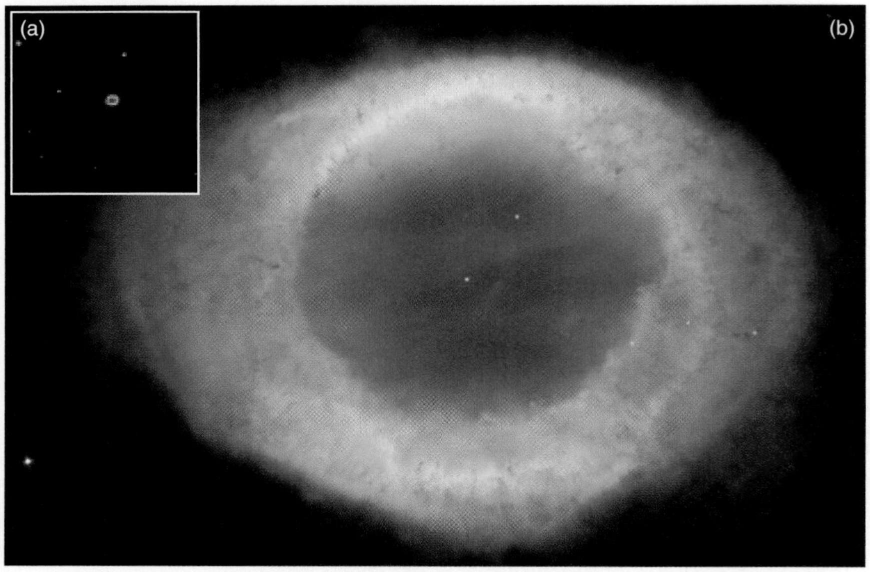

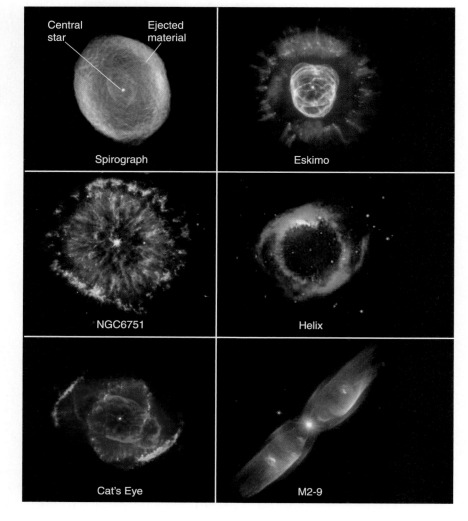

FIGURE 16.11 At the end of its life, a low-mass star ejects its outer layers and may form a "planetary nebula" consisting of an expanding shell of gas surrounding the white-hot remnant of the star. (a) This picture of a famous planetary nebula called the Ring Nebula, as it appears through a small amateur telescope, shows why astronomers thought these objects looked like planets. (b) However, an HST image of the ring shows the remarkable and complex structure of this expanding shell of gas.

FIGURE 16.12 Planetary nebulae are not all simple spherical shells around their parent stars. These HST images of planetary nebulae show a wealth of structure resulting from the complex processes by which low-mass stars eject their outer layers.

CONNECTIONS 16.2

Seeding the Universe with New Chemical Elements

One of the grand themes of 20th and 21st century astronomy is the origin of the chemical elements. When we discuss the properties of the early universe, we will see that the only chemical elements that were formed in abundance when the universe was young were hydrogen and helium, along with trace amounts of lithium, beryllium, and boron.

Astronomers studying red giant and AGB stars see direct evidence that more massive chemical elements formed deep within the interiors of low-mass stars find their way to the "outside" world. In Chapter 14 we learned that if the change in temperature within a star becomes too steep, convection will begin. Like the boiling motion in a pot of water, convection is a process of mixing. Material from the inner parts of the star is carried upward and mixed with material in the outer parts of the star. As a star ascends first the red giant and later the asymptotic giant branch of the H-R diagram, the core of the star grows hotter and hotter, and the temperature gradient within the star grows steeper. Convection spreads through more and more of the star. Under certain circumstances convection can spread so deep into a star that it is able to dredge up chemical elements formed by nuclear burning within the star and carry them to the star's surface. Many AGB stars, known as **carbon stars**, show an over-

abundance of carbon and other by-products of nuclear burning in their spectra. This extra carbon originated in the star's helium-burning shell and then was carried to its surface by convection.

Mass loss from giant stars carries the chemical elements enriching the stars' outer layers off into interstellar space. Planetary nebulae—the ejected atmospheres of low-mass stars—often show an overabundance of elements such as carbon, nitrogen, and oxygen, which are by-products of nuclear burning. Once this chemically enriched material leaves the star, it mixes with the interstellar gas there, enriching the chemical diversity of the universe. Novae and Type I supernovae, discussed at the end of this chapter, also do their part. The chemical burning that occurs in these explosive events goes far beyond the formation of elements such as carbon. Novae and Type I supernovae help to seed the universe with much more massive atoms such as iron and nickel.

As we continue to discuss the evolution of stars and the universe in the chapters to come, we will learn more about the origin of the chemical elements. Today when scientists study the patterns of the abundance of different chemical elements on Earth and in our Solar System, they recognize the patterns they see. Those patterns are clear signatures of nuclear reactions in stars.

The Star Becomes a White Dwarf

Within 50,000 years or so, a post-AGB star burns all of the fuel remaining on its surface, leaving nothing behind but a cinder—a nonburning ball of carbon with a mass of about $1\ M_\odot$. In the process the star plummets down the left side of the H-R diagram, becoming smaller and fainter. Within

---

A newly formed white dwarf is hot but tiny.

---

a few thousand years the burned-out core shrinks to about the size of Earth, at which point it has become fully degenerate and so can shrink no further. The degenerate stellar cinder is now referred to as a **white dwarf**. The white dwarf continues to radiate energy away into space, and as it does

so it cools, just as the filament of a lightbulb cools when the switch is turned off. Because the white dwarf is electron degenerate, its size does not change much as it cools, so it moves down and to the right on the H-R diagram, following a line of constant radius. Even though the white dwarf may remain very hot for 10 million years or so, its luminosity may now be only 1/1,000 that of a main sequence star like our Sun. Many white dwarfs are known, but all are much too faint to be seen without the aid of a telescope. Yet when you look at Sirius, you are also looking at a white dwarf. The brightest star in Earth's sky has a faint white dwarf as a binary companion.

Here we have the fate of our Sun, some 5 billion years from now. It will become a white dwarf that, bereft of all sources of nuclear fuel, will shine ever more feebly as it

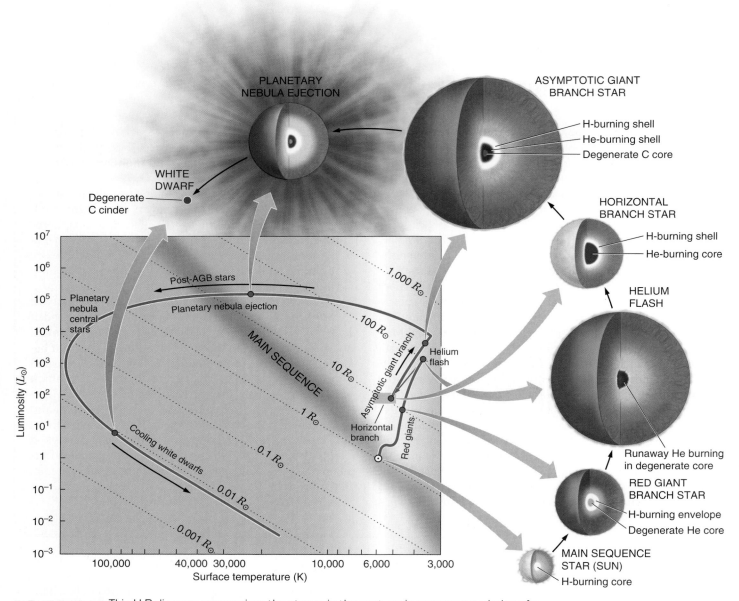

FIGURE 16.13 This H-R diagram summarizes the stages in the post–main sequence evolution of a one-solar-mass star.

continues to cool, radiating its thermal energy away into space. It is amazing to recall that this superdense ball—with a density of a ton per teaspoonful—actually began its life billions of years earlier as a cloud of interstellar gas billions of times more tenuous than the vacuum in the best vacuum chamber on Earth.

Figure 16.13 recaps the evolution of a solar-type 1 $M_\odot$ main sequence star through to its final existence as a 0.6 $M_\odot$ white dwarf. We point out once again that what we have described is representative of the fate of low-mass stars.

Although all low-mass stars form white dwarfs at the end points of their evolution, the exact path a low-mass star follows from core hydrogen burning on the main sequence to white dwarf depends on many details particular to the star. Some stars less massive than the Sun may become white dwarfs composed largely of helium rather than carbon. Temperatures in the cores of evolved 2 $M_\odot$ to 3 $M_\odot$ stars are high enough to allow additional nuclear reactions to occur, leading to the formation of somewhat more massive white dwarfs composed of materials such as oxygen, neon, and magnesium. Differences in chemical composition of a star can also lead to dramatic differences in its post–main sequence evolution. Finally, we note that studies of the evolution of stars whose main sequence masses are less than about 0.8 $M_\odot$ or so are pure theory. These stars are able to burn hydrogen in their cores for longer than the age of the universe. No star with a mass lower than that has ever evolved beyond the main sequence except in the virtual world of a computer's calculations.

In our discussion of stellar evolution we have focused on what happens after a star leaves the main sequence. It is important, however, that we remind ourselves that the spectacle of a red giant or AGB star is ephemeral. The Sun will travel the path from subgiant to white dwarf in less than a 10th of the time that it spends on the main sequence steadily burning hydrogen to helium in its core. Stars spend most of their luminous lifetimes on the main sequence, which is why most of the stars that we see in the sky are main sequence stars. (Were it otherwise, the "main sequence" would not be "main" at all!) Yet in the end it is white dwarfs that will carry the day. Too faint to be noticed as we look at the sky on a clear summer evening, white dwarfs still constitute the final resting place for the vast majority of stars that have been or ever will be formed.

16.6 Many Stars Evolve as Pairs

Possibly the most significant complication in our picture of the evolution of low-mass stars arises from the fact that many stars in the sky are members of binary systems. Binary stars were quite useful to us when we were uncovering the properties of stars. We used them as a sort of celestial bathroom scale, applying Kepler's laws to the orbits of two stars about each other to determine their masses. While the members of a binary pair are on the main sequence, they probably have little effect on each other. However, when one of the stars evolves off the main sequence, it can swell to hundreds of times its original size, and its outer layers can literally cross the line between what belongs to the evolved star and what belongs to its companion. We will shortly see the consequences of this.

Mass Flows from an Evolving Star onto Its Companion

Think for a moment about what would happen if you were to travel in a spacecraft from Earth toward the Moon. When you are still near Earth, the force of Earth's gravity is far

stronger than that of the Moon. There is no question that you "belong" to Earth, just as there is no question that you "belong" to Earth as you sit in your easy chair reading this book. Yet as you move away from Earth and closer to the Moon, this terrestrial hegemony becomes less secure: The gravitational attraction of Earth weakens, and the gravitational attraction of the Moon becomes stronger. There comes a point when you reach an interplanetary zone where neither body has the upper hand. If you continue beyond this point, the lunar gravity begins to successfully assert itself until you find yourself firmly in the grip of the Moon.

Exactly the same situation exists between two stars. Gas near each star clearly "belongs" to that star. But what happens if, as a star swells up, its outer layers cross that gravitational dividing line separating the star from its companion? Any material that crosses this line no longer belongs to the first star, but instead can be pulled toward the companion.

An evolving star loses mass to its companion.

A star reaches this point when it fills up its portion of an imaginary figure-8-shaped volume of space shown in **Figure 16.14**. These regions surrounding the two stars are referred to as the **Roche lobes** of the system (see **Connections 16.3**). Once the star has expanded to fill its Roche lobe, material begins to pour through the "neck" of the figure 8 and fall toward the other star. Astronomers refer to this exchange of material between the two stars as **mass transfer**.

Evolution of a Binary System

The best way to understand how mass transfer affects the evolution of stars in a binary system is to apply what we have learned from the evolution of single low-mass stars. Figure 16.14(a) shows a binary system consisting of two low-mass stars. We will rather unimaginatively call the more massive of the two stars star 1 and the less massive of the two star 2. This is an ordinary binary system, and each of these stars is an ordinary main sequence star for most of the system's lifetime. However, when one of the stars begins to evolve, things start to get interesting.

The more massive a main sequence star is, the more rapidly it evolves. Therefore star 1, which is more massive, will be the first to use up the hydrogen at its center and begin to evolve off the main sequence (Figure 16.14(b)). If the two

The more massive star evolves first.

stars are close enough together, star 1 will eventually grow to fill its Roche lobe, and the overflow from star 1 will start falling onto star 2 (Figure 16.14(c)). A number of interesting things can happen at this point. For example, the transfer

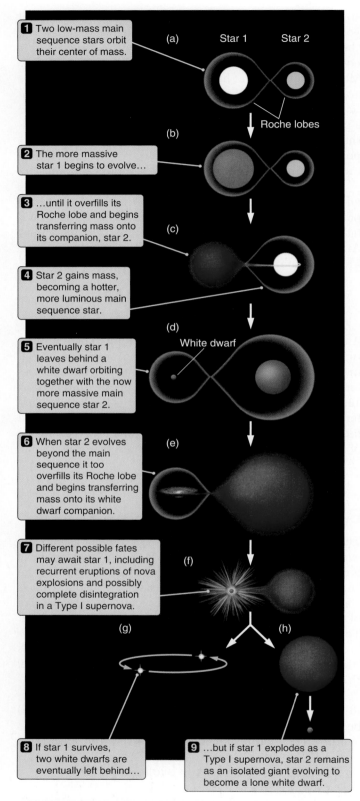

1 Two low-mass main sequence stars orbit their center of mass.

(a)   Star 1   Star 2

Roche lobes

(b)

2 The more massive star 1 begins to evolve...

3 ...until it overfills its Roche lobe and begins transferring mass onto its companion, star 2.

(c)

4 Star 2 gains mass, becoming a hotter, more luminous main sequence star.

(d)   White dwarf

5 Eventually star 1 leaves behind a white dwarf orbiting together with the now more massive main sequence star 2.

6 When star 2 evolves beyond the main sequence it too overfills its Roche lobe and begins transferring mass onto its white dwarf companion.

(e)

7 Different possible fates may await star 1, including recurrent eruptions of nova explosions and possibly complete disintegration in a Type I supernova.

(f)

(g)   (h)

8 If star 1 survives, two white dwarfs are eventually left behind...

9 ...but if star 1 explodes as a Type I supernova, star 2 remains as an isolated giant evolving to become a lone white dwarf.

FIGURE 16.14 A close binary system consisting of two low-mass stars passes through a sequence of stages as the stars evolve and mass is transferred back and forth. The evolution of the binary system can lead to novae or even to a Type I supernova that destroys the initially more massive star 1.

of mass between the two stars can result in a sort of "drag" that causes the orbits of the two stars to shrink, bringing the stars closer together and further enhancing mass loss. The two stars can even reach the point where they are effectively two cores sharing the same extended envelope of material.

Despite these complexities, star 2 probably remains a basically normal main sequence star throughout this process, burning hydrogen in its core. However, it does not remain the same main sequence star that it started out to be. The mass of star 2 increases, courtesy of the largesse of its companion. As it does so, the structure of star 2 must change to accommodate its new status as a higher-mass star. If we could watch how star 2 moves on the H-R diagram during this period, we would see it move up and to the left along the main sequence, becoming larger, hotter, and more luminous.

While star 2 gains from the interaction, star 1 suffers. Star 1 never gets to experience the full glory of being an isolated red giant or AGB star because star 1 can never grow larger than its Roche lobe. The presence of star 2 prevents star 1 from swelling to the size of the behemoths that populate the top of the H-R diagram. Yet star 1 continues to

> **The first star becomes a white dwarf orbiting a main sequence star.**

evolve, spending its allotted time burning helium in its core on the horizontal branch, proceeding through a stage of helium shell burning, and finally losing its outer layers and leaving behind a white dwarf. Figure 16.14(d) shows the binary system after star 1 has completed its evolution. All that remains of star 1 is a white dwarf, orbiting about its bolstered main sequence companion, star 2.

Fireworks Occur When the Second Star Evolves

The stage is now set for a sequence of events that can lead to one of nature's most spectacular displays. Figure 16.14(e) picks up the evolution of the binary system as star 2 begins to evolve off the main sequence. Like star 1 before it, star 2 grows to fill its Roche lobe; as it does, material from star 2 begins to pour through the "neck" connecting the Roche lobes of the two stars. However, this time the mass is not being added to a normal star but is drawn toward the tiny white dwarf left behind by star 1. Because the white dwarf is so small, the orbit of the infalling material generally causes it to miss. Instead of landing directly on the white dwarf, the infalling mass forms an **accretion disk** around the white dwarf, similar in some ways to the accretion disk that forms around a protostar. As in the process of star formation, the accretion disk serves as a way station for material that is destined to find its way onto the white dwarf but

CONNECTIONS 16.3

Roche Limits and Roche Lobes

You may recall the name *Roche* from our earlier discussion of the tidal disruption of moons and comets leading to the formation of planetary rings. The Roche limit determines how close a small object might come to a planet before being torn apart by the planet's tides. In this chapter we discuss Roche lobes in connection with the transfer of mass from one member of a binary star system to the other. The two processes are actually similar. Mass transfer between stars in a binary can be thought of as the "tidal stripping" of one star by the other. The physical principles at work in the two situations are much the same, and Edouard Roche is responsible for early calculations of both.

that starts out with too much angular momentum to hit the white dwarf directly.

We have already seen that a white dwarf has a mass comparable to that of the Sun but is closer to the size of Earth. Large mass and small radius mean strong gravity. In Chapter 6 we saw how the gravitational energy of material falling toward a forming protostar is converted into thermal energy when the material hits the accretion disk around the protostar. The same principle applies here, but with a

---

Material from the second star gets hot as it falls onto the white dwarf.

---

vengeance. A kilogram of material falling from space onto the surface of a white dwarf releases a hundred times more energy than a kilogram of material falling from the outer Solar System onto the surface of the Sun. The material streaming toward the white dwarf in the binary system falls down into an incredibly deep gravitational "well." All of this energy has to go somewhere, and it goes into thermal energy. The spot where the stream of material from star 2 hits the accretion disk can be heated to millions of kelvins, where it glows in the far ultraviolet and X-ray parts of the electromagnetic spectrum.

Eventually the infalling material accumulates on the surface of the white dwarf (**Figure 16.15(a)**), where it is compressed by the enormous gravitational pull of the white dwarf to a density close to that of the white dwarf itself. As more and more material builds up on the surface of the white dwarf, the white dwarf shrinks (just as the core of the red giant shrinks as it grows more massive). The density gets greater and greater, and at the same time the release of gravitational energy drives the temperature of the white dwarf higher and higher. Now recall that the infalling material is from the outer, unburned layers of star 2, so it is mostly composed of hydrogen. We have a dangerous situation here. Hydrogen, the best nuclear fuel around, is being compressed to higher and higher densities and heated to higher and higher temperatures on the surface of the white dwarf.

The mental picture you should have at this point is of gasoline pooling on the floor of a match factory. Once the temperature at the base of the layer of hydrogen reaches about 10 million K, hydrogen begins to burn. But this is not the contained hydrogen burning like that taking place in the center of the Sun. Rather, this is explosive hydrogen burning in a degenerate gas. Energy released by hydrogen burning drives up the temperature, and the higher temperature drives up the rate of hydrogen burning. This runaway thermonuclear reaction is much like the runaway helium burning that takes place during the helium flash, except now there are no overlying layers of a star to keep things

---

Runaway burning of hydrogen on a white dwarf causes a nova.

---

contained. The result is a tremendous explosion—a **nova** (plural *novae*)—that blows part of the layer covering the white dwarf out into space at speeds of thousands of kilometers per second (see Figure 16.14(f) and Figure 16.15(b)).

About 50 novae are thought to occur in the Milky Way Galaxy each year, but because our view is restricted by interstellar extinction, we can only see a few of these, typically two or three each year. Novae get bright in a hurry —typically reaching their peak brightness in only a few hours—and for a brief time can be almost half a million times more luminous than the Sun. Although the brightness of a nova sharply declines in the weeks following the outburst, it can sometimes still be seen for years. During this time the glow from the expanding cloud of material ejected by the explosion is powered by the decay of radioactive isotopes created in the explosion.

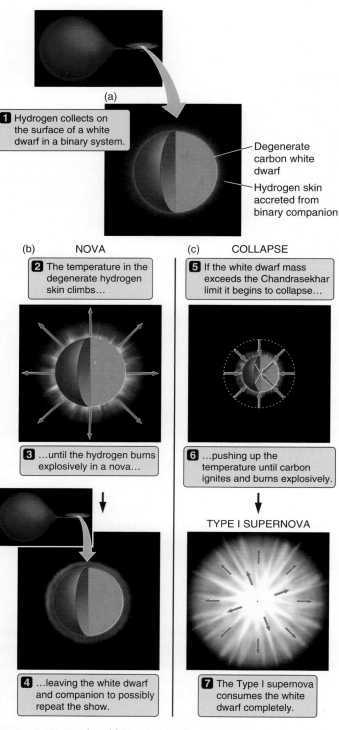

(a)

1 Hydrogen collects on the surface of a white dwarf in a binary system.

Degenerate carbon white dwarf

Hydrogen skin accreted from binary companion

(b)   NOVA

2 The temperature in the degenerate hydrogen skin climbs…

3 …until the hydrogen burns explosively in a nova…

4 …leaving the white dwarf and companion to possibly repeat the show.

(c)   COLLAPSE

5 If the white dwarf mass exceeds the Chandrasekhar limit it begins to collapse…

6 …pushing up the temperature until carbon ignites and burns explosively.

TYPE I SUPERNOVA

7 The Type I supernova consumes the white dwarf completely.

FIGURE 16.15 In a binary system in which mass is transferred onto a white dwarf, a skin of hydrogen builds up on the surface of the degenerate white dwarf. If hydrogen burning ignites on the surface of the white dwarf, the result is a nova. If enough hydrogen accumulates to force the white dwarf to begin to collapse, carbon ignites and the result is a Type I supernova.

The explosion of a nova does not destroy the underlying white dwarf star. In fact, much of the material that had built up on the white dwarf can remain behind after the explosion. The nova leaves the binary system in much the same configuration it was in previously—the configuration shown in Figure 16.15(a)—with material from star 2 still pouring onto the white dwarf. This cycle can repeat itself many times, with material building up and igniting over and over again on the surface of the white dwarf. In most cases outbursts are probably separated by thousands of years, so most novae have been seen only once in historical times. However, some novae are known to be recurrent, erupting anew every decade or so.

A Stellar Cataclysm May Await the White Dwarf in a Binary System

It is possible that eventually star 2 will simply go on to form a white dwarf, leaving behind a binary system consisting of two white dwarfs, as in Figure 16.14(g). There is another possibility, however. Novae may be spectacular events, but they pale in comparison with the fate that may eventually await the white dwarf in a binary system. Through millions of years of mass transfer from star 2 onto the white dwarf, and possibly through countless nova outbursts, the mass of the white dwarf slowly increases. As it does, the white dwarf shrinks, and the grip of self-gravity on this already extremely dense object gets tighter and tighter. The mass of a white dwarf can be pushed only so far. There comes a point

The mass of a white dwarf cannot exceed 1.4 $M_\odot$.

when even the pressure supplied by degenerate electrons is no longer enough to balance gravity. This precipice occurs at a mass of about 1.4 $M_\odot$, a value referred to as the **Chandrasekhar limit**. It is impossible for a star to increase in mass above 1.4 $M_\odot$ and remain a white dwarf. Raise its mass above the Chandrasekhar limit, and it will begin to collapse into even more extreme forms of matter (see Figure 16.15(c)).

As the white dwarf in our binary system approaches this limit, it grows smaller and smaller and hotter and hotter. As the star crosses the Chandrasekhar limit it begins to collapse, and the conversion of gravitational energy into thermal energy quickly drives the temperature in the white dwarf past the 6×10^8 K needed to slam carbon nuclei together with enough force to overcome their electric repulsion and fuse them. Once again the onset of nuclear burning in a degenerate gas leads to a thermonuclear runaway, with ever higher temperatures and ever faster nuclear burning feeding off each other. But whereas the thermonuclear runaway in a nova involves only a thin shell of material on the surface of the white dwarf, runaway carbon burning involves the entire white dwarf. Within about a second the entire white dwarf is consumed in the resulting conflagration. In this single instant 100 times more energy is liberated than will

---

The carbon in a white dwarf can ignite in a Type I supernova.

---

be given off by the Sun over its entire 10 billion-year lifetime on the main sequence. Runaway fusion reactions convert a large fraction of the mass of the star into elements such as iron and nickel, and the explosion blasts the shards of the white dwarf into space at top speeds in excess of 20,000 km/s. The explosion completely destroys star 1, leaving star 2 be-

hind as a lone giant to continue its evolution toward becoming a white dwarf (Figure 16.14(h)).

Explosive carbon burning in a white dwarf is a leading theory used to explain colossal events called **Type I supernovae**. Type I supernovae[5] occur in a galaxy the size of the Milky Way about once a century. For a brief time they can shine with a luminosity 10 billion times that of our Sun, possibly outshining the galaxy itself.

---

[5] Strictly speaking, we should refer here to Type Ia supernovae. Technically a "Type I" supernova is a supernova that shows no emission lines from hydrogen. This includes some variants of the supernovae from massive stars that will be discussed in the next chapter. We hope the purist will forgive our simplification of referring to all supernovae from white dwarfs as Type I and all supernovae from massive stars as Type II.

Summary

- All stars eventually exhaust their nuclear fuel.
- The more massive a star, the shorter will be its lifetime.
- Helium accumulates like ash in the cores of main sequence stars.
- After exhausting its hydrogen, a low-mass star leaves the main sequence and swells to become a red giant.
- A red giant burns helium via the triple-alpha process.
- In their dying stages some stars form planetary nebulae.
- All low-mass stars eventually become white dwarfs.
- A white dwarf is very hot but very small, and cannot exceed 1.4 $M_\odot$.
- White dwarfs cool to become cold, dark cinders—the stellar graveyard.
- Transfer of mass within a binary system can lead to novae or supernovae.

Seeing the Forest through the Trees

Stars are born of the tenuous matter of interstellar space, then settle down onto the main sequence as stable and steady cosmic citizens. But the universe is a place of constant change and evolution. Stars consume the fuel

they were born with, and when that fuel is gone, they must die.

In the world of humans it is wealth and power and birthright that determine our lifestyle. In the lives of stars, birthright is spelled "mass." Mass determines the weight bearing down on the interior of a star, which in turn controls the rate of the nuclear reactions taking place there. High-mass stars are spectacular but ephemeral, burning out the fuel in their cores in only a few million years. Our Sun is typical instead of the much longer-lived low-mass stars. Yet eventually even these steady performers must exhaust the hydrogen at their centers.

In this chapter we have followed the evolution of a low-mass star like our Sun as it evolves beyond the main sequence, passing through a fascinating series of evolutionary stages. Some of these stages are dramatic and violent, such as the runaway thermonuclear reactions of the helium flash or the ejection of a planetary nebula that carries the chemical legacy of nuclear burning deep within the dying AGB star. Other phases are far more quiescent, such as the star's time on the horizontal branch burning helium to carbon in its core and hydrogen to helium in a surrounding shell. Eventually even an undistinguished star like our Sun may reach a luminosity rivaling all but the most luminous main sequence O stars, and swell to a size that would swallow much of our inner Solar System. Curiously, such a behemoth is supported by frenzied nuclear reactions in a region surrounding the tiniest of cores—a superdense inert cinder crushed by the weight of the overlying star to a size little larger than Earth. At a density of a thousand kilograms per cubic centimeter, the electron degenerate core defies our normal notions of matter.

All of these stages are signposts along the star's inexorable march toward its eventual fate, in which the outer layers of the star will have been ejected into space and nothing will remain but a tiny cooling cinder—a degenerate white dwarf that will eventually fade from the notice of the rest of the universe. Yet if it happens to orbit about a binary companion, the white dwarf may live again. Unburned nuclear fuel pouring onto the surface of the white dwarf may trigger the episodic nuclear outbursts of novae or even a cataclysmic Type I supernova.

We have never witnessed the life cycle of a low-mass star from birth to death, nor will we. Not even a hundred million human lifetimes would be enough to see the saga through from beginning to end. Yet we are confident of most of what we have reported here, just as we are confident of our knowledge of the internal structure of the Sun and Earth. Physics works, and the closing decades of the 20th century saw steady improvements in the quality of our calculations of the structure and evolution of stars. And although we cannot witness the life cycle of a single star, when we look at the sky we find many examples of red giants, AGB stars, white dwarfs, and the other members of this evolutionary menagerie. Later on our journey we will also see how clusters of stars that form together provide snapshots that allow us to test our theories of stellar evolution in exquisite detail.

What we have seen is representative of the fate awaiting the vast majority of stars in the universe. In the next chapter we turn our attention to the evolution of the small fraction of stars that fall into the category of high-mass stars. There we will find that the extreme and counterintuitive behavior of white dwarfs presages a whole zoo of bizarre and exotic objects. Where low-mass stars bent our notions about matter, high-mass stars will shatter those notions, leading us to consider forms of matter so extreme that they cease to be matter at all.

Key Terms

low-mass star, p. 458
main sequence lifetime, p. 459
electron degenerate, p. 463
hydrogen shell burning, p. 463
subgiant, p. 465
subgiant branch, p. 466
red giant branch, p. 466
triple-alpha process, p. 468
alpha particle, p. 468
helium flash, p. 469
horizontal branch, p. 470
asymptotic giant branch (AGB), p. 471
stellar mass loss, p. 472
planetary nebula, p. 472
white dwarf, p. 475

Roche lobe, p. 477
mass transfer, p. 477
accretion disk, p. 478
nova, p. 479
Chandrasekhar limit, p. 480
Type 1 supernova, p. 481

Student Questions

THINKING ABOUT THE CONCEPTS

1. Why do most nearby stars turn out to be low-mass, low-luminosity stars?

2. The Sun is slowly becoming more luminous as it ages on the main sequence. What effect will this have on terrestrial life?

3. What is the primary reason that the most massive stars have the shortest lifetimes? (The answer can be expressed in just a few words.)

4. Consider a hypothetical star having the same mass as our Sun being one of the first stars that formed when the universe was still very young. Would you expect it to be surrounded by planets? Explain your answer.

5. We typically say that the mass of a newly formed star determines its destiny from birth to death. However, there is a frequent environmental circumstance in which this statement is not true. Explain why the birth mass of a star might not explain its full destiny.

6. Suppose Jupiter were not a planet, but instead were a G5 main sequence star with a mass of 0.8 $M_\odot$.
   a. What effect, if any, do you think this would have on terrestrial life?
   b. How would this affect the Sun as it came to the end of its life?

7. Suppose you were an astronomer making a survey of the observable stars in our galaxy. What would be your chances of seeing a star undergoing the helium flash? Explain your answer.

8. The intersection of the Roche lobes in a binary system is the equilibrium point between the two stars where the gravitational attraction from both stars is equally strong and opposite in direction. Is this an example of stable or unstable equilibrium? Explain.

9. In Latin *nova* means "new." Novae, as we now know, are not new stars. Explain how novae might have gotten their name and why they are really not new stars.

10. T Corona Borealis is a well-known recurrent nova.
    a. Is it a single star or a binary system? Explain.

b. What mechanism causes a nova to flare up?

c. How can a nova flare-up happen more than once?

APPLYING THE CONCEPTS

11. The escape velocity ($v_{esc} = \sqrt{2\ GM_\odot/r_\odot}$) from the Sun's surface today is 618 km/s.

   a. What will the escape velocity be when the Sun becomes a red giant with a radius 50 times greater and a mass only 0.9 times that of today?

   b. What will it be when the Sun becomes an AGB star with a radius 200 times greater and a mass only 0.7 times that of today?

   c. How would these changes in escape velocity affect mass loss from the surface of the Sun as a red giant, and later as an AGB star?

12. Each form of energy generation in stars depends on temperature.

   a. The rate of hydrogen fusion (proton–proton chain) near 10^7 K increases with temperature as T^5. If you raise the temperature of the hydrogen-burning core by 10 percent, how much does the hydrogen fusion energy increase?

   b. Helium fusion (triple-alpha) at 10^8 K increases with an increase in temperature at a rate of T^{38}. If you raise the temperature of the helium-burning core by 10 percent, how much does the helium fusion energy increase?

13. After leaving the main sequence, low-mass stars "seed" the interstellar medium with heavy elements such as carbon, nitrogen, and oxygen. Assume that the universe is approximately 1.4×10^{10} (14 billion) years old and that the Sun is 4.6 billion years old. Refer to Table 16.1 for the masses and main sequence lifetimes of stars. How many generations of each type of star preceded the birth of our Sun and Solar System?

14. Assume that there are twice as many F0 stars as A0 stars, and that each type contributes heavy elements to the interstellar medium in approximate proportion to its main sequence mass. What was the relative contribution from the two types of star before the Sun was born?

15. A white dwarf has a density of approximately 10^9 kg/m^3. Earth has an average density of 5,500 kg/m^3 and a diameter of 12,700 km. If Earth were compressed to the same density as a white dwarf, how large would it be?

16. Recall from Chapter 4 that the luminosity of a spherical object at temperature T is given by $L = 4\pi R^2\ \sigma T^4$, where R is its radius. If the Sun becomes a white dwarf with a radius of 10^7 m, what would be its luminosity at temperatures of 10^8 K, 10^6 K, 10^4 K, and 10^2 K?

StudySpace

wwnorton.com/astro21

provides a Study Plan for each chapter that includes a reading outline, animations, keyword flash cards, and gradebook-enabled multiple-choice quizzes. From StudySpace you can also access premium content in the ebook and SmartWork.

It does not do to leave a live dragon out of
your calculations, if you live near him.

J. R. R. TOLKIEN (1892–1973)
THE HOBBIT

The inner Crab Nebula
seen in X-ray (blue) and
visible (red) light.

CHAPTER 17

Live Fast, Die Young

17.1 There Be Dragons...

So far in our discussion of the lives of stars we have concentrated on what happens with low-mass stars like our Sun. These are stars that steadily burn hydrogen to helium in their cores and produce a fairly constant output of energy for billions of years—long enough for our Earth to cool from a molten ball of rock and for life to evolve to the point of contemplating its place in the universe. High-mass stars—stars with masses greater than about 8 $M_\odot$—are very different beasts. These are objects that burn with luminosities of thousands or even millions of times that of our Sun and squander their generous allotments of nuclear fuel in less time than mammals have walked on Earth.

The difference between the ways in which high-mass stars and low-mass stars evolve lies in the same relationship between gravity, pressure, and the rate of nuclear burning that determines the balance in a star like our Sun. The litany should be familiar by now. More mass means stronger gravity and therefore more weight bearing down on the inner parts of the star. This means higher pressure; higher pressure means faster reaction rates; and faster reaction rates mean greater luminosity. There are many differences in the evolution of low- and high-mass stars, but in the end all of these trace back to the greater gravitational force bearing down on the interior of a high-mass star.

KEY CONCEPTS

Our Sun is but a pale ember compared with the brilliance of more massive stars. These stars consume their fuel in a cosmic blink of the eye, then in their death dazzle us with brilliant fireworks, seed the universe with the elements of life, and force us to once again alter our concept of space and time. As our exploration of stars draws to a close, we will discover

- How the structure of high-mass stars differs from that of low-mass stars.

- The stages of nuclear burning that evolving high-mass stars experience as they form progressively more massive chemical elements.

- How H-R diagrams allow us to measure the ages of stars and test our theories of stellar evolution.

- The death of massive stars in Type II supernovae.

- The origin of the most massive chemical elements.

- Degenerate neutron stars the size of small cities, with masses several times that of the Sun.

- A menagerie of bizarre objects, including pulsars and X-ray binary stars.

- That gravity is a consequence of the way that mass distorts the very shape of spacetime.

- The bottomless pits in spacetime that we call black holes.

17.2 High-Mass Stars Follow Their Own Path

Recall from Chapter 14 that the first step in the hydrogen-burning proton–proton chain is the collision and fusion of two protons. Protons have only a single positive charge, so they are the easiest atomic nuclei to force close enough together to fuse. This is a great advantage for low-mass stars, where hydrogen burns at temperatures as low as a few million kelvins. However, the proton–proton chain suffers from a large disadvantage as well. The "affinity" between two protons is not very great. Even when two protons are slammed together hard enough for the strong nuclear force to act on them, the likelihood that they will fuse is low. The lack of vigor of this first step in the proton–proton chain limits how rapidly this process can run.

At the much higher temperatures at the center of a high-mass star, additional nuclear reactions become possible. In particular, hydrogen nuclei are able to interact with the nuclei of more massive elements such as carbon. It takes a lot of energy to get past the electric barrier that repels a carbon nucleus, with its six protons, from a hydrogen nucleus. But if this barrier can be overcome, the strong nuclear force interaction between the carbon and the hydrogen is much more favorable than the interaction between two hydrogen nuclei. This reaction, in which a ^{12}C nucleus and a proton combine to form a ^{13}N nucleus consisting of seven protons and six neutrons, is the first reaction in a chain of events that is referred to as the **carbon–nitrogen–oxygen (CNO) cycle**. The remaining steps in the CNO cycle are shown in **Figure 17.1**. Note that carbon is not consumed by the CNO cycle but instead is a catalyst. The ^{12}C nucleus that started

The CNO cycle burns hydrogen in massive stars.

the cycle is there again at the end of the cycle, but now instead of the four hydrogen nuclei that took part in the chain of reactions, we have a helium nucleus. **Figure 17.2** shows that the CNO cycle becomes far more efficient than the proton–proton chain in stars more massive than 1.5 $M_\odot$.

The different ways that hydrogen burning takes place in high- and low-mass stars are reflected in the different structures of their cores. The temperature gradient in the

Convection stirs the core of a high-mass star.

core of a high-mass star is so steep that convection sets in within the core itself, "stirring" the core like the water in a boiling pot. Compare **Figure 17.3** with the similar figure in Chapter 16 (Figure 16.2). Rather than building up from the center outward, helium ash is spread uniformly through-

---

EXCURSIONS 17.1

Stars Are Unique

The distinction made in this book between low-mass and high-mass stars is a useful, convenient way to think about stellar evolution. After all, there is a fundamental and qualitative difference between stars that end their lives quietly as dying cinders and those that die in spectacular explosions. Even so, this distinction is an oversimplification. In reality, each star is unique, with its own individual mass and composition and its own particular circumstances.

Stars with masses between about 3 $M_\odot$ and 8 $M_\odot$ exist in something of a gray area between the low-mass stars discussed in Chapter 16 and the high-mass stars discussed in this chapter. While they are on the main sequence, these **intermediate-mass stars** burn hydrogen via the CNO cycle like massive stars. These stars also leave the main sequence as massive stars do, burning helium in their cores immediately after their hydrogen is exhausted and skipping the red giant and helium flash phases of low-mass star evolution. However, at the completion of the He-core-burning phase, the temperature at the center of an intermediate-mass star is too low for carbon to burn. From this point on, the star evolves more like a low-mass star, ascending the asymptotic giant branch, burning helium and hydrogen in shells around a degenerate core, then ejecting its outer layers and leaving behind a white dwarf. Yet even at the end, the star carries the signature of its early "high-mass" years. The chemical compositions of planetary nebulae and white dwarfs left behind by intermediate-mass stars can be quite distinct from those of truly low-mass stars.

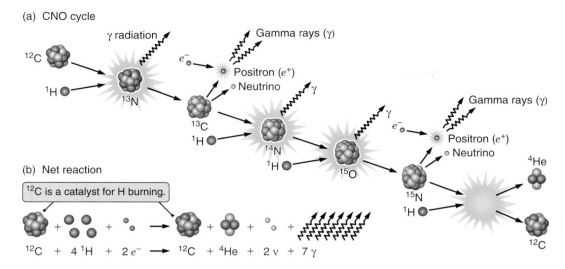

(a) CNO cycle

(b) Net reaction

^{12}C is a catalyst for H burning.

$$^{12}\text{C} + 4\,^1\text{H} + 2\,e^- \longrightarrow\ ^{12}\text{C} + \,^4\text{He} + 2\,\nu + 7\,\gamma$$

FIGURE 17.1 In high-mass stars, carbon serves as a catalyst for fusion of hydrogen to helium. This process is called the carbon–nitrogen–oxygen (or CNO) cycle.

out the core of a high-mass star as the star consumes its hydrogen.

The High-Mass Star Leaves the Main Sequence

These differences between high- and low-mass stars might seem a bit esoteric, but as a star's life on the main sequence comes to an end, the visible differences in the structure and evolution of the star become far more pronounced. As the high-mass star runs out of hydrogen in its core, the weight of the overlying star compresses the core, just as it did in a low-mass star. Yet long before the core of the high-mass star becomes electron degenerate, the pressure and temperature in the core reach the 10^8 K point needed for helium burning to begin. There will be no growing degen-

erate core in the high-mass star. There will be no accelerating ascent up the red giant and asymptotic giant branches on the H-R diagram. Instead the star makes a fairly smooth transition from hydrogen burning to helium burning. The overall structure of the star responds to the changes taking place in its interior, but its luminosity will change relatively little. When a low-mass star leaves the main sequence, the path that it follows on the H-R diagram is largely vertical, going to higher and higher luminosities. As the high-mass star leaves the main sequence, on the

High-mass stars "skip" the red giant phase.

other hand, it grows in size while its surface temperature falls, so it moves mostly off to the right on the H-R diagram (**Figure 17.4**). The massive star now has the same structure as a low-mass horizontal branch star, burning helium in

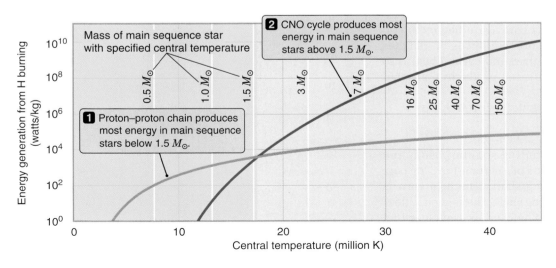

FIGURE 17.2 Plots of the rate of energy generation as a function of temperature for the proton–proton chain and the CNO cycle. At the higher central temperatures of stars more massive than 1.5 $M_\odot$, the CNO cycle more efficiently fuses hydrogen into helium.

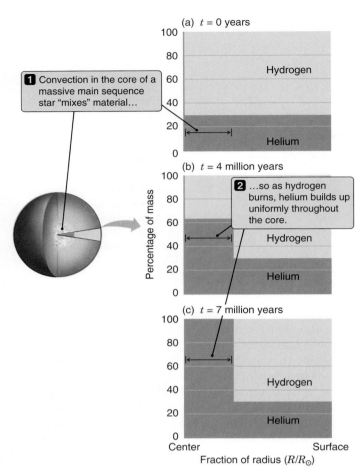

FIGURE 17.3 Convection keeps the core of a high-mass main sequence star well mixed, so the composition remains uniform throughout the evolving core. (Evolution times are for a 25 $M_\odot$ star.)

its core and hydrogen in a surrounding shell, but it has skipped the red giant phase of low-mass star evolution.

The next stage in the evolution of a high-mass star has no analog in low-mass stars. When the high-mass star exhausts the helium in its core, its core again begins to collapse, but this time as the core collapses it reaches temperatures of 8×10^8 K or higher, and carbon begins to burn (see **Table 17.1**). Carbon burning produces a number

---

Progressively more massive elements burn as massive stars evolve.

---

of more massive elements, including sodium, neon, and magnesium. The star now consists of a carbon-burning core surrounded by a helium-burning shell, which in turn is surrounded by a hydrogen-burning shell. The sequence does not end here. When carbon is exhausted as a nuclear fuel at the center of the star, neon burning picks up the slack; and when neon is exhausted, oxygen begins to burn. The

structure of the evolving high-mass star, shown in **Figure 17.5**, is reminiscent of an onion, with layer surrounding layer surrounding layer. As we move inward toward the center of the star, we pass through a layer of hydrogen burning, then a layer of helium burning, then a layer of carbon burning, and so on. Finally we reach the most advanced stages of nuclear burning in the core of the star.

Not All Stars Are Stable

When we discussed the structure of the Sun in Chapter 14 and a low-mass star in Chapter 16, we asserted that a single stable solution always exists. Given one solar mass with a mix of 70 percent hydrogen and 30 percent helium, that mass *will* form a stable star with the structure of our Sun. As we "what-if" the Sun, considering the consequences of forcing the Sun to be larger or smaller than it is, we find that changes in the star would cause it to settle back into this unique balance between the inward pull of gravity and the outward push of pressure. This remains true for high-mass stars on the main sequence as well. However, it is *untrue* for many evolved stars. As a star undergoes post–main

FIGURE 17.4 When massive stars leave the main sequence, they move horizontally across the H-R diagram.

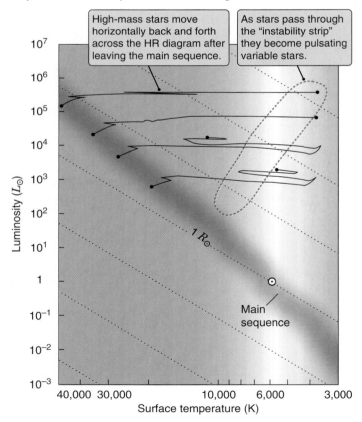

TABLE 17.1

Burning Stages in High-Mass Stars

Core Burning Stage	9 $M_\odot$ Star	25 $M_\odot$ Star	Typical Core Temperatures
H burning	20 million years	7 million years	$(3–10) \times 10^7$ K
He burning	2 million years	700,000 years	$(1–7.5) \times 10^8$ K
C burning	380 years	160 years	$(0.8–1.4) \times 10^9$ K
Ne burning	1.1 years	1 year	$(1.4–1.7) \times 10^9$ K
O burning	8 months	6 months	$(1.8–2.8) \times 10^9$ K
Si burning	4 days	1 day	$(2.8–4) \times 10^9$ K

Evolved stars pulsate while passing through the instability strip.

sequence evolution, it may make one or more passes through a region of the H-R diagram known as the **instability strip**, shown in Figure 17.4. Rather than finding a steady balance, stars in the instability strip pulsate, alternately growing larger and smaller.

Stars lying within the instability strip of the H-R diagram are heat engines, powering their pulsation by tapping into the thermal energy flowing outward through them. Changes in the ionization state of the gas within the star alternately trap thermal energy, forcing the pressure in the star up and causing it to expand, and then release this energy, allowing the star to contract again. These changes serve the role of the valves in an engine: They rob the star of pressure support at one part of its cycle, allowing it to shrink too far, and then pump the pressure in the star back up at a different point in its cycle, puffing up the star to a larger size than it can support. The process is illustrated in **Figure 17.6**. Rather than settling at a constant radius, like the Sun, the star alternately expands and shrinks, moving out and in like the piston of a steam engine. The pulsations in the outer parts of the star have very little effect on the nuclear burning in the star's interior. However, the pulsations do affect the light escaping from the star. Both the luminosity and the color of the star change as the star expands and shrinks. The star is at its brightest and bluest while it expands through its equilibrium size, and at its faintest and reddest while it falls back inward. Such stars are referred to as **pulsating variable stars**.

The highest-mass and most luminous pulsating variable stars are named **Cepheid variables**, after Delta Cephei, the first recognized member of this class. A Cepheid variable completes one cycle of its pulsation in anywhere from about 1 to 100 days, depending on its luminosity. The more luminous the star, the longer it takes it to complete its cycle. This **period–luminosity relationship** for Cepheid variables, first discovered experimentally by **Henrietta Leavitt** (1868–1921) in 1912, is the basis for the use of Cepheid variables as indicators of the distances to galaxies beyond our own.

Cepheid variables are not the only type of **variable star**. The horizontal branch of the evolutionary tracks of low-mass stars (see Figure 16.8) may pass through the instability strip as well. These unstable horizontal branch stars are known as **RR Lyrae variables** after their prototype. They pulsate by the same mechanism as Cepheid variables but are typically hundreds of times less luminous. The instability strip also intersects the main sequence around spectral type A, and many A stars do show significant variability (although it is much less pronounced than the variability in Cepheids or RR Lyrae variables). A number of other kinds of variable stars are seen elsewhere in the H-R diagram, each driven by its own variety of "engine."

As low-mass stars evolve beyond the main sequence, they expel significant amounts of mass back into interstellar space. For a high-mass star, violent mass loss is a fact of life throughout its existence. Even while on the main

Massive stars generate high-velocity winds.

sequence, massive O and B stars drive away tenuous winds at velocities that can reach 3,000 km/s. These winds are pushed outward by the pressure of the radiation from the star. We do not normally think of light as "pushing" on something, but the pressure of the intense radiation at the

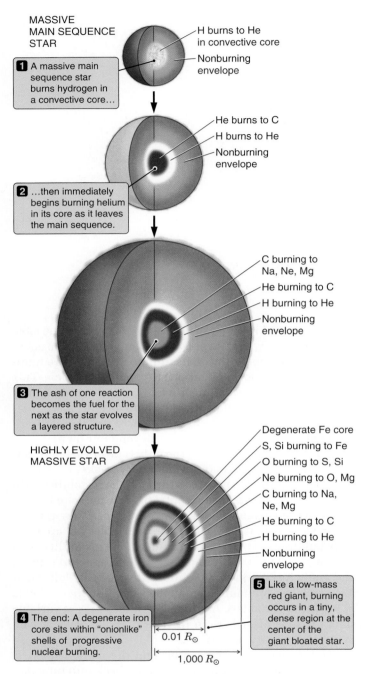

MASSIVE
MAIN SEQUENCE
STAR

H burns to He
in convective core

Nonburning
envelope

1 A massive main sequence star burns hydrogen in a convective core…

He burns to C

H burns to He

Nonburning
envelope

2 …then immediately begins burning helium in its core as it leaves the main sequence.

C burning to
Na, Ne, Mg

He burning to C

H burning to He

Nonburning
envelope

3 The ash of one reaction becomes the fuel for the next as the star evolves a layered structure.

HIGHLY EVOLVED
MASSIVE STAR

Degenerate Fe core

S, Si burning to Fe

O burning to S, Si

Ne burning to O, Mg

C burning to Na,
Ne, Mg

He burning to C

H burning to He

Nonburning
envelope

5 Like a low-mass red giant, burning occurs in a tiny, dense region at the center of the giant bloated star.

4 The end: A degenerate iron core sits within "onionlike" shells of progressive nuclear burning.

$0.01\ R_\odot$

$1,000\ R_\odot$

FIGURE 17.5 As a high-mass star evolves, it builds up a layered structure like an onion, with progressively more advanced stages of nuclear burning found deeper and deeper within the star. Note the change in scale in the bottom figure.

surface of a massive star is enough to overcome the star's gravity and drive away material in its outermost layers. Main sequence O and B stars lose mass at rates ranging from about 10^{-7} to $10^{-5}\ M_\odot$ of material per year, with the greatest mass loss occurring in the most massive stars. These numbers may sound tiny, but added up over millions of years

they mean that mass loss plays a prominent role in the evolution of high-mass stars. O stars with masses of 20 $M_\odot$ or more may lose around 20 percent of their mass while on the main sequence, and possibly more than 50 percent of their mass over their entire lifetimes. Even an 8 $M_\odot$ star may lose 5 percent to 10 percent of its mass. An extreme example of mass loss in a massive star is seen in **Eta Carinae** (**Figure 17.7**), a 100 $M_\odot$ star with a luminosity of 3 million suns. Currently Eta Carinae is losing mass at a rate of about $10^{-3}\ M_\odot$ per year (or 1 $M_\odot$ every 1,000 years). However, during a 19th century eruption, when Eta Carinae became the second brightest star in the sky, its mass loss must have reached the amazing level of 0.1 $M_\odot$ per year, shedding 2 $M_\odot$ of material over a mere 20 years.

17.3 High-Mass Stars Go Out with a Bang

A low-mass star approaches the end of its life relatively slowly and gently as the outer parts of the star are ejected into nearby space (sometimes forming a planetary nebula), leaving behind the degenerate core of the star. In stark contrast, for a high-mass star the end comes suddenly and amid considerable fury. An evolving high-mass star builds up its onionlike structure (Figure 17.5) as nuclear burning in its interior proceeds to more and more advanced stages. Hydrogen burns to helium, helium burns to carbon and

Iron is the most massive element formed by fusion.

oxygen, carbon burns to magnesium, oxygen burns to sulfur and silicon, and then silicon and sulfur burn to iron. There are many different types of nuclear reactions that occur up to this point, forming almost all of the different stable isotopes of elements less massive than iron. But the essential point is this: *With iron, the chain of nuclear fusion stops.*

Why does hydrogen burn while iron does not? More generally, what must be true for a material to serve as fuel? Gasoline burns because when it combines with oxygen, energy is released. This extra energy pushes up the temperature, which then causes the chemical reaction between gasoline and oxygen to go faster. The reaction is *self-sustaining,* which means that the reaction itself is the source of thermal energy needed to cause the reaction to go. The same is true of nuclear fusion reactions in the interiors of stars. When four hydrogen atoms combine to form a helium atom, the resulting helium atom has less energy than the four hydrogen atoms had separately. This difference in energy gets converted to thermal energy, which maintains the temperature of the gas at the high levels needed to sustain the reaction.

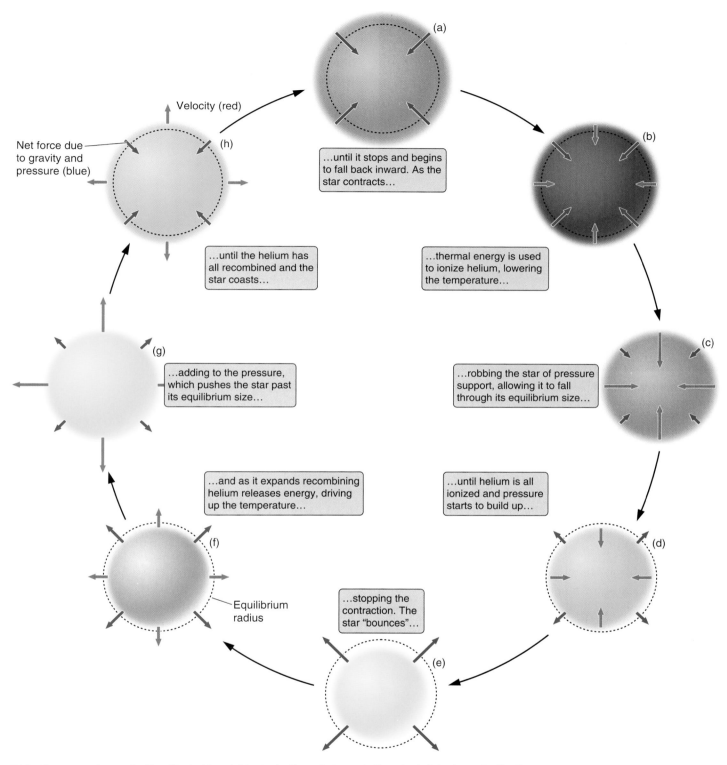

Velocity (red)

Net force due to gravity and pressure (blue)

(h)

(a) ...until it stops and begins to fall back inward. As the star contracts...

(b) ...thermal energy is used to ionize helium, lowering the temperature...

...until the helium has all recombined and the star coasts...

(g) ...adding to the pressure, which pushes the star past its equilibrium size...

(c) ...robbing the star of pressure support, allowing it to fall through its equilibrium size...

...and as it expands recombining helium releases energy, driving up the temperature...

...until helium is all ionized and pressure starts to build up...

(f)

Equilibrium radius

(d)

...stopping the contraction. The star "bounces"...

(e)

FIGURE 17.6 In a pulsating Cepheid variable, ionization of atoms in the star's interior acts like the valves in a steam engine, alternately allowing the surface of the star to fall inward, then pushing it back out again. (Color changes shown here are greatly exaggerated.)

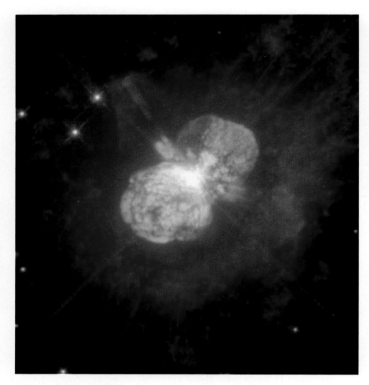

FIGURE 17.7 This Hubble Space Telescope image shows an expanding cloud of dusty material ejected by the luminous blue variable star Eta Carinae. The star itself, which is largely hidden by the surrounding dust, has a luminosity 3 million times that of the Sun and a mass probably in excess of 100 $M_\odot$. Dust is created when volatile material ejected from the star condenses.

How much energy is available from a nuclear reaction of combining, say, three helium nuclei to form a ^{12}C nucleus (the triple-alpha process)? We can answer the question in steps. First, how much energy would it take to break the three helium nuclei into their constituent six neutrons and six protons? Next, how much energy would be released if these six protons and neutrons combined to form a ^{12}C nucleus? The net energy produced by the reaction is just the difference between these two amounts.

The energy that it would take to break up an atomic nucleus into its constituent parts is called the **binding energy**

Different nuclei have different binding energies.

---

of the nucleus. The net energy released by a nuclear reaction is the difference between the binding energy of the products and the binding energy of the reactants:

$$\text{Net energy} = \frac{\text{Binding energy}}{\text{of products}} - \frac{\text{Binding energy}}{\text{of reactants}}$$

Figure 17.8 shows a plot of the nuclear binding energy per kilogram for a number of different atomic nuclei.

It is worth taking a brief digression to see how this works in practice. Using these values to work through the previous example of the triple-alpha process, we find that the binding energy of a helium nucleus is 4.53×10^{-12} joules, or 6.822×10^{14} J per kilogram of helium. The binding energy of a ^{12}C nucleus is 1.477×10^{-11} J, or 7.410×10^{14} J per kilogram of carbon. To find the amount of energy available from fusing a kilogram of helium nuclei into carbon, take the difference between the two numbers:

$$\frac{\text{Net energy from}}{\text{burning 1 kg He}} = \frac{\text{Binding energy}}{\text{of C formed}} - \frac{\text{Binding energy}}{\text{of He burned}}$$

$$= 7.410 \times 10^{14}\,\text{J} - 6.822 \times 10^{14}\,\text{J}$$

$$= 5.84 \times 10^{13}\,\text{J}$$

So helium is a good nuclear fuel. We can see this directly by looking at Figure 17.8. Moving from helium to carbon on the plot increases the binding energy of each nucleon,[1] so fusing helium to carbon releases energy. But what if we instead try to fuse iron to form more massive elements? Because iron is at the peak of the binding energy curve, the products of iron burning will have *less* binding energy than the reactants (going from iron to more massive elements takes us down on the binding energy curve in Figure 17.8), so the net energy in the reaction will be *negative*. Rather than producing energy, fusion of iron *uses* energy. Iron does not burn.[2]

The Final Days in the Life of a Massive Star

The nuclear reactions following hydrogen burning are energetically much less favorable than conversion of hydrogen to helium. A look at Figure 17.8 shows that conversion of 1 kg of helium to carbon produces less than $\frac{1}{10}$ as much energy as conversion of 1 kg of hydrogen to helium. In order to support the star against gravity, this less efficient nuclear fuel must be consumed more rapidly. Although conversion of hydrogen into helium can provide the energy needed to support the high-mass star against the force of gravity for millions of years, helium burning can support the star for only a few hundred thousand years.

Following helium burning, the nature of the balance within the star becomes qualitatively different. There is almost as much energy available from burning a kilogram of carbon, neon, oxygen, or silicon as there is from burning

---

[1] A nucleon is a constituent of an atomic nucleus—that is, a proton or a neutron.

[2] Iron does not burn in a nuclear sense, but iron does burn *chemically*. Combining iron and oxygen increases the *chemical* binding energy of the atoms involved in the reaction. Of course, chemical burning plays no role in the interior of a star.

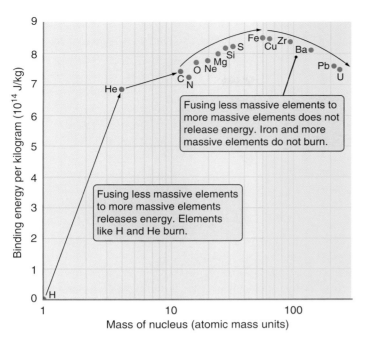

FIGURE 17.8 The nuclear binding energy of a kilogram of material is plotted against the mass of the atomic nuclei. This is the energy that it would take to break a kilogram of the material apart into protons and neutrons. Energy is released by nuclear fusion only if the binding energy of the products is greater than the binding energy of the reactants.

air pouring out through a huge rent in the side of a balloon, neutrinos produced in the interior of the star stream through the overlying layers of the star as if they were not even there, carrying the energy from the stellar interior out into space. Keeping the balloon inflated becomes difficult indeed. As thermal energy pours out of the interior of the star, the outer layers of the star push inward, driving up the density and temperature, and forcing nuclear reactions to run at the furious rate necessary to replace the energy escaping in the form of neutrinos.

Once this process of **neutrino cooling** becomes significant, the star begins evolving much more rapidly. Carbon burning is capable of supporting the star only for something less than about a thousand years. Oxygen burning holds the star up for only about a year. Once a massive star reaches the point that it begins to burn silicon, it is within a few days of the end. The luminosity of a silicon-burning star in electromagnetic radiation may be little greater than it was when the star was burning helium in its core. But if our eyes could sense neutrinos, we would see that a star burning silicon is actually giving off about 200 million times more energy per second than its former self!

Following silicon burning, we come to the end of the line. Once a star forms its iron core, no source of nuclear energy remains to replenish the energy that is being taken away by escaping neutrinos. The high-mass star's life as a balancing act between gravity and a controlled thermonuclear furnace is over. Gravity will have its way.

Stages of burning after helium burning are progressively shorter-lived.

a kilogram of helium. We might expect, then, that each of these stages would last about as long as helium burning. Yet when we look at Table 17.1, we see that the star proceeds from helium burning through to the end of its life in a cosmic blink of the eye. So where is the catch?

You can think of the balance in a star as something like trying to keep a leaky balloon inflated. The larger the leak, the more rapidly you have to pump air into the balloon. A star that is burning hydrogen or helium is like a balloon with a slow leak. At the temperatures generated by hydrogen or helium burning, energy leaks out of the interior of a star primarily by radiation and convection. Neither of these processes is very efficient because the outer layers of the star act like a thick warm blanket. Nuclear fuels need to burn at only a relatively modest rate to support the weight of the outer layers of the star while keeping up with the energy escaping outward. Beginning with carbon burning, this balance shifts in a dramatic and fundamental way. Rather than being carried by radiation and convection, energy now begins to escape from the core primarily in the form of neutrinos produced by the many nuclear reactions occurring there. When this happens, the floodgates are opened. Like

The Core Collapses and the Star Explodes

Throughout the story of stellar evolution, we have seen that although gravity can be held off for a time—perhaps a very long time—it is infinitely patient, and in the end it generally wins the battle. Bereft of support from thermonuclear fusion, the iron core of the massive star begins to collapse. Refer to **Figure 17.9** as we follow the dramatic sequence of events that comes next.

The early stages of the collapse of the iron ash core of an evolved massive star are much the same as in the collapse of a nonburning core in a low-mass star. As the core collapses, its density and temperature skyrocket, and the force of gravity becomes even stronger. The gas in the core is once again compressed beyond the realm of normal matter, becoming electron degenerate when it reaches about the size of Earth. Unlike the electron degenerate core of a low-mass red giant, however, the weight bearing down on the interior of the iron ash core is too great to be held up by electron degeneracy. This time gravity is too strong even for the quantum mechanical rules that limit how tightly electrons can be packed together. As the collapse continues, the core reaches temperatures of 10 billion K (10^{10} K) and

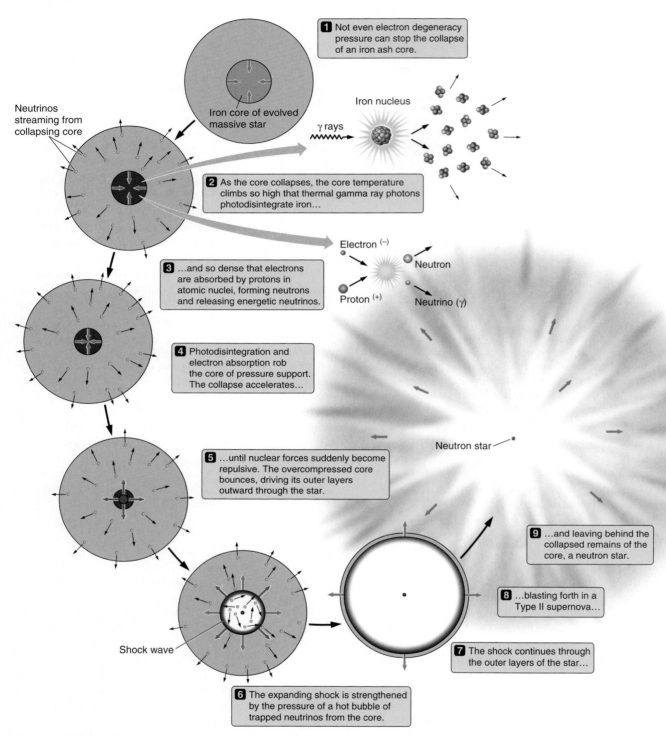

Neutrinos streaming from collapsing core

1 Not even electron degeneracy pressure can stop the collapse of an iron ash core.

Iron core of evolved massive star

Iron nucleus

γ rays

2 As the core collapses, the core temperature climbs so high that thermal gamma ray photons photodisintegrate iron...

Electron (−)

Neutron

Proton (+)

Neutrino (γ)

3 ...and so dense that electrons are absorbed by protons in atomic nuclei, forming neutrons and releasing energetic neutrinos.

4 Photodisintegration and electron absorption rob the core of pressure support. The collapse accelerates...

Neutron star

5 ...until nuclear forces suddenly become repulsive. The overcompressed core bounces, driving its outer layers outward through the star.

9 ...and leaving behind the collapsed remains of the core, a neutron star.

8 ...blasting forth in a Type II supernova...

7 The shock continues through the outer layers of the star...

Shock wave

6 The expanding shock is strengthened by the pressure of a hot bubble of trapped neutrinos from the core.

FIGURE 17.9 Shown here are the stages that a high-mass star goes through at the end of its life as its core collapses and the star explodes as a Type II supernova.

higher, while the density exceeds 10 metric tons per cubic centimeter, or 10 times the density of an electron degenerate white dwarf.

These phenomenal temperatures and pressures trigger fundamental changes in the makeup of the core. The laws describing thermal radiation say that at these temperatures the nucleus of the star will be awash in extremely energetic thermal radiation. This radiation is so energetic that thermal gamma ray photons are produced with enough energy to break iron nuclei apart into helium nuclei. This process, known as **photodisintegration**, literally begins undoing the results of nuclear fusion—a task that uses up a tremendous amount of the thermal energy of the core. At the same time the density of the core is so great that electrons are squeezed into atomic nuclei, where they combine with protons to produce neutron-rich isotopes in the core of the star. This process uses up thermal energy as well, robbing the core of even more of its pressure support. All the while neutrinos continue to pour out of the core of the dying star. These events take place within a second! The collapse of the core accelerates, reaching velocities of 70,000 km/s, or almost a quarter of the speed of light, on its inward fall.

The next hurdle standing in the way of the collapsing core is the force that holds atomic nuclei together. As material in the collapsing core reaches and exceeds the density of an atomic nucleus, the strong nuclear force actually becomes repulsive. Computer models say that about half of the collapsing core suddenly slows its inward fall. The remaining half slams into the innermost part of the star at a significant fraction of the speed of light and "bounces," sending a tremendous shock wave back out through the star.

Under the extreme conditions in the center of the star, neutrinos are being produced at an enormous rate. Over the next second or so almost a fifth of the mass of the material in the core is converted into neutrinos and their energy. Most of these neutrinos pour outward through the star; but at the phenomenal densities found in the collapsing core of the massive star, not even neutrinos pass with complete freedom. A few tenths of a percent of the energy of the neutrinos

> **The star explodes as a tremendous Type II supernova.**

streaming out of the core of the dying star are trapped by the dense material behind the expanding shock wave. The energy of these trapped neutrinos drives the pressure and temperature in this region ever higher, inflating a bubble of extremely hot gas and intense radiation around the core of the star. The pressure of this bubble adds to the strength of the shock wave moving outward through the star. Within about a minute the shock wave has pushed its way out through the helium shell within the star. Within a few hours it reaches the surface of the star itself, heating the stellar surface to 500,000 K and blasting material outward at velocities of up to about 30,000 km/s. Our evolved massive star has exploded in an event referred to as a **Type II supernova**.

The death of a star as a Type I or Type II supernova should mean more to us than just another example of nature's spectacular fireworks. Were it not for supernovae, it is highly unlikely that either we or our planet would exist, as explained in **Connections 17.1**.

17.4 The Spectacle and Legacy of Supernovae

A Type II supernova is comparable in its spectacle to a Type I supernova. For a brief time a Type II supernova can shine with the light of 100 billion suns; yet the energy that comes out of the supernova in the form of light is but a tiny fraction of the energy of the explosion itself. About a hundred times more energy comes out as the kinetic energy of the outer parts of the star that are blasted into space by the explosion. This ejected material contains about 10^{47} J of kinetic energy—enough energy to accelerate the mass of our Sun to a speed of 10,000 km/s. It is the kinetic energy of the ejecta from supernovae of both Type I and Type II that is responsible for heating the hottest phases of the interstellar medium and pushing around the clouds in the interstellar medium. Yet even this amount of energy is a pittance in comparison with the energy carried away from the supernova explosion by neutrinos—an amount of energy at least hundred times larger still!

One of the most important astronomical events in the last part of the 20th century was the explosion of a massive star in the small companion galaxy to the Milky Way known as the Large Magellanic Cloud. Even at a distance of 160,000 light-years, Supernova 1987A was so bright that it dazzled sky gazers in the Southern Hemisphere (see **Figure 17.10**). While astronomers working in all parts of the electromagnetic spectrum scrambled to point their telescopes at the new supernova, solar astronomers had already quietly and unknowingly captured one of the true scientific prizes of SN1987A. Solar neutrino telescopes recorded a burst of neutrinos passing through Earth—neutrinos that originated not in the Sun but in the tremendous stellar explosion that occurred beyond the bounds of our galaxy itself. The detection of neutrinos from SN1987A provided us with a rare and crucial glimpse of the very heart of a massive star at the moment of its death, confirming a fundamental prediction of our theories about the collapse of the core and its effects.

The Energetic and Chemical Legacy of Supernovae

Type II supernova explosions leave a rich and varied legacy to the universe. We have already mentioned their energetic

Origins—The Chemistry of Life

When we speak about the chemistry of life, what we really mean is life itself. All living organisms are composed of a more or less common suite of chemicals, and very complex ones at that. We begin by looking at ourselves. Approximately two-thirds of the atoms in our bodies are hydrogen (H), about a quarter are oxygen (O), a tenth are carbon (C), and a few hundredths are nitrogen (N). The remaining atomic elements, and there are several dozen of them, make up only 0.2 percent of our total inventory. It turns out that all living creatures—at least those we know of—are an assemblage of molecules composed almost entirely of these four atomic elements (sometimes called CHON), along with small amounts of phosphorus and sulfur. Some of these molecules are enormous. Consider deoxyribonucleic acid—called DNA for short—which is responsible for our genetic code. DNA is made entirely from only *five* atomic elements: CHON and phosphorus. But the DNA in each cell of our bodies is composed of combinations of *tens of billions* of atoms of these same five elements. Then there are proteins, the huge molecules responsible for the structure and function of living organisms. Proteins are long chains of smaller molecules called amino acids. Terrestrial life employs 20 specific amino acids, which also contain no more than five atomic elements—in this case CHON plus sulfur instead of phosphorus. Can this be all there is to our body chemistry? No, the chemistry of life is far too complex to get by with a mere half dozen atomic species. Many others are present in smaller amounts but are essential to the complicated chemical processes that make living organisms tick. They include sodium, chlorine, potassium, calcium, magnesium, iron, manganese, and iodine. Finally there are the so-called *trace elements*, such as copper, zinc, selenium, and cobalt.

They also play a crucial role in life chemistry but are needed in only tiny amounts.

We know the infant universe was composed basically of hydrogen and helium and very little else. Well enough. But fast forward now to a later time—some 9 billion years later. All the chemical elements essential to life were present and available in the molecular cloud that gave birth to our Solar System. So, we might ask, how and where were these heavier atoms created, and how did they find their way into our molecular cloud? The answers can be found right here in this chapter and the previous one. Many of these heavier elements were forged in the nuclear furnaces of low- and high-mass stars and were then dispersed into space through sometimes passive, sometimes violent, mass loss (see Connections 16.2). As dying red giants, low-mass stars lose their gravitational grip on their overly extended atmospheres. Along with hydrogen and helium, the newly created heavy elements are blown off into space, eventually finding their way into molecular clouds.

Most of the trace elements essential to biology are more massive than iron and must have a different origin because elements heavier than iron cannot be formed in stellar interiors by nuclear fusion. They are instead created within a matter of minutes during the violent supernova explosions that mark the death of high-mass stars.

Although we have focused here on the chemistry of biology, we should not forget that life needs an accommodating planet such as our own on which to get a start and then evolve. Earth is made up largely of silicates—a combination of silicon and oxygen—and silicon is yet another element created within the cores of stars. So the picture is now clear. We (and not incidentally our own planet Earth and our entire Solar System) exist courtesy of the generations of low- and high-mass stars that came before us.

legacy: The energy from these explosions is responsible for driving much of the dynamics of the interstellar medium. When we look at the sky over a wide range of wavelengths, we see huge expanding bubbles of million-kelvin gas, like that in **Figure 17.11(a)**, aglow in X-rays and driving visible shock waves (**Figure 17.11(b)**) into the surrounding interstellar medium. These bubbles are the still-powerful blast

waves of supernova explosions that took place thousands of years ago. In fact, in many cases supernova explosions are thought to be responsible for compressing nearby clouds (like the one in **Figure 17.11(c)**) enough to trigger their collapse toward the formation of new generations of stars.

Perhaps even more important to us is the chemical legacy left behind by supernova explosions. Later in our journey

FIGURE 17.10 SN1987A was a supernova that exploded in a small companion galaxy of the Milky Way called the Large Magellanic Cloud (LMC). These images show the LMC before the explosion, then while the supernova was near its peak.

we will turn our attention to the earliest moments after the birth of the universe; we will find that the only chemical elements that formed at that time were the least massive elements—hydrogen, helium, and trace amounts of lithium, beryllium, and boron. All of the rest of the chemical elements, including a large fraction of the atoms of which we

Supernovae eject newly formed massive elements into interstellar space.

are made, were formed in the hearts of stars, then returned to the interstellar medium. This process of progressive chemical enrichment of the universe is called **nucleosynthesis**. We introduced the idea of nucleosynthesis in our discussion of planetary nebulae in Chapter 16. But although

mass loss from both low- and high-mass stars enriches the interstellar medium with massive elements formed in their interiors, Type I and II supernovae are the true champions of nucleosynthesis. A look at Figure 17.5 should offer the first clue as to why this is so. Whereas low-mass stars are able to form elements only as massive as carbon and oxygen, nuclear burning in high-mass stars produces elements as massive as iron. Yet this is not the whole story. A look at a table of elements that occur in nature shows that there are many elements, up through uranium, that are far more massive than iron. If iron is the most massive element that can be formed by nuclear burning, then where do these more massive elements come from?

Answering this question takes us back to the reason why high temperatures are necessary for fusion to occur. Under

FIGURE 17.11 The Cygnus Loop is a supernova remnant—an expanding interstellar blast wave caused by the explosion of a massive star. (a) Gas in the interior with a temperature of millions of kelvins glows in X-rays, while (b) visible light comes from locations where the expanding blast wave pushes through denser gas in the interstellar medium. (c) A Hubble Space Telescope image of a location where the blast wave is hitting an interstellar cloud.

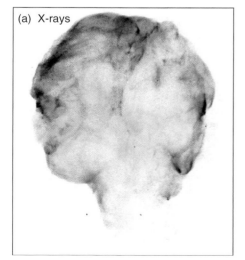
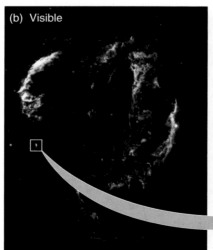
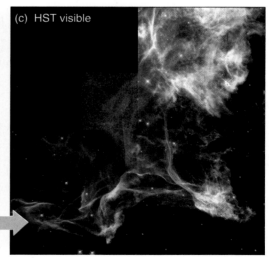

(a) X-rays

(b) Visible

(c) HST visible

normal circumstances electric repulsion keeps positively charged atomic nuclei far apart. Extreme temperatures are needed to slam nuclei together hard enough to overcome this electric repulsion. Free neutrons, on the other hand, are not subject to these ground rules: They have no net electric charge, so there is no electric repulsion to prevent them from simply running into an atomic nucleus, regardless of how many protons that nucleus contains. Under normal circumstances in nature, free neutrons are very rare. However, in the interiors of evolved stars a number of nuclear reactions produce free neutrons, and under some circumstances—including those shortly before and during a Type II supernova—free neutrons can be produced in very large numbers. Free neutrons are easily captured by atomic nuclei and later decay to become protons. In this way elements with higher and higher mass are formed. The most massive element that is stable enough to last for long periods of time is uranium. As discussed in **Foundations 17.1**, stellar nucleosynthesis correctly predicts the patterns in the abundances of atoms of different kinds found in the very Earth beneath our feet.

We have seen the fate of the outer parts of the star, which are blasted back into interstellar space by the explosion of a Type II supernova. But what remains of the core that was

---

A neutron degenerate core is left behind.

---

left behind? Picking up our story where we left off, the material at the center of the massive star has collapsed to the point where it has about the same density of matter as the nucleus of an atom. As long as the mass of the core left behind by the explosion is no more than about 3 $M_\odot$, this col-

---

FOUNDATIONS 17.1

The Chemical Composition of the Universe: Comparison of Observation and Theory

Calculations of nucleosynthesis in stars make clear predictions about which nuclei should be formed in abundance and which should not. These same patterns are found in measurements of the abundances of nuclei on Earth, in meteorite material, and in the atmospheres of stars. While some of these patterns are subtle, others can be appreciated by comparing what we have learned about nucleosynthesis in this book with a plot of the relative abundance of elements found in nature **(Figure 17.12)**. Notice first of all that less massive elements are far more abundant than more massive elements—a consequence of the way more massive elements are progressively built up from less massive elements. An exception to this is the dip in the abundances of the light elements lithium, beryllium, and boron. These light elements are easily destroyed by nuclear burning, but their production is mostly bypassed by the main reactions involved in burning H and He. Conversely, carbon, nitrogen, and oxygen are big winners in the CNO cycle of hydrogen burning and the triple-alpha process of helium burning, and their high abundances reflect this fact. The spike in the abundances of the "iron peak" elements is evi-

dence of the nucleosynthetic processes, including those in Type I and Type II supernovae that favor these most tightly bound of all nuclei. Even the sawtooth pattern in the abundances of even- and odd-numbered elements can be understood as a consequence of the process of stellar nucleosynthesis. (To complete our discussion of nucleosynthesis, we will need to consider the formation of elements that took place not in stars but during the formation of the universe itself. We will take up this discussion in Chapter 20.)

Once again we find connections and insights in places that prior to our study of the universe we might never have imagined to look. Our understanding of the processes at work within the interiors of dying stars is being confirmed by analysis of the chemical composition of Earth under our feet. At the same time we can make a pretty good stab at saying what kinds of stars are responsible for forming the atoms that make up our own bodies. Our growing understanding of the chemical evolution of the universe and our connection to it represents one of the triumphs of modern astronomy.

lapse will be halted by quantum mechanical rules similar to those responsible for holding up a white dwarf. Only now instead of electrons it is neutrons that are forced together as tightly as the rules of quantum mechanics allow. The neutron degenerate core left behind by the explosion of a Type II supernova is referred to as a **neutron star**. It has a radius of perhaps 10 km, making it roughly the size of a small city; but into that volume is packed a mass more than 1.4 times that of our Sun. At a density of around a billion metric tons per cubic centimeter, the neutron star is a billion times more dense than a white dwarf and a thousand trillion (10^{15}) times more dense than water! That density is roughly what we would get by crushing the entire Earth down to an object the size of a football stadium.

As if neutron stars were not extraordinary enough in their own right, they also form the hearts of a number of other exotic objects. If the massive star responsible for the formation of a neutron star is part of a binary system, then the neutron star will be left with a binary companion. Such a system might remind you of the white dwarf binary systems responsible for novae and Type I supernovae discussed in Chapter 16. As the second star in such

X-ray binaries arise from the accretion of mass onto neutron stars.

a binary system evolves and overfills its Roche lobe, matter plummets down the deep gravitational well of the tiny but massive neutron star. This matter slams into the accretion disk around the neutron star with enough energy to heat the disk to temperatures of millions of kelvins, and the accretion disk glows brightly in X-rays. Such an

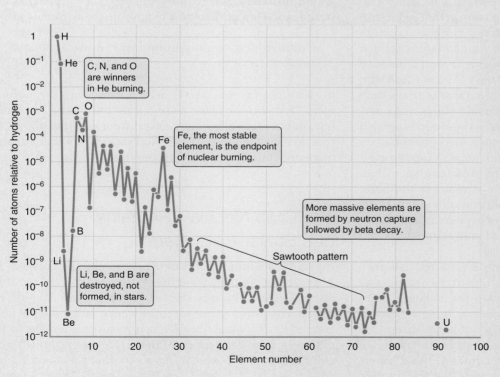

FIGURE 17.12 The relative abundances of different elements on Earth are plotted against the mass of the nucleus. This pattern can be understood as a result of the process of nucleosynthesis in stars.

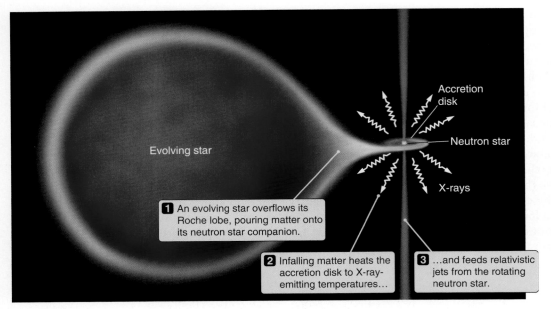

FIGURE 17.13 X-ray binaries are systems consisting of a white dwarf, neutron star, or black hole and a normal evolving star. As the evolving star overflows its Roche lobe, mass falls toward the collapsed object. The gravitational well of the collapsed object is so deep that when the material hits the accretion disk, it is heated to such high temperatures that it radiates away most of its energy as X-rays.

Labels in figure:
- Evolving star
- Accretion disk
- Neutron star
- X-rays
- **1** An evolving star overflows its Roche lobe, pouring matter onto its neutron star companion.
- **2** Infalling matter heats the accretion disk to X-ray-emitting temperatures…
- **3** …and feeds relativistic jets from the rotating neutron star.

object, illustrated in **Figure 17.13**, is known as an **X-ray binary.** Many fascinating phenomena occur in X-ray binaries, including the formation of powerful jets that blast away from the neutron star in directions perpendicular to its accretion disk and at speeds approaching the speed of light.

Besides its phenomenal density, a neutron star has a number of other extraordinary properties. The same principle of conservation of angular momentum that requires a collapsing molecular cloud to spin faster as it grows smaller also says that as the core of a massive star collapses, it must spin faster as well. As a main sequence O star, a massive star rotates perhaps once every few days. As a neutron star, it might instead rotate tens or even hundreds of times each second! We also saw in our discussion of a collapsing interstellar cloud how the magnetic field in the cloud is carried along and concentrated by the collapse. This phenomenon also occurs in the collapse of the star, amplifying the magnetic field to values that are trillions of times greater than the magnetic field at Earth's surface. A neutron star has a magnetosphere, just as Earth and several other planets do, except that the neutron star's magnetosphere is unimaginably stronger and is whipped around many times a second by the spinning neutron star.

Energetic subatomic particles such as electrons and **positrons** move along the magnetic field lines of the neutron star and are "funneled" by the field toward the magnetic poles of the system. Conditions there produce intense electromagnetic radiation, which is beamed out in the directions away from the magnetic poles of the neutron star as shown in **Figure 17.14**. As the neutron star rotates, these beams of radiation sweep through space much like the rotating beams of a lighthouse. When we are located in the paths of these beams, we see what sailors coming into harbor at night would see looking at a lighthouse. The neutron star appears to flash on and off with a regular period equal to the period of rotation of the star (or half the rotation period if we see both beams).

Rapidly pulsing objects were first discovered by observers working at radio wavelengths in 1967. These objects, which blinked like regularly ticking clocks, puzzled astronomers. One of the early tongue-in-cheek names given

---

Pulsars are rapidly spinning magnetized neutron stars.

---

to these objects was *LGMs*, which stood for "little green men." Today these objects are referred to by the less flamboyant but more accurately descriptive term **pulsar.** As of this writing well over a thousand pulsars are known, and more are being discovered all the time.

The Crab Nebula—Remains of a Stellar Cataclysm

In A.D. 1054 Chinese astronomers recorded the presence of a "guest star" in the part of the sky that we call the constellation Taurus. The new star was so bright that it could be seen during the daytime for three weeks, and it did not fade from visibility altogether for many months. On the basis of the Chinese description of the changing brightness and color of the object, we can say today that the guest star of 1054 was a fairly typical Type II supernova. When we look at this spot in the sky today we see an expanding cloud of

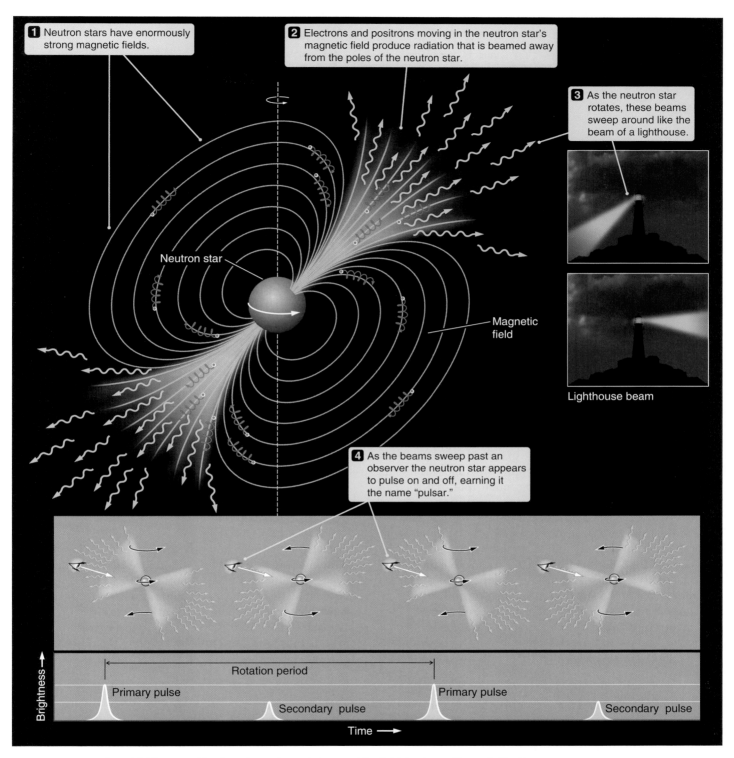

1 Neutron stars have enormously strong magnetic fields.

2 Electrons and positrons moving in the neutron star's magnetic field produce radiation that is beamed away from the poles of the neutron star.

3 As the neutron star rotates, these beams sweep around like the beam of a lighthouse.

Neutron star

Magnetic field

Lighthouse beam

4 As the beams sweep past an observer the neutron star appears to pulse on and off, earning it the name "pulsar."

Brightness →

Rotation period

Primary pulse

Secondary pulse

Primary pulse

Secondary pulse

Time →

FIGURE 17.14 As a highly magnetized neutron star rotates rapidly, light is given off, much like the beams from a rotating lighthouse lamp. From our perspective, as these beams sweep past us, the star will appear to pulse on and off, earning it the name *pulsar*.

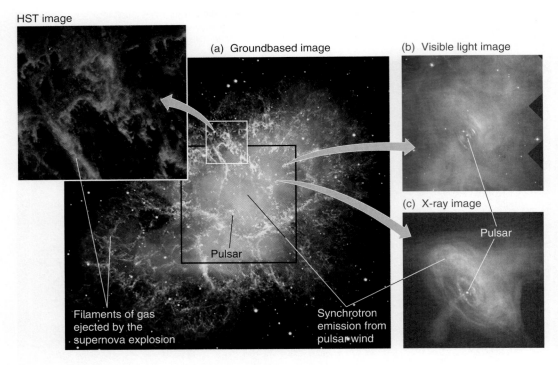

HST image

(a) Groundbased image

(b) Visible light image

(c) X-ray image

Pulsar

Pulsar

Filaments of gas ejected by the supernova explosion

Synchrotron emission from pulsar wind

FIGURE 17.15 The Crab Nebula is the remnant of a supernova explosion witnessed by Chinese astronomers in A.D. 1054. (a) The object we see today is an expanding cloud of "shrapnel" from that earlier cataclysm. The spinning pulsar at the heart of the Crab Nebula sends off a "wind" of electrons and positrons moving at close to the speed of light. The synchrotron radiation from these is shown (b) in visible light and (c) in X-rays.

debris from this explosion—an extraordinary object called the **Crab Nebula (Figure 17.15)**.

The Crab Nebula consists of several components. Images of the Crab Nebula taken in the light of nebular emission lines show filaments of glowing gas. Doppler shift measurements of these filaments reveal a pattern much like that seen in planetary nebulae—the hallmark of an expanding shell. But whereas planetary nebulae are expanding at 20 to 30 km/s, the shell of the Crab is expanding at closer to 1,500 km/s. Studies of the spectra of these filaments show that they contain anomalously high abundances of helium and other more massive chemical elements—the products of the nucleosynthesis that took place in the supernova and its progenitor star.

There is a pulsar at the center of the Crab Nebula. This was the first pulsar to be seen at visible wavelengths as it flashed on and off 30 times a second. Actually the Crab pulsar flashes 60 times a second with a main pulse associated with one of the lighthouse beams, then a fainter secondary pulse associated with the other lighthouse beam. Today the Crab Nebula has been observed in all parts of the electromagnetic spectrum, from low-energy radio waves to high-energy X-rays and even higher-energy gamma rays (see the chapter opening photograph).

As the Crab pulsar spins 30 times a second, it whips its powerful magnetosphere around with it. At a distance from the pulsar about equal to the radius of the Moon, material in the magnetosphere must move at almost the speed of light to keep up with this rotation. Like a tremendous slingshot, the rotating pulsar magnetosphere flings elementary particles—probably mostly electrons and **antimatter** electrons

called positrons—away from the neutron star in a powerful wind moving at nearly the speed of light. Material from this wind fills the space between the pulsar and the expanding shell. The Crab Nebula is almost like a big balloon; but instead of being filled with hot air, it is filled with a mix of

A pulsar powers the Crab's eerie synchrotron glow.

relativistic particles and strong magnetic fields—an environment more like what we might find in a physicist's particle accelerator than what we normally think of as interstellar space. Images of the Crab Nebula (Figure 17.15(b) and (c)) show this bizarre bubble as an eerie glow. This glow is synchrotron radiation from the relativistic electrons and positrons as they spiral around the magnetic field in the Crab.

17.5 Star Clusters Are Snapshots of Stellar Evolution

Over the course of this and the previous chapters we have told a remarkable tale about the evolution of stars of different masses, presenting the story as a well-corroborated theory. In other words, we have presented the story of stellar evolution as fact. Yet even the most massive stars take hundreds of thousands of years to evolve, which is far longer than the handful of decades we have spent studying their ways. Upon what testable predictions of our theories of stellar evolution do we base such bold claims of knowledge?

As we learned in Chapter 15, when an interstellar cloud collapses, it fragments into pieces, forming not one star but

Stars in clusters formed together at about the same time.

many stars of different masses. We see many such *star clusters* around us today, containing anywhere from a few dozen

to millions of stars. The fact that all of the stars in a cluster formed together at nearly the same time means that clusters are snapshots of stellar evolution. A look at a cluster that is 10 million years old shows us what stars of all different masses evolve into during the first 10 million years after they are formed. A look at a cluster 10 *billion* years after it formed shows us what has become of stars of different masses after 10 billion years have passed.

This basic result—the fact that high-mass stars in a cluster evolve more rapidly than low-mass stars that formed at the same time—provides the key to our knowledge of stellar evolution. **Figure 17.16** shows the H-R diagram of a simulated cluster of 40,000 stars as it would appear at

FIGURE 17.16 H-R diagrams of star clusters are snapshots of stellar evolution. These are H-R diagrams of a simulated cluster of 40,000 stars of solar composition seen at different times following the birth of the cluster. Note the progression of the main sequence turnoff to lower and lower masses.

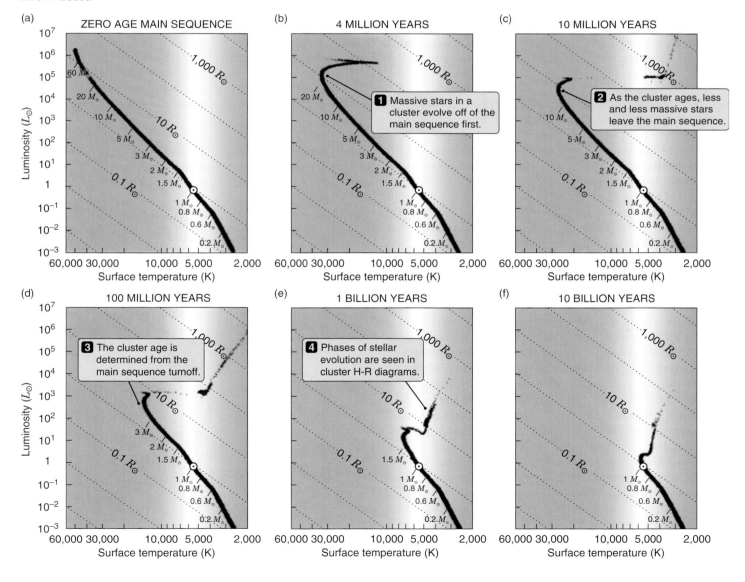

several different ages. In Figure 17.16(a) stars of all masses are located on the zero-age main sequence, showing where they begin their lives as main sequence stars. The increasing masses of stars along the main sequence are indicated. We would never expect to see a cluster H-R diagram that looks like the one in Figure 17.16(a), however, for the simple reason that the stars in a cluster do not all reach the main sequence at exactly the same time. Star formation in a molecular cloud is spread out over several million years, and it takes a considerable time for lower-mass stars to contract to reach the main sequence. The H-R diagram of a very young cluster normally shows many lower-mass stars located well above the main sequence, still descending their Hayashi tracks.

The more massive a star is, the shorter its life on the main sequence will be. After only 4 million years (Figure 17.16(b)), all stars with masses greater than about 20 $M_\odot$ have evolved off the main sequence and are now spread out across the top of the H-R diagram. The most massive stars have already disappeared from the H-R diagram entirely, having vanished in supernovae.

As time goes on, lower- and lower-mass stars evolve off the main sequence, and the turnoff point moves toward the bottom right in the H-R diagram. As in Figure 17.16(c), by

---

The main sequence turnoff shifts to lower-mass stars as a cluster grows older.

---

the time the cluster is 10 million years old, only stars with masses less than about 15 $M_\odot$ remain on the main sequence. The cluster H-R diagram looks as if we grabbed the top of the band of cluster stars and gradually peeled it away from its original location along the main sequence. The location of the most massive star that is still on the main sequence is called the **main sequence turnoff**. As the cluster ages, the main sequence turnoff moves farther and farther down the main sequence to stars of lower and lower mass.

As a cluster ages (Figures 17.16(d) and (e)) we see more than the motion of the main sequence turnoff to lower and lower masses; we see the details of all stages of stellar evolution. By the time the star cluster is 10 *billion* years old (Figure 17.16(f)), stars with masses of only 1 $M_\odot$ are beginning to pull away from the main sequence. Stars slightly more massive than this are seen as giant stars of various types. In this cluster the horizontal branch appears as a knot of stars sometimes referred to as the "red clump." Note how few giant stars are present in any of the cluster H-R diagrams. The giant, horizontal, and asymptotic giant branch phases in the evolution of low- and intermediate-mass stars pass so quickly in comparison with a star's main sequence lifetime that even though the cluster started with 40,000 stars, only a handful of stars are seen in these phases of evolution at any given time. Similarly, it takes a newly formed white dwarf only a few tens of millions of years to cool to the point that it disappears off the bottom of these figures. Even though the

majority of evolved stars in an old cluster are white dwarfs, all but a few of these stars will have cooled and faded into obscurity at any given time.

The cluster H-R diagrams in Figure 17.16 are theoretical calculations of what clusters of different ages *should* look like. The crucial point is that this is what H-R diagrams of real clusters *must* look like if our theories of stellar evolution are correct. Fortunately, this is also what H-R diagrams of real clusters *do* look like. **Figure 17.17** shows the observed H-R diagram for the cluster 47 Tucanae, along with a theoretical calculation of the H-R diagram for a 12-billion-year-old cluster. The quality of the agreement speaks for itself. The fact that observed star clusters have H-R diagrams that agree so well with the predictions of models is strong support for our theories of stellar evolution.

Armed with confidence in the reliability of our ideas about stellar evolution, we can turn cluster evolution models into a powerful tool for studying the history of star formation. When we observe a star cluster, the location of the main sequence turnoff immediately tells us the age of the cluster. If shown any of the cluster H-R diagrams in Figure 17.16, we should have no difficulty estimating the age of the cluster. **Figure 17.18** traces the observed H-R diagrams for several real star clusters. Once we know what to look for, the difference between young and old clusters is obvious. NGC 2362 is clearly a young cluster. Its complement

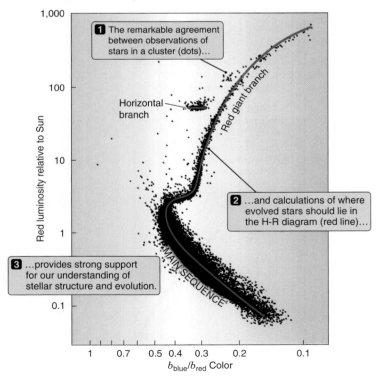

FIGURE 17.17 The observed H-R diagram of the cluster 47 Tucanae agrees remarkably well with the theoretical calculation *(solid line)* of the H-R diagram of a 12-billion-year-old cluster.

1 The remarkable agreement between observations of stars in a cluster (dots)…

Horizontal branch

Red giant branch

2 …and calculations of where evolved stars should lie in the H-R diagram (red line)…

3 …provides strong support for our understanding of stellar structure and evolution.

MAIN SEQUENCE

Red luminosity relative to Sun

b_{blue}/b_{red} Color

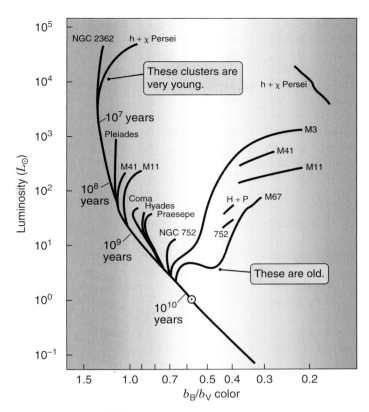

FIGURE 17.18 H-R diagrams for clusters having a range of different ages. The ages associated with the different main sequence turnoffs are indicated.

while stars with lower abundances of massive *elements* in their atmospheres often look significantly bluer than their more chemically enriched counterparts. Even so, we can usually get some idea about the properties of a group of stars by looking at its overall color. We will put this knowledge

Colors reveal the ages of stellar populations.

to good use in the chapters that follow as we turn our attention to the large collections of stars called galaxies. A group of stars with similar ages and other characteristics is referred to as a **stellar population**. If the color of a galaxy or a part of a galaxy is especially bluish, it often signifies that the galaxy contains a young stellar population that still includes hot, luminous, blue stars that must have formed recently. In contrast, if a galaxy or part of a galaxy has a reddish color, we expect that it is composed primarily of an old stellar population.

17.6 Gravity Is a Distortion of Spacetime

Having discussed the evolution of clusters of stars of all masses, we now return to the stellar remnants left behind by the evolution of the most massive stars. You might imagine that, surely, the neutron star represents the final extreme of stellar evolution; but there is one more step a star can take. The physics of a neutron star is much like the physics of a white dwarf, except that it is neutrons rather than electrons that cause it to be degenerate. A white dwarf can have a mass of no more than about 1.4 $M_\odot$. This is the Chandrasekhar limit, discussed in Chapter 16. If the mass of the object exceeds this limit, then gravity will be able to overcome electron degeneracy, and the white dwarf will begin collapsing again.

Just as there is a Chandrasekhar limit for white dwarfs, there is a Chandrasekhar limit for the mass of neutron stars as well. If the mass of a neutron star exceeds about 3 $M_\odot$, then gravity will begin to win out over matter once again. The neutron star grows smaller, and gravity becomes stronger and stronger at an ever accelerating pace. However, this time there is no force in nature powerful enough

If a neutron star's mass exceeds 3 $M_\odot$, it will collapse to a black hole.

to prevent gravity's final victory. The collapsing object quickly crosses a threshold where the escape velocity from its surface exceeds the speed of light, and not even light can escape its gravity. From this point on, nothing can

of massive, young stars shows it to be only a few million years old. In contrast, 47 Tucanae (Figure 17.17) has a main sequence turnoff of around 0.85 $M_\odot$, signifying a cluster age of around 12 billion years. (We should note that the evolution of stars depends on their chemical composition as well as their masses. Detailed comparisons between models and observed cluster H-R diagrams must account for the abundances of massive elements in the atmospheres of cluster stars, as well as the cluster age.)

We can apply our understanding of stellar evolution even when groups of stars are so far away that we cannot see each star individually. Although many fewer high-mass stars form in a cluster than low-mass stars, higher-mass stars are *far* more luminous than their lower-mass siblings. Likewise giant, evolved low-mass stars are far more luminous than less massive stars that remain on the main sequence. As a result, the most massive, most luminous stars present will dominate the light from a star cluster. If the cluster is young, then most of the light we see will usually come from luminous hot, blue stars. If the cluster is old, then the light from the cluster will have the color of red giants and relatively cool, low-mass stars.

As always, there are caveats to such general statements. Young clusters may contain very luminous red supergiants, escape from the collapsing object and find its way back into the universe of which it was once a part. The object is now a **black hole**. A black hole will form if the stellar core left behind by a Type II supernova exceeds about 3 $M_\odot$. Alternatively, a neutron star will collapse to become a black hole if it accretes enough matter from a binary companion to push it over the 3 $M_\odot$ limit. Regardless of how it formed, any collapsed object with a mass greater than 3 $M_\odot$ must be a black hole.

Free Fall Is the Same as Free Float

In Chapter 4 our exploration of special relativity began with the observation that the speed of light is always the same (regardless of the motion of an observer or the source), and ended by shattering our everyday notions of space and time. Now as we confront the properties of black holes—indeed, of all massive objects in the universe—our concepts of space and time will be pulled even further from the comfortable

---

Mass warps the fabric of spacetime.

---

absolutes of Newtonian physics. First we learned that what we traditionally called (three-dimensional) space and time are actually just a result of our particular, limited perspective on a four-dimensional spacetime that is different for each observer. Now we discover that this four-dimensional spacetime is warped and distorted by the masses it contains. One of the consequences of this deformation is the gravity that holds you to Earth. This realization, called the **general theory of relativity**, is one of Albert Einstein's great contributions to science.

A crucial clue to the fundamental connection between gravity and spacetime has been with us since Chapter 3, where we found that the *inertial mass* of an object—the mass appearing in Newton's $F = ma$—is *exactly* the same as the object's gravitational mass. Another clue is that, left on their

---

Inertial mass and gravitational mass are the same.

---

own, any two objects at the same location and moving with the same velocity will follow the same path through spacetime, regardless of their masses. The space shuttle astronaut falls around Earth, moving in lockstep with the space shuttle itself. A feather dropped by an *Apollo* astronaut standing on the Moon falls toward the surface of the Moon at exactly the same rate as a dropped hammer. In some sense, rather than thinking of gravity as a "force" that "acts on" objects, it is more accurate to think of it as a consequence of the path through spacetime that objects will follow in the absence of other forces. *Gravitation is the result of the shape of the spacetime terrain through which objects move.*

The essence of special relativity is that any inertial reference frame is as good as any other. There is no experiment you can do to distinguish between sitting in an enclosed spaceship floating stationary in deep space (**Figure 17.19(a)**) and sitting in an enclosed spaceship traveling through our galaxy at 0.99999 times the speed of light (**Figure 17.19(b)**). You cannot tell any difference between these two cases —they do not *feel* any different—because there *is* no difference between them. Each of these reference frames is an equally valid inertial reference frame. As long as nothing acts to *change* the motion—that is, nothing pushes on either spaceship—the laws of physics are exactly the same inside both spacecraft.

General relativity begins by applying this same idea to an astronaut inside the space shuttle orbiting Earth, as shown in **Figure 17.19(c)**. So long as we restrict our atten-

---

A freely falling object defines an inertial reference frame.

---

tion to a small enough volume of space and a short enough period of time that we can ignore changes in the strength and direction of gravity from place to place, our astronaut again has no way to tell the difference between being inside the space shuttle as it falls around Earth and being inside a spaceship coasting through interstellar space. Close your eyes and jump off a diving board. For the brief time that you are falling freely through Earth's gravitational field, the sensation you feel is exactly the same as the sensation that you would feel adrift in the abyss of interstellar space! The implications of this result are somewhat startling. Even though its velocity is constantly changing as it falls, *the inside of a space shuttle orbiting Earth is as good an inertial frame of reference as that of an object drifting along a straight line through interstellar space.* This principle— which can be simply stated as "free fall is the same as free float"—is called the **equivalence principle**.

The equivalence principle says that a falling object is simply following its "natural" path through spacetime—it is going where its inertia carries it—every bit as much as an object that drifts along a straight line at a constant speed through deep space. The natural path that an object will follow through spacetime in the absence of other forces is referred to as the object's "world line" or **geodesic**. In the absence of a gravitational field, the geodesic of an object is a straight line—hence Newton's statement of inertia that,

---

Falling objects follow straight lines through curved spacetime.

---

unless acted on by an unbalanced external force, an object will move at a constant speed in a constant direction. However, in the presence of mass the shape of spacetime becomes distorted, so an object's geodesic becomes curved.

The equivalence principle gives us a way of understanding why gravitational mass and inertial mass are one and

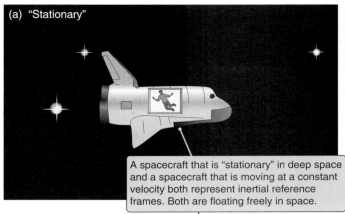

(a) "Stationary"

A spacecraft that is "stationary" in deep space and a spacecraft that is moving at a constant velocity both represent inertial reference frames. Both are floating freely in space.

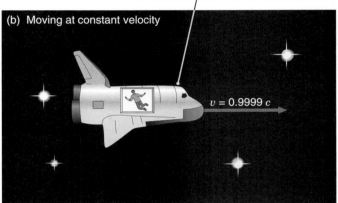

(b) Moving at constant velocity

$v = 0.9999\ c$

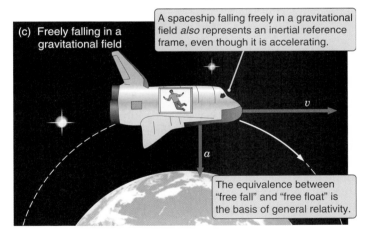

(c) Freely falling in a gravitational field

A spaceship falling freely in a gravitational field *also* represents an inertial reference frame, even though it is accelerating.

v

a

The equivalence between "free fall" and "free float" is the basis of general relativity.

FIGURE 17.19 Special relativity says that there is no difference between (a) a reference frame that is floating "stationary" in space and (b) one that is moving through the galaxy at constant velocity. General relativity adds that there is no difference between these inertial reference frames and (c) an inertial reference frame that is falling freely in a gravitational field. "Free fall" is the same as "free float" as far as the laws of physics are concerned.

the same thing. When we discussed Newton's law, $F = ma$, we found it useful to state it instead as $a = F/m$. When you apply a force F to an object you move it away from its natural path, and its inertial mass m tells you how strongly it resists the change. General relativity says that when you are standing still on the surface of Earth, your natural path through spacetime—your geodesic—is actually a path falling inward toward the center of Earth. In the absence of any external forces, this is what you would do. Of course, the surface of Earth gets in your way. Put another way, the surface of Earth exerts an external force on your feet, and that force causes you to accelerate continuously away from your natural path through spacetime.

This leads to a different, equally valid, way of stating the equivalence principle. Another thought experiment, shown in **Figure 17.20**, demonstrates the point. Imagine you are in a box inside a rocket ship that is accelerating through deep space at a rate of 9.8 m/s² in the direction of the arrow, as shown in Figure 17.20(b). The floor of the box exerts enough of a force on you to overcome your inertia and cause you to accelerate at 9.8 m/s², so you feel as though you are being pushed into the floor of the box.

> Being stationary in a gravitational field is the same as being in an accelerated reference frame.

Now imagine instead that you are sitting in a closed box on the surface of Earth. Again the floor of the box exerts enough upward force on you to overcome your inertia, causing you to accelerate at 9.8 m/s². You feel as though you are being pushed into the floor of the box. According to the equivalence principle, *the two cases are identical.* There is no difference between sitting in an armchair in a rocket ship traveling through deep space with an acceleration of 9.8 m/s² and sitting in an armchair on the surface of Earth reading this book. In the first case, the force of the rocket ship is pushing you away from your "floating" straight-line geodesic through spacetime. In the second case, Earth's surface is pushing you away from your curved "falling" geodesic through a spacetime that has been distorted by the mass of Earth. An acceleration is an acceleration, regardless of whether you are being accelerated off a straight-line geodesic through deep space or being accelerated off a "falling" geodesic in the gravitational field of Earth. And in all cases, it is the same mass —the mass that gives an object inertia—that resists the change. Gravitational mass and inertial mass are the same thing!

There is an important caveat to the equivalence principle. In an accelerated reference frame such as an accelerating rocket ship, the *same* acceleration is experienced *everywhere.* In contrast, the curvature of space by a massive object changes from place to place. Tides are one result of changes in the curvature of space from one place to another. A more careful statement of the equivalence principle is that the effects of gravity and acceleration are indistinguishable *locally*—that is, so long as we restrict our attention to small enough volumes of space that changes in gravity can be ignored.

FIGURE 17.20 (a) According to the equivalence principle, an object falling freely in a gravitational field is in an inertial reference frame, whereas (b) an object at rest in a gravitational field is in an accelerated frame of reference.

Spacetime as a Rubber Sheet

General relativity is a *geometrical* theory. It describes how mass distorts the *geometry* of spacetime. You can get a sense for how mass distorts spacetime by imagining the surface of a tightly stretched rubber sheet. The rubber sheet is flat. If you roll a marble across the sheet, it will roll in a straight line. All of the Euclidean geometry that you learned in high

---
Mass distorts the geometry of spacetime.
---

school applies on the surface of the sheet as well: The angles in a triangle on the sheet add up to 180°. Right triangles obey the Pythagorean theorem. If you draw a circle on the sheet, you will find that the circumference of the circle is equal to 2π times its radius. If you draw a line on this sheet and a point off to one side, there will be exactly one line that passes through that point but never intersects the first line.

But now think about what happens if you place a bowling ball in the middle of the rubber sheet as in **Figure 17.21**. The surface of the sheet will be stretched and distorted. Now if you roll a marble across the sheet, its path will dip and curve (Figure 17.21(a)). You might even find that you can roll

the marble so that it moves around and around the bowling ball, like a planet orbiting about the Sun. Next you might revisit the relationship between the radius of a circle and its circumference. If you draw a circle around the bowling ball, measure the distance around that circle, and then compare that distance with the distance from the circle to its center along the surface of the sheet (Figure 17.21(b)), you will find that the circumference of the circle is less than $2\pi r$. Finally, you might try to draw a triangle on the surface of the sheet, connecting three points around the bowling ball with the straightest and shortest lines you can draw on the surface of the sheet, as in Figure 17.21(c). If you do this and then look at the sheet from above, you will be amazed to see that rather than adding to 180°, the angles in this new triangle sum to more than 180°. The surface of the sheet is no longer flat, and Euclid's geometry no longer applies. (That is why Euclid's geometry is called "plane geometry.")

Mass has an effect on the fabric of spacetime that is analogous to the effect of the bowling ball on the fabric of the rubber sheet. The bowling ball stretches the sheet, changing the distances between any two points on the surface of the sheet. (Think of the deep depression in the rubber sheet as a well. We will frequently use this language.)

Similarly, mass distorts the shape of spacetime, changing the "distance" between any two locations or events in that spacetime.

It is easy to understand how the two-dimensional surface of the sheet is distorted by the bowling ball because we can visualize how the sheet is stretched through a third spatial dimension. It is virtually impossible for us to "see" in our mind's eye what a curved four-dimensional spacetime would "look like." Once again we have run into a limitation in how our brains are wired. Yet there are experiments we can perform, much like those done on the surface of the rubber sheet, that demonstrate that the geometry of our four-dimensional spacetime is distorted much like the rubber sheet.

Before going any further in our discussion of general relativity, however, it is important to point out that general relativity does not mean that Newton's law of gravitation is "wrong." See **Connections 17.2** for a discussion of what happens when one physical law supplants another.

The Observable Consequences of General Relativity

There are many observable consequences of curved spacetime. You can imagine, at least in principle, stretching a rope all the way around the circumference of Earth's orbit

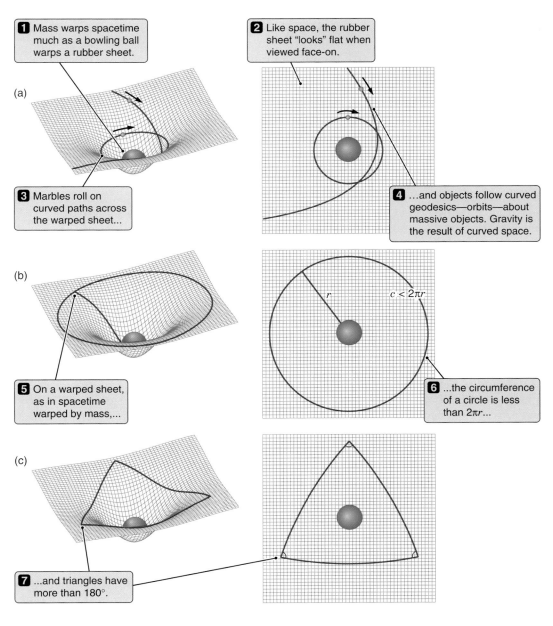

1 Mass warps spacetime much as a bowling ball warps a rubber sheet.

2 Like space, the rubber sheet "looks" flat when viewed face-on.

(a)

3 Marbles roll on curved paths across the warped sheet...

4 ...and objects follow curved geodesics—orbits—about massive objects. Gravity is the result of curved space.

(b)

5 On a warped sheet, as in spacetime warped by mass,...

r $c < 2\pi r$

6 ...the circumference of a circle is less than $2\pi r$...

(c)

7 ...and triangles have more than 180°.

FIGURE 17.21 Mass warps the geometry of spacetime in much the same way that a bowling ball warps the surface of a stretched rubber sheet. This distortion of spacetime has many consequences, such as these: (a) Objects follow curved paths or geodesics through curved spacetime, (b) the circumference of a circle around a massive object is less than 2π times the radius of the circle, and (c) angles in triangles need not sum to exactly 180°.

CONNECTIONS 17.2

Gravitation: When One Physical Law Supplants Another

If you have been reading this chapter closely, you might feel some justifiable annoyance with this business of gravity. Throughout the book until this point we have described gravity as a force that obeys Newton's universal law of gravitation: $F = Gm_1m_2/r^2$. Now we suddenly introduce the ideas of general relativity and in the process ask you to totally change the way you think about gravity. So which is the real deal? If general relativity is right, does that not imply that Newton's formulation of gravity is wrong? And if so, then why have we continued to use Newton's law?

The answers to these questions go to the heart of how science progresses and how our conception of the universe evolves. Under most circumstances there is virtually no difference between the predictions made using general relativity and the predictions made using Newton's law. So long as a gravitational field is not too strong, Newton's law of gravitation is a very close *approximation* to the results of a calculation using general relativity. The meaning of "too strong" in this context is relative. For example, in most ways the enormous gravitational field near the core of a massive main sequence star would be considered "weak." Had we used a general relativistic formulation of gravity rather than Newton's laws to compute the structure of a main sequence star, it would have made virtually no difference in the results of the calculation. Similarly, even though spacetime is curved by gravity, this curvature near Earth is very slight, so over small regions it can be ignored entirely. The flat Euclidean geometry is a good "local" approximation even to curved spacetime—and is a lot easier to use. This is exactly the kind of approximation we use when, despite the curvature of Earth, we navigate a city using a flat road map.

This is not the first time we have run into the idea that the physics of our everyday experience is only an approximation to the more general rules governing the behavior of matter and energy. Newton's laws of motion are one of the great triumphs of the human intellect, and they still form the basis of several years of study for students of physics. Yet we now know that Newton's laws of motion are actually *approximations* to the more generally applicable rules of special relativity and quantum mechanics. In fact, we can *derive* Newton's laws from special relativity and quantum mechanics by making the "everyday" assumptions that speeds of objects are much less than the speed of light and that the sizes of objects are much larger than the subatomic particles from which atoms are made. We use Newton's laws of motion and gravitation most of the time because they are far easier to apply than the relativistically and quantum mechanically "correct" laws, and because any inaccuracies we introduce by using Newtonian approximations are usually far too tiny to matter. It is only in conditions very different from those of our everyday lives (such as the behavior of an electron in an atom or the gravitational field of a black hole), or in special cases when very high accuracy is needed (such as the precise timing used by the global positioning system satellite network), that the more general laws must be used.

This is a general feature of new scientific theories. If a new theory is to replace an earlier, highly successful scientific theory, the new theory must normally "hold the old theory within it"—it must be able to reproduce the successes of the earlier theory—just as general relativity holds within it the successful Newtonian description of gravity that we have used throughout the book.

about the Sun, and then comparing the length of that rope with the length of a rope taken from the orbit of Earth to the center of the Sun. Having taken geometry in high school, you might expect to find that the circumference of Earth's orbit is equal to 2π times the radius of Earth's orbit, just like a circle drawn on a flat piece of paper. However,

if you could carry this experiment out, you would find instead that *the rope around the circumference of Earth's orbit was shorter by 10 km than 2π times the length of the rope stretched from Earth to the center of the Sun*—just as the circumference of the circle on the stretched rubber sheet is less than 2π times the radius of the circle. It is not practical to stretch a rope from Earth to the Sun. But we *can* do an experiment that is conceptually much the same thing. The long axis of Mercury's elliptical orbit about the Sun is slowly *precessing*—that is, the axis of Mercury's orbit is slowly changing its direction. Even after allowing for the

The consequences of curved spacetime include precession of Mercury's orbit.

perturbations caused by the gravity of the other planets, this precession is not predicted by Newton's inverse square law of gravity, working in a flat Euclidean space. However, it is *just* what is predicted for the path of a planet that is moving alternately deeper and outward within the stretched-out non-Euclidean fabric of spacetime that has been warped by the Sun.

The real-life equivalent of the triangle with more than 180° is probably easier to understand. A straight line in space is *defined* by the path followed by a beam of light. This is the shortest distance between any two points. A

Gravitational lensing can displace and distort an object's image.

beam of light moving through the distorted spacetime around a massive object is bent by gravity, just as the lines in Figure 17.21(c) are bent by the curvature of the sheet. This phenomenon is called **gravitational lensing** because the curvature of spacetime bends the path of light something like the lens in a pair of eyeglasses.

The first measurement of gravitational lensing came during the solar eclipse of 1919. Prior to the eclipse, Sir Arthur Eddington measured the positions of a number of stars in the part of the sky where the eclipse would occur. Eddington then repeated his measurement during the solar eclipse and found that the apparent positions of the stars had been deflected outward by the presence of the Sun. The light from the stars followed a bent path through the curved spacetime around the Sun, causing the stars to appear farther apart in Eddington's measurement (see **Figure 17.22**). During the eclipse the triangle formed by Earth and the two stars contained more than 180°—just like the triangle on the surface of our rubber sheet. The results of Eddington's measurements were just as predicted by Einstein's theory. Eddington's result was the first experimental test of a prediction of general relativity and is considered to be one of the landmark experiments of 20th century physics. More recently, gravitational lensing has been used to search for unseen massive objects adrift in space. These objects do not give off light, but their gravity can distort the light from background stars that they happen to pass in front of. (See Excursions 19.2.)

General relativity also affects spacetime in ways that have no direct comparison with a rubber sheet because mass distorts not only the geometry of space, but the geometry of time as well. The deeper we descend into the gravitational field of a massive object, the more slowly our clocks appear to run from the perspective of a distant observer. This effect is called **general relativistic time dilation**. To understand one consequence of general relativistic time dilation,

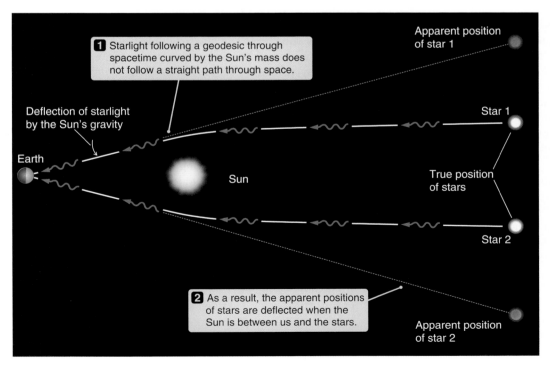

FIGURE 17.22 Measurements obtained by Sir Arthur Eddington during the total solar eclipse of 1919 found that the gravity of the Sun bends the light from distant stars by the amount predicted by Einstein's general theory of relativity. This is an example of gravitational lensing. Note that the "triangle" formed by Earth and the two stars contains more than 180°, just like the triangle in Figure 17.21(c).

suppose a light is attached to a clock sitting on the surface of a neutron star. The light is timed so that it flashes once a second. However, because time near the surface of the star is dilated—stretched out—to an observer far from the neutron star it seems that the light is pulsing less frequently

Time runs more slowly near massive objects.

than once a second. The frequency of the flashing is lowered. Now suppose that we have an emission line source on the surface of the neutron star. Because time is running slowly on the surface of the neutron star, at least from our distant perspective, the light that reaches us will have a lower frequency as well. Remember that lower frequency means longer wavelength. So the light from the source will be seen at a longer, redder wavelength than the wavelength at which it was emitted.

This phenomenon, shown in **Figure 17.23**, is called the **gravitational redshift** because the wavelengths of light from objects deep within a gravitational well are shifted to longer wavelengths. Gravitational redshift is similar in its effect to the **Doppler redshift** we saw earlier. In fact, there is no way to tell the difference between light that is redshifted by gravity and the Doppler-shifted light from an object moving away from us. Astronomers often describe the gravitational redshift of an object as an "equivalent velocity." The gravitational redshift of lines formed on the surface of the Sun is equivalent to a Doppler shift of 0.6 km/s. The gravitational redshift of light from the surface of a white dwarf is equivalent to a Doppler shift of about 50 km/s. The gravitational redshift from the surface of a neutron star is equivalent to a Doppler shift of about a tenth the speed of light. Sometimes astronomers get sloppy and talk about the gravitational redshift as if it truly were a Doppler shift. We might say, for example, that the "gravitational redshift of the surface of a particular white dwarf is 57.1 km/s." However, this does not mean that the surface of the white dwarf is moving away from us at 57.1 km/s. It means that time is running slowly enough on the surface of the white dwarf that the light reaching us from the white dwarf *looks like* it is coming from an object moving away from us at 57.1 km/s.

Bringing this a bit closer to home, a clock on the top of Mount Everest gains about 80 nanoseconds a day compared with a clock at sea level. The difference between an object on the surface of Earth and an object in orbit is much greater. A global positioning system (GPS) receiver uses the results of sophisticated calculations of the effects of general relativistic time dilation to help you accurately find your position on the surface of Earth. Even after allowing for slowing due to special relativity, the clocks on the satellites that make up the GPS run faster than clocks on the surface of Earth. If the satellite clocks *and* your GPS receiver did not correct

for this and other effects of general relativity, then the position your GPS receiver would report would be in error by up to half a kilometer. The fact that the GPS system works is actually a strong experimental confirmation of a number of predictions of general relativity, including general relativistic time dilation.

We could easily fill the rest of this book with fascinating tales about general relativity. It is pretty heady stuff if you think about it—discussing the fabric of the universe itself as though it were a substance in a test tube to be poked and prodded (see **Connections 17.3**). One final phenomenon that we should mention before moving on is the phenomenon of **gravity waves**. If you thump the surface of our rubber sheet, waves will move away from where you thump it, something like ripples spreading out over the surface of a pond. Similarly, the equations of general relativity predict

Gravity waves travel through the fabric of spacetime.

that if you "thump" the fabric of spacetime (for example, with the catastrophic collapse of a high-mass star or the formation of a black hole), then ripples in spacetime will move outward at the speed of light. These gravity waves are like electromagnetic waves in some respects. Accelerating an electrically charged particle gives rise to an electromagnetic wave. Accelerating a massive object gives rise to gravity waves.

Gravity waves have never been observed in a laboratory or anywhere else, but there is strong circumstantial evidence for their reality. In 1974 astronomers discovered a binary system consisting of two neutron stars, one of which is an observable pulsar. By using the pulsar as a precise clock, astronomers are able to very accurately measure the orbits of the stars. The stars themselves are 2.8 solar radii apart, but their orbits are gradually decaying, which means that they are losing energy *somewhere*. Calculations show that the energy being lost by the system is just what general relativity predicts the system should be losing in the form of gravity waves. More recently astronomers discovered a similar binary pair separated by only 1.0 solar radii, and they found once again that the orbital energy loss was consistent with the radiation of gravity waves. Still, both systems provide only indirect evidence. These binary systems suggest a very strong probability for, but do not prove the existence of, gravity waves.

Let's stop for a minute and think about this. Here we are talking about a phenomenon that has *never been observed*, at least not directly. At the moment gravity waves exist only as a scientific theory. Is this so different from the pseudoscientific theory of "intelligent design," which claims that life on Earth was deliberately designed by some intelligent agent? Yes, very much so. The prediction that gravity waves exist is *falsifiable*. By this we mean that the

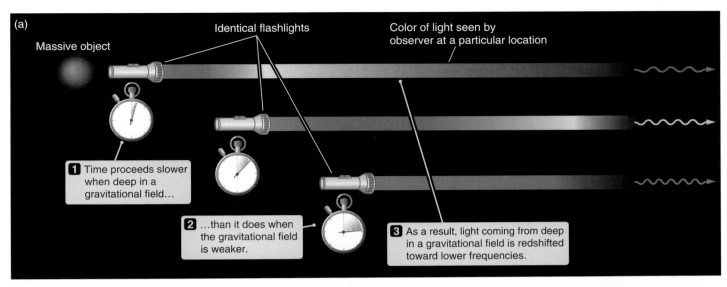

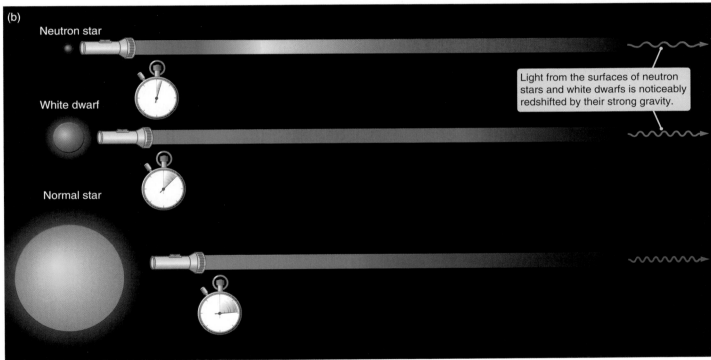

FIGURE 17.23 Time passes more slowly near massive objects because of the curvature of spacetime. As a result, to a distant observer, light from near a massive object will have a lower frequency and longer wavelength. (a) The closer to the object the source of radiation is, or (b) the more massive and compact the object is, the greater the gravitational redshift.

gravity wave theory *can be shown to be false*. We have already noted that the theory predicts that certain events, such as the catastrophic collapse of a high-mass star or the formation of a black hole, will generate gravity waves. As we learned in Chapter 5, physicists will soon have a new kind of "telescope" that will be able to detect the pre-

dicted gravity waves emanating from such events. When this happens, either we will see them or we will not. The theory that predicts gravity waves *can* be tested because it is falsifiable! Pseudoscientific theories on the other hand are *not* falsifiable: They cannot be put to a straightforward "yes or no" test.

CONNECTIONS 17.3

General Relativity and the Structure of the Universe

Our discussion of gravity as the warping of spacetime by the mass it contains was motivated by the events that accompany the collapse of a massive star. Were it only for black holes, it might not have been appropriate to devote so much effort in this chapter to a discussion of general relativity. However, this is not the last time we will run across general relativity. Many of the same physical processes we have seen at work here also shape events in the larger universe. In our discussion of galaxies, for example, we will find that some galaxies contain supermassive black holes at their centers—objects with masses millions of times that of our Sun, which grow by consuming entire stars and interstellar clouds. Material falling into such monsters is a leading candidate to explain the powerful radiation from *quasars*—beacons at the edge of the observable universe—which we will explore in greater detail in Chapter 18. We will also find that just as the Sun bends the path of light passing near it, as observed by Eddington during the total eclipse of 1919, the curved spacetime around distant galaxies can act as a lens, magnifying and distorting the appearance of even more distant objects behind them.

The grandest application of general relativity will come as we consider the history and fate of the universe itself. We live in an expanding universe—one that is the result of a singular event that took place around 14 billion years ago. This event, called the *Big Bang*, was not the result of mass coming into existence *within* the spacetime of the universe. Rather, *it was the coming into existence of the spacetime of the universe itself.* When we say that the universe is expanding, we do not mean that the stars and the galaxies in the universe are flying away from each other through space. Instead, the spacetime of the universe itself is expanding with time. Just as the spacetime around a black hole has a shape, so does the spacetime of the universe, and that shape is determined at least in part by the mass that our universe contains. What was the universe like when it was very young? Will the universe expand forever, or is there enough mass in the universe to warp spacetime into a closed shape that will eventually collapse back in on itself? General relativity gives astronomers the tools they need to address such basic questions about the nature of existence itself.

Back to Black Holes

We began our digression into general relativity when we encountered black holes, and we return to the nature of black holes now. When we placed an object on the surface of our rubber sheet, it caused a funnel-shaped distortion that is analogous to the distortion of spacetime by a mass.

Black holes are bottomless pits in spacetime.

Now imagine such a funnel-shaped distortion in the rubber sheet that is *infinitely* deep—a funnel that keeps getting narrower and narrower as we go deeper and deeper, but that has no bottom. This is the rubber-sheet analog to a black hole. A black hole is a place where the mathematics describing the shape of spacetime fails in the same way that the

mathematical expression $1/x$ fails when $x = 0$. Such a mathematical anomaly is called a **singularity**. Black holes are singularities in spacetime (see **Figure 17.24**).

A black hole has only three properties—mass, electric charge, and angular momentum. The amount of mass that falls into a black hole determines the extent of its distortion of spacetime. The electric charge of a black hole is the net electric charge of the matter that fell into it. The angular momentum of a black hole causes the spacetime around the black hole to be twisted around much like the water around an eddy in a river. Apart from these three properties, all information about the material that fell into the black hole is lost. Nothing of its former composition, structure, or history survives.

We can never actually "see" the singularity at the center of a black hole. The closer an object is to a black hole, the greater is its escape velocity (that is, the speed at which it

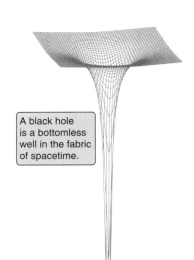

FIGURE 17.24 A black hole is a singularity in the curvature of spacetime. It is a gravitational well with no bottom.

A black hole is a bottomless well in the fabric of spacetime.

would have to move to escape from the gravity of the black hole). There is a radial distance from the black hole at which the escape velocity reaches the speed of light. This point of no return, beyond which even light is trapped by the black hole, is called its **event horizon**. The radius of the event horizon of a black hole is proportional to the black hole's mass. A black hole with a mass of 1 $M_\odot$ has an event horizon radius of about 3 km. A 2 $M_\odot$ black hole has a corresponding radius of about 6 km. The radius for a 3 $M_\odot$ black hole is about 9 km. A black hole with a mass equivalent to that of Earth would have an event horizon radius of only about a centimeter—the mass of our planet compressed into a volume equal to that of a pecan!

The event horizon is the boundary of no return.

Let us consider what would happen if an adventurer were willing to journey into a black hole (**Figure 17.25**). From our perspective outside the black hole, we would see our adventurer fall toward the event horizon, but as she did her watch would run more and more slowly, and her progress toward the event horizon would get slower and slower as well. Like Achilles in Zeno's famous paradox, even though our adventurer got closer and closer to the event horizon, she would never quite make it, from our perspective. The event horizon is where the gravitational redshift becomes infinite and where clocks stop altogether. Yet our adventurer's own experience would be very different. From her standpoint there would be nothing special about the event horizon at all. She would fall past the event horizon and on, deeper into the black hole's gravitational well. How-

ever, she would now have entered a region of spacetime that was cut off from the rest of the universe. The event horizon is like a one-way door: Once our adventurer has passed through, she can never again pass back into the larger universe of which she was once a part.

Actually, we have overlooked a rather crucial fact. Our intrepid explorer would have been torn to shreds long before she reached the black hole. Near the event horizon of a 3 $M_\odot$ black hole, the difference in gravitational acceleration between our explorer's feet and her head—the tidal "force" pulling her apart—would be about a billion times her weight on the surface of Earth. Obviously this is not an experiment we would ever want to perform! Although scientific theories *must* produce testable predictions, it is not required that *all* individual predictions be directly testable.

"Seeing" Black Holes

In 1974 the British physicist **Stephen Hawking** (1942–) realized that black holes should actually be *sources* of radiation. In the ordinary vacuum of empty space, quantum theory says that particles and their antiparticle "mates" spontaneously spring into existence and then quickly annihilate each other and disappear. These particle pairs typically live

FIGURE 17.25 A trip into a black hole.

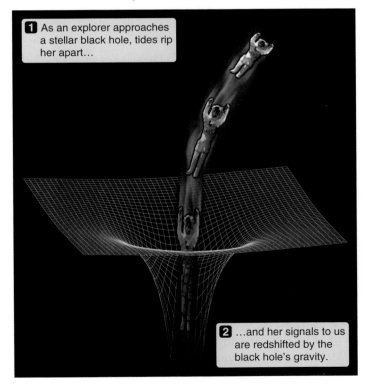

1 As an explorer approaches a stellar black hole, tides rip her apart…

2 …and her signals to us are redshifted by the black hole's gravity.

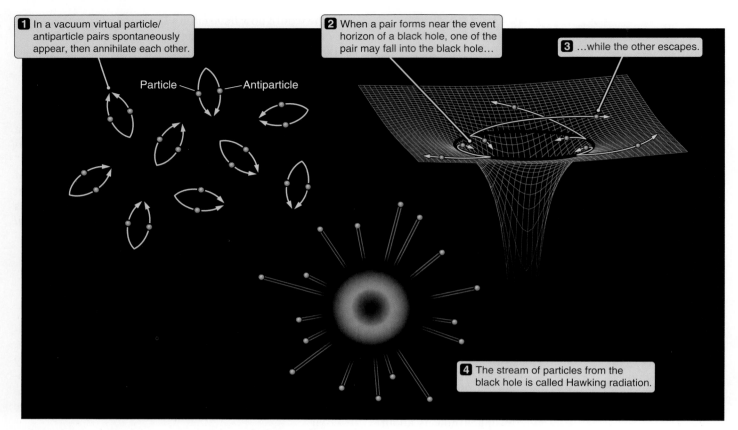

FIGURE 17.26 In the vacuum, particles and antiparticles are constantly being created and then annihilating each other. However, near the event horizon of a black hole, one particle may cross the horizon before it recombines with its partner. The remaining particle leaves the black hole as Hawking radiation.

for less than about 10^{-21} second, but their effects are seen in sensitive measurements of atomic transitions. If such a pair of **virtual particles** comes into existence near the event horizon of a very small black hole, as shown in **Figure 17.26**, then one of the particles might wind up falling into the black hole while the other particle is able to escape. Some of the gravitational energy of the black hole will have been used up in making one of the pair of virtual particles real. When all of the esoteric physics is taken into account, Hawking was able to show that a black hole should actually emit a Planck spectrum, and that the effective temperature of this spectrum would increase as the black hole became smaller. Although this phenomenon, called **Hawking radiation**, is of considerable interest to physicists and astronomers, in a practical sense it is usually negligible. A 3 $M_\odot$ black hole should emit radiation at a whopping temperature of only 2×10^{-8} K, which means that the black hole should radiate with a power of 1.6×10^{-29} watts—very feeble indeed.

Hawking radiation is hardly a useful way to "see" a black hole. For all intents and purposes black holes remain true to their name. Nonetheless, by the end of the 20th century astronomers had found strong circumstantial evidence for black holes in two very different kinds of systems. In Chapter 18 we will find that there is strong evidence for supermassive black holes at the very centers of galaxies; but the first and perhaps strongest evidence for black holes comes from X-ray binary stars in our own galaxy. In 1972 astronomers did not yet have a good understanding of the X-ray emission from stars. In that year the Uhuru X-ray satellite

Black holes are found through the effects of their gravity.

made a puzzling discovery. The brightest X-ray source in the constellation Cygnus was found to be rapidly flickering. We now know that the brightness of the X-ray emission from this object, called **Cygnus X-1**, can change in as little as 0.01 second. For reasons we will cover in more detail in our discussion of quasars in Chapter 18, this means that the source of the X-rays must be smaller than the distance light travels in 0.01 second, or 3,000 km. Thus the source of X-rays in Cygnus X-1 must be smaller than Earth!

When astronomers began to study this object in other parts of the electromagnetic spectrum, Cygnus X-1 was identified with both a radio star and with an already cataloged optical star called HD226868. The spectrum of HD226868 shows that it is a normal B0 supergiant star with a mass of about 30 $M_\odot$ Such a star is far too cool to explain the X-ray emission from Cygnus X-1. But HD226868 was also discovered to be part of a binary system. The wavelengths of absorption lines in the spectrum of HD226868 are Doppler shifted back and forth with a period of 5.6 days. Using the same techniques we used to measure the masses of stars in Chapter 13 (namely, analyzing the orbits of binaries), astronomers found that the mass of the unseen compact companion of HD226868 must be at least 6 $M_\odot$. (Only a lower limit can be determined because the tilt of the orbit of the binary is not known.) The companion to HD226868 is too compact to be a normal star, yet it is much more massive than the Chandrasekhar limit for a white dwarf or a neutron star. According to our understanding of the laws of physics, such an object can only be a black hole.

Since 1972 a number of other good candidates for stellar-mass black holes have been discovered. One such object is a rapidly varying X-ray source in our companion galaxy, the Large Magellanic Cloud. Called LMC X-3, this X-ray source orbits a B3 main sequence star every 1.7 days, and the data show that the compact source must have a mass of at least 9 $M_\odot$. Although the evidence that these systems contain black holes is circumstantial, the arguments that lead to this conclusion seem airtight. With a dozen or so compelling examples of such objects on the books, the evidence is in. Black holes, once regarded as nothing more than a bizarre quirk of the mathematics describing gravitation and spacetime, exist in nature!

Summary

- The CNO cycle burns hydrogen in massive stars.

- More massive elements are created in successive burning stages.

- Iron is the most massive element formed by fusion.

- Massive stars eventually explode as supernovae, leaving behind a neutron star.

- Supernovae eject newly formed massive elements into interstellar space.

- Neutron stars contain more than 1.4 $M_\odot$ packed into a 10-km carbon sphere.

- Pulsars are rapidly spinning magnetized neutron stars.

- Inertial mass and gravitational mass are the same.

- Mass warps the fabric of spacetime.

- Objects deep in a gravitational well appear redshifted.

- Time runs more slowly near massive objects.

- Black holes are bottomless pits in spacetime.

Seeing the Forest through the Trees

The story of low-mass stars like our Sun is by and large a story of longevity and stability. No one has ever seen a star less massive than about 0.8 $M_\odot$ evolve off the main sequence because all the time in the universe has literally not been enough for this to happen even once. What a contrast on this leg of our journey to look instead at stars that inhabit the high-mass end of the family of stars. Rather than stability, these stars offer spectacle. Living for only a short while, they blaze with the light of thousands or even millions of suns. Such stars are few and far between, but the role they play in the life (and in our understanding) of the universe is significant far beyond their numbers. Indeed, when we move on to study galaxies and the universe, massive stars such as Cepheid variables will provide the signposts that we use to gauge the scale of our universe.

Contemplating the lives of massive stars has also taken us even further afield into the extremes to which matter, space, and time can be molded. White dwarfs may have seemed incomprehensibly bizarre when we encountered them in our discussion of low-mass stars. Objects with densities of a ton per teaspoonful are so far beyond our everyday experience as to be all but unimaginable. Now we have encountered neutron stars—objects a billion times more dense yet. The matter in a neutron star is to the electron degenerate matter in a white dwarf as the matter in a white dwarf is to the air in a summer breeze. But even the neutron star is but a way station on our journey to the limits of matter and mind. As we contemplate massive stars, we also come face to face with the most extreme object of all—if *object* is even an appropriate term for the bottomless pit in the fabric of the universe that we call a black hole.

This part of our journey has not only taken us into the realm of tortured spacetime but has also shown us the answers to immediate questions about our own existence. We have watched as high-mass stars begin their lives by tapping the same reservoir of nuclear fuel as their low-mass counterparts, fusing hydrogen into helium and helium into elements such as carbon. But the alchemy of high-mass stars does not end with a cinder of carbon and oxygen. Instead it goes on, fusing massive elements into more massive elements, and those elements into more massive elements yet.

And finally we have seen the spectacle of a Type II supernova. As gravity has its final victory, pulling the inner parts of a massive star down into a neutron star or black hole, the outer layers of the star blaze forth, casting into the universe the seeds of future generations of planets and, in all likelihood, life.

We now come to the end of this leg of our journey, having filled out the family album of stars, following them from their origins in the vast reaches of interstellar space to their final places of rest. But we are not done with stars—far from it. All that we have seen has happened on the stage of galaxies, and the universe is the grand hall in which that stage resides. Knowledge of stars and the interstellar medium is the astronomers' starting point as we endeavor to tease from galaxies and the universe their secrets. At the same time, knowledge of the theater will give us further insight into the players on the stage.

And so we again head outward on the last major segment of our journey: an investigation that will begin with the properties of galaxies and end with a consideration of the origin and nature of the universe itself.

Key Terms

CNO cycle, p. 486
intermediate-mass stars, p. 486
instability strip, p. 489
variable star, p. 489
pulsating variable star, p. 489
Cepheid variable, p. 489
RR Lyrae variable, p. 489
binding energy, p. 492
neutrino cooling, p. 493
photodisintegration, p. 495
Type II supernova, p. 495
nucleosynthesis, p. 497
neutron star, p. 499
X-ray binary, p. 500
positron, p. 500
antimatter, p. 502
main sequence turnoff, p. 504
stellar population, p. 505
black hole, p. 506
general theory of relativity, p. 506
gravitational lensing, p. 511

general relativistic time dilation, p. 511
gravitational redshift, p. 512
gravity waves, p. 512
singularity, p. 514
event horizon, p. 515
Hawking radiation, p. 516

Student Questions

THINKING ABOUT THE CONCEPTS

1. Explain the differences between the way that hydrogen is converted to helium in a low-mass star (proton–proton chain) and in a high-mass star (CNO cycle). What is the catalyst in the CNO cycle, and how does it take part in the reaction?

2. What are the two reasons why each post-helium-burning cycle for high-mass stars (carbon, neon, oxygen, silicon, and sulfur) becomes shorter than the preceding cycle?

3. Cepheids are highly luminous variable stars in which the period of variability is directly related to luminosity. Explain why Cepheids are good indicators for determining stellar distances that lie beyond the limits of accurate parallax measurements.

4. List and explain two important ways in which supernovae influence the formation and evolution of new stars.

5. Why can the accretion disk around a neutron star release so much more energy than the accretion disk around a white dwarf even though both stars have approximately the same mass?

6. An experienced astronomer can take one look at the H-R diagram of a star cluster and immediately estimate its age. How is this possible?

7. What do we mean by the *binding energy* of an atomic nucleus? How does this quantity help us to calculate the energy given off in nuclear fusion reactions?

8. An astronomer sees a redshift in the spectrum of an object. With no other information available, can she determine whether this is an extremely dense object (gravitational redshift) or one that is receding from us (Doppler redshift)? Explain your answer.

9. Of the four forces in nature (strong nuclear, electromagnetic, weak nuclear, and gravity), gravity is by far the weakest. Why then is gravity such a dominant force

in stellar evolution? *Note:* Although not explicitly discussed so far in this text, the weak nuclear force is involved in certain decay processes within the nucleus.

10. If you could watch a star falling into a black hole, how would the color of the star change as it approached the event horizon?

APPLYING THE CONCEPTS

11. In our galaxy there are about 50,000 stars of average mass (0.5 $M_\odot$) for every main sequence star of 20 $M_\odot$. But stars with 20 $M_\odot$ are about 10^4 times more luminous than the Sun, and 0.5 $M_\odot$ stars are only 0.08 times as luminous as the Sun.
    a. How much more luminous is a single massive star than the total luminosity of the 50,000 less massive stars?
    b. How much mass is in the lower-mass stars compared to the single high-mass star?
    c. What does this tell you about which stars contain the most mass in the galaxy and which stars produce the most light?

12. In 1841 the 150 $M_\odot$ star Eta Carinae was losing mass at the rate of 0.1 $M_\odot$ per year. Let's put that into perspective.
    a. The mass of the Sun is 2×10^{30} kg. How much mass (in kg) did Eta Carinae lose each minute?
    b. The mass of the Moon is 7.35×10^{22} kg. How does Eta Carinae's mass loss per minute compare with the mass of the Moon?

13. The approximate relationship between the luminosity and the period of Cepheid variables is L_{star} ($L_\odot$ units) = 335 P (days). Delta Cephei has a cycle period of 5.4 days and a parallax of 0.0033 arcseconds. A more distant Cepheid variable appears 1/1,000 as bright as Delta Cephei and has a period of 54 days.
    a. How far away (in parsecs, or pc) is the more distant Cepheid variable?

    b. Could the distance of the more distant Cepheid variable be measured by parallax? Explain.

14. If the Crab Nebula has been expanding at an average velocity of 3,000 km/s since A.D. 1054, what was its average radius in the year 2006? (There are approximately 3×10^7 seconds in a year.)

15. We know that pulsars are rotating neutron stars. For a pulsar that rotates 30 times per second, at what radius in the pulsar's equatorial plane would a co-rotating satellite (rotating about the pulsar 30 times per second) have to be moving at the speed of light? Compare this to the pulsar radius of 1 km.

16. The Moon has a mass equal to 3.74×10^{-8} $M_\odot$. Suppose the Moon suddenly collapsed into a black hole.
    a. What would be the radius of the event horizon (the "point of no return") around the black hole Moon?
    b. What effect would this have on tides raised by the Moon on Earth? Explain.
    c. Do you think this event would generate gravity waves? Explain.

StudySpace
wwnorton.com/astro21
provides a Study Plan for each chapter that includes a reading outline, animations, keyword flash cards, and gradebook-enabled multiple-choice quizzes. From StudySpace you can also access premium content in the ebook and SmartWork.

PART IV

Galaxies, the Universe, and Cosmology

... it may not be amiss to point out some other very remarkable Nebulae which cannot well be less, but are probably much larger than our own system; and being also extended, the inhabitants of the planets that attend the stars which compose them must likewise perceive the same phenomena. For which reason they may also be called milky ways. . . .

SIR WILLIAM HERSCHEL (1738–1822)

A large barred spiral galaxy, 70 million light-years away in the constellation Eridanus.

CHAPTER 18

Galaxies

18.1 Twentieth Century Astronomers Discovered the Universe of Galaxies

We have come a long way on our journey of discovery, building an understanding of stars and the planetary systems that surround them. Yet cosmically speaking, everything we have come across so far is in our own backyard. A deep-space image of a piece of "dark" sky, such as the Hubble Space Telescope image shown in **Figure 18.1**, reveals myriad faint smudges of light filling the gaps among a sparse smattering of nearby stars. It has long been known that the sky contains faint, misty patches of light. These objects were originally referred to as "nebulae" because of their nebulous appearance. Prior to the 1780s only about 100 of these smudges of light had been found by eye with telescopes. In 1784 the French comet hunter **Charles Messier** (1730–1817) published a catalog of 103 nebulous objects, mostly as a warning to other comet hunters not to waste their time on these objects. Yet 20 years later, courtesy of the remarkable observations of William Herschel and his sister Caroline, that number jumped to 2,500. From this time on, astronomers were aware of systematic differences in the appearance of nebulae. While some of the Herschels' nebulae looked diffuse and amorphous, most were round or elliptical or resembled spiraling whirlpools. These distinctions formed the original three categories—diffuse, elliptical, and spiral—used to classify nebulae.

Discovering the existence of nebulae was one thing, but uncovering their true nature was quite another. Speculations about the nature of these objects abounded for the next

KEY CONCEPTS

As humans it is difficult for us to comprehend the scale of even a single star such as our Sun; yet there are as many stars in the universe as there are grains of sand on all the world's beaches. Stars are not spread uniformly through space, but are instead grouped into what Kant referred to as "island universes" and what we refer to today as "galaxies." As we look beyond stars to the galaxies of which they are a part, we will find that

- Galaxies are classified into different morphological types that are given their shapes by the properties of the orbits of the stars they contain.

- The arms of spiral galaxies, which form whenever the disk of a spiral galaxy is disturbed, are sites of star formation.

- Stars and gas account for only a small fraction of the mass of a galaxy; galaxies are mostly composed of an unknown form of "dark matter."

- Most, and perhaps all, large galaxies have supermassive black holes at their centers.

- When these supermassive black holes are fed, as during encounters between galaxies, they may blaze forth with the light of thousands of normal galaxies coming from an active galactic nucleus no larger than our own Solar System.

FIGURE 18.1 A deep image of a section of "blank" sky made with the Hubble Space Telescope. When we look hard enough, the sky seems literally to be covered with faint galaxies. Nearly every object in this image is a galaxy. The faintest smudges are images of galaxies being formed more than 10 billion years ago.

140 years. Prior to the 1920s many astronomers thought that the sum of existence—the universe—consisted solely of the swarm of stars to which our Sun belongs. It was even suggested that spiral nebulae might be planetary systems in various stages of formation. The great 18th century philosopher **Immanuel Kant** (1724–1804) had a very different idea. He speculated that nebulae were instead "island universes" —realms of existence separate from our own. Herschel himself shared this belief but realized that no telescope of his day would ever be able to resolve the issue. It was not until the first third of the 20th century that the technological tools became available to turn philosophical musing into scientific knowledge.

Today we know that Kant was correct. Our Milky Way is only one of Kant's island universes, which were renamed **galaxies** to reflect this change in understanding. (The term **universe** is now used to refer to the full expanse of space and all that it contains.) Whereas most diffuse nebulae are nearby clouds of gas and dust in our own Milky Way, the elliptical

and spiral nebulae are instead galaxies located far beyond the bounds of our local galaxy. Each tiny smudge in Figure 18.1 is such a galaxy. The universe contains billions upon billions of galaxies—possibly more galaxies than stars in our

There are hundreds of billions of galaxies in the universe.

own galaxy! Each galaxy is a collection ranging from millions to hundreds of billions of stars, thereby rivaling our own cosmic home. Most of these galaxies are located at such astonishing distances from us that they appear too small and faint to see with any but the most powerful telescopes.

As has often been the case in astronomy, the nature of galaxies was, at its heart, a question about size and distance. Early attempts to understand the size of the Milky Way were confounded by interstellar dust, which blocks the passage of visible light, limiting our view. Early astronomers, not knowing of the existence or consequences of dust, assumed that what they could see in visible light was all that there is. Unable to see past this obscuring shroud, they concluded that we live in a system of stars some 6,000 light-years across. It was not until the mid-1910s that **Harlow Shapley** (1885–1972), of the Harvard College Observatory, found that our galaxy is over 300,000 light-years in size.

In an interesting historical twist, the same insight that brought Shapley to the correct conclusion about the size of the Milky Way also led him to an erroneous conclusion about the nature of the spiral and elliptical nebulae. In 1920 Shapley met Lick Observatory's **Heber D. Curtis** (1872–1942) in Washington, D.C., to publicly debate these issues. Historians call this meeting astronomy's Great Debate. When the

> The Great Debate focused attention on the size and distance of nebulae.

Great Debate was held, Curtis defended the earlier, smaller model of our galaxy, but the tide against that picture was already turning. However, the question about the nature of what we now call **spiral galaxies** was still wide open. In Shapley's opinion, his far larger Milky Way was ample to encompass everything in the universe. (Having worked to show that the galaxy is 50 times larger than previously thought, Shapley balked at the idea that the whole universe was hugely larger still.) Curtis, on the other hand, favored the idea that the spiral nebulae were really galaxies separate from our own and that the universe was, indeed, far larger than our own galaxy.

Unlike questions of politics and law, scientific questions are not resolved by the rhetorical skills of partisans. Instead they are settled by the results of well-crafted and carefully conducted experiments and observations. However, scientific debates do help bring issues into sharper focus, leading scientists to concentrate their attention and efforts on key questions. The reason we call that 1920 meeting the Great Debate is because it clearly marked the final steps that would lead to a correct understanding of nebulae. The 1920 debate set the stage and pointed the direction for the work of **Edwin P. Hubble** (1889–1953), whose name was to become forever entwined with our modern understanding of the universe.

Using the newly finished 100-inch telescope on Mt. Wilson, high above the then-small city of Los Angeles, Hubble was able to find some variable stars in the large neighboring galaxy of Andromeda (see the opening photograph for Part

> The nondebater Hubble settled the Great Debate by measuring the distance to galaxies.

IV). He recognized that these stars were very similar to the Cepheid variable stars studied by Henrietta Leavitt, but these stars were much fainter in appearance. Using the period–luminosity relation for Cepheid variable stars discussed in Chapter 17, Hubble turned his observations of these stars into

measurements of the distances to these objects. The results showed that these nebulae are far more distant than even Shapley's measurement of the size of our galaxy. No doubt remained: Spiral and elliptical nebulae are really galaxies in their own right, similar in size to our own galaxy but located at truly immense distances. Hubble may not have shared the stage during the Great Debate, but when he spoke through his results, those earlier questions were answered once and for all. Shapley's Milky Way, itself vast beyond comprehension, is but a speck adrift in a universe full of galaxies.

18.2 Galaxies Are Classified Based on Their Appearance

Imagine what you would see if you were to take a handful of coins and throw them into the air as shown in **Figure 18.2**. You know that all of these objects are very much the same: dimes, pennies, nickels, quarters—all flat and circular. When you look at the objects falling through the air, however, they do not all appear to be the same. Some coins you see face-on, appearing circular. Some coins are instead seen edge-on, appearing as nothing but thin lines. Most

FIGURE 18.2 Galaxies are like a handful of coins when thrown in the air. We see some face-on, some edge-on, and most somewhere in between.

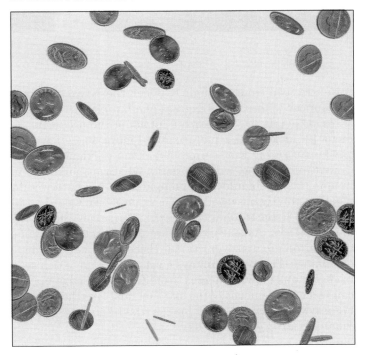

FIGURE 18.3 Disk galaxies seen from various perspectives or angles. The variety of angles we see for galaxies corresponds to the range of perspectives for the coins in Figure 18.2.

coins are seen from an angle between these two extremes and appear with various degrees of *ellipticity*.

In principle we can learn a lot about the properties of coins by looking at a picture like Figure 18.2. If we begin by assuming that coins have some particular three-dimensional shape, we can predict what we should see in Figure 18.2. We could then compare our prediction with what is actually observed. In this example we would probably have little trouble convincing ourselves that coins must be disks. As astronomers we play exactly this game in our efforts to discover the true three-dimensional shape of galaxies. **Figure 18.3** shows a set of spiral galaxies seen from various viewing angles, from face-on to edge-on. We can infer from images of the sky that, just like the coins in Figure 18.2, galaxies often have a disklike shape and are randomly oriented on the sky.

A quick look at a group of galaxies, like that in **Figure 18.4**, shows that galaxies come in a wide range of sizes and shapes. The first step in understanding galaxies came by sorting these different shapes into categories. The classifications we use today date back to the 1930s, when Edwin Hubble devised a scheme much like that shown in **Figure 18.5**. On the bottom (or "handle") of this **tuning fork diagram** sit oval-shaped objects called **elliptical galaxies**. On the two "tines" of the fork are the spirals (as well as a class we will describe shortly, the *S0 galaxies*). Galaxies that fall into none of these classes are called **irregular galaxies**. Originally Hubble thought that his tuning fork diagram might do for galaxies what the H-R diagram did for stars. This hope turned out to be incorrect, but his classification scheme did succeed in bringing order to the study of these objects.

FIGURE 18.4 A Hubble Space Telescope image of a small group of galaxies called Hickson Compact Group 87. This image shows something of the range of shapes and sizes found among galaxies.

Stellar Motions Give Galaxies Their Shapes

Galaxies are not solid objects like coins, but collections of stars, gas, and dust orbiting under the influence of the galaxy's overall gravitational field. In any region in an elliptical galaxy, some stars are falling in while others are climbing out—in fact, there are stars moving in all possible directions. Unlike planets, which move on simple elliptical orbits about the Sun, stars in an elliptical galaxy follow orbits with a wide range of different shapes, as shown in **Figure 18.6**. These orbits are more complex than the orbits of planets because the gravitational field within an elliptical galaxy does not come from a single central object.

All of these stellar orbits, taken together, are what give an elliptical galaxy its shape. The faster the stars are moving, the more spread out the galaxy is. (After all, if the stars

> The collective orbits of all its stars give an elliptical galaxy its shape.

were not moving at all, they would all clump together at the center of the galaxy.) If the stars in an elliptical galaxy are moving in random directions, then the galaxy will have a spherical shape. However, if stars tend to move faster in one direction than in others, then the galaxy will be more spread

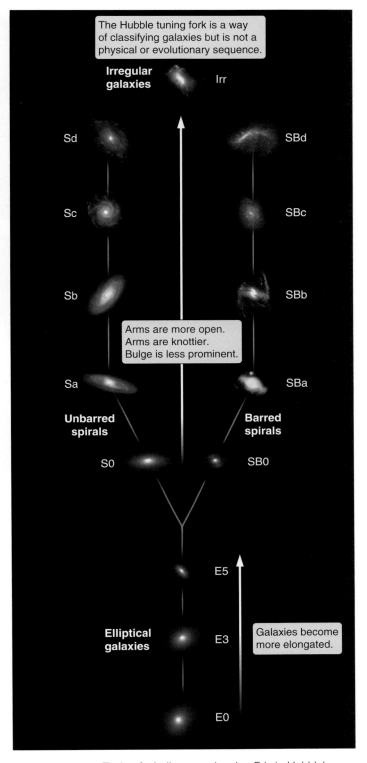

FIGURE 18.5 Tuning fork diagram showing Edwin Hubble's scheme for classifying galaxies based on their appearance. Elliptical galaxies form the "handle" of this tuning fork, shown here at the bottom of the diagram. Unbarred and barred spiral and S0 galaxies lie along the left and right tines of the fork, respectively. Irregular galaxies are not placed on the tuning fork.

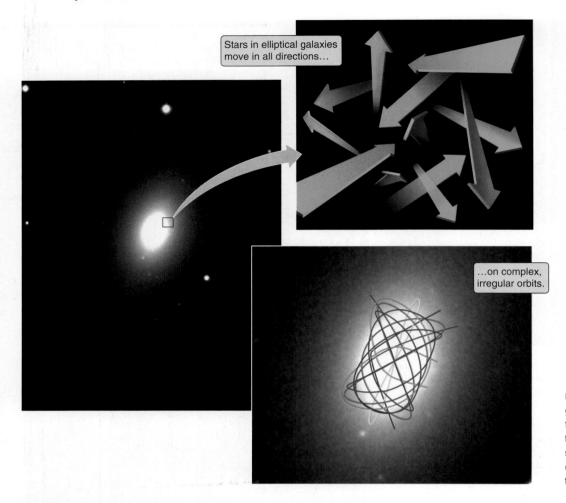

Stars in elliptical galaxies move in all directions…

…on complex, irregular orbits.

FIGURE 18.6 Elliptical galaxies take their shape from the orbits of the stars they contain. Shown here is a sample of stellar orbits in an elliptical galaxy superposed on the galaxy itself.

out in that direction, giving it an elongated shape. Hubble noted that some elliptical galaxies (those on the bottom of the tuning fork handle—see Figure 18.5) are round while others (those on the top of the handle) are elongated. One of the most frustrating problems faced by astronomers studying elliptical galaxies, however, is that their appearance in the sky does not necessarily tell us their true shape. For example, a galaxy might actually be shaped like an American football, but if we happen to see it end-on, it will look instead like a baseball.

The orbits of stars in spiral galaxies are quite different from those of stars in elliptical galaxies. The components of a spiral galaxy are shown in **Figure 18.7**. The defining

Spiral galaxies have a rotating disk and a central bulge.

feature of a spiral galaxy is a flattened, rotating disk. Like the planets of our Solar System, most of the stars in the disk of a spiral galaxy follow nearly circular orbits in the same direction about the center of the galaxy. Spiral galaxies also contain the spiral arms that give these galaxies their name.

In addition to disks and arms, spiral galaxies have central **bulges**, which look like elliptical galaxies. This similarity is more than appearance: Like elliptical galaxies, the bulges of spiral galaxies get their shapes from the range of orbits of the stars in the bulges.

Hubble noticed that the bulges of roughly half of the spiral galaxies are bar-shaped (see the opening photograph of this chapter). He called these **barred spirals** and placed them along the right tine of the tuning fork in Figure 18.5. Spirals that lack a barlike bulge lie along the left tine. Placement of spiral galaxies vertically along the tines of the fork is based on the prominence of the central bulge and how tightly the spiral arms are wound. The criteria Hubble used to classify galaxies are summarized in **Table 18.1**.

There is another big difference between spiral and elliptical galaxies. Most spiral galaxies contain large amounts of molecular gas and dust concentrated in the midplanes of their disks. Just as the dust in the disk of our own galaxy can be seen on a clear summer night as a dark band slicing the Milky Way in two (see Chapter 15, Figure 15.1), the dust in an edge-on spiral galaxy appears as a dark, obscuring band running down the midplane of the disk (**Figure 18.8**).

Ellipticals contain mostly hot gas, and spirals contain mostly cold gas.

The molecular gas that accompanies the dust can also be seen in radio observations of spiral galaxies. In contrast, elliptical galaxies contain large amounts of very hot gas that we see primarily by observing the X-rays it emits. The difference in shape between elliptical and spiral galaxies offers some insight into why the gas in ellipticals is hot, while spirals contain a large amount of cold dense gas. Just as gas settles into a disk around a forming star, so too does conservation of angular momentum cause cold gas to settle into the disk of a spiral galaxy. In contrast, the only place in an elliptical galaxy where cold gas could collect is at the center. However, the density of stars in elliptical galaxies is so high that Type I supernovae continually reheat this gas, preventing most of it from cooling off and forming cold clouds.

Hubble recognized that the distinction between spiral and elliptical galaxies is not always clear-cut. Some galaxies seem to be a cross between the two types, having stellar disks but no spiral arms. Hubble called these **S0 galaxies** and placed them near the junction of his tuning fork. Today the distinction between elliptical and S0 galaxies is even more blurred. Based on better observations, it now seems that many, if not most, elliptical galaxies contain small rotating disks. In addition to this similarity, we note that both elliptical and S0 galaxies have currently stopped producing any new stars.

Other Differences among Galaxies

Stars form from dense clouds of cold molecular gas. Spiral galaxies contain large amounts of such gas, whereas ellipticals and S0 galaxies do not. From these two observations we would predict that many spiral galaxies are actively forming stars today, while star formation is quite rare in

Stars form in spiral galaxies but not in elliptical galaxies.

ellipticals and S0 galaxies. One consequence of this prediction is that the disks of spiral galaxies should contain both young and old stars, while elliptical and S0 galaxies should mostly contain only a much older population of stars. These predictions have proven to be correct. The differences in stellar population show up as differences in the colors of the two types of galaxies. The disks of spiral galaxies tend

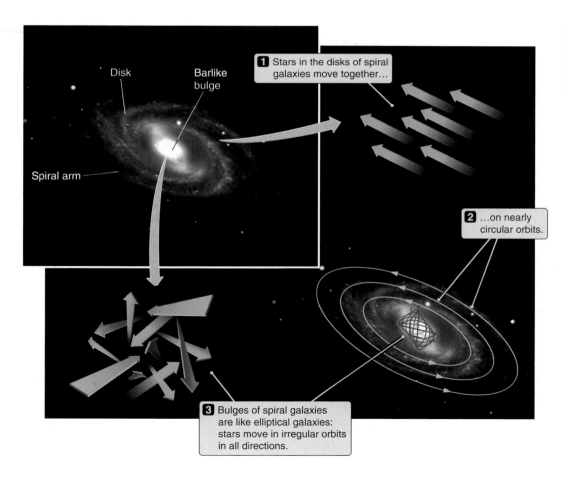

FIGURE 18.7 The components of a barred spiral galaxy. The orbits of stars in the rotating disk and the ellipselike bulge are indicated.

Disk | Barlike bulge

1 Stars in the disks of spiral galaxies move together…

Spiral arm

2 …on nearly circular orbits.

3 Bulges of spiral galaxies are like elliptical galaxies: stars move in irregular orbits in all directions.

to be fairly blue in color, reflecting the fact that their light is dominated by the luminous, young, massive, hot stars that they contain. In contrast, elliptical galaxies are much redder, reflecting the fact that they contain only an older population of lower-mass stars.

Turning this around, the colors of spiral and elliptical galaxies tell us a great deal about their star formation histories. The red colors of elliptical and S0 galaxies tell us that we are looking at old stellar populations and that there has been little or no star formation for quite some time. The blue colors of the disks of spiral galaxies, on the other hand, tell us that we are looking at regions of ongoing star formation where massive young stars are being born. Even though *most* of the stars in a spiral disk are old, the massive young stars are so luminous that their blue light dominates what we see. When it comes to star formation, most irregular galaxies are like spiral galaxies. Some irregular galaxies are currently forming stars at prodigious rates, given their relatively small sizes.

Galaxies range in luminosity from around a million solar luminosities up to a million million solar luminosities (10^6–10^{12} $L_\odot$) and in size from around 1,000 light-years up

Galaxies come in a wide range of sizes for all types.

to hundreds of thousands of light-years. There is no strict size difference between elliptical and spiral galaxies. Although it is true that the most luminous elliptical galaxies are more luminous than the most luminous spiral galaxies,

TABLE 18.1

The Hubble Sequence of Galaxies*

Category/Criteria	Abbreviation	Sequence Range of Features		
Ellipticals	E0	Rounder		
	E1	↕		
Mostly bulge	E2			
Old, red stellar population	E3			
Smooth appearing	E4			
	E5	Flatter		
S-zeros (unbarred/barred)	S0/SB0	Smooth disk and bulge		
Bulge and disk with no arms				
Bulge and disk contain mostly old, red stars				
Spirals (unbarred/barred)	Sa/SBa	More bulge	Tightly wound arms	Smooth arms
Bulge and disk with arms	Sb/SBb	↕	↕	↕
Bulge has old, red stars	Sc/SBc			
Disk has both old, red stars and young, blue stars	Sd/SBd	Little bulge	Open arms	Knotty arms
Spirals (S) have roundish bulges				
Barred spirals (SB) have elongated or barred bulges				
Irregulars	Irr			
No arms				
No bulge				
Some old stars, but mostly young stars, giving a knotty appearance				

*A morphological classification scheme based on the gross properties of galaxies.

FIGURE 18.8 The dust in the plane of an edge-on spiral galaxy is seen as a dark obscuring band in the midplane of the galaxy. Compare this image with Figure 15.1, which shows the dust in the plane of the Milky Way.

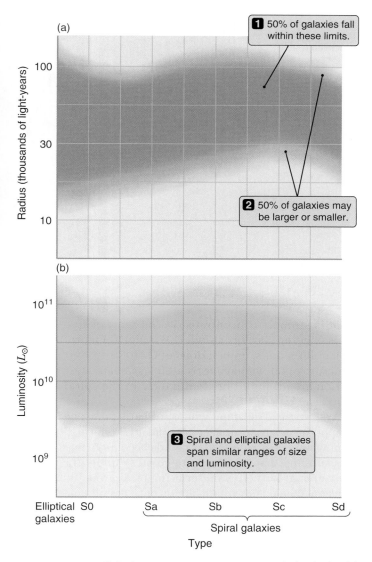

FIGURE 18.9 Galaxies span an enormous range in both size (a) and luminosity (b). Note the great overlap in luminosity and size among galaxies of different Hubble types.

there is considerable overlap in the range of sizes and luminosities among all Hubble types (**Figure 18.9**).

Earlier in our journey we found that mass is the single most important parameter in determining the properties and evolution of a star. In contrast, differences in mass and size do not lead to such obvious differences in the appearance of galaxies. While slight differences in color and concentration exist between large and small galaxies, these differences are more subtle. Even when a smaller, nearby spiral galaxy is seen next to a larger, distant spiral (**Figure 18.10**), it can be difficult to tell which is which. Still, astronomers prefer to call galaxies that are of relatively low luminosity (less than 1 billion solar luminosities) **dwarf galaxies** and galaxies more luminous than this **giant galaxies**. Only elliptical and irregular galaxies come in both types. In fact, among spiral and S0 galaxies, we find only giants. It is relatively easy to tell the difference between a dwarf elliptical galaxy and a giant elliptical galaxy (as shown in **Figure 18.11**). Giant elliptical galaxies have a much higher density of stars than dwarf ellipticals.

18.3 Stars Form in the Spiral Arms of a Galaxy's Disk

Spiral galaxies take their name from the spiral arms they contain, but what is a spiral arm? From pictures of spiral galaxies outside of our own, we might have guessed that stars in the disk of a spiral galaxy are concentrated in the spiral arms. This turns out not to be the case. **Figure 18.12** shows images of the same spiral galaxy taken in ultravio-

let light (Figure 18.12(a)) and in red light (Figure 18.12(b)). Notice that whereas the spiral arms are relatively prominent in the UV image, they are much less prominent when viewed in red light. If we carefully trace the actual numbers of stars rather than just their brightness, we find that although stars are slightly concentrated in spiral arms, this concentration is not strong—certainly not strong enough to account for the prominence of the spiral arms we see. In fact, the concentration of stars in the disks of spiral galaxies varies quite smoothly as it decreases outward from the center of the disk to the edge of the galaxy.

Spiral arms look so prominent when viewed in blue or UV light because that is where we find significant concentrations of young, massive, luminous stars. In other words,

Which galaxy is 4 times farther away and 10 times as luminous as the other?

Size does not determine a galaxy's appearance.

FIGURE 18.10 The mass or size of a spiral galaxy does not determine its appearance. Here a larger, more distant galaxy (left) looks much the same as a smaller, closer galaxy.

what is strongly concentrated in the arms of spiral galaxies is ongoing star formation. H II regions (Figure 18.12(c)), molecular clouds, associations of O and B stars, and other structures that we have learned to associate with star formation are all found predominantly in the spiral arms of galaxies.

We can say a lot about what spiral arms must be like just by applying what we already know about star formation. Stars form when dense interstellar clouds become so massive and concentrated that they begin to collapse under the force of their own gravity. If stars form in spiral arms, then spiral arms must be places where clouds of interstellar gas pile up

and are compressed. Such is indeed the case. There are many ways to trace the presence of gas in the spiral arms of galaxies. Pictures of face-on spiral galaxies, like that in **Figure 18.13(a)**, show dark lanes where clouds of dust block starlight.

Gas, dust, and young stars are concentrated in spiral arms.

These lanes provide one of the best tracers of spiral arms. Spiral arms also show up in other tracers of concentrations of gas, such as 21-cm radiation from neutral hydrogen or radio emission from carbon monoxide (**Figure 18.13(b)**).

FIGURE 18.11 Dwarf elliptical galaxies, as illustrated on the left, differ in appearance from giant elliptical galaxies, such as the one shown on the right.

In ultraviolet light we see massive, young hot stars concentrated in spiral arms.

Older stars seen in red light are less concentrated in arms.

Glowing clouds of hydrogen trace massive stars forming in spiral arms.

(a) Ultraviolet light

(b) Red light

(c) Hydrogen emission

FIGURE 18.12 Images of the barred spiral galaxy UGC 12343, taken in (a) ultraviolet light, (b) red light, and (c) emission from glowing interstellar clouds of ionized hydrogen gas. Note that the spiral arms are most prominent in the image in ultraviolet light, which is dominated by young hot stars, and in emission from interstellar clouds that are ionized by the radiation from young hot stars. The spiral arms are much less prominent in the image taken in red light, in which the smooth underlying disk of old stars accounts for more of what we see.

When a Galaxy's Disk Is "Kicked," Spiral Structure Forms

Spiral arms are concentrations of gas where stars form, but why do spiral arms exist at all? Part of the answer is that any disturbance in the disk of a spiral galaxy will naturally be made into a spiral pattern by the disk's rotation. Material closer to the center takes less time to complete a revolution around the galaxy than material farther out in the galaxy. **Figure 18.14** illustrates the point. We begin with a single lin-

ear arm through the center of a model galaxy, then watch what happens as the model galaxy rotates. In the time that it takes for the inner part of the galaxy to complete several ro-

> **Rotation in a disk galaxy naturally produces spiral structure.**

tations, the outer parts of the galaxy may not have completed even a single revolution. In the process, the originally straight arms are slowly made into the spiral structure shown.

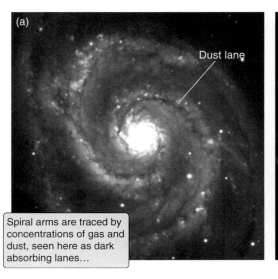

(a)

Dust lane

Spiral arms are traced by concentrations of gas and dust, seen here as dark absorbing lanes…

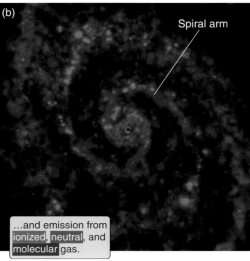

(b)

Spiral arm

…and emission from ionized, neutral, and molecular gas.

FIGURE 18.13 Two images of a face-on spiral galaxy showing the spiral arms. (a) An image in visible light, which also shows dust absorption. (b) 21-cm emission, which shows the distribution of neutral interstellar hydrogen CO emission from cold molecular clouds and Hα emission from ionized gas.

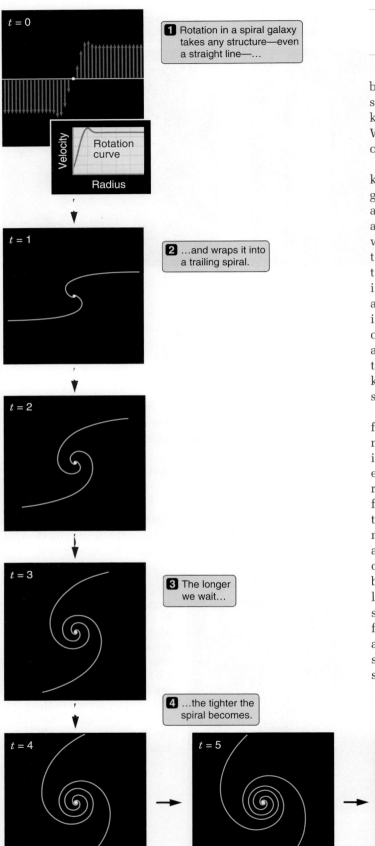

1 Rotation in a spiral galaxy takes any structure—even a straight line—...

Velocity

Rotation curve

Radius

$t = 0$

$t = 1$

2 ...and wraps it into a trailing spiral.

$t = 2$

$t = 3$

3 The longer we wait...

4 ...the tighter the spiral becomes.

$t = 4$

$t = 5$

$t = 20$

Gravitational interactions and star formation are processes that kick disks.

The way disk galaxies rotate implies that any disturbances or "kicks" to a galaxy's disk will naturally lead to spiral structure in that disk. One way that a galaxy can be kicked is by gravitational interactions with other galaxies. We will see much more of these interactions later in this chapter.

There are several ways in which a spiral galaxy can feel a kick. Just giving a galaxy disk a single kick (whether through gravitational interactions or star formation) will not produce a stable spiral-arm pattern. For the same reason that spiral arms form at all, those produced from a "one-shot" kick will wind themselves up completely in two or three rotations of the disk, then disappear. However, it turns out that some types of kicks are repetitive and so are capable of sustaining spiral structure indefinitely. Some come from within a galaxy itself. If the bulge in the center of a spiral galaxy is not spherically symmetric (as indeed seems to be the case for most spiral galaxies), then the bulge will produce a gravitational disturbance in the disk. As the disk rotates through this disturbance, it is subjected to the repetitive kick needed to trigger star formation and the formation of spiral structure.

Another spur to the formation of spiral structure comes from the process of star formation itself. Regions of star formation dump considerable energy into their surroundings in the form of UV radiation, stellar winds, and supernova explosions. All of this energy drives up the pressure in the region, compressing clouds of gas and triggering more star formation. Because many massive stars typically form in the same region at about the same time, their combined mass outflows and supernova explosions will occur one after another in the same region of space over the course of only a few million years. The result can be large, expanding bubbles of hot gas that sweep out cavities in the interstellar medium and concentrate the swept-up gas into dense star-forming clouds, much like the snow that piles up in front of a snowplow. In this way star formation can actually propagate through the disk of a galaxy. The resulting strings of star-forming regions can then be swept into spiral structures by rotation.

FIGURE 18.14 The differential rotation of a spiral galaxy will naturally take even an originally linear structure and wrap it into a progressively tighter spiral as time goes by.

Many galaxies show clear evidence of a relationship between the shapes of their bulges and the structure of their spiral arms. Barred spirals, for example, have a characteristic two-armed spiral pattern that is tied to the elongated bulge (see Figure 18.12). Even the bulges of galaxies that are

Regular disturbances lead to two-armed spirals.

not obviously barred may be nonspherical enough to contribute to the formation of two-armed spiral structure. Smaller galaxies in orbit about larger galaxies can also give rise to a periodic kick, triggering the same sort of two-armed structure.

Regular disturbances in the disks of spiral galaxies are called **spiral density waves**. We call them density waves because they are regions of greater mass density and increased pressure in the interstellar medium that move around a disk in the pattern of a two-armed spiral. Spiral density waves act like the spiral-shaped blade in a blender. Models of how spiral density waves form tell us that this spiral-shaped two-armed wave pattern does not necessarily rotate at the same rate as the rest of the galaxy. This means that as material in the disk orbits about, it passes through the spiral density waves.

The motions of stars are little affected as they pass through the spiral density wave. However, for gas it is a very different story. Consider what happens when you turn on the tap in your kitchen sink. The water hits the bottom of the sink and spreads out in a thin, rapidly moving layer. A few inches out, depending on the rate at which water is flowing, there is a sudden increase in the depth of the water

Spiral density waves compress gas, triggering star formation.

called a "hydrostatic jump." Spiral arms in galaxies work in much the same way. Gas flows into the spiral density wave and piles up like water in a hydrostatic jump. Stars form in the resulting compressed gas. Massive stars have such short lives (typically 10 million years or so) that they never get the chance to drift far from the spiral arms where they were born, and so that is where we see them. Less massive stars, on the other hand, have plenty of time to move away from their places of birth, forming a smooth underlying disk.

18.4 Galaxies Are Mostly Dark Matter

As we found earlier, although mass may play the dominant role in determining the properties of stars, the same is not true for galaxies. Even so, efforts to measure the masses of galaxies during the last decades of the 20th century led to some of the most remarkable and surprising findings in the history of astronomy. To understand this work, we first need to ask how to go about measuring the mass of a galaxy. One way is to look at the amount of light it gives off. We can use the spectrum of starlight from a galaxy to determine the types of stars the galaxy contains. We then use our knowledge of stellar evolution to turn the luminosity of the galaxy into an estimate of the total mass in stars. Finally, our knowledge of the physics of radiation from interstellar gas at X-ray, infrared, and radio wavelengths allows us to estimate the mass of these other components. Together the stars, gas, and dust in a galaxy are called **luminous matter**, or simply **normal matter**, because this matter emits electromagnetic radiation.

Looking at the light from a galaxy is not a sure way to determine its mass, however. Imagine, for example, if we were to replace the Sun with a black hole of the same mass. We would see no light coming from this black hole, and we would not include its mass in our estimate based on the luminosity of starlight from our galaxy. Yet the planets would continue on their orbits, moving under the influence of its gravity. At every point in this book we have relied on

Gravity and Kepler's laws are the sure way to measure galaxy mass.

measuring the effect of gravity on motion as the one sure way of determining the masses of objects. This time will be no different. The disks of spiral galaxies are rotating, which means that the stars in those disks are following orbits that are much like the Keplerian orbits of planets around their parent stars and binary stars around each other. To measure the mass of a spiral galaxy, all we need do is apply Kepler's laws, just as we did for those other systems.

We might begin our study of the rotation of spiral galaxies with the hypothesis that the mass in a galaxy is distributed in the same way as its light. That is, we could begin by assuming that the luminous mass in these galaxies is all the mass that there is. On the basis of this hypothesis, we would make a prediction. The light of all galaxies, including spiral galaxies, is highly concentrated toward their centers. If all the mass of a spiral galaxy were contained in its centrally concentrated stars, gas, and dust, we would predict that the orbital velocities of the gas and stars in the disk should behave like the orbital velocities of the planets in our Solar System. We should see fast orbital velocities close in and slower orbital velocities farther out.

Figure 18.15 shows the prediction of how the orbital velocities of material in the disk of a spiral galaxy should change with distance from the center of the spiral, assuming that the mass is distributed like the light. To test this prediction, we use the Doppler effect to measure orbital motions. There are several ways to do this. We might measure the velocities of stars from observations of absorption lines in their spectra, or we might measure the velocities

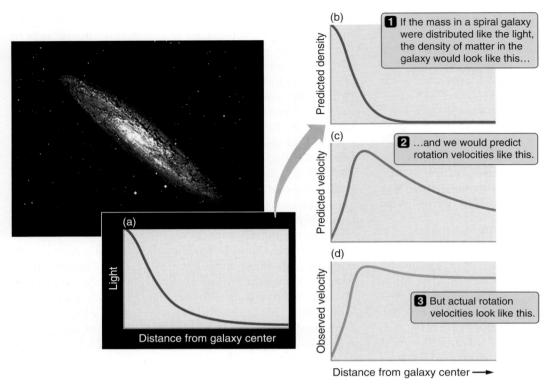

(b)

Predicted density

1 If the mass in a spiral galaxy were distributed like the light, the density of matter in the galaxy would look like this…

(c)

Predicted velocity

2 …and we would predict rotation velocities like this.

(d)

Observed velocity

3 But actual rotation velocities look like this.

Distance from galaxy center ⟶

(a)

Light

Distance from galaxy center

FIGURE 18.15 (a) The profile of the light in a typical spiral galaxy. (b) The mass density of stars and gas located at a given distance from the galaxy's center. If stars and gas accounted for all of the mass of the galaxy, then the galaxy's rotation curve should be as shown in (c), but real galaxies have observed rotation curves more like that shown in (d).

of interstellar gas using emission lines such as Hα emission or 21-cm emission from neutral hydrogen.

A graph that shows how orbital velocity in a galaxy varies with distance from the galaxy's center is called a **rotation curve**. Astronomers had to wait until the mid-1970s before telescope instrumentation permitted such rotation curves to be measured reliably outside the inner, bright regions of galaxies. Much to their surprise, when the observations were made, the prediction that mass in galaxies is distributed like the light from galaxies was falsified! Rather

Rotation curves of spiral galaxies are remarkably flat.

than finding rotation curves that fall off to lower and lower velocities in the outer parts of spiral galaxies, as predicted, observations instead showed that spiral galaxy rotation velocities remain about the same out to the most distant measured parts of the galaxies. As shown in Figure 18.15(d), the rotation curves of spiral galaxies appear level, or "flat," in their outer parts. These are naturally referred to as **flat rotation curves**. Observations of 21-cm radiation from neutral hydrogen even show that the rotation curves remain flat well outside the extent of the visible disks. This discovery came as a shock. It meant that long-held ideas about the distribution of mass in galaxies were wrong.

The rotation curve of a spiral galaxy allows us to directly determine how the mass in that galaxy is distributed.

All we need to do is apply Kepler's laws to these rotation curves and ask, How much mass must be present inside a given radius to account for the orbital velocity we measure at that radius? (Recall from Chapter 10 that only the mass inside a given radius contributes to the net gravitational force felt by an object. Strictly speaking, this is true only for spherically symmetric objects, but a spiral galaxy is symmetric enough for this to be a good approximation.) **Figure 18.16** shows the result of such a calculation. In addition to the centrally concentrated luminous matter, there must be a second component to these galaxies consisting of matter that does not show up in our census of stars, gas, and dust. This material, which reveals itself only by the influence of its gravity, is called **dark matter**.

We see the presence of dark matter in spiral galaxies because it is distributed differently from the starlight. Although both the starlight and the dark matter are concentrated toward the center of a galaxy, this is less true for dark matter than for luminous matter. The rotation curves of the inner parts of spiral galaxies match fairly well what would

Most of the mass comprising spiral galaxies is dark matter.

be predicted by their luminous matter, indicating that normal luminous matter dominates the inner part of spiral galaxies. Within the part of a galaxy that can be seen in visual light, the mix of dark and luminous matter is about half and half. However, rotation curves measured using 21-cm radiation from neutral hydrogen show that the outer parts of spiral galaxies are mostly dark matter. It is currently estimated that up to 95 percent of the total mass in a spiral galaxy consists of a greatly extended **dark matter halo**, far larger than the visible spiral portion of the galaxy located at its center. This is a startling statement. A spiral galaxy illuminates only the inner part of a much larger distribution of mass that is dominated by some type of matter we cannot see!

We measure the luminous matter in elliptical galaxies the same way we measure the luminous matter in spiral galaxies. However, elliptical galaxies do not rotate, so we need a different approach to measure their masses. Our earlier

---

Ellipticals' dark matter enables them to hold onto their hot gases.

---

discussion of planetary atmospheres in Chapter 8 provides just the tool we need. A planet's ability to hold onto its atmosphere depends on its mass. In like fashion, an elliptical galaxy's ability to hold onto its hot, X-ray-emitting gas depends on its mass. When the masses of elliptical galaxies are

inferred from X-ray images, such as the one in **Figure 18.17**, we discover the same thing that we found from the rotation curves of spirals. Elliptical galaxies contain up to 20 times as much mass as can be accounted for by their stars and gas alone, so they too are dominated by dark matter. As with spirals, the luminous matter in ellipticals is more centrally concentrated than is the dark matter.

The transition from the inner part of galaxies (where luminous matter dominates) to the outer parts of galaxies (which are dominated by dark matter) is remarkably smooth. This seamless combination of dark and luminous matter is an important piece of evidence that any successful theory of galaxy formation must explain.

So what is this dark matter of which galaxies are mostly made? We do not yet know. A number of suggestions have been made over the years, ranging from space filled with Jupiter-like objects, to swarms of black holes, to copious numbers of white dwarf stars, to exotic unknown elementary particles. (The last of these is currently the favored explanation.) If galaxies are formed mostly of dark matter, then have we been wasting our time paying so much attention to the small fraction of mass tied up in stars and the interstellar medium? No. For one thing, stars and gas are the only

FIGURE 18.16 The flat rotation curve of the spiral galaxy NGC 3198, along with the total mass within a given radius that can be accounted for by stars and gas, and the extra "dark" mass needed to explain the rotation curve. In addition to matter we can see, galaxies must be surrounded by halos containing a large amount of dark matter.

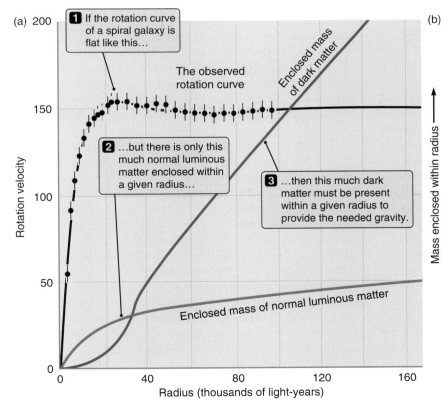
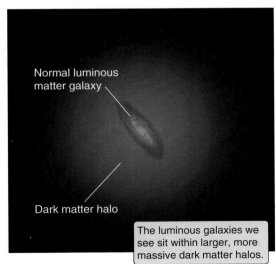

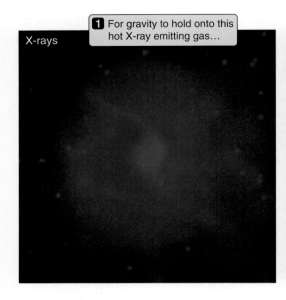

X-rays

1 For gravity to hold onto this hot X-ray emitting gas…

Visible light

2 …an elliptical galaxy must be much more massive than the combination of all stars we see.

3 Elliptical galaxies are mostly dark matter.

FIGURE 18.17 The X-ray emission from hot gas around an elliptical galaxy extends well beyond the region of visible light from stars.

parts of galaxies that we can see directly. Stars are also of undeniable importance from our human perspective. Stars formed the atoms in our bodies and are the kernels around which planetary systems form. The star we call our Sun supplies the energy that makes life on Earth possible. The attention we have paid to stars is well placed indeed!

18.5 There Is a Beast at the Centers of Galaxies

Galaxies are remarkable objects, each shining with the light of hundreds of billions of stars. However, galaxies themselves pale in comparison with the most brilliant beacons of all—**quasars**. *Quasar* is short for "quasi-stellar radio source," so named because they were first observed as unresolved points at radio wavelengths. Quasars are phenomenally powerful, pouring forth the luminosity of a trillion to a thousand trillion (10^{12}–10^{15}) Suns! Quasars are objects

Quasars are phenomenally luminous.

of the distant universe. The nearest quasar to us is approximately 1 billion light-years away. There are literally billions of galaxies that are closer to us than the nearest quasar. Because light travels at a finite speed, the distance to an object also tells us the amount of time that has passed since the light from that object left its source. The fact that the nearest quasar is seen as it existed about a billion years ago tells us that quasars are quite rare in the universe today. Quasars were once much more common. The discovery of quasars

(see **Excursions 18.1**) in the distant and therefore earlier universe provided one of the first pieces of evidence that the universe has evolved over time.

Quasars are not isolated beacons, but instead are centers of violent activity in the hearts of large galaxies, as seen in **Figure 18.18**. Quasars are now recognized as only the most extreme form of activity that can occur in the nuclei of galaxies. In today's universe approximately 3 percent of galaxies contain brilliant points of light in their centers that may outshine all of the stars in the galaxies that host them. Together quasars and their less luminous but still active cousins are called **active galactic nuclei**, or simply **AGNs**.

Seyfert galaxies, named after **Carl Seyfert** (1911–1960), who discovered them in 1943, are spiral galaxies that contain AGNs discernible in visible light at their centers. The luminosity of a typical Seyfert nucleus can be 10 billion to

There are several types of active galactic nuclei.

100 billion $L_\odot$, comparable to the luminosity of the rest of the galaxy as a whole. The luminosities of AGNs found in elliptical galaxies are similar to those of Seyfert nuclei (10 billion to 100 billion $L_\odot$). Unlike Seyfert nuclei, however, AGNs in elliptical galaxies are usually most prominent in the radio portion of the electromagnetic spectrum, earning them the name **radio galaxies**. Radio galaxies, and their distant, extremely luminous cousins the quasars, are often the sources of thin jets that extend outward millions of light-years from the galaxy, powering twin lobes of radio emission, such as those seen in **Figure 18.19**.

What could power such phenomenal cosmic beacons as quasars, Seyfert galaxies, and radio galaxies? Clues lie in the light they emit. **Figure 18.20** shows the electromagnetic spectrum of the well-studied quasar 3C273 from radio frequencies to high-energy gamma rays. This quasar is about 100 times as luminous as our entire Milky Way Galaxy and exhibits much stronger radio and gamma ray emission than is typical for AGNs. Much of the light from AGNs is synchrotron radiation. This is the same type of radiation that

---

AGNs emit synchrotron radiation.

---

we first encountered coming from Jupiter's magnetosphere and later saw again in such extreme environments as the Crab Nebula. Recall from Foundations 9.1 that synchrotron radiation comes from relativistic charged particles spiraling around the direction of a magnetic field (see Figure 9.17). The fact that AGNs accelerate large amounts of material to close to the speed of light indicates that they are very violent objects indeed. In addition to synchrotron radiation, the spectra of many quasars and Seyfert nuclei show emission lines that are smeared out by the Doppler effect across a wide range of wavelengths. This implies that gas in AGNs

is swirling around the centers of these galaxies at speeds of thousands or even tens of thousands of kilometers per second.

AGNs Are as Small as the Solar System

The enormous radiated power and mechanical energy of AGNs are mind-boggling on their own, but they are made even more spectacular by the fact that all this power emerges from a region that can be no larger than a light-day or so across. That is comparable in size to our own Solar System! How can we even make such a claim? Quasars and other AGNs appear only as unresolved points of light even in our most powerful telescopes. What information is there in the light we receive from AGNs that tells us they are such compact objects? For an answer we turn not to the sky but to the halftime show at a local football game.

EXCURSIONS 18.1

Quasars—When Conventional Thinking Failed

The story of the discovery of quasars provides an interesting insight into the discovery of new phenomena in astronomy and how conventional thinking can sometimes hinder progress. In the late 1950s radio surveys had detected a number of bright, compact objects that at first seemed to have no optical counterparts. Eventually, improved positions revealed that the radio sources coincided with faint, very blue, stellarlike objects. Astronomers, unaware of their true nature, called them "radio stars." Obtaining spectra of the first two radio stars was a laborious task, requiring 10-hour exposures with the only recording technique available in those days: slow photographic plates (see Chapter 5). Astronomers, hardly prepared for what they would see in these painstakingly obtained spectra, were greatly puzzled by the results. Rather than displaying the expected absorption lines that are characteristic of blue stellar objects, the spectra showed only a single pair of emission lines that were broad—indicating very rapid motions within these objects—and that did not seem to

correspond to the lines of any known substances. Puzzling indeed! For several years astronomers believed they had discovered a new type of star. After all, "if it looks like a star, it must be a star." Finally one astronomer, Martin Schmidt, realized what no one else had considered: that these broad spectral lines were, in fact, the highly redshifted lines of ordinary hydrogen. The implications were astounding! These "stars" were not stars. They were extraordinarily luminous objects at enormous distances.

Other "quasars," as they came to be known, were soon found by the same techniques. Many were relatively easy to identify because of their unusual blue color. As still more were found, astronomers did what astronomers always do: They began cataloging them. The young daughter of two of the catalogers misunderstood what her parents termed "quasi-stellar objects," hearing it instead as "crazy-stellar objects"—a not altogether unfitting description of this strange new phenomenon, and a source of much amusement in that household!

Figure 18.21 illustrates the unalterable bane of every marching band director. When a band is all together in a tight formation at the center of the field, the notes you hear in the stand are clear and crisp. The band plays together beautifully. But as the band spreads out across the field, its sound begins to get mushy. This is not because the marchers are poor musicians. Rather, it is a consequence of the fact that sound travels at a finite speed. The speed of sound on a cold, dry December day is around 330 m/s. At this speed it takes sound approximately a third of a second to

---

A marching band cannot play a clean note.

---

travel from one end of the football field to the other. Even if every musician on the field played a note at exactly the same instant in response to the director's cue, in the stands you would hear the musicians close to you first but would have to wait longer for the sound from the far end of the field to arrive.

If the band is spread from one end of the field to the other, then the beginning of a note will be smeared out over about a third of a second, or the difference in sound travel time from the near side of the field to the far side. If the band were spread out over two football fields, it would take about two-thirds of a second for the sound from the most distant musicians to arrive at your ear. If our marching band were spread out over a kilometer, then it would take roughly three seconds—the time it takes sound to travel a kilometer—for us to hear a sharply played note start and stop. Even with

our eyes closed, it would be easy to tell whether the band was in a tight group or spread out across the field.

Exactly the same principle works for AGNs; but here we are working with the speed of light, not the speed of sound. Quasars and other AGNs can change their brightness dramatically over the course of only a day or two. The

---

AGNs vary rapidly and so must be relatively small.

AGN powerhouse must therefore be no more than a light-day or so across because, if the powerhouse were larger, what we see could not possibly change in a day or two. Here is the image that should come to mind when you think of quasars: *the light of 10,000 galaxies pouring out of a region of space that would come close to fitting within the orbit of Pluto!*

Supermassive Black Holes and Accretion Disks Run Amok

When AGNs were first discovered, a variety of ideas were put forward to explain them. However, as their tiny sizes and incredible energy densities became clear, only one answer seemed to make sense. AGNs are powered by violent accretion disks surrounding **supermassive black holes** with masses from thousands to tens of billions of solar masses.

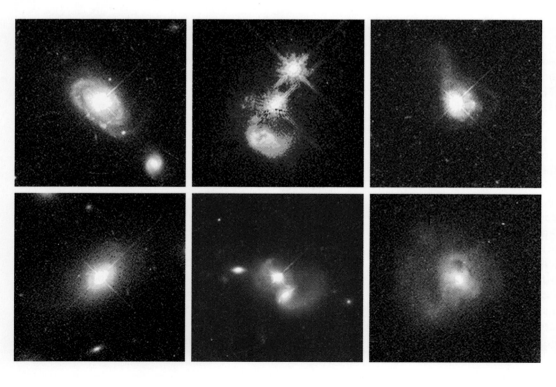

FIGURE 18.18 HST images of the environments around quasars. Quasars are found in the centers of galaxies. Those galaxies often show evidence of interactions with other galaxies.

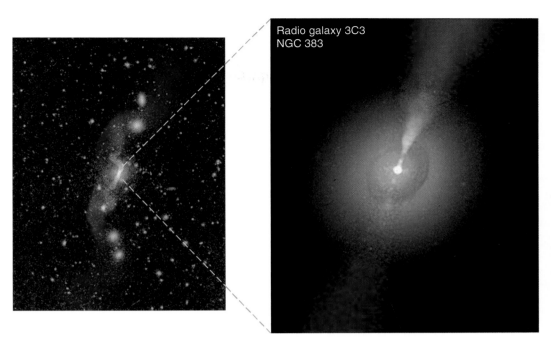

Radio galaxy 3C3
NGC 383

FIGURE 18.19 Radio emission from a double-lobed radio galaxy (shown in red) is superposed over an image of visible starlight from the galaxy (shown in blue). The lobes, powered by a relativistic jet streaming outward from the nucleus of the galaxy, are over 3 million light-years from the galaxy.

We have run across accretion disks on a number of occasions during our journey. Accretion disks surround young stars, providing the raw material for solar systems. Accretion disks around white dwarfs, fueled by material torn from their bloated evolving companions, lead to novae and Type I supernovae. Accretion disks around neutron stars and star-sized black holes a few kilometers across are seen

AGNs are powered by accretion onto supermassive black holes.

as X-ray binary stars. Take this mental picture and scale it up to a black hole with a mass of a billion solar masses and a radius comparable in size to the orbit of Uranus. Rather than small amounts of material being siphoned off a star, imagine an accretion disk fed by substantial fractions of entire galaxies. *That* is an active galactic nucleus.

This basic picture of a supermassive black hole surrounded by an accretion disk has been developed into a more complete physical description called the **unified model of AGNs**. The unified model takes its name from the fact that it attempts to unify our understanding of all AGNs —quasars, Seyfert galaxies, and radio galaxies—within the same framework of understanding.

Figure 18.22(a) shows a diagram of the various components of the unified model of AGNs. In the unified model, a supermassive black hole is surrounded by an accretion disk. Much farther out lies a large torus (doughnut) of gas and dust consisting of material that is feeding the central engine. Each of the different components of the unified model accounts for different observed properties of AGNs.

In our discussion of star formation, we learned how gravitational energy is converted to thermal energy as material moves inward toward the growing protostar. As material moves inward toward a supermassive black hole, conversion of gravitational energy heats the disk to hun-

FIGURE 18.20 Electromagnetic energy spectrum of the well-studied and bright quasar 3C273 from the low-frequency radio to high-energy gamma rays. Most of the energy emitted by this quasar lies between optical and gamma ray frequencies.

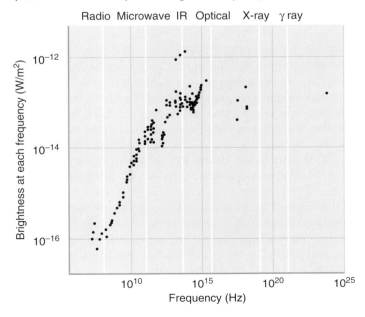

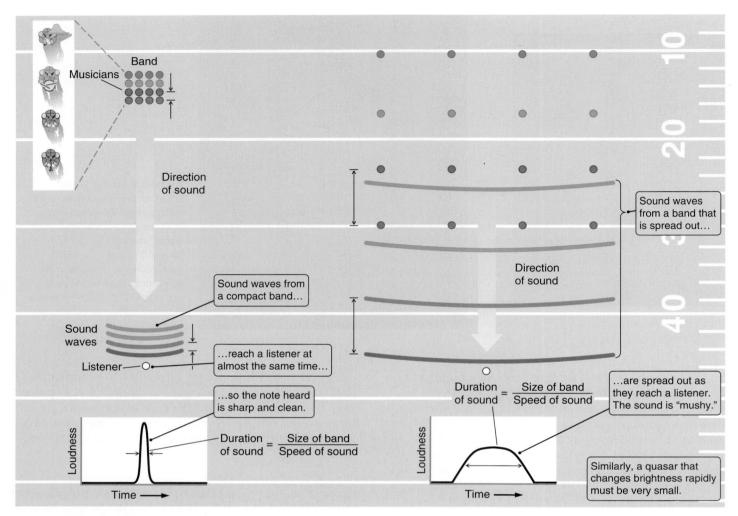

FIGURE 18.21 A marching band spread out across a field cannot play a clean note. Similarly, AGNs must be very small to explain their rapid variability.

dreds of thousands of kelvins, causing it to glow brightly in visible and ultraviolet light. Conversion of gravitational energy to thermal energy as material falls onto the accretion disk is also a source of X-rays, UV radiation, and other energetic emission. When we discussed the Sun, we marveled at the efficiency of fusion, which converts 0.7 percent

AGN accretion disks are phenomenally efficient at converting mass to energy.

of the mass of hydrogen into energy. In contrast, approximately 50 percent of the mass of infalling material around a supermassive black hole is converted to luminous energy. The rest of that mass is pulled into the black hole itself, causing it to grow even more massive.

The interaction of the accretion disk with the black hole gives rise to powerful radio jets—superpowerful analogs to the Herbig-Haro jets formed by accretion disks around young stellar objects (see Chapter 15). Throughout, twisted magnetic fields accelerate charged particles such as electrons and protons to relativistic speeds, accounting for the observed synchrotron emission. Gas in the accretion disk or in nearby clouds orbiting the central black hole at high speeds gives off emission lines that are smeared out by the Doppler effect into the broad lines seen in AGN spectra. This is the "central engine" that leads to AGNs.

The outer torus plays a somewhat different role in the unified model. Located far from the inner turmoil of the accretion disk, and far larger than the central engine, some of the outer torus is ionized by UV light from the AGN much as an H II region is ionized by the UV light from O stars. This provides one possible source for H II region–like emission lines seen in many AGNs. And of course, the outer torus *must* be there in some form if material is continuing to fall inward toward the accretion disk. Yet the most important conceptual role of the outer torus is that it allows a single model to explain many of the differences observed among various AGNs. The outer torus obscures our view of

> **The outer torus determines what we can see.**

the central engine in different ways depending on the angle from which we happen to see it. By invoking variations in the viewing angle, the mass of the black hole, and the rate at which it is being fed, the unified model of AGNs can account for a wide range of AGN properties, including the average spectra of AGNs, an example of which is shown in Figure 18.20.

What AGN We See Depends on Our Perspective

When we view the unified model edge-on, we see emission lines from the surrounding torus and other surrounding gas. We can also sometimes see the torus in absorption

> **Viewed edge-on, an AGN's accretion disk is not visible.**

against the background of the galaxy. **Figure 18.22(b)** is a Hubble Space Telescope image of the inner part of the galaxy NGC 7052, showing what appears to be just such a

FIGURE 18.22 (a) An illustration of the unified model of active galactic nuclei. The appearance of the object changes when the disk and torus are viewed (b) from the edge, (c) at higher inclination, and (d) close to face-on. Viewing angle, the mass of the central black hole, and the rate at which it is being fed are thought to determine the properties of any AGN that we see.

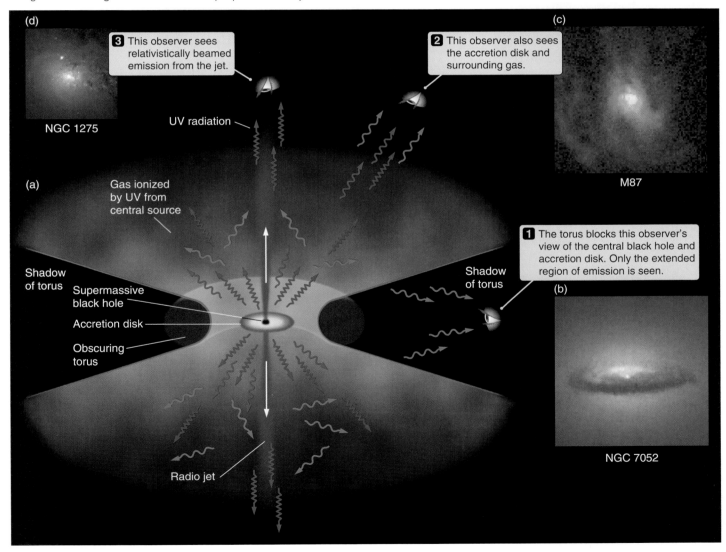

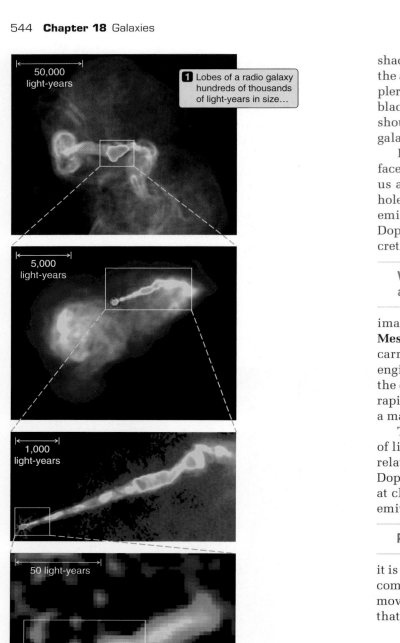

1 Lobes of a radio galaxy hundreds of thousands of light-years in size...

50,000 light-years

5,000 light-years

1,000 light-years

50 light-years

10 light-years

1 light-year

0.1 light-year

shadow. When viewing from this orientation, we cannot see the accretion disk itself, so we do not expect to see the Doppler-smeared lines that originate closer to the supermassive black hole. However, if jets are present in the AGN, we should be able to see these emerging from the center of the galaxy.

If we look at the inner accretion disk somewhat more face-on, then we can see over the edge of the torus, giving us a more direct look at the accretion disk and the black hole. In this case we should see more of the synchrotron emission from the region around the black hole and the Doppler-broadened lines produced in and around the accretion disk. **Figure 18.22(c)** shows a Hubble Space Telescope

When we view it more face-on, we see the accretion disk.

image of one such object called M87 (object number 87 in **Messier's catalog**). M87 is a source of powerful jets that carry on for millions of light-years but originate in the tiny engine at the heart of the galaxy (**Figure 18.23**). Spectra of the disk at the center of this galaxy (**Figure 18.24**) show the rapid rotation of material around a central black hole with a mass of 3 billion $M_\odot$.

The material in an AGN jet travels very close to the speed of light. As a result, what we see is strongly influenced by relativistic effects. One of these is an extreme form of the Doppler effect called **relativistic beaming**. Matter traveling at close to the speed of light concentrates any radiation it emits into a tight beam pointed in the direction in which

Relativistic effects influence what we see.

it is moving. As a result of relativistic beaming, an AGN jet coming toward us will look much brighter than its twin moving away from us. The unified model clearly predicts that AGN jets should be two-sided, but most of the time

FIGURE 18.23 The visible jet from the galaxy M87 originates in a tiny volume at the heart of the galaxy but extends over hundreds of thousands of light-years.

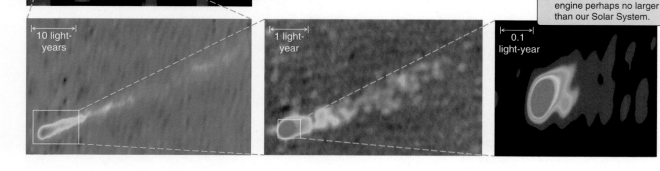

2 ...originate in a central engine perhaps no larger than our Solar System.

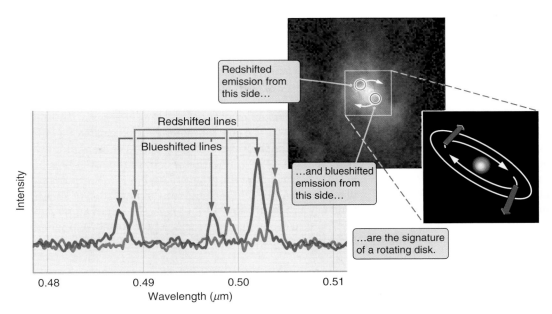

Redshifted emission from this side…

Redshifted lines

Blueshifted lines

…and blueshifted emission from this side…

…are the signature of a rotating disk.

FIGURE 18.24 An HST measurement of the rotation velocities of gas near the center of the nearby radio galaxy M87. The high velocities observed provide direct evidence of rotation about a supermassive black hole at this galaxy's center.

what we actually see appear to be one-sided jets. This is because we see only the portion of the jet that is being beamed toward us. The emission from the other portion of the jet always exists. We infer this from the fact that the radio lobes of radio galaxies are always two-sided. We do not see the jet moving away from us because it is beaming its radiation in the other direction.

In rare cases we happen to see the accretion disk in a quasar or radio galaxy almost directly face-on. When this happens, relativistic beaming dominates what we see. Rather than emission lines and other light coming from hot gas in the accretion disk, we are blinded by the bright glare of jet emission beamed directly at us (Figure 18.22(d)).

Relativistic beaming is not the only thing that makes appearances deceiving when we look down the barrel of an AGN jet. The material in an AGN jet is moving so close to the speed of light that the radiation it emits is barely able to outrun its source. As observers, we see all of the light emitted by the jet over thousands of years arrive at our telescopes over the course of only a few years. Time appears to be compressed. From our perspective the jet seems to travel great distances in brief periods of time. In extreme cases, such as the jet in M87 (**Figure 18.25**), features in the jet *appear* to be moving across the sky faster than the speed of light. We stress the word "appear" because this phenomenon, referred to as **superluminal motion**, is an optical illusion. Despite the name, nothing in these jets is actually traveling through space faster than the speed of light. Einstein's theory of special relativity remains safe.

It is worth noting that we have used the same galaxy, M87, as an example of several of the phenomena we have discussed. The presence of many different phenomena

FIGURE 18.25 Because the jet material in M87 is moving toward us at relativistic speeds, we see an optical illusion: The radio blobs appear to be moving apart at more than six times the speed of light.

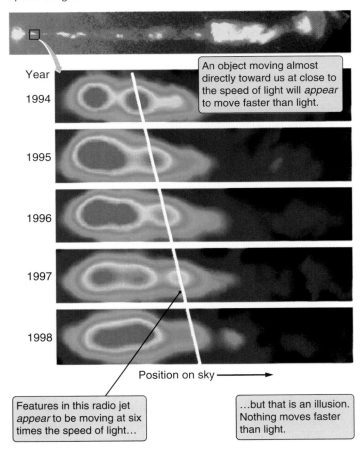

Year

1994

1995

1996

1997

1998

An object moving almost directly toward us at close to the speed of light will *appear* to move faster than light.

Position on sky ⟶

Features in this radio jet *appear* to be moving at six times the speed of light…

…but that is an illusion. Nothing moves faster than light.

predicted by the unified model in the same object is important support for the view that the model truly is *unified*.

Normal Galaxies and AGNs— A Question of Feeding the Beast

The unified model of AGNs has been around in some form since the 1980s. This model, constructed to explain existing observations of AGNs, has been modified frequently to account for new and better data. It is comforting that the unified model of AGNs has been able to keep up. No observations have come along that have forced us to abandon the model altogether. *On the other hand, the model was invented to explain observations of AGNs.* The unified model, therefore, cannot help but be consistent with those observations. However, scientific knowledge is not based on whether a theory is consistent with its own assumptions, but on experimental confirmation of a theory's testable predictions. What testable predictions does the unified model make, and how have those predictions fared in the light of new observational tests?

The essential elements of the unified model are a central engine (an accretion disk surrounding a supermassive black hole) and a source of fuel (gas and stars flowing onto the accretion disk). We commonly speak of the infall of material onto the accretion disk as "feeding the beast." If we were to shut off this infall—if we were to stop feeding the beast —the supermassive black hole would remain, absent all the fireworks. Without a source of matter falling onto the black hole, an AGN would no longer be active. If we were to look at such an object, we would see a normal (not active) galaxy with a supermassive black hole sitting in its center.

Should such objects be common? As we have noted, only about 3 percent of present-day galaxies contain AGNs. However, when we look at more distant galaxies (and therefore

> The unified model predicts that normal galaxies contain supermassive black holes.

look back in time), the percentage is much larger. Our observations show that when the universe was younger, there were many more AGNs than there are today. If the unified model of AGNs is correct, then all the supermassive black holes that powered those dead AGNs should still be around. If we combine what we know of the number of AGNs in the past with ideas about how long a given galaxy remains in an active phase, we are led by the unified model to predict that many—perhaps even *most*—normal galaxies today contain supermassive black holes!

This is a somewhat startling prediction. Quiescent collections of stars, gas, and dust like our own Milky Way should have slumbering beasts at their centers. It is a bit like suggesting that all of our dignified, soft-spoken grandmothers were once members of biker gangs. Yet here is a prediction of the unified model that can be tested.

If supermassive black holes are present in the centers of normal galaxies, they should reveal themselves in a number of ways. For one thing, the presence of such a concentration of mass at the center of a galaxy should draw surrounding stars close to it. The central region of such a galaxy would be much brighter than could be explained if stars alone were responsible for the gravitational field in the inner part of the galaxy. Stars feeling the gravitational pull of a supermassive black hole in the center of a galaxy should also orbit at very high velocities. We should therefore see large Doppler shifts in the light from stars near the centers of normal galaxies. Evidence of this sort has now been found in every normal

> Supermassive black holes have been discovered in the nuclei of many nearby galaxies.

galaxy with a substantial bulge in which a careful search has been conducted. The masses inferred for these black holes range from 10,000 $M_\odot$ (for a "small" black hole) to 5 billion $M_\odot$ (a "gargantuan" one)! The mass of the supermassive black hole seems to be related to the mass of the elliptical galaxy or spiral galaxy bulge in which it is found. At the beginning of the 21st century we have come to accept that all large galaxies probably contain supermassive black holes. These observations confirm the most fundamental prediction of the unified model. They also tell us something remarkable about the structure and history of normal galaxies.

Apparently the only difference between a normal galaxy and an active galaxy is whether the supermassive black hole at its center is being fed at the time we see that galaxy. The fact that only 3 percent of present-day galaxies have AGNs does not indicate which galaxies have the potential for AGN activity. Rather, it indicates which galaxy centers are being lit up at the moment. If we were to drop a large amount of gas and dust directly into the center of any large galaxy, this material would fall inward toward the central black hole, forming an accretion disk and a surrounding torus. The predicted result of this process is that the nucleus of this galaxy would change into an AGN.

Mergers and Interactions Make the Difference

Galaxies do not exist in isolation. Even our own Milky Way has several neighbors. In Chapter 10 we discussed tidal interactions between galaxies. **Figure 18.26** shows the havoc that such interactions can cause, pulling interacting galaxies into distorted shapes in which stars and gas are drawn out into sweeping arcs and tidal tails (also see Figures 5.31 and 10.14). Sometimes when galaxies interact, they pass by each other and go their separate ways. Such interactions can trigger the formation of spiral structure, and they can also slam clouds of interstellar gas together at high speeds, triggering additional star formation. Tidally distorted or interacting galaxies often contain regions of vigorous ongoing star formation. Sometimes when two galaxies interact, they merge to form a single, larger galaxy. This turns out to be an important part of the process of galaxy formation, as we will learn in Chapter 21.

Interactions and mergers must have been much more prevalent in the past to account for the many large galaxies that we see today. The prevalence of interacting galaxies when the universe was younger explains the large number

Galaxy–galaxy interactions fuel AGN activity.

of AGNs that existed in the past. Computer models show that galaxy–galaxy interactions can cause gas located tens of thousands of light-years from the center of a galaxy to fall inward toward the center, where it can provide fuel for an AGN. During mergers, a significant fraction of a cannibalized galaxy might wind up being fed to the beast. HST images of quasars, such as those in Figure 18.18, often show that quasar host galaxies are tidally distorted or are surrounded by other visible matter that is probably still falling into the galaxies. The most violent forms of AGN activity were likely most common in the early universe because that was the time when galaxies were forming, and large amounts of matter were constantly being drawn in by the gravity of newly formed galaxies. This process is still at work today. Galaxies that show evidence of recent interactions with other galaxies are far more likely than average to house AGNs in their centers.

There are still many puzzles. For example, the unified model has not been developed to the point that it predicts

FIGURE 18.26 These tidally interacting galaxies show severe distortions, including stars and gas drawn into long tidal tails.

how long an outburst of AGN activity will last, or how often galaxies will undergo episodes of AGN activity. To answer these questions the unified model of AGNs will have to be combined with much better models of galaxy formation and evolution than we currently have. There are also some observations that the unified model of AGNs does not yet explain. For instance, the unified model does not account for why one quasar can be a very powerful radio source while another, identical in all other respects, is radio-quiet, even when observed with the most sensitive radio telescopes.

Our understanding of AGNs is far from complete. Even so, the unified model has had many successes—enough for us to say with confidence that any large galaxy, including our own, is only a chance encounter away from becoming an AGN. How different our own sky might be if, a few tens of millions or hundreds of millions of years from now, our descendants look toward the center of our galaxy and see the brilliant light of a powerful Seyfert nucleus blazing forth.

Summary

- There are at least hundreds of billions of galaxies in our own observable universe.

- The shapes of galaxies and the types of orbits for their stars determine their morphological type or their Hubble classification.

- Star formation is currently occurring in the disks of spiral galaxies but not in ellipticals or S0 galaxies.

- Spiral arms are regions of intense star formation, and the arms are visible because of the concentration of bright young stars.

- Most of the mass in galaxies does not reside in gas, dust, or stars; rather about 90 percent of a galaxy's mass is in the form of dark matter, which does not emit or absorb light to any significant degree.

- Most, and perhaps all, large galaxies have supermassive black holes at their centers.

- When gas accretes onto one of these supermassive black holes, the center of the galaxy becomes an active galactic nucleus, which can emit as much as a thousand times the light of the whole galaxy, all coming from a region the size of our solar system.

Seeing the Forest through the Trees

A century is but a blink of an eye compared with the age of Earth or even the age of our species. Even so, a century can seem like forever when viewed from the perspective of the changes that a century brings. Such is the story of 20th century astronomy. A hundred years ago, humankind knew virtually nothing of the true answers to the most basic questions we might ask about the universe. Virtually all that we know of these matters we have learned within living memory. The discovery and study of galaxies is one of the major threads running throughout the story of that century.

From the time that Copernicus first dislodged Earth from the center of creation, each new discovery has pushed us further and further from the human-centered conceptions that shaped our view of the world for millennia (and persist in many ways to this day). The discovery of the almost unthinkable size of our own Milky Way and the realization that the universe contains countless more galaxies comparable to our own were huge steps in humanity's expanding awareness of the universe. The work of Shapley, Hubble, and others did far more than merely answer a few arcane and esoteric scientific questions. These giants of 20th century science tore away some of the last vestiges of a curtain that had hidden the reality of existence from the human mind. These scientists forever changed the way we must view everything, including ourselves.

The scale and distance of galaxies may have shattered preconceptions about the extent of the universe, but galaxies themselves seemed at first to be composed of familiar objects. The zoo grew far larger, but the animals themselves seemed the same. Stars in distant galaxies shine in accordance with the same laws that govern the Sun. Galaxies are held together by the same force of gravity that set the course of the cannonball in Newton's famous thought experiment. But as our understanding grew and new observations were made, observations of galaxies forced astronomers to push our knowledge of physics in extreme and unexpected ways.

When we first learned of black holes, we were faced with the thought of several solar masses compressed into a region only a few kilometers across, and we had to confront ideas of general relativity that turn our everyday notions of space and time inside out. Now we discover that such star-sized black holes are the merest of specks compared with the monsters residing in the centers of large galaxies. We have witnessed the consequences when these behemoths, with masses as great as billions of times that of the Sun, are fed with gas supplied by collisions between galaxies. Radio observations of the sky show us relativistic jets stretching across millions of light-years of intergalactic space. Yet the source of such a jet is tiny. The light of a quasar outshines a thousand galaxies—a beacon that can be seen from the very edge of the universe—yet originates from a region no larger than our Solar System. As we continue our journey we will find that such a monster lurks at the heart of our own mundane Milky Way, waiting (perhaps forever) for its next meal.

Active galaxies stir the imagination, but one of the most startling and fundamental results of our study of galaxies comes from the simple application of Newton's derivation of Kepler's laws to the motions of stars in galaxies. We have used this technique over and over on our journey to measure the mass of planets and stars. When we apply this comfortable tool to galaxies, however, the results are shocking. The matter we see in the stars, gas, and dust is but the tip of a much larger iceberg, the rest of which is composed of a substance known only as dark matter. Most of the matter in the universe consists of we know not what.

The 20th century saw dramatic progress in building a physical understanding of the formation, evolution, and death of stars. We can use the laws of physics to pull back the layers of a star, peer into the heart of a supernova, or run the clock forward and see our Sun's ultimate fate. No such understanding exists for galaxies. This is nothing to apologize for. Remember how little time has passed since galaxies were even recognized for what they are. It is important to remember that the study of galaxies is still in its infancy. Many galactic astronomers still spend their lives comparing and contrasting morphological traits, much as Linnaeus sorted living things into kingdoms without real understanding of how those kingdoms arose or what they signified. It is unlikely there will ever be a simple scheme that does for galaxies what the H-R diagram did for stars. On the other hand, astronomers are beginning to better understand what the "ecology" within galaxies is like and to piece together something of how they form and evolve. In the next chapter we take a step down this road by looking in more detail at the galaxy we know best—our own Milky Way.

Key Terms

galaxy, p. 524
universe, p. 524
spiral galaxy, p. 525
elliptical galaxy, p. 526
irregular galaxy, p. 526
bulge, p. 528
S0 galaxy, p. 529
dwarf galaxy, p. 531
giant galaxy, p. 531
spiral density wave, p. 535
luminous matter, p. 535
normal matter, p. 535
rotation curve, p. 536
flat rotation curve, p. 536
dark matter, p. 536
dark matter halo, p. 537
quasar, p. 538
active galactic nucleus (AGN), p. 538
Seyfert galaxy, p. 538
radio galaxy, p. 538
supermassive black hole, p. 540
superluminal motion, p. 545

Student Questions

THINKING ABOUT THE CONCEPTS

1. Name and describe some objects that appear so dissimilar when viewed from different angles that you might think from these different views that they are actually different objects.

2. What are the principal differences between spiral and elliptical galaxies?

3. What determines the shape of an elliptical galaxy?

4. Explain why star formation in spiral galaxies takes place mostly in the spiral arms.

5. Some galaxies have regions that are relatively blue in color, while other regions appear redder. Aside from color, what can you say about the differences between these regions?

6. What evidence do we have that most galaxies are composed largely of dark matter?

7. It is likely that most galaxies contain supermassive black holes, yet in many galaxies there is no obvious evidence for their existence. Why do some black holes reveal their presence while others do not?

8. What distinguishes a "normal" galaxy from one we call "active"?

9. The nearest quasar is about a billion light-years away. Why do we not see any that are closer?

10. Describe what must be happening at the centers of galaxies that contain AGNs.

APPLYING THE CONCEPTS

11. Assume the Sun is located 27,000 light-years (2.6×10^{17} km) from the center of the Milky Way Galaxy and is moving along a circular orbit at a speed of 220 km/s. How long does it take our Solar System to make one complete circuit around our galaxy?

12. A disk star has an orbital period of 3×10^8 years at an average distance of 2.4×10^4 light-years (which equals 1.5×10^9 AU) from the center of its galaxy. Remember that Newton's explanation of Kepler's third law has mass $(M_\odot) = [A(\text{AU})]^3/[P(\text{yr})]^2$.
    a. What is the mass of the galaxy, in units of solar masses, that lies inside the orbit of this star?
    b. If luminous energy implies 2.5×10^9 solar masses of visible matter inside the star's orbit, what is the ratio of dark matter to visible matter?

13. Assume that there are 1 trillion (10^{12}) galaxies in the universe, that the average galaxy has a mass equivalent to 100 billion (10^{11}) average stars, and that an average star has a mass of 10^{30} kg.
    a. Ignoring dark matter, how much mass (in units of kg) does the universe contain?
    b. If the mass of an average particle of normal matter is 10^{-27} kg, how many particles are there in the entire universe?

14. A quasar has the same brightness as a foreground galaxy that happens to be 6 million light-years distant. If the quasar is 1 million times more luminous than the galaxy, what is the distance of the quasar?

15. You read in the newspaper that astronomers have discovered a "new" cosmological object that appears to be flickering with a period of 83 minutes. Having read *21st Century Astronomy,* you are able to quickly estimate the maximum size of this object. How large can it be?

16. A solar-type star (mass $= 2 \times 10^{30}$ kg), accompanied by its retinue of planets, approaches a supermassive black hole. As it crosses the event horizon, half of its mass falls into the black hole, while the other half is completely converted to luminous energy.

a. As it signals its demise in a burst of electromagnetic radiation, how much energy (in units of joules) does the dying solar system send out to the rest of the universe?

b. This is one possible way that quasars emit energy. If a luminous quasar has a luminosity of 2×10^{41} joules/second, how many solar masses per year does this quasar consume to maintain its average energy output?

17. Say that the number density of galaxies in the universe is, on average, 3×10^{-68} galaxies/m³. If astronomers could observe all galaxies out to a distance of 10^{10} light-years, how many galaxies would they find?

StudySpace
wwnorton.com/astro21
provides a Study Plan for each chapter that includes a reading outline, animations, keyword flash cards, and gradebook-enabled multiple-choice quizzes. From StudySpace you can also access premium content in the ebook and SmartWork.

O Milky Way, sister in whiteness
To Canaan's rivers and the bright
Bodies of lovers drowned,
Can we follow toilsomely
Your path to other nebulae?

GUILLAUME APOLLINAIRE (1880–1918)

An all-sky view of
the Milky Way and
Comet Hyakutake.

CHAPTER 19

The Milky Way—
A Normal Spiral Galaxy

19.1 We Look Up and See Our Galaxy

We live in a universe full of galaxies of many sizes and types, visible in our most powerful telescopes all the way to the edge of the observable universe. Yet when we go outside at night away from city lights and look up, it is not this universe of galaxies that we see. Rather, the night sky is filled with a single galaxy—our home, the galaxy we call the Milky Way. With what we know of galaxies from the previous chapter, there is a great deal that the appearance of the night sky can tell us about our local galaxy. **Figure 19.1(a)** shows a complete picture of what the sky looks like

We live in a barred spiral galaxy called the Milky Way.

as seen from Earth, while **Figure 19.1(b)** shows an image of an edge-on spiral galaxy. Comparison of the two can leave little doubt that we live in the disk of such a spiral. The flattened disk of the Milky Way is obvious when we look at the sky, once we appreciate what we are looking at. We can even see dark bands where clouds of interstellar gas and dust obscure much of the central plane of our galaxy. This single look tells us a great deal about our galaxy, but there is much that it does not tell us. There are many things that we know about galaxies only from our experience with the Milky Way. We see it from a much closer perspective than we do any other galaxy. We see it from the inside!

At the same time, there are real disadvantages in the perspective that we have on the Milky Way Galaxy. Buried

KEY CONCEPTS

Of the hundreds of billions of galaxies in the universe, the one that means the most to us is our cosmic home, the Milky Way. The Milky Way may be just another galaxy, but it is the only galaxy we can study at close range. As we focus our attention on our galaxy we will learn

- How variable stars in globular clusters are used as standard candles, allowing us to measure the size of the Milky Way.

- How Doppler-shifted radio emission from gas throughout the rotating disk of the Milky Way allows us to map the galaxy's structure.

- That the Milky Way is a typical giant barred spiral galaxy, and like all such galaxies, it is composed mostly of dark matter.

- How the chemical composition of the Milky Way has evolved with time.

- What differences in age and chemical composition of groups of stars tell us about the history of star formation in our galaxy.

- About the environment within the disk of the Milky Way, and the halo of stars, globular clusters, and dark matter that surrounds our galaxy.

- About the black hole at the center of our galaxy.

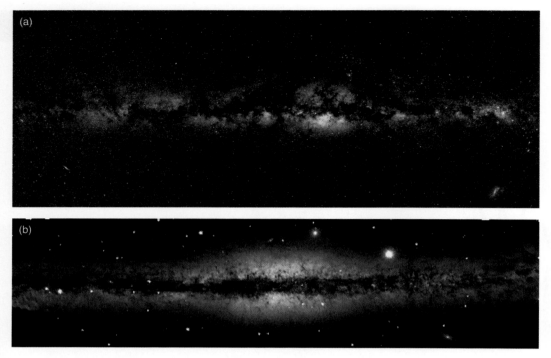

FIGURE 19.1 (a) The night sky as viewed from Earth. (b) The edge-on spiral galaxy NGC 891, whose disk greatly resembles the Milky Way.

within the disk of the Milky Way, we lack a bird's-eye view of our galactic home. The situation is further complicated by the dust in the surrounding interstellar medium that limits our view. **Figure 19.2** shows a model of the structure of the Milky Way proposed in the early 1800s by William Herschel. Herschel did not know about the interstellar ex-

---

Dust obscures our view of much of our own galaxy.

---

tinction of starlight, the properties of stars, or the relationship between our galaxy and the "spiral nebulae" that he studied throughout his life. His model, which bears little resemblance to our modern understanding of the Milky Way, was based simply on counting the number and brightness of stars seen in different directions.

Early models of our galaxy are of interest today mostly as historical curiosities. Even so, today's astronomers must cope with the same suite of difficulties that stood in the way of 19th and early 20th century astronomers. What type of spiral galaxy do we live in? Are the spiral arms prominent? Is the bulge barlike, and is it large or small relative to the disk? The questions that are so easy to ask and answer for distant galaxies are far more difficult to address for our own galaxy. Yet at this stage in our journey, we have already learned far more about our galaxy than about any other. In fact, everything that we have discussed in this book up until the previous chapter falls in the category of "things we know about the Milky Way." Stars, planets, and the interstellar medium—almost all that we know about

these we have learned within the context of our own galactic home. It is now time to make this home the focus of our study. It is time to merge the perspective of our look outward at the universe of other galaxies with our knowledge of our own locality to better understand our Milky Way as a spiral galaxy.

19.2 Measuring the Milky Way

One of the more difficult practical issues faced by astronomers is determining the distances to objects in the sky. In Chapter 13 we learned how geometrical parallax is used to measure the distances to nearby stars, but this got us out to distances of only a few hundred light-years. In the previous chapter we mentioned that Harlow Shapley determined the size of the Milky Way and the effect his measurement had on astronomy. We also mentioned Hubble's discovery of Cepheid variables in nearby galaxies, and we discussed galaxies that are many millions or even billions of light-years distant. Yet for the most part the important question of how we actually *know* these distances was temporarily swept under the rug. The difficult question of measuring distances in the universe will be a theme in much of the rest of our journey, and it is an issue that must now come to the foreground.

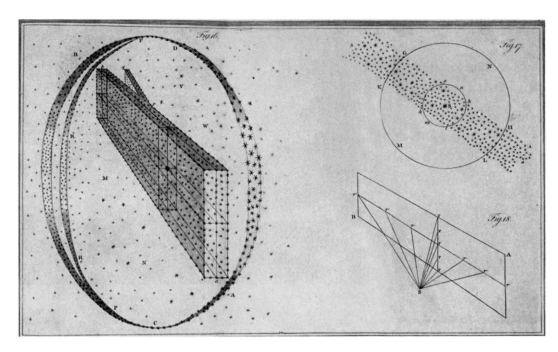

FIGURE 19.2 Two models of the Milky Way proposed by William Herschel in the late 1700s and early 1800s.

Globular Clusters and the Size of the Milky Way

The key to discovering the size of the Milky Way Galaxy turned out to be the presence of **globular clusters** in our galaxy. A globular cluster, such as the one in **Figure 19.3**, is a large spheroidal group of stars held together by gravity. Many clusters can be seen through small telescopes. At first glance they look something like small elliptical galaxies, and the analogy is not a bad one. The motions of stars within a globular cluster are much like the motions of stars within an elliptical galaxy. However, globular clusters are quite different from elliptical galaxies in size and in concentration of stars. There are more than 150 cataloged globular clusters in our galaxy (and likely many more—dust in the disk of our galaxy may hide them from view).

> Globular clusters are very luminous and easy to identify at great distances within the galaxy.

The known globular clusters have luminosities ranging from a low of 400 $L_\odot$ to a high of around 1 million $L_\odot$. A typical globular cluster consists of 500,000 stars packed into a volume of space with a radius of only 15 light-years. In contrast, we find only about 50 stars within that distance from our own Sun. Globular clusters are therefore much denser concentrations of stars than occur on average throughout our galaxy. On the other hand, our galaxy has over 100,000 times as many stars as a typical globular cluster.

About a quarter of the globular clusters in the Milky Way reside in or near the disk of our galaxy. The rest of them occupy a large volume of space surrounding the disk and bulge, referred to as the **halo** of the Milky Way. Globular

FIGURE 19.3 A Hubble Space Telescope image of the globular cluster M80.

clusters offer two distinct advantages that make them relatively easy to study. First, they are very luminous and so can be easily seen to great distances. Second, because many globular clusters lie outside the disk of the Milky Way, we can see them at great distances without encountering much absorption due to the obscuring dust within the disk. However, simply being able to see globular clusters does not necessarily put us any closer to measuring their distances. To do that we need to look at the properties of the stars they contain.

This is not the first time we have been confronted with measuring the distance to remote objects. If we know the luminosity of a star and can measure its brightness, then we can use the inverse square law of radiation to determine its distance. We discussed the inverse square law at length in Chapter 4. Rearranging that law for our present purpose, we have

$$d = \sqrt{\frac{L_{\text{star}}}{4\pi b_{\text{star}}}}.$$

In words, the distance of a star is proportional to the square root of the ratio of its luminosity (L_{star}) to its brightness (b_{star}) as seen from Earth.

Some types of stars are especially useful for determining distances. These are stars that are both very luminous (so that they can be seen at great distances) and have *known*

We use stars of known luminosity as standard candles to measure distances.

---

luminosities. Such stars are referred to as **standard candles**.[1] The story of measuring distances to remote objects has largely involved the search for better, more luminous standard candles.

Do globular clusters contain good standard candles? **Figure 19.4** shows the H-R diagram for the stars in globular cluster M92. The main sequence turnoff in this cluster's H-R diagram occurs for stars with masses of around 0.8 $M_\odot$, which (for the low abundance of heavy elements in the stars in this globular cluster) corresponds to a main sequence lifetime of close to 13 billion years. The age of this globular cluster is similar to those of many others, making them the oldest objects known in our galaxy or in any nearby galaxy. Globular clusters must have formed when the universe and our galaxy were very young. Compared to globular cluster stars, our Sun at 5 billion years old is a relatively young member of our galaxy.

In Chapter 17 we found that there is a region in the H-R diagram, called the *instability strip,* in which stars pulsate. As they do so, their luminosity changes. In an old cluster like a globular cluster, the horizontal branch of the H-R

---

[1] This term is borrowed from an old unit of light intensity that is actually based on candles.

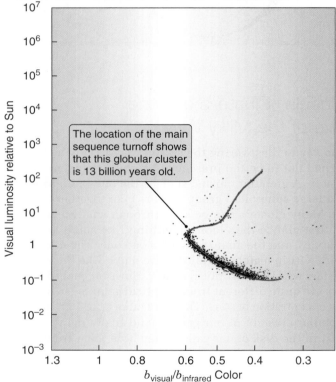

The location of the main sequence turnoff shows that this globular cluster is 13 billion years old.

FIGURE 19.4 The globular cluster M92 and an H-R diagram of the stars it contains. (In this instance the color plotted is based on the ratio of visible to infrared light.) The main sequence turnoff at 0.8 $M_\odot$ indicates that the cluster is about 13 billion years old.

diagram crosses the instability strip. Recall that horizontal branch stars that lie in the instability strip are variable stars called *RR Lyrae stars.* RR Lyrae stars in globular clusters are easy to spot because they are very luminous (horizontal branch stars are giant stars) and because their periodic changes in brightness signal their identity. As

Nightfall

The famous science and science fiction writer Isaac Asimov (1920–1992) imagined what might happen to a civilization on a planet orbiting within a system of six stars, located in the heart of a giant globular cluster. His short story *Nightfall* has become one of the more famous works of science fiction. On Asimov's fictional planet Lagash, at least one of its six stars is almost always above the horizon. "Nightfall" occurs on Lagash only once every 2,049 years. The story tells of the great madness that afflicts the inhabitants on this one night as they recoil in fear from a sky filled with hundreds of thousands of bright stars.

Asimov's story leads us to consider how our perspectives as human beings are formed by the circumstances in which we live. Yet we need not go to science fiction to ask these kinds of questions. Our perspective on the universe changes with time. At the moment we are traversing an open, rather dust-free part of the Milky Way. On a moonless night we see a dark sky and gaze with our telescopes into a universe of galaxies, but that will not always be the case. Just "down the road," astronomically speaking, are dark interstellar clouds and star-forming regions filled with glowing gas and obscuring dust through which the Sun and its entourage of planets will occasionally pass. How different would our view of the universe be if, instead of a dark sky, we looked up each night and saw a sky filled with a soft green glow, punctuated by a few points of intense light? How much different would our history be if at some point during the rise of our civilization we had suddenly emerged over the course of just a few years from within a molecular cloud and gotten our first look at the larger universe?

Globular clusters contain RR Lyrae standard candles.

with Cepheid variables, the time it takes for an RR Lyrae star to undergo one pulsation is related to the star's luminosity. Harlow Shapley used Henrietta Leavitt's determination of this period–luminosity relationship to determine the luminosities of RR Lyrae stars in globular clusters, then used the inverse square law of radiation to combine these

Harlow Shapley used globular clusters to determine the size of our galaxy.

luminosities with measured brightnesses to determine the distances to globular clusters. Shapley then cross-checked himself by noting that more distant clusters (as measured with his standard candle) also tended to appear smaller in the sky, as expected.

Knowing both the distances to globular clusters and where they appear in the sky, Shapley was able to make a three-dimensional map of the distribution of globular clusters in space. This map showed that globular clusters occupy a roughly spherical region of space with a radius of about 300,000 light-years! These globular clusters trace out the halo of the Milky Way Galaxy, as shown in **Figure 19.5**, which reflects the modern view of the globular cluster distribution.

The globular clusters around the Milky Way are moving about under the gravitational influence of the galaxy, just as stars in an elliptical galaxy move about under the gravitational influence of that kind of galaxy. Symmetry therefore requires that the center of the distribution of globular clusters coincide with the gravitational center of the galaxy itself. Shapley realized that because he could determine the distance to the center of this distribution, he had actually determined the size of the Milky Way itself.

Figure 19.6 identifies the disk, bulge, and halo of the Milky Way. Modern determinations indicate that the Sun is located about 27,000 light-years from the center of the galaxy, or roughly halfway out toward the edge of the disk. Armed with strong evidence, Shapley confidently presented the then-new, greatly expanded galaxy in 1915. It took all of human history up to 1610 to go from an Earth-centered universe to a Sun-centered Solar System. It took 305 years to go from a small galaxy to a large galaxy. Yet it was barely a decade later that Edwin Hubble, using Cepheid stars and the period–luminosity relationship, proved that Shapley's enormous Milky Way is but one of billions of galaxies in the universe.

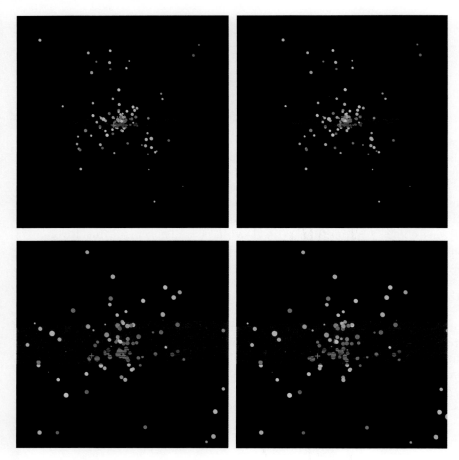

FIGURE 19.5 Stereoscopic views of the distribution of globular clusters in the Milky Way Galaxy. Colors represent the average colors of stars in each cluster. The green cross shows the location of the Sun. The light gray area represents the plane of the galaxy. (See Figure 13.1 for viewing instructions.)

center of the galaxy we see that on one side hydrogen clouds are moving toward us, while on the other side clouds are moving away from us. This is a pattern we have seen before—it is the pattern of the rotation velocity of gas in a disk. The only difference is that instead of looking at it from outside, we see our own galaxy's rotation curve from a vantage point located within—and rotating with—the galaxy. In other directions the velocities we see are complicated by our moving vantage point within the disk and so are more difficult to interpret at a glance. Even so, observed velocities of neutral hydrogen allow us to measure our galaxy's rotation curve and even determine the structure present throughout the disk of our galaxy.

Figure 19.8 shows a reconstruction of the Milky Way based on 21-cm data and other observations. Our galaxy has a modest bulge, which is barlike, as indicated by infrared studies of the distribution and motions of stars toward the center of our galaxy. Spiral arms sweep through the galaxy's disk, just like the arms we saw in external spiral galaxies. If we put all of the available information together, we conclude that the Milky Way is a middle-of-the-road giant barred spiral. From outside our galaxy probably looks much like the galaxy shown in **Figure 19.9**, and it would be placed about halfway along the right tine of the Hubble tuning fork diagram (see Figure 18.5). In fact, we would classify the Milky Way as an SBbc galaxy.

Rotation of the Galaxy Is Measured Using 21-cm Radiation

Clearly there is much we would like to know about the Milky Way, apart from its size and the fact that it is a spiral galaxy. However, as noted many times, there is a problem: We live inside the dusty disk of the galaxy, which badly obscures the visible-light view of our own galaxy. If you go out at night and look in the direction of the center of our galaxy (located in the constellation Sagittarius), instead of a bright spot you will see a dark lane of dusty clouds. To probe the structure of our galaxy, we must therefore use long-wavelength infrared and radio radiation that can penetrate the disk without being affected much by dust. The most powerful tool for this work is the same 21-cm line from neutral interstellar hydrogen that we used in the previous chapter to measure the rotation of other galaxies.

Figure 19.7 is a map (another name for a plot or graph) of the velocities of interstellar hydrogen measured from

> 21-cm Doppler velocities show that we live inside a rotating galaxy.

21-cm radiation, shown as a function of the direction in which we are looking. Looking in the region around the

The Milky Way Is Mostly Dark Matter

Figure 19.10 shows the rotation curve of the Milky Way as inferred primarily from 21-cm observations. The outermost

> Our galaxy has a flat rotation curve.

point in the rotation curve, at a distance of roughly 160,000 light-years from the center of the galaxy, is obtained from the orbital motion of the nearby dwarf galaxy called the Large Magellanic Cloud. As we have come to expect for spiral galaxies, the Milky Way has a fairly flat rotation curve.

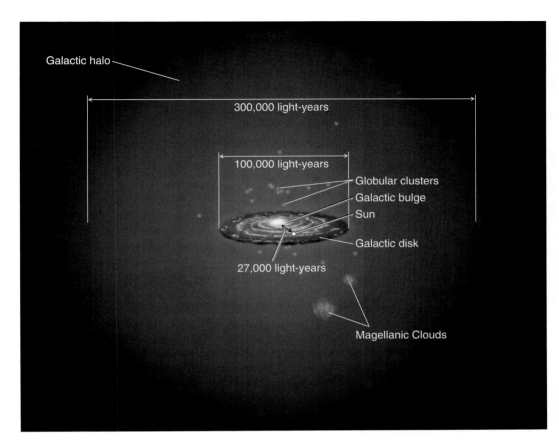

FIGURE 19.6 A diagram of the disk, bulge, and halo of the Milky Way Galaxy showing the location of the Sun.

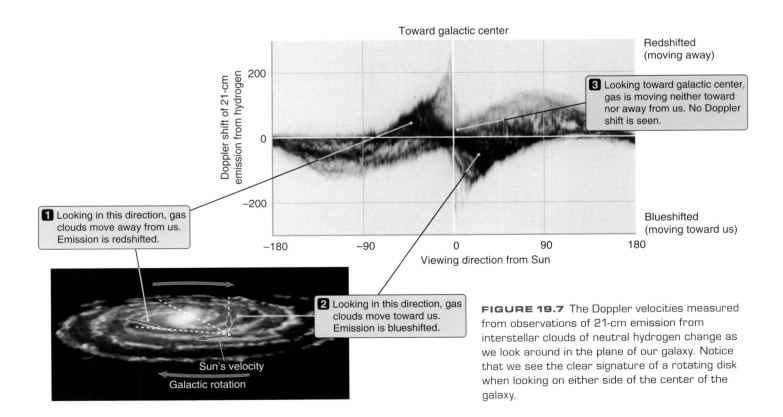

FIGURE 19.7 The Doppler velocities measured from observations of 21-cm emission from interstellar clouds of neutral hydrogen change as we look around in the plane of our galaxy. Notice that we see the clear signature of a rotating disk when looking on either side of the center of the galaxy.

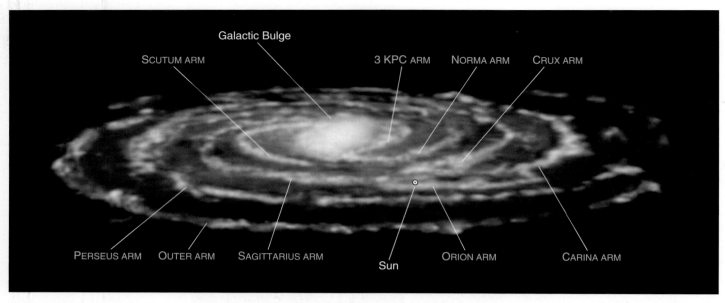

FIGURE 19.8 The Milky Way Galaxy as reconstructed from 21-cm radiation from neutral hydrogen and other observations.

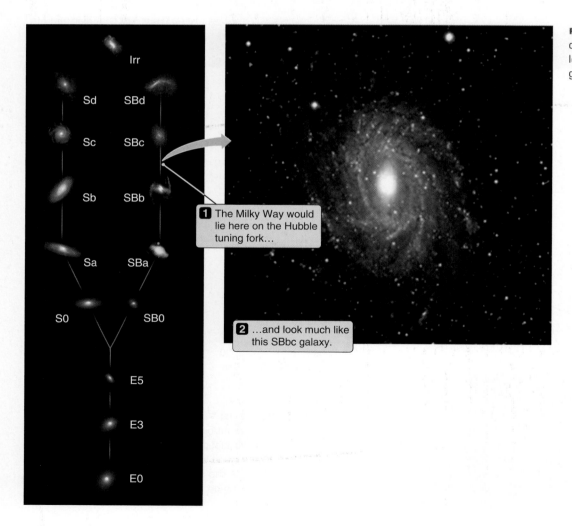

FIGURE 19.9 From the outside, the Milky Way would look much like this barred spiral galaxy, NGC 6744.

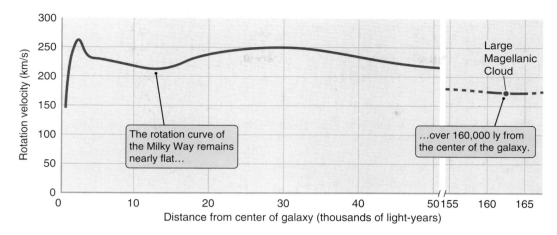

FIGURE 19.10 A plot showing rotation velocity versus distance from the center for the Milky Way. The most distant point comes from measurements of the orbit of the Large Magellanic Cloud.

The most important result from our analysis of galactic rotation curves in the previous chapter was the use of such curves to measure the distribution of mass within galaxies. By applying these techniques to the rotation curve of the Milky Way, we infer that the galaxy's mass must be about 6×10^{11} $M_\odot$. However, if we instead estimate the mass of the Milky Way by measuring its infrared luminosity (using infrared to see through the dust), we find a much lower value. Like other spiral galaxies, the Milky Way is mostly dark matter. The spatial distribution of dark and normal matter within the Milky Way is also much like what we have seen in other galaxies. The inner part of our galaxy is dominated by visible matter. Its outer parts, at least to a distance of 150,000 light-years from the center of our galaxy, are dominated by dark matter (see **Excursions 19.2**).

19.3 Studying the Milky Way Galaxy Up Close and Personal

Earlier in the chapter we concentrated on the disadvantages of our perspective that make it hard to see the Milky Way as a galaxy. However, our perspective also has advantages. For example, we can study the stellar content of the Milky Way at very close range, star by star, looking at subtle aspects of the populations of stars that give us direct clues about how spiral galaxies like ours form. It is much more difficult to glean such information by observing other spiral galaxies.

As with other galaxies, the shapes of the different parts of our galaxy are determined by the shapes of the orbits of the stars they contain. The stars in the disk rotate about the center of the galaxy, just like the gas and dust in the disk. The stars in the halo move in orbits similar to those of stars in elliptical galaxies. The bar shape of the bulge of our galaxy is defined primarily by stars and gas moving both in highly elongated orbits up and down the long axis of the bar and in short orbits aligned perpendicular to the bar.

The Sun is a middle-aged disk star, located mostly among stars that are middle-aged as well. Yet also near the Sun are stars that are part of the halo, passing through the disk in their orbits. As a result we can study stars both in the disk and in the halo just by studying the ages, chemical abundances, and motions of stars near our Sun.

Stars of Different Age and Chemical Composition

The most fundamental categories into which populations of stars can be grouped are based on their ages and the abundances of massive elements found in their atmospheres. Conveniently, some stars come prepackaged into groups that split up along just these two lines. There are two different varieties of star clusters in the Milky Way. Globular clusters—the densely packed collections of millions of stars studied by Shapley—are mostly halo objects. With ages of up to 13 billion years, they are among the oldest objects

Globular and open clusters differ in age, location, and chemical abundance.

known. In contrast, **open clusters**, like the one in **Figure 19.12(a)**, are much less tightly bound collections of a few dozen to a few thousand stars that are found orbiting in the disk of our galaxy. As with globular clusters, the stars in an open cluster all formed in the same region at about the same time. When we study the H-R diagrams of open clusters

EXCURSIONS 19.2

Searching for Dark Matter in the Halo

There is a very clever way to search for dark matter within the halo of our own galaxy, if the dark matter consists of compact objects such as low-mass stars, planets, white dwarfs, neutron stars, or black holes. We refer to such dark matter candidates as **MACHOs**, which stands for "massive compact halo objects."[2] If the dark matter in our halo consists of MACHOs, there would have to be a lot of these objects, and they would each exert gravitational force but not emit much light.

How might we detect MACHOs? Because of their gravity, MACHOs can gravitationally deflect light according to Einstein's general theory of relativity. If we are ob-

serving a distant star and if a MACHO were to pass between us and the star, the star's light would be deflected and perhaps amplified by the intervening MACHO as it passed across our line of sight, as illustrated in **Figure 19.11(a)**.

We would be remarkably lucky if such an event occurred just as we were observing a single distant star. However, astronomers have monitored the stars in the Large and Small Magellanic Clouds (two of the small satellite galaxies of the Milky Way Galaxy), observing tens of millions of stars for several years. They found a number of examples of events of the sort shown in **Figure 19.11(b)**, but not nearly enough to account for the amount of dark matter in the halo of our galaxy. Thus it was concluded that the dark matter in our galaxy is probably *not* primarily composed of MACHOs.

[2]MACHOs are not to be confused with another dark matter candidate called "weakly interacting massive particles," or WIMPs. Who says astronomers have no sense of humor?

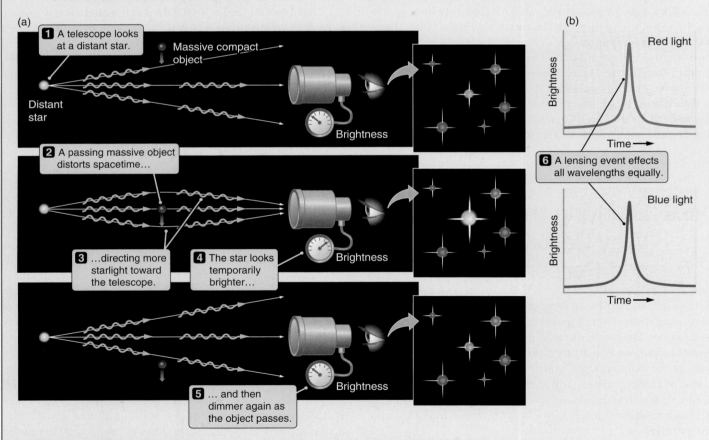

(a)

1 A telescope looks at a distant star.

Massive compact object

Distant star

Brightness

2 A passing massive object distorts spacetime…

3 …directing more starlight toward the telescope.

4 The star looks temporarily brighter…

Brightness

5 … and then dimmer again as the object passes.

Brightness

(b)

Red light

Brightness

Time

6 A lensing event effects all wavelengths equally.

Blue light

Brightness

Time

FIGURE 19.11 (a) The light from a distant star is affected by a compact object crossing our line of sight. Because gravity affects all wavelengths equally, such "lensing events" should look the same in all colors. (b) The observed light curves of a real star experiencing a lensing event.

(**Figure 19.12(b)**), we find a wide range of ages. Some open clusters contain the very youngest stars known. Other open clusters contain stars that are somewhat older than the Sun. There is no overlap in age between open clusters and globular clusters, however. Even the youngest globular clusters are several billion years older than the oldest open clusters. Open clusters do not survive long in the disk of our galaxy because they are loosely bound together and easily disrupted by the gravitational tug from nearby objects.

The differences in ages between globular and open clusters immediately tell us something interesting about the history of star formation in our galaxy. Stars in the halo formed first, but this epoch of star formation did not last long. No young globular clusters are seen. In the disk of the galaxy, star formation seems not to have gotten started until later, but it has been continuing ever since. The process that formed the stars in the massive, compact globular clusters must have also been much different from the more sedate process responsible for creating stars in the less massive, more scattered open clusters.

It is obvious why we would be so interested in grouping stars according to their ages, but why would we place emphasis on the chemical composition of stars? We have discussed the chemical evolution of the universe on a number of occasions. When the universe was very young, only the least massive of elements existed. All elements more massive than boron must have formed by nucleosynthesis in

---

Abundances of massive elements record the cumulative history of star formation.

---

stars. For this reason, the abundance of massive elements in the interstellar medium provides a record of the cumulative amount of star formation that has taken place up until the present time. Gas that shows large abundances of massive elements must have gone through a great deal of stellar processing, whereas gas with low abundances of massive elements is more pristine.

In turn, the abundance of massive elements in the atmosphere of a star provides a snapshot of the chemical composition of the interstellar medium *at the time that star formed!* (In main sequence stars, material from the core does not mix with material in the atmosphere, so the abundances of chemical elements inferred from the spectra of a star are the same as the abundances in the interstellar gas from which the star formed.) As illustrated in **Figure 19.13**, the chemical composition of a star's atmosphere is a record of the cumulative amount of star formation up until the time that star formed.

If our ideas about the chemical evolution of the universe are correct, we would expect to see large differences in massive element abundances between globular and open clusters. Stars in globular clusters, being among the earliest stars to form, should contain only very small amounts of massive elements. That is exactly what we see. Some globular cluster stars contain only 0.5 percent as much of these massive ele-

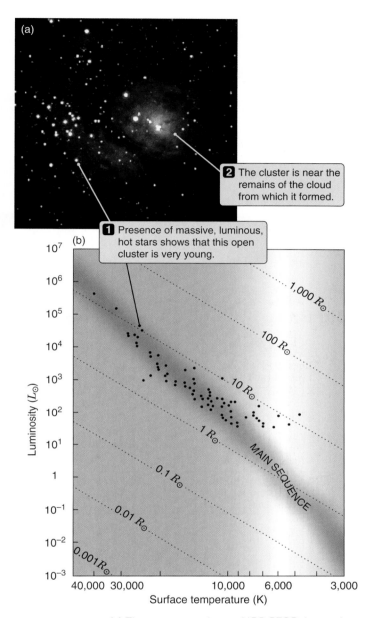

FIGURE 19.12 (a) The open star cluster NGC 6530, located 5,200 light-years away, in the disk of our galaxy. (b) The H-R diagram of stars in this cluster shows that it has an age of a few million years or less.

ments as does our Sun. This relationship between age and abundances of massive elements is seen throughout much of the galaxy. The chemical evolution of the Milky Way has

---

Younger stars typically have higher massive element abundances than older stars.

---

continued within the disk as generation after generation of disk stars has further enriched the interstellar medium with the products of their nucleosynthesis. Within the disk,

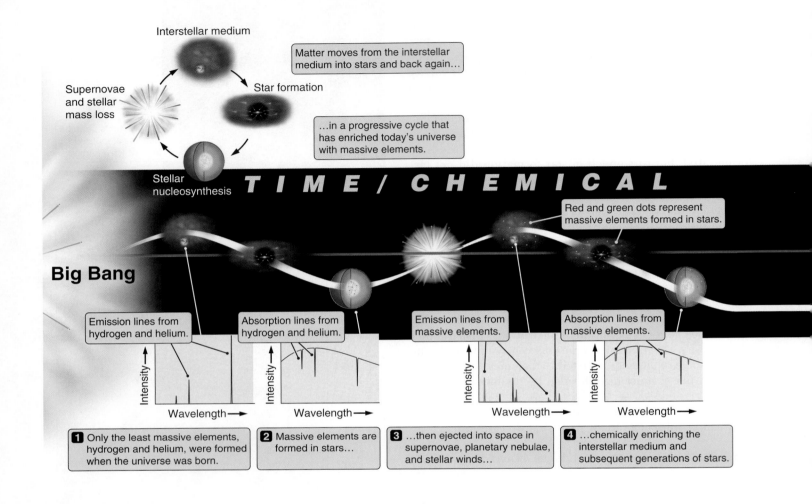

Interstellar medium

Matter moves from the interstellar medium into stars and back again...

Supernovae and stellar mass loss

Star formation

...in a progressive cycle that has enriched today's universe with massive elements.

Stellar nucleosynthesis

TIME / CHEMICAL

Red and green dots represent massive elements formed in stars.

Big Bang

Emission lines from hydrogen and helium.

Absorption lines from hydrogen and helium.

Emission lines from massive elements.

Absorption lines from massive elements.

Intensity — Wavelength→

Intensity — Wavelength→

Intensity — Wavelength→

Intensity — Wavelength→

1 Only the least massive elements, hydrogen and helium, were formed when the universe was born.

2 Massive elements are formed in stars...

3 ...then ejected into space in supernovae, planetary nebulae, and stellar winds...

4 ...chemically enriching the interstellar medium and subsequent generations of stars.

younger stars typically have higher abundances of massive elements than do older stars. Similarly, older stars in the outer parts of our galaxy's bulge have lower massive element abundances than young stars in the disk. This is the case not only for globular cluster stars but also for all the stars in our galaxy's halo, where globular cluster stars comprise only a minority among the total number of stars in the galactic halo.

Within the galaxy's disk, we can even see differences in abundances of massive elements from place to place that are related to the rate of star formation in different regions. Star formation is generally more active in the inner part of the Milky Way than in the outer parts. This is a result of the denser concentrations of gas that are found in the inner galaxy. If this has continued throughout the history of our galaxy, we might predict massive elements to be more abundant in the inner parts of our galaxy than in the outer parts. Observations of chemical abundances in the interstellar medium, based both on interstellar absorption lines in the spectra of stars and on emission lines in glowing

H II regions, confirm this prediction. As expected, there is a smooth decline in abundances of massive elements from the inner to the outer parts of the disk. Similar trends are often seen in other galaxies. These trends can be seen in stars as well. Relatively old stars near the center of a galaxy often have massive element abundances that are greater than those of young stars in the outer parts of the disk.

Our basic idea about higher massive element abundances following the more prodigious star formation in the inner galaxy seems correct; but as always, the full picture is not this simple. The chemical composition of the interstellar medium at any location depends on a wealth of factors. New material falling into the galaxy might affect interstellar chemical abundances. Chemical elements produced in the inner disk might be blasted into the halo in great fountains powered by the energy of massive stars, only to fall back onto the disk elsewhere. Past interactions with other galaxies might have stirred the Milky Way's interstellar medium, mixing gas from those other galaxies in with our

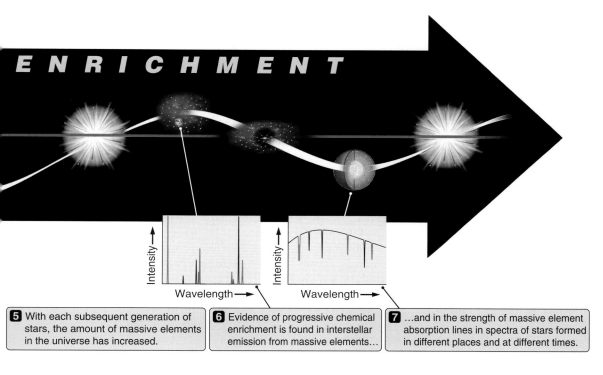

FIGURE 19.13 As subsequent generations of stars formed, lived, and died, they enriched the interstellar medium with massive elements —the products of stellar nucleosynthesis. The chemical evolution of our galaxy and other galaxies can be traced in many ways, including by the strength of interstellar emission lines and stellar absorption lines.

own. At this moment two small companions of the Milky Way, the Sagittarius Dwarf and the Ursa Major Dwarf, are plowing through the disk of the galaxy on the other side of the bulge. How chemical abundances vary from place to place within the Milky Way and other galaxies and what these variations tell us about the history of star formation and nucleosynthesis are active topics of research at the beginning of the 21st century.

Although the details are complex, there are several clear and important lessons to be learned from patterns in massive element abundances in the galaxy. The first is that even the

> Generations of stars must have formed even before globular clusters did.

very oldest globular cluster stars contain *some* amount of massive chemical elements. The implication is clear: Globular cluster stars and other halo stars were not the first stars in our galaxy to form. *There must have been at least one generation of massive stars that lived and died, ejecting newly synthesized massive elements into space, before even the oldest globular clusters formed.* Further, every star less massive than about 0.8 $M_\odot$ that ever formed is still around today. Even so, we find *no* disk stars with exceptionally low massive element abundances. We would have found these stars by now if they existed. The gas that wound up in the plane of the Milky Way must have seen a significant amount of star formation *before* it settled into the disk of the galaxy and made stars.

We have placed great emphasis on variations in chemical abundances from place to place. These tell us a lot about the history of our galaxy and a lot about the origin of the material that we are made from. It is important to remember, however, that even a chemically "rich" star like the Sun, which is made of gas processed through approximately 9 billion years of previous generations of stars, is still composed of less than 2 percent massive elements. Luminous matter in the universe is still dominated by hydrogen and helium formed long before the first stars.

A Cross Section through the Disk

The youngest stars in our galaxy are most strongly concentrated in the plane of our galaxy, defining a disk about 500 light-years thick (but over 100,000 light-years across—thin indeed). The older population of disk stars, distinguishable by lower abundances of massive elements, has a much "thicker" distribution (about 4,000 light-years thick). **Figure 19.14** illustrates how the population of stars changes with

There are thin and thick parts of our galactic disk.

distance from the galactic plane. Not too surprisingly, the youngest stars are concentrated closest to the plane of the galaxy for the simple reason that this is where the molecular clouds are. Older stars make up the thicker parts of the disk. There are two hypotheses for the origin of this thicker disk. One suggests that these stars formed in the midplane of the disk long ago but have since been kicked up out of the plane of the galaxy, primarily by gravitational interactions with massive molecular clouds. The other hypothesis suggests that these stars were acquired from the merging process that formed our galaxy.

When we discussed the formation of accretion disks in Chapters 6 and 15, we found that a rotating cloud of gas could do nothing other than collapse into a thin disk. This was a consequence of the fact that gas falling from one direction ran into gas falling from the other direction. Clouds of gas cannot pass through each other, so the gas had no

choice but to settle into a disk. The same thing applies to clouds of gas that are pulled by gravity toward the midplane of the disk of a spiral galaxy. Although stars are free to pass back and forth from one side of the disk to the other,

Gas tends to concentrate into thin sheets.

cold, dense clouds of interstellar gas settle down into the central plane of the disk. These clouds are seen as the concentrated dust lanes that slice the disks of spiral galaxies (see Figure 19.1(b)) like a layer of bologna in a very thin, flat sandwich. This thin lane slicing through the midplane of the disk is both where new stars form and where new stars are found.

We mentioned earlier that energy from regions of star formation can impose interesting structure on the interstellar medium, clearing out large regions of gas in the disk of a galaxy. Many massive stars forming in the same region can blow "chimneys" out through the disk of the galaxy via a combination of supernova explosions and strong stellar winds. If enough massive stars are formed together, sufficient energy may be deposited to blast holes all the way through the plane of the galaxy. In the process, dense interstellar gas can be thrown high above the plane of the galaxy (**Figure 19.15**). Maps of the 21-cm emission from neutral hydrogen in our galaxy and visible-light images of hydrogen emission from some edge-on external galaxies (**Figure 19.16**) show a wealth of vertical structure in the interstellar medium of disk galaxies. These vertical structures are often interpreted as the "walls" of these chimneys.

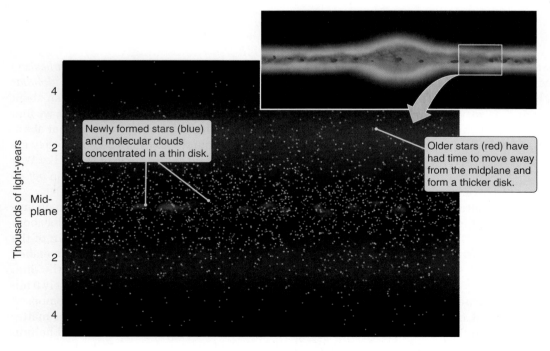

FIGURE 19.14 A vertical profile of the disk of the Milky Way Galaxy. Gas and young stars are concentrated in a thin layer in the center of the disk, but older populations of stars define ever-thickening portions of the disk.

Newly formed stars (blue) and molecular clouds concentrated in a thin disk.

Older stars (red) have had time to move away from the midplane and form a thicker disk.

Thousands of light-years

Mid-plane

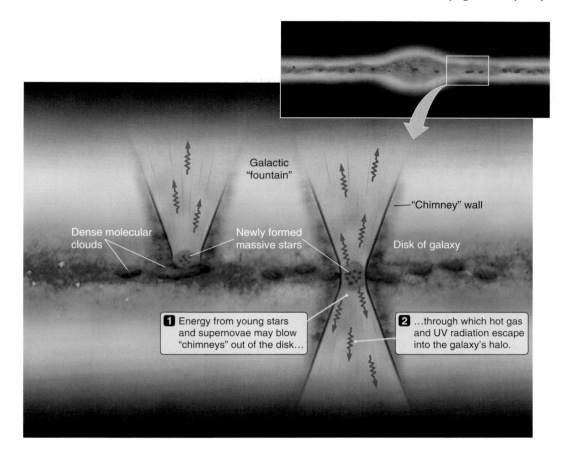

Galactic "fountain"

"Chimney" wall

Dense molecular clouds

Newly formed massive stars

Disk of galaxy

1 Energy from young stars and supernovae may blow "chimneys" out of the disk...

2 ...through which hot gas and UV radiation escape into the galaxy's halo.

FIGURE 19.15 In the "galactic fountain" model of the disk of a spiral galaxy, gas is pushed away from the plane of the galaxy by energy released by young stars and supernovae, then falls back onto the disk.

The Halo Is More Than Globular Clusters

Earlier, when discussing the halo of our galaxy, we concentrated our attention on its globular clusters. These are very important to astronomy, in part because they tell us a great deal about the history of star formation in the halo. Yet globular clusters account for only about 1 percent of the total mass of stars in the halo. As halo stars fall through the disk of the Milky Way, some pass close to the Sun, giving us a sample of the halo that we can study at close range.

Because most of the stars near our Sun are, like the Sun, disk stars, you might be wondering how we can tell them apart from nearby halo stars. They are distinguishable in two ways. First, most halo stars have much lower abundances of massive elements than disk stars. Second, halo stars appear to be whizzing by us at high velocities. Actually, it is often not the halo stars that are moving fast, but we who are moving fast relative to them. This is because halo stars do not rotate about the center of the galaxy in the same way that disk stars do, but rather are in the same kind of odd-shaped orbits as stars in elliptical gal-axies. In contrast, the disk stars near the Sun are moving, mostly together, in 220 km/s rotation about the center of our galaxy. Just as it is easy to tell the difference between a person sitting beside you on a bus and the people who (in your frame of reference) are whizzing by outside the window, it is easy to tell the difference between disk stars that share our motion and halo stars that, to us, are just passing through.

Reflecting once again the bias of the observer, astronomers call the halo stars **high-velocity stars**, even though it is usually the disk stars that are moving faster relative to the galaxy as a whole! Enough halo stars are near us that we can measure their distances and map out their orbits, giving us a very detailed look at the kinds of orbits that stars in a halo (or in an elliptical galaxy) have. The picture we get is one of confusion. About half the halo stars are orbiting in the same direction as the disk rotation, while the other half are moving in the *opposite* direction. The motions of the halo stars near us indicate that, in their full orbits, these halo stars fill a volume of space similar to that occupied by the globular clusters in the halo. Halo stars and globular clusters both paint the same picture of the halo of our galaxy.

(a) 21-cm image of part of Milky Way disk

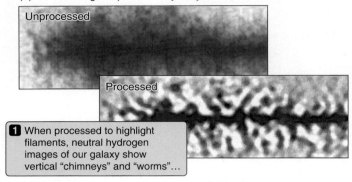

1 When processed to highlight filaments, neutral hydrogen images of our galaxy show vertical "chimneys" and "worms"...

(b) Processed Hα image of edge-on spiral galaxy

2 ...much like those seen in the disks of other galaxies.

FIGURE 19.16 Observations of emission from (a) neutral hydrogen in our own galaxy and (b) ionized hydrogen in other galaxies, such as NGC 891, show "chimneys" extending up out of the disks of these galaxies. These chimneys are understood to be the walls of cavities blown by energy released from young stars and supernovae.

Magnetic Fields and Cosmic Rays Fill the Galaxy

Almost all that we know about the galaxy we learn from analyzing the electromagnetic radiation it emits—almost, but not quite all. An important component of the galaxy's disk, whose energy we can measure directly on Earth, is **cosmic rays**. Despite their name, cosmic rays are not a form of electromagnetic radiation. (They were named before their true nature was known.) Rather, they are charged particles moving at close to the speed of light. Cosmic rays are continually hitting Earth.

Most cosmic ray particles are protons, but some are nuclei of helium, carbon, and all of the other elements produced by nucleosynthesis. What makes cosmic rays especially fascinating is that they span an enormous range in particle energy. We can observe the lowest-energy cosmic rays using interplanetary spacecraft. These cosmic rays have energies as low as about 10^{-11} joules, which corresponds to the energy of a proton moving at a velocity of a few tenths the speed of light. In contrast, the most energetic cosmic rays are 10 trillion (10^{13}) times more energetic than the lowest-energy cosmic rays. These high-energy cosmic rays

are known from the showers of elementary particles that they cause when crashing through Earth's atmosphere.

It is thought that cosmic rays are accelerated to these incredible energies in the shock waves produced in supernova explosions. The highest-energy cosmic rays are more difficult to explain. These are as much as a billion times more energetic than any particle ever produced in a particle accelerator on Earth. These cosmic rays have energies up to

Some cosmic rays have incredibly high energies.

100 J, corresponding to the energy of a proton moving at a velocity that is 99.99999999999999999999 percent of the speed of light. A better way to visualize this is to realize that if you were to drop your copy of *21st Century Astronomy* (this irreplaceable textbook) from the ceiling, the energy it would have when it hit the floor is about the same as the energy concentrated into a *single* high-energy cosmic ray proton!

The interstellar medium of the Milky Way is laced with remarkably strong magnetic fields that are wound up and strengthened by the rotation of the galaxy's disk. When we discussed Earth's magnetosphere in Chapter 8, we saw that charged particles are able to move freely along a magnetic

Cosmic rays are trapped by the Milky Way's magnetic field.

field, but are unable to move across it. The other side of this coin is that magnetic fields cannot freely escape from a cloud of gas containing charged particles. Earlier we mentioned that dense clouds of interstellar gas are concentrated by gravity in the midplane of the Milky Way. The weight of these clouds anchors the galaxy's magnetic field to the disk, as shown in **Figure 19.17**.

The disk of our galaxy glows from synchrotron radiation produced by cosmic rays (mostly electrons) looping around the direction of the galaxy's magnetic field. Such synchrotron emission is seen in the disks of other spiral galaxies as well, telling us that they, too, have magnetic fields and populations of energetic cosmic rays. Even so, the very highest-energy cosmic rays are moving much too fast to be confined to our galaxy. Any such cosmic rays formed in the Milky Way soon stream away from the galaxy into intergalactic space. It is also quite possible that some fraction of the energetic cosmic rays reaching Earth originated in energetic events outside our galaxy.

The total energy of all of the cosmic rays in the galactic disk can be estimated from the energy of the cosmic rays reaching Earth. The strength of the interstellar magnetic field can be measured in a variety of ways, including the

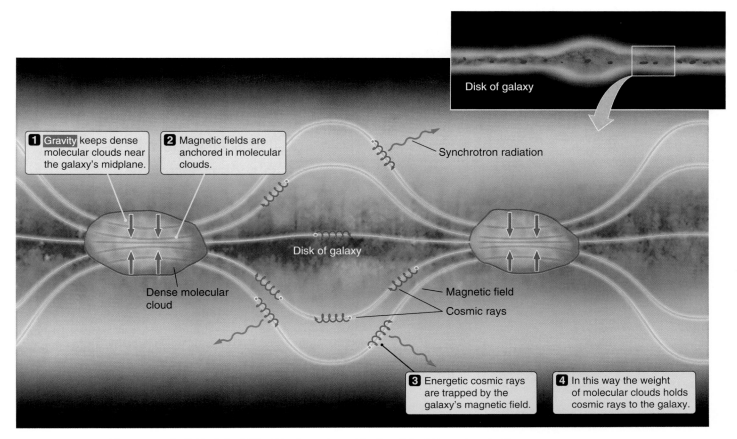

FIGURE 19.17 The magnetic field of the Milky Way is anchored to the disk of the galaxy by the weight of interstellar clouds. The magnetic field in turn traps the galaxy's cosmic rays, much as a planetary magnetosphere traps charged particles.

effect that it has on the properties of radio waves passing through the interstellar medium. These measurements indicate that in our galaxy, the magnetic field energy and the cosmic ray energy are about equal to each other. Both are comparable to the energy present in other energetic components of the galaxy, including the motions of interstellar gas and the total energy of electromagnetic radiation within the galaxy. Magnetic fields and cosmic rays are *not* bit players in the production we call the Milky Way.

The Milky Way Hosts a Supermassive Black Hole

In the previous chapter we discussed evidence that most galaxies with bulges have supermassive black holes at their centers. The Milky Way Galaxy is no exception. Observations of our galaxy's rotation curve show rapid rotation velocities very close to its center. This is the same sort of

The center of our galaxy shows rapid rotation and AGN-like emission.

evidence that led us to surmise the presence of supermassive black holes in other galaxies. We estimate the black hole at the center of our own galaxy is a relative lightweight, having a mass of "only" $4 \times 10^6 \, M_\odot$.

Radio observations of the inner part of our galaxy (**Figure 19.18**) show synchrotron radiation from fascinating wisps and loops of material throughout the region. These are reminiscent of the synchrotron emission seen from AGNs, albeit at far lower levels. Without a source of large amounts of material to feed the beast that resides at the center of our galaxy, the Milky Way is quiescent at the moment. However, like steam rising from the cauldron of a dormant volcano, the synchrotron emission from the inner Milky Way is a reminder that our galaxy was likely "active" at some time in the past, and, like an episodic volcano, could become active again in the future.

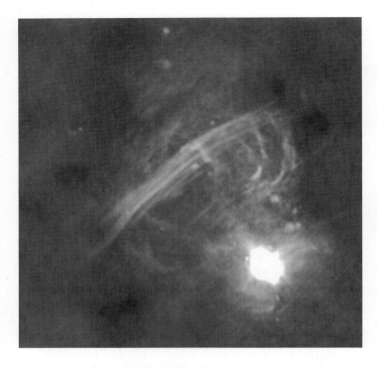

FIGURE 19.18 Radio observations of the center of the Milky Way showing strong synchrotron emission. The galactic center is also a source of strong X-rays and gamma rays, evidence of the "beast" at the center of our own galaxy.

19.4 The Milky Way Offers Clues about How Galaxies Form

One of the fundamental goals of stellar astronomy is to understand the life cycle of stars, including how stars form from clouds of interstellar gas. In Chapter 15 we were able to tell a fairly complete story of this process, at least as it occurs today, and tie this story strongly to observations of our galactic neighborhood. Galactic astronomy has the same basic goal. That is, astronomers would like very much to have a complete and well-tested theory of how our galaxy formed. Unfortunately, such a complete theory is not yet in hand. Even so, what we have seen so far offers us many clues.

Important among these clues are the properties of globular clusters and high-velocity stars in the halo of the galaxy. For reasons discussed earlier, these objects must have been among the first stars formed that still exist today. The fact that they are not concentrated in the disk or bulge of the galaxy says that they formed from clouds of gas well before those clouds had settled into the galaxy's disk. The observations that globular clusters are very old and that the youngest cluster is older than the oldest disk stars agree

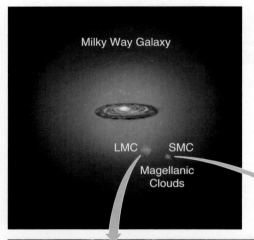

FIGURE 19.19 The Large and Small Magellanic Clouds were named for Ferdinand Magellan, who headed the first European expedition to venture far enough into the Southern Hemisphere to see these two dwarf companions of the Milky Way.

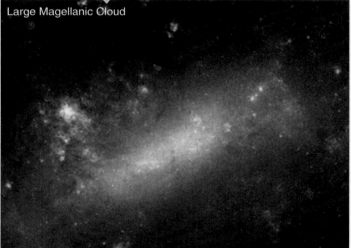

with this surmise. The presence of small amounts of massive elements in the atmospheres of halo stars also tells us that there must have been at least one generation of stars

At least one generation of stars had to exist before today's halo stars were formed.

that lived and died *before* the formation of the halo stars we see today. We have yet to find any stars from that first generation in our galaxy today.

Based on these and other clues, we have come to understand that our galaxy must have formed when the gas within a large "clump" of dark matter collapsed into a large number of small protogalaxies. Some of these smaller clumps are still around today in the form of small dwarf galaxies near our own, including the Sagittarius and Ursa Major Dwarfs and the Large and Small Magellanic Clouds, shown in **Figure 19.19**. The remainder of these protogalaxies merged to form the barred spiral galaxy we call the Milky Way. In this process, stars were formed in the halo, in the bulge, and in the disk. The first stars to form ended up in the halo —some in globular clusters, but many not. Gas that settled into the disk of the Milky Way quickly formed several generations of stars. In this process a seed somehow formed, which quickly grew into the supermassive black hole at the

Our galaxy formed from the mergers of many smaller protogalaxies.

center of the galaxy. The details of this process are sketchy, but some calculations indicate that so much mass was concentrated in this small region that almost any sequence of events would have led to the formation of a massive black hole.

We need to be careful not to get ahead of ourselves. The Milky Way offers many clues about the way galaxies form, but much of what we know of the process comes from looking beyond our local system. Images of distant galaxies (which we see as they existed billions of years ago), as well as observations of the glow left behind by the formation of the universe itself, provide equally important pieces of the puzzle. We will have to leave the story of galaxy formation as we currently understand it unfinished, for now, as we turn our attention to the immensely larger structure of the universe as a whole.

Summary

- We live in the disk of a barred spiral SBbc galaxy called the Milky Way about 27,000 light-years from its center.

- We use variable stars of known luminosity to find the distances to globular clusters, which allows us to measure the size of the Milky Way.

- The Doppler velocities of radio lines show that the rotation curve of the Milky Way is flat, like those of other galaxies, and that the mass of our galaxy is mostly in the form of dark matter.

- The chemical composition of the Milky Way has evolved with time, and there must have been a generation of stars before the oldest halo and globular cluster stars we see today formed.

- Star formation is actively occurring in the disk of our galaxy, leading to complex structures within the disk.

- At the center of the galaxy is a massive black hole, which produces rapid rotational velocities nearby.

Seeing the Forest through the Trees

Many books of puzzles contain optical illusions in which familiar patterns lie hidden in jumbles of lines and shapes. When looking at such a picture, at first we see nothing but confusion. Then in a flash of insight we suddenly see the pattern! Once the pattern is recognized, our perception shifts: A moment ago all we saw was a mess, but now when we look at the picture, the "totally obvious" pattern jumps off the page at us. Such is the case with observing our galaxy. Go outside on a dark, moonless summer night, away from the city lights, and stare at the sky. In addition to the dazzling jewels of individual stars that fill your view, you will also see a faint, patchy river of light running from horizon to horizon. This river—this "Milky Way"—has been known since earliest prehistory. Even so, its nature was unknown until a remarkably short time ago. As recently as the early 20th century, it was unclear how this river of light fit into models of our stellar system based on counts of individual stars. Yet once the answer is known, there it is for all to see—how could we ever have been so blind as to miss the obvious? Stretched across our sky lies an unmistakable spiral galaxy, as viewed edge-on from within the disk.

Our conclusion from this flash of insight is confirmed by the results of countless observational studies. Whether we look at the distribution of globular clusters, 21-cm observations of the motions of hydrogen gas, or any number of other tracers of structure, the answer is always the same. The Milky Way is a giant "plain vanilla," dark matter–dominated, barred spiral galaxy, with a supermassive black hole at its center, just like countless others. The Sun and Earth are located in a relatively open and clear region of interstellar space part way out in the galaxy's disk.

Our vantage point within the Milky Way is hardly the best from which to answer questions about global properties of our galaxy. If ever there was a problem in astronomy of seeing the forest for the trees, this is it. For example, we can determine the morphological classification of the Andromeda Galaxy by glancing at a single picture. In contrast, obtaining a reliable and complete classification of our own Milky Way is a task that has taken decades of detailed study. The classification of the Milky Way as an SBbc galaxy was difficult to obtain and remains at least somewhat uncertain today.

On the other hand, the Milky Way gives us the ultimate insider's view. For example, stars in the Milky Way are not just part of a large blur, but are instead individuals that can be studied as such. Here is where we learn about the stars in globular and open clusters. These collections of stars not only tell us about the Milky Way as a galaxy, but also provide the data we use to test and refine our ideas about the way stars everywhere evolve. In the Milky Way we see stars in the bulge, stars in the thin and thicker portions of the disk, and stars in the halo that are falling through the disk near us. We measure their properties—luminosity, size, temperature, and abundance of the chemical elements of which they are made—and we learn what each variety of stars in a spiral galaxy is like. Spiral arms in the Milky Way are not just bright swirls against the background of a spiral disk. Instead they are collections of young stars, old stars, and clouds of gas and dust that can be studied at close range. We know that stars form in the arms of other spiral galaxies because we see the brightest of those stars, along with the generally blue color of the arms. We know that stars form in the spiral arms of the Milky Way because that is where we see dense molecular cloud cores containing individual young stars surrounded by their accretion disks.

So it is difficult to study the Milky Way as a galaxy, but it is worth the investment. We are guided in our study of the grand patterns within the Milky Way by what we know of other galaxies; conversely, in the Milky Way we can learn the details of processes that are at work shaping all of those other islands of activity spread out across the expanse of the universe. These details are beginning

to help us clarify our understanding of some of the "big" questions about the way galaxies form and evolve.

Our location in the Milky Way may not give us the best perspective from which to study our galaxy, but it is far from the worst. Fate has been kind. Since the Sun and Earth formed, they have made numerous trips around the galaxy, passing back and forth through the disk and encountering a wide range of environments. What if our species had become aware at a moment when Earth was in the depths of a molecular cloud, and the sky held nothing of interest other than the Sun and planets? That would bring a whole new and very literal meaning to the notion of the "dark ages."

In some sense this portion of our journey has been about context. We have learned to see stars within the context of the history of the Milky Way. We have seen star-forming molecular clouds within the context of the sweep of spiral arms. We have learned to see the Milky Way within the context of its similarities with other galaxies. On the next leg of our journey we will pose the ultimate question of context as we learn to see galaxies, including our Milky Way, within the context of the universe itself.

Key Terms

globular cluster, p. 555
halo, p. 555
standard candle, p. 556
open cluster, p. 561
MACHO, p. 562
high-velocity star, p. 567
cosmic ray, p. 568

Student Questions

THINKING ABOUT THE CONCEPTS

1. What do astronomers mean by a "standard candle"?

2. Explain the observational evidence that shows we live in a spiral galaxy, not an elliptical galaxy.

3. What can you see for yourself in the night sky (unaided by a telescope) that suggests that we live in a spiral galaxy? Explain your answer.

4. Why do we need to use 21-cm radio observations to probe the structure of our galaxy?

5. Discuss reasons for the differences between globular clusters and open clusters, with respect to their relative

ages, abundance of massive elements, numbers of stars, and distribution throughout the galaxy.

6. Why are the youngest stars concentrated close to the plane of our galaxy?

7. Stars in many globular clusters tend to have a lower abundance of massive elements than stars in the disk of our galaxy. Explain why this is so.

8. How do we know that the stars in globular clusters are the oldest stars in our galaxy?

9. To observers in Earth's Southern Hemisphere, the Large and Small Magellanic Clouds look like detached pieces of the Milky Way. What are these "clouds," and why is it not surprising that they look so much like pieces of the Milky Way?

10. Describe how our environment would be different and how our skies might appear if our Sun and Solar System were located
   a. Near the center of the galaxy.
   b. Near the center of a large globular cluster.
   c. Near the center of a large, dense molecular cloud.

11. How do we know there is a supermassive black hole at the center of the Milky Way?

APPLYING THE CONCEPTS

12. The Sun completes one trip around the center of the galaxy in approximately 230 million years. How many times has our Solar System made the circuit since its formation 4.6 billion years ago?

13. The Sun is located about 27,000 light-years from the center of the galaxy, but the disk probably extends another 30,000 light-years farther out from the center. Assume that the Sun's orbit takes 230 million years to complete.
   a. Assuming a truly flat rotation curve, how long would it take a globular cluster located near the edge of the disk to complete one trip around the center of the galaxy?
   b. How many times has that globular cluster made the circuit since its formation 13 billion years ago?

14. Parallax measurements of the variable star RR Lyrae indicate that it is located 750 light-years from the Sun. A similar star observed in a globular cluster located far above the galactic plane has the same period of variability but appears 160,000 times fainter than RR Lyrae.
   a. How far from the Sun is this globular cluster?
   b. What does your answer to (a) tell you about the size of the galaxy's halo compared to the size of its disk?

15. The flat rotation curve indicates that the total mass of our galaxy is $6 \times 10^{11}\ M_\odot$. However, electromagnetic radiation associated with normal matter suggests a total mass of only $3 \times 10^{10}\ M_\odot$. Given this information, calculate the fraction of our galaxy's mass that is made up of dark matter.

16. A cosmic ray proton is traveling at nearly the speed of light (3×10^8 m/s).
   a. Using Einstein's familiar relationship between mass and energy ($E = mc^2$), show how much energy (in joules) the cosmic ray proton would have if m is based only on the proton's rest mass (1.7×10^{-27} kg).
   b. The actual measured energy of the cosmic ray proton, however, is 100 J. What then is the relativistic mass of the cosmic ray proton?
   c. How much greater is the relativistic mass of this cosmic ray proton than the mass of a proton at rest?

StudySpace
wwnorton.com/astro21
provides a Study Plan for each chapter that includes a reading outline, animations, keyword flash cards, and gradebook-enabled multiple-choice quizzes. From StudySpace you can also access premium content in the ebook and SmartWork.

A man said to the universe, "Sir, I exist!" "However," replied the universe, "The fact has not created in me a sense of obligation."

STEPHEN CRANE (1871–1900)

A hemispheric sky map showing tiny fluctuations in the cosmic background radiation (CBR).

CHAPTER 20

Our Expanding Universe

20.1 The Cosmological Principle Shapes Our View of the Universe

As we look back on our journey of the mind—and on the history of our species—it is remarkable how far we have come. Beginning with Copernicus's early realization that Earth is not the center of all things, we have shared the insights of Galileo and Newton and their intellectual heirs as they tore down the conceptual barriers separating terrestrial existence from that of the heavens. Casting aside the aura of mysticism and magic, we can now go out and look at the night sky and begin to see it for what it is. Earth sits within an enormous spiral galaxy consisting of hundreds of billions of stars. This galaxy in turn is but one of hundreds of billions of galaxies that fill a universe vastly larger than our ancestors might have imagined. Through it all, the cosmological principle has been at the center of our conceptual understanding. No progress has been possible without the enabling realization that the rules that apply to one part of our universe apply everywhere.

The time has come to put the cosmological principle to work in its namesake field—the field of **cosmology**. The science of cosmology is the study of the universe itself, including its structure, history, origins, and fate. As we stressed before, the cosmological principle is not an article of faith. Rather, it is a testable scientific theory. An important prediction of the cosmological principle is that the conclusions we reach about our universe should be more or less the same, regardless of whether we live in the Milky Way or in a galaxy billions of light-years away at the limits of

KEY CONCEPTS

Just as stars make up the structure of our own galaxy, so too do galaxies make up the structure of the universe itself. Observations of the motions of galaxies led to one of the most remarkable discoveries in the history of our species: The universe is expanding! As we investigate this discovery we will learn

- About the discovery of Hubble's Law relating the redshift of a galaxy to its distance.

- That galaxies are not flying apart through space, but rather "ride along" as space itself expands in accordance with the general theory of relativity.

- How Hubble's Law is used to map the universe in space and look back in time.

- That spacetime itself was born approximately 13.7 billion years ago in an event called the Big Bang.

- That each of the major predictions of Big Bang theory, from the glow of the cosmic background radiation to the types of atoms that fill the universe, is confirmed by observation.

- How the history, shape, and fate of the universe are determined by the amount of mass it contains and by a poorly understood cosmological constant.

- What observations of the cosmic background tell us about the properties of the young universe.

the observable universe. In other words, we predict that if the cosmological principle is correct, then our universe will be **homogeneous**.

According to one popular dictionary, the word *homogeneous* means "having at all points the same composition and properties." Clearly the universe is not truly homogeneous. The conditions we encounter at the surface of Earth are very

---

Observers everywhere should see the same universe.

---

different from those we would encounter in deep space or the heart of the Sun. (Even homogenized milk varies at the molecular level from place to place.) When we speak as cosmologists of homogeneity of the universe, we mean instead that stars and galaxies in our part of the universe are much the same, and behave in the same manner, as stars and galaxies in remote corners of the universe. We also mean that

stars and galaxies everywhere are distributed in space in much the same way as they are in our cosmic neighborhood, and that observers in those galaxies see the same properties for our universe that we do. In other words, when we speak of a homogeneous universe, we are using the term *homogeneity* in a very "broad brush" sense.

It is not easy to verify the prediction of homogeneity directly. We do not have the luxury of traveling from our galaxy to a galaxy in the remote universe to see whether conditions are the same. However, we can observe light arriving from the distant universe and see in what ways features look the same or different. For example, we can look at the way galaxies are distributed in space (**Figure 20.1**) and ask whether that distribution is homogeneous.

In addition to predicting that the universe is homogeneous, the cosmological principle requires that all observers (including us) have the same impression of the universe, regardless of the *direction* in which they are looking. This

FIGURE 20.1 Homogeneity and isotropy in four universes. (a) The distribution of galaxies is uniform, so this universe is both homogeneous and isotropic. (b) The density of galaxies is decreasing in one direction, so this universe is neither homogeneous nor isotropic. (c) The bands of galaxies define a unique direction, making this universe anisotropic. (d) The distribution of galaxies is uniform, but galaxies move along only one direction, so this universe, too, is anisotropic.

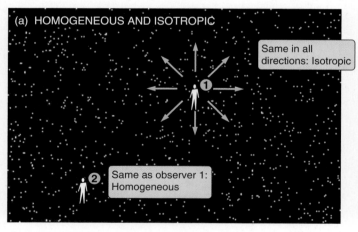

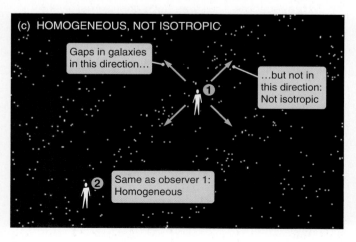

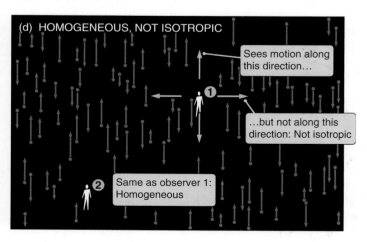

prediction of the cosmological principle is much easier to test directly than homogeneity. For example, if galaxies were lined up in rows—a violation of the cosmological principle (Figure 20.1(c))—we would get very different impressions depending on the direction in which we looked. If something is the same in all directions, then it is **isotropic**. In most instances isotropy goes hand in hand with homogeneity, and the cosmological principle requires them both.

The isotropy and homogeneity of the distribution of galaxies in the universe are predictions of the cosmological principle that we can go to the telescope and test directly. As you read this, you can benefit from the experience of

The distribution of galaxies is isotropic and homogeneous, as predicted.

decades of astronomers who have used the world's most powerful telescopes to test these predictions. These observations show that the properties of the universe are basically the same, regardless of the direction in which we look. If we look on very large scales, the universe appears homogeneous as well. Had these observations turned out otherwise, they would have proven the cosmological principle false, but the cosmological principle withstood the test. As has been true over and over again on our journey, more of the predictions of the cosmological principle are found to be correct.

20.2 We Live in an Expanding Universe

One of the great pioneers of cosmology was Edwin Hubble, who is shown on the cover of a 1948 edition of *Time* magazine in **Figure 20.2**. In the 1920s Hubble and his coworkers were studying the properties of a large collection of galaxies. In addition to taking images, a colleague of Hubble's named **Vesto Slipher** (1875–1969) was using the facilities at Lowell Observatory to obtain spectra of those galaxies. One

Emission from distant galaxies is redshifted to longer wavelengths.

aspect of what Slipher's spectra showed was no surprise at all. Galaxy spectra look like the spectra of ensembles of stars but with a bit of glowing interstellar gas mixed in for good measure. The surprise, however, was that the emission and absorption lines in the spectra of galaxies that Slipher observed were seldom seen at the same wavelengths at which these features appeared in laboratory-generated spectra. The lines were almost always shifted to longer wavelengths, as shown in **Figure 20.3**.

FIGURE 20.2 Edwin Hubble, as he appeared on the cover of the February 9, 1948, issue of *Time* magazine.

Slipher characterized the observed shifts in galaxy spectra as **redshifts** because almost all galaxies have spectral lines shifted to longer (in other words, redder) wavelengths. The wavelength at which a line is observed in an object that is stationary relative to the observer is called the rest wavelength of the line, written λ_{rest}. The redshift of a galaxy, written z, is defined as the difference between the observed wavelength, $\lambda_{observed}$, and the rest wavelength, divided by the rest wavelength:

$$z = \frac{\lambda_{observed} - \lambda_{rest}}{\lambda_{rest}}$$

Note that the redshift of a galaxy is the same, regardless of the wavelength of the line used to measure it. Also note that this expression is very similar to the equation for the Doppler shift that we studied in Chapter 4; in fact, this equation is identical to the Doppler formula if we take z to be the speed of an object moving away from us (v_r) divided by the speed of light. Following this reasoning, Hubble interpreted Slipher's redshifts as Doppler shifts. By setting

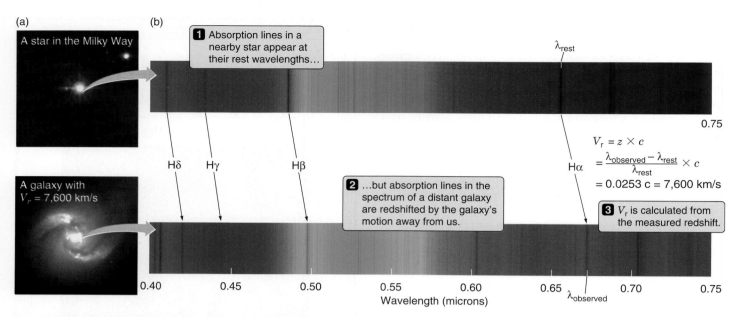

FIGURE 20.3 (a) A distant galaxy seen together with a star in our galaxy. (b) The spectrum of each, shown on the same scale. Note that lines in the galaxy spectrum are redshifted to longer wavelengths.

$z = v_r/c$, he concluded that almost all of the galaxies in the universe are moving away from the Milky Way.

When Hubble combined these measurements of galaxy recession velocities with his own estimates of the distances to these galaxies, he made one of the greatest discoveries in the history of astronomy. Hubble found that distant galax-

---

Hubble's Law: A galaxy's recession velocity is proportional to its distance.

---

ies are moving away from us more rapidly than nearby galaxies. Specifically, *the velocity at which a galaxy is moving away from us is proportional to the distance of that galaxy.* This simple relationship between distance and recession velocity has become known as **Hubble's Law**.

When talking about the distances of remote galaxies in the universe, we need a suitable yardstick. The yardstick we choose is the **mega-light-year (Mly)**, equal to a million

---

Hubble's constant relates a galaxy's recession velocity to its distance.

---

light-years. According to Hubble's Law, a galaxy located 100 Mly away is, on average, moving away from us at twice the speed of a galaxy at a distance of 50 Mly.

We usually write Hubble's Law as

$$v_r = H_0 \times d_G,$$

where d_G is the distance to the galaxy and v_r is the recession velocity of the galaxy. H_0 is a constant of proportionality and

is called the **Hubble constant**. When thinking of Hubble's Law, just remember that if you double the distance d_G you also double the recession velocity v_r.

Any Observer Sees the Same Hubble Expansion

Hubble's Law is a remarkable observation about the universe that has far-reaching implications. For one thing, Hubble's Law helps us test the prediction that the universe is homogeneous and isotropic. When we look at galaxies in one direction in the sky, we find they obey the same Hubble Law as galaxies observed in other directions in the sky. However, while Hubble's Law corroborates the prediction that our view of the universe is isotropic, the law appears at first glance to contradict the prediction of the cosmological principle that the universe is homogeneous. Hubble's Law might seem to imply that we are sitting in a very special place—at the *center* of a tremendous explosion, with everything else in the universe streaming away from us. However, this initial impression is incorrect. *Hubble's Law actually says that we are sitting in a uniformly expanding universe and that*

Hubble's Law says the universe is expanding uniformly.

---

the expansion looks the same, regardless of the location of the galaxy from which we view it! To understand why this is the case, we now turn to a useful toy model that you can build for yourself with materials you can probably find in your desk.

Figure 20.4 shows a long rubber band with paper clips attached along its length. If you stretch the rubber band, the paper clips, which represent galaxies in an expanding universe, get farther and farther apart. Imagine what this expansion would look like if you were an ant riding on paper clip A. As the rubber band is stretched, you notice that all of the paper clips are moving away from you. When you look at paper clip B, which is the next paper clip over, you

see it moving away slowly. When you look at paper clip C, located twice as far down the line from you as B, you see it moving away twice as fast as B. By the time you work your way down the line to paper clip E (located four times as far away as paper clip B), you see it moving away four times as fast as B. In other words, from the perspective of an ant riding on paper clip A, all of the other paper clips on the rubber band are moving away with a speed that is proportional to their distance. The paper clips located along the rubber band obey a Hubble Law.

FIGURE 20.4 Imagine a stretched rubber band with paper clips evenly spaced along its length. An ant riding on paper clip A will observe a Hubble Law along the rubber band, with paper clip C moving away twice as fast as paper clip B. Similarly, an ant riding on paper clip E will see a Hubble Law, with paper clip C moving away twice as fast as paper clip D. Any ant would see the same Hubble Law, regardless of which paper clip it was riding.

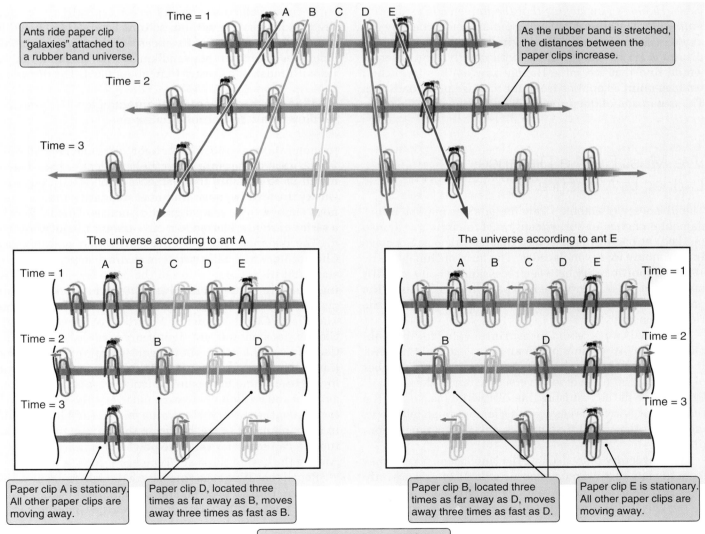

This is a handy result, but the key insight comes from realizing that there is nothing special about paper clip A. If instead you were riding on paper clip E, then it would have been paper clip D that was moving away slowly, and paper clip A that was moving away four times as fast. Repeat this experiment for *any* paper clip along the rubber band, and you will arrive at the same result: The speed at which other paper clips are moving away from you is proportional to their distance. Bringing our cosmological terminology to bear, the stretching rubber band is homogeneous. We would see the *same* Hubble Law, regardless of the paper clip that we choose as our vantage point.

The observation that nearby paper clips move away slowly and distant paper clips move away more rapidly does not say that we are at the center of anything. Instead, it says that the rubber band is being stretched uniformly along its length. In like fashion, *Hubble's Law for galaxies does not mean that our galaxy is at the center of an expanding universe. Hubble's Law means that the universe is expanding uniformly.* Any observer viewing our universe from any galaxy will see nearby galaxies moving away slowly and more distant galaxies moving away more rapidly. Any observer would find that the same Hubble Law applies from their vantage point as applies from our vantage point on Earth. The expansion of the universe is homogeneous.

We Must Build a Distance Ladder to Measure H_0

The discovery of Hubble's Law for galaxies marked a fundamental change in our perception of the universe. It also marked the beginning of a quest that has taken center stage in astronomy for seven decades. The form of Hubble's Law —as a proportionality between recession velocity and distance—tells us our universe is expanding. But to know the present *rate* of that expansion, we need a good value for the Hubble constant H_0.

How shall we go about measuring a value for the Hubble constant? In principle, the answer is straightforward. If we can measure the redshifts and distances of a number of galaxies, and plot velocity versus distance, then H_0 will be the slope of the resulting line. Yet while measuring the redshifts of galaxies is easy, given a large enough telescope, measuring the actual distances to galaxies is much more difficult.

The difficulty in measuring the Hubble constant comes from the fact that we must measure the distances not only to nearby galaxies, but to galaxies that are very far away. To see why, think about the motion of the water in a river. All of the water in a flowing river moves downstream, but even very uniform and steady rivers contain eddies and cross currents that disturb the uniform flow. If you want

to get a good overall picture of river flow, you need to look at a large portion of the river, not just the motion of a single leaf or two drifting downstream.

Likewise, there are eddies and cross currents in the motions of galaxies that make up our universe. The overall motion of galaxies in accord with Hubble's Law is often referred to as *Hubble flow*. Departures from a smooth Hubble flow, referred to as **peculiar velocities**, are the result of gravitational interactions that cause galaxies to fall toward their neighbors or toward large concentrations of mass in the universe. If we look only at nearby galaxies, their motions due to Hubble's Law will be small, so most of the motion that we see will instead be due to their peculiar velocities. If we want to measure the Hubble flow itself to obtain a reliable value for H_0, we need to study galaxies that are far enough away that the part of their velocities due to Hubble flow is much greater than any peculiar velocities they might have.

Peculiar velocities of galaxies are typically a few hundred kilometers per second, so to measure H_0 we need to accurately measure the distances to galaxies with Hubble velocities of several thousand km/s or more (that is, galaxies with redshift z greater than about 0.01). The distances

---

Measuring H_0 requires measuring distances to remote galaxies.

---

to such galaxies—150 million light-years and beyond—are far too great to measure using the same techniques that allowed us to measure the distances of objects in our own galaxy. Instead we must find new standard candles luminous enough to be seen at great distances. This is done in a series of steps, referred to as the **distance ladder**.

The distance ladder begins with stellar parallax (see Chapter 13), which allows us to measure distances to nearby stars and trace the position of the main sequence. By using the main sequence itself as a standard candle, we can measure the distance to very nearby galaxies such as the Magellanic Clouds, shown in Figure 19.19. The Magellanic Clouds, located 160,000 light-years away, contain many Cepheid variable stars, allowing us to improve the calibration of these standard candles. Cepheid variables in turn allow us to accurately measure distances to galaxies as far out as about 100 million light-years, but even this is not enough for us to measure a reliable value for the Hubble constant. In that volume of space, however, there are many galaxies that we can scour for yet more powerful standard candles. Among the best of these are Type I supernovae.[1]

Recall that Type I supernovae are thought to occur when gas flows from an evolved star onto its white dwarf companion, pushing the white dwarf over the Chandrasekhar

---

[1] Formally, a specialist in the field would refer to these as Type Ia supernovae, as noted previously in Chapter 16.

limit for the mass of an electron degenerate object. When this happens, the overburdened white dwarf first begins to collapse, and then explodes. Because all Type I supernovae occur in white dwarfs of the same mass, we might

Type I supernovae are very luminous standard candles.

expect all such explosions to have about the same luminosity. This is borne out by observations of Type I supernovae in galaxies with known distances. With a peak luminosity that outshines a hundred billion suns (**Figure 20.5**), Type I supernovae can be seen and measured using current telescopes reaching much of the way to the edge of the observable universe.

It is often said that a chain is no stronger than its weakest link. When constructing the cosmic distance ladder, the situation is even worse: The final ladder is no better than the *combination* of the weaknesses in each of its rungs. For the last few decades of the 20th century, astronomers working to measure H_0 were largely split into two camps. One group favored a Hubble constant of around 18 km/(s Mly), while a second camp viewed the data as supporting a value of around 35 km/(s Mly). A virtual war raged between these two groups for years, until HST observations in the mid-1990s began to converge on an intermediate value of about 22 km/(s Mly).[2] The most recent (2006) analysis of satellite data using other measures of distance has yielded results completely consistent with that value.

Figure 20.6 plots the measured recession velocities of galaxies against their measured distances. Notice how well the universe follows Hubble's Law. Today most astronomers

Current measurements give H_0 at 22 km/(s Mly).

believe we have measured the Hubble constant to an accuracy of 10 percent and perhaps better. Therefore, we can be confident that it lies between 20 and 25 km/(s Mly) and that it is likely to be further refined in the years to come.

Hubble's Law Maps the Universe in Space and Time

So why all the fuss about Hubble's Law and the Hubble constant? What do these tell us about the universe? First, Hubble's Law gives us a practical tool for measuring distances to remote objects. Once we know the value of H_0,

BEFORE · AFTER · Supernova

FIGURE 20.5 The same galaxy before and after the explosion of a Type I supernova. Type I supernovae are extremely luminous standard candles.

Hubble's Law allows us to turn a straightforward measurement of the redshift of a (relatively nearby) galaxy into knowledge of its distance. For example, a galaxy with a

Redshift tells us a galaxy's distance.

redshift of 0.1 is 1.4 billion light-years away; a galaxy with a redshift of 0.2 is twice that distance away. In short, once we know H_0, Hubble's Law makes the once-difficult task of measuring distances in the universe relatively easy, providing us with a tool to literally map the structure of the universe. We will put this tool to good use in the next chapter when we turn our attention to the large-scale structure of the universe.

Hubble's Law does more than place galaxies in space. It also places galaxies in time. Light travels at a huge but finite speed. When we look at the Sun, we see it as it existed 8 minutes ago. When we look at Alpha Centauri, the nearest stellar system beyond the Sun, we see it as it existed 4.3

When observing the universe, elsewhere is "elsewhen."

years ago. When we look at the center of our galaxy, the picture that we see is 27,000 years old. When looking at a distant object, we speak of its **look-back time**—the time it has taken for the light from that object to reach our telescope. As we look into the distant universe, look-back times become very great indeed. The distance to a galaxy where $z = 0.1$ is 1.4 billion light-years (assuming $H_0 = 22$ km/(s Mly)), so the look-back time to that galaxy is 1.4 billion years. The look-back time to a galaxy where $z = 0.2$ is 2.7 billion years. As we look at objects with greater and greater redshifts, we are seeing increasingly younger versions of our universe.

---

[2] Astronomers normally express H_0 in units of km/s per million parsecs rather than km/s per million light-years. In these units, 22 km/(s Mly) corresponds to 72 km/(s Mpc).

(a)

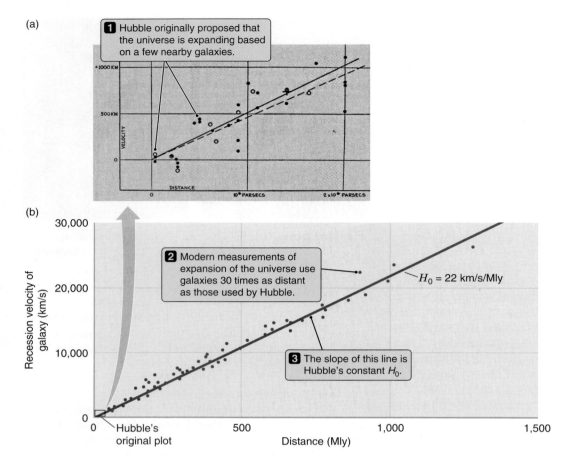

1 Hubble originally proposed that the universe is expanding based on a few nearby galaxies.

(b)

2 Modern measurements of expansion of the universe use galaxies 30 times as distant as those used by Hubble.

$H_0 = 22$ km/s/Mly

3 The slope of this line is Hubble's constant H_0.

Recession velocity of galaxy (km/s)

Hubble's original plot

Distance (Mly)

FIGURE 20.6 (a) Hubble's original figure illustrating that more distant galaxies are receding faster than less distant galaxies. (b) Modern data on galaxies up to 30 times farther away than those studied by Hubble show that recession velocity is proportional to distance.

20.3 The Universe Began in the Big Bang

Hubble's Law provides a very powerful and practical tool for mapping the distribution of galaxies throughout the universe. Yet the most significant aspect of Hubble's Law is what it tells us about the structure of the universe itself. We know that all galaxies in the universe are moving away from each other, but if we could run the movie backward in time, we would find the galaxies getting closer and closer together.

Figure 20.7 shows two galaxies located 100 Mly ($d_G = 9.5 \times 10^{20}$ km) away from each other. If these two galaxies are flying apart from each other, then at some point in the past, they must have been together in the same place at the same time. According to Hubble's Law, the distance between these two galaxies is increasing at the rate $v_r = H_0 \times d_G = 2,200$ km/s, assuming that $H_0 = 22$ km/(s Mly). Traveling at this speed, the two galaxies would have taken about 13.6 billion years to travel the 100 Mly that separates them. [Time = Distance/Speed = $(9.5 \times 10^{20}$ km)/(2,200 km/s) = 4.32×10^{17} s.] In other words, *if* expansion of the universe

has been constant, two galaxies that today are 100 Mly apart started out at the same place 13.6 billion years ago.

Now we do the same calculation with two galaxies that are twice as far (200 Mly) apart (see Figure 20.7). These two galaxies are twice as far apart, but the distance between them is increasing twice as rapidly: $v_r = H_0 \times d_G = 4,400$ km/s. We again calculate time equals distance divided by speed (twice the distance divided by twice the speed) to find that

> Hubble's Law shows that galaxies have been traveling apart for 13 billion years.

these galaxies *also* took 13.6 billion years to reach their current locations. We find that galaxies beginning 300 Mly apart were in the same place 13.6 billion years ago, as well. We could do this calculation again and again for any pair of galaxies in the universe today. The farther apart the two galaxies are, the faster they are moving. The thing that is the *same* for all galaxies is the *time* that it took them to get to where they are today.

A look at the following math makes this clear. Velocity equals Hubble's constant times distance, so when we calculate a time by saying "time equals distance divided by

$1/H_0$ is one measure of the age of the universe.

velocity," the distance factors on top and bottom cancel out. Writing this out as an equation, we get

$$\text{Time} = \frac{\text{Distance}}{\text{Velocity}} = \frac{\text{Distance}}{H_0 \times \text{Distance}} = \frac{1}{H_0}.$$

The startling implications of this result are illustrated in **Figure 20.8**. About 6.8 billion years ago, when the universe was half its present age, all of the galaxies in the universe were separated from each other by half their present distances. Twelve billion years ago, all of the galaxies in the universe must have been separated from each other by about a tenth of their present distances. Assuming that gal-

Expansion started in a Big Bang.

axies have been moving apart at the same speed that we see today, then 13.6 billion years ago (a time equal to $1/H_0$), *all the stars and galaxies that make up today's universe must have been concentrated together at the same location!* The value of 1 divided by the Hubble constant is referred to as the **Hubble time**. If our line of reasoning is correct, then today's universe is hurtling outward from a tremendous explosion that took place approximately 13 to 14 billion years ago. This colossal event, which marked the beginning of our universe, is referred to as the **Big Bang**.

The idea of the Big Bang greatly troubled many astronomers in the early and middle years of the 20th century. Several different suggestions were put forward to explain the observed fact of Hubble expansion without resorting to the idea that the universe came into existence in an extraordinarily dense fireball billions of years ago. However, as more and more observations have come in, and more discoveries about the structure of the universe have been made, the Big Bang theory has only grown stronger. Today only a very few astronomers working seriously in this field doubt whether the Big Bang took place. Virtually all the major predictions of the Big Bang theory (expansion of the universe being but one of them) have proven to be correct. As we will see, the Big Bang theory for the origin of our universe is now such a well-corroborated theory that most astronomers would probably say it has crossed into the realm of scientific fact.

The implications of Hubble's Law are striking. This single discovery forever changed our concept of the origin, history, and possible future of the universe in which we live. At the same time Hubble's Law has pointed to many new questions about the universe. To address them we next need to consider exactly what we mean by the term *expanding universe*.

Galaxies Are *Not* Flying Apart through Space

The mental picture that you have of the expanding universe at this point in our discussion is probably one of a cloud of debris from an explosion flying outward through space. One of the first questions students usually ask about the Big Bang is, "Where did the explosion take place?" The answer to this question, amazing as it seems, is that the explosion took

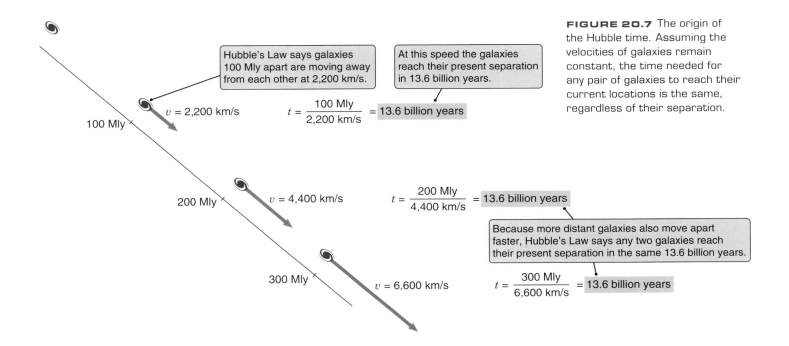

FIGURE 20.7 The origin of the Hubble time. Assuming the velocities of galaxies remain constant, the time needed for any pair of galaxies to reach their current locations is the same, regardless of their separation.

Hubble's Law says galaxies 100 Mly apart are moving away from each other at 2,200 km/s.

At this speed the galaxies reach their present separation in 13.6 billion years.

100 Mly

$v = 2,200$ km/s

$t = \dfrac{100 \text{ Mly}}{2,200 \text{ km/s}} = 13.6$ billion years

200 Mly

$v = 4,400$ km/s

$t = \dfrac{200 \text{ Mly}}{4,400 \text{ km/s}} = 13.6$ billion years

Because more distant galaxies also move apart faster, Hubble's Law says any two galaxies reach their present separation in the same 13.6 billion years.

300 Mly

$v = 6,600$ km/s

$t = \dfrac{300 \text{ Mly}}{6,600 \text{ km/s}} = 13.6$ billion years

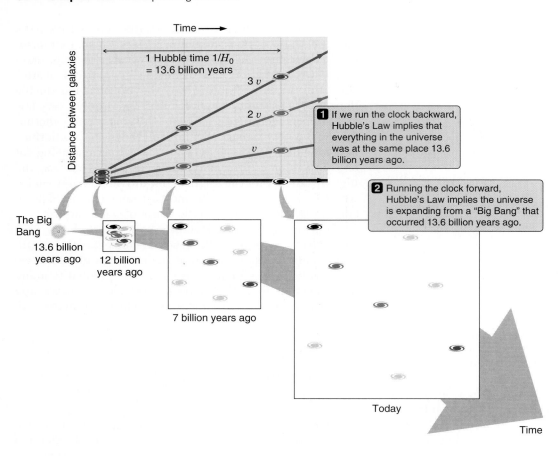

Time ⟶

Distance between galaxies

1 Hubble time $1/H_0$
= 13.6 billion years

3 v

2 v

v

1 If we run the clock backward, Hubble's Law implies that everything in the universe was at the same place 13.6 billion years ago.

2 Running the clock forward, Hubble's Law implies the universe is expanding from a "Big Bang" that occurred 13.6 billion years ago.

The Big Bang
13.6 billion years ago

12 billion years ago

7 billion years ago

Today

Time

FIGURE 20.8 Looking backward in time, the distance between any two galaxies is smaller and smaller, until all matter in the universe is concentrated together at the same point, the Big Bang.

place *everywhere.* Wherever you are in the universe today, you are sitting at the site of the Big Bang. The reason for this is that galaxies are not flying apart through space at all.

The Big Bang happened everywhere.

Rather, it is *space itself* that is expanding, carrying the stars and galaxies that populate the universe along with it.

This may seem an incredible notion, but we have already dealt with the basic ideas that allow us to understand the expansion of space. In our discussion of neutron stars and black holes back in Chapter 17, we encountered Einstein's general theory of relativity. General relativity says that space is distorted by the presence of mass, and that the consequence of this distortion is gravity. For example, the mass

Space itself is expanding similar to a two-dimensional rubber sheet.

of the Sun, like any object, distorts the geometry of space-time around it, so that Earth, coasting along in its inertial frame of reference, follows a curved path around the Sun. We illustrated this with the analogy of a ball placed on a stretched rubber sheet, showing how the ball distorted the surface of the sheet.

There are other ways to distort the surface of a rubber sheet as well. Imagine a number of coins placed on a rubber sheet, as shown in **Figure 20.9**. Suppose we grab the edges of the sheet and begin pulling them outward. As the rubber sheet stretches, each coin remains at the same location on the surface of the sheet, but the distances between the coins increase. Two coins sitting close to each other move apart only slowly, while coins farther apart move away from each other more rapidly. In other words, the distances and relative motions of the coins on the surface of a rubber sheet would obey a Hubble-type relationship as the sheet is stretched.

This is what is happening in the universe, with galaxies taking the place of the coins and space itself taking the place of the rubber sheet. Obviously, in the case of coins on a rubber sheet, there is a limit to how far we can stretch the sheet before it breaks. With space and the real universe, there is no such limit. The fabric of space can, in principle, go on expanding forever. Hubble's Law is the observational consequence of the fact that the space making up the universe is expanding.

Expansion Is Described with a Scale Factor

When astronomers discuss the expansion of the universe, they talk in terms of the **scale factor** of the universe. To understand this concept, we return to our analogy of the rubber sheet. Suppose we place a ruler on the surface of the sheet and draw a tick mark every centimeter, as in **Figure 20.10(a)**. If we want to know the distance between two points on the sheet, all we need to do is count the marks between the two points and multiply by 1 cm per tick mark.

But as the sheet is stretched, the distance between the tick marks does not remain 1 cm. When the sheet is stretched to 150 percent of the size that it had when we drew our ruler, each tick mark is separated from its neighbors by one and one half times their original distance, or 1.5 cm. If we wanted to know the distance between two points, we could still count the marks, but we would have to *scale up* the distance in tick marks by 1.5 to find the distance in

centimeters. The scale factor of the sheet is now 1.5. And when the sheet is twice the size that it was when we drew the ruler (**Figure 20.10(b)**), each mark would correspond to 2 cm of actual distance; the scale factor of the sheet would now be 2. The scale factor tells us the size of the sheet relative to its size at the time when we drew our ruler. The scale factor also tells us how much the distance between points on the sheet has changed.

We can apply this same idea to the universe. Suppose we choose today to lay out a "cosmic ruler" on the fabric of space, placing an imaginary tick mark every 10 Mly. We

---

The scale factor R_U increases as the universe expands.

---

define the scale factor of the universe at this time to be 1. In the past, when the universe was smaller, distances between the points in space marked by our cosmic ruler would have been less than 10 Mly apart. The scale factor of that

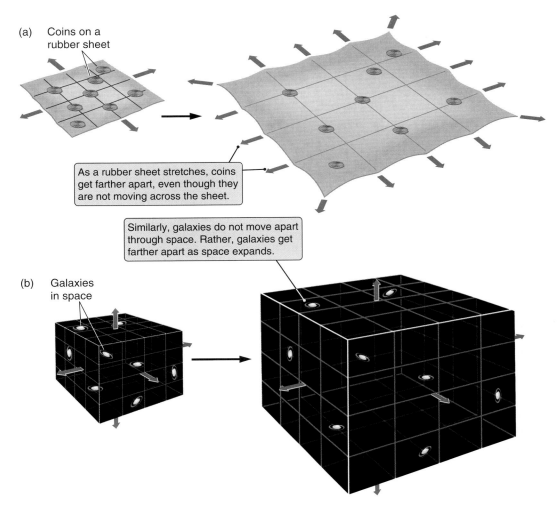

(a) Coins on a rubber sheet

As a rubber sheet stretches, coins get farther apart, even though they are not moving across the sheet.

Similarly, galaxies do not move apart through space. Rather, galaxies get farther apart as space expands.

(b) Galaxies in space

FIGURE 20.9 (a) As a rubber sheet is stretched, coins on its surface move farther apart even though they are not moving with respect to the sheet itself. Any coin on the surface of the sheet will observe a Hubble Law in every direction. (b) In analogous fashion, galaxies in an expanding universe are not flying apart through space. Rather, space itself is stretching.

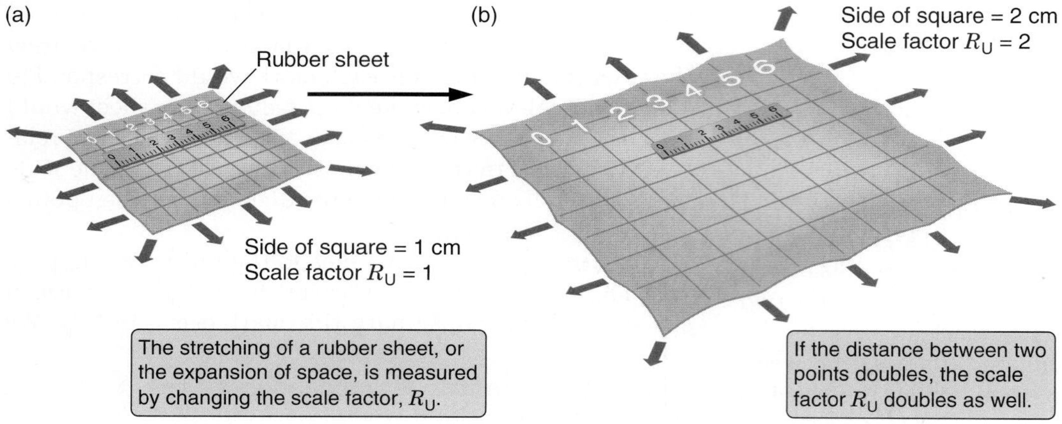

(a)

Rubber sheet

Side of square = 1 cm
Scale factor $R_U = 1$

The stretching of a rubber sheet, or the expansion of space, is measured by changing the scale factor, R_U.

(b)

Side of square = 2 cm
Scale factor $R_U = 2$

If the distance between two points doubles, the scale factor R_U doubles as well.

FIGURE 20.10 Tick marks are drawn on a rubber sheet at a distance of 1 cm apart (a). As the sheet is stretched, the tick marks move farther apart. When the spacing between the tick marks is 2 cm, or twice the original value, we say that the scale factor of the sheet, R, has doubled (b). A similar scale factor, R_U, is used to describe the expansion of the universe.

younger, smaller universe would have been less than 1 compared with today. In the future, as the universe continues to expand, the distances between the tick marks on our cosmic ruler will grow to more than 10 Mly, and the scale factor of the universe will be greater than 1. In this way we can use the scale factor, usually written as R_U, to keep track of the changing scale of the universe.

It is important to remember, when thinking about the expanding universe, that the laws of physics are themselves unchanged by the changing scale factor. For example, when

Expansion does not affect local physics— stars, atoms, or anything else.

we stretched out the rubber sheet, we did not change the properties of the coins on its surface. In like fashion, as the universe expands, the sizes and other physical properties of atoms, stars, and galaxies also remain unchanged.

We return to the question of locating the center of expansion. Looking back in time, the scale factor of the universe gets smaller and smaller, approaching zero as we get closer and closer to the Big Bang. The fabric of space that today spans billions of light-years spanned much smaller distances when the universe was young. When the universe was only a day old, all of the space that we see today amounted to a region only a few times the size of our Solar System. When the universe was a 50th of a second old, the vast expanse of space that makes up today's observable universe (and all the matter in it) occupied a volume only the size of today's Earth. As we continue to approach the Big Bang itself going backward in time, the space that makes up today's observable universe becomes smaller and smaller—the size of a grapefruit, a marble, an atom, a proton.... Every point in the fabric of space that makes up today's universe was right there at the beginning, a part of that unimaginably tiny, dense universe that emerged from the Big Bang.

These points bear repetition: Where is the center of the Big Bang? There is no center. The Big Bang did not occur at a specific point in space because space itself came into existence with the Big Bang. Where did the Big Bang happen? It happened everywhere, including right where you are sitting. This is an important result. If there were a particular point in today's universe that marked the site of the Big Bang, that would be a very special point indeed. But there is no such point. The Big Bang happened everywhere. A Big Bang universe is homogeneous and isotropic, consistent with the cosmological principle.

Redshift Is Due to the Changing Scale Factor of the Universe

General relativity gives us a powerful tool for interpreting Hubble's great discovery. It also forces us to rethink just what we mean when we talk about the redshift of distant galaxies. Although it is true that the distance between galaxies is increasing as a result of the expansion of the universe, and that we can use the equation for Doppler shifts to measure the redshifts of galaxies, these redshifts are not due to Doppler shifts at all! As light comes toward us from distant galaxies, the scale factor of the space through which the light travels is constantly increasing; and as it does so, the distance between adjacent wave crests increases as well. The light is "stretched out" as the space it travels through expands. (See **Foundations 20.1**.)

Let's return to our rubber sheet analogy. If we were to draw a series of bands on the rubber sheet to represent the crests of an electromagnetic wave, as in **Figure 20.12**, we could watch what happens to the wave as the sheet is stretched out. By the time the sheet is stretched to twice its original size—that is, by the time the scale factor of the sheet is 2—the distance between wave crests has doubled.

FOUNDATIONS 20.1

When Redshift Exceeds 1

In our discussion of Doppler shift, we found that $(\lambda_{observed} - \lambda_{rest})/\lambda_{rest}$ is equal to the velocity of an object away from us, divided by the speed of light. In this chapter we see how Edwin Hubble used this result to interpret the observed redshifts of galaxies as evidence that galaxies throughout the universe are moving away from us. Einstein's special theory of relativity says that nothing can move faster than the speed of light. Hubble's initial assumption was that redshifts are due to the Doppler effect. The resulting relation $z = v_r/c$ would then seem to imply that no object can have a redshift z greater than 1. Yet that is not the case. Astronomers routinely observe redshifts significantly in excess of 1. As of this writing, the most distant objects known have redshifts of over 6! What is wrong here? It is worth taking a brief digression to consider how redshifts can exceed 1.

The first thing to note is that, to arrive at the expression for the Doppler effect—$v_r/c = (\lambda_{observed} - \lambda_{rest})/\lambda_{rest}$—we have to *assume* that v_r is much less than c. If that were not the case—if v_r were close to c—then we would have to take into account more than just the fact that the waves from an object are stretched out by the object's motion away from us. We would also have to consider relativistic effects, including the fact that moving clocks run slowly. When combining these effects, we would find that as the speed of an object approaches the speed of light, its redshift becomes arbitrarily large (**Figure 20.11**).

A second source of redshift is the *gravitational redshift* discussed in Chapter 17. As light escapes from deep within a gravitational well, it loses energy, so photons are shifted to longer and longer wavelengths. If the gravitational well is deep enough, then the observed redshift of this radiation can be boundlessly large. In fact, the event horizon of a black hole—that is, the surface around the black hole from which not even light can escape—is where the gravitational redshift becomes infinite.

Cosmological redshift, which is most relevant to this chapter, results from the amount of "stretching" space has undergone during the time the light from its original source has been en route to us. The amount of stretching that has occurred is given by the factor $1 + z$. When we look at a distant galaxy whose redshift $z = 1$, we are seeing the universe at a time when it was half the size that it is today. When we see light from a galaxy with $z = 2$, we are seeing the universe when it was one-third its current size. Nearby, this means that distance and look-back time are proportional to z. As we look back closer and closer to the Big Bang, however, redshift climbs more and more rapidly.

Doppler's original formula is essentially correct—as long as we are looking at shifts due to an object's motion and the velocities we measure are far less than the speed of light. In this case, $v_r/c = (\lambda_{observed} - \lambda_{rest})/\lambda_{rest}$. When we look at the motions of orbiting binary stars or the peculiar velocities of galaxies relative to the Hubble flow, this equation works just fine. But anytime you have a redshift of 1 or greater, you should immediately recognize that you are now in the realm of relativity.

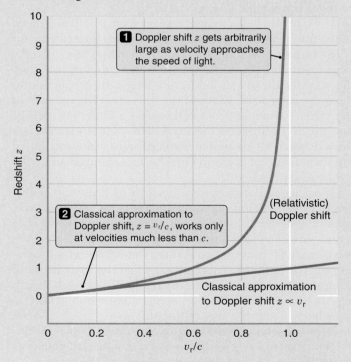

FIGURE 20.11 Plot of the redshift z of an object versus its recession velocity v_r as a fraction of the speed of light. According to special relativity, as v_r approaches c, the redshift becomes large without limit.

1 Doppler shift z gets arbitrarily large as velocity approaches the speed of light.

2 Classical approximation to Doppler shift, $z = v_r/c$, works only at velocities much less than c.

(Relativistic) Doppler shift

Classical approximation to Doppler shift $z \propto v_r$

Redshift z

v_r/c

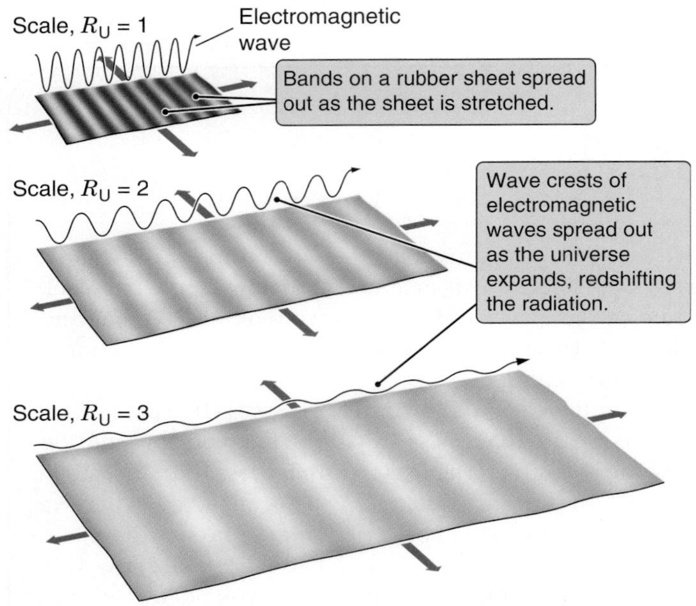

FIGURE 20.12 Bands drawn on a rubber sheet represent the positions of the crests of an electromagnetic wave in space. As the rubber sheet is stretched—that is, as the universe expands—the wave crests get farther apart. The light is redshifted.

When the sheet has been stretched to three times its original size (a scale factor of 3), the wavelength of the wave will be three times what it was originally.

We apply this idea to light coming from a distant galaxy. When the light left the galaxy of its origin, the scale factor of the universe was smaller than it is today. The universe expanded while the light was in transit, and as it did so, the wavelength of the light grew longer in proportion to the increasing scale factor of the universe. The redshift of light from distant galaxies is therefore a direct measure of how much the universe has expanded since the time when the radiation left its source. Redshift measures how much the scale factor of the universe, R_U, has changed since the light was emitted.

If we see radiation with a redshift of 1, then its wavelength is twice as long as when that radiation left its source. When the light left its source, R_U was equal to ½ compared with today. If we see radiation with a redshift of 2, then the wavelength of the radiation is three times its original wavelength. The radiation was emitted when the scale factor of the universe was ⅓. This wonderfully direct relationship lets us convert from the observed redshift of a galaxy to knowledge about the size of the universe at the look-back time to that galaxy. Written as an equation, the scale factor of the universe we see when looking at a distant galaxy is equal to 1 divided by 1 plus the redshift of the galaxy:

$$R_U(z) = \frac{1}{1+z}.$$

20.4 The Major Predictions of the Big Bang Theory Are Resoundingly Confirmed

The questions we are grappling with when discussing ideas like the origin of the universe are some of the most fundamental questions humankind can ask about the universe. Throughout human history, answers to these same questions have been among the most prized goals of philosophers and theologians. It is quite remarkable that we live in a time when we are finding real, testable answers to these questions by appealing not to divine inspiration or the blind logic of philosophy, but rather to the empirical methods of science. It is essential, then, that we place extraordinary demands on the quality of the evidence that we rely on to support the theory of the Big Bang. What evidence is there, apart from the observed expansion of the universe itself, that requires us to accept that the Big Bang actually took place?

We See Radiation Left Over from the Big Bang

The story of the single most important confirmation of the Big Bang theory begins in the mid-1940s, when **George Gamow** (1904–1968) was thinking, along with his student **Ralph Alpher** (1921–), about the implications of Hubble expansion. When a gas is compressed, it grows hotter. Similarly, as a gas expands, it becomes cooler. So, reasoned Gamow and Alpher, since the universe is expanding, it must also be cooling. Consequently, when the universe was very young and small, it must have consisted of an extraordinarily hot, dense gas. As with any hot, dense gas, this early universe would have been awash in the same kind of radiation that we have encountered so many times before on our journey of discovery: radiation from a blackbody, which exhibits a Planck spectrum.

Gamow and Alpher took this idea a step further. As the universe expanded, they reasoned, this radiation would have been redshifted to longer and longer wavelengths. Recall Wien's Law, which states that the temperature associated with Planck radiation is inversely proportional to the peak wavelength: $T = (2{,}900\ \mu m)/\lambda_{peak}$. Shifting the wavelength of Planck radiation to longer and longer wavelengths is therefore the equivalent of shifting the characteristic temperature of the radiation to lower and lower values. As illustrated in **Figure 20.13**, doubling the wavelength of the photons in a Planck spectrum by doubling the scale factor of the universe is equivalent to cutting the temperature of the Planck spectrum in half.

On April Fools' Day, 1948, Alpher, **Hans Bethe** (1906–2005), and Gamow published a paper asserting that this radiation should still be visible today and should have a Planck spectrum with a temperature in the neighborhood of 5 to 10 K. (The paper actually listed as authors not only Alpher and Gamow, but also Hans Bethe, who, though an esteemed

Alpher, Bethe, and Gamow predicted the glow from a young hot universe in 1948.

physicist in his own right, did not actively participate in this research. Gamow, being a renowned jokester, added Bethe's name between his and Alpher's. "Alpher, Bethe, and Gamow" is a play on the first three letters of the Greek and Hebrew alphabets—an appropriate authorship for a paper making predictions about the very early universe!)

This prediction languished until the early 1960s, when two physicists from Bell Laboratories, **Arno Penzias** (1933–) and **Robert Wilson** (1936–), were trying to bounce radio signals off the newly launched ECHO satellites. This hardly seems much of a feat today, when we routinely use handheld units that communicate directly with satellites. Yet at the time it pushed radio technology to its limits. Penzias and Wilson needed a very sensitive microwave telescope for their work. Any spurious signals coming from the telescope itself might wash out the faint signals bounced off a satellite. Penzias and Wilson, shown in **Figure 20.14** along with their radio telescope, worked tirelessly to eliminate all possible sources of interference originating from within their instrument. This work included such endless and menial tasks as keeping the telescope free of bird droppings and other

extraneous material. Even so, Penzias and Wilson found that no matter how hard they tried to eliminate sources of extraneous noise, they could still detect a faint microwave signal when they pointed the telescope at the sky. Eventually they came to accept that the signal they were detecting was real. The sky faintly glows in microwaves.

In the meantime, Robert Dicke (1916–1997) and his colleagues at Princeton University had also predicted a hot early universe, arriving independently at the same basic conclusions that Alpher and Gamow had reached two decades earlier. When Dicke and colleagues heard of the signal that Penzias and Wilson had found, they interpreted it as the radiation left behind by the hot early universe. The strength of the detected signal was consistent with the glow

Penzias and Wilson discovered the cosmic background radiation.

from a blackbody with a temperature of about 3 K, very close to the predicted value. Their results, published in 1965, reported the discovery of the glow left behind by the Big Bang. Penzias and Wilson shared the 1979 Nobel Prize in physics for their remarkable discovery. (It is worth noting that timing can be everything in science. Alpher, who first predicted the existence of a faint glow from the Big Bang, searched unsuccessfully for the signal 10 years before Penzias and Wilson made their discovery. Unfortunately, however, the technology of the late 1940s and early 1950s was simply not up to the task.)

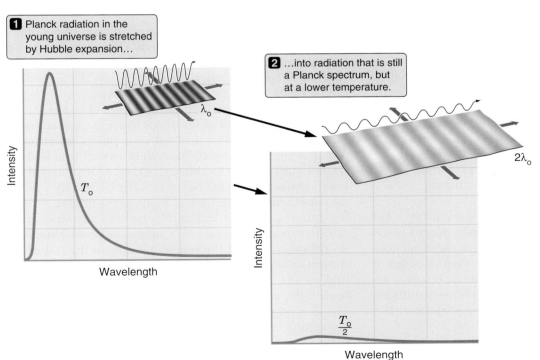

1 Planck radiation in the young universe is stretched by Hubble expansion…

2 …into radiation that is still a Planck spectrum, but at a lower temperature.

λ_0

$2\lambda_0$

Intensity

Wavelength

T_0

Intensity

Wavelength

$\dfrac{T_0}{2}$

FIGURE 20.13 As the universe expanded, Planck radiation left over from the hot young universe was redshifted to longer wavelengths. Redshifting a Planck spectrum is equivalent to lowering its temperature.

FIGURE 20.14 Penzias and Wilson next to the horn telescope with which they discovered the cosmic background radiation.

This radiation left over from the early universe is called the **cosmic background radiation (CBR).** Today the cosmic background radiation and the conditions in the early universe are much better understood than they were in the early 1960s. The origin of the CBR is illustrated in **Figure 20.15.**

> **The CBR is thermal radiation that arose when the universe was hot and ionized.**

When the universe was young, it was hot enough that all of the atoms in the universe were ions. In our discussion of the structure of the Sun and stars, we found that radiation does not travel well through an ionized plasma. Free electrons in a plasma interact strongly with the radiation, blocking its progress. At this time in the early universe, the conditions within the universe were much like the conditions within a star: The universe was an opaque blackbody.

As the universe expanded, the gas filling the universe cooled. By the time the universe was about a thousandth of its current size, the temperature had dropped to a few

thousand kelvins, so protons and electrons were able to combine to form hydrogen atoms. This event, called the **recombination** of the universe, occurred when the universe was several hundred thousand years old.

Hydrogen atoms are much less effective at blocking radiation than free electrons are; so when recombination occurred, the universe suddenly became transparent to radiation. Since that time the radiation left behind from the Big Bang has been able to travel largely unimpeded throughout the universe. At the time of recombination, when the

> **Since recombination, the CBR has traveled freely and cooled by a factor of 1,000.**

temperature of the universe was a few thousand kelvins, the wavelength of this radiation peaked at around 1 μm according to Wien's Law. As the universe expanded, this radiation was redshifted to longer and longer wavelengths. Today the scale of the universe has increased a thousand-fold since recombination, and the peak wavelength of the cosmic background radiation has increased by a thousand-fold as well to a value close to 1 mm. The spectrum of the CBR still has the shape of a Planck spectrum, but with a characteristic temperature of 2.73 K—only a thousandth what it was at the time of recombination.

COBE Removed Any Reasonable Doubt That the CBR Is Real

The presence of cosmic background radiation with a Planck spectrum is a very strong prediction of the Big Bang theory. Penzias and Wilson had confirmed that a signal with the correct strength was there, but they could not say for certain whether the signal they saw had the spectral shape of a Planck spectrum. From the remainder of the 1960s to the 1980s, most experiments at different wavelengths supported these same conclusions. Yet it was not until the end of the 1980s that the predictions of Big Bang cosmology for the CBR were put to the ultimate test. The year 1989 saw the launch of a satellite called the Cosmic Background Explorer, or COBE. COBE carried on-board instruments capable of making extremely precise measurements of the CBR at many wavelengths, from a few micrometers out to 1 cm. In January 1990 hundreds of astronomers gathered in a large conference room in Washington, D.C., at the winter meeting of the American Astronomical Society to hear the COBE team present its first results. Security surrounding the new findings had been tight, so the atmosphere in the room was electric. The tension did not last for long: Presentation of a single viewgraph brought the room's occupants to their feet in a spontaneous ovation.

The data shown on that viewgraph are reproduced in **Figure 20.16**. The small circles in the figure are the COBE

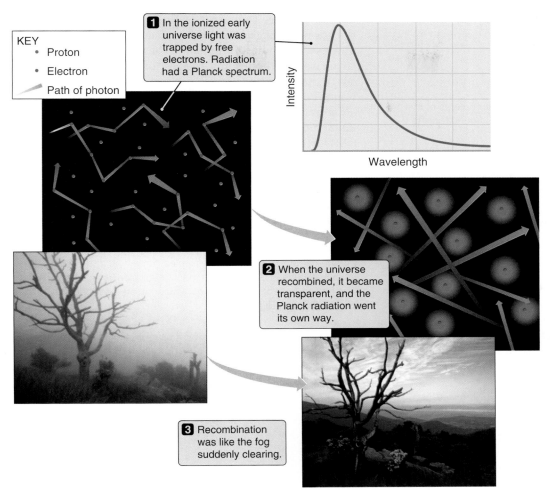

1 In the ionized early universe light was trapped by free electrons. Radiation had a Planck spectrum.

Intensity

Wavelength

2 When the universe recombined, it became transparent, and the Planck radiation went its own way.

KEY
• Proton
• Electron
Path of photon

3 Recombination was like the fog suddenly clearing.

FIGURE 20.15 The origin of the cosmic background radiation. Prior to recombination the universe was like a foggy day. Radiation interacted strongly with free electrons and so could not travel far. The trapped radiation had a Planck spectrum. When the universe recombined, the fog cleared, and this radiation was free to go on its way.

measurements of the CBR at different frequencies. The uncertainty in each measurement is far less than the size of each dot. The line in the figure, which runs perfectly through

COBE unambiguously showed the CBR to have a Planck spectrum at 2.73 K.

the data points, is a Planck spectrum with a temperature of 2.73 K. The agreement between theoretical prediction and observation is truly remarkable. The observed spectrum so perfectly matches the one predicted by Big Bang cosmology that there can be no real doubt we are seeing the residual radiation left behind from the primordial fireball of the early universe.

The CBR Measures Earth's Motion Relative to the Universe Itself

COBE provided us with much more than a measurement of the spectrum of the cosmic background radiation. **Figure**

20.17(a) shows a map obtained by COBE of the CBR from the entire sky. The different colors in the map correspond to variations in the temperature of the CBR. The differences are not as extreme as the colors might suggest, however. The range in temperature in the map corresponds to a variation of only about 0.1 percent in the temperature of the CBR. Most of this range of temperature is present because one side of the sky looks slightly warmer than the opposite side of the sky. This difference has nothing to do with the large-scale structure of the universe itself, but rather is the result of the motion of Earth with respect to the CBR.

Time and again on our journey we have stressed that there is no preferred frame of reference. The laws of physics are the same in *any* inertial reference frame, so none is

Our motion makes the CBR slightly hotter in the direction we are moving toward and slightly cooler behind us.

better than any other. Yet in a certain sense there *is* a preferred frame of reference at any point in the universe. This is the frame of reference that is at rest with respect to the

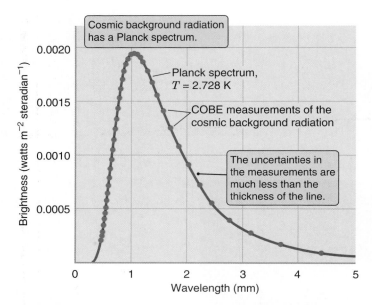

FIGURE 20.16 The spectrum of the cosmic background radiation as measured by the Cosmic Background Explorer (COBE) satellite (red dots). The uncertainty in the measurement at each wavelength is much less than the size of a dot. The line running through the data is a Planck spectrum with a temperature of 2.728 K.

expansion of the universe and in which the CBR is isotropic. The COBE map shows that one side of the sky is slightly hotter than the other because our Sun and our Earth are moving at a velocity of 368 km/s in the direction of the constellation of Crater, relative to this cosmic reference frame. Radiation coming from the direction in which we are moving is slightly blueshifted (shifted to a higher characteristic temperature) by our motion, whereas radiation coming from the opposite direction is Doppler-shifted toward the red (or cooler temperatures). As we shall see in Chapter 21, our motion is due to a combination of factors, including the motion of our Sun around the Milky Way and the motion of the galaxy relative to the CBR.

If we subtract from the COBE map this asymmetry in the CBR caused by the motion of Earth, only slight variations in the CBR remain, as shown in **Figure 20.17(b)**. The slight variations seen in this map have an amplitude that

> **COBE found, and WMAP further revealed, tiny variations in the CBR resulting from the formation of structure during early times.**

is only about 1/100,000 the brightness of the CBR. This means that the brighter parts of this image are only about 1.00001 times brighter than the fainter parts. These slight variations might not seem like much, but they are actually of crucial importance in the history of the universe. These

fluctuations are the result of the gravitational redshift of the cosmic background radiation caused by concentrations of mass in the early universe. These concentrations later gave rise to galaxies and the rest of the structure that we see in the universe today. Subsequent observations from the South Pole and from instruments carried aloft by balloons support the COBE findings. More recently, beginning in 2001, more precise measurements of the variations of the CBR have been carried out by a satellite called the Wilkin-

FIGURE 20.17 (a) The COBE map of the cosmic background radiation. The CBR is slightly hotter (by about 0.003 K) in one direction in the sky than in the other direction. This is due to Earth's motion relative to the CBR. (b) The COBE map with Earth's motion removed, showing tiny ripples remaining in the CBR. (c) WMAP has provided the highest resolution yet of the CBR. The radiation seen in this image was emitted less than 400,000 years after the Big Bang.

(a)

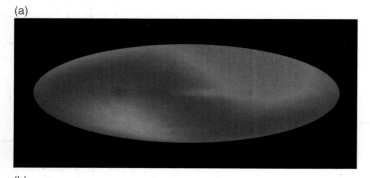

(b)

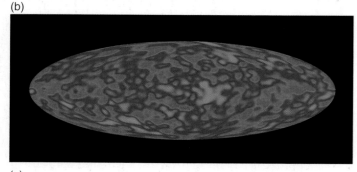

(c)

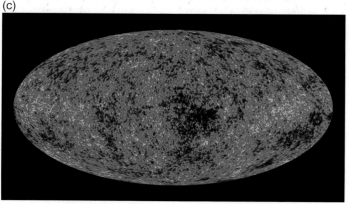

son Microwave Anisotropy Probe, or WMAP. In **Figure 20.17(c)** we show the ripples measured by WMAP with much higher resolution than could be detected by COBE. The much higher-resolution maps obtained by WMAP have profound implications for our understanding of the origin of structure in the universe and allow us to determine several cosmological parameters. For example, the value of the Hubble constant we use in this book is identical to the value inferred from the WMAP experiment.

The Big Bang Theory Correctly Predicts the Abundance of the Least Massive Elements

The next confrontation between the predictions of the Big Bang theory and observations of the universe came from a very different direction. When the universe was only a few minutes old, the temperature and density in the universe were high enough for nuclear reactions to take place. Just as we can use our knowledge of nuclear physics to calculate the nuclear reactions occurring in the interiors of stars, we can

use this knowledge to calculate the nuclear reactions that took place in the early universe. Collisions between protons in the early universe built up low-mass nuclei, including deuterium (heavy hydrogen) and isotopes of helium, lithium, beryllium, and boron. The formation of new elements in this early nuclear brew, called **Big Bang nucleosynthesis**, determined the final chemical composition of the matter that emerged from the hot phase of the Big Bang.

We have previously discussed how differences in the abundances of the products of *stellar* nucleosynthesis help us track the chemical evolution of the universe and the history of star formation. These ideas played an important role in our discussion of the Milky Way in Chapter 19, and they will come to the fore again in our later discussion of galaxy formation. Here we focus on the products of Big Bang nucleosynthesis.

The amounts of various elements that formed from Big Bang nucleosynthesis depended in detail on the temperature and density of normal matter in the early universe. **Figure 20.18** shows the calculated predictions of Big Bang nucleosynthesis plotted as a function of the present-day density of normal (luminous) matter in the universe. The first thing we see from this figure is that about 24 percent of the mass of the normal matter formed in the early universe should have

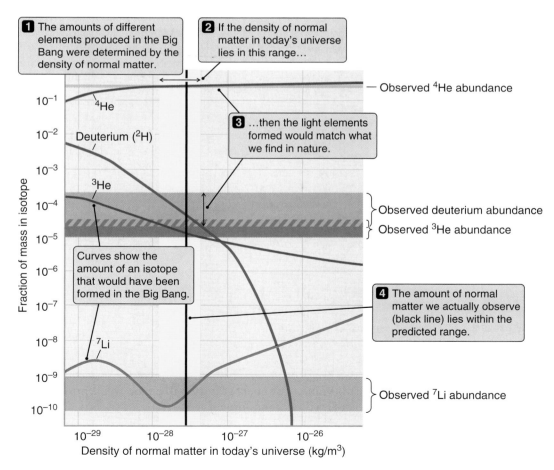

1 The amounts of different elements produced in the Big Bang were determined by the density of normal matter.

2 If the density of normal matter in today's universe lies in this range…

3 …then the light elements formed would match what we find in nature.

Curves show the amount of an isotope that would have been formed in the Big Bang.

4 The amount of normal matter we actually observe (black line) lies within the predicted range.

— Observed ^{4}He abundance

Observed deuterium abundance

Observed ^{3}He abundance

Observed ^{7}Li abundance

^{4}He

Deuterium (^{2}H)

^{3}He

^{7}Li

Fraction of mass in isotope

10^{-1}
10^{-2}
10^{-3}
10^{-4}
10^{-5}
10^{-6}
10^{-7}
10^{-8}
10^{-9}
10^{-10}

10^{-29} 10^{-28} 10^{-27} 10^{-26}

Density of normal matter in today's universe (kg/m^3)

FIGURE 20.18 Calculations of the abundances of the products of Big Bang nucleosynthesis, plotted against the density of normal matter in today's universe. Big Bang nucleosynthesis correctly predicts the amounts of these isotopes found in the universe today.

ended up in the form of the very stable isotope ^{4}He, regardless of exactly what the density of matter in the universe was. Indeed, when we look about us in the universe today, we find that about 24 percent of the mass of normal matter in the universe is in the form of ^{4}He, in good agreement with the prediction of Big Bang nucleosynthesis.

Unlike helium, the abundances of most of the isotopes formed in the Big Bang depended sensitively on the density of normal matter in the universe. Beginning with the amounts of isotopes such as deuterium (^{2}H) and ^{3}He found in the universe (shown as horizontal bands in Figure 20.18), we can ask what the density of normal matter in today's universe must be for these isotopes to have been formed in the

Big Bang. This prediction is shown as the yellow vertical band in the figure. We have talked a great deal about the ways astronomers measure the amount of normal matter in the universe. These measurements give a value of about 3×10^{-28} kg/m^3 for the average density of normal matter in the universe today. This value, shown as a vertical black line in Figure 20.18, lies well within the predicted range. Once again, the agreement is remarkable. Turning this around, we can begin with an observation of the amount of normal matter in and around galaxies, then use our understanding of the Big Bang to calculate what the chemical composition emerging from the Big Bang should have been. When we do this, the answer we get agrees remarkably well with the amounts of these elements we actually find in nature.

Two other points are worth noting here. The first is that no elements more massive than boron could have been formed in the Big Bang. Reactions such as the triple-alpha process, which forms carbon in the interior of stars, simply would not work under the conditions existing in the early universe. That is how we know that all the more massive elements in the universe, including the atoms making up

the bulk of our planet and ourselves, must have formed in subsequent generations of stars. The second point is that the agreement between the observed density of normal matter in the universe and the abundances of light elements also provides a powerful constraint on the nature of the dark matter dominating the mass in the universe. Dark mat-

ter (see Chapter 18) *cannot* consist of normal matter made up of neutrons and protons; if it did, the density of neutrons and protons in the early universe would have been much higher, and the resulting abundances of light elements in the universe would have been much different from what we actually observe.

20.5 The Universe Has a Destiny and a Shape

We live in an expanding universe, but will that expansion continue forever? This is clearly one of the great questions of modern cosmology. What is the fate of the universe? The answer depends in part on the amount of distributed mass the universe contains on very large scales. This distributed matter gravitationally affects how the universe evolves.

Think back on our discussion of escape velocity in Chapter 3. The fate of a projectile fired straight up from the surface of the Moon depends on its speed. As long as the speed is less than the *escape velocity* from the Moon (2.4 km/s), gravity will eventually stop the rise of the projectile and pull it back to the Moon's surface. However, if the speed of the projectile is greater than the Moon's escape velocity, then gravity will lose. Although the projectile will slow, it will never stop. It will escape from the Moon entirely.

Just as the gravity of the Moon pulls on a projectile, slowing its climb, so the gravity arising from the mass contained in the universe acts to slow its expansion. If there

is enough mass in the universe, then gravity will be strong enough to stop the expansion. If that is the case, then the universe will slow, stop, and eventually collapse in on itself in a catastrophic "Big Crunch." On the other hand, if there is not enough mass, then the expansion of the universe may slow, but it will never stop. The universe will expand forever.

The escape velocity from a planet is determined by the planet's mass and radius. The "escape velocity" of the universe is also determined by its mass and size—specifically, its average *density*. If the universe is denser on average than some particular value, called the **critical mass density**, then we expect gravity will be strong enough to eventually stop and reverse the expansion. If the universe is less dense than this value, we expect gravity will be too weak, and the universe will expand forever.

The faster the universe is expanding, the more mass is needed to turn that expansion around. For that reason, the critical mass density depends on the value of H_0. Assuming

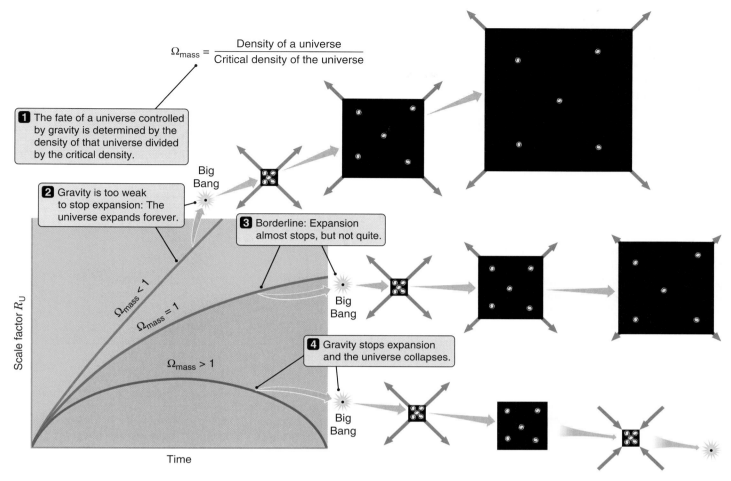

$$\Omega_{\text{mass}} = \frac{\text{Density of a universe}}{\text{Critical density of the universe}}$$

1 The fate of a universe controlled by gravity is determined by the density of that universe divided by the critical density.

2 Gravity is too weak to stop expansion: The universe expands forever.

3 Borderline: Expansion almost stops, but not quite.

4 Gravity stops expansion and the universe collapses.

Big Bang

$\Omega_{\text{mass}} < 1$

$\Omega_{\text{mass}} = 1$

$\Omega_{\text{mass}} > 1$

Scale factor R_{U}

Time

Big Bang

Big Bang

FIGURE 20.19 We would expect gravity to slow the expansion of the universe with time. These plots show the changing scale factor of the universe as a function of time for three possible scenarios, depending on the density of mass in the universe. (These plots ignore any cosmological constant.)

that $H_0 = 22$ km/(s Mly) and that gravity is the only thing we have to worry about, the critical density has a value of 8×10^{-27} kg/m³. Rather than trying to keep track of such awkward numbers, we will instead talk about the *ratio* of the actual density of the universe divided by its critical density. We call this ratio Ω_{mass} (pronounced "omega sub mass"). Note that Ω_{mass} is dimensionless (has no units).

We can follow the expansion of different possible universes in **Figure 20.19**, which shows a plot of the scale factor R_{U} versus time for different values of Ω_{mass}. (In these plots

Ω_{mass} **determines the fate of a universe governed exclusively by gravity.**

we have assumed that gravity alone controls the fate of the universe, an assumption that we will revisit in the following section.) If Ω_{mass} in a universe controlled by gravity alone is greater than 1, then gravity is strong enough to turn the

expansion around. Like a projectile fired from the Moon at less than the escape velocity, the expansion of such a universe will slow and eventually stop, then the universe will fall back in on itself. Conversely, if Ω_{mass} in a possible universe is less than 1, that universe will expand forever. The dividing line, where Ω_{mass} equals 1, corresponds to a universe that expands more and more slowly forever, but never quite turns around.

Until the closing years of the 20th century, most astronomers thought that was all there was to the question of expansion and collapse. Great efforts were focused on carefully measuring the mass of galaxies and assemblages of galaxies. As we have seen time and again, measuring masses of objects in the universe is tricky. When we calculate Ω_{mass} using just the luminous matter we see in galaxies and groups of galaxies, we get a value for Ω_{mass} of about 0.02. There is about 10 times as much dark matter as normal matter in galaxies, so adding in the dark matter in galaxies pushes the value of Ω_{mass} up to about 0.2. Finally, when we include the mass of dark matter *between* galaxies (a subject we will return to in the next chapter), Ω_{mass} could increase to 0.3 or higher. In other words, by our current accounting, there is perhaps about a third as much mass in the universe as we would expect it should take to stop its expansion.

Reviving Einstein's "Great Blunder"

As they were closing in on a good value for Ω_{mass}, it seemed to astronomers that our understanding of the expansion of the universe was almost complete. Then the other shoe dropped.

If the expansion of the universe has been slowing with time, as we might expect, then when the universe was young, it must have been expanding more rapidly than it is today. That is a prediction that we can go to telescopes and check. Objects that are very far away (so that we see them as they were long ago) should have larger velocities than we would expect if the Hubble expansion were carrying on at the same pace since the universe was young.

In the closing years of the 20th century, some groups of astronomers began using tools such as the Hubble Space Telescope and the giant Keck telescopes in Hawaii to test this prediction. They measured the brightnesses of standard

---

There is evidence that the expansion of the universe is accelerating.

---

candles (in particular, Type I supernovae) in very distant galaxies, and compared those brightnesses with expected brightness based on the redshifts of those galaxies. The early findings of these studies have set the astronomical community abuzz. Rather than showing that the expansion of the universe has slowed down over time, the data indicate that it is *speeding up*!

Figure 20.20 shows the data indicating that the expansion of the universe is speeding up. The observations on which this conclusion is based are extremely difficult to carry out, and many different factors have to be taken into account when interpreting the data. Even so, the fact that similar findings have been obtained independently by different groups of astronomers lends credence to the results. In addition, results obtained early in the 21st century by the WMAP experiment provide independent confirmation of this increasing rate of expansion of the universe.

How could it be that the rate of expansion of the universe has *increased* over time? For this to be the case, there would have to be some force other than gravity at work to push the universe outward. Physicists have some ideas about how such a hypothetical repulsive force might originate. These are related to theories from particle physics concerning the nature of what we call the **vacuum**. Normally we think of a

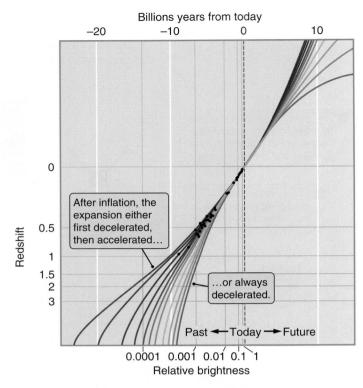

FIGURE 20.20 Observed brightness of Type Ia supernovae, plotted as a function of their redshift, *z*. Surprisingly, the observations suggest that the redshifts are too small for their distance—evidence that the universe is expanding faster today than in the past.

vacuum as "empty space," but it turns out that even empty space has some very interesting physical properties. Discussion of these ideas (which are related to Hawking radiation from black holes, discussed in Chapter 17) will have to wait until Chapter 21, where we will take up the question of the origin of physical law. For now we turn our attention to the effects that such a force would have.

The idea of a repulsive force opposing the attractive force of gravity is not new. When Einstein used his newly formulated equations of general relativity to calculate the structure of spacetime in the universe, he was greatly troubled. The equations clearly indicated that any universe containing mass could not be static, any more than a ball can hang motionless in the air. However, this was over a decade before Hubble did his epic work on the expansion of the universe, and the conventional wisdom at that time was that the universe is static.

In order to force his new theory of general relativity to allow for a static universe, Einstein inserted a "fudge factor" into his equations. He called this fudge factor the **cosmological constant**, which we write as Ω_Λ.[3] The cosmo-

---

[3] Einstein probably never wrote Ω_Λ, referring instead to the related quantity Λ.

Einstein invented a cosmological constant
to oppose gravity in a static universe.

logical constant acts as a repulsive force in the equations, opposing gravity and allowing galaxies to remain stationary despite their mutual gravitational attraction. The cosmological constant was Einstein's version of the magician's trick that allows a subject to hang apparently unsupported in midair.

When Hubble announced his discovery that the universe is expanding, Einstein realized his mistake. The unmodified equations of general relativity demand that the structure of the universe be dynamic. Instead of doctoring them with the inclusion of Ω_Λ, Einstein realized that he should have *predicted* that the universe must either be expanding or contracting with time. What a coup it would have been for his new theory to successfully predict such an amazing and previously unsuspected result. He called the introduction of his fudge factor, the cosmological constant, the "greatest blunder" of his career as a scientist. It is ironic that with the new results on the brightness of Type I supernovae, "Einstein's greatest blunder" has returned to center stage. The repulsive force represented by the infamous Ω_Λ in Einstein's equations is just what is needed to describe a universe that is expanding at an ever-accelerating rate.

How does the possibility of a nonzero value for Ω_Λ affect the possible fates that await our universe? If Ω_Λ is not zero, the fate of the universe is no longer controlled exclusively by Ω_{mass}. If there is something effectively pushing outward from within the universe, adding to its expansion, then gravity will have a harder time turning the expansion around. In that case, the mass needed to halt the expansion of the universe would be greater than the critical mass we already discussed.

Figure 20.21 shows plots of scale factor versus time, similar to those shown in Figure 20.19; but now we have included the effects of a nonzero cosmological constant. The evolution of a universe that collapses back on itself (below the horizontal line in **Figure 20.22**) is similar, regardless of whether or not Ω_Λ is zero. In contrast, if a universe with a nonzero cosmological constant expands forever, its evolution will look drastically different from any universe in which Ω_Λ is zero. As a universe expands, gravity gets weaker and weaker because the mass is more and more spread out. Unlike gravity, however, the effect of the cosmological constant becomes *greater*. While a universe is young and compact, gravity is strong enough to dominate the effect of the cosmological constant. Unless gravity is able to turn the expansion around, however, the cosmological constant wins out in the end, causing the expansion to continue accelerating forever.

So which of these fates awaits our universe? Figure 20.22 shows the range of values for Ω_{mass} and Ω_Λ allowed by current observations. Assuming that these observations are

correct, the values for Ω_{mass} and Ω_Λ are about 0.3 and 0.7, respectively. It appears that the expansion of our universe is *already* accelerating under the dominant effect of the cosmological constant.

The Age of the Universe

Our values for Ω_{mass} and Ω_Λ not only affect our predictions for the future of the universe. They also influence our interpretation of the past. **Figure 20.23** shows plots of the scale factor of the universe versus time. Measurement of the Hubble constant H_0 tells us how fast the universe is expanding *today*. That is, it tells us the *slope* of the curves in Figure 20.23 *at the current time*. As we saw earlier, if the expansion of the universe has not changed in time, then the plot of R_U versus time is the straight red line in Figure 20.23. The age of the universe in this case is equal

FIGURE 20.21 Plots of scale factor R_U versus time for cosmologies with and without a cosmological constant, Ω_Λ. If there is enough mass in a universe, gravity may still overcome the cosmological constant and cause that universe to collapse. Any universe without enough mass to eventually collapse will instead end up expanding at an ever-increasing rate.

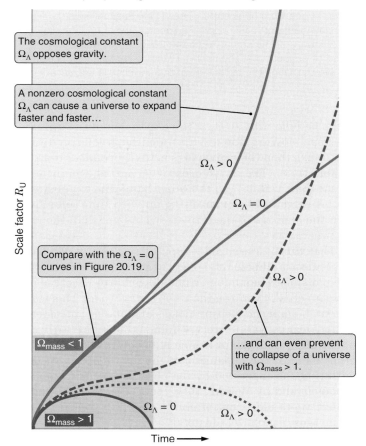

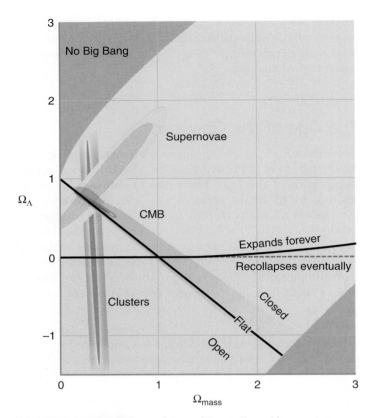

FIGURE 20.22 Values of Ω_Λ and Ω_{mass} allowed by current observations from different sources (Type Ia supernovae, measurements of mass in galaxies and clusters, and detailed observations of the structure of the CBR). These sources together suggest that the best current estimate for Ω_Λ is about 0.7, which means the expansion of the universe is accelerating.

to the Hubble time, $1/H_0$. If the expansion of the universe has been slowing down with time (the green line in Figure 20.23), then the universe is actually *younger* than the Hubble time. (The curve crosses $R_U = 0$ at a point more recent than $1/H_0$.) If, on the other hand, the expansion of the universe has been speeding up with time (blue line), then the true age of the universe is greater than the Hubble time.

The current measured value for H_0 (22 km/(s Mly)) corresponds to a Hubble time ($1/H_0$) of about 13.6 billion years. If expansion of the universe has slowed over time, the universe is actually younger than 13.6 billion years. This would be a problem if the measured ages of globular clusters—13 billion years—were correct. (Globular clusters clearly cannot be older than the universe that contains them!) On the other hand, if the expansion of the universe has sped up with time, as suggested by the observations of Type I supernovae and by WMAP, then the universe is around 14 billion years old—comfortably older than measurements of the ages of globular clusters.

The Universe Has a Shape

By this time the idea that spacetime is described by general relativity is getting to be a familiar friend. Space is a "rubber sheet" that has stretched outward from the Big Bang. In Chapter 17 we saw that the rubber sheet of space is also curved by the presence of mass. In that chapter we saw how the shape of space around a massive object can be detected through changes in simple geometrical relationships, such as the ratio of the circumference of a circle to its radius, or the sum of the angles in a triangle. If the mass of a star, planet, or black hole causes a distortion in the shape of space, then should not the mass of all of the galaxies and dark matter in the universe also distort the shape of the universe *as a whole?* The answer is yes.

There are three basic shapes that our universe might have. Which shape actually describes the universe is determined by the sum of Ω_{mass} and Ω_Λ. Continuing with the rubber sheet analogy, the first possibility, corresponding to $\Omega_{mass} + \Omega_\Lambda = 1$, is that we live in a **flat universe**. A flat universe is described overall by the rules of Euclidean geometry. As shown in **Figure 20.24(a)**, circles in a flat universe have a circumference of 2π times their radius, and triangles contain angles whose sum is 180°. A flat universe stretches on forever.

The second possibility is that the geometry of the universe is shaped something like the surface of a saddle (**Fig-**

FIGURE 20.23 A plot of the scale factor R_U versus time for three possible universes. If the universe has expanded at a constant rate, then its age is equal to the Hubble time, $1/H_0$. If the expansion of the universe has slowed with time, it is younger than the Hubble time. If the expansion of the universe has sped up with time, then the universe is older than $1/H_0$.

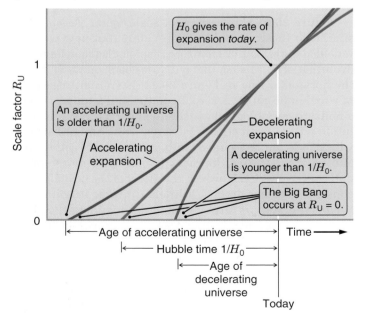

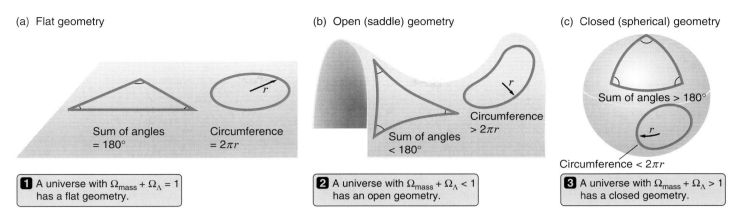

FIGURE 20.24 Two-dimensional representations of the possible geometries that space can have in a universe. (a) In a flat universe, Euclidean geometry holds. Triangles have angles that sum to 180°, and the circumference of a circle equals 2π times the radius. In (b), an open universe, or (c), a closed universe, these relationships are no longer correct over very large distances.

ure 20.24(b)). This type of universe, in which $\Omega_{mass} + \Omega_\Lambda < 1$, is also infinite and is referred to as an **open universe**. In this type of universe the circumference of a circle is greater than 2π times its radius, and triangles contain less than 180°.

The final possibility, in which $\Omega_{mass} + \Omega_\Lambda > 1$, is a universe with a geometry shaped like the surface of a sphere (**Figure 20.24(c)**). The geometric relationships on a sphere are similar to those in the vicinity of a massive object, as discussed in Chapter 17. The circumferences of circles on a sphere are less than 2π times their radii, and triangles contain more than 180°. This possibility is called a **closed universe** because space is finite and closes back on itself.

Again we face the question, Which of these shapes describes the universe in which we live? The measurements

> Recent evidence suggests that our universe is remarkably flat.

are difficult. Even so, as seen in Figure 20.22, $\Omega_{mass} + \Omega_\Lambda$ is close to 1 (0.3 + 0.7), meaning that our universe is very nearly flat.

20.6 Problems Lead to New Understanding

It is remarkable that Big Bang cosmology makes so many correct predictions about the properties of the universe in which we live. A century ago astronomers were struggling just to get a handle on the size of the universe. Today we have a comprehensive theory that ties together many diverse facts about nature, ranging from the constancy of the

speed of light, to the properties of gravity, to the motions of galaxies, to the origins of the very atoms of which we are made. The case for the Big Bang is compelling. Even so, as our knowledge of the expansion of the universe has grown and our observations of the cosmic background radiation have improved, a number of puzzles have arisen. The solution to these puzzles has forced us to consider some remarkable ideas about how our universe expanded when it was very young.

The Universe Is Much Too Flat

The first problem that we run into when observing our universe is that the universe is too flat. In fact, the universe is much too close to being exactly flat for this to have happened by chance. To see why this is a problem, imagine a model universe in which $\Omega_\Lambda = 0$. (This is a reasonable approximation for the very early days of *any* possible universe. When a universe is very young, it is also very dense, and Ω_{mass} is all that matters.) As our model universe expands out of its own version of the Big Bang, its density falls. At the same time, because the universe is expanding more slowly, the critical density needed to eventually stop the expansion falls as well. If Ω_{mass} is *exactly* 1, the decline in the actual density and the decline in the critical density go hand in hand: The ratio between the two, Ω_{mass}, remains 1 for all time, as shown by the middle curve in **Figure 20.25**. A universe that starts out perfectly flat *stays* perfectly flat.

On the other hand, a universe that does *not* start out perfectly flat has a very different fate. If a universe started out with Ω_{mass} even slightly greater than 1, its expansion would slow more rapidly than that of the flat universe, meaning that less and less density would be required to stop the

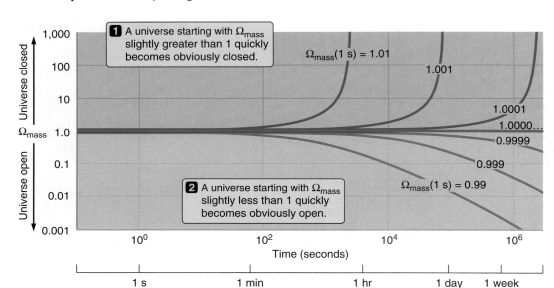

FIGURE 20.25 Universes in which Ω_{mass} has slightly different values at an age of 1 second. Notice that even tiny differences from $\Omega_{mass} = 1$ are rapidly amplified as the model universe expands. To have a value of Ω so close to 1 today, our universe must have started out with a value of Ω exquisitely close to 1. (The value of Ω_Λ does not affect the calculations shown.)

expansion. At the same time the actual density would be falling less rapidly than in the flat universe. This disparity between the actual density of the universe and the critical density of the universe would increase, causing the ratio between the two, Ω_{mass}, to skyrocket. This is shown by the curves that climb toward the top of Figure 20.25. A universe that starts out even *slightly* closed rapidly becomes *obviously* closed.

Conversely, if a universe started with Ω_{mass} even a tiny bit less than 1, the expansion would slow *less* rapidly than in a flat universe. As time passed, more and more mass would be required for gravity to stop the too rapidly expanding universe. At the same time the actual density of the universe would be dropping faster than in a flat universe. In this case Ω_{mass} (the ratio between the actual density and the critical density) would plummet, leading to the curves that dive toward the bottom of Figure 20.25.

Adding Ω_Λ to the picture makes the math a bit more complex, but it does not change the basic results. Try balancing a razor blade on its edge. If the blade is tipped just a tiny bit in one direction, it quickly falls that way. If the blade is tipped just a tiny bit in the other direction, it quickly falls in the other direction instead. By all rights we would expect our universe to be either *obviously* open or *obviously* closed —analogous to the tipped razor blade. We find ourselves instead in a universe in which $\Omega_{mass} + \Omega_\Lambda$ is so close to 1 that we have difficulty telling which way the razor blade is tipped at all! Discovering that $\Omega_{mass} + \Omega_\Lambda$ is extremely close to 1 after 13 billion years or more is like balancing a razor blade on its edge and coming back 10 years later to find that it still has not tipped over!

For the present-day value of $\Omega_{mass} + \Omega_\Lambda$ to be as close to 1 as it is, $\Omega_{mass} + \Omega_\Lambda$ could not have differed from 1 by more than 1 part in 100,000 when the universe was 2,000 years old. When the universe was 1 second old, it had to be flat to

1 part in 10 billion. At even earlier times, it had to be much flatter still. This is simply too special a situation to be due to chance—a fact referred to in cosmology as the **flatness problem**. *Something* about the early universe must have *forced* $\Omega_{mass} + \Omega_\Lambda$ to have a value that was incredibly close to 1.

The Cosmic Background Radiation Is Much Too Smooth

The second problem faced by our cosmological models is that the cosmic background radiation is uncomfortably smooth. Following the discovery of the CBR in the 1960s, many observers turned their attention to mapping this background glow. At first they were reassured as result after result showed that the temperature of the CBR was remarkably constant, regardless of where one looked in the sky. Yet over time this strong confirmation of Big Bang cosmology turned instead into a puzzle that challenged our view of the early universe. Once we remove our motion relative to the CBR from the picture, the CBR is not just smooth—it is *too* smooth.

Why should we expect the CBR to be less uniform than it is? To understand the answer, we need to shift our attention from the very large to the very small. In Chapter 4 we discovered the bizarre world of quantum mechanics that

---

We expect nonuniformities in the early universe.

---

shapes the world of atoms, light, and elementary particles. When the universe was extremely young, it was so small that quantum mechanical effects played a role in shaping the structure of the universe as a whole. In particular, the early universe was subject to the quantum mechanical uncertainty principle. The uncertainty principle says that as we look at a system at smaller and smaller scales, the properties of that system become less and less well determined. This applies whether we are talking about the properties of an electron in an atomic orbital, or about our entire universe at the time when it would fit within the size of an atom.

As this chapter is being written, one of your authors is sitting on the beach looking out across the ocean. Looking off into the distance, he sees the surface of the ocean as smooth and flat. The horizon looks almost like a geometrical line. Yet the apparent smoothness of the ocean as a whole hides the tumultuous structure present at smaller scales, where waves and ripples upon waves fluctuate dramatically from place to place. In similar fashion, quantum mechanics says that as we look at smaller and smaller scales in the universe, conditions *must* fluctuate in unpredictable ways. In particular, quantum mechanics says that the smaller the universe we consider (that is, the earlier in the history of our universe that we go), the more dramatic those fluctuations become. When the universe was young, it could *not* have been smooth. There must have been dramatic variations (that is, "ripples") in the density and temperature of the universe from place to place.

If the universe had expanded slowly, those ripples would have smoothed themselves out. But the universe expanded much too rapidly for this. Different parts of the universe could not have "communicated" with each other (telling

The CBR is smoother than the early universe should have been.

each other to smooth out the ripples) rapidly enough to smooth these ripples out. So when we look at the universe today, we should see the fingerprint of those early ripples imprinted on the cosmic background radiation—but we do not. The fact that the CBR is so smooth is referred to as the **horizon problem** in cosmology: Different parts of the universe are too much like other parts of the universe that should have been "over their horizon" and beyond the reach of any signals that might have smoothed out the early quantum fluctuations.

"Inflation" Solves the Problems

In the early 1980s **Alan Guth** (1947–) offered a solution to the flatness and horizon problems of cosmology. Guth suggested that for a brief time the young universe underwent a period of **inflation** during which the universe itself expanded at a rate *far* in excess of the speed of light. Like so many things we have seen on our journey, the numbers used to describe inflation are far beyond our ability to grasp intuitively. Between the ages of about 10^{-35} s and 10^{-33} s, the

scale factor R_U of the universe increased by a factor of at least 10^{30} and perhaps much more. In that incomprehensibly brief instant, the size of the observable universe grew from 10 trillionths the size of the nucleus of an atom to a region

Inflation was a period of intense expansion of the universe.

about 3 meters across. That is like a grain of very fine sand growing to the size of today's *entire universe*—all in a billionth the time that it takes light to cross the nucleus of an atom! (At this point you may well be saying to yourself, "Wait a minute! What happened to all that business about nothing traveling faster than the speed of light?" Inflation does not violate the rule that no signal can travel *through* space at greater than the speed of light. During inflation space *itself* expanded so rapidly that the distances between points in space increased faster than the speed of light.)

To understand how inflation solves the flatness and horizon problems of cosmology, imagine that you are an ant living in the *two-dimensional* universe defined by the surface of a golf ball, as shown in **Figure 20.26**. Your universe would have two very apparent characteristics. First, it would be obviously curved. If you were to walk around the circumference of a circle in your two-dimensional universe and then measure the radius of the circle, you would find the circumference to be less than 2π times the radius. If you were to draw a triangle in your universe, the sum of its angles would be greater than 180°. The second obvious characteristic would be the dimples, approximately a millimeter deep, on the surface of the golf ball.

Now imagine how this would change if your golf ball universe suddenly grew to the size of Earth. (Granted, this change is nowhere near comparable to the inflation experienced by the real universe, but you can get the idea.) First,

Inflating the size of a dimpled sphere makes it seem flatter.

the curvature of your universe would no longer be apparent. An ant walking along the surface of Earth would be hard pressed to tell that Earth is not flat. The circumference of a circle would be 2π times its radius, and there would be 180° in a triangle. In fact, it took us most of our history as a species to realize that Earth really is round. In the case of inflationary cosmology, the universe after inflation would be extraordinarily flat (that is, with $\Omega_{mass} + \Omega_\Lambda$ extraordinarily close to 1) *regardless* of what the geometry of the universe was before inflation. Because the universe was inflated by a factor of at least 10^{30}, $\Omega_{mass} + \Omega_\Lambda$ immediately after inflation must have been 1 to within 1 part in 10^{60}, which is flat enough for $\Omega_{mass} + \Omega_\Lambda$ to remain close to 1 today. Today's universe is not flat by chance. It is flat because *any* universe that underwent inflation would emerge with a value for $\Omega_{mass} + \Omega_\Lambda$ that was within a gnat's eyelash of 1.

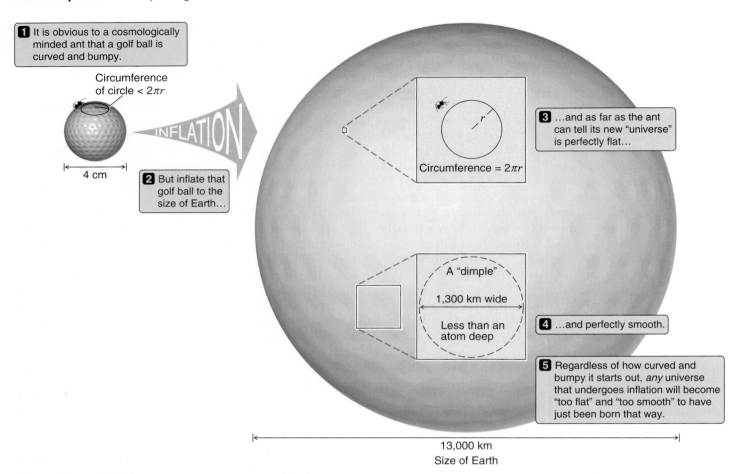

1 It is obvious to a cosmologically minded ant that a golf ball is curved and bumpy.

Circumference of circle < $2\pi r$

4 cm

INFLATION

2 But inflate that golf ball to the size of Earth...

Circumference = $2\pi r$

3 ...and as far as the ant can tell its new "universe" is perfectly flat...

A "dimple"

1,300 km wide

Less than an atom deep

4 ...and perfectly smooth.

5 Regardless of how curved and bumpy it starts out, *any* universe that undergoes inflation will become "too flat" and "too smooth" to have just been born that way.

13,000 km
Size of Earth

FIGURE 20.26 If a round lumpy golf ball were suddenly inflated to the size of Earth, it would seem extraordinarily flat and smooth to an ant on its surface. Similarly, following inflation, any universe would be both extremely flat and extremely smooth, regardless of the exact geometry and irregularities it started out with.

So much for the flatness problem. What about the horizon problem? When our golf ball universe inflated to the size of Earth, the dimples that covered the surface of the golf ball were stretched out as well. Instead of being a millimeter or so deep and a few millimeters across, these dim-

Huge expansion also smoothes out inhomogeneities.

ples now are only an atom deep but are hundreds of kilometers across. Again, our ant would be hard pressed to detect any dimples at all. In the case of the real universe, inflation took the large fluctuations in conditions caused by quantum uncertainty in the preinflationary universe and stretched them out so much that they are unmeasurable in today's postinflationary universe. The slight irregularities that we see in the CBR are the faint ghosts of quantum fluctuations that occurred as the universe inflated.

An early era of inflation in the history of the universe offers a handy way of solving the horizon and flatness problems, but why would the real universe have the temerity to do such a thing? It seems quite remarkable that the universe should undergo a period during which it expanded at such an "astronomical" rate. But there are many remarkable things about the universe. It is remarkable that spacetime has a shape. It is remarkable that light is both a wave and a particle. It is remarkable that galaxies and stars and planets exist at all.

We live in a universe of structure. Expansion, inflation, and the physical laws that govern everything are all part of this structure, as is the distribution of stars and galaxies that we see around us today. The fundamental question we face at this stage of our journey is *not* simply "Why did the universe undergo inflation?" but rather "How did structure in the universe arise?" It is to this question that we now turn.

Summary

- The cosmological principle requires that our universe be homogeneous and isotropic, which it does indeed seem to be.

- The expansion of the universe is governed by Hubble's Law: A galaxy's recession velocity is proportional to its distance.

- The universe is expanding uniformly from a Big Bang, which occurred nearly 14 billion years ago.

- Expansion of the universe produces the observed redshifts, but it *does not* affect the local physics or structure of objects.

- We observe thermal radiation at 2.73 K, which is the cooled remnant of the radiation present in the universe when it was much smaller and hotter.

- The CBR allows us to measure our velocity with respect to the background radiation, and it also shows evidence of the ripples that grew to become large-scale structure in the universe.

- Nuclear reactions of normal matter in the hot early universe produced all the helium we see today as well as trace amounts of other light elements.

- Both gravity and the cosmological constant (or vacuum energy) determine the fate of the universe. Observations suggest that rather than slowing down, the expansion of the universe is accelerating.

- The very early universe may have gone through a brief but dramatic period of exceptionally rapid expansion, called inflation. If true, this would explain both the flatness and the homogeneity of the universe we see today.

Seeing the Forest through the Trees

The journey that we have taken in *21st Century Astronomy* has involved many amazing discoveries. We have peered into the heart of the Sun, watched as planetary systems formed, and witnessed the violent death throes of massive stars. We have seen our ideas of cause and effect shattered by the quantum mechanical world of atoms, and have come to accept that space and time are joined in a four-dimensional fabric of spacetime that is bent and even torn by the presence of mass. Yet even when witnessed within the context of such a remarkable journey of discovery, what we see when we look at the structure of the universe itself is truly mind-boggling.

You never know when you get up in the morning if this will be one of those rare days when you happen upon some extraordinary experience that leaves you looking at the world in a different way. So it was for Edwin Hubble. Hubble, who also was the first to prove the true nature of galaxies, was certainly at the right place at the right time in his career. He was doing as scientists do, using the cutting-edge tools of his day and following his instincts about what questions might turn out to have interesting answers. His investigation into the basic properties of the newly recognized class of objects called *galaxies* took an amazing and unexpected turn. With one graph, plotting the redshift of galaxies against his estimates of their distances, he forever changed our view of the universe. What must it have felt like to be Edwin Hubble, looking at his data and realizing for the first time what the implications of those data were? What would it mean to be the first human to glimpse the true history of the universe?

We live in a universe in which the spacetime described by general relativity has been expanding outward for about 14 billion years, an expansion that can be traced back to a single moment when spacetime and all that it contains came into existence. The Big Bang is a theory, but one of the lessons we have learned along the way is that a well-tested and well-corroborated scientific theory is the closest humans can ever come to certain knowledge. The predictions of the Big Bang have been borne out time and again, in everything from observations of the cosmic background radiation, to the predictions the theory makes about the abundances of chemical elements, to the fact of the expansion of the universe itself. A full accounting of the experimentally verified predictions of the Big Bang theory could easily fill a huge book. By any reasonable definition of the term, the Big Bang is a fact.

Again and again during our exploration we have stopped to marvel at how vast the universe is. We have pondered the thousands of stars in the night sky that can be seen with the naked eye, realizing they are but the tiniest fraction of the hundreds of billions of stars that make up our Milky Way. Then we faced the task of comprehending a universe in which our Milky Way is itself but one of hundreds of billions of galaxies spread out across a universe so large that even light takes nearly 30 billion years to cross it. Now we are forced to confront the idea that at one time everything we see in that vast universe, including space itself, was contained in a volume that was as small relative to the nucleus of an atom as we are as compared with the enormity of today's universe. All we need to do to be reminded of the reality of that hot, dense early universe is look at the faint microwave glow that fills the sky. Looking in the other direction in time, we have considered the fate of the universe and found it equally startling. Although the jury is still out, the best models of the universe at the beginning of the 21st century tell of a future in which the expansion of the universe continues forever at an ever-accelerating pace.

Seldom have novelists, theologians, or philosophers dared dream of anything so extraordinary as the universe of modern cosmology. Yet we arrive at our conclusion not by fantasy or conjecture, but by following a path laid down by hard-won observation and well-tested physical law.

In this chapter we have set the stage. We have discovered the shape of the canvas on which our existence is painted. Yet our world is not defined by spacetime alone. The trees and planets and stars are still there for us to see. We are part of a universe that is filled with structure. How did that structure arise? What are the acts in the grand play that continues on the cosmic stage? The time has come to tell the story from the beginning—to start as close to the Big Bang as our physics will take us —and run the clock forward, watching as the structure and complexity of today's universe emerge.

Hold onto your hats. We have found the rabbit hole. Now we will see how deep it goes.

Key Terms

homogeneous, p. 576
isotropic, p. 577
redshift, p. 577
Hubble's Law, p. 578
mega-light-year (Mly), p. 578
Hubble constant, p. 578
peculiar velocity, p. 580
distance ladder, p. 580
look-back time, p. 581
Hubble time, p. 583
Big Bang, p. 583
cosmological redshift, p. 587
cosmic background radiation (CBR) , p. 590
Big Bang nucleosynthesis, p. 593
critical mass density, p. 594
vacuum, p. 596
cosmological constant, p. 596
flat universe, p. 598
open universe, p. 599
closed universe, p. 599
flatness problem, p. 600
horizon problem, p. 601
inflation, p. 601

Student Questions

THINKING ABOUT THE CONCEPTS

1. Why are we not able to use the measured radial velocities of galaxies in the Local Group to evaluate the Hubble constant (H_0)?

2. Imagine you are standing in the middle of a dense fog. Would you describe your environment as isotropic? Would you describe it as homogeneous? Explain your answers.

3. Knowing that you are studying astronomy, a curious friend asks where the center of the universe is located. You smile and answer "right here and everywhere." Explain in detail why you would give this answer to your friend.

4. The general relationship between radial velocity (v_r) and redshift (z) is $v_r = cz$. This simple relationship fails, however, for very distant galaxies with large redshifts. Explain why.

5. Why is Earth not expanding together with the rest of the universe?

6. As astronomers extend their distance ladder beyond 100 Mly, they change their standard candle from Cepheid variable stars to Type I supernovae. Explain why this is necessary.

7. Very distant galaxies have redshifts indicating recession velocities of 100,000 km/s or more, yet their true speeds are probably no more than a few hundred kilometers per second. Explain.

8. If new observations suggested that the mass fraction of helium is still 24 percent but that the mass fraction of deuterium (D) is really 10^{-6}, how would that affect our estimates of the density of normal matter in the universe?

9. Describe the observational evidence suggesting that Einstein's "cosmological constant" (a repulsive force) may be needed to explain the historical expansion of the universe.

10. If you could accurately measure a triangle and a circle drawn on the surface of Earth, you would find that the sum of the triangle's interior angles is more than 180° and the circle's circumference is less than $2\pi r$. This is contrary to what you probably learned about triangles and circles in your introductory geometry class. Explain why this is so and how it relates to a "spherical universe."

11. During the period of inflation, the universe may have briefly expanded at 10^{25} (10 trillion trillion) times the speed of light. Why did this not violate Einstein's spe-

cial theory of relativity, which says that neither matter nor communication can travel faster than the speed of light?

12. As the sensitivity of our instrumentation increases, we are able to look ever farther into space and, therefore, ever farther back in time. When we reach the era of recombination, however, we run into a wall and can see no farther back in time. Explain why.

APPLYING THE CONCEPTS

13. Hubble time $(1/H_0)$ represents the age of a universe that has been expanding at a constant rate since the Big Bang. Assuming a H_0 value of 22 km/(s Mly) and a constant rate of expansion, calculate the age of the universe in years. Note that one year $= 3.16 \times 10^7$ s and one light-year $= 9.46 \times 10^{12}$ km.

14. One of the most distant known quasars has a redshift $z = 5.82$ and a recession velocity of 287,000 km/s or about 96 percent of the speed of light.
    a. If $H_0 = 22$ km/(s Mly) and the rate of expansion of the universe is constant, how far away is this quasar?
    b. Assuming a Hubble time of 13 billion years, how old was the universe at the look-back time of this quasar?
    c. What was the scale factor (R_U) of the universe at that time?

15. The spectrum of a distant galaxy shows the H_α line of hydrogen $(\lambda_{rest} = 0.65628$ microns$)$ at a wavelength of 0.98442 microns. Assume that $H_0 = 22$ km/(s Mly).
    a. What is the redshift (z) of this galaxy?
    b. What is its recession velocity (in km/s)?
    c. What is the distance of the galaxy (in Mly)?

16. The rest wavelength of the Ly_α line of hydrogen is in the extreme ultraviolet region of the spectrum at 0.1216 microns.

a. What would be the wavelength of this line in the spectrum of a quasar with a redshift $z = 6.4$?
b. In what region of the spectrum would this redshifted line be located?

17. Suppose we observe two galaxies, one at a distance 35 Mly with a radial velocity of 580 km/s, and another at a distance of 1,100 Mly with radial velocity of 25,400 km/s.
    a. Calculate the Hubble constant for each of these two observations.
    b. Which of the two calculations would you consider to be more trustworthy? Why?
    c. Estimate the peculiar velocity of the closer galaxy.
    d. If the more distant galaxy had this same peculiar velocity, how would that change your calculated value of the Hubble constant?

18. The universe today has an average density $\rho_o = 3 \times 10^{-28}$ kg/m³. Assuming that the average density depends on the scale factor as $\rho = \rho_o/R_U{}^3$, what was the scale factor of the universe when its average density was about the same as Earth's atmosphere at sea level $(\rho = 1.23$ kg/m³$)$?

There is a theory which states that if ever anyone discovers exactly what the universe is for and why it is here, it will instantly disappear and be replaced by something even more bizarre and inexplicable.

There is another which states that this has already happened.

The search for our origins is one of the grand themes of modern science.

CHAPTER 21

The Origin of Structure

21.1 Whence Structure?

Throughout the ages there have been as many different ideas about the origin of the universe as there have been cultural traditions. Thoughts about what the universe was like once upon a time have been part of the mythologies and traditions of all great civilizations. We live at a remarkable moment in history, when the nature of the early universe has moved from philosophical speculation to the realm of scientific fact. All that we have to do to see the early universe is point our microwave telescopes at the sky. The glow of the cosmic background radiation is there for anyone with the appropriate technology to see.

The early universe was an extraordinary place—an expanding fireball that was far more uniform than the blue of the bluest sky on the clearest day. How different that universe was from the universe we see about us today! Today's universe is one of stars and galaxies, viewed from the surface of a planet with oceans and mountains and uncountable species of living things. The contrast between these two realities cries out for explanation. How did we get from there to here? We come to one of the most philosophically intriguing questions in the history of human thought: What is the origin of structure?

KEY CONCEPTS

The universe that emerged from the Big Bang was incredibly uniform, wholly unlike today's universe of galaxies, stars, and planets. As our journey comes to its end, we tackle face-on the question that has been with us all along: Where does structure in the universe come from? Here we find that complex structure is a natural, unavoidable consequence of the action of physical law in an evolving universe:

- Matter and the fundamental forces of nature "froze out" of the uniformity of the expanding and cooling universe moments after the Big Bang.

- Galaxies formed as slight ripples in the dark matter emerging from the Big Bang, which then collapsed under the force of gravity, pulling in normal matter as well.

- Galaxies were drawn together by gravity to form the large groupings of galaxies we see today.

- Much lies ahead for Earth and for the universe.

- Like planets, stars, and galaxies, life is another form of structure that evolved through the action of the physical processes that shape the universe—the universe may teem with life.

- Evolution of structure through the action of physical law is the lynchpin that ties all of modern science together and brings sense to what we see.

21.2 Galaxies Form Groups, Clusters, and Larger Structures

Just as stars and clouds of glowing gas show us the structure of our galaxy, it is the distribution of galaxies themselves that shows us the structure of our universe. And just as gravity holds galaxies together, giving them their shape, it is gravity that shapes the universe itself. No galaxy exists in utter isolation. The vast majority of galaxies are parts of gravitationally bound collections of galaxies. The smallest and most common of these are called **galaxy groups**. A galaxy group is an irregular collection of a few dozen galaxies, most of them dwarf galaxies. Our Milky Way is a member of the **Local Group**, which consists of two giant spirals (the Milky Way and the Andromeda Galaxy), along with more than 30 smaller dwarf galaxies in a volume of space about 4 Mly in diameter (see **Figure 21.1**). Close to 98 percent of the galaxy mass in the Local Group resides in just its two giant galaxies.

Larger systems of galaxies, which can consist of hundreds of galaxies and often have a more regular structure than groups, are called **galaxy clusters**. Galaxy clusters are larger than groups, typically occupying a volume of space 10 to 15 Mly across; but clusters and groups are similar in composition. Giant galaxies dominate the mass of the far more numerous dwarf galaxies. One interesting difference among galaxy groups and clusters is that while spiral galaxies are common in most systems, elliptical galaxies are common in only about a quarter of groups and clusters. The Virgo cluster (**Figure 21.2(a)** and Figure 21.1), located 53 Mly from Earth, is an example of the first type of cluster, whereas the Coma cluster (**Figure 21.2(b)** and Figure 21.1) is an example of a cluster that is dominated by giant elliptical and S0 galaxies.

Clusters and groups of galaxies themselves bunch together to form enormous **superclusters**, which contain tens of thousands or even hundreds of thousands of galaxies and span regions of space typically more than 100 Mly in size. Our Local Group is part of the Virgo supercluster, which also includes the Virgo cluster.

Hubble's Law is a powerful tool for mapping the distribution of galaxies in space. Using Hubble's Law, all that we need to determine the distance to a galaxy is a single spec-

---

Galaxy redshift surveys measure distances to large numbers of galaxies.

---

trum from which we can measure the galaxy's redshift. Although it is easy in principle to measure the redshift of a galaxy, in practice this can be a time-consuming business. The first redshifts were measured from spectra recorded on photographic plates. Exposures of several hours were required to capture the feeble signal, and results rolled in at

the breakneck pace of one or two redshifts per night of painstaking observation. As of 1975 redshifts had been measured for only about 1,000 of the hundred billion or so galaxies we can see. Since that time there has been a happy marriage of large telescopes, new instruments (such as CCD detectors and spectrographs capable of observing many galaxies at once), and powerful computers that can process large amounts of data in rapid, automated fashion. By 1990 the redshifts of over 10,000 galaxies had been measured; and by the time you read this book, that number will be well over 1 million galaxy redshifts.

The first large redshift survey was conducted by the Harvard Center for Astrophysics, which in 1986 presented

FIGURE 21.1 The Local Group of galaxies and the Virgo and Coma clusters of galaxies.

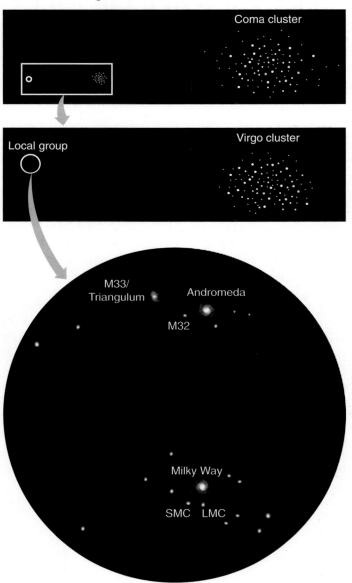

FIGURE 21.2 (a) The Virgo cluster contains a large fraction of spiral galaxies. (b) The larger, richer Coma cluster is dominated by elliptical galaxies.

the astronomical community with a "slice of the universe." **Figure 21.3(a)** shows what this slice looked like. The observations show that the clusters and superclusters, rather than being scattered randomly through space, are linked together in an intricate network of "filaments" and "walls." The concentrations of galaxies in turn surround large voids, regions of space that are largely devoid of galaxies. These voids represent some of the largest "structures" seen in the universe. Though the voids may seem like "a whole lot of nothing," we do not know that they are empty of *matter*—only

that they are largely empty of galaxies. Clusters and superclusters are located within the walls and filaments. This structure is not peculiar to the "nearby" universe. Subsequent surveys have looked at much larger volumes of space.

Large-scale structure fills the universe.

Figure 21.3(b) shows the results of one such survey conducted using the Anglo-Australian Telescope in Siding Spring, Australia. A more recent survey, the Sloan Digital Sky Survey (SDSS), is shown in **Figure 21.3(c)**. Out as far as our observatories can currently measure, the universe has a porous structure reminiscent of a sponge. Together, galaxies and the larger groupings in which they are found are referred to as **large-scale structure**.

21.3 Gravity Forms Large-Scale Structure

Large-scale structure cries out for explanation. Clearly it is the consequence of well-defined physical processes, but what processes? A number of ideas have been proposed. For example, early on it was suggested that voids were the result of huge expanding blast waves from tremendous explosions that might have occurred in the early universe. The correct answer has turned out to be less fanciful, but far more satisfying. Large-scale structure is the fingerprint of our old friend gravity.

At the close of the previous chapter we learned that the slight ripples in the glow of the cosmic background radiation probably result from quantum mechanical variations

Ripples in the universe were the seeds of structure.

that imprinted structure on the early universe at the time of inflation. These variations provided the seeds from which galaxies and collections of galaxies grew. In our discussion of star formation in Chapter 15 we learned about gravitational instabilities. As illustrated in Figure 15.14, we begin with a molecular cloud with clumps inside it, and gravity then causes those clumps to collapse faster than their surroundings. If conditions are right, then what began as relatively minor variations in the density of a cloud are turned by gravity into stars. If you replace "molecular cloud" with "universe" and "star" with "galaxy," you will have a good starting point for understanding the way galaxies formed from slight inhomogeneities in the distribution of matter following recombination.

It is one thing to say that galaxies and larger structures formed from gravitational instabilities that began with slight

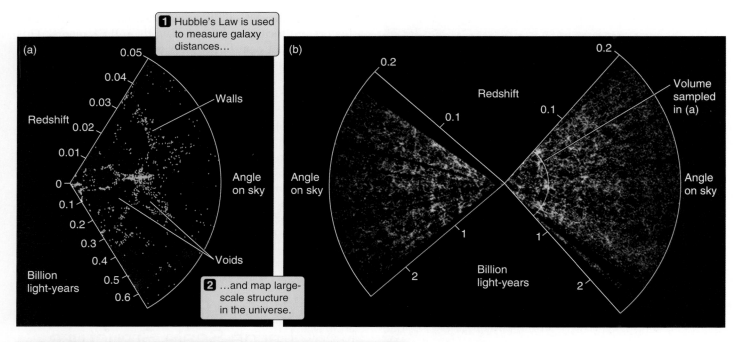

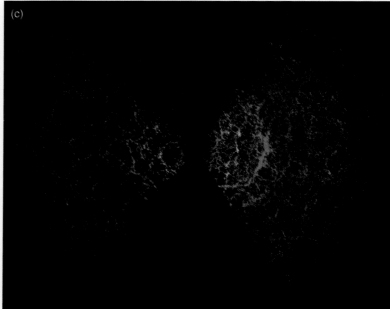

FIGURE 21.3 Redshift surveys use Hubble's Law to map the universe. (a) The Harvard-Smithsonian Center for Astrophysics redshift survey, called "A Slice of the Universe," was the first to show that clusters and superclusters of galaxies are part of even larger-scale structures. (b) The 2dF Galaxy Redshift Survey shows similar structures at even larger distances. (c) The Sloan Digital Sky Survey map of the universe extends outward to a distance of about 2 billion light-years. Shown here is a sample of 67,000 galaxies that lie near the plane of the Earth's equator.

irregularities in the early universe. It is quite another to turn this statement into a real scientific theory with testable predictions. To make that step, we return to the same basic

Models show how gravity turns early seeds into large-scale structure.

technique we have used to look into the centers of planets and stars and to answer many other questions about things we cannot observe directly. We combine the ideas we want to test with the laws of physics, construct a model, and then

compare the predictions of that model with observations of the universe.

To build a model of the formation of large-scale structure, we have to begin by making three key choices. First we have to decide what universe we are going to model. That is, what values of Ω_{mass} and Ω_Λ (see Chapter 20) are we going to assume? These are important in part because they determine how rapidly the universe expands and therefore how difficult it is for gravity to overcome this expansion in some region. The more rapidly a universe is expanding, or the less mass it contains, the more difficult it will be for gravity to pull material together into galaxies and larger-scale structures.

The second thing we need to know to construct our model is what the early bumps in the density of the universe looked like. How large and how concentrated were these early bumps? There are several ways to approach this question. One such method uses observations of variations in the CBR. Both COBE and WMAP (see Figure 20.17 and the opening photograph for Chapter 20) provide a good picture of structure in the CBR, from which we can infer what the early inhomogeneities in the universe must have looked like. Future observations such as those from the European Space Agency's proposed Planck mission will do even better.

A different way to approach this question is to look at models of inflation, which make predictions about the structure that will emerge following this episode of rapid expansion. These are among the predictions of the inflation model that will be tested in the years to come. These are especially important predictions to test because they tie together the large-scale structure of today's universe with our most basic

> **Smaller structures form first. Larger structures form later.**

ideas about what the universe was like in the briefest instant after the Big Bang. While we await final answers about the details of structure, we know enough to say that the early universe was more irregular on smaller (galaxy-sized) spatial scales than it was on the scales of the clusters, superclusters, filaments, and voids seen in Figure 21.3. An immediate consequence of this fact is that smaller structures (such as galaxies and even subgalactic clumps) formed first, whereas larger structures took more time to form. The idea of small structure forming first and larger structure forming later is referred to as **hierarchical clustering**. Hierarchical clustering has become one of the most important themes in our growing understanding of how structure in the universe formed.

The third thing we need if we are to model the formation of large-scale structure is a complete list of the types and amounts of ingredients the early universe was made of. Specifically, we need to know the balance between radiation, normal matter, and dark matter. We also need to make some choices about the nature of the dark matter that we use in our model. We'll discuss this further in a bit.

Once we have made these three choices—the shape of the universe we are modeling, the way matter inhomogeneities developed, and the nature and mixture of different forms of matter we are using—the rest is physics and calculations: We simply need to "turn the crank," as the expression goes. As we discuss later, there are some real difficulties in carrying out this strategy. For example, our uncertainties about how gas turns into stars make it difficult to compare our models with the real universe. Despite technical dif-

ficulties, we are aiming to answer this key question: What choices lead to a model universe that most resembles the real universe in which we live?

Galaxies Formed Because of Dark Matter

Observations of nonuniformity in the CBR made with COBE and WMAP show variations of about 1 part in 100,000. Models clearly show that such tiny variations at the time of recombination (when the universe was about half a million years old) are far too slight to explain the structure we see in today's universe. Gravity is simply not strong enough to grow galaxies and clusters of galaxies from such poor "seeds." These models indicate that for ripples in the density of the universe to collapse to form today's galaxies, the density of those ripples must have been at least 0.2 percent greater than the average density of the universe at the time of recombination.

If normal luminous matter in the early universe had been clumped at this level, then the variations in the CBR today would be at least 30 times larger than what is observed. At first glance this discrepancy might seem to be a horrible problem with our understanding of the origin of structure in the universe, but it has turned out instead to be a crucial result that ties together a number of pieces of the puzzle. The theme of this story is "dark matter."

Dark matter first appeared on our journey back in Chapter 18 as an ad hoc construction—an annoyance, really—used by astronomers to explain the oddly flat rotation curves of spiral galaxies. When we turn our attention to clusters of galaxies, we find that dark matter dominates normal matter on those scales as well. Now we find that irregularities in normal matter in the early universe were too slight to form galaxies. So where are we going to turn for an answer to the question of how structure in the universe *did* form? You guessed it: dark matter.

In our discussion of primordial nucleosynthesis we learned that dark matter cannot be made of normal matter consisting of neutrons, protons, and electrons. If it were, then it would have affected the formation of chemical elements in the early universe. The abundances of several isotopes of the least massive elements would be quite different

> **Dark matter provides the seeds for observed structure.**

from what we find in nature. Dark matter must be something else—something that has no electric charge (so it does not interact with electromagnetic radiation) and that interacts only feebly with normal matter. Clumps of such dark matter in the early universe would not interact with radiation or normal matter, so they would not be seen directly when looking at the CBR. This unseen dark matter saves the day for modeling the formation of galaxies and clusters of galaxies.

What is so different about the behavior of dark matter and normal matter in the early universe? First, pressure waves and radiation did a very good job of smoothing out ripples in the distribution of normal matter in the early universe. Feebly interacting dark matter would be immune to these processes, so clumps of dark matter survived long after bumps in the normal matter had been smoothed out, as illustrated in **Figure 21.4**. Second, the dark matter in these clumps does not glow, so we do not see it directly when we look at the cosmic background radiation. Although these clumps of dark matter do cause slight gravitational redshifts in the light coming from normal matter, the resulting variations fit well with current observations of the CBR. Thus clumps of dark matter could have (and, according to our theories of the very early universe, *should* have) existed in the early universe, even though we are unable to see evidence of them directly.

Just because we cannot see clumps of dark matter in the early universe does not mean they had no effect. In the same way that dark matter dominates the gravitational fields of today's galaxies and clusters, so too did the mass of dark matter control the growth of gravitational instabilities in the early universe.

There Are Two Classes of Dark Matter

Here is the story of galaxy formation in a nutshell. Dark matter in the early universe was much more strongly clumped than normal matter. Within a few million years after recombination, these dark matter clumps pulled in the surrounding normal matter. Later, gravitational instabilities caused these clumps to collapse. The normal matter in the clumps went on to form visible galaxies. This story seems plausible enough, but the details of how this happened depend greatly on the properties of dark matter itself. Even though we do not yet know exactly what dark matter in the universe is made of, we can talk about two broad classes of dark matter based on how it behaves.

One possibility is that dark matter consists of feebly interacting particles that are moving about relatively slowly, like the slowly moving atoms and molecules in a cold gas.

Cold dark matter consists of relatively massive, slowly moving particles.

For obvious reasons this type of dark matter is called **cold dark matter**. There are several candidates for cold dark matter. It is possible that cold dark matter consists of tiny black holes that might have been produced in the early universe. Few physicists and cosmologists favor this idea. Most think

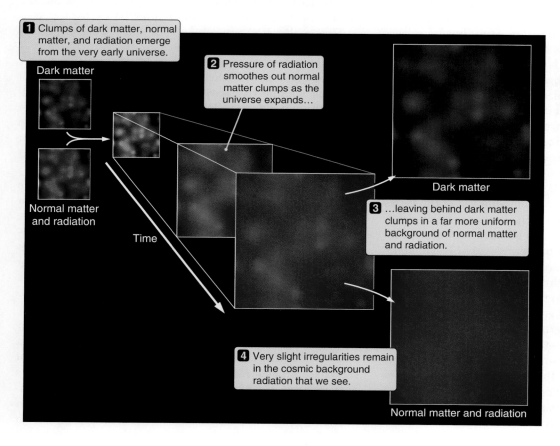

① Clumps of dark matter, normal matter, and radiation emerge from the very early universe.

Dark matter

Normal matter and radiation

Time

② Pressure of radiation smoothes out normal matter clumps as the universe expands...

Dark matter

③ ...leaving behind dark matter clumps in a far more uniform background of normal matter and radiation.

④ Very slight irregularities remain in the cosmic background radiation that we see.

Normal matter and radiation

FIGURE 21.4 Radiation pressure and other processes in the early universe smoothed out variations in normal matter, but irregularities in the dark matter survived to become the seeds of galaxy formation.

instead that cold dark matter consists of some unknown elementary particle. One candidate is the **axion**, a hypothetical particle first proposed to explain some observed properties of neutrons. Even though axions would have very low mass, they would have been produced in great abundance in the Big Bang. Another candidate is the **photino**, an elementary particle related to the photon. Some theories of particle physics predict that these particles exist and have a mass of around 10,000 times that of the proton. Our state of knowledge might soon change quickly: Photinos might be detected in current particle accelerators, and experiments are under way to search for axions and photinos that are trapped in the dark matter halo of our galaxy.

If the first class of dark matter is called cold dark matter, then the other class of dark matter is called—are you ready?—**hot dark matter**. Hot dark matter consists of particles that are moving very rapidly. Neutrinos are one example of hot dark matter that we know exists. We have seen that

> Hot dark matter consists of less massive, rapidly moving particles.

neutrinos interact with matter so feebly that they are able to flow freely outward from the center of the Sun, passing through the dense overlying layers of matter as if they were not there. There is no question that the universe is filled with neutrinos. Calculations indicate that around 300 million of these cosmic relics of the Big Bang fill each cubic meter of space. Preliminary measurements of the neutrino mass obtained at the Super Kamiokande detector in Japan and the Sudbury Neutrino Observatory in Canada suggest that neutrinos might account for as much as 5 percent of the mass of the universe. Although this percentage is not enough to account for all of the dark matter in the universe, it may be enough to have had a noticeable effect on the formation of structure.

The different effects of cold and hot dark matter on structure formation have to do with differences in the way the two types of dark matter cluster in a gravitational field. The faster an object is moving, the harder it is for gravity to hold onto it. This is the same basic idea that we used to explain why hydrogen and helium are able to escape Earth's atmo-

> Cold dark matter forms galaxies.

sphere while molecules such as oxygen and nitrogen are not. It is also the same idea we used to explain why the hot gas in elliptical galaxies extends far beyond their visible boundaries. Slow-moving particles are more easily corralled by gravity than fast-moving particles, so particles of cold dark matter clump together more easily into galaxy-sized structures than do particles of hot dark matter. The result of the calculations of galaxy formation is clear: To account for the formation of today's galaxies, we need cold dark matter.

A Galaxy Forms within a Collapsing Clump of Dark Matter

We can best see how models of galaxy formation work by following the events predicted by the models step by step. Let us consider the case of a universe made up of 90 percent cold dark matter and 10 percent ordinary matter, clumped together in a manner consistent with observations of the cosmic background radiation. On the scale of an individual galaxy, the effect of the cosmological constant is so small it can be ignored.

Figure 21.5(a) shows the state of affairs of one clump of dark matter at the time of recombination. The dark matter is less uniformly distributed than normal matter, but overall the distribution of matter is still remarkably uniform. By a few million years after recombination (**Figure 21.5(b)**), the universe of our model calculation has expanded severalfold. Spacetime is expanding, so the clump of dark matter is also expanding. However, the clump of dark matter is not expanding as rapidly as its surroundings because its self-gravity has slowed down its expansion. The clump now stands out more with respect to its surroundings. Another change that has taken place is that the gravity of the dark matter clump has begun to pull in normal matter as well. By the stage shown in **Figure 21.5(c)**, normal matter is clumped in much the same way as dark matter.

A ball thrown in the air will slow, stop, and then fall back to Earth. In like fashion, the clump of dark matter will stop expanding when its own self-gravity slows and then stops its initial expansion. By the time the universe is around a billion years old, the clump of dark matter has

> Clumps of dark matter can collapse only so far.

reached its maximum size and is beginning to collapse (Figure 21.5(c)). The collapse of the dark matter clump stops when the clump is about half its maximum size, however, because the particles making up the cold dark matter are moving too rapidly to be pulled in any closer (**Figure 21.5(d)**). The clump of cold dark matter is now given its shape by the orbits of its particles, in the same way that an elliptical galaxy is given its shape by the orbits of the stars it contains.

Unlike dark matter (which cannot emit radiation), the normal matter in the clump is able to radiate away energy and cool. As the gas cools, it loses pressure; and as it loses pressure, it collapses. Models show that small concentrations of normal matter within the dark matter clump collapse under their own gravity to form clumps of normal matter that range from the sizes of globular clusters to the sizes of dwarf galaxies. These clumps of normal matter then fall inward toward the center of the dark matter clump, as shown in **Figure 21.5(e)**. We discussed a similar chain of events in Chapter 15: the collapse of a molecular cloud core on its way to becoming a star. It is important to note that,

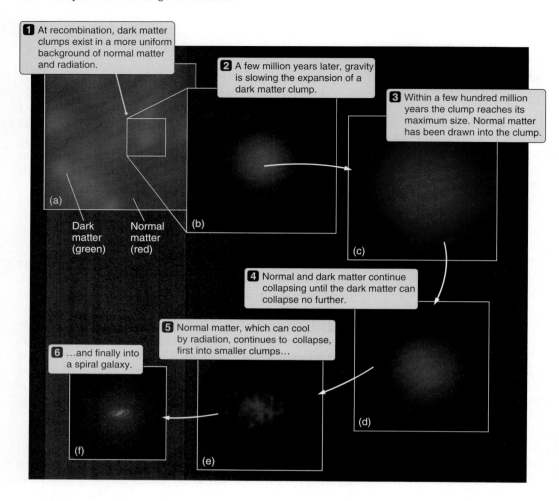

1 At recombination, dark matter clumps exist in a more uniform background of normal matter and radiation.

2 A few million years later, gravity is slowing the expansion of a dark matter clump.

3 Within a few hundred million years the clump reaches its maximum size. Normal matter has been drawn into the clump.

4 Normal and dark matter continue collapsing until the dark matter can collapse no further.

5 Normal matter, which can cool by radiation, continues to collapse, first into smaller clumps…

6 …and finally into a spiral galaxy.

Dark matter (green) — Normal matter (red)

(a) (b) (c) (d) (e) (f)

FIGURE 21.5 Stages in the formation of a spiral galaxy from the collapse of a clump of cold dark matter.

according to models, gas in our universe can cool quickly enough to fall in toward the center of the dark matter clump only if the clump has a mass of 10^8 to 10^{12} $M_\odot$. This is just

> **Normal matter cools and falls toward the center of the dark matter halo.**

the range of masses of observed galaxies! This agreement between theory and observations is an important success of the cold dark matter theory of galaxy formation.

As discussed in **Connections 21.1**, there are many similarities between the collapse of a molecular cloud to form stars and the collapse of a clump in the early universe to form galaxies. Another process that we saw at work in star formation also comes into play at this point. The clumps of matter collapsing to form galaxies do not exist in isolation. They have been tugged on by the gravity of neighboring clumps and have been pushed around by the pressure waves that ran through the young universe, smoothing out its structure. As a result, each protogalactic clump has a little bit of rotation when it begins its collapse. As normal

matter falls inward toward the center of the dark matter clump, this rotation forces much of the gas to settle into a rotating disk (**Figure 21.5(f)**), just as the collapsing cloud that was to become our Sun settled first into an accretion disk. And just as the accretion disk around the Sun provided the raw materials for the planets of our Solar System, the disk formed by the collapse of each protogalactic clump becomes the disk of a spiral galaxy.

Searching for Signs of Dark Matter

Dark matter has come to play a very important role in our understanding of the universe. We might reasonably be uncomfortable having our models of galaxy formation rely so heavily on something we cannot see directly. Fortunately, dark matter shows itself in a variety of ways. The flat rotation curves of spiral galaxies, for example, are compelling evidence for the existence of extended halos comprised of dark matter. Similarly, the ability of elliptical galaxies to hold

CONNECTIONS 21.1

Parallels between Galaxy and Star Formation

As you read about galaxy formation, you might find it enlightening to think back to our discussion of star formation in Chapter 15. Both processes involve the gravitational collapse of vast clouds to form denser, more concentrated structures. To help you with the comparison, we list here a few of the similarities and differences between the two:

Gravitational instability: In both star and galaxy formation, the collapse begins with a gravitational instability. Regions only slightly denser than their surroundings are pulled together by their own self-gravity. As the matter in these regions becomes more compact, gravity becomes stronger, and the collapse process snowballs. One key difference is that for a galaxy to form, the dark matter clump must collapse rapidly enough to counteract the overall expansion of the universe itself.

Fragmentation: In both cases the original cloud separates into smaller pieces as a result of the gravitational instability. However, the order of fragmentation differs between star and galaxy formation. In molecular clouds large regions begin to collapse first, then fragment further to form individual stars. In contrast to this "top-down" process, galaxy formation is "bottom-up": Smaller structures collapse first and then merge together to form galaxies and eventually assemblages of galaxies.

Compression, heating, and thermal support: As an interstellar molecular cloud collapses, its temperature climbs and the pressure in the cloud increases. The higher pressure would eventually be enough to prevent further collapse except that the cloud core is able to radiate away thermal energy. That is the bright infrared radiation that allows us to see star-forming cores. Compare this with galaxy formation: As a dark matter clump collapses, its temperature also climbs, and it too reaches a point where there is a balance between gravity and the thermal motions of the dark matter. However, dark matter is *not* able to radiate away energy, so once this balance is reached, the collapse of the dark matter is over. Only the normal matter within the cloud of dark matter is able to radiate away thermal energy and continue collapsing. That is why normal matter collapses to form galaxies, while the dark matter remains in much larger dark matter halos.

As galaxies form, dark matter remains in extended halos. Dark matter is too hot to settle into galaxy disks. It is far too hot to become concentrated into even smaller structures such as molecular clouds or to take part in the formation of stars. Dark matter may be the dominant form of matter in the universe, and it may determine the structure of galaxies; but dark matter can never collapse enough to play a role in the processes that shape stars, planets, or the interstellar medium.

Angular momentum and the formation of disks: Conservation of angular momentum is responsible for the formation of disk galaxies, just as conservation of angular momentum is responsible for the formation of the accretion disks around young stars. The Milky Way and the Solar System are both flat for the same reasons. The origin of the angular momentum is different, though. Turbulent motions within the molecular cloud precursors of stars produce the net angular momentum for a stellar disk, whereas gravitational interactions with nearby clumps lead to the angular momentum of the Milky Way.

The end product: Once a stellar accretion disk forms, most of the matter moves inward and is collected into a star. In contrast, much of the matter in a spiral galaxy remains in the disk.

onto hot gas convincingly demonstrates that they contain far more mass than can be accounted for by their stars alone.

We can use a similar approach to look for evidence of dark matter on larger scales. Galaxy clusters are filled with extremely hot (10 million to 100 million K) gas, which occupies the space between galaxies (**Figure 21.6**). Even though this gas is *extremely* tenuous, the volume of space that it occupies is enormous: The mass of this hot gas can be up to five times the mass of all the stars in that cluster. X-ray spectra show that this gas contains significant amounts of massive elements that must have formed in stars. This chemically enriched gas has been either blown out of galaxies in winds driven by the energy of massive stars, or stripped from galaxies during encounters with neighboring galaxies. Were it not for the gravity of the dark matter filling the volume of the cluster, this hot gas would have dispersed long ago.

We can also look at the motions of the cluster's galaxies (which behave like very large "atoms" in this context) and again ask, How strong must gravity be to hold this cluster together? Again, the answer is that the total mass of the clusters, including dark matter, must be about 10 times greater than the normal matter they contain.

A third way to look for dark matter relies on the predictions of Einstein's general theory of relativity, which states that mass distorts the geometry of spacetime, causing even light to bend near a massive object. In particular, light from a distant object is bent by the gravity of a galaxy or cluster of galaxies, so that images of the distant object can be seen magnified on either side of the intervening galaxy or cluster. The result is a **gravitational lens (Figure 21.7(a))**.

We have already encountered gravitational lensing in our discussion of MACHOs in Excursions 19.2, where we found that lenses can make background objects appear brighter. When considering the more complex three-dimensional geometry of gravitational lenses, we find that

Gravitational lenses are one way to measure the masses of galaxy clusters.

lenses can show multiple images of background objects, and that these magnified images are often drawn out into an arclike appearance. The greater the gravitational lensing, the greater the mass that must be in the cluster. **Figure 21.7(b)** shows an image of a galaxy cluster that is acting as a gravitational lens for a number of background galaxies. Analysis of such images allows us to determine the mass of the lensing cluster.

Regardless of how we measure the masses of galaxy clusters—by looking at the motions of their galaxies, by

Dark matter dominates the mass of galaxy groups and clusters.

measuring their hot gas, or by using them as gravitational lenses—the results are the same. By mass, galaxy clusters, like the galaxies they are made of, are dominated by the dark matter they contain.

(a) X-rays

1 For gravity to hold onto this hot X-ray emitting gas…

1 Mly

(b) Visible light

2 …this galaxy cluster must be more massive than the galaxies we see.

3 Galaxy clusters are mostly dark matter.

FIGURE 21.6 (a) An X-ray image of a cluster of galaxies shows that the cluster is full of hot tenuous gas. (b) An image of the same field of view taken in visible light. (The two brightest objects are foreground stars.)

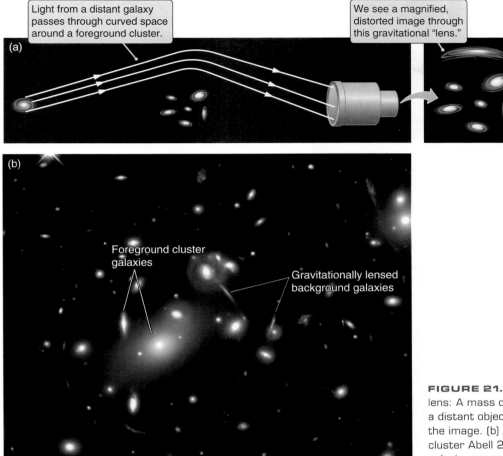

FIGURE 21.7 (a) The geometry of a gravitational lens: A mass can gravitationally focus the light from a distant object, thereby magnifying and distorting the image. (b) A Hubble Space Telescope image of the cluster Abell 2218 showing many gravitationally lensed galaxies, seen as arcs.

Mergers Play a Large Role in Galaxy Formation

The picture of galaxy formation given in the previous sections is much cleaner and more idealized than what happened in reality. In our hierarchical clustering picture, smaller structures collapsed first. If there was substructure within the original galaxy-sized dark matter clump, that substructure would itself have collapsed into subgalaxy-sized objects before the galaxy as a whole finished forming. This gives us a way to understand the existence of such objects as the globular star clusters and dwarf companion galaxies near the Milky Way and other galaxies. These objects formed inside the larger, cold dark matter clump when the luminous matter in the clump was still settling toward what would become the disk of the galaxy.

It is also likely that the same larger clump often produced more than one galaxy, and that these galaxies later interacted or even merged. When two spiral (or protospiral) galaxies merged, all sorts of commotion were likely. The tidal interactions between the galaxies and the collisions between gas clouds in the galaxies probably triggered many regions of star formation throughout the combined system.

The merging of galaxies also answers another puzzle. We have seen how spiral galaxies could form in the early universe, but what about the formation of elliptical galax-

Ellipticals form from the mergers of spirals.

ies? Ellipticals are now thought to result from the merger of two or more spiral galaxies. **Figure 21.8** shows a computer simulation of such a merger. If the merging galaxies were not originally spinning in the same direction, then the resulting merged galaxy loses its disklike character. The dark matter halos of the galaxies merge, and the stars eventually settle down into the bloblike shape of an elliptical galaxy. As we might expect on the basis of this picture, elliptical galaxies are known to be more common in dense clusters where mergers are likely to have been more frequent.

Such events also played a major role in feeding supermassive black holes, which themselves must have formed

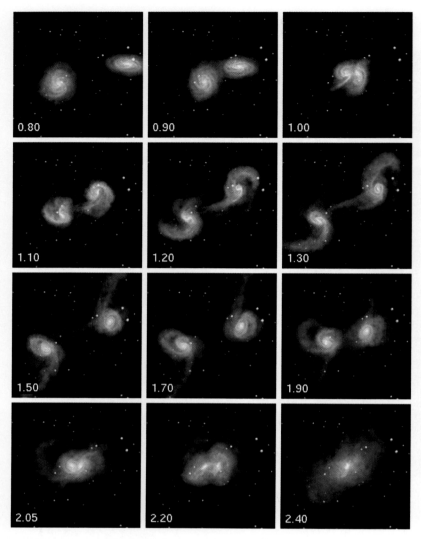

When we look at the bright, active galaxies that populate the early universe, it is important to remember that what we actually see is only the tip of the iceberg. The luminous galaxies are small in comparison with the dark matter clumps that surround them. These clumps are still found in today's universe as the invisible dark matter halos responsible for the flat rotation curves of spiral galaxies.

Observations Help Fill Gaps in the Models

We would like to be able to turn our models of collapsing cold dark matter clumps into detailed models of what galaxies in the young universe should look like, but we have not yet reached such a level of sophistication. One major problem is that we lack a good understanding of how stars form in young galaxies. We know that most of the normal matter in a galaxy winds up as stars; but when and where do those stars form, and how long does it take? Although we can say a great deal about how individual stars form in our galaxy, as we did in Chapter 15, we do not yet have theories or models that predict such important pieces of the puzzle as what fractions of stars should have what masses. Nor do we clearly understand the differences between star formation in the early universe and the star formation going on around us today. In a universe devoid of massive elements there were no dense, dusty molecular clouds from which stars might form, and no spiral disks in which such clouds might congregate. Instead the first stars must have formed from the collapse of clouds of hydrogen and helium gas within the overall clump of matter of which they were a part.

Although a detailed understanding of star formation in a forming galaxy is in the future, we can still make a number of factual statements about when and where stars form in a collapsing galaxy. These clues come from the study of our

Star formation began very early on in collapsing galaxies.

own galaxy. The fact that the atmospheres of even the oldest halo stars in the Milky Way contain some amount of massive elements tells us that a significant amount of star formation must have taken place *very* early during the collapse of the Milky Way. If we could closely inspect a collapsing proto-

very early on in the collapse of galaxies and protogalaxies. The same mergers and interactions that formed the giant galaxies that we see today also provided the fuel to power the quasars and other AGNs that were common in the early universe.

With galaxies crashing together, supermassive black holes forming, quasars flooding the young universe with intense radiation and powerful jets, and star formation running amok, galaxy formation in the young universe must have been a violent, messy business. This conclusion has clear consequences for what we should expect to see when looking back to a time in the history of the universe when galaxies were actively forming. Rather than the well-formed spirals and ellipticals that dominate today's universe, the early universe should have contained many clumpy, irregular objects. As expected, images of the young universe such as the Hubble Space Telescope image shown in Figure 18.1 are filled with lumpy, distorted objects that are still in the process of settling in.

galactic clump in the stages depicted in Figure 21.5(d) and (e), we would expect to see stars already forming and supernovae exploding from place to place in the collapsing halo. The halo stars that we see today must have formed while the Milky Way was still collapsing, before the gas they formed from coalesced into the disk. In contrast, disk stars (which all have relatively high abundances of massive elements) formed from gas that was enriched by early generations of halo stars before it settled into the galaxy's disk.

Thus while we cannot say exactly how stars formed in the early universe, we can say that star formation must have been going on throughout the process of galaxy formation. The pattern of stellar ages and massive element abundances evident in our galaxy today fits naturally with our current models in which galaxies formed from collapsing clumps of cold dark matter.

Uncertainties in our understanding of star formation complicate quantitative comparisons between models of galaxy formation and observations of the early universe. Fortunately, though, when we talk about structure on scales much larger than galaxies, star formation becomes less of an issue. To understand the formation of clusters and larger structures, we only need worry about the way that galaxy-sized clumps of matter themselves fall together under the force of gravity.

Figure 21.9 shows the results of one computer simulation that follows the motions of millions of clumps of dark matter as they fall through space under their mutual gravitational attraction. The scale of this simulation is so large that we cannot follow the outcome of individual galaxies,

Clusters, filaments, and voids form after galaxies form.

but instead are looking at the overall distribution of dark matter. The simulation shows that the smaller-scale structures develop first. Over the first few billion years, dark matter falls together into structures comparable in size to today's clusters of galaxies. Only later do the spongelike filaments, walls, and voids become well defined.

The similarities between the results of these models and observations of large-scale structure, such as those in Figure 21.3, are quite remarkable. Not any model will do, however. Only model universes with certain combinations of shape, mass, nature of ripples, type of dark matter, and values for the cosmological constant will produce structure similar to what we actually see. It is a very important result that models beginning with assumptions consistent with our knowledge of the early universe wind up predicting the formation of large-scale structure similar to what we see in today's universe.

FIGURE 21.9 A computer simulation of the formation of very large-scale structure in a universe filled with cold dark matter.

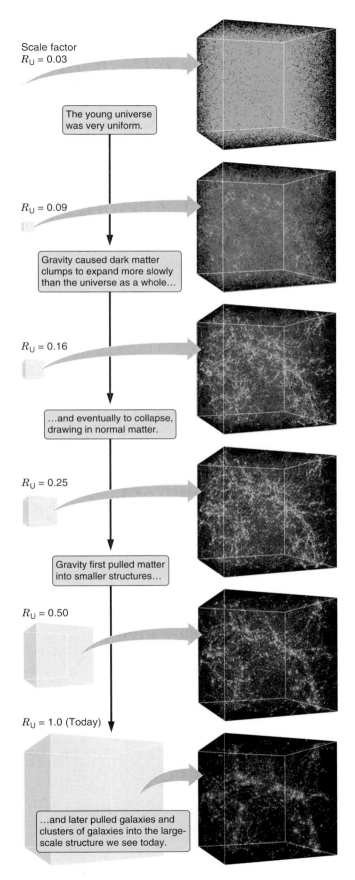

Scale factor
$R_U = 0.03$

The young universe was very uniform.

$R_U = 0.09$

Gravity caused dark matter clumps to expand more slowly than the universe as a whole…

$R_U = 0.16$

…and eventually to collapse, drawing in normal matter.

$R_U = 0.25$

Gravity first pulled matter into smaller structures…

$R_U = 0.50$

$R_U = 1.0$ (Today)

…and later pulled galaxies and clusters of galaxies into the large-scale structure we see today.

Peculiar Velocities Trace the Distribution of Mass

In our comparison between models and observations, there is an important caveat. The models show where the *mass* is, whereas our observations show where the *light* is. We have

---

The distribution of light may not be the same as the distribution of mass.

---

already noted that our understanding of star formation in the early universe is incomplete at best. What if a dark matter clump formed and collapsed, but for some reason the normal matter associated with that clump did not produce stars? Such clumps could still contain a significant amount of mass and could even affect the geometry of the universe, but would themselves not be seen in images of the sky.

There is a way to get at the question of the overall distribution of dark matter more directly. If the clumping of galaxies is a result of the action of gravity, then we should be able to look at how galaxies are moving and see them falling together. In fact, measurements of the way galaxies fall together should allow us to infer the gravitational field they are experiencing, and hence the overall distribution of dark matter.

In the previous chapter we learned that the cosmic background radiation is blueshifted in one direction in space and redshifted in the opposite direction, telling us of our peculiar velocity relative to the CBR. Peculiar velocities of galaxies other than our own are difficult to measure. They require us to accurately determine the distances to galaxies using standard candles, and then to use those distances along with Hubble's Law to predict what the redshifts of those galaxies should be. Comparison of observed redshifts with redshifts predicted on the basis of Hubble expansion then tells us how fast these galaxies are moving with re-

spect to the cosmic background radiation (at least along our line of sight).

Figure 21.10(a) shows one such map of the peculiar velocities inferred for galaxies in our part of space. If we use observations of peculiar velocities to map concentrations of mass in this part of the universe, we get a picture that looks like two mountains surrounded by foothills and valleys, as shown in **Figure 21.10(b)**. Unfortunately, these two "mountains" of mass, which exert the greatest gravitational pulls on our own galaxy, are located in regions of the sky that are heavily obscured by the Milky Way's dusty disk.

Peculiar velocities tell us how dark matter is distributed.

The largest pull that we are experiencing comes from a region dubbed the Great Attractor, so-called because of its gravitational tug.

The picture painted in this chapter of the formation of large-scale structure is almost certainly correct in its broad outlines. Even so, uncertainties in star formation, our understanding of the exact nature of dark matter, and our knowledge of the shape of the universe limit the amount of detail in any comparisons we can make between models and observations. As the 21st century opens, this is a field of very active research. Several projects such as the Sloan Digital Sky Survey are currently collecting large amounts of data on the distribution and redshifts of galaxies, and satellites such as WMAP and the ESA's proposed Planck mission can dramatically improve our knowledge of the structure in the CBR. Larger and more sophisticated models are being built all the time, while studies of star-forming regions in our galaxy and in galaxies with lower massive element abundances are continuing to improve our understanding of this important process. There is also hope that new theories and laboratory experiments might identify

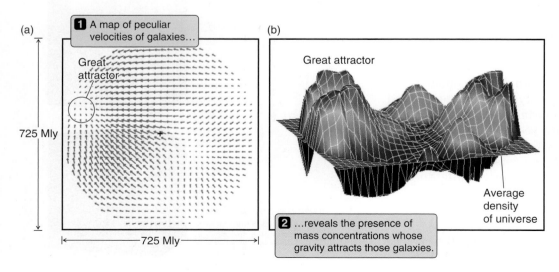

FIGURE 21.10 (a) A map of peculiar velocities of galaxies in our neighborhood. Arrows show the velocities of material lying in the plane of our own local supercluster. Motions seem to be converging onto the position of the Great Attractor. The black cross marks the location of the Milky Way Galaxy. (b) A map of the mass distribution inferred from observations of peculiar velocities.

the particles that make up dark matter. A more complete understanding of the formation of galaxies and large-scale structure in the universe will not likely come in a single "eureka" moment, but will instead arrive in bits and pieces throughout the first part of the 21st century as a result of advances in all of these different areas.

21.4 The Earliest Moments

It is easy to look at galaxies and stars and say, "Aha, structure!" But there is more to the origin of structure than simply the clumping of matter. The forces that govern the behavior of matter and energy in the universe are themselves a kind of structure. There are four fundamental forces in nature, and everything in the universe is a result of their action. Chemistry and light are products of the *electromagnetic force* acting between protons and electrons in atoms and molecules. The energy produced in fusion

> There are four fundamental forces in nature.

reactions in the heart of the Sun comes from the *strong nuclear force* that binds together the protons and neutrons in the nuclei of atoms. Beta decay of nuclei, in which a neutron decays into a proton, an electron, and an antineutrino, is governed by the **weak nuclear force**. Finally there is our old friend *gravity,* which has played such a major role at every point along our journey. How these forces— these physical laws—came into being is part of the history of the universe as well.

May the Forces Be with You

In Chapter 4 we spoke of electromagnetism using the electric and magnetic fields, but we also spoke of the quantum mechanical description of light as a stream of particles called photons. Because there is only one reality, both of these descriptions of electromagnetism have to coexist. The branch of physics that deals with this reconciliation is called **quantum electrodynamics** or **QED**.

QED treats charged particles almost as if they were baseball players engaged in an endless game of catch. As the baseball players throw and catch baseballs, they experience forces. Similarly, in QED, charged particles "throw" and "catch" an endless stream of virtual photons, as illustrated in **Figure 21.11**. Earlier we grappled with the idea that quantum mechanics is a science of possibilities rather than certainties. QED describes the electromagnetic interaction between two charged particles by averaging over all of the

possible ways that the particles could throw photons back and forth. The result is a force that, over large scales, acts like the classical electric and magnetic fields described by Maxwell's equations. Physicists speak of the electromag-

> In QED, photons carry the electromagnetic force.

netic force being "mediated by the exchange of photons." As is always the case with quantum mechanics, the world described by QED is hard to picture. Even so, QED is one of the most accurate, well-tested, and precise branches of physics. As of this writing, not even the tiniest measurable difference between the predictions of the theory and the outcome of an actual experiment has been found.

FIGURE 21.11 (a) The classical view of an electron being deflected from its course by the electric field from a proton. (b) According to quantum electrodynamics, the interaction is properly viewed as an ongoing exchange of photons between the two particles.

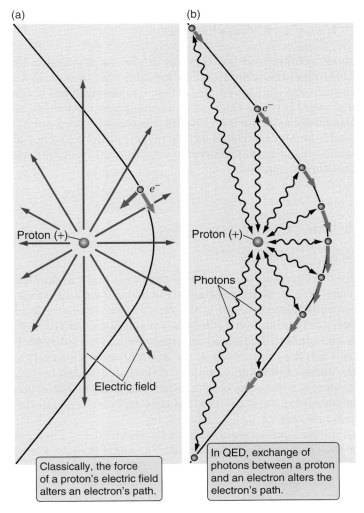

(a)

Proton (+)

e^-

Electric field

Classically, the force of a proton's electric field alters an electron's path.

(b)

e^-

Proton (+)

Photons

In QED, exchange of photons between a proton and an electron alters the electron's path.

The central idea of QED—forces mediated by the exchange of carrier particles—provides a template for understanding two of the other three fundamental forces in

The weak nuclear and electromagnetic forces combine in electroweak theory.

nature. The electromagnetic and weak nuclear forces have been combined into a single theory called **electroweak theory**. This theory predicts the existence of three particles —labeled the W^+, W^-, and Z^0—that mediate the weak nuclear force. The essential predictions of electroweak theory were confirmed in the 1980s when these particles were identified in laboratory experiments.

The strong nuclear force is described by a third theory, called **quantum chromodynamics** or **QCD**. In this theory particles such as protons and neutrons are composed of more fundamental building blocks, called **quarks**, that are bound together by the exchange of another type of carrier particle dubbed **gluons**. Together electroweak theory and QCD are referred to as the **standard model** of particle physics. A deeper investigation of the standard model must

Electroweak theory + QCD = The standard model of particle physics.

await another journey. Here we leave the discussion by pointing out that the standard model is able to explain all the currently observed properties of matter and has made many predictions that were subsequently confirmed by laboratory experiments. The standard model does not exclude the neutrino from having mass, but the existence of neutrino mass would have an explanation lying outside the standard model.

A Universe of Particles and Antiparticles

For modern theories of particle physics to make any sense, every type of particle in nature must also have an alter ego —an *antiparticle*—that is the opposite of that particle in every way described by quantum mechanics. For the electron there is the antielectron, otherwise known as the positron. For the proton there is the antiproton, for the neutron the antineutron, and so on down the list. One fascinating property of these particle–antiparticle pairs is that if you bring such a pair together, the two particles will annihilate each other.

When a particle–antiparticle pair annihilates, the mass of the two particles is converted into energy in accord with Einstein's special theory of relativity ($E = mc^2$). For example, in **Figure 21.12(a)** an electron and a positron annihilate each other, and the energy is carried away by a pair of photons.

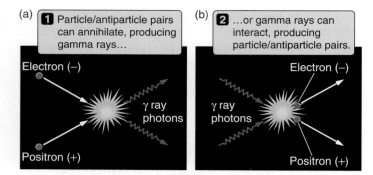

FIGURE 21.12 (a) An electron and a positron annihilate, creating two gamma ray photons that carry away the energy of the particles. (b) In the reverse process, pair creation, two gamma ray photons collide to create an electron–positron pair.

(This is the idea behind *Star Trek*'s "antimatter" engines.) When we run time backward (as we are allowed to do in particle physics), we see two high-energy photons colliding with each other, as in **Figure 21.12(b)**, creating in their place an electron–positron pair. This is an example of how an energetic event can create a particle and its corresponding antiparticle—a process called **pair creation**.

In principle, *any* type of particle and its antiparticle can be created in this way. The only limitation comes when there is not enough energy available to supply the rest mass of the particles being created. A specific example helps to show how this works. An electron (or positron) has a mass of $m_e = 9.11 \times 10^{-31}$ kg, which corresponds to an energy ($E = m_e c^2$) of 8.20×10^{-14} joules. If two gamma ray photons with a combined energy greater than 16.40×10^{-14} J ($= 2 \times m_e c^2$) collide, then the two photons may disappear and leave an electron–positron pair behind in their place. If the photons have more than the necessary energy, then the extra energy goes into the kinetic energy of the two newly formed particles.

Now we apply this idea to a hot universe awash in a bath of Planck radiation. Using Wien's Law from Chapter 4 [$\lambda_{peak} = (2{,}900\ \mu\text{m K})/T$] and the expression for photon energy ($E = hc/\lambda$), we can show that when the universe was less than about 100 seconds old and had a temperature greater than a billion kelvins, it was filled with photons that had enough energy to create electron–positron pairs. Under

The early universe was filled with photons, electrons, and positrons.

these conditions, photons were constantly colliding, creating electron–positron pairs; and electron–positron pairs were constantly annihilating each other, creating pairs of gamma ray photons. The whole process reached an equilibrium, determined strictly by temperature, in which pair creation and pair annihilation exactly balanced each other. Rather than being filled only with a swarm of photons, at this time the universe was filled with a swarm of photons, electrons, and positrons, as illustrated in **Figure 21.13(a)**.

The Frontiers of Physics

Physics has given us the tools to understand all the structures we have seen so far on our journey. There are still many gaps in what we know, but we are confident that everything from planets to stars to galaxies can be understood by applying our current knowledge of the four fundamental forces. At the turn of the 21st century, most of astronomy is a struggle with the complexity of the universe, rather than a shortcoming of our understanding of the fundamental laws governing matter, energy, and spacetime. However, as we push further back toward the Big Bang itself, the nature of the game changes. We need new physical theories.

In Foundations 10.1 we explored the power of symmetry—the idea that we can learn a lot about nature just by thinking about the way one part of something matches up with another. There we were talking about the gravitational forces that hold planets together, but other kinds of symmetry exist as well. In the process of pair creation there is a symmetry between matter and antimatter. For every particle created, its antiparticle is created as well.

As the universe cooled, there was no longer enough energy to support the creation of particle pairs, so the swarm of particles and antiparticles that filled the early universe annihilated each other and were not replaced. When this happened, every electron should have been annihilated by a positron. Every proton should have been annihilated by an antiproton. This was almost the case, but not quite. For every electron in the universe today, there were 10 billion

> For every 10 billion positrons there were 10 billion and one electrons.

and one electrons in the early universe, but only 10 billion positrons. This one part in 10 billion excess of electrons over positrons meant that when electron–positron pairs finished annihilating each other, some electrons were left over—enough to account for all the electrons in all the atoms in the universe today (**Figure 21.13(b)**).

If the standard model of particle physics were a complete description of nature, then the one-part-in-10-billion imbalance between matter and antimatter would not have been there in the early universe. The symmetry between matter and antimatter would have been complete. No matter at all would have survived into today's universe, and we would not exist. The fact that you are reading this page demonstrates that something more needs to be added to the model.

According to current ideas, the symmetry between matter and antimatter may be broken in a theory that joins the

> GUTs unify the strong and electroweak interactions.

electroweak and strong nuclear forces together in much the same way that electroweak theory unified our understanding

FIGURE 21.13 (a) A swarm of electrons, positrons, and photons in the very early universe. For every 10 billion positrons, there were really 10 billion and one electrons. (b) After these particles annihilated, only the one electron was left.

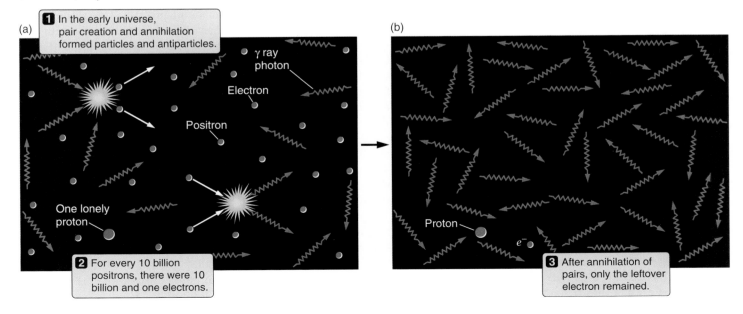

(a)

1 In the early universe, pair creation and annihilation formed particles and antiparticles.

γ ray photon

Electron

Positron

One lonely proton

2 For every 10 billion positrons, there were 10 billion and one electrons.

(b)

Proton

e^-

3 After annihilation of pairs, only the leftover electron remained.

of electromagnetism and the weak nuclear force. Such a theory, which combines three of the four fundamental forces into a single grand, unified force, is referred to as a **grand unified theory** or **GUT**.

Many possible grand unified theories exist, and they make many predictions about the world. The problem is that the particles that mediate GUTs are so massive that it takes enormous amounts of energy to bring them into existence—roughly a trillion times as much energy as can be achieved in today's particle accelerators! Even so, some predictions of GUTs are testable with current technology. For

---

GUTs predict that even the proton will decay.

---

example, GUTs predict that protons should be unstable particles that, given enough time, will decay into other types of elementary particles. This is a *very* slow process. Over the course of your life, GUTs predict that there is about a 1 percent chance that *one* of the 10^{28} or so protons in your body will decay. As of this writing, proton decay has yet to be observed. As we speak, however, large arrays of detectors are peering into huge tanks of water, waiting to see the signature of such an event. Perhaps your local news stand holds a story heralding the confirmation of this central prediction of GUTs.

The particles that mediate GUTs may be beyond the reach of today's high-energy physics labs; but when the universe was *very* young (younger than about 10^{-35} second) and *very* hot (hotter than about 10^{27} K), there was enough energy available for these particles to be freely created. During this time, the distinction between the electromagnetic, weak, and strong nuclear forces had not yet come into being. There was only the one grand, unified force. Welcome to the era of GUTs.

During this era of GUTs, the apparent size of the entire universe was less than a trillionth the size of a single proton. This may seem virtually incomprehensible, yet the basic ideas needed to understand this universe are in place. However, as we move backward in time, there is one threshold we have yet to cross. The story of advances in our understanding of physical law has been a story of unification—of the electromagnetic and weak forces into the electroweak theory, and then of these and the strong nuclear force into a GUT. But this program is incomplete. How does gravity fit into this scheme?

General relativity provides a beautifully successful description of gravity that correctly predicts the orbits of planets, describes the ultimate collapse of stars, and even allows

---

Gravity does not fit into the GUT picture.

---

us to calculate the structure of the universe. Yet general relativity's description of gravity "looks" very different from our theories of the other three forces. Rather than talking

about the exchange of photons or gluons or other carrier particles, general relativity talks instead about the smooth, continuous canvas of spacetime upon which events are painted.

We might be tempted to say, "Oh well. Gravity works one way and the other forces work another way, and that is how the universe happens to be." In practice that is exactly what we do when we call on quantum mechanics to tell us about the properties of atoms, then use relativity to describe the passage of time or the expansion of the universe. Even the era of GUTs is described perfectly well by treating gravity as a separate force. However, as we push back even closer to the moment of the Big Bang, this happy coexistence between relativity and quantum mechanics turns instead into a brutal confrontation.

Toward a Theory of Everything

When the universe was younger than about 10^{-42} second old, the density of the universe was incomprehensibly high. The universe was so small that 10^{60} universes would have fit into the volume of a single proton! Under these extreme conditions, quantum mechanical fluctuations in the matter and radiation making up the universe involved immense amounts of mass—so much mass that quantum fluctuations made mincemeat out of spacetime. Rather than a smooth sheet, spacetime was a quantum mechanical froth. General relativity fails to describe this early universe in ways that are reminiscent of the failure of Newtonian mechanics to describe the structure of atoms. An electron in an atom

---

In the Planck era the whole universe was a quantum mechanical froth.

---

must be thought of in terms of probabilities rather than certainties. Similarly, there is no unique history for the earliest moments after the Big Bang. This era in the history of the universe is referred to as the **Planck era**, signifying that we can understand the structure of the universe itself during this period only by using the ideas of quantum mechanics.

The conflict between the continuous and the discrete —between general relativity and quantum mechanics— brings us to the current limits of human knowledge. The physics that we know can take us back to a time when the universe was a millionth of a trillionth of a trillionth of a trillionth of a second old, but to push back any further we need something new. We need a theory that combines general relativity and quantum mechanics into a single theoretical framework unifying all four of the fundamental forces. Here we have reached the holy grail of modern physics. To understand the earliest moments of the universe, we need a **theory of everything (TOE)**.

A successful theory of everything would do more than unify general relativity with quantum mechanics. It would tell us which of the possible GUTs is correct and would provide an answer for the nature of dark matter. A successful theory of everything would also necessarily answer several outstanding issues in cosmology, including the how,

Superstring theory is a possible theory of everything.

when, and why of inflation, and the physics underlying the possible cosmological constant adding to the expansion of the universe. Physicists are currently grappling with what a TOE might look like. The current contender for the title is **superstring theory**. Here elementary particles are viewed not as points but as tiny loops called *strings*. A guitar string vibrates in one way to play an F, another way to play a G, and yet another way to play an A. According to superstring theory, different types of elementary particles are like different "notes" played by vibrating loops of string.

Superstring theory in principle provides a way to reconcile general relativity and quantum mechanics, but there is a price to pay for this success. To make superstring theory work, we have to imagine that these tiny loops of string are vibrating in a universe with 10 spatial dimensions! (Adding time to the list would make our universe 11-dimensional.)

Superstring theory predicts 11 dimensions, but we experience only 4 of them.

How can that be, when we clearly experience only three spatial dimensions? Whereas the three spatial dimensions that we know spread out across the vastness of our universe, the other seven dimensions predicted by string theory wrap tightly around themselves (**Figure 21.14**), extending no further today than they did a brief instant after the Big Bang.

To better appreciate how this bizarre notion works, imagine what it would be like to live in a three-dimensional universe (like the one we experience) in which one of those dimensions extended for only a tiny distance. Living in such a universe would be like living within a thin sheet of paper that extended on for billions of light-years in two directions but was far smaller than an atom in the third. In such a universe we would easily be aware of length and width—we could move in those directions at will. In contrast, we would have no freedom to move in the third dimension at all, and might not even recognize that the third dimension existed. Perhaps our only inkling of the true nature of space would come from the fact that in order to explain the results of particle physics experiments, we would have to assume that particles extended into a third, unseen dimension. In like fashion, if superstring theory is correct, we see three spatial dimensions extending on possibly forever, but we are unaware of the fact that each point in our three-dimensional space actu-

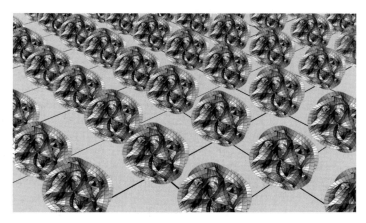

FIGURE 21.14 It is difficult to visualize what seven spatial dimensions wrapped up into structures far smaller than the nucleus of an atom would be like. Here such geometries are projected onto the two-dimensional plane of the paper.

ally has a tiny but finite extent in seven other dimensions at the same time!

Superstring theory is only a pale shadow of the sort of well-tested theories that we have made use of throughout this book. In some respects superstring theory is no more than a promising idea providing direction to theorists searching for a TOE. It is worth noting that we will probably never be able to build particle accelerators that allow us to directly search for the most fundamental particles predicted by a TOE. The energies required are simply too high. Fortunately, however, nature has provided us with the ultimate particle accelerator—the Big Bang itself! The structure of the universe that we see around us today is the observable result of that grand experiment.

At first glance particle physics and cosmology might seem to have almost nothing in common. Particle physics is the study of the quantum mechanical world that exists on the tiniest scales imaginable, whereas cosmology is the study of the changing structure of a universe that extends for billions of light-years and probably much farther. Yet the last quarter of the 20th century saw the boundary between these two fields fade and eventually disappear as cosmologists and particle physicists came to realize that the structure of the universe and the fundamental nature of matter are two sides of the same scientific coin.

One of the most intriguing questions that a successful TOE may answer is whether the universe could have been different. Do we live in only one of many possible universes,

Is ours the only possible universe?

each with different physical laws? Or are the physical laws that govern our universe the *only* consistent set of physical laws that could exist? We do not know the answer yet, but possibly within your lifetime we will.

Order "Froze Out" of the Cooling Universe

We started our discussion of the fundamental forces of nature by pointing out that these forces themselves represent a type of structure in the universe. Just as galaxies and stars are structure that condensed out of the uniformity of the Big Bang, so too are the four fundamental forces structure that condensed out of the uniformity of the theory of everything.

Figure 21.15 illustrates the origin of structure in the evolving universe. The first 10^{-43} second after the Big Bang is described by the TOE, when the physics of elementary particles and the physics of spacetime were one and the same. As the universe expanded and cooled, gravity parted ways with the forces described by the GUT. Spacetime took on the properties described by general relativity. Inflation may also have been taking place at this time.

As the universe continued to expand and its temperature fell further, less and less energy was available for the creation of particle–antiparticle pairs. When the particles responsible for GUT interactions could no longer form, the

---

First, atomic nuclei and then atoms formed as the universe cooled.

---

strong force split off from the electroweak force. One might speak of this transition as the strong force "freezing out" of the GUT because this and other similar transitions are reminiscent of the phase change that occurs as water changes to ice, and molecules become more constrained in their motions. Somewhere along the line, as the unity of the original TOE was lost, the symmetry between matter and antimatter was broken. As a result, the universe ended up with more matter than antimatter.

The next big change took place when the particles responsible for unifying the electromagnetic and weak nuclear forces froze out, leaving these two forces independent of each other. The four fundamental forces of nature that govern today's universe had now come into their own. The temperature of the universe had fallen to a chilly 10^{16} K, and a 10-trillionth of a second had ticked off the cosmic clock. It was a full minute or two before the universe cooled to the billion-kelvin mark, below which not even pairs of electrons and positrons could form.

Although the universe was now too cool to form additional particles and their antiparticles, it was still hot enough for the thermal motions of protons to overcome the electric barriers between them, allowing nuclear reactions to take place. These reactions formed the least massive elements, including helium, lithium, beryllium, and boron, but could not create more massive elements.

Big Bang nucleosynthesis came to an end by the time the universe was 5 minutes old, and the temperature of the universe had dropped below about 800 million K. The density

of the universe at this point had fallen to only about a 10th that of water. Normal matter in the universe now consisted of atomic nuclei and electrons, awash in a bath of radiation. So the universe remained for the next several hundred thousand years, until finally the temperature dropped so far that electrons were able to combine with atomic nuclei to form neutral atoms. We have encountered this event before. This was the era of recombination, which we see directly when we look at the cosmic background radiation.

21.5 Life Is Another Form of Structure

As we near the end of our journey, let's briefly appreciate the distance we have covered. We have followed the origin of structure in the universe from the earliest instants after the Big Bang, when the fundamental forces of nature came to be, through to formation of the galaxies and other large-scale structure that is visible today. Earlier on our journey, we watched as stars (including our Sun) formed from clouds of gas and dust within these galaxies, and planets (including our Earth) formed around those stars. We learned of the geological processes that shaped early Earth into the planet we see around us today. In short, we have traced the origin of structure from the instant the universe came into existence up until the modern day. To help put this into perspective, **Excursions 21.1** shows how the major events in the history of the universe would fit into a single 24-hour day that began with the Big Bang and ended today. Yet there is one piece that we have left out. Although the focus of our journey is 21st century *astronomy,* no discussion of how structure evolved in the universe would be complete without some consideration of the origin of the particular type of structure we refer to as *life* (**Figure 21.16**).

Evolution Is Unavoidable

Imagine that just once during the first few hundred million years after the formation of Earth, a single molecule formed by chance somewhere in Earth's oceans. That molecule had a very special property: Chemical reactions between that

---

To create life, only one self-replicating molecule needed to form by chance.

---

molecule and other molecules in the surrounding water resulted in that molecule making a copy of itself. Now there were two such molecules. Being duplicates of the original, chemical reactions would produce copies of each of these molecules as well, making four. Four became 8, 8 became 16,

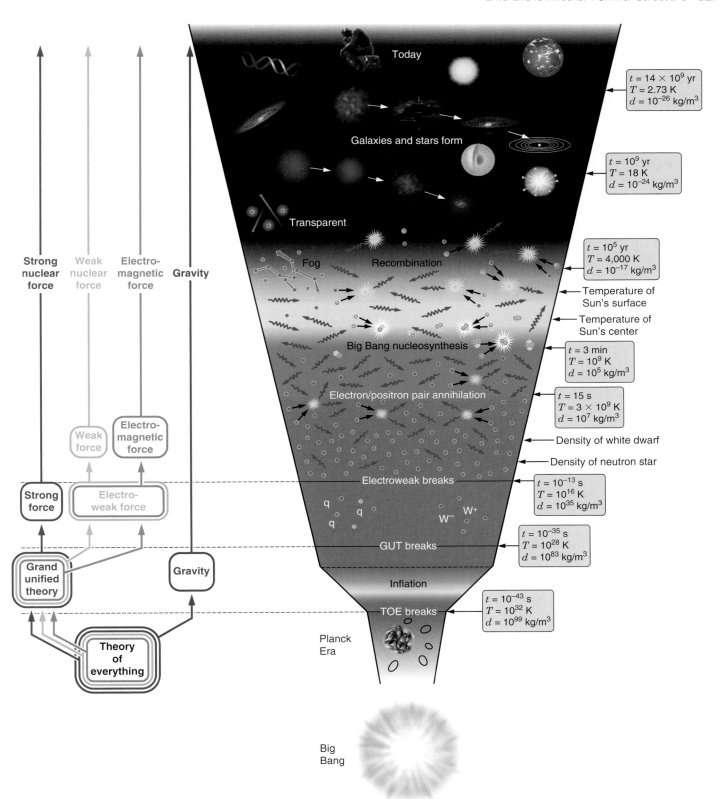

FIGURE 21.15 Eras in the evolution of the universe. As the universe expanded and cooled following the Big Bang, it went through a series of phases determined by what types of particles could be created freely at that temperature. Later the structure of the universe was set by the gravitational collapse of material to form galaxies and stars, and by the chemistry made possible by elements formed in stars.

EXCURSIONS 21.1

Forever in a Day

We have seen that the events that ultimately led to our existence here as inhabitants of planet Earth stretched out over many billions of years, intervals that even astronomers have difficulty visualizing. We can sometimes get a better grasp of such enormous spans of time by compressing them into much shorter intervals with which we have day-to-day experience. Try then to imagine the age of the universe and those important events associated with our origins as if they were all taking place within a single day. Our cosmic day begins at the stroke of midnight:

12:00:00 A.M. The embryonic universe is a mixed broth, minute specks of matter suspended in a vast soup of radiation. The entire universe exists only as an extraordinarily hot bath of photons and a zoo of elementary particles.

12:00:02 A.M. It is now just 2 seconds after midnight. All of the early eras in the history of the universe have passed. The fundamental forces have frozen out, Big Bang nucleosynthesis has formed the universe's original complement of atomic nuclei, and things have cooled down enough for atomic nuclei to combine with electrons to produce neutral atoms. Both normal and dark matter are now available to create galaxies and stars, but that process will take some time.

12:40 A.M. The first stars, quasars and galaxies appear. At some point—we are not sure exactly when—our own galaxy becomes visible as star formation begins. Throughout the cosmic day, stars will continue to form. The more massive stars each go through their brief life cycles in only 5 to 10 seconds of our imaginary 24-hour clock. Each massive star shines briefly, creates its heavy elements, and then disperses this material throughout interstellar space as it dies in a violent supernova explosion. Stars similar to our Sun go through less dramatic life cycles, each lasting about 16 cosmic hours. Stars with mass less than 0.8 $M_\odot$ will last for several dozen cosmic hours and will survive beyond the end of our cosmic day.

4:00 P.M. Our Solar System forms out of a giant cloud of gas and dust. Collapse of the cloud's protostellar core, followed by the appearance of the Sun and the planets—including Earth—all take place within the span of a single cosmic minute.

4:05 P.M. A Mars-size planetesimal crashes into Earth, forming the Moon.

---

16 became 32,... and so on. By the time the original molecule had copied itself just 100 times, over a *million trillion trillion* (10^{30}) of these molecules existed. That is 100 million times more of these molecules than there are stars in the observable universe!

Chemical reactions are never perfect. Sometimes when a copy is attempted, it is not an exact duplicate of its predecessor. The imperfection in the attempted copy is called a **mutation**. Most of the time such an error would be devas-

Success breeds success.

tating, leading to a molecule that could no longer reproduce at all; but occasionally a mutation would actually help. It would lead to a molecule that was *more* successful in duplicating itself than the original. Even if imperfections in the copying process cropped up only once every 100,000 times a molecule reproduced itself, and even if only 1 out of 100,000 of these errors turned out to be beneficial, that still meant that after only 100 generations there were a hundred million trillion (10^{20}) errors that, by blind luck, improved on the original design. Copies of each of these improved molecules would inherit the change. These molecules would have happened upon a form of **heredity**—the ability of one generation of structure to pass on its characteristics to future generations.

In this way, as these molecules continued to interact with their surroundings and make copies of themselves, they split into many different varieties. Eventually these descendants of our original molecule became so numerous that the building blocks they needed in order to reproduce became scarce. Faced with this scarcity of resources, varieties of molecules that were more successful than others in reproducing themselves became more numerous. Varieties that could break down other varieties of self-replicating molecules and use them as raw material were especially

5:20 P.M. The first primitive life appears on Earth. It evolves into the simplest life forms: unicellular organisms such as bacteria, cyanobacteria, and archaebacteria.

8:40 P.M. More complex single-cell organisms appear, making it possible for multicellular life to develop.

9:30 P.M. The first multicellular organisms (fungi) appear on dry land.

11:00 P.M. Multicellular organisms become abundant. This paves the way for still larger and more complex life forms.

11:20 P.M. The first animals (fish) make the transition from ocean to dry land, a major step toward our existence.

11:35 P.M. The first dinosaurs make their appearance. Various small animals appear as well, but they remain dominated by larger life forms.

11:53:10 P.M. A large comet or asteroid crashes into Mexico's Yucatán Peninsula. Seventy percent of all species on Earth (including the dinosaurs) suddenly vanish. In the minutes that follow, the mammals,

being more adaptable to the changed environment, gain prominence.

11:59:25 P.M. Our earliest human ancestors finally appear on the plains of Africa just 35 seconds before the end of our cosmic day.

11:59:59.8 P.M. Modern humans arrive with only a fifth of a second to spare—a fraction of a heartbeat before the day's end.

12:00:00 A.M. Just as our cosmic day draws to a close, will a worldwide catastrophe occur? Will 21st century humans permanently scar the face of the planet, driving many of its life forms into oblivion by polluting the atmosphere and poisoning the land, the rivers, and the oceans? Or will 21st century humans finally break free of their gravitational bondage to their planetary home and begin to claim the rest of the Solar System as their own? In comparison to all that came before, what will happen in the next century will occur in a blur—less than a blink of the eye.

successful in this world of limited resources. Competition and predation had entered the picture. After a few generations, certain varieties of molecules came to dominate the mix, while less successful varieties became less and less common. This competition, in which better-adapted molecules thrive and less well-adapted molecules die out, is referred to as **natural selection**.

Four billion years is a long time—enough time for the combined effects of heredity and natural selection to shape the descendants of that early self-copying molecule into a huge variety of complex, competitive, successful structures. Geological processes on Earth, such as sedimentation, have preserved a fossil record of the history of these structures, as illustrated in **Figure 21.17**. Among these descendants are "structures" capable of thinking about their own existence and unraveling the mysteries of the stars.

The molecules of DNA (deoxyribonucleic acid) that make up the chromosomes in the nuclei of the cells throughout

your body are direct descendants of those early self-duplicating molecules that flourished in the oceans of a young Earth. Although the game played by the molecules

Evolution is inevitable.

of DNA in your body is far more elaborate than the game played by those early molecules in Earth's oceans, the fundamental rules remain the same. We now realize that this process is inevitable: Any system that combines the elements of heredity, mutation, and natural selection *must* and *will* evolve.

In *The Selfish Gene* (1976), Richard Dawkins points out that, when viewed from a purely utilitarian perspective, a human being is a machine whose "purpose" is to produce copies of its genetic material. The remarkable thing about humanity is that our intelligence (itself a very powerful tool for survival in a competitive world) has also allowed us to

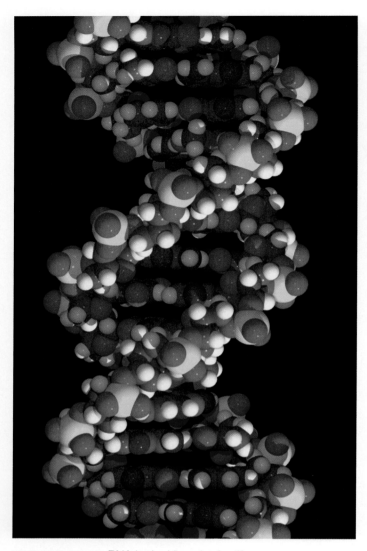

FIGURE 21.16 DNA is the blueprint for life.

develop more noble pursuits (for example, astronomy) than those dictated by our basic biochemical imperative.

Are We Alone?

The story of the evolution of life cannot be separated from the story of 21st century astronomy. We know that we live in a universe full of stars and that systems of planets orbit many and probably most of those stars. We also know that there is nothing mysterious about the origin of life or the processes that cause it to evolve. In fact, as discussed in **Connections 21.2**, the evolution of life on Earth is but one of the many examples that we have encountered of the emergence of structure in an evolving universe. This leads naturally to one of the more profound questions that we ask about the universe. Has life arisen elsewhere, and is it intelligent? Are we alone?

Humans have already made preliminary efforts to "reach out and touch someone." The *Pioneer 11* spacecraft, which will probably spend eternity drifting through interstellar space, carries the plaque shown in **Figure 21.19** describing ourselves to any future interstellar traveler who might happen to find it. This may not be the most efficient way to make contact with the universe, but it is a significant gesture nonetheless. A somewhat more practical effort was made in 1974 when the kilometer-wide dish of the Arecibo radio telescope was used to beam the message shown in **Figure 21.20** toward the star cluster M13. (If someone from M13 answers, we will know in 48,000 years!)

The first serious effort to search for intelligent extraterrestrial life was made by astronomer **Frank Drake** (1930–) in 1960. Drake used what was then astronomy's most powerful radio telescope to listen for signals from two nearby stars. Although his search revealed nothing unusual, it prompted him to develop an equation that still bears his name. The **Drake equation** is a prescription for calculating the likelihood that intelligent civilizations exist beyond our own Solar System.

The Drake equation estimates the number of advanced civilizations in the Milky Way.

In its basic form, the Drake equation is a prescription for calculating N, the number of technologically advanced civilizations in our galaxy today:

$$N = F f_p n_{pm} f_l f_c L$$

The six factors on the right side of the equation relate to the conditions that must be met for a civilization to exist:

- F is the total number of stars in our galaxy. We know this to be several hundred billion stars.

- f_p is the fraction of stars that form planetary systems. Questions remain, but we do know that planets form as a natural by-product of star formation and that many, perhaps most, other stars have planets. For this calculation we will assume that f_p is between 0.5 and 1.

- n_{pm} is the average number of planets and moons in each planetary system capable of supporting life. If such planets are like Earth, they will have to be of the terrestrial type and orbit within a "life zone," where temperatures are neither too hot nor too cold. If our own Solar System is a good example, there would be one or two in each planetary system. (In addition to Earth, Mars appears capable of supporting life, whether or not it presently exists there.) Although we now know that many planetary systems do not resemble our own, it is possible that moons orbiting giant planets near stars could also harbor life. For purposes of calculation, we put the value of n_{pm} between 0.1 and 2.

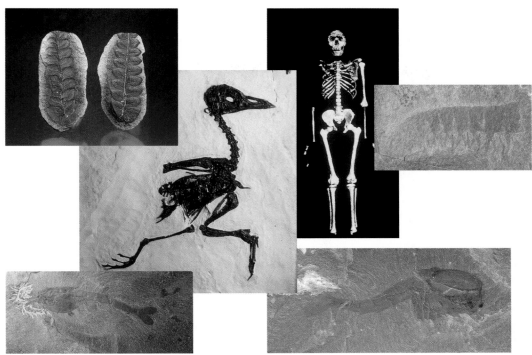

FIGURE 21.17 Fossils record the history of the evolution of life on Earth.

- f_l is the fraction of planets and moons capable of supporting life on which life actually arises. Remember that just a single self-replicating molecule may be enough to get the ball rolling. Many biochemists now believe that if the right chemical and environmental conditions are present, then life *will* develop. If they are correct, then f_l is close to 1. We will use a range for f_l of 0.01 to 1.

- f_c is the fraction of those planets harboring life that eventually develop technologically advanced civilizations. With only one example of a technological civilization to work with, f_c is hard to estimate. Intelligence is certainly the kind of survival trait that might often be strongly favored by natural selection. On the other hand, on Earth it took 4 billion years—half the expected lifetime of our star—to evolve tool-building intelligence. The correct value for f_c might be close to 1, or it might be closer to 1 in 1,000.

- L is the likelihood that any such civilization exists *today*. This is certainly the most difficult factor of all to estimate because it depends on the long-term stability of advanced civilizations. We have had a technological civilization on Earth for around 100 years, and during that time have developed, deployed, and used weapons with the potential to eradicate our civilization and render Earth hostile to life for many years to come. We have also so degraded our planet's ecosystem that many respectable biologists and climatologists wonder if we are nearing the brink. Do all technological civilizations destroy themselves within a thousand years? A thousand years is only one 10-millionth of the lifetime of our star.

On the other hand, if most technological civilizations learn to use their technology for survival rather than self-destruction, perhaps they instead will survive for hundreds of millions of years. For our calculation we will use a range for L of between 10^{-7} (for civilizations that live 1,000 years) and 10^{-2} (for civilizations that live 100 million years).

As illustrated in **Figure 21.21**, the conclusions we draw based on the Drake equation depend a great deal on the assumptions we make. Using the most pessimistic of our estimates, the Drake equation sets the likelihood of finding a technological civilization in our galaxy at 1 percent, or 1 chance in 100. If this is correct, then we are in all likelihood the *only* technological civilization in the Milky Way.

The nearest advanced civilization may be as far away as 30 Mly...

In fact, this would suggest that only 1 in 100 galaxies would contain a technological civilization *at all*. Such a universe would still be full of intelligent life. With a hundred billion galaxies in the observable universe, even these pessimistic assumptions would mean that there are a billion technological civilizations out there somewhere. On the other hand, we would have to go a *very* long way (30 Mly or so) to find our nearest neighbors.

At the other extreme, what if we take the most optimistic numbers, assuming that intelligent life arises and survives everywhere it gets the chance? The Drake equation then says that there should be 40 million technological

CONNECTIONS 21.2

Origins—Life, the Universe, and Everything

Popular discussions of the origin of life almost always wind up struggling with the question of how it is that complex, highly ordered structures such as living things could have emerged from a simpler, more disordered past. Place a drop of ink in a glass of water and watch what happens. The ink spreads out, diffusing through the water, until the only sign that the ink is there is the fact that the water is a different color. The order represented by the discrete drop of ink naturally fades away as the ink spreads out through the water. Yet no matter how long we watch, we will never see that drop of ink spontaneously reassemble itself.

Physicists discuss the degree of order of a system using the concept of **entropy**, which is a measure of the number of different ways a system could be rearranged and still appear the same. A neatly ordered system (such as the drop of ink and the glass of clear water) has low entropy, whereas a disordered system (the glass of inky water, which can be stirred or turned and still look the same) has higher entropy. The **second law of thermodynamics** says that, left on its own, an isolated system will always move toward higher entropy—that is, from order toward disorder. This is common sense. As time goes by we are more likely to find a system in a state that is, well, more likely.

In light of this inescapable march toward disorder dictated by the second law, how can ordered structure emerge spontaneously? Creationist claims that the origin of life flies in the face of the second law of thermodynamics are about as common as reports of Elvis sightings in supermarket tabloids. Yet such claims make a crucial

mistake: They focus attention on one player while ignoring the rest of the game.

We have all seen what happens when we set a glass of ice water out on a hot, humid day (**Figure 21.18**). Water vapor from the surrounding air condenses into drops of liquid water on the surface of the cold glass. This is amazing: We have just watched as ordered structure (drops of water) spontaneously emerged from disorder (water vapor in the air). It is almost as if we saw the drop of ink reassemble itself. This simple, everyday event appears to violate the second law of thermodynamics... but it does not.

To understand why, we need to step back and look at more than just the drops of water on the glass. When the drops condensed, they released a small amount of thermal energy that slightly heated both the glass and the surrounding air. Heating something increases its entropy. The decrease in entropy due to the formation of the drops is more than made up for by the increase in the entropy of their surroundings. Ordered structure spontaneously emerged, but *overall,* disorder increased.

It is important to note that energy can be used to reduce entropy at one location while increasing entropy somewhere else. An air conditioner is a good example. Electric energy produced at a power plant is used to pump thermal energy from inside a home and dump that thermal energy outside. When an air conditioner is run in reverse to provide heating, it is referred to as a "heat pump," but "entropy pump" might be a more accurate description. As you sit in your armchair on a summer day, all you immediately notice is that when the air

civilizations in our galaxy alone! In this case, our nearest neighbors may be "only" 40 or 50 light-years away. If scientists in that civilization are listening to the universe with

...or as near as 40 light-years.

their own radio telescopes, hoping to answer the question of life in the universe for themselves, then as you read this page they may be puzzling over an episode of *I Love Lucy.*

It is fascinating to speculate about the implications of the Drake equation. If there are civilizations around every corner, cosmically speaking, then why have we not heard from them? Perhaps they are not interested in talking to the new kid on the block, or perhaps the fact that we know of no other civilizations simply means that there are none nearby.

If we did run across another technologically advanced civilization, what would it be like? Looking back at the conditioner comes on, the temperature drops. Entropy decreases inside your house. If you look at the system as a whole, on the other hand, including the heating of air by the outside coils and the entropy produced by the burning of coal or natural gas during the production of electricity, then you see that turning on the air conditioning causes an overall *increase* in entropy.

This idea is especially important when we consider the origin of life. A living thing, whether an amoeba or a human being, represents a huge local increase in order. However, no violation of the second law of thermodynamics is involved. In our everyday lives, the food we eat gives us the energy we need to stave off the relentless advance of entropy. Viewed even more broadly, the evolution of life on Earth was powered primarily by energy striking Earth in the form of sunlight. A local increase in order on Earth (such as you) is "paid for" in the end by the much greater decrease in order that accompanies thermonuclear fusion in the heart of the Sun. Order emerges in localized regions within a system, but the second law of thermodynamics is obeyed overall.

The unifying theme of this chapter and, in many ways, this entire book has been understanding the origin of structure. The answer we have found is clear. Whether we are discussing the freezing out of matter and the fundamental forces in the young universe; the gravitational collapse of stars, planets, and galaxies; the evolution of life; or water beading up on the outside of a cold glass—ordered structure *does* emerge spontaneously as an unavoidable consequence of the action of physical law.

FIGURE 21.18 On a humid day, water condenses into droplets on the surface of a cold glass. The second law of thermodynamics does not mean that ordered structure cannot spontaneously emerge.

Drake equation, it is highly unlikely that we have neighbors nearby unless civilizations typically live for many thousands or even millions of years. In this case, any civilization that we encountered would almost certainly have been around for much longer than we have. Having survived that

Any civilization we discover will almost certainly be advanced.

long, would its members have learned the value of peace, or would they have developed strategies for controlling pesky neighbors? Movie theaters and the science fiction shelves of libraries and bookstores are filled with amusing and thoughtful stories that explore what life in the universe might be like (**Figure 21.22**). For the moment let's set aside such speculation to look at the question as scientists. Given this fascinating question, how might we go about finding a real answer?

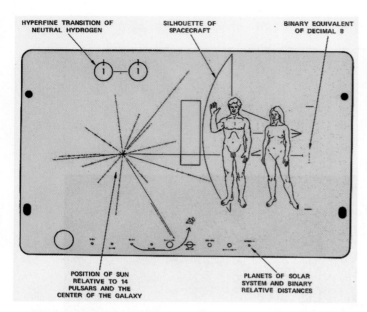

HYPERFINE TRANSITION OF NEUTRAL HYDROGEN

SILHOUETTE OF SPACECRAFT

BINARY EQUIVALENT OF DECIMAL 8

POSITION OF SUN RELATIVE TO 14 PULSARS AND THE CENTER OF THE GALAXY

PLANETS OF SOLAR SYSTEM AND BINARY RELATIVE DISTANCES

FIGURE 21.19 The plaque included with the *Pioneer 11* probe, which left the Solar System to travel through the millennia in interstellar space.

The Search for Signs of Intelligent Life in the Universe

One question we may be able to answer using observations of our own Solar System is, What is the likelihood of life originating at all? When the *Viking* landers failed to discover life on Mars in 1976, hopes faded for the presence of other life on worlds orbiting the Sun. Since that time, however, optimism has been renewed. A better understanding of the history of Mars indicates that at one time the planet was wet and warm, and many scientists believe that fossil life or even living microbes may be buried under the planet's surface.

Even more exciting are discoveries in the outer Solar System. There is strong evidence that life on Earth may have originated deep under the oceans, where geothermal vents (**Figure 21.23**) provided the thermal and chemical energy needed for life to gain a toehold. If we are correct that geological activity churns the floor of a deep ocean on Jupiter's moon Europa, for example, then this could be an excellent place to find life. One other example is all we need. If life arose independently *twice* in the same planetary system, then we would have no choice but to conclude that f_l in the Drake equation is close to 1 and that life is ubiquitous throughout the universe.

Another way to search for intelligent life is to simply turn our ears to the sky and listen. Drake's original project of listening for radio signals from intelligent life around two nearby stars has grown over the years into a much more elaborate program that is referred to as the Search for Extraterrestrial Intelligence, or **SETI**. Scientists from around

the world have thought carefully about what strategies might be useful for finding life in the universe. Most of these have focused on the idea of using radio telescopes to listen for signals from space that bear an unambiguous signature of an intelligent source. Some have listened intently at certain

SETI listens for radio signals from other civilizations.

"magic" frequencies, such as the frequency of the interstellar 21-cm line from hydrogen gas. The assumption behind this technique is that if a civilization wanted to be heard, they would tune their broadcasts to a channel that astronomers throughout the galaxy should be listening to. More recent searches have made use of advances in technology to record as broad a range of radio signals from space as possible, and then use computers to search these databases for types of regularity in the signals that might suggest that they are intelligent in origin.

Unlike much astronomical research, SETI is funded mostly through private rather than governmental means, and SETI researchers have been very ingenious at finding ways to accomplish as much as possible with limited resources. One especially clever idea, coming out of the SETI Institute in Mountain View, California, uses the underutilized computing power of thousands of personal computers around the world to analyze the institute's data. SETI screen saver programs installed on worldwide desktops download radio observations from the SETI Institute over the Internet, analyze these data while the computers' owners are off living their lives, and then report the results of their searches back to the institute. It is fun to think that the first sign of intelligent life in the universe might be found by a computer sitting on a table in the corner of your living room!

A number of SETI projects are on the drawing board. One is called the Allen Telescope Array (ATA), named after the cofounder of Microsoft, Paul Allen, who provided much of the initial financing for the project. The ATA (see **Figure 21.24**) will consist of a "farm" of hundreds of small, inexpensive radio dishes like those used to capture sig-

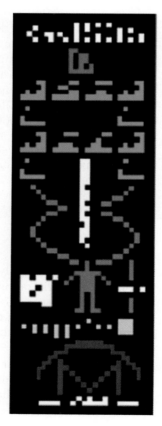

FIGURE 21.20 The message we beamed toward the star cluster M13. A reply may be forthcoming in 48,000 years.

nals from orbiting communication satellites. The ATA, a joint venture between the SETI Institute and the University of California, will use sensitive modern receiver technology to search the sky 24 hours a day, seven days a week, for signs of intelligent life. Just as your brain can sort out sounds coming from different directions, this array of radio telescopes will be able to determine the direction a signal is coming from, allowing it to listen to many stars at the same time. Over several years' time the ATA is expected to survey as many as a million stars, hoping to find a civilization that has sent a signal in our direction. If reality is anything like the more optimistic of the assumptions we used in evaluating the Drake equation, this project will stand a good chance of success.

If SETI finds even *one* nearby civilization in our galaxy, then the message will be clear. If we find a *second* technological civilization in our own small corner of the universe, then it means that the universe as a whole must literally be teeming with intelligent life. SETI may not be in the mainstream of astronomy, and the likelihood of its success may be difficult to predict, but its potential payoff is enormous. Few discoveries would do more to change our understanding of ourselves than certain knowledge that we are not alone.

21.6 The Future, Near and Far

We have used our understanding of physics and cosmology to look back through time and watch as structure formed throughout the universe. Now, as our journey nears its end, we look toward the future and contemplate the fate that awaits Earth, humanity, and the universe as a whole. Beginning close to home, around 5 billion years from now, the Sun will end its long period of relative stability. Shedding its identity as the passive, benevolent star that has nurtured life on Earth for nearly 4 billion years, the Sun will expand to become a red giant and later an asymptotic giant branch (AGB) star, swelling to hundreds of times its present size. The giant planets, orbiting outside the extended red giant atmosphere, should survive the Sun's cranky old age in some form. Even so, they will suffer the blistering radiation from a Sun grown thousands of times more luminous than it is today.

The terrestrial planets will not be so lucky. Some and perhaps all of the worlds of the inner Solar System will be engulfed by the expanding Sun. Just as an artificial satellite is slowed by drag in Earth's tenuous outer atmosphere and eventually falls to the ground in a dazzling streak of white-hot light, so too will a terrestrial planet caught in the Sun's atmosphere be consumed by the burgeoning star. If this is Earth's fate, our home world will leave no trace other than

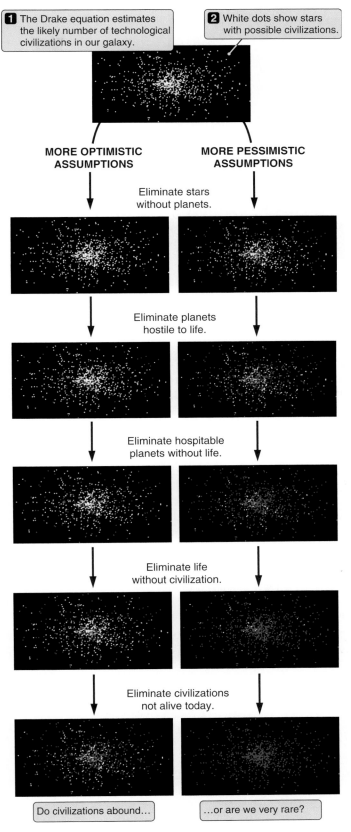

1 The Drake equation estimates the likely number of technological civilizations in our galaxy.

2 White dots show stars with possible civilizations.

MORE OPTIMISTIC ASSUMPTIONS   MORE PESSIMISTIC ASSUMPTIONS

Eliminate stars without planets.

Eliminate planets hostile to life.

Eliminate hospitable planets without life.

Eliminate life without civilization.

Eliminate civilizations not alive today.

Do civilizations abound...   ...or are we very rare?

FIGURE 21.21 Simulation of the Drake equation for optimistic and pessimistic assumptions about the factors affecting the prevalence of intelligent life in the universe.

FIGURE 21.22 The classic 1951 film *The Day the Earth Stood Still* portrayed intelligent extraterrestrials who had no interest in our internal affairs, but promised destruction of Earth if we carried our violent ways into space.

Far future Earth will be consumed by the Sun or left as an icy cinder.

a slight increase in the amount of massive elements in the Sun's atmosphere. As the Sun loses more and more of its atmosphere in an AGB wind, our atoms may be expelled back into the reaches of interstellar space from which they came.

Another fate is possible, however. As the red giant Sun loses mass in a powerful wind, its gravitational grasp on the planets will weaken, and the orbits of both the inner and outer planets will spiral outward. If Earth moves out far enough, it may survive as a seared cinder, orbiting the white dwarf that the Sun will become. Barely larger than Earth and with its nuclear fuel exhausted, the white dwarf

Sun will slowly cool, eventually becoming a cold, inert sphere of degenerate carbon, orbited by what remains of its retinue of planets. The ultimate outcome for our Earth —consumed in the heart of the Sun or left behind as a frigid burned rock orbiting a long-dead white dwarf—is not yet certain. In either case Earth's status as a garden spot will be at an end.

The Fate of Life on Earth

Life on our planet will not survive long enough to witness the Sun's departure from the main sequence. Well before that cataclysmic event takes place, the luminosity of our heretofore benevolent star will begin to rise. As solar luminosity increases, so will temperatures on all the planets, including our own. Eventually Earth's temperatures will climb so high that all animal and plant life will perish. Models of the Sun's evolution are still too imprecise to predict with certainty when that fatal event will occur, but the end of all terrestrial life may be only 1 or 2 billion years away. That is, of course, a comfortably distant time from now, but it is well short of the Sun's departure from the main sequence.

Yet it is far from certain that the descendants of today's humanity will even be around a billion years from now. Some of the threats that await us come from beyond Earth. For the remainder of the Sun's life, the terrestrial planets, including Earth, will continue to be bombarded by asteroids and comets. Perhaps a hundred or more of these impacts will involve kilometer-sized objects, capable of causing the kind of devastation that eradicated the dinosaurs (and most other species) 65 million years ago. Although these events may create new surface scars, they will have little effect on

FIGURE 21.23 Life on Earth may have arisen near ocean geothermal vents like the one at left. Similar environments might exist elsewhere in the Solar System. Life around vents (right) is powered by geothermal rather than solar energy.

FIGURE 21.24 When complete, the Allen Telescope Array, seen here in an artist's view, will listen for evidence of intelligent life from as many as a million stellar systems.

the integrity of Earth itself. Earth's geological record is filled with such events, and each time they happen, life manages to recover and reorganize.

It seems likely, then, that some form of life will survive to see the Sun begin its march toward becoming a red giant. On the other hand, individual species do not necessarily fare so well when faced with cosmic cataclysm. If the descendants of humankind survive, it will be because we chose to become players in the game by changing the odds of such planetwide biological upheavals. We are rapidly developing technology that could allow us to detect most threatening asteroids and modify their orbits well before they can strike Earth. Comets are more difficult to guard against because long-period comets appear from the outer Solar System with little warning. To offer protection, defense capabilities would have to be in place, ready to be

To survive, humanity must learn to manage the threat of impacts.

used on very short notice. We have been slow to take such threats seriously. Although impacts from kilometer-sized objects are infrequent, objects a few dozen meters in size, carrying the punch of a several-megaton bomb, strike Earth about once every 100 years. Perhaps an explosion like the 1908 Tunguska blast occurring over New York or Paris would be enough to convince us that such precautions are worthwhile.

We might protect ourselves from the fate of the dinosaurs, but in the long run the descendants of humanity will either leave this world or die out. Planetary systems surround other stars, and all that we know tells us that many

other Earth-like planets should exist throughout our galaxy. Colonizing other planets is currently the stuff of science fiction, but if our descendants are ultimately to survive the death of our home planet, off-Earth colonization must become science fact at some point in the future.

But though humankind may soon be capable of protecting Earth from life-threatening comet and asteroid impacts, in other ways we are our own worst enemy. We are poison-

Humanity is the worst threat to its own survival.

ing the atmosphere, the water, and the land that form the habitat for all terrestrial life. As our population grows unchecked, we are occupying more and more of Earth's land and consuming more and more of its resources, while sending thousands of species of plants and animals to their extinction each year. At the same time, human activities are dramatically affecting the balances of atmospheric gases. The climate and ecosystem of Earth constitute a finely balanced, complex system capable of exhibiting chaotic behavior. The fossil record shows that Earth has undergone sudden and dramatic climatic changes in response to even minor perturbations. Such drastic changes in the overall balance of nature would certainly have consequences for our own survival. When politics is added to the mix, even more immediate dangers await. For the first time in human history, we possess the means to unleash nuclear or biological disasters that could threaten the very survival of our species. In the end, the fate of humanity will depend more than anything on whether we accept stewardship of ourselves and of our planet.

The Future of Our Expanding Universe

So much for our planet. What about the universe itself? As we have seen, if the mass of the universe is large enough and the cosmological constant small enough, gravity will win in the end. Hubble expansion will eventually reverse, and the universe will collapse back into a state resembling that of the young, hot universe. Galaxies, stars, planets, molecules, atoms, and subatomic particles—all might cease to exist as matter is replaced by pure energy. Perhaps from such a state, a new universe would emerge. At the dawn of the 21st century, however, few cosmologists see such a "Big Crunch" in the future of the universe. It appears that our universe will expand forever, perhaps even at an ever-accelerating pace. Does this mean the universe will go on without end? Yes, but not in the form that we see today.

In 1997 Fred Adams and Gregory Laughlin of the University of Michigan published their calculations of the great eras, past and future, in the history of the universe. During the first era—the first 500,000 years after the Big Bang and before

---

Other eras await our universe in the far future.

---

recombination—the universe was a swarm of radiation and elementary particles. Today we live during the second era, the Era of Stars, but this too will end. Some 100 trillion (10^{14}) years from now the last molecular cloud will collapse to form stars, and a mere 10 trillion years later the least massive of these stars will evolve to form white dwarfs.

Following the Era of Stars, most of the normal matter in the universe will be locked up in degenerate stellar objects: brown dwarfs, white dwarfs, and neutron stars. During this Era of Degeneracy, the occasional star will still flare up as ancient substellar brown dwarfs collide, merging to form low-mass stars that burn out in a short trillion years or so. However, the main source of energy during this era will come from the decay of protons and neutrons and the annihilation of particles of dark matter. Even these processes will eventually run out of fuel. In 10^{39} years white dwarfs will have been destroyed by proton decay, and neutron stars will have been destroyed by the beta decay of neutrons.

As the Era of Degeneracy comes to an end, the only significant concentrations of mass left will be black holes. These will range from black holes with the masses of single stars to greedy monsters that grew during the Era of Degeneracy to have masses as large as those of galaxy clusters. During the period that follows, the Era of Black Holes, these black holes will slowly evaporate into elementary particles through the emission of Hawking radiation. A black hole

with a mass of a few solar masses will evaporate into elementary particles in 10^{65} years, whereas galaxy-sized black holes will evaporate in around 10^{98} years. By the time the universe reaches an age of 10^{100} years, even the largest of the black holes will be gone. A universe vastly larger than ours will contain little but photons with colossal wavelengths, neutrinos, electrons, positrons, and other waste products of black hole evaporation. The Dark Era will have arrived as the universe continues to expand forever into the long, cold, dark night of eternity. This may be the final victory of **entropy**—the **heat death** of the universe.

From our perspective, the universe of the far distant future sounds like an extremely dull and lifeless place—but will that necessarily be so? Imagine for a moment that in-

---

We cannot predict what structure might arise in the far future.

---

telligent life evolved amid the swarm of free quarks and gluons that filled the universe immediately after inflation. Such organisms would have been far smaller than today's atoms and perhaps would have lived out their lives in 10^{-40} second or so. To such organisms the universe—all three meters of it—would have seemed incomprehensibly vast. These creatures and their entire civilization would have had to evolve, live, and die out in a millionth of a trillionth of a trillionth of the time that it takes for a single synapse in our brains to fire. If such creatures ever calculated the conditions in *today's* universe, they would have recoiled in horror. They would have imagined a frozen time when the temperature of the universe was only a thousandth of a trillionth of a trillionth of what they knew—a time when most matter had ceased to exist altogether, and the tiny fraction that remained was spread out over a volume of space 10^{75} times greater than that of their universe. In short, such creatures would have looked forward and seen *our* universe as the frozen, desolate future. It seems doubtful they could have foreseen the existence of stars, galaxies, planets, and intelligent creatures for whom a single thought took longer than a billion trillion times the entire history of the universe they knew.

Now turn the tables, and think *forward* to a time when the universe is 10^{50} times older than it is today. Who can say that there will not be life then as well? Perhaps they will be organisms of magnetic fields and tenuous electron plasmas, spread across countless trillions of light-years of space, whose lives unfold over untold eons of time. For such organisms, if they ever exist, it will be *we* who are the impossible creatures, alive for the briefest of instants, still immersed in the momentary fireball of the Big Bang.

Summary

- Galaxies reside in groups, clusters, and larger structures, all of which formed after galaxies did.

- Structure formed in the universe through the gravitational collapse of cold dark matter inhomogeneities arising from the early universe.

- Observed galaxies come from complex mergers, with the visible gas in galaxies cooling and falling inward to form the visible stars, which are surrounded by a dark matter halo.

- Understanding the very earliest moments in the universe requires that we also understand how the four fundamental forces of nature were all unified into one basic phenomenon at early times.

- Life may be an inevitable consequence of the formation of a single molecule that evolves through natural selection to more complex structures.

- Other advanced civilizations may be as close as 40 light-years or as far away as 30 million light-years, and we are searching for signals from such civilizations using SETI.

Seeing the Forest through the Trees

As our journey nears its end, we have reached high ground, a vantage point from which we can look back and survey the terrain we have covered. From this perspective it becomes clear that from the beginning, our journey has been guided by a single underlying quest: to understand how we came to be. Whether we are talking about the formation of galaxies, the origin and evolution of stars, the history of our Solar System, the geology of our own world, the evolution of life, the changes that shaped the universe itself during its earliest moments, or the eras of the far future, 21st century astronomy (the science, as well as the book) is organized around the desire to better understand the origin of structure in the universe.

The past century has seen remarkable strides toward this goal, and along the way a pattern has emerged. Look back on the quotation opening Chapter 16. What statement remains true in all circumstances? "And this, too, shall pass away." Structure is ephemeral. The universe is not about destination. The universe is a place of process and change. The things that "matter" are the things that happen along the way!

When you see your breath fog up on a cold winter day, think about elementary particles freezing out of the early universe, giving rise not only to matter but to the fundamental forces of nature. As you watch the clouds build before a thunderstorm, think of galaxies coalescing within halos of dark matter. Glance at a crystalline snowflake as it lands on the sleeve of your coat, and in your mind's eye see a star that condensed out of clouds of interstellar gas and dust. The origin of structure is everywhere around us and is no less remarkable for its familiarity.

As it is for galaxies and stars and planets, so too is it for ourselves. A recurring theme on our journey of discovery has been the systematic dismantling of conceptual walls that in our minds separated us from the larger universe. Humanity does not stand outside the processes that shape the universe. We are instead one more variety of the structure to which the universe has given birth—a way station on a long road of evolving structure stretching back 14 billion years. Few single words are capable of eliciting as much emotional reaction from some people as the word *evolution.* Yet the public controversy surrounding evolution cannot change the scientific standing of this theory as one of the best tested and most successful theories in all of science. As stated by Stephen J. Gould, "The theory of evolution is in about as much trouble as the theory that Earth revolves around the Sun." By any reasonable standards of scientific knowledge, evolution is a fact.

Opponents of evolutionary theory often act as if evolution were a tiny piece of science that can simply be cast aside without doing violence to the rest. Having traveled the journey of *21st Century Astronomy,* you should see clearly how absurd such a claim is. Modern astronomy would simply cease to be were it stripped of our understanding of the origin and evolution of galaxies, stars, planets, and every other component of the universe that we observe. Everything we see says that the universe formed 14 billion years ago, not 6,000. Cosmology is the ultimate evolutionary science—the science of the origin and evolution of the universe itself. During the 20th century geology became the science of the evolution of the surface of Earth, while planetary science applied our understanding of terrestrial geology to understanding the evolution of other planets and their moons. The most fundamental questions in physics concern the origin and evolution of physical laws. So too is the case in modern biology, which simply makes no sense until it is organized around the theme of evolution by natural selection.

Evolution is anything but a scientific appendix that can be harmlessly removed. It is the backbone and central nervous system of modern science. In *Darwin's Dangerous Idea* (1995) Daniel Dennett speaks of the concept of evolution as "universal acid: it eats through just about every traditional concept, and leaves in its wake a revolutionized worldview." As the 21st century gets under way, evolution of structure is *the* unifying theme that ties the breadth of modern science together into a beautiful, powerful, comprehensive whole. If we tried to pull the thread of evolution out of this tapestry, the whole cloth would unravel before our eyes.

Key Terms

galaxy group, p. 608
Local Group, p. 608
galaxy cluster, p. 608
supercluster, p. 608
large-scale structure, p. 609
hierarchical clustering, p. 611
cold dark matter, p. 612
photino, p. 613
hot dark matter, p. 613
weak nuclear force, p. 621
quantum electrodynamics (QED) , p. 621
electroweak theory, p. 622
quantum chromodynamics (QCD) , p. 622
quark, p. 622
gluon, p. 622
standard model, p. 622
pair creation, p. 622
grand unified theory (GUT) , p. 624
Planck era, p. 624
theory of everything (TOE) , p. 624
superstring theory, p. 625
mutation, p. 628
heredity, p. 628
Drake equation, p. 630
entropy, p. 632
second law of thermodynamics, p. 632
heat death, p. 638

Student Questions

THINKING ABOUT THE CONCEPTS

1. Of the four fundamental forces in nature, which does not depend on electric charge?

2. Suppose you could view the early universe at a time when galaxies were first forming. How would it be different from the universe we see today?

3. What are the basic differences between a grand unified theory (GUT) and the theory of everything (TOE)?

4. As clumps containing cold dark matter and normal matter collapse, they heat up. When a clump collapses to about half its maximum size, the increased thermal motion of particles tends to inhibit further collapse. Whereas normal matter can overcome this effect and continue to collapse, dark matter cannot. Explain the reason for this difference.

5. How do astronomers use the following to measure the amount of dark matter contained in a cluster of galaxies?
   a. Motions of individual members of the cluster.
   b. Extremely hot gas that fills the intergalactic space within the cluster.
   c. Gravitational lensing by the cluster.

6. Imagine that there are galaxies in the universe composed mostly of dark matter with relatively few stars or other luminous normal matter. If this were true, how might we learn of the existence of such galaxies?

7. Why is dark matter so essential to the galaxy formation process?

8. If we should eventually find life on Europa, what would this tell you about the probability of finding life on worlds around other stars?

9. A few scientists believe we may be the only advanced life in the galaxy today. If this were indeed the case, which factors in the Drake equation would have to be extremely small?

10. The second law of thermodynamics says that the entropy (a measure of disorder) of the universe is always increasing. Yet living organisms exist by creating order from disorder. Why does this not violate the second law of thermodynamics?

APPLYING THE CONCEPTS

11. The proton and antiproton each have the same mass, $m_p = 1.67 \times 10^{-27}$ kg. What is the energy (in joules) of each of the two gamma rays created in a proton–antiproton annihilation?

12. Suppose you brought together a gram of ordinary matter hydrogen atoms (each composed of a proton and an electron) and a gram of antimatter hydrogen atoms (each composed of an antiproton and a positron). Keeping in mind that two grams is less than the mass of a dime:
    a. Calculate how much energy (J) would be released as the ordinary matter and antimatter hydrogen atoms annihilated one another.

b. Compare this with the energy released by a one-megaton hydrogen bomb (1.6×10^{14} J).

13. Excursions 21.1 ("Forever in a Day") takes events spread out over enormous intervals of time and compresses them into the more comprehensible interval of a single 24-hour day. Make your own "Life in a Day" by compressing all the important events of your lifetime into a single day, starting with your birth at the stroke of midnight and continuing to the present at the end of the day.

14. Consider an organism Beta that, because of a genetic mutation, has a 5 percent greater probability of survival than its nonmutated form, Alpha. Alpha has only a 95 percent probability (p_r) of reproducing itself compared to Beta. After n generations, Alpha's population within the species would be $S_p = (p_r)^n$ compared to Beta's. Calculate Alpha's relative population after 100 generations. (You may need a scientific calculator or help from your instructor to evaluate the quantity 0.95^{100}.)

15. To fully appreciate the power of heredity, mutation, and natural selection, consider Alpha's relative population (from Question 14) after 5,000 generations in a case where Beta has a mere 0.1 percent survivability advantage over Alpha.

16. Make your own best guess at the six factors on the right side of the Drake equation to estimate the number of technologically advanced civilizations in our galaxy.

17. The Drake equation estimates the number of technologically advanced civilizations in our galaxy. Instead estimate the most optimistic and most pessimistic number of systems containing life (whether advanced or not) in our galaxy.

18. The kinetic energy of an object is given by $KE = mv^2/2$. A 100-meter radius spherical rock having a mass of 10^{10} kg strikes the Earth with a speed of 20 km/sec. How much energy does it deposit on Earth compared to a one-megaton hydrogen bomb (1.6×10^{14} J)?

StudySpace
wwnorton.com/astro21

provides a Study Plan for each chapter that includes a reading outline, animations, keyword flash cards, and gradebook-enabled multiple-choice quizzes. From StudySpace you can also access premium content in the ebook and SmartWork.

Sometimes the light's all shining on me
Other times I can barely see
Lately it occurs to me
What a long strange trip it's been.

ROBERT HUNTER (1938–)

Epilogue: We Are Stardust in Human Form

The Long and Winding Road

Go out at night and look at the stars, and feel the same sense of wonder and awe that our kind has always felt at the sight. Take it in, be amazed at the majestic canopy overhead, and let your imagination roam, just as our ancestors have done for thousands of years. As you do, drift back down the road we have followed. Reflect on all we have come to know about the universe, and on how much more magnificent the heavens are than our ancestors ever could have imagined. Our journey has been more than a description of the universe and what it contains. It has been a travelogue of the struggle and triumph of the human mind and spirit. Ever since humans first recognized that the patterns shaping our lives are echoed in the sky, we have searched for the threads connecting us to the cosmos and have sought to understand our place in it. We have the privilege of being among the first generations of humans to find those threads and to learn real answers to those age-old questions.

We have no way to count how many different stories have been told about the heavens, but we do know that for thousands of years most of those stories shared a common foundation. One cornerstone was the belief that Earth occupies a special place in the scheme of things. In our minds we were at the center of Creation, fixed and immovable, and all that we saw was present only to give meaning to our existence. A second cornerstone of this traditional worldview was the belief that the heavens are fundamentally different from Earth. To our ancestors our world was the realm of the ordinary and mundane—a terrestrial existence built from Aristotle's earth, wind, fire, and water. In contrast, the heavens had their own separate reality. There we saw a realm of gods and angels, of mysticism and magic, of the perfect and unchanging fifth element.

So it remained for thousands of years until, at the dawn of the Renaissance, a Polish monk dared to challenge the wisdom of the ages and to think the unthinkable. Reviving a notion that had been discarded long before by the Greeks, Nicolaus Copernicus allowed himself to imagine that perhaps it was the motion of Earth, rather than the motions of the Sun and stars, that shaped the passage of the days and the years. In so doing he not only conceptually dislodged Earth from its moorings, but he also broke the shackles that had for so long constrained the human mind. What began as a crack in the foundation of our preconceptions would in the end turn that old view of the world to rubble. In its place we would construct an edifice of knowledge that has given us dominion over our world and carried our thoughts to the edges of the universe.

Copernicus was one of a succession of people with great minds who could not rest without first picking at the loose threads of the ideas in which they had been raised to believe. Water runs into the cracks in a slab of granite and freezes, expanding and pushing the cracks open, exposing the flaws in the rock. As the seasons come and go, the imposing boulder stands no chance in the face of this persistent onslaught. In like fashion, the persistent questioning and probing of great minds would eventually shatter the reign of enforced ignorance and entrenched authority. We have met a few of these great minds on our journey; there were many others. They were of different nationalities, different upbringings, and different dispositions and beliefs. But their work shared a common theme: The answers to questions about the world come not from the pronouncements of authority or the prejudices taught in childhood, but from observing nature itself and thinking carefully, honestly, and openly about what we see.

In Newton's famous thought experiment, a ball fired from an imaginary cannon moves rapidly enough that it falls around the world in a circle. Such a "cannon" could not be built with the technology available in Newton's day. That would have to await the launch of *Sputnik* hundreds of years later. Yet although Newton could not make his cannon reality, he did not have to. All he had to do was look at the sky and watch as the Moon traced out its monthly path. Newton realized that the force holding the Moon in its orbit about Earth is the same force that gives us weight and guides the path of a ball thrown into the air. In fact, all Newton had to do to see his cannonball was look out across the English countryside—all of us ride Newton's cannonball as the force of the Sun's gravity holds Earth in its yearly orbit.

This thought experiment was an important step in Newton's work. Eventually it led him to invent calculus, which he used to calculate the motions of the planets, making predictions that were confirmed by Kepler's empirical laws. The philosophical and scientific significance of Newton's thought experiment goes far deeper, however: It signifies the final collapse of the barriers that humankind had placed between Earth and the heavens. With Newton's insight we came to see the heavens as part of the world around us—made of the same substance and shaped by the same physical laws. It is ironic that for knowledge to progress, we had to turn the early cornerstones of our thinking upside down. The modern foundation of our understanding of the universe, the cosmological principle, is the literal negation of those early beliefs: Not only is Earth *not* the center of the universe, but Earth occupies no special place in the universe *at all.* The heavens are *not* "the other," but are instead "the same." We can know the heavens by going into terrestrial laboratories and learning about the nature of matter and energy and radiation, then applying this knowledge to careful observations of a universe of stars, planets, and galaxies that is governed by physical law.

In our journey we have followed the trail of discovery that grew from this profound change in our understanding. We have watched as stars and planets formed, as stars lived out their lives and died, and as galaxies coalesced out of the primordial fireball of the Big Bang. We have followed our physics back to the very briefest of instants after the event that brought space and time into existence, and we have seen the hints of theories that may in our lifetimes carry us the final step. There is no doubting the wonder of what we have seen. Yet for you there is another aspect to our journey that, in a practical sense, should be even more significant. While learning *about* the universe, you have also come to

better appreciate *how we know* those—things and in so doing have found a powerful, workable definition for what it means "to know."

According to Richard Paul, director of the Foundation for Critical Thinking in Dillon Beach, California, the three most common standards that people apply to knowledge even today are "it is true because I believe it," "it is true because we believe it," *and* "it is true because I want to believe it." Of course none of these has anything to do with what really *is* true. To learn about the universe and our world, we have had to set aside these notions, which blur the line between reality and fantasy, and replace them with a tough, unforgiving, and very different standard: "It is *provisionally* true because we have worked very hard to show that it is false, but so far have failed." This standard alone places reality itself in front of our parochial ideas and beliefs. It puts what *is* true ahead of what we would *like* to be true. Only by testing the falsifiable predictions of our theories about the world have we learned to push aside the comfortable notions that for so long prevented us from truly seeing our world and our universe.

We Are Stardust in Human Form

As witnessed by the obelisks of Stonehenge or the ruins of a Mayan pyramid, humans have always built temples to the stars. We still build temples to the stars today. They are seen as an array of radio telescopes spread across the high desert of New Mexico, or a city of domes atop the summit of a dormant Hawaiian volcano, or a satellite telescope carried into orbit and subsequently repaired by space shuttle astronauts, or a tiny rover crawling across the surface of Mars. These modern temples are the legacy of insights by Copernicus, Kepler, Galileo, Newton, Einstein, Hubble, and countless others.

The discoveries that pour forth from these modern-day temples stretch the mind and stir the imagination. We have walked on the Moon and have come to see the planets not as points of light in the sky, but as worlds as rich and complex as our own. We have looked at the remnants of stars that exploded long ago and have peered into eerie columns of glowing interstellar gas within which new stars are being born. We have gazed back in time at galaxies forming when the universe was young, and we have even learned to recognize the birth of the universe itself in the faint glow of the cosmic background radiation. As we contemplate those wonders, the words from act I, scene V, of Shakespeare's *Hamlet* seem almost frighteningly appropriate. As Hamlet faces the challenges and revelations brought by the ghost of his father, Horatio cries out:

"O day and night, but this is wondrous strange!"

To this comes Hamlet's immortal reply:

> And therefore as a stranger give it welcome.
> There are more things in heaven and earth, Horatio,
> Than are dreamt of in your philosophy.
> There are more things in heaven and earth, Horatio, than
> are dreamt of in your philosophy

Here is a message to shout back through the ages. At the start of our journey we asked the most basic question about what we see in the sky—"How big is it?"—and the answers were enough to expose the comedy of humanity's ancient conceits. Traveling at the speed of a modern jetliner, it would take us over 5 million years to cross the distance to even the nearest star beyond our Sun. Even so, we live in a galaxy containing hundreds of billions of such stars, which are themselves outnumbered by other galaxies filling a universe that may stretch on forever. Using the speed of light as our yardstick, we have come to realize that Earth—the world of our birth and the stage on which all of human history has been played—is to the expanse of the observable universe as a single snap of our fingers is to the aeons that have transpired since time itself came into existence, roughly 14 billion years ago.

As we stare at the images that have come to symbolize 21st century astronomy and consider what they show, it is easy to understand why some people recoil from these insights. "Wouldn't it be nice," they say, "if we could just go back to imagining that Earth is only 6,000 years old, and that humanity occupies a special place at the pinnacle of Creation?" Indeed, if this were where our story ended—with the fact of our seeming insignificance in the universe—we might *all* long for an excuse to retreat into ignorance. Fortunately, this is not where our story ends. Rather, this is where our true journey of discovery begins. Although modern science may have shattered our egotistical views about our exalted place in the scheme of things, it has also offered us a wonderful new appreciation and understanding of ourselves to fill that void.

When we look at distant galaxies, the light we see is produced by stars like our Sun. Each of those stars formed when a cloud of interstellar gas and dust collapsed under the same force of gravity that guided Newton's cannonball. As each of those clouds collapsed, it spun faster and faster, obeying the same laws of motion that accelerate the spin of an Olympic skater as she pulls her arms and leg ever more tightly to her body. This spin prevented those clouds from collapsing directly into stars, forcing them instead to settle into flat rotating disks. We see such disks today when we look at the newest generation of stars. Inside those disks, grains of dust stick together to make larger grains, which stick together to make still larger grains—the beginning of a bottom-up process that culminates with planetesimals crashing together to make worlds. Our Earth is one such world. We have come to view our Sun, Earth, and Solar System as products of natural processes still going on around us today. As we watch new generations of stars form and search for the planets that surround them, we are witnessing a replay of the birth of our own world, 4.6 billion years ago. We have found our roots in the stars.

Go out at sunset on an evening when a waxing crescent Moon hangs low above the western horizon, and several planets stretch out across the darkening sky. The plane of the ecliptic is there in front of you, and your mind's eye might even envision the flat, rotating accretion disk from which our Solar System formed. Once you realize what you are looking at, the cradle of our world and ourselves hangs there in the night sky for all to see.

Looked at in this way, the sunset takes on a whole new significance. It will never be the same. Yet even the majesty of the planets spread across the sky fails to capture the intimacy of our connection with the universe. In the most basic sense, the question "What are we?" is easily answered. Humans and all other terrestrial life are an organized assemblage of various organic molecules, most of which are very complex. Counting the numbers of atoms in our bodies, we are approximately 60 percent hydrogen, 26 percent oxygen, 11 percent carbon, and 2 percent nitrogen, with a small fraction of metals and other heavy elements mixed in. Over the course of our journey we have witnessed the history of those atoms. A very long time ago—roughly 14 billion years—something wonderful happened. The universe came into being, and time began. From an infinitesimally small volume of concentrated energy, the universe expanded, growing in size at the speed of light. Within the first few minutes, particles of solid matter, including protons and electrons, condensed out of this dense, primordial ball of energy. Nuclear reactions caused some of the protons to fuse into other light nuclei. Several hundred thousand years later, when the universe had cooled to a temperature of a few thousand kelvins, those nuclei combined with electrons to form atoms. Of those atoms, roughly 90 percent were hydrogen atoms and 10 percent were helium atoms. There were traces of lithium, beryllium, and boron as well, but that was all that existed in the way of normal luminous matter as the universe expanded past the threshold of recombination.

The hydrogen atoms in our bodies date back to this early time in the history of the universe, but what of the rest? Having taken our journey of discovery, you know the answers. As the universe emerged from the Big Bang, clumps of dark matter began to collapse under the force of gravity, pulling normal matter along with it. Within these collapsing protogalaxies the first generations of stars formed—nuclear furnaces powered by the fusion of those original hydrogen atoms into increasingly massive elements. Carbon, oxygen, silicon, sulfur—elements all the way up to iron and nickel were formed in those stellar infernos. As those first generations of stars ended their lives, they blasted this nuclear ash back into the reaches of interstellar space. Nucleosynthesis did not end with fusion, however. In the extreme environments of supernovae, free neutrons were captured by the products of fusion, building more massive elements still. Atoms of copper, zinc, tin, silver, and gold—all the way up to the most massive naturally occurring element, uranium—were formed and expelled into space. Here were the chemical elements to fill the periodic table and to build the compounds of life. As early protogalaxies merged to form early galaxies, more generations of stars continued to enrich the universe with the fruits of their alchemy. As early galaxies settled into the well-ordered ellipticals and spirals of today's universe, still more generations of stars came and went, adding to the chemical richness of the universe. When our Solar System appeared on the scene 4.6 billion years ago, it formed from interstellar material that carried the chemical building blocks of planets and of life—atoms produced both in the Big Bang and in the hearts of generations of stars that had lived and died during the 9 billion years that had transpired since the universe began. "What are we, and how did we get here?" We are stardust in human form.

The Future Arrives Every Day

While traveling the highways and back roads of 21st century astronomy, we have come to see our world and ourselves in a very different light. Even so, nothing we have seen has changed the most basic circumstances of our day-to-day existence. Earth remains our world, our home. The hopes and dreams we humans feel are no less real today than they were a thousand years ago. In 1968 Stanley Kubrick and Arthur C. Clarke collaborated to make the film *2001: A Space Odyssey*. This provocative piece of speculative fiction captured the imagination of a generation, and the year 2001 came to signify the future. That future is now here: yet little of our modern-day life is recognizable in those cinematic prognostications dating from only three decades ago. It is ironic that while we can forecast the future of the Sun with great accuracy and can even calculate the fate of the universe itself, our vision of our own future is so much less certain.

These words were first drafted on the afternoon of December 31, 2000. At that moment in the lives of your authors, roughly half the surface of Earth remained in the 20th century, while the other half of the world had witnessed the beginning not only of a new century but also of a new millennium. As the dividing line between the two swept across Europe and out into the Atlantic Ocean, it was just another moment in the long dance of Earth as it spins on its axis and falls around the Sun. The Sun took no notice of the moment as it continued its orbit around the center of the Milky Way Galaxy, which itself is but a speck in an expanding universe. Yet while that December afternoon had no objective physical significance, it was loaded with symbolic meaning for our species. The year of Kubrick and Clarke's mind-bending tale was at hand. Though the details of their vision have turned out to be incorrect, there is no question that we live at an amazing moment in history—a moment when science has for the first time allowed us to truly see the universe beyond ourselves and offers the promise of showing us the universe within ourselves, as well.

There is no denying the fact that we live in an evolving universe—a place of ongoing and unending change that continues to shape humanity as certainly as it shapes the cosmos itself. The grim prospects mentioned in the closing sections of Chapter 21 are real. Saying that they do not exist or choosing to push them from our minds will not make them go away. Yet at the same time modern medicine, food production, transportation, and a thousand other technologies that push back the ancient scourges of humanity are equally real. Differences among people remain, but modern communication offers the hope of spreading understanding. Meanwhile, a picture of Earthrise above the lunar horizon taken by the *Apollo 8* astronauts (**Figure E.1**) hangs forever as part of the human experience and perspective, showing us as nothing else could that Earth is a tiny fragile island to be cherished. Whether we like it or not, humanity is a single, interrelated, interdependent global village that will face the future together—or not at all.

At the dawn of the 21st century, science has shown us the wonders of the universe and at the same time has given us the knowledge and power to shape our world and choose our future. The future of humanity may depend entirely on how well we treat Earth and ourselves over the few decades and centuries ahead. If 21st century astronomy does nothing else, it forces us to change our perspective. As we study the laws that govern the workings of atoms, we learn something about the conditions of our own lives: The future is not yet written. As we use our telescopes to stare at galaxies 10 billion light-years away, collecting light from stars that died billions of years before our Sun was even born, manifest destiny seems a pretty silly concept. There are no guarantees that things will work out in the end for one tiny world or for the species to which it gave birth. If we choose to destroy our world, either through direct action or simple neglect, so be it. The universe as a whole will carry on in sublime indifference to our fate.

This is not, however, a message of despair. Instead it is a message of hope, responsibility, and maturity. We have the power to make our Earth a paradise or to leave our children's children to cope with a world choking from our shortsighted excess. The choices are ours, whether we want them or not. Borrowing from the book of Genesis, we have truly tasted of the tree of knowledge. As we stand at the beginning of the 21st century, we face a future filled with choices: but one choice we are not allowed is refusing to acknowledge responsibility for our own destiny.

As you read these words your authors' moment of introspection at the threshold between the second and third millennia has long passed. Even so, the moment when you read these words is not so different from that New Year's Eve of the year A.D. 2000. Although the symbolism of your moment may not be as palpable as that of an instant when the world passes a major milestone in humanity's accounting of events, your moment is a milestone nonetheless. It is one of a stream of moments that define your life as the possibilities of the future cross, inexorably, into the unchangeable reality of the past. That moment is there for you to use as you will. The course of the future is yours to shape.

With that thought, we come to the end of our journey. We, the authors, hope this journey has helped to open your eyes to the wonders of the world and the universe around you. Even more, we hope this journey has given you pause to reflect on who we are as humans and on our place in the larger reality in which we find ourselves. Finally, we hope this journey has changed the shape of the way you think—not only about the sights you see in the night sky, but also about the events of your daily life. If any or all of these hopes are fulfilled, then the journey will have been worth taking—worthy of your time and thought and of ours.

FIGURE E.1 This image of Earth taken from lunar orbit by the *Apollo 8* astronauts forever changed our understanding of our Earth and ourselves.

Mathematical Tools

Working with Proportionalities

Most of the mathematics in *21st Century Astronomy* involves proportionalities—statements about the way that one physical quantity changes when another quantity changes. Here we offer a practical guide to working with proportionalities.

To use a statement of proportionality to compare two objects, begin by turning the proportionality into a ratio. For example, the price of a bag of apples is proportional to the weight of the bag:

$$\text{Price} \propto \text{Weight}.$$

Here the symbol "$\propto$" is pronounced "is proportional to." What this means is that the ratio of the prices of two bags of apples is equal to the ratio of the weights of the two bags:

$$\text{Price} \propto \text{Weight} \quad means \quad \frac{\text{Price of A}}{\text{Price of B}} = \frac{\text{Weight of A}}{\text{Weight of B}}.$$

Work a specific example: Suppose bag A weighs two pounds, while bag B weighs one pound. That of course means that bag A will cost twice as much as bag B. We can turn our proportionality into this equation:

$$\frac{\text{Price of A}}{\text{Price of B}} = \frac{\text{Weight of A}}{\text{Weight of B}} = \frac{2\ \text{lb}}{1\ \text{lb}} = 2.$$

So the price of bag A is two times the price of bag B.

Let's work another, more complicated, example. In Chapter 13 we discuss the relationship between the luminosity, brightness, and distance of stars. The luminosity of a star—the total energy the star radiates each second—is proportional to the star's brightness times the square of its distance:

$$\text{Luminosity} \propto \text{Brightness} \times \text{Distance}^2.$$

What this proportionality means is that if we have two stars—call them A and B—then

$$\frac{\text{Luminosity of A}}{\text{Luminosity of B}} = \frac{\text{Brightness of A}}{\text{Brightness of B}} \times \left(\frac{\text{Distance of A}}{\text{Distance of B}}\right)^2.$$

If we use the symbols L, b, and d to represent luminosity, brightness, and distance, respectively, this becomes

$$\frac{L_A}{L_B} = \frac{b_A}{b_B} \times \left(\frac{d_A}{d_B}\right)^2.$$

As an example, suppose Star A appears twice as bright in the sky as Star B, but Star A is located 10 times as far away as Star B. Compare the luminosities of the two stars. Because we know that

$$\text{Luminosity} \propto \text{Brightness} \times \text{Distance}^2$$

we write

$$\frac{\text{Luminosity of A}}{\text{Luminosity of A}} = \frac{\text{Brightness of A}}{\text{Brightness of B}} \times \left(\frac{\text{Distance of A}}{\text{Distance of B}}\right)^2$$

$$= \frac{2}{1} \times \left(\frac{10}{1}\right)^2 = 200.$$

Star A is 200 times as luminous as Star B.

A final note: In our original example we said that the price of a bag of apples is proportional to the weight of the bag, which allowed us to say that a two-pound bag of apples costs twice as much as a one-pound bag of apples. This is a statement about the way the world *works*. A two-pound bag of apples costs *more* than a one-pound bag, not *less* than a one-pound bag. To actually figure out how much a bag of apples will cost, we need another piece of information: the price per pound. The price per pound is an example of a *constant of proportionality*. Wrapped up in the price per pound is all sorts of information about the cost of growing apples, the cost of transporting them, how the market is at the moment, the profit the grocer needs to make, and so forth. In other words, this constant of proportionality is a statement about the way the world *is*.

Proportionalities help us understand how the world works and let us compare one object to another. Constants of proportionality allow us to calculate real values for things. In *21st Century Astronomy* it is usually the "understanding"—the proportionality itself—that we care about.

Scientific Notation

Astronomy is a science both of the very large and the very small. The mass of an electron, for example, is

$$0.0000000000000000000000000000009109 \text{ kg}$$

whereas the distance to a galaxy far, far away might be around

$$100,000,000,000,000,000,000,000,000 \text{ m}.$$

All it takes is a quick glance at these two numbers to see why astronomers, like most scientists, make heavy use of scientific notation to express numbers.

Powers of 10

Our number system is based on powers of 10. Going to the left of the decimal place,

$$10 = 10 \times 1,$$

$$100 = 10 \times 10 \times 1,$$

$$1,000 = 10 \times 10 \times 10 \times 1,$$

and so on. Going to the right of the decimal place.

$$0.1 = \frac{1}{10} \times 1,$$

$$0.01 = \frac{1}{10} \times \frac{1}{10} \times 1,$$

$$0.001 = \frac{1}{10} \times \frac{1}{10} \times \frac{1}{10} \times 1,$$

and so on for as long as we care to continue. In other words, each place to the right or left of the decimal place in a number represents a power of 10. For example, 1 million can be written

$$\begin{aligned}1 \text{ million} &= 1,000,000 \\ &= 1 \times 10 \times 10 \times 10 \times 10 \times 10 \times 10.\end{aligned}$$

That is to say, 1 million is 1 times six factors of 10. Scientific notation combines these factors of 10 in convenient short-hand. Rather than writing out all six factors of 10, instead we combine them in easy shorthand using an exponent:

$$1 \text{ million} = 1 \times 10^6$$

which means 1 times six factors of 10.

Moving to the right of the decimal place, each step we take *removes* a power of 10 from the number. One millionth can be written

$$1 \text{ millionth} = 1 \times \frac{1}{10} \times \frac{1}{10} \times \frac{1}{10} \times \frac{1}{10} \times \frac{1}{10} \times \frac{1}{10}.$$

We are removing powers of 10, so we express this using a negative exponent. We write

$$\frac{1}{10} = 10^{-1}$$

and

$$1 \text{ millionth} = 1 \times 10^{-1} \times 10^{-1} \times 10^{-1} \times 10^{-1} \times 10^{-1} \times 10^{-1}$$

$$= 1 \times 10^{-6}$$

Returning to our earlier examples, the mass of an electron is 9.109×10^{-31} kg, whereas the distant galaxy is located 1×10^{26} away: these are *much* more convenient ways of writing this information. *Notice that the exponent in scientific notation gives you a feeling for the size of a number at a glance.* The exponent of 10 in the electron mass is −31, which tells us instantly that it is a very small number. The exponent of 10 in the distance to a remote galaxy, +26, tells us immediately that it is a very large number. This exponent is often called the *order of magnitude* of a number. When you see a number written in scientific notation while reading *21st Century Astronomy* (or elsewhere), just remember to look at the exponent to better understand what the number is telling you.

Scientific notation is also convenient because it makes multiplying and dividing numbers easier. Again, let's look at an example. Two billion times eight thousandths can be written

$$2,000,000,000 \times 0.008$$

but it is more convenient to write these two numbers using scientific notation as

$$(2 \times 10^9) \times (8 \times 10^{-3}).$$

This can be regrouped as

$$(2 \times 8) \times (10^9 \times 10^{-3})$$

The first part of the problem is just $2 \times 8 = 16$. The more interesting part of the problem is the multiplication in the right parentheses. The first number, 10^9, is just shorthand for $10 \times 10 \times 10 \ldots$ nine times. That is, it represents nine factors of 10. The second number stands for three factors of $1/10$—or removing three factors of 10 if you prefer to think of it that way. All together, that makes $9 - 3 = 6$ factors of 10. In other words,

$$10^9 \times 10^{-3} = 10^{9-3} = 10^6.$$

Putting the problem together,

$$(2 \times 10^9) \times (8 \times 10^{-3}) = (2 \times 8) \times (10^9 \times 10^{-3})$$

$$= 16 \times 10^6.$$

By convention, when a number is written in scientific notation, only one digit is placed to the left of the decimal point. In this case, there are two. However, 16 is 1.6×10, so we can add this two factor of 10 to the exponent at right, making the final answer

$$1.6 \times 10^7.$$

Dividing is just the inverse of multiplication. Dividing by 10^3 means removing three factors of 10 from a number. Using the previous number.

$$(1.6 \times 10^7) \div (2 \times 10^3) = (1.6 \div 2) \times (10^7 \div 10^3)$$
$$= 0.8 \times 10^{7-3}$$
$$= 0.8 \times 10^4.$$

This time we have only a zero to the left of the decimal point. To get the number into proper form, say that $0.8 = 8 \times 10^{-1}$, giving

$$0.8 \times 10^4 = (8 \times 10^{-1}) \times 10^4 = 8 \times 10^3.$$

Adding and subtracting numbers in scientific notation is somewhat more difficult because each number has to be written as a value times the *same* power of 10 before they can be added or subtracted. However, almost all modern calculators have scientific notation built in. They keep up with the powers of 10 for you. If you do not have such a calculator you may want to buy one and learn to use it before tackling the mathematical problems in this book. There are more examples on our Web site that you can use to better learn how to work with scientific notation.

Significant Figures

In the previous example we actually broke some rules in the interest of explaining how powers of 10 are treated in scientific notation. The rules that we broke involve the *precision* of the numbers we are expressing. In everyday speech we might say, "The store is a kilometer away," by which we probably mean that the store is *roughly* a kilometer away. If it turned out to be 0.8 km, or 1.2 km, it is unlikely that the recipient of our directions would quibble. But when expressing quantities in science, it is extremely important to know not only of the value of a number, but also how precise that value is.

The most complete way to keep track of the precision of numbers is to actually write down the uncertainty in the number. For example, if we know the distance to the store (call it d) is between 0.8 km and 1.2 km, we can write

$$d = 1.0 \pm 0.2 \text{ km}$$

where the symbol "±" is pronounced "plus or minus." In this example, d is between $1.0 - 0.2 = 0.8$ km and $1.0 + 0.2 = 1.2$ km. This is an unambiguous statement about the limitations on our knowledge of the value of d: but carrying along the formal errors with every number we write would be cumbersome at best. Instead we keep track of the approximate precision of a number by using *significant figures*.

The convention for significant figures works like this: We assume the number we write has been rounded from a number that had one additional digit to the right of the decimal point. If we say that some quantity d, which might represent the distance to the store, is "1.", what we mean is that d is close to 1. It is likely not as small as 0., and it is likely not as large as 2. If we say instead

$$d = 1.0$$

then we mean that d is likely not 0.9 and is likely not 1.1. It is roughly 1.0 to the nearest tenth. The greater the number of significant figures, the more precisely the number is being specified. For example, 1.00000 is *not* the same number as 1.00. The first number, 1.00000, represents a value that is probably not as small as 0.99999 and is probably not as large as 1.00001. The second number, 1.00, represents a value that is probably not as small as 0.99 nor as large as 1.01.

When we carry out mathematical operations, significant figures are important. For example, $2.0 \times 1.6 = 3.2$. It does *not* equal 3.20000000000. The product of two numbers cannot be known to any greater accuracy than the numbers themselves! As a general rule, when we multiply and divide, the answer should have the same number of significant figures as the less precise of the numbers being multiplied or divided. In other words, $2.0 \times 1.602583475 = 3.2$. Because all we know is that the first factor is probably closer to 2.0 than to 1.9 or 2.1, all we know about the product is that it is between about 3.0 and 3.4. It is 3.2. It is not 3.205166950 (*even if that is the answer your calculator gives!*). The rest of the digits to the right of 3.2 just do not mean anything.

When we add and subtract, the rules are a bit different. If one number has a significant figure with a particular place value but another number does not, their sum or difference cannot have a significant figure in that place value. In other words,

$$1,045.$$

$$\underline{+1.34567}$$

$$1,046.$$

The answer is 1,046., *not* 1,046.34567. Again, the extra digits to the right of the decimal place have no meaning because 1,045. is not known to that accuracy.

There is always a fly in the ointment, and this is no exception. What is the precision of the number 1,000,000? As it is written, the answer is unclear. Are all those zeros really significant, or are they placeholders? If we write the number in scientific notation, on the other hand, there is never a question. Instead of 1,000,000 we write 1.0×10^6 for a number that is known to the nearest hundred thousand or so: or we write 1.00000×10^6 for a number that is 1 million to the nearest 10.

So our earlier example would have been more correct had we said

$$(2.0 \times 10^9) \times (8.0 \times 10^{-3}) = 1.6 \times 10^7.$$

Algebra

There are many branches of mathematics. The branch that tells us about the relationships between quantities is called *algebra*. If you are reading this book, you have almost certainly taken an algebra class: but you are not alone if you feel a little review is in order. Basically, algebra begins by using symbols to represent quantities. For example, we might write the distance you travel in a day as d. As it stands, d has no value. It might be 10,000 miles. It might be 30 feet. It does, however, have *units*—in this case the units of distance.

The average speed at which you travel is equal to the distance you travel divided by the time you take. If we use the symbol v to represent your average speed and the symbol t to represent the time you take, then instead of writing out, "Your average speed is equal to the distance you travel divided by the time taken," we can write

$$v = \frac{d}{t}.$$

The meaning of this algebraic expression is exactly the same as the sentence quoted before it, but it is much more concise. As it stands, v, d, and t still have no specific values. There are no numbers assigned to them yet. However, this expression tells us what the relationship between those numbers will be when we *do* look at a specific example. For example, if

you go 500 km ($d = 500$ km) in 10 hours ($t = 10$ hours), this expression tells you that your average speed is

$$v = \frac{d}{t} = \frac{500 \text{ km}}{10 \text{ hours}} = 50 \text{ km/hour.}$$

Notice that the units in this expression act exactly like the numerical values. They are just multiplicative factors. When we say "500 km" what we really mean is "500 × kilometers." Likewise, 10 hours means "10 × hours." When we divide the two we find that the units of v are kilometers divided by hours, or km/hr (pronounced kilometers per hour).

We introduced algebra as shorthand for expressing relations between quantities, but it is far more powerful than that. Algebra provides rules for manipulating the symbols used to represent quantities. We begin with a bit of notation for *powers* and *roots*. When we talk about raising a quantity to a power, we mean multiplying the quantity by itself some number of times. For example, if S is a symbol for something (anything), then S^2 (pronounced "S squared" or "S to the second power") means $S \times S$, and S^3 (pronounced "S cubed" or "S to the third power") means $S \times S \times S$. Suppose S represents the length of the side of a square. The area of the square is given by

$$\text{Area} = S \times S = S^2.$$

If $S = 3$ m, then the area of the square is

$$S^2 = 3 \text{ m} \times 3 \text{ m} = 9 \text{ m}^2$$

(pronounced 9 square meters). It should be obvious why raising a quantity to the second power is called "squaring" the quantity. We could have done the same thing for the sides of a cube and found that the volume of the cube is

$$\text{Volume} = S \times S \times S = S^3$$

If $S = 3$ m, then the volume of the cube is

$$S^3 = 3 \text{ m} \times 3 \text{ m} \times 3 \text{ m} = 27 \text{ m}^3$$

(pronounced 27 cubic meters). Again, it is clear why raising a quantity to the third power is called "cubing" the quantity.

Roots are the reverse of this process. The square root of a quantity is the value that, when squared, gives the original quantity. The square root of 4 is 2, which means that $2 \times 2 = 4$. The square root of 9 is 3, which means that $3 \times 3 = 9$. Similarly, the cube root of a quantity is the value that, when cubed, gives the original quantity. The cube root of 8 is 2, which means that $2 \times 2 \times 2 = 8$. Roots are written with the symbol $\sqrt{}$. For example, we write

$$\sqrt{9} = 3$$

for the square root of 9 and

$$\sqrt[3]{8} = 2$$

for the cube root of 8. If the volume of a cube is $V = S^3$, we can also write

$$S = \sqrt[3]{V} = \sqrt[3]{S^3}.$$

Roots can also be written as powers. Powers and roots behave exactly like the exponents of 10 in our discussion of scientific notation. (They had better: The exponents used in scientific notation are just powers of 10.) For example, if a, n, and m are all algebraic quantities, then

$$a^n \times a^m = a^{n+m}, \quad \text{and} \quad \frac{a^n}{a^m} = a^{n-m}.$$

(To see if you understand all this, explain why the square root of a can also be written $a^{\frac{1}{2}}$ and the cubed root of a can be written $a^{\frac{1}{3}}$.)

Some of the rules of algebra are summarized next. These are really no more than the rules of arithmetic applied to the symbolic quantities of algebra. The important thing is this: So long as we apply the rules of algebra properly, then the relationships among symbols we arrive at through our algebraic manipulations remain true for the physical quantities those symbols represent.

Here we summarize a few algebraic rules and relationships. In this summary, a, b, c, n, m, r, x, and y are all algebraic quantities:

Associative rule:

$$a \times b \times c = (a \times b) \times c = a \times (b \times c)$$

Commutative rule:

$$a \times b = b \times a$$

Distributive rule:

$$a \times (b + c) = (a \times b) + (a \times c)$$

Cross multiplication:

If $\frac{a}{b} = \frac{c}{d}$, then $ad = bc$.

Working with exponents:

$$\frac{1}{a^n} = a^{-n} \qquad a^n a^m = a^{n+m}$$

$$\frac{a^n}{a^m} = a^{n-m} \qquad (a^n)^m = a^{n \times m} \qquad \left(\frac{a}{b}\right)^n = \frac{a^n}{b^n}$$

Equation of a line with slope m and y-intercept b:

$$y = mx + b$$

Equation of a circle with radius r centered at $x = 0$, $y = 0$:

$$x^2 + y^2 = r^2$$

Angles and Distances

The farther away something is, the smaller it appears. This is common sense and everyday experience. In astronomy, where we seldom get to walk up to the object we are studying and measure it with a meterstick, our knowledge about the sizes of things usually depends on knowing the relationship between the size of an object, its distance, and the angle it covers in the sky.

The natural way to measure angles is using a unit called *radians*. As shown in **Figure A1.1(a)**, the size of an angle in radians is just the length of the arc subtending the angle divided by the radius of the circle. In the figure, the angle $x = S/r$ radians.

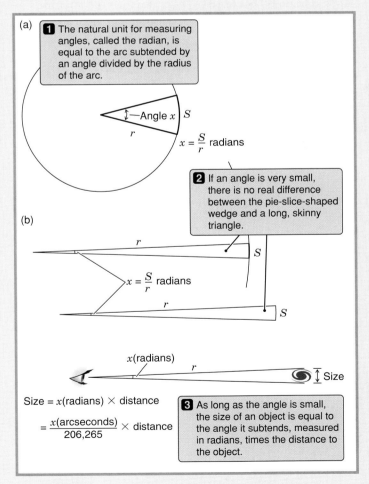

(a)

1 The natural unit for measuring angles, called the radian, is equal to the arc subtended by an angle divided by the radius of the arc.

Angle x S

r

$x = \frac{S}{r}$ radians

2 If an angle is very small, there is no real difference between the pie-slice-shaped wedge and a long, skinny triangle.

(b)

r S

$x = \frac{S}{r}$ radians

r

S

x(radians)

r

Size

Size $= x$(radians) $\times$ distance

$= \dfrac{x(\text{arcseconds})}{206{,}265} \times$ distance

3 As long as the angle is small, the size of an object is equal to the angle it subtends, measured in radians, times the distance to the object.

FIGURE A1.1

Because the circumference of a circle is 2π times the radius, $C = 2\pi r$, a complete circle has an angular measure of $(2\pi r)/r = 2\pi$ radians. In more conventional angular measure, a complete circle is 360°: so we can say that

$$360° = 2\pi \text{ radians}$$

or that

$$1 \text{ radian} = \frac{360°}{2\pi} = 57.2958°$$

When talking about stars and galaxies, we often use seconds of arc to measure angles. A degree is broken into 60 minutes of arc, each of which is broken into 60 seconds of arc—so there are 3,600 seconds of arc in a degree. That means

$$3{,}600 \, \frac{\text{arcseconds}}{\text{degree}} \times 57.2958 \, \frac{\text{degrees}}{\text{radian}} = 206{,}625 \, \frac{\text{arcseconds}}{\text{radian}}$$

If the angle is small enough (which it usually is in astronomy), there is precious little difference between the pie slice just described and a long skinny triangle with a short side of length S (see **Figure A1.1(b)**). So, if we know the distance d to an object, and we can measure the angular size x of the object, then the size of the object is just

$$S = x \text{ (in radians)} \times d = \frac{x \text{ (in degrees)}}{57.2958 \text{ degrees/radian}} \times d$$

$$= \frac{x \text{ (in arcseconds)}}{206{,}265 \text{ arcseconds/radian}} \times d,$$

which is all we need to turn our knowledge of the angular size and the distance to an object into a measurement of the object's physical size. (Our Web site has plenty of examples of such calculations.)

Circles and Spheres

To round out our mathematical tools, here are a few useful formulas for circles and spheres. The circle or sphere in each case has a radius r.

$$\text{Circumference}_{\text{circle}} = 2\pi r$$

$$\text{Area}_{\text{circle}} = \pi r^2$$

$$\text{Area}_{\text{circle}} = 4\pi r^2$$

$$\text{Volume}_{\text{sphere}} = \frac{4}{3}\pi r^3$$

Physical Constants and Units

Fundamental Physical Constants

Constant	Symbol	Value
Speed of light in a vacuum	c	2.99792×10^8 m/s
Universal gravitational constant	G	6.673×10^{-11} N m^2/kg^2
Planck constant	h	6.62607×10^{-34} J s
Electric charge of electron or proton	e	1.60218×10^{-19} C
Boltzmann constant	k	1.38065×10^{-23} J/k
Stefan-Boltzmann constant	σ	5.67040×10^{-8} W/(m^2 k^4)
Mass of electron	m_e	9.10938×10^{-31} kg
Mass of proton	m_p	1.67262×10^{-27} kg

Source: National Institute of Standards and Technology (http://physics.nist.gov).

Unit Prefixes

Prefix[a]	Name	Factor[b]
n	nano-	10^{-9}
μ	micro-	10^{-6}
m	milli-	10^{-3}
k	kilo-	10^3
M	mega-	10^6
G	giga-	10^9
T	tera-	10^{12}

These prefixes ([a]), when appended to a unit, change the size of the unit by the factor ([b]) given. For example, 1 km (kilometer) is 10^3 meters.

Units and Values

Quantity	Fundamental Unit	Values
Length	meter (m)	Radius of Sun ($R_\odot$) = 6.96265×10^8 m astronomical unit (AU) = 1.49598×10^{11} m 1 astronomical = 149,598,000 km Light-year (ly) = 9.4605×10^{15} m 1 light-year = 6.324×10^4 AU 1 parsec (pc) = 3.261 ly = 3.0857×10^{16} m 1 m = 3.281 feet
Volume	meters³ (m³)	1 m³ = 1,000 liters = 264.2 gallons
Mass	kilogram (kg)	1 kg = 1,000 g Mass of Earth ($M_\oplus$) = 5.9736×10^{24} kg Mass of Sun ($M_\odot$) = 1.9891×10^{30} kg
Time	seconds (s)	1 hour (h) = 60 minutes (min) = 3,600 s Solar day (noon to noon) = 86,400 s Sidereal day (Earth rotation period) = 86,164.1 s Tropical year (equinox to equinox) = 365.24219 days = 3.15569×10^7 s Sidereal year (Earth orbital period) = 365.25636 days = 3.15581×10^7 s
Speed	meters/second (m/s)	1 m/s = 2.236 miles/h 1 km/s = 1,000 m/s = 3,600 km/h c = 3.00×10^8 m/s = 300,000 km/s
Acceleration	meters/second² (m/s²)	Gravitational acceleration on Earth (g) = 9.78 m/s²
Energy	joules (J)	1 J = 1 kg m²/s² 1 megaton = 4.19×10^{15} J
Power	watt (W)	1 W = 1 J/s Solar luminosity ($L_\odot$) = 3.827×10^{26} W
Force	newton (N)	1 N = 1 kg m/s² 1 pound (lbs) = 4.448 N 1 N = 0.22481 pounds (lbs)
Pressure	newtons/meter² (N/m²)	Atmospheric pressure at sea level = 1.013×10^5 N/m² = 1.013 bar
Temperature	kelvins (K)	Absolute zero = 0 K = −273.15°C = −459.67°F

Sources: National Space Science Data Center (2002); *Observer's Handbook 2002*, Rajiv Gupta (Royal Astronomical Society of Canada, 2001); National Institute of Standards and Technology (2002).

Periodic Table of the Elements

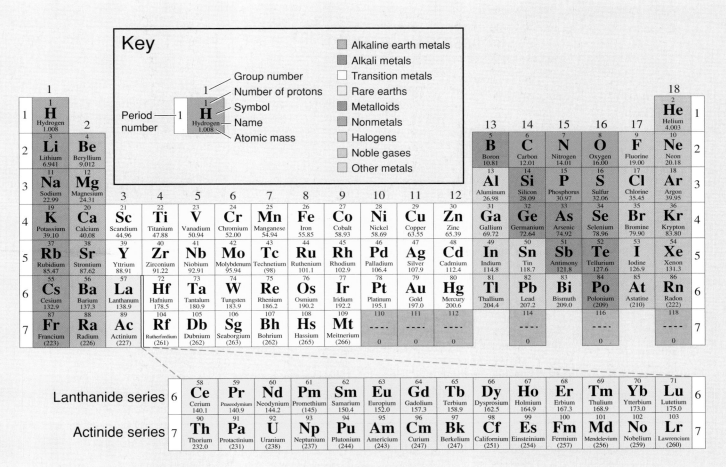

Key

Group number
Number of protons
Symbol
Period number — Name
Atomic mass

- Alkaline earth metals
- Alkali metals
- Transition metals
- Rare earths
- Metalloids
- Nonmetals
- Halogens
- Noble gases
- Other metals

Sources: Los Alamos National Laboratory; National Institute of Standards and Technology.

Properties of Planets, Dwarf Planets, and Moons

Physical Data for Planets and Dwarf Planets

Planet	Equatorial Radius km	R/R$_\odot$	Mass kg	M/M$_\odot$	Average Density (relative to water[a])	Rotation Period (days)	Tilt of Rotation Axis (relative to orbit)	Surface Gravity (relative to Earth[b])	Escape Velocity (km/s)	Average Surface Temperature (K)
Mercury	2,440	0.383	3.30×10^{23}	0.055	5.427	58.64	0.01°	0.378	4.3	440 (100 725)[d]
Venus	6,052	0.949	4.87×10^{24}	0.815	5.243	243.02[c]	177.36°	0.907	10.36	737
Earth	6,378	1.000	5.97×10^{24}	1.000	5.515	1.000	23.45°	1.000	11.19	288 (183 331)[d]
Mars	3,397	0.533	6.42×10^{23}	0.107	3.933	1.0260	25.19°	0.377	5.03	210 (133 293)[d]
Ceres	950	0.075	9.60×10^{20}	0.0002	2.100	9.075	3.0°	0.27	0.51	200
Jupiter	71,492	11.209	1.90×10^{27}	317.83	1.326	0.4136	3.13°	2.364	59.5	165
Saturn	60,268	9.449	5.68×10^{26}	95.16	0.687	0.4440	26.73°	0.916	35.5	134
Uranus	25,559	4.007	8.68×10^{25}	14.537	1.270	0.7183[c]	97.77°	0.889	21.3	76
Neptune	24,764	3.883	1.02×10^{26}	17.147	1.638	0.6713	28.32°	1.12	23.5	58
Pluto	1,195	0.187	1.25×10^{22}	0.0021	1.750	6.387[c]	122.53°	0.059	1.1	40
Eris	1,200	0.188	1.5×10^{22} (est.)	0.0025 (est.)	?	8.?	?	?	?	30

[a]The density of water is 1,000 kg/m³.
[b]The surface gravity of Earth is 9.78 m/s².
[c]Venus, Uranus, and Pluto rotate opposite to the directions of their orbits. Their north poles are south of their orbital planes.
[d]Where given, values in parentheses give extremes of recorded temperatures.

Orbital Data for Planets and Dwarf Planets

Planet	Mean Distance from Sun (A[a]) 10^6 km	AU	Orbital Period (P) (sidereal years)	Eccentricity	Inclination (relative to ecliptic)	Average Speed (km/s)
Mercury	57.91	0.387	0.2408	0.2056	7.005°	47.87
Venus	108.2	0.723	0.6152	0.0067	3.395°	35.02
Earth	149.6	1.000	1.000	0.0167	0.000°	29.78
Mars	227.9	1.524	1.8809	0.0935	1.850°	24.13
Ceres	413.9	2.767	4.6027	0.097	9.73°	17.88
Jupiter	778.6	5.204	11.8618	0.0489	1.304°	13.07
Saturn	1,433.5	9.582	29.4566	0.0565	2.485°	9.69
Uranus	2,872.5	19.201	84.0106	0.0457	0.772°	6.81
Neptune	4,495.1	30.047	164.7856	0.0113	1.769°	5.43
Pluto	5,869.7	39.236	247.6753	0.2488	17.142°	4.72
Eris	10,123	67.668	557.	0.4418	44.187°	3.44

[a]A is the semimajor axis of the planet's elliptical orbit.
Sources: National Space Science Data Center, *Astronomical Almanac.*

Properties of Selected Moons[a]

Planet	Moon	Orbital Properties P (days)	A (10^3 km)	Physical Properties R (km)	M (10^{20} kg)	Density[b] (water = 1)
Earth (1 moon)	Moon	27.32	384.4	1,737.4	735	3.34
Mars (2 moons)	Phobos	0.32	9.38	13.5 × 10.8 × 9.4	0.0001	2.0
	Deimos	1.26	23.46	7.5 × 6.1 × 5.5	0.00002	1.7
Jupiter (63 known moons)	Metis	0.29	127.97	20	0.00096	2.8
	Amalthea	0.50	181.30	131 × 73 × 67	0.0717	1.8
	Io	1.77	421.60	1,815	894	3.55
	Europa	3.55	670.90	1,569	480	3.01
	Ganymede	7.16	1,070	2,631	1,480	1.94
	Callisto	16.69	1,883	2,403	1,080	1.86
	Himalia	250.57	11,480	93	0.0956	2.8
	Pasiphae	735[c]	23,500	25	0.0019	2.9
	Callirrhoe	759[c]	24,100	4.3	0.00001	2.6

(continued)

Properties of Selected Moons[a]

Planet	Moon	Orbital Properties		Physical Properties		
		P (days)	A (10^3 km)	R (km)	M (10^{20} kg)	Density[b] (water = 1)
Saturn (56 known moons)	Pan	0.58	133.58	20	0.00003	—
	Prometheus	0.61	139.35	72.5 × 42.5 × 32.5	0.0027	0.7
	Pandora	0.63	141.70	57 × 42 × 31	0.0022	0.7
	Mimas	0.94	185.52	196	0.38	1.17
	Enceladus	1.37	238.02	250	0.84	1.24
	Tethys	1.89	294.66	530	7.55	1.21
	Dione	2.74	377.40	560	10.5	1.43
	Rhea	4.52	527.04	765	24.9	1.33
	Titan	15.95	1,222	2,575	1,350	1.88
	Hyperion	21.28	1,481	205 × 130 × 110	0.177	1.4
	Iapetus	79.33	3,561	735	18.8	1.21
	Phoebe	550.48[c]	12,952	107	0.04	0.7
	Paaliaq	686.9	15.200	11	0.0001	2.3
Uranus (27 known moons)	Cordelia	0.34	49.75	20	0.0004	1.3
	Miranda	1.41	129.78	236	0.64	1.15
	Ariel	2.52	191.24	579	12.7	1.56
	Umbriel	4.14	265.97	585	12.7	1.52
	Titania	8.71	435.84	789	34.9	1.70
	Oberon	13.46	582.60	761	30.3	1.64
	Setebos	2.225[c]	17,418	15	0.0002	1.5
Neptune (16 known moons)	Naiad	0.29	48.0	48 × 30 × 26	0.002	1.3
	Larissa	0.55	73.6	108 × 102 × 84	0.05	1.3
	Proteus	1.12	117.6	210 × 208 × 202	0.5	1.3
	Triton	5.88[c]	354.8	1,353	214	2.07
	Nereid	360.14	5,513.40	170	0.3	1.5
Pluto (3 moons)	Charon	6.39	19.60	593	16.2	1.85
Eris	Dysnomia	14	36	?	?	?

[a]Innermost, outermost, largest, and/or a few other moons for each planet.
[b]The density of water is 1,000 kg/m³.
[c]Irregular moon (has retrograde orbit).

Nearest and Brightest Stars

Stars within 15 Light-Years of Earth

Name[a]	Visual[b] Brightness (Sirius = 1,000,000)	Distance (ly)	Spectral Type[c]	Visual[b] Luminosity (Sun = 1.000)
Sun	1.29×10^{16}	1.58×10^{-5}	G2V	1.000
Alpha Centauri C (Proxima Centauri)	10.3	4.23	M5.5V	0.00006
Alpha Centauri B	76,600	4.40	K1V	0.5
Alpha Centauri A	263,000	4.40	G2V	1.6
Barnard's star (dbl)[d]	40	6.0	M4Ve	0.0004
CN Leonis	1.1	7.8	M5.5	0.00002
BD +36-2147	260	8.3	M2	0.0055
Sirius A	1,000,000	8.6	A1V	23
Sirius B	110	8.6	wdDA	0.0025
BL Ceti A	2.6	8.8	M5.5	0.00006
BL Ceti B	1.7	8.8	M5.5	0.00004
V1216 Sagittarii	19	9.7	M3.5V	0.0005
HH Andromedae	3.2	10.3	M5var	0.0001
Epsilon Eridani (dbl)	8,500	10.5	K2V	0.28
Lacaille 9352	300	10.7	M0.5	0.010
FI Virginis	9.3	10.9	M4	0.0003
EZ Aquarii	3.1	11.1	M5.5	0.0001
61 Cygni A	2,200	11.4	K5V	0.084
61 Cygni B	990	11.4	K7V	0.039
Procyon A	184,000	11.4	F5IV-V	7.4
Procyon B	14	11.4	wdDA	0.00054

(continued)

Stars within 15 Light-Years of Earth

(continued) Name[a]	Visual[b] Brightness (Sirius = 1,000,000)	Distance (ly)	Spectral Type[c]	Visual[b] Luminosity (Sun = 1.000)
Gleise 227-046B	34	11.5	M3.5	0.0014
Gleise 227-046A	69	11.7	M3V	0.0028
Groombridge 34 A	150	11.7	M1	0.0062
Groombridge 34 B	9.5	11.7	M3.5	0.00039
DX Cancri	0.31	11.8	M6	0.00001
Epsilon Indi (dbl)	3,500	11.8	K4.5V	0.15
Tau Ceti	10,500	11.9	G8V	0.45
YZ Ceti	3.8	12.1	M4.5V	0.00017
Luyten's star	30	12.4	M3.5	0.0014
Kapteyn's star	75	12.8	M0	0.0037
AX Microscopii	550	12.9	M1/M2V	0.027
Kruger 60A	38	13.1	M2V	0.0020
Kruger 60B	7.9	13.1	M6V	0.0004
V577 Monocerotis A	9.3	13.4	M4.5V	0.0005
V577 Monocerotis B	0.38	13.4	—	0.00002
CD 25 10553A (dbl)	5.4	13.9	M1	0.0003
Gliese 153-058	24	13.9	M3.5	0.0014
FL Virginis A	1.6	14.2	M5	0.0001
FL Virginis B	1.1	14.2	M7	0.00007
Gliese 267-025	98	14.2	M1.5	0.0060
HIP 15689 (dbl)	3.6	14.4	—	0.0002
Van Maanen 2	2.9	14.4	wdDG	0.0002
TZ Arietis	3.3	14.6	M4.5	0.0002
LHS 288	0.74	14.7	M	0.00005
CD 25 10553B	3.9	14.7	M1.5	0.0003
LP 731-58	0.15	14.8	M6.5	0.00001
Gliese 240-063 (dbl)	57	14.8	M3	0.0037
CD 46 11540	46	14.8	M2.5	0.0030

[a]Stars may carry many names, including common names (such as Sirius), names based on their prominence within a constellation (such as Alpha Canis Majoris, another name for Sirius), or names based on their inclusion in a catalog (such as BD + 36 – 2147). Addition of letters A, B, and so on, or superscripts indicates membership in a multiple-star system.

[b]Brightness and luminosity in these tables refer only to radiation in "visual" light.

[c]Spectral types such as M3 are discussed in Chapter 13. Other letters or numbers provide additional information. For example, V after the spectral type indicates a main sequence star, whereas III indicates a giant star.

[d](dbl) means an unresolved double star.

The 52 Brightest Stars in the Sky

Name	Common Name	Visual Brightness (Sirius = 1,000,000)	Distance (ly)	Spectral Type	Visual Luminosity (Sun = 1.000)
Sun	Sun	1.29×10^{16}	1.58×10^{-5}	G2V	1.000
Alpha Canis Majoris	Sirius	1,000,000	8.60	A1V	23
Alpha Carinae	Canopus	506,000	310	F0II	15,000
Alpha Bootis	Arcturus	270,000	36.7	K1.5IIIFe-0.5	110
Alpha^1 Centauri	Rigel Kentaurus	263,000	4.40	G2V	1.6
Alpha Lyrae	Vega	254,000	25.3	A0Va	50
Alpha Aurigae	Capella	242,000	42	G5IIIe+G0III	130
Beta Orionis	Rigel	233,000	770	B8Ia:	43,000
Alpha^1 Canis Minoris	Procyon A	184,000	11.4	F5IV-V	7.4
Alpha Eridani	Achernar	171,000	144	B3Vpe	1,100
Alpha Orionis	Betelgeuse	164,000	430	M1-2Ia-Iab	9,300
Beta Centauri	Hadar	149,000	530	B1III	13,000
Alpha Aquilae	Altair	128,000	16.8	A7V	11
Alpha Tauri	Aldebaran	119,000	65	K5+III	160
Alpha Scorpii	Antares	108,000	600	M1.5Iab-Ib+B4Ve	12,000
Alpha Virginis	Spica	106,000	260	B1III-IV+B2V	2,200
Beta Geminorum	Pollux	91,200	33.7	K0IIIb	32
Alpha Piscis Austrinus	Fomalhaut	89,500	25.1	A3V	17
Beta Crucis	Becrux	82,400	350	B0.5III	3,200
Alpha Cygni	Deneb	82,400	3,000	A2Ia	270,000
Alpha^1 Crucis	Acrux A	76,600	320	B0.5IV	2,400
Alpha^2 Centauri	Alpha Centauri B	76,600	4.40	K1V	0.5
Alpha Leonis	Regulus	75,200	77	B7V	140
Epsilon Canis Majoris	Adhara	65,500	430	B2II	3,800
Gamma Crucis	Gacrux	58,100	88	M3.5III	140
Lambda Scopii	Shaula	58,100	700	B2IV+B	8,900
Gamma Orionis	Bellatrix	57,500	240	B2III	1,100
Beta Tauri	El Nath	57,000	131	B7III	300
Beta Carinae	Miaplacidus	55,500	111	A2IV	210
Epsilon Orionis	Alnilam	54,500	1,300	B0Ia	30,000
Alpha^2 Crucis	Acrux B	53,000	800	B1V	11,000

(continued)

The 52 Brightest Stars in the Sky

(continued) Name	Common Name	Visual Brightness (Sirius = 1,000,000)	Distance (ly)	Spectral Type	Visual Luminosity (Sun = 1.000)
Alpha Gruis	Al Na'ir	52,500	101	B7IV	170
Epsilon Ursae Majoris	Alioth	51,000	81	A0pCr	100
Gamma2 Velorum	Suhail	50,600	840	WC8+O9I	11,000
Alpha Persei	Mirfak	50,100	590	F5Ib	5,400
Alpha Ursae Majoris	Dubhe	50,100	124	K0IIIa	240
Delta Canis Majoris	Wezen	47,900	1,800	F8Ia	47,500
Epsilon Sagittarii	Kaus Australis	47,400	145	B9.5III	300
Epsilon Carinae	Avior	47,000	630	K3III+B2:V	5,800
Eta Ursae Majoris	Benetnasch	47,000	101	B3V	150
Theta Scorpii	Sargas	46,600	270	F1II	1,100
Beta Aurigae	Menkalinan	45,300	82	A2IV	94
Alpha Trianguli Australis	Atria	44,500	415	K2IIb-IIIa	2,400
Gamma Geminorum	Alhena	44,100	105	A0IV	150
Alpha Pavonis	Peacock	43,700	183	B2IV	450
Delta Velorum	—	42,900	80	A1V	84
Beta Canis Majoris	Murzim	42,100	500	B1II-III	3,200
Alpha Geminorum	Castor	42,100	52	A1V	35
Alpha Hydrae	Alphard	42,100	177	K3II-III	410
Alpha Arietis	Hamal	41,300	66	K2-IIICa-1	55
Alpha Ursae Minoris	Polaris (Pole Star)	40,600	430	F7:Ib-II	2,300
Sigma Sagittarii	Nunki	40,600	224	B2.5V	630

Sources for Appendix 5: *The Hipparcos and Tycho Catalogues*, 1997, European Space Agency SP-1200; *Bright Star Catalogue*, 5th rev. ed. (Hoffleit, 1991), NSSDC Astronomical Data Center; *Burnham's Celestial Handbook* (Dover, 1983).

Observing the Sky

The purpose of this appendix is to provide enough information so you can make sense of a star chart or list of astronomical objects, as well as find a few objects in the sky.

Celestial Coordinates

In Chapter 2 we discuss the celestial sphere—the imaginary sphere with Earth at its center upon which celestial objects appear to lie. A number of different coordinate systems are used to specify the positions of objects on the celestial sphere. The simplest of these is the *altitude–azimuth coordinate system*. The altitude–azimuth coordinate system is based on the "map" direction to an object (the object's azimuth, with north = 0°, east = 90°, south = 180°, and west = 270°) combined with how high the object is above the horizon (the object's altitude, with the horizon at 0° and the zenith at 90°). For example, an object that is 10° above the eastern horizon has an altitude of 10° and an azimuth of 90°. An object that is 45° above the horizon in the southwest is at altitude 45°, azimuth 225°.

The altitude–azimuth coordinate system is the simplest way to tell a friend where in the sky to look at the moment, but it is not a good coordinate system for cataloging the positions of objects. The altitude and azimuth of an object are different for each observer, depending on their position on Earth, and are constantly changing as Earth rotates on its axis. If we need to specify the direction to an object in a way that is the same for everyone, we need a coordinate system that is fixed relative to the celestial sphere. The most common such coordinates are called *celestial coordinates*.

Celestial coordinates are illustrated in **Figure A6.1.** Celestial coordinates are much like the traditional system of latitude and longitude used on the surface of Earth. On Earth, latitude specifies how far you are from Earth's equator, as discussed in Chapter 2. If you are on Earth's equator, your latitude is 0°. If you are at Earth's North Pole, your latitude is 90° north. If you are at Earth's South Pole, your latitude is 90° south.

The latitudelike coordinate on the celestial sphere is called *declination,* often signified with the Greek letter δ (delta). The celestial equator has $\delta = 0°$. The north celestial pole has $\delta = +90°$. The south celestial pole has $\delta = -90°$. (See Chapter 2 if you need to refresh your memory of the celestial equator or celestial poles.) Declination is usually expressed in degrees, minutes of arc, and seconds of arc. For example, Sirius, the brightest star in the sky, has $\delta = -16° \, 42' \, 58''$, meaning that it is located not quite 17° south of the celestial equator.

On Earth, east–west position is specified by longitude. Lines of constant longitude run north–south from one pole to the other. Unlike latitude, for which the equator provides a natural place to call "zero," there is no natural starting point for longitude, so we just have to invent one. By arbitrary convention, the Royal Greenwich Observatory in Greenwich, England, is defined to lie at a longitude of 0°. On the celestial sphere the longitudelike coordinate is called *right ascension,* often signified with the Greek letter α (alpha). Unlike longitude, there *is* a natural point on the celestial sphere to use as the starting point for right ascension—the vernal equinox, or the point at which the ecliptic crosses the celestial equator with the Sun moving from the southern sky into the northern sky. The vernal equinox defines the line of right ascension at which $\alpha = 0°$. The autumnal equinox, located on the opposite side of the sky, is at $\alpha = 180°$.

Normally right ascension is measured in units of time rather than in degrees. It takes Earth 24 hours (of sidereal time) to rotate on its axis, so the celestial sphere is broken into 24 hours of right ascension, with each hour of right ascension corresponding to 15°. Hours of right ascension are then subdivided into minutes and seconds of time. Right ascension increases going to the east. The right ascension of Sirius, for example, is $\alpha = 06^h \, 45^m \, 08.9^s$, meaning that Sirius is about 101° (that is, $06^h \, 45^m$) east of the vernal equinox. Time is a natural unit for measuring right ascension because time naturally tracks the motion of objects due to Earth's rotation on its axis. If stars on the meridian at a certain time have 6^h, then an hour later the stars

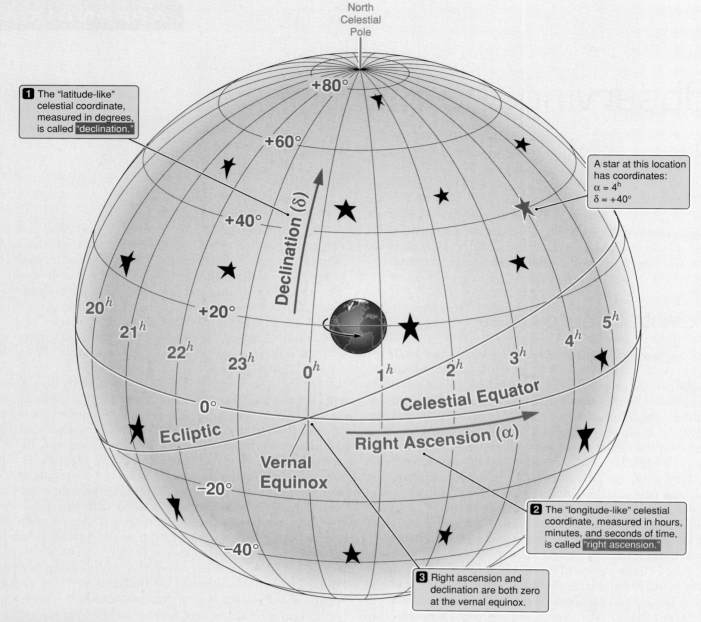

North
Celestial
Pole

1 The "latitude-like" celestial coordinate, measured in degrees, is called "declination."

+80°

+60°

A star at this location has coordinates:
$\alpha = 4^h$
$\delta = +40°$

+40°

Declination (δ)

+20°

20^h

21^h

22^h

23^h

0^h

1^h

2^h

3^h

4^h

5^h

0°

Celestial Equator

Ecliptic

Right Ascension (α)

Vernal
Equinox

−20°

2 The "longitude-like" celestial coordinate, measured in hours, minutes, and seconds of time, is called "right ascension."

−40°

3 Right ascension and declination are both zero at the vernal equinox.

FIGURE A6.1

on the meridian will have 7^h, and an hour after that they will have $\alpha = 8^h$. The *local sidereal time*, or "star time," at your location right now is equal to the right ascension of the stars that are on your meridian at the moment. Because of Earth's motion around the Sun, a sidereal day is about 4 minutes shorter than a solar day: so local sidereal time constantly gains on solar time. At midnight on September 23 the local sidereal time is 0^h. By midnight on December 22 local sidereal time has advanced to 6^h. On March 21

local sidereal time at midnight is 12^h. Local sidereal time at midnight on June 21 is 18^h.

Putting this all together, right ascension and declination provide a convenient way to specify the location of any object on the celestial sphere. Sirius is located at $\alpha = 06^h$ 45^m 08.9^s, $= -16° \, 42' \, 58''$, which means that at midnight on December 22 (local sidereal time = 6^h) you will find Sirius about 45^m east of the meridian, not quite 17° south of the celestial equator.

There is just one final caveat. As we discussed in Chapter 2, the directions of the celestial equator, celestial poles, and vernal equinox are constantly changing as Earth's axis wobbles like the axis of a spinning top. In Chapter 2 we called this 26,000-year wobble the "precession of the equinoxes," meaning that the location of the equinoxes is slowly advancing along the ecliptic. So when we specify the celestial coordinates of an object, we need to specify the date at which the positions of the vernal equinox and celestial poles were measured. By convention, coordinates are usually referred to with the position of the vernal equinox on January 1, 2000. A complete, formal specification of the coordinates of Sirius would then be $\alpha(2000)06^h\ 45^m\ 08.9^s$, $\delta(2000)-16°\ 42'\ 58''$, where the "2000" in parentheses refers to the equinox of the coordinates.

Constellations and Names

Although it is certainly possible to specify exactly any location on the surface of Earth by giving its latitude and longitude, it is usually convenient to use a more descriptive address. We might say, for example, that one of the coauthors of this book works near latitude 37° north, longitude 122° west: but it would probably mean a lot more to you were we to say that George Blumenthal works in Santa Cruz, California.

Just as the surface of Earth is divided into nations and states, the celestial sphere is divided into 88 *constellations,* the names of which are often used to refer to objects within their boundaries (see the star charts in **Figure A6.2,** starting on the next page). The brightest stars within the boundaries of a constellation are referred to using a Greek letter combined with the name of the constellation. For example, the star Sirius is the brightest star in the constellation Canis Major (the Great Dog) and so is referred to as α *Canis Majoris.* The bright red star in the northeastern corner of the constellation of Orion is referred to as α *Orionis,* also known as Betelgeuse. Rigel, the bright blue star in the southwest corner of Orion, is also called β *Orionis.*

Astronomical objects can take on a bewildering range of names. For example, the bright southern star Canopus, also known as α *Carinae* (the brightest star in the constellation of Carina) has no fewer than 34 different names, most of which are about as memorable as "SAO 234480" (number 234,480 in the Smithsonian Astrophysical Observatory catalog of stars).

You may have noticed a slight diffrence in the way a constellation is spelled when it becomes part of a star's name. Thus we see that Sirius is called α *Canis Majoris,* not α *Canis Major*; Rigel is referred to as α *Orionis,* not α *Orion*; and Canopus becomes α *Carinae,* not α *Carina.* This is because we use the Latin genitive or possessive case with star names where, for example, *Orionis* means "of *Orion."*

Astronomical Magnitudes

Throughout the text we refer to the brightness of objects: but when discussing the appearance of an object in the sky, astronomers normally speak instead of the object's *magnitude.* The system of astronomical magnitudes dates back to the Greek astronomer Hipparchus, who when classifying stars ordered them according to their "rank." First–rank stars were the brightest stars in the sky, second–rank stars were the next brightest, and so on. The faintest stars that could be seen were called sixth–rank stars. When methods were devised to actually measure the brightness of stars, it was discovered that first–rank stars were typically about 100 times as bright as sixth–rank stars and that the steps from one rank to the next were *logarithmic.* That is a first–rank star was typically about 2.51 *times* as bright as a second–rank star, whereas a second–rank star was typically about 2.51 *times* as bright as a third–rank star.

This rough classification of stellar "rank" was formalized into the modern system of astronomical magnitudes. A difference of five magnitudes between the brightness of two stars (say a star with $m = 6$ and a star with $m = 1$), corresponds to a hundredfold difference in brightness. Notice that the magnitude scale is backward—the *greater the magnitude, the fainter the object.*

If five steps in magnitude correspond to a factor of 100 in brightness, then one step in magnitude must correspond to a factor of $100^{1/5} = 10^{2/5} = 2.512\ldots$ in brightness ($100^{1/5} \times 100^{1/5} \times 100^{1/5} \times 100^{1/5} \times 100^{1/5} = 100$). The relationship between brightness and magnitude is most easily written using common or base-10 logarithms. If star 1 has a brightness of b_1 and star 2 has a brightness of b_2, then the difference in magnitude $m_2 - m_1$ between the two stars is

$$m_2 - m_1 = -2.5\ log_{10} \frac{b_2}{b_1}.$$

To convert from magnitude differences to brightness ratios, divide by −2.5 (that is, multiply by −0.4) and raise 10 to the resulting power:

$$\frac{b_2}{b_1} = 10^{-0.4 \times (m_2 - m_1)}$$

The last thing we need to set the magnitude scale is an object that we define to have a magnitude of zero. Many different magnitude scales exist, but perhaps the most common scale for visual magnitudes defines the star Vega as having a magnitude of 0.

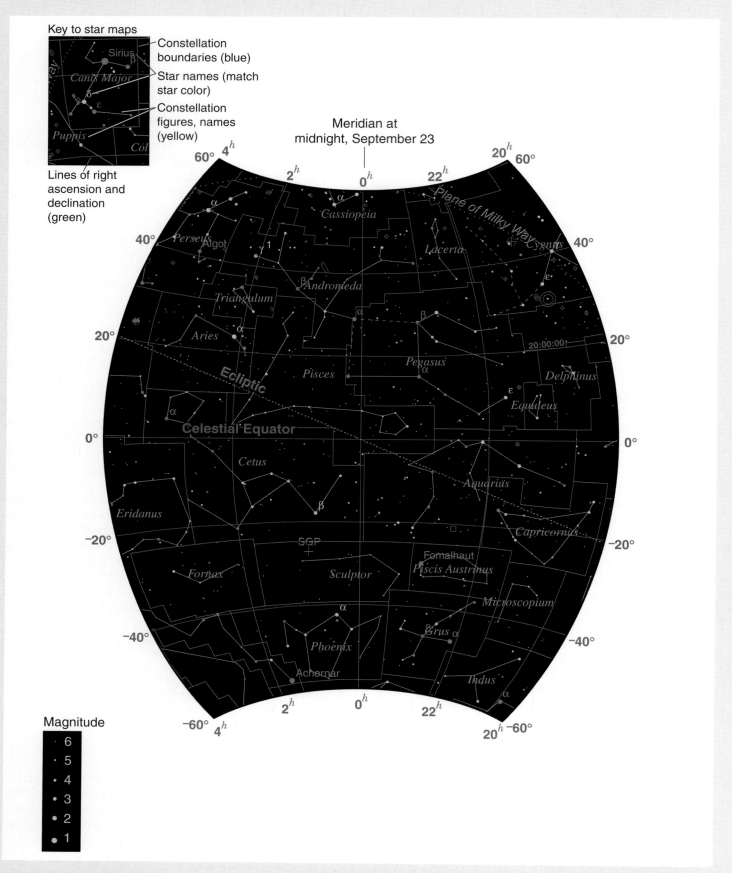

Key to star maps

Constellation boundaries (blue)

Star names (match star color)

Constellation figures, names (yellow)

Lines of right ascension and declination (green)

Meridian at midnight, September 23

Magnitude
6
5
4
3
2
1

FIGURE A6.2(A) The sky from right ascension 20ʰ to 4ʰ and declination −60° to +60°.

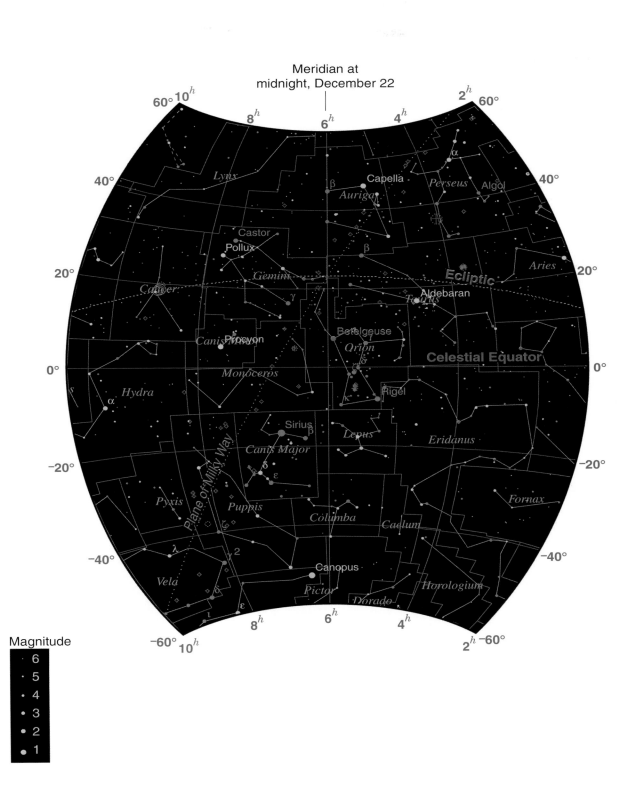

FIGURE A6.2(B) The sky from right ascension 2ʰ to 10ʰ and declination –60° to +60°.

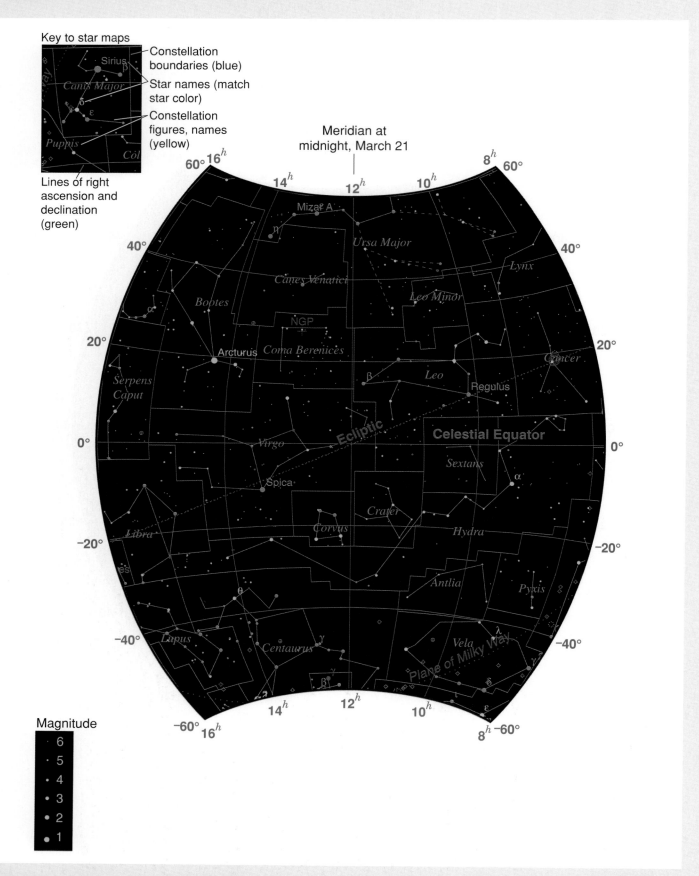

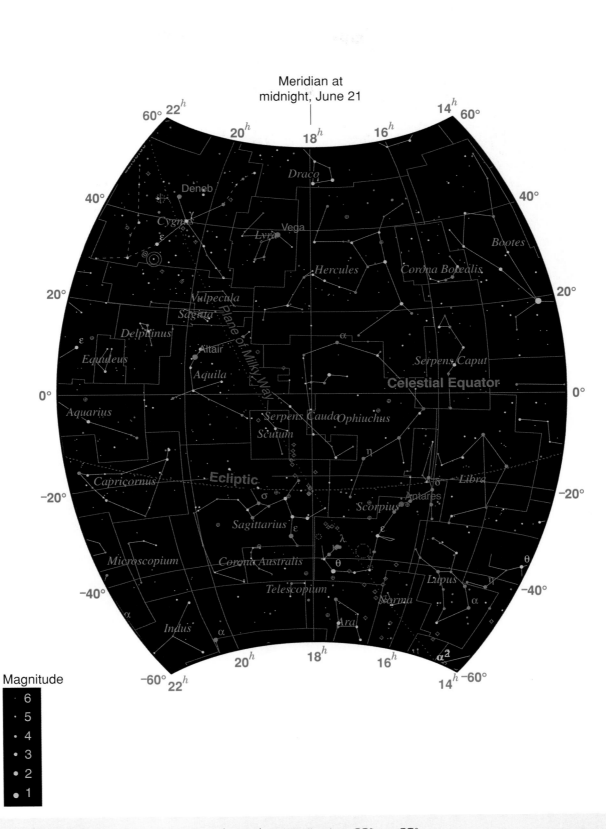

FIGURE A6.2(D) The sky from right ascension 14ʰ to 22ʰ and declination −60° to +60°.

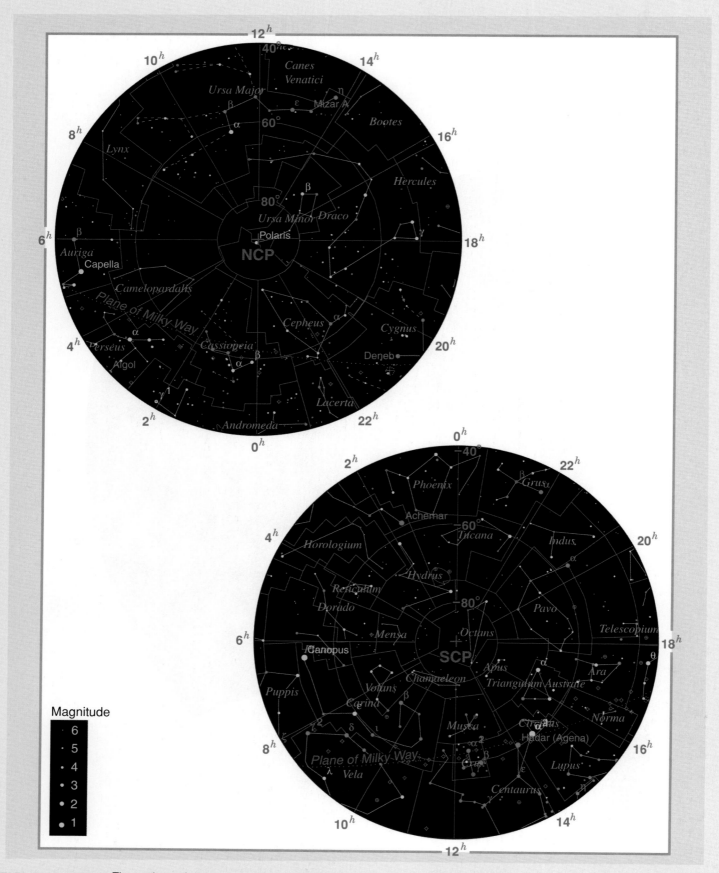

FIGURE A6.2(E) The regions of the sky north of declination +40° and south of declination −40°.

A final note: In Chapters 13 and onward, we used "colors" based on the ratio of the brightness of a star as seen in two different parts of the spectrum. The "b_B/b_V color," for example, was just the ratio of the brightness of a star seen through a blue filter, divided by the brightness of a star seen through a red filter. Normally astronomers instead discuss the "B-V color" of a star, which is equal to the difference between a star's blue magnitude and its visual magnitude. We can use the expression previous for a difference in magnitude to write

$$\text{B-V color} = m_B - m_V = -2.5 \, log_{10} \frac{b_B}{b_V}.$$

So a star with a b_B/b_V color of 1 has a B-V color of 0. A star with a b_B/b_V color of 1.4 has a B-V color of −0.37. Notice that, as with magnitudes, B-V colors are "backward." The bluer a star the greater its b_B/b_V color, but the less its B-V color.

Although the system of astronomical magnitudes and colors is actually convenient in many ways—which is why astronomers continue to use it—it can also be confusing to new students, especially students for whom math is a foreign language. Having seen something of how astronomical magnitudes work, we think you may agree with our choice to discuss the brightness of stars rather than their magnitudes throughout the text! However, if you do need to use the magnitude system (for example, to read star charts, make sense of a popular astronomy article, or complete a lab assignment), just remember three things and you will probably get by:

1. The greater the magnitude, the *fainter* the object.

2. A magnitude less means about two and a half times brighter.

3. The brightest stars in the sky have magnitudes of less than 1, while the faintest stars that can be seen with the naked eye on a dark night have magnitudes of about 6.

Uniform Circular Motion and Circular Orbits

In Chapter 3 (see Section 3.5 and Figure 3.15) we discuss the motion of an object moving in a circle at a constant speed. This motion, called uniform circular motion, is the result of centripetal force always acting toward the center of the circle. The key question when thinking about uniform circular motion is "how hard do I have to pull to keep the object moving in a circle?" Part of the answer to this question is pretty obvious—the more massive an object is, the harder it will be to keep it moving on its circular path. According to Newton's Second Law, $F = ma$, or in this case, the centripetal force equals the mass times the centripetal acceleration. The larger the mass, the greater the force required to keep it moving in its circle.

The centripetal force needed to keep an object moving in constant circular motion also depends on two other quantities: the speed of the object and the size of the circle. The faster an object is moving, the more rapidly it has to change direction to stay on a circle of a given size. The second quantity that influences the needed acceleration is the radius of the circle. The smaller the circle, the greater the pull needed to keep it on track. You can understand this by looking at the motion. A small circle requires a continuous "hard" turn, whereas a larger circle requires a more gentle change in direction. It takes more force to keep an object moving faster in a smaller circle than it does to keep the same object moving more slowly in a larger circle. (To get a better feeling for how this works, think about the difference between riding in a car that is taking a tight curve at high speed and a car that is moving slowly around a gentle curve.)

To arrive at the circular velocity and other results discussed in Chapter 3, we need to turn these intuitive ideas about uniform circular motion into a quantitative expression for exactly how much centripetal acceleration is needed to keep an object moving in a circle with radius r at speed v. **Figure A7.1** shows a ball moving around a circle of radius r at a constant speed v at two different times. The centripetal acceleration that is keeping the ball on the circle is a. Remember that the acceleration is always directed toward the center of the circle, whereas the velocity of the ball is always perpendicular to the acceleration. The ball's velocity and its acceleration are always at right angles to each other. As the object moves around the circle, the direction of motion and the direction of the acceleration change together in lockstep.

In the figure we have drawn two triangles. In triangle 1 we show the velocity (speed and direction) at each of the two times. The arrow labeled Δv connecting the heads of the two velocity arrows shows how much the velocity changed between time 1 and time 2. This change is the effect of the centripetal acceleration. If we imagine that points 1 and 2 are very close together—so close that the direction of the centripetal acceleration does not change by much between the two—then we can say that the centripetal acceleration equals the change in the velocity divided by the time between the two, $\Delta t = t_2 - t_1$. So we have $\Delta v = a\Delta t$.

In triangle 2 we do a similar thing. Here the arrow labeled Δr indicates the change in the position of the ball between time 1 and time 2. Again, if we imagine that the time between the two points is very short, we can say that Δr is equal to the velocity times the time, or $\Delta r = v\Delta tv$.

The line between the center of the circle and the ball is always perpendicular to the velocity of the ball. So if the direction of the ball's velocity changes by an angle α, then the direction of the line between the ball and the center of the circle must also change by the same angle α. In other words, triangles 1 and 2 are *similar triangles*. They have the same *shape*. If the triangles are the same shape, the ratio of two sides of triangle 1 must equal the ratio of the two corresponding sides of triangle 2. Using this fact we can write

$$\frac{a\Delta t}{v} = \frac{v\Delta t}{r}.$$

If we divide out the Δt from both sides of the equation, then cross-multiply, we obtain

$$ar = v^2,$$

which after dividing both sides of the equation by r becomes

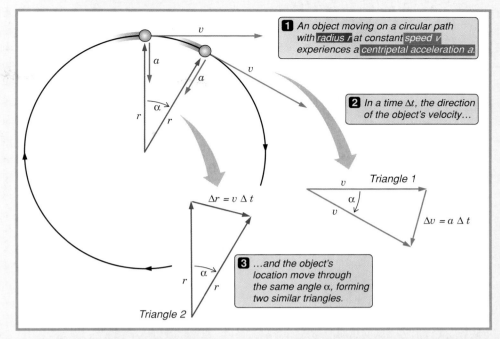

FIGURE A7.1 *Similar triangles used to find the centripetal force needed to keep an object moving at a constant speed on a circular path.*

$$a_{\text{centripetal}} = \frac{v^2}{r}.$$

We have added the subscript "centripetal" to a to signify that this is the centripetal acceleration needed to keep the object moving in a circle of radius r at speed v. The centripetal force required to keep an object of mass m moving on such a circle is then

$$F_{\text{centripetal}} = ma_{\text{centripetal}} = \frac{mv^2}{r}.$$

Circular Orbits

In the case of an object moving in a circular orbit there is no string to hold the ball on its circular path. Instead this force is provided by gravity.

Think about an object with mass m in orbit about a much larger object with mass M. The orbit is circular, and the distance between the two objects is given by r. The force needed to keep the smaller object moving at speed v in a circle with radius r is given by the previous expression for $F_{\text{centripetal}}$. The force actually provided by gravity (see Chapter 3) is

$$F_{\text{gravity}} = G\frac{Mm}{r^2}.$$

If gravity is responsible for holding the mass in its circular motion, then it should be true that $F_{\text{gravity}} = F_{\text{centripetal}}$. That is, if mass m is moving in a circle under the force of gravity, the force provided by gravity *must* equal the centripetal force needed to explain that circular motion. Setting our two expressions for $F_{\text{centripetal}}$ and F_{gravity} equal to each other gives

$$\frac{mv^2}{r} = G\frac{Mm}{r^2}.$$

All that remains is a bit of algebra. Dividing out the m on both sides of the equation and multiplying both sides by r gives

$$v^2 = G\frac{M}{r}.$$

Taking the square root of both sides then brings us to the desired result:

$$v_{\text{circular}} = \sqrt{\frac{GM}{r}}$$

This is the "circular velocity" we presented in Chapter 3. It is the velocity at which an object in a circular orbit *must* be moving. If the object were not moving at this velocity, then gravity would not be providing the needed centripetal force, and the object would not move in a circle.

IAU 2006 Resolutions: Definition of a Planet in the Solar System, and Pluto

24 August 2006, Prague
SOURCE: International Astronomical Union

Resolutions

Resolution 5 is the principal definition for the IAU usage of "planet" and related terms.

Resolution 6 creates for IAU usage a new class of objects, for which Pluto is the prototype. The IAU will set up a process to name these objects.

Resolution 5

Definition of a Planet in the Solar System

Contemporary observations are changing our understanding of planetary systems, and it is important that our nomenclature for objects reflect our current understanding. This applies, in particular, to the designation "planets." The word "planet" originally described "wanderers" that were known only as moving lights in the sky. Recent discoveries lead us to create a new definition, which we can make using currently available scientific information.

The IAU therefore resolves that planets and other bodies, except satellites, in our Solar System be defined into three distinct categories in the following way:

(1) A "planet"[1] is a celestial body that (a) is in orbit around the Sun, (b) has sufficient mass for its self-gravity to overcome rigid body forces so that it assumes a hydrostatic equilibrium (nearly round) shape, and (c) has cleared the neighbourhood around its orbit.

(2) A "dwarf planet" is a celestial body that (a) is in orbit around the Sun, (b) has sufficient mass for its self-gravity to overcome rigid body forces so that it assumes a hydrostatic equilibrium (nearly round) shape[2], (c) has not cleared the neighbourhood around its orbit, and (d) is not a satellite.

(3) All other objects,[3] except satellites, orbiting the Sun shall be referred to collectively as "Small Solar-System Bodies."

Resolution 6

Pluto

The IAU further resolves:

Pluto is a "dwarf planet" by the above definition and is recognized as the prototype of a new category of Trans-Neptunian Objects.[4]

[1] The eight planets are: Mercury, Venus, Earth, Mars, Jupiter, Saturn, Uranus, and Neptune.

[2] An IAU process will be established to assign borderline objects into either dwarf planet and other categories.

[3] These currently include most of the Solar System asteroids, most Trans-Neptunian Objects (TNOs), comets, and other small bodies.

[4] An IAU process will be established to select a name for this category.

Glossary

A

aberration of starlight The apparent displacement in the position of a star due to the finite speed of light and Earth's orbital motion around the Sun.

absolute zero The temperature at which thermal motions cease. The lowest possible temperature. Zero on the kelvin temperature scale.

absorption The capture of electromagnetic radiation by matter.

absorption line An intensity minimum in a spectrum due to absorption of electromagnetic radiation at a specific wavelength determined by the energy levels of an atom or molecule.

acceleration The rate at which the speed and/or direction of an object's motion is changing.

accretion The process by which smaller bodies coalesce to form larger ones.

accretion disk A flat, rotating disk of gas and dust surrounding a condensed mass, such as a young stellar object, a forming planet, or a collapsed star in a binary pair.

achondrite A stony meteorite that does not contain chondrules.

active comet A comet nucleus that approaches close enough to the Sun to show signs of activity, such as the production of a coma and tail.

active galactic nucleus (AGN) A highly luminous, compact galactic nucleus whose luminosity may exceed that of the rest of the galaxy.

adaptive optics Electro-optical systems that largely compensate for image distortion caused by Earth's atmosphere.

AGB star A star on the asymptotic giant branch.

AGN See *active galactic nucleus.*

albedo The fraction of electromagnetic radiation incident on a surface that is reflected by the surface.

alpha particle A He^4 nucleus consisting of two protons and two neutrons. Alpha particles are given off in the type of radioactive decay referred to as *alpha decay*—hence the name.

Alpher, Ralph (1921–) The physicist who worked with George Gamow to predict the existence of the cosmic background radiation.

Amors A Family of asteroids with orbits that cross the orbit of Mars but not that of Earth.

amplitude In a wave, the maximum excursion from equilibrium. For example, in a water wave the amplitude is the vertical distance from crests to the undisturbed water level.

angular momentum A conserved property of a rotating or revolving system whose value depends on the velocity and distribution of its mass.

annular solar eclipse A solar eclipse that occurs when the apparent diameter of the Moon is less than that of the Sun, leaving a visible ring of light (annulus) surrounding the dark disk of the Moon.

Antarctic Circle The circle on Earth with latitude 66.5° south, marking the limit of the Antarctic region.

antimatter Matter made from antiparticles.

antiparticle An elementary particle of antimatter identical in mass but opposite in charge and all other properties to its corresponding ordinary matter particle.

aperture The clear diameter of a telescope's objective lens or primary mirror.

aphelion (pl. **aphelia**) The point in a solar orbit that is farthest from the Sun.

Apollos A group of asteroids whose orbits cross the orbits of both Earth and Mars.

arcminute See *minute of arc.*

arcsecond See *second of arc.*

Arctic Circle The circle on Earth at latitude 66.5° north, marking the limit of the Arctic region.

Aristotle (384–322 B.C.) The ancient Greek philosopher whose observations of the natural world later became the basis of modern science.

asteroid A primitive rocky or metallic body (planetesimal) that has survived planetary accretion. Asteroids are parent bodies of meteoroids. Asteroids are also known as minor planets.

asteroid belt The region between the orbits of Mars and Jupiter that contains most of the asteroids in our Solar System.

astrobiology An interdisciplinary science combining astronomy, biology, and geology to study life in the cosmos.

astrology The belief that the positions and aspects of stars and planets influence human affairs and characteristics as well as terrestrial events.

astronomical seeing A measurement of the degree to which Earth's atmosphere degrades the resolution of a telescope's view of astronomical objects.

astronomical unit (AU) The average distance from the Sun to Earth: approximately 150 million kilometers.

astronomy The scientific study of planets, stars, galaxies, and the universe as a whole.

astrophysics The application of physical laws to the understanding of planets, stars, galaxies, and the universe as a whole.

asymptotic giant branch (AGB) In the H-R diagram, a separate "branch" that goes from the horizontal branch toward higher luminosities and lower temperatures, asymptotically approaching and then rising above the red giant branch.

Atens A group of asteroids whose orbits cross Earth's orbit, but not the orbit of Mars.

atmosphere The gravitationally bound, outer gaseous envelope surrounding a planet, moon, or star.

atmospheric greenhouse effect A warming of planetary surfaces produced by atmospheric gases that transmit optical solar radiation but partially trap infrared radiation. Compare with *greenhouse effect.*

atmospheric windows   A regions of the electromagnetic spectrum in which radiation is able to penetrate a planet's atmosphere.

atom   The smallest piece of an element that retains the properties of that element. Each atom is composed of a nucleus (neutrons and protons) surrounded by a cloud of electrons.

aurora   Emission in the upper atmosphere of a planet from atoms that have been excited by collisions with energetic particles from the planet's magnetosphere.

autumnal equinox   (a) The point where the ecliptic crosses the celestial equator, with the Sun moving from north to south. (b) The day, around September 23, on which the Sun appears at this location, which is the first day of autumn (fall) in the Northern Hemisphere.

axion   A hypothetical elementary particle first proposed to explain certain properties of the neutron and now considered a candidate for *cold dark matter.*

B

backlighting   Illumination from behind a subject as seen by an observer. Fine material such as human hair and dust in planetary rings stands out best when viewed under backlighting conditions.

bar   A unit of pressure. One bar is equivalent to 10^5 newtons per square meter—approximately equal to Earth's atmospheric pressure at sea level.

barred spiral   A spiral galaxy with a bulge having an elongated, barlike shape.

basalt   Gray to black volcanic rock, rich in iron and magnesium.

b_B/b_V color   Measurement of the color of an object on the basis of the ratio of its brightness in blue light (b_B) to its brightness in "visual" (or yellow-green) light (b_V).

beta decay   The decay of a neutron into a proton by emission of an electron (beta ray) and an antineutrino, or the decay of a proton into a neutron by emission of a positron and a neutrino.

Bethe, Hans (1906–2005)   The physicist who won the Nobel Prize for his work on the source of stellar energy.

Big Bang   The event that occurred about 14 billion years ago that marks the beginning of time and the universe.

Big Bang nucleosynthesis   The formation of low-mass nuclei (H, He, Li, Be, B) during the first few minutes following the Big Bang.

binary star   A system in which two stars are in gravitationally bound orbits about their common center of mass.

binding energy   The minimum energy required to separate an atomic nucleus into its component protons and neutrons.

bipolar outflow   Material streaming away in opposite directions from either side of the accretion disk of a young star.

blackbody   An object that absorbs and can reemit all electromagnetic energy it receives.

blackbody spectrum   See *Planck spectrum.*

black hole   An object so dense its escape velocity exceeds the speed of light: a *singularity* in spacetime.

blueshift   The Doppler shift toward shorter wavelengths of light from an approaching object. Compare with *redshift.*

Bohr model   A model of the atom, proposed by Neils Bohr in 1913, in which a small positively charged nucleus is surrounded by orbiting electrons, similar to a miniature solar system.

Bohr, Neils (1885–1962)   The Danish theoretical physicist best known for his discovery of the quantization of atomic states.

bolide   A very bright, exploding meteor.

Boltzmann's constant (k)   The constant that relates the temperature of a gas to the average kinetic energy of the molecules of the gas.

bound orbit   A closed orbit in which the velocity is less than the escape velocity.

bow shock   (a) The boundary at which the speed of the *solar wind* abruptly drops from supersonic to subsonic in its approach to a planet's *magnetosphere;* the boundary between the region dominated by the solar wind and the region dominated by a planet's magnetosphere. (b) The interface between strong collimated gas and dust outflow from a star and the interstellar medium.

Brahe, Tycho (1546–1601)   The foremost astronomical observer before the invention of the telescope. Tycho provided the data that permitted Kepler to deduce the motions of the planets around the Sun.

brightness   The apparent intensity of light from a luminous object. Brightness depends on both the luminosity of a source and its distance. Units at the detector: watts per square meter (W/m^2).

brown dwarf   A "failed" star that is not massive enough to cause hydrogen fusion in its core.

bulge   The central region of a spiral galaxy that is similar in appearance to a small elliptical galaxy.

C

Cannon, Annie Jump (1863–1941)   The American astronomer who systematically classified the spectral types of stars in the Henry Draper Catalog, the first catalog to list stellar spectral types for 250,000 stars.

carbonaceous chondrite   A primitive stony meteorite that contains chondrules and is rich in carbon and volatile materials.

carbon–nitrogen–oxygen cycle   See *CNO cycle.*

carbon star   A cool *red giant* or AGB star that has an excess of carbon in its atmosphere.

Cassini Division   The largest gap in Saturn's rings, discovered by Jean-Dominique Cassini in 1675.

Cassini, Jean-Dominique (1625–1712)   The Italian astronomer (nationalized as French in 1673) who discovered four satellites of Saturn as well as the division within Saturn's ring that bears his name.

catalyst   An atomic and molecular structure that permits or encourages chemical and nuclear reactions but does not change its own chemical or nuclear properties.

CBR   See *cosmic background radiation.*

CCD   See *charge-coupled device.*

celestial equator   The imaginary great circle that is the projection of Earth's equator onto the celestial sphere.

celestial sphere   An imaginary sphere with celestial objects on its inner surface and Earth at its center. The celestial sphere has no physical existence but is a convenient tool for picturing the directions in which celestial objects are seen from the surface of Earth.

Celsius (C)   The arbitrary temperature scale with 0°C at the freezing point of water and 100°C at the boiling point of water at sea level. Defined by Anders Celsius (1701–1744). Also known as the Centigrade scale.

center of mass   The location associated with an object system at which we may regard the entire mass of the system as being concentrated. The point in any isolated system that moves according to Newton's first law of motion.

centripetal acceleration   The acceleration of an object directed toward the center of curvature of its motion.

centripetal force   A force directed toward the center of curvature of an object's curved path.

Cepheid variable   An evolved high-mass star with an atmosphere that is pulsating, leading to variability in the star's luminosity and color.

Chandrasekhar limit The upper limit on the mass of an object supported by electron degeneracy pressure—approximately 1.4 $M_\odot$.

chaos Behavior in complex, interrelated systems in which tiny differences in the initial configuration of a system result in dramatic differences in the system's later evolution.

charge-coupled device (CCD) A common type of solid-state detector of electromagnetic radiation that transforms the intensity of light directly into electric signals.

chemistry The study of the composition, structure, and properties of substances

chondrite A stony meteorite containing chondrules.

chondrule A spherical inclusion of rapidly cooled melt found inside some meteorites.

chromatic aberration A detrimental property of a lens in which rays of different wavelengths are brought to different focal distances from the lens.

chromosphere The region in the Sun's atmosphere located between the *photosphere* and the *corona.*

circular velocity The orbital velocity needed to keep an object moving in a circular orbit.

circumpolar That part of the sky, near either celestial pole, that can always be seen above the horizon from a specific location on Earth.

classical mechanics The science of applying Newton's laws to the motion of objects.

climate The state of an atmosphere averaged over an extended time.

closed universe A finite universe with a curved spatial structure such that the angles of a triangle always exceed 180°.

CNO cycle (carbon–nitrogen–oxygen cycle) One of the ways in which hydrogen burning (the fusion of four hydrogen atoms to form one helium atom) can take place. See also *triple-alpha process.*

cold dark matter Dark matter particles that move slowly enough to be gravitationally bound even in the smallest galaxies.

coma (pl. **comae**) The nearly spherical cloud of gas and dust surrounding the nucleus of an active comet.

comet A complex object consisting of a small solid, icy nucleus, an atmospheric halo, and a tail of gas and dust.

comet nucleus (pl. **nuclei**) A primitive planetesimal, composed of ices and *refractory materials* that has survived

planetary accretion. The "heart" of a comet, containing nearly the entire mass of the comet. A "dirty snowball."

comparative planetology The study of planets through comparison of their chemical and physical properties.

complex system An interrelated system capable of exhibiting chaotic behavior. See also *chaos.*

composite volcano A large, cone-shaped volcano.

compound lens A lens made up of two or more elements of differing refractive index, the purpose of which is to minimize *chromatic aberration.*

conservation law A physical law stating that the amount of some physical quantity (such as *energy or angular momentum*) of an isolated system does not change over time.

conservation of angular momentum The physical law stating that the amount of angular momentum of an isolated system does not change over time.

conservation of energy The conservation law stating that in an isolated, closed system, the total energy does not change.

constant of proportionality The multiplicative factor by which one quantity is related to another.

constellation An imaginary image formed by patterns of stars; any of 88 defined areas on the celestial sphere used by astronomers to locate celestial objects.

constructive interference A state in which the amplitudes of two intersecting waves reinforce one another. See also *destructive interference, interference.*

continental drift The slow motion (centimeters per year) of Earth's continents relative to each other and to Earth's mantle.

continuous radiation Electromagnetic radiation with intensity that varies smoothly over some range of wavelengths.

convection The transport of thermal energy from the lower (hotter) to the higher (cooler) layers of a fluid by motions within the fluid driven by variations in buoyancy.

convective zone A region within a star in which energy is transported outward by convection.

Copernicus, Nicolaus (1473–1543) The Polish monk who first proposed the model of a Sun-centered Solar System to the Western world.

core (a) The innermost region of a planetary interior. (b) The innermost part of a star.

core accretion A process for forming giant planets, whereby large quantities of surrounding hydrogen and helium are gravitationally captured onto a massive rocky core.

Coriolis effect The apparent displacement of objects in a direction perpendicular to their true motion as viewed from a rotating frame of reference. On a rotating planet, different latitudes rotating at different speeds cause this effect.

corona The hot, outermost part of the Sun's atmosphere.

coronal hole A lower-density region in the solar corona containing "open" magnetic field lines along which coronal material is free to stream into interplanetary space.

coronal mass ejection An eruption on the Sun that ejects hot gas and energetic particles at much higher speeds than are typical in the solar wind.

cosmic background radiation (CBR) Isotropic microwave radiation from the celestial sphere having a 2.73 K Planck spectrum. The CBR is understood as residual radiation from the Big Bang.

cosmic ray A very fast-moving particle (usually an atomic nucleus) that resides in the disk of our galaxy.

cosmological constant A constant introduced into general relativity by Einstein that characterizes an extra, repulsive force in the universe due to the vacuum of space itself.

cosmological principle The (testable) assumption that the same physical laws that apply here and now also apply everywhere and at all times, and that there are no special locations or directions in the universe.

cosmological redshift (z) The redshift that results from the expansion of the universe rather than from the motions of galaxies or gravity (see *gravitational redshift*).

cosmology The study of the large-scale structure and evolution of the universe as a whole.

Crab Nebula The remnant of the Type II supernova explosion witnessed by Chinese astronomers in A.D. 1054.

crescent Phases of the Moon, Mercury or Venus in which the object appears less than half illuminated by the Sun.

Cretaceous period The interval in Earth's history from 146 million to 65 million years ago.

Cretaceous–Tertiary (K-T) boundary The boundary between the Cretaceous and Tertiary periods in Earth's history;

corresponds to the time of the impact of an asteroid or comet and the extinction of the dinosaurs.

critical mass density The value of the mass density of the universe that, ignoring any cosmological constant, is just barely capable of halting expansion of the universe.

crust The relatively thin, outermost, hard layer of a planet, which is chemically distinct from the interior.

cryovolcanism Low temperature volcanism in which the *magmas* are composed of motter *ices* rather than rocky material. See also *volcanism*.

C-type asteroid An asteroid made of material that has largely been unmodified since the formation of the Solar System; the most primitive type of asteroid.

cumulonimbus A towering cumulus-type cloud associated with thunderstorms, also known as a *thunderhead*.

cumulus A large puffy cloud with a flat base.

Curtis, Heber D. (1872–1942) The Lick Observatory astronomer who argued, in the Great Debate of 1920, that galaxies are island universes. See also *Shapley, Harlow*.

cyclonic motion The rotation of a weather system resulting from the Coriolis effect as air moves toward a region of low atmospheric pressure.

Cygnus X-1 A binary X-ray source and probable black hole.

D

dark matter The matter existing in galaxies and in groups and clusters of galaxies that does not emit or absorb electromagnetic radiation; thought to comprise most of the mass in the universe.

dark matter halo The centrally condensed, greatly extended dark matter component of a galaxy that contains up to 95 percent of the galaxy's mass.

Darwin, Charles (1809–1882) The English naturalist whose publication, *The Origin of Species*, opened the way to the modern theory of evolution.

daughter element An element resulting from radioactive decay of a more massive parent element.

decay (a) The process of a radioactive nucleus changing into its daughter element. (b) An atom or molecule dropping from a higher-energy state to a lower-energy state.

density The measure of an object's mass per unit of volume. Units: kg/m^3.

destructive interference A state in which the amplitudes of two intersecting waves cancel one another. See also *constructive interference, interference*.

differential rotation Rotation of different parts of a system at different rates.

differentiation The process by which materials of higher density sink toward the center of a molten or fluid planetary interior.

diffraction The spreading of a wave after it passes through an opening or past the edge of an object.

diffraction limit The limit of a telescope's angular resolution caused by diffraction.

dispersion The separation of rays of light into their component wavelengths.

distance ladder A sequence of techniques for measuring cosmic distances; each method is calibrated using the results from other methods that have been applied to closer objects.

Doppler effect The change in wavelength of sound or light due to the relative motion of the source toward or away from the observer.

Doppler redshift See *redshift*.

Doppler shift The change in observed wavelength in a wave from a source moving toward or away from an observer.

Drake equation A prescription for estimating the number of intelligent civilizations existing elsewhere.

Drake, Frank (1930–) The American astronomer best known for his formulation of the *Drake equation*.

dust devil A small tornadolike column of air containing dust or sand.

dust tail A type of comet tail consisting of dust particles that are pushed away from the comet's head by radiation pressure from the Sun.

dwarf galaxy A small galaxy with a luminosity ranging from 1 million to 1 billion solar luminosities.

dwarf planet A body with characteristics similar to that of a classical planet except that it has not cleared smaller bodies from the neighboring regions around its orbit. Compare with *planet*.

dynamic equilibrium A situation in which the configuration of a system does not change even though the components of that system are in motion.

dynamo A device that converts mechanical energy into electrical energy in the form of electric currents and magnetic fields. The *dynamo effect* is thought to create magnetic fields in planets and stars by electrically charged currents of material flowing within their cores.

E

eccentricity (e) A measure of the departure of an ellipse from circularity; the ratio of the distance between the two foci of an ellipse to its major axis.

eclipse season Times during the year when the line of nodes is sufficiently close to the Sun for eclipses to occur.

eclipsing binary A binary system in which the orbital plane is oriented such that the two stars appear to pass in front of one another as seen from Earth.

ecliptic (a) The apparent annual path of the Sun against the background of stars. (b) The projection of Earth's orbital plane onto the celestial sphere.

effective temperature The temperature at which a black body, such as a star, appears to radiate.

Einstein, Albert (1879–1955) The eminent 20th century German physicist whose theories about relativity and the properties of light changed how we view the physical universe.

ejecta (a) Material thrown outward by the impact of an asteroid or comet on a planetary surface, leaving a crater behind. (b) Material thrown outward by a stellar explosion.

electric field A field that is able to exert a force on a charged object, whether at rest or moving.

electric force The force exerted on a charged particle by an electric field.

electromagnetic force The force, including both electric and magnetic forces, that acts on electrically charged particles. One of four fundamental forces of nature. The force mediated by photons.

electromagnetic radiation A traveling disturbance in the electric and magnetic fields caused by accelerating electric charges. In quantum mechanics, a stream of photons. Light.

electromagnetic spectrum The spectrum made up of all possible frequencies or wavelengths of electromagnetic radiation, ranging from gamma rays through radio waves and including the portion our eyes can use.

electromagnetic wave A wave consisting of oscillations in the electric field strength and the magnetic field strength.

electron (e^-) A fundamental particle having a negative charge of 1.6×10^{-19} coulomb, a rest mass of 9.1×10^{-31} kg, and mass-equivalent energy of 8×10^{-14} joules.

electron degenerate Describes the state of material compressed to the point where electron density reaches the limit imposed by the rules of quantum mechanics.

electroweak theory The quantum theory that combines descriptions of both the electromagnetic force and the weak nuclear force.

element One of 92 naturally occurring substances (such as hydrogen, oxygen, and uranium) and more than 20 human-made ones (such as plutonium). Each element is chemically defined by the specific number of protons in the nuclei of its atoms.

elementary particle One of the basic building blocks of nature, such as the *electron* and the *quark*.

ellipse A conic section produced by the intersection of a plane with a cone by passing the plane through the cone at some angle to the axis other than 0° or 90°.

elliptical galaxy A galaxy with a circular to elliptical outline on the sky containing almost no disk and a population of old stars (abbreviated E galaxy).

emission The release of electromagnetic energy when an atom, molecule, or particle drops from a higher-energy state to a lower-energy state.

emission line An intensity peak in a spectrum due to sharply defined emission of electromagnetic radiation in a narrow range of wavelengths.

empirical science Based primarily on observations and experimental data. Descriptive rather than based on theoretical inference.

energy The conserved quantity that gives objects and systems the ability to do work. Units: joules (J).

energy transport The transport of energy from one location to another. In stars, energy transport is mostly carried out by radiation or convection.

entropy A measure of the disorder of a system related to the number of ways a system can be rearranged without affecting its appearance.

equator The great circle on the surface of a body midway between its poles. The equatorial plane passes through the center of the body and is perpendicular to its rotation axis.

equilibrium The state of an object in which physical processes balance each other so that its properties or conditions remain constant.

equinox ("equal night") (a) One of two positions on the ecliptic where it intersects the celestial equator. (b) Either of the two times of year at which the Sun is at one of these two positions. At this time, night

and day are of the same length everywhere on Earth. See also *autumnal equinox, vernal equinox.*

equivalence principle The principle stating that there is no difference between a reference frame that is freely floating through space and one that is freely falling within a gravitational field.

escape velocity The minimum velocity needed for an object to achieve a parabolic trajectory and thus permanently leave the gravitational grasp of another mass.

Eta Carinae An extremely massive luminous star that underwent an episode of great mass loss during the 1840s.

event A particular location in *spacetime.*

event horizon The effective "surface" of a *black hole.* Nothing inside this surface—not even light—can escape from a black hole.

evolutionary track The path that a star follows across the H-R diagram as it evolves through its lifetime.

excited state An energy level of a particular atom, molecule, or particle, higher than its ground state.

extrasolar planet A planet orbiting a star other than the Sun.

F

Fahrenheit (F) The arbitrary temperature scale with 32°F at the melting point of water and 212°F at the boiling point of water at sea level. Defined by Gabriel Fahrenheit (1686–1736).

fault A fracture in the crust of a planet or moon along which blocks of material can slide.

filter An instrument element that transmits a limited wavelength range of electromagnetic radiation. For the optical range such elements are typically made of different kinds of glass and take on the hue of the light they transmit.

first quarter Moon The phase of the Moon as seen from Earth in which only the western half of the Moon is illuminated by the Sun; occurs about a week after a new Moon.

fissure A fracture in the planetary lithosphere from which magma emerges.

flatness problem The surprising result that the sum of Ω_{mass} plus Ω_Λ is extremely close to unity in the present-day universe; equivalent to saying that it is surprising the universe is so flat.

flat rotation curve A rotation curve of a spiral galaxy in which rotation rates do

not decline in the outer part of the galaxy, but remain relatively constant to the outermost points.

flat universe An infinite universe whose spatial structure obeys Euclidean geometry, such that the sum of the angles of a triangle always equals 180°.

flux The total amount of energy passing through each square meter of a surface each second. Units: watts per square meter (W/m²).

flux tube Tubelike structures in a plasma, such as the atmosphere of the Sun, or the interaction of Io and Jupiter, that are "bundled" together by magnetic fields.

flyby A spacecraft that first approaches and then continues on past a planet or moon. Flybys can visit multiple objects, but they remain in the vicinity of their targets only briefly. See also *orbiter.*

focal length The optical distance between a telescope's objective lens or primary mirror and the plane (called the focal plane) on which the light from a distant object is focused.

focal plane The plane, perpendicular to the optical axis of a lens or mirror, in which an image is formed.

focus (pl. **foci**) (a) One of two points that define an ellipse. (b) A point in the *focal plane* of a telescope.

fold A location where rock is bent or warped.

Foucault, Jean-Bernard-Leon (1819–1868) The French physicist who was the first to demonstrate that Earth rotates using the periodic motions of a pendulum. A pendulum that demonstrates Earth's rotation is called a *Foucault pendulum* in his honor.

Foucault pendulum A pendulum used to demonstrate the rotation of Earth.

frame of reference A frame against which an observer measures positions, motions, and the like.

free fall The motion of an object when the only force acting on it is gravity.

frequency The number of times per second that a periodic process occurs. Unit: hertz (Hz), 1/s.

full Moon The phase of the Moon as seen from Earth in which the near side of the Moon is fully illuminated by the Sun; occurs about two weeks after a new Moon.

G

galactic cluster See *open cluster.*

galaxy A gravitationally bound system that consists of stars and star clusters, gas, dust, and dark matter; typically greater than 1,000 light-years across; and recognizable as a discrete, single object.

galaxy cluster A large, gravitationally bound collection of galaxies containing hundreds to thousands of members; typically 10 to 15 Mly across.

galaxy group A small, gravitationally bound collection of galaxies containing from several to a hundred members; typically 4 to 6 Mly across.

Galileo Galilei (1564–1642) The first person to use a telescope to make useful discoveries about the heavens and to develop the concept of inertia.

gamma ray Electromagnetic radiation with higher frequency, higher photon energy, and shorter wavelength than all other types of electromagnetic radiation.

Gamow, George (1904–1968) The Ukrainian-American theoretical physicist who, with his student Ralph Alpher, first predicted the existence of the *cosmic background radiation.*

gas giant A giant planet formed mostly of hydrogen and helium.

general relativistic time dilation The verified prediction that time passes more slowly in a gravitational field than in the absence of a gravitational field. See *time dilation.*

general relativity See *general theory of relativity.*

general theory of relativity Einstein's theory explaining gravity as the distortion of spacetime by massive objects; also known simply as *general relativity.* This theory deals with all types of motion; for contrast, see *special relativity.*

geodesic The path an object will follow through *spacetime* in the absence of external forces.

geology The study of the composition, behavior, and history of solid bodies, including Earth, the terrestrial planets, and the moons of the Solar System.

giant galaxy A galaxy with luminosity greater than about a billion solar luminosities.

giant molecular cloud An interstellar cloud composed primarily of molecular gas and dust, having hundreds of thousands of solar masses.

giant planet One of the largest planets in the Solar System, typically 10 times the size and many times the mass of the terrestrial planets and lacking a solid surface.

gibbous Phases of the Moon, Mercury, or Venus in which the object appears more than half illuminated by the Sun.

global circulation The overall, planet-wide circulation pattern of a planet's atmosphere.

globular cluster A spherically symmetric, highly condensed cluster of stars, containing tens of thousands to a million members.

gluon The particle that carries (or, equivalently, mediates) interactions due to the *strong nuclear force.*

gradation The leveling of a planet's surface through weathering, erosion, transpiration, and deposition of rock debris by water, wind, and gravity.

grand unified theory (GUT) A unified quantum theory that combines the *strong nuclear, weak nuclear,* and *electromagnetic forces* but does not include *gravity.*

granite Rock that is cooled from magma and is relatively rich in silicon and oxygen.

grating A *diffraction* grating. An optical surface containing many narrow, closely and equally spaced parallel grooves or slits that create *dispersion* of reflected or transmitted light.

gravitational lens A massive object that gravitationally focuses the light of a more distant object to produce multiple brighter, magnified, possibly distorted images.

gravitational lensing The bending of light by gravity.

gravitational potential energy The potential energy that an object has solely due to its position within a gravitational field.

gravitational redshift The shifting to longer wavelengths of the wavelength of radiation from an object deep within a gravitational well.

gravity (a) The mutually attractive force between massive objects. (b) An effect arising from the bending of *spacetime* by massive objects. (c) One of four fundamental forces of nature.

gravity wave Wave in the fabric of *spacetime* emitted by accelerating masses.

great circle Any circle on a sphere that has as its center the center of the sphere. The celestial equator, the meridian, and the ecliptic are all great circles on the sphere of the sky, as is any circle drawn through the zenith.

Great Red Spot (GRS) The giant, oval, brick-red anticyclone seen in Jupiter's southern hemisphere.

greenhouse effect The solar heating of air in an enclosed space, such as a closed building or car, resulting primarily from the inability of the hot air to escape. Compare with *atmospheric greenhouse effect.*

greenhouse molecule One of a group of atmospheric molecules such as carbon dioxide that are transparent to visible radiation but absorb infrared radiation.

Gregorian calendar The modern calendar. A modification of the Julian calendar decreed by Pope Gregory XIII in 1582. By this time the less accurate Julian calendar had developed an error of 10 days over the 13 centuries since its inception.

ground state The lowest possible energy state for a system or part of a system, such as an atom, molecule, or particle.

GUT See *grand unified theory.*

Guth, Alan (1947–) The physicist who first proposed that the universe underwent an early, rapid *inflation,* which resolves both the flatness and horizon problems.

H

Hadley circulation A simplified, and therefore uncommon, atmospheric global circulation that carries thermal energy directly from the equator to the polar regions of a planet.

half-life The time it takes half a sample of a particular radioactive element to decay to a daughter element.

halo The spherically symmetric, low-density distribution of stars and dark matter that defines the outermost regions of a galaxy.

harmonic law Another name for *Kepler's third law* of planetary motion.

Hawking, Stephen (1942–) The British physicist/astrophysicist who has done much to further our understanding of the universe and the phenomena surrounding black holes.

Hawking radiation Radiation from a black hole.

Hayashi track The path that a protostar follows on the H-R diagram as it contracts toward the main sequence.

head The part of a comet that includes both the nucleus and the inner part of the coma.

heat death The possible eventual fate of an open universe, in which entropy has triumphed and all energy- and structure-producing processes have come to an end.

heavy element See *massive element.*

Heisenberg uncertainty principle The physical limitation that the *product of the position and the momentum* of a particle cannot be smaller than a well-defined value, *Planck's constant* (h).

Heisenberg, Werner (1901-1976) The German theoretical physicist and Nobel laureate who was one of the founders of quantum mechanics. He is perhaps best known for his discovery of the *Heisenberg uncertainty principle.*

helioseismology The use of solar oscillations to study the interior of the Sun.

helium flash The runaway explosive burning of helium in the degenerate helium core of a *red giant* star.

Herbig-Haro (HH) object A glowing, rapidly moving knot of gas and dust that is excited by bipolar outflows in very young stars.

heredity The process by which one generation passes on its characteristics to future generations.

hertz (Hz) A unit of frequency equivalent to cycles per second.

Hertz, Heinrich (1857–1894) The 19th century physicist who supplied the first experimental data confirming Maxwell's predictions about electromagnetic radiation.

Hertzsprung, Einar (1873–1967) The Danish astronomer who, along with Henry Norris Russell, defined the main diagnostic tool for studying the properties of stars, in which we graph stellar luminosity versus stellar spectral type (or temperature, or color). The *Hertzsprung-Russell (H-R) diagram* is named in his honor.

Hertzsprung-Russell diagram See *H-R diagram.*

HH object See *Herbig-Haro object.*

hierarchical clustering The "bottom-up" process of forming large-scale structure. Small-scale structure first produces groups of galaxies, which in turn form clusters, which then form superclusters.

high-velocity star A star belonging to the halo found near the Sun, distinguished from disk stars by moving far faster and often in the direction opposite to the rotation of the disk and its stars.

homogeneous In cosmology, describes a universe in which observers in any location would observe the same properties.

horizon The boundary that separates the sky from the ground.

horizon problem The puzzling observation that the cosmic background radiation is so uniform in all directions, despite the fact that widely separated regions should have been "over the horizon" from each other in the early universe.

horizontal branch A region on the H-R diagram defined by stars burning helium to carbon in a stable core.

hot dark matter Particles of dark matter that move so fast that gravity cannot confine them to the volume occupied by a galaxy's normal luminous matter.

hot Jupiter A large, Jovian-type *extrasolar planet* located very close to its parent star.

hot spot A place where hot plumes of mantle material rise near the surface of a planet.

H-R diagram The Hertzsprung-Russell diagram, which is a plot of the luminosities versus the surface temperatures of stars. The evolving properties of stars are plotted as tracks across the H-R diagram.

H II region A region of interstellar gas that has been ionized by UV radiation from nearby hot massive stars.

Hubble constant *(H_0)* The constant of proportionality relating the recession velocities of galaxies to their distances. See also *Hubble time.*

Hubble, Edwin P. (1889–1953) The Ammerican astronomer who discovered the true nature of galaxies and that the universe is expanding. The *Hubble Space Telescope* is named after him.

Hubble's law The law stating that the speed at which a galaxy is moving away from us is proportional to the distance of that galaxy.

Hubble Space Telescope (HST) A 2.4-m telescope placed in Earth orbit in 1990.

Hubble time An estimate of the age of the universe from the inverse of the Hubble constant, $1/H_0$.

hurricane A large tropical cyclonic system circulating counterclockwise in the Northern Hemisphere and clockwise in the Southern Hemisphere. Hurricanes can extend outward from their center to more than 600 km and generate winds in excess of 300 km/h. Hurricanes are also known as *cyclones* or *typhoons.*

Huygens, Christiaan (1619–1695) Among the founders of mechanics and optics; the Dutch physicist first to recognize that Saturn has rings.

hydrogen burning The release of energy from the nuclear fusion of four hydrogen atoms into a single helium atom.

hydrogen shell burning Fusion of hydrogen in a shell surrounding a stellar core that may be either degenerate or fusing more massive elements.

hydrosphere The portion of Earth that is largely liquid water.

hydrostatic equilibrium The condition in which the weight bearing down at some point within an object is balanced by the pressure within the object.

hypothesis A well-thought-out idea, based on scientific principles and knowledge, that makes testable predictions.

hypothetical Proposed but not yet tested.

I

ice The solid form of a volatile material; sometimes the volatile material itself, regardless of its physical form.

ice giant A giant planet formed mostly of the liquid form of volatile substances.

ideal gas A gas in which all collisions between individual atoms or molecules are like collisions between billiard balls or marbles.

ideal gas law The relationship between pressure (P), number density of particles (n), and temperature (T) expressed as $P = nkT$, where k is Boltzmann's constant.

igneous activity The formation and action of molten rock or magma.

impact crater The scar of the impact left on a solid planetary or moon surface by collision with another object.

index or refraction (n) The ratio of the speed of light in a vacuum, c, to the speed of light in an optical medium.

inert gas A gaseous element that combines with other elements only under conditions of extreme temperature and pressure—such as helium, neon, and argon.

inertia The tendency for objects to retain their state of motion.

inertial frame of reference A reference frame that is not accelerating. In general relativity, a reference frame that is falling freely in a gravitational field.

inflation An extremely brief phase of ultra-rapid expansion of the very early universe. Following inflation, the standard Big Bang models of expansion apply.

infrared (IR) radiation Electromagnetic radiation occurring in the spectral region between those of visible light and microwaves.

instability strip A region of the H-R diagram containing stars that pulsate with a periodic variation in luminosity.

integration time The time interval over which photons are collected and added up in a detecting device.

intensity Amount per second per unit area. In the case of electromagnetic radiation: W/m².

intercloud gas   A low-density region of the interstellar medium that fills the space between interstellar clouds.

interference   The interaction of two sets of waves pruducing high and low intensity, depending on whether their amplitudes reinforce or cancel. See also *constructive interference* and *destructive interference*.

interferometer   A group or array of separate but linked optical or radio telescopes whose overall separation determines the angular resolution of the system.

interferometric array   See *interferometer*.

intermediate-mass star   Star with mass between about 3 $M_\odot$ and 8 $M_\odot$.

interstellar cloud   A discrete, high-density region of the interstellar medium made up mostly of atomic or molecular hydrogen and dust.

interstellar dust   Small particles or grains (0.01 to 10 μm) of matter, primarily carbon and silicates, distributed throughout interstellar space.

interstellar extinction   The dimming of visible and ultraviolet light by interstellar dust.

interstellar medium   The gas and dust that fill the space between the stars within a galaxy.

inverse square law   The rule that a quantity or effect diminishes with the square of the distance from the source.

ion   An *atom* or *molecule* that has lost or gained one or more *electrons*.

ionize   The process by which electrons are stripped free from an *atom* or *molecule*, resulting in free *electrons* and a positively charged atom or molecule.

ionosphere   A layer high in Earth's atmosphere in which most of the atoms are ionized by solar radiation.

ion tail   A type of comet tail consisting of ionized gas. Particles in the ion tail are pushed directly away from the comet's head in the antisolar direction at high speeds by the solar wind.

iron meteorite   A metallic meteorite composed mostly of iron–nickel alloys.

irregular galaxy   A galaxy without regular or symmetric appearance.

irregular moon   A moon that has been captured by a planet. Some revolve in the opposite direction from the rotation of the planet, and many are in distant, unstable orbits.

isotopes   Forms of the same element with differing numbers of neutrons.

isotropic   In cosmology, a universe whose properties observers find to be the same in all directions.

J

jansky (*Jy*)   The basic unit of flux density. Units: W/m²/Hz.

Jansky, Karl (1905-1950)   The American physicist and radio engineer whose discovery of radio emissions from the *Milky Way Galaxy* led to the birth of radio astronomy.

jet   (a) A stream of gas and dust ejected from a comet nucleus by solar heating. (b) A collimated linear feature of bright emission extending from a *protostar or active galactic nucleus*.

joule (J)   A unit of energy or work. 1 J = 1 newton meter.

K

Kant, Immanuel (1724–1804)   The 18th century German philosopher who hypothesized that the nebulae seen by Messier and Herschel were "island universes," separate from our own.

kelvin (K)   The temperature scale using Celsius-sized degrees, but with 0 K defined as *absolute zero* instead of the melting point of water. Defined by William Thompson, better known as Lord Kelvin (1824–1907).

Kepler, Johannes (1571–1630)   The discoverer of the elliptical shape of planetary orbits and how orbital period is related to average distance from the Sun.

Kepler's first law   The law stating that planets move in orbits of elliptical shapes with the Sun at one focus.

Kepler's laws   The three rules of planetary motion inferred by Johannes Kepler from the data acquired by Tycho Brahe.

Kepler's second law   The law stating that a line drawn from the Sun to a planet sweeps out equal areas in equal times as the planet orbits the Sun. Also called the *law of equal areas*.

Kepler's third law   The relationship between the period of a planet's orbit and its distance from the Sun. Also called the *harmonic law*.

kinetic energy   The energy of an object due to its motions. KE = 1/2 mv^2. Units: joules (J).

Kirkwood, Daniel (1814-1895)   The American astronomer who discovered resonant gaps in the asteroid belt, now known as *Kirkwood gaps*.

Kirkwood gap   A gap in the main asteroid belt related to orbital resonances with Jupiter.

Kuiper Belt   A disk-shaped population of comet nuclei extending from Neptune's orbit to perhaps several thousand AU from the Sun.

Kuiper Belt object (KBO)   An icy planetesimal (comet nucleus) that orbits within the Kuiper Belt beyond the orbit of Neptune. Also known as a *Trans-Neptunian Object* (TNO).

Kuiper, Gerard Peter (1905-1973)   The Dutch-American planetary astronomer who discovered CO_2 in the atmosphere of Mars and pioneered airborne infrared observing. Kuiper is well known for his suggestion that a belt of trans-Neptunian objects, now known as the *Kuiper Belt,* is the source of *short-period comets*.

L

Lagrange, Joseph (1736–1813)   Italian-French mathematician who contributed to the theory of celestial mechanics.

Lagrangian equilibrium point   One of five points of equilibrium in a system consisting of two massive objects in nearly circular orbit around a common center of mass. Only two (L_4 and L_5) represent stable equilibrium. A third smaller body located at one of the five points will move in lockstep with the center of mass of the larger bodies.

lambda peak (λ_{peak})   The wavelength at which the thermal radiation from an object is most intense. Literally, the peak of the Planck spectrum as specified by Wien's Law.

lander   An instrumented spacecraft designed to land on a planet or moon. See also *rover*.

large-scale structure   Observable aggregates on the largest scales in the universe, including galaxy groups, clusters, and superclusters.

latitude   The angular distance north (+) or south (–) from the equatorial plane of a nearly spherical body.

law of conservation of angular momentum   The physical law stating that a system's total angular momentum remains constant, unless acted on by an external torque.

law of equal areas   See *Kepler's second law*.

law of gravitation   See *universal law of gravitation*.

leap year   A year that contains 366 days. Leap years occur every four years when the year is divisible by four, correcting for the accumulated excess time in a normal year, which is approximately 365¼ days long.

Leavitt, Henrietta (1868–1921)   The American astronomer who discovered that periods of variation of Cepheid stars are related to their luminosities by studying the variable stars in the Large and Small Magellanic Clouds.

Leonids   A November meteor shower associated with the dust debris left by comet Tempel-Tuttle.

libration   The apparent wobble of an orbiting body that is tidally locked to its companion (such as Earth's Moon) resulting from the fact that its orbit is elliptical rather than circular.

light   All electromagnetic radiation, which comprises the entire *electromagnetic spectrum.*

light-year (ly)   The distance light travels in one year—about 9 trillion kilometers.

limb darkening   The darker appearance caused by increased atmospheric absorption near the limb of a planet or star.

line of nodes   (a) A line defined by the intersection of two orbital planes. (b) The line defined by the intersection of Earth's equatorial plane and the plane of the ecliptic.

Lippershey, Hans (1570-1619)   The German-Flemish spectacle maker who is generally credited with the discovery of the optical telescope.

lithosphere   The solid, brittle part of Earth (or any planet or moon), including the crust and the upper part of the mantle.

lithospheric plate   A separate piece of Earth's lithosphere capable of moving independently. See *continental drift, plate tectonics.*

Local Group   The small group of galaxies of which the Milky Way and the Andromeda Galaxy are members.

longitudinal wave   A wave that oscillates parallel to the direction of the wave's propagation.

long-period comet   A comet with an orbital period greater than 200 years.

look-back time   The time that it has taken the light from an astronomical object to reach Earth.

low-mass star   A star with a main sequence mass of less than about 3 $M_\odot$.

luminosity   The total flux emitted by an object. Unit: watts (W).

luminosity–temperature–radius relationship   A relationship among these three properties of stars indicating that if any two are known, the third can be calculated.

luminous matter   Matter in galaxies—including stars, gas, and dust—that emits electromagnetic radiation.

lunar eclipse   An eclipse that occurs when the Moon is partially or entirely in Earth's shadow.

lunar tide   The component of Earth's tides due to the Moon.

M

MACHO   "Massive compact halo object," such as brown dwarfs, white dwarfs, and black holes, which are candidates for being considered dark matter. See also *WIMP.*

magma   Molten rock often containing dissolved gases and solid minerals.

magnetic field   A field that is able to exert a force on a moving electric charge. See *electromagnetic force.*

magnetic force   A force associated with, or caused by, the relative motion of charges.

magnetosphere   The region surrounding a planet filled with relatively intense magnetic fields and plasmas.

magnitude   A system used by astronomers to describe the brightness or luminosity of stars. The brighter the star, the smaller its magnitude.

main asteroid belt   See *asteroid belt.*

main sequence   The strip on the H-R diagram where most stars are found. Main sequence stars are fusing hydrogen to helium in their cores.

main sequence lifetime   The amount of time a star spends on the main sequence, fusing hydrogen into helium in its core.

main sequence turnoff   The location on the H-R diagram of a single-aged stellar population (such as a star cluster) where stars have just evolved off the main sequence. The position of the main sequence turnoff is determined by the age of the stellar population.

mantle   The solid portion of a rocky planet that lies between the crust and the core.

mare (pl. **maria**)   A dark region on the Moon, composed of basaltic lava flows.

mass   (a) Inertial mass is the property of matter that resists changes in motion. (b) Gravitational mass is the property of matter defined by its attractive force on other objects. According to general relativity the two are equivalent.

massive element   (a) In astronomy, refers to all elements more massive than helium. (b) In other sciences (and sometimes also

in astronomy), refers to the most massive elements in the periodic table, such as uranium and plutonium. Also called heavy element.

mass transfer   The transfer of mass from one member of a binary star system to its companion. Mass transfer occurs when one of the stars evolves to the point that it overfills its *Roche lobe,* so that its outer layers are pulled toward its binary companion.

mathematics   The science and language of patterns; the language of science. Mathematics provides the tools used by scientists to understand, describe, and predict the patterns found in nature.

matter   Objects made of particles that have mass, such as protons, neutrons, and electrons; anything that occupies space and has mass.

Maunder minimum   The period 1645–1715, when there were few sunspots.

Maxwell, James Clerk (1831–1879)   The 19th century Scottish physicist who summarized all of classical electromagnetism, including electromagnetic radiation, with a set of four equations.

megabar   A unit of pressure equal to 1 million *bars.*

mega-light-year (Mly)   A unit of distance equal to 1 million *light-years.*

meridian   The imaginary arc in the sky running from the horizon at due north through the zenith to the horizon at due south. The meridian divides the observer's sky into eastern and western halves.

mesosphere   A portion of Earth's atmosphere that lies above the stratosphere.

Messier, Charles (1730–1817)   An early French comet hunter; published the first catalog of bright, diffuse celestial objects to prevent their being confused with comets.

Messier's Catalog   The catalog made by Charles Messier in the late 18th century in which he identifies 103 bright, diffuse-appearing objects visible from France, all of them either star clusters, planetary nebulae, H II regions, or galaxies; the catalog was later expanded to 110 objects.

meteor   The incandescent trail produced by a small piece of interplanetary debris as it travels through the atmosphere at very high speeds.

meteorite   A meteoroid that survives to reach a planet's surface.

meteoroid   A small cometary or asteroidal fragment ranging in size from 100 μm to 100 m. When entering a planetary atmosphere, the meteoroid creates a *meteor,* which is an atmospheric phenomenon.

meteor shower A larger than normal display of meteors, occurring when Earth passes through the orbit of a disintegrating comet, sweeping up its debris.

micrometer (μm) 10^{-6} meter; a unit of length used for measuring the wavelength of electromagnetic radiation. Also called *micron.*

micron See *micrometer.*

microwave radiation Electromagnetic radiation occurring in the spectral region between those of infrared radiation and radio waves.

Milky Way Galaxy The *galaxy* in which our Sun and Solar System reside.

minute of arc (′) A unit for measuring angles. A minute of arc is 1/60 of a degree of arc. Also called *arcminute.*

Mly See *mega-light-year.*

modern physics A term usually used to refer to those physical principles, including relativity and quantum mechanics, developed since Maxwell's equations were published.

molecular cloud An interstellar cloud composed primarily of molecular hydrogen.

molecular cloud core A dense clump within a molecular cloud that forms as the cloud collapses and fragments. Protostars form from molecular cloud cores.

molecule Generally, the smallest particle of a substance that retains its chemical properties and is composed of two or more atoms. A very few types of molecules, such as helium, are composed of single atoms.

momentum The product of the *mass* and *velocity* of a particle. Units: kg m/s.

moon A less massive satellite orbiting a more massive object. Moons are found around planets, dwarf planets, asteroids, and KBOs.

M-type asteroid An asteroid once part of the metallic core of a larger, differentiated body that has since been broken into pieces; mostly made of iron and nickel.

mutation In biology, an imperfect reproduction of self-replicating material.

N

natural selection The process by which forms of structure, ranging from molecules to whole organisms, that are best adapted to their environment become more common than less well-adapted forms. See also *Theory of Evolution.*

nature The word frequently used by scientists to denote the physical universe. Note the difference from other common uses.

neap tide An especially weak tide that occurs around the time of the first or third quarter Moon when the gravitational forces of the Moon and the Sun on Earth are at right angles to each other, thus producing the least pronounced tides.

near-Earth asteroid An asteroid whose orbit brings it close to the orbit of Earth. See also *near-Earth object.*

near-Earth object (NEO) An asteroid, comet, or large meteoroid whose orbit intersects Earth's orbit.

nebula (pl. nebulae) A cloud of interstellar gas and dust, either illuminated by stars (bright nebula) or seen in silhouette against a brighter background (dark nebula).

neutrino A very low-mass, electrically neutral particle emitted during beta decay. (Neutrinos interact with matter only very feebly and so can penetrate through great quantities of matter.)

neutrino cooling The process in which thermal energy is carried out of the center of a star by neutrinos rather than by electromagnetic radiation or convection.

neutron (n^0) A subatomic particle having no net electric charge, and a rest mass and rest energy nearly equal to that of the proton.

neutron star The neutron degenerate remnant left behind by a Type II supernova.

new Moon The phase of the Moon that occurs when the Moon is between Earth and the Sun, and we see only the side of the Moon not being illuminated by the Sun.

newton (N) The force required to accelerate a 1-kg mass at a rate of 1 m/s². Units: kg m/s².

Newton's first law of motion The law stating that an object will remain at rest or will continue moving along a straight line at a constant speed until an unbalanced force acts on it.

Newton, Sir Isaac (1642–1727) The English discoverer and codifier of the laws describing the motion of objects and of the law of gravity; in many respects, the founding father of modern science.

Newton's laws See *Newton's first law of motion, Newton's second law of motion,* and *Newton's third law of motion.*

Newton's second law of motion The law stating that if an unbalanced force acts on a body, the body will have an accelera-

tion proportional to the unbalanced force and inversely proportional to the object's mass. The acceleration will be in the direction of the unbalanced force.

Newton's third law of motion The law stating that for every force, there is an equal and opposite force.

nonthermal radiation Any form of electromagnetic radiation, such as *synchrotron radiation,* that does not originate from thermal energy.

normal matter Matter in galaxies that emits and absorbs electromagnetic radiation. Compare with *dark matter.*

north celestial pole The northward projection of Earth's rotation axis onto the celestial sphere.

North Pole The point in the Northern Hemisphere where Earth's rotation axis intersects the surface of Earth.

north star See *Polaris.*

nova (pl. novae) A stellar explosion that results from runaway *nuclear fusion* in a layer of material on the surface of a white dwarf in a binary system.

nuclear fusion The combination of two less massive atomic nuclei into a single more massive atomic nucleus. Release of energy by fusion of low-mass elements is often referred to as *nuclear burning.*

nuclear burning Release of energy by fusion of low-mass elements.

nucleosynthesis The formation of more massive atomic nuclei from less massive nuclei, either in the Big Bang (Big Bang nucleosynthesis) or in the interiors of stars (stellar nucleosynthesis).

nucleus (pl. nuclei) (a) The dense, central part of an atom. (b) The central core of a galaxy, comet, or other diffuse object.

O

oblateness The flattening of an otherwise spherical planet or star caused by its rapid rotation.

obliquity The inclination of a celestial body's equator to its orbital plane.

observational uncertainty The fact that real measurements are never perfect; all observations are uncertain by some amount.

Occam's razor The principle that the simplest hypothesis is the most likely; named after William of Occam (circa 1285–1349), the medieval English cleric to whom the idea is attributed.

Oort Cloud A *spherical* distribution of comet nuclei stretching from beyond the

Kuiper Belt to more than 50,000 AU from the Sun.

Oort, Jan (1900–1992) The Dutch radio astronomer who first suggested that a distant, spherical swarm of objects, now known as the *Oort Cloud*, is the source of *long-period comets.*

opacity A measure of how effectively a material blocks the radiation going through it.

open cluster Group of a few dozen to a few thousand stars that formed together in the disk of a spiral galaxy. Also called *galactic clusters.*

open universe An infinite universe with a negatively curved spatial structure (much like the surface of a saddle) such that the sum of the angles of a triangle is always less than 180°.

orbit The path taken by one object moving around another object under the influence of their mutual gravitational or electric attraction.

orbital resonance A situation in which the orbital periods of two objects are related by a ratio of small integers.

orbiter A spacecraft placed in orbit around a planet or moon. See also *flyby.*

organic A substance, not necessarily of biological origin, containing the element carbon.

origins The subject relating to the genesis of the universe and life.

P

pair creation The production of a particle–antiparticle pair from a source of electromagnetic energy.

paleomagnetism The record of Earth's magnetic field as preserved in rocks.

palimpsest A flat, circular feature in icy lithospheres, believed to be a scar of an ancient impact.

parallax The apparent shift in the position of one object relative to another object, caused by the changing perspective of the observer. In astronomy, the displacement in the apparent position of a nearby star caused by the changing location of Earth in its orbit.

parent element A radioactive element. See *daughter element.*

parsec (pc) The distance to a star with a parallax of 1 arcsecond using a base of 1 AU. One parsec is approximately 3.26 light-years.

partial solar eclipse Any solar eclipse in which the observer is outside the path of totality.

peculiar velocity The motion of a galaxy relative to the overall expansion of the universe.

pendulum A mass supported by a string, wire, or rod that is free to swing back and forth in a gravitational field.

penumbra (pl. **penumbrae**) (a) The outer part of a shadow, where the source of light is only partially blocked. (b) The region surrounding the umbra of a sunspot. The penumbra is cooler and darker than the surrounding surface of the Sun but not as cool or dark as the umbra of the sunspot.

penumbral lunar eclipse A lunar eclipse in which the Moon passes through the penumbra of Earth's shadow.

Penzias, Arno (1933–) The Bell Laboratories physicist who, together with Robert Wilson, in the early 1960s, discovered the cosmic background radiation, for which they won the Nobel Prize in physics.

perihelion (pl. **perihelia**) The closest point to the Sun on a solar orbit.

period The time it takes for a regularly repetitive process to complete one cycle.

period–luminosity relationship The relationship between the period of variability of a pulsating variable star, such as a Cepheid or RR Lyrae variable, and the luminosity of the star. Longer-period Cepheid or RR Lyrae variables are more luminous than their shorter-period cousins.

Perseids A prominent August meteor shower associated with the dust debris left by comet Swift-Tuttle.

phase One of the various appearances of the sunlit surface of the Moon or a planet caused by the change in viewing location of Earth relative to both the Sun and the object.

photino An elementary particle related to the *photon*. One of the leading candidates for the cold dark matter.

photochemical A chemical reaction driven by the absorption of electromagnetic radiation.

photodisintegration The disintegration of atomic nuclei by absorption of high energy radiation. For example, the splitting of iron into helium by the absorption of gamma rays.

photodissociation The breaking apart of molecules into smaller fragments or individual atoms by the action of photons.

photoelectric effect An effect whereby electrons are emitted from a substance illuminated by photons above a certain critical frequency.

photon A discrete unit or particle of electromagnetic radiation; a *quantum of*

light. The energy of a photon is equal to Planck's constant (h) times the frequency (f) of its electromagnetic radiation: $E_{photon} = h \times f$. The particle that mediates the electromagnetic force.

photosphere The apparent surface of the Sun as seen in visible light.

physical laws Broad statements that predict some aspect of how the physical universe behaves and that are supported by many empirical tests.

physics The scientific study of the fundamental principles that govern the behavior of matter and energy, and the application of those principles to understanding and predicting natural phenomena.

pixel The smallest picture element in a digital image array.

Planck era The early time, just after the Big Bang, when the universe as a whole must be described with quantum mechanics.

Planck, Max (1858–1947) The German physicist who first postulated the quantum nature of electromagnetic radiation.

Planck's constant (h) The constant of proportionality between the energy of a photon and the frequency of the photon. This constant defines how much energy a single photon of a given frequency or wavelength has. Value: $h = 6.63 \times 10^{-34}$ joule-second.

Planck spectrum The spectrum of electromagnetic energy emitted by a blackbody per unit area per second, which is determined only by the temperature of the object. Also called *blackbody spectrum.*

planet (a) A large body that orbits the Sun or other star that shines only by light reflected from the Sun or star. (b) In the Solar System, a body that orbits the Sun, has sufficient mass for self-gravity to overcome rigid body forces so that it assumes a spherical shape, and has cleared smaller bodies from the neighborhood around its orbit. Compare with *dwarf planet.*

planetary nebula The expanding shell of material ejected by a dying AGB star. A planetary nebula glows from flourescence caused by intense ultraviolet light coming from the hot, stellar remnant at its center.

planetary system A system of planets and other smaller objects in orbit around a star.

planetesimal A primitive body of rock and ice, 100 m or more in diameter, that combines with others to form planets.

plasma A gas composed largely of charged particles but that also may include some neutral atoms.

plate tectonics The geological theory concerning the motions of lithospheric plates, which in turn provides the theoretical basis for continental drift.

Polaris The star that currently is the closest bright star to the north celestial pole.

positron A positively charged subatomic particle; the *antiparticle* of the electron.

power The rate at which work is done or at which energy is delivered. Unit: watts or joules/second (W or J/s).

precession of the equinoxes The slow change in orientation between the ecliptic plane and the celestial equator caused by the wobbling of Earth's axis.

pressure Force per unit area. Units: newtons per square meter (N/m^2) or bars.

primary atmosphere An atmosphere, composed mostly hydrogen and helium, that forms at the same time as its host planet.

primary mirror The principal optical mirror in a reflecting telescope, which determines the telescope's light-gathering power and resolution.

primary wave A longitudinal seismic wave in which the oscillations involve compression and decompression parallel to the direction of travel (that is, a pressure wave).

principle A general idea or sense about how the universe is that guides us in constructing new scientific theories. Principles can be testable theories.

prograde (a) Describes rotational or orbital motion of a moon that is in the same sense as the planet it orbits. (b) Describes the counterclockwise orbital motion of Solar System objects as seen from above Earth's orbital plane. See also *retrograde*.

prominence An archlike projection above the solar photosphere often associated with a sunspot.

proportional Two things are proportional if their ratio is a constant.

proton (p or p^+) A fundamental particle having a positive electric charge of 1.6×10^{-19} coulomb, a mass of 1.67×10^{-27} kg, and a rest energy of 1.5×10^{-10} joules.

proton–proton chain One of the ways in which hydrogen burning—the fusion of four hydrogen atoms to form a helium atom—can take place. This is the most important path for hydrogen burning in low-mass stars such as the Sun.

protoplanetary disk The remains of the accretion disk around a young star from which a planetary system may form.

protostar A young stellar object that derives its luminosity from the conversion of gravitational energy to thermal energy,

rather than from nuclear reactions in its core.

protostellar disk The accretion disk that forms as an interstellar cloud collapses on its way to becoming a star.

pulsar A rapidly rotating neutron star that beams radiation into space in two searchlight-like beams. To a distant observer, the star appears to flash on and off, earning its name.

pulsating variable star A variable star that undergoes periodic radial pulsations.

Q

QCD See *quantum chromodynamics*.

QED See *quantum electrodynamics*.

quantized Describes a quantity that exists as discrete, irreducible units.

quantum chromodynamics (QCD) The quantum mechanical theory describing the strong nuclear force and its mediation by gluons.

quantum efficiency The fraction of photons falling on a detector that actually produces a response in the detector.

quantum electrodynamics (QED) The quantum theory that describes the electromagnetic force and its mediation by photons.

quantum mechanics The branch of physics that deals with the quantized and probabilistic behavior of atoms and subatomic particles.

quantum of light The discrete particle of light we call a *photon*.

quark The building block of protons and neutrons.

quasar The most luminous of the active galactic nuclei (AGNs), seen only at great distances from our galaxy; shortened from "quasi-stellar radio" source.

Quaternary Period The period in Earth's history from 1.8 million years ago through today.

R

radial velocity The component of velocity that is directed toward or away from the observer.

radian The angle at the center of a circle subtended by an arc equal to the length of the circle's radius. Therefore, 2π radians equals 360° and 1 radian equals approximately 59°.296.

radiant The direction in the sky from which the meteors in a meteor shower seem to come.

radiation belt A toroidal ring of high-energy particles surrounding a planet.

radiative transfer The transport of energy from one location to another by electromagnetic radiation.

radiative zone A region in the interior of a star through which energy is transported outward by radiation.

radio galaxy An elliptical galaxy having very strong emission (10^{35} to 10^{38} watts) in the radio part of the electromagnetic spectrum.

radiometric dating Use of radioactive decay to measure the ages of materials such as minerals.

radio telescope An instrument for detecting and measuring radio-frequency emissions from celestial sources.

radio wave Electromagnetic radiation in the extreme long-wavelength region of spectrum, beyond the region of microwaves.

ray (a) A beam of electromagnetic radiation. (b) A bright streak emanating from a young impact crater.

recombination (a) The combining of ions and electrons to form neutral atoms. (b) An event early in the evolution of the universe in which hydrogen and helium nuclei combined with electrons to form neutral atoms. The removal of electrons caused the universe to become transparent to electromagnetic radiation.

reddening The effect by which stars and other objects, when viewed through interstellar dust, appear redder than they actually are. Reddening is caused by blue light being more strongly absorbed and scattered than red light.

red giant A low-mass star that has evolved beyond the main sequence and is now fusing hydrogen in a shell surrounding a degenerate helium core.

red giant branch A region on the H-R diagram defined by low-mass stars evolving from the main sequence toward the *horizontal branch*.

redshift The shifting of the wavelength of light to longer wavelengths by any of several effects including Doppler shifts, gravitational redshift, or cosmological redshift. Compare with *blueshift*.

reflecting telescope A telescope that uses mirrors for collecting and focusing incoming electromagnetic radiation to form an image in their focal planes. The size of a reflecting telescope is defined by the diameter of the first mirror from which incoming light is reflected (called the *primary mirror*).

reflection The redirection of a beam of light incident on the surface between two media having different refractive indices. If the surface is flat and smooth, the angle of incidence equals the angle of reflection.

refracting telescope A telescope that uses objective lenses to collect and focus light.

refraction The bending of a beam of light when it crosses the boundary between two media having different refractive indices.

refractory material Material that remains solid at high temperatures. Compare with *volatile material*.

regular moon A moon that formed together with the planet it orbits.

relative humidity The amount of water vapor held by a volume of air at a given temperature compared (stated as a percentage) to the total amount of water that could be held by the same volume of air at the same temperature.

relative motion The difference in motion between two individual frames of reference.

relativistic Pertaining to physical processes that take place in systems traveling at nearly the speed of light or located in the vicinity of very strong gravitational fields.

relativistic beaming The effect created when material moving at nearly the speed of light beams the radiation it emits in the direction of its motion.

remote sensing The use of images, spectra, radar, or other techniques to measure the properties of an object from a distance.

resolution The ability of a telescope to separate two point sources of light. Resolution is determined by the telescope's *aperture* and the *wavelength* of light it receives.

rest wavelength The wavelength of light we see coming from an object at rest with respect to the observer.

retrograde (a) Describes rotation or orbital motion of a moon that is in the opposite sense to the rotation of the planet it orbits. (b) Describes the clockwise orbital motion of Solar System objects as seen from above Earth's orbital plane. See also *prograde*.

ring An aggregation of small particles orbiting a planet or star. The rings of the four giant planets of the Solar System are composed variously of silicates, organic materials, and ices.

ring arc A discontinuous region of higher density within an otherwise continuous, narrow ring.

ringlet A narrow subdivision within a larger feature called a *ring*.

Roche, Edouard A. (1820–1883) The French astronomer who systematized the effects of tidal stress on the structure of moons and planets.

Roche limit The distance at which a planet's tidal forces exceed the self-gravity of a smaller object, such as a moon, asteroid, or comet, causing the object to break apart.

Roche lobe The hourglass or figure eight–shaped volume of space surrounding two stars, which constrains material that is gravitationally bound by one or the other.

Roentgen, W. C. (1845–1923) The German discoverer of X-rays, for which he won the first Nobel Prize in physics in 1901.

Rømer, Ole (1644–1719) The Danish astronomer who first estimated the speed of light by timing eclipses of Jupiter's moons when Earth was at different distances from Jupiter.

rotation curve A plot showing how the orbital velocity of stars and gas in a galaxy changes with radial distance from the galaxy's center.

rover A remotely controlled instrumented vehicle designed to traverse and explore the surface of a terrestrial planet or moon. See also *lander*.

RR Lyrae variable A variable giant star whose regularly timed pulsations are good predictors of its luminosity. RR Lyrae stars are used for distance measurements to globular clusters.

Russell, Henry Norris (1877–1957) The American astronomer who, together with Einar Hertzsprung, defined the main diagnostic tool for studying the properties of stars. The *Hertzsprung-Russell (H-R) diagram* is named in his honor, as is the *Vogt-Russell theorem*.

Rutherford, Ernest (1871–1937) The New Zealand nuclear physicist, often considered the "father" of nuclear physics, who pioneered the orbital theory of the atom.

S

satellite (a) An object in orbit about a more massive body. (b) A moon.

saturation An atmosphere reaching the point at which it can hold no more water (or other substance) in the gas phase.

scale factor (R_U) A dimensionless number proportional to the distance between two points in space. The scale factor increases as the universe expands.

scattering The random change in the direction of travel of photons, caused by their interactions with molecules or dust particles.

science The search for the rules that govern the behavior of the universe and the explanation of observed phenomena in terms of those rules.

scientific method The formal procedure—including hypothesis, prediction, and experiment or observation—used to test (attempt to falsify) the validity of scientific hypotheses and theories.

scientific notation The standard expression of numbers with one digit (which can be zero) to the left of the decimal point and multiplied by 10 to the exponent required to give the number its correct value. Example: $2.99 \times 10^8 = 299{,}000{,}000$.

secondary atmosphere A planetary atmosphere that formed—as a result of volcanism, comet impacts, or some other process—sometime after the planet formed.

secondary crater A crater formed from ejecta by a primary impact.

secondary mirror A small mirror placed on the optical axis of a reflecting telescope that returns the beam back through a small hole in the primary mirror, thereby shortening the mechanical length of the telescope.

secondary wave A transverse seismic wave.

second law of thermodynamics The law stating that the entropy or disorder of an *isolated* system always increases as the system evolves.

second of arc (") A unit used for measuring very small angles. A second of arc is 1/60 of a minute of arc, or 1/3,600 of a degree. Also called *arcsecond*.

seismic wave A vibration due to earthquakes, large explosions, or impacts on the surface that travels through a planet's interior.

seismometer An instrument that measures the amplitude and frequency of seismic waves.

self-gravity The gravitational attraction between all the parts of the same object.

semimajor axis Half of the longer axis of an ellipse.

SETI The Search for Extraterrestrial Intelligence project, which uses advanced technology combined with radio telescopes to search for evidence of intelligent life elsewhere in the universe.

Seyfert, Carl (1911–1960) The American astronomer who discovered AGNs in spiral galaxies. *Seyfert galaxies* are named in his honor.

Seyfert galaxy A type of spiral galaxy with an active galactic nucleus (AGN) at its center, first discovered in 1943 by Carl Seyfert.

Shapley, Harlow (1885–1972) The Harvard College Observatory astronomer who first showed that our galaxy is nearly 300,000 light-years in size. Shapley was the participant in the Great Debate who argued that our galaxy alone encompassed the entire universe. See also *Curtis, Heber*.

shepherd moon A moon that orbits close to rings and gravitationally confines the orbits of the ring particles.

shield volcano A volcano formed by very fluid lava flowing from a single source, spreading out from that source.

short-period comet A comet with an orbital period of less than 200 years.

sidereal The orbital or rotational period of an orbit measured with respect to the stars.

sidereal time Time based on the true rotation period of Earth. The 24-hour sidereal day used by astronomers is approximately four minutes shorter than a solar day.

silicate One of the family of minerals composed of silicon and oxygen in combination with other elements.

singularity (a) The point where a mathematical expression or equation becomes meaningless, such as the denominator of a fraction approaching zero. (b) A black hole.

Slipher, Vesto (1875–1969) The Lowell Observatory astronomer who first made measurements of the redshifts of galaxies, supplying most of the data used by Edwin Hubble to discover that our universe is expanding.

small solar system body A collective term that includes most asteroids, comets, and most objects orbiting beyond Neptune.

solar abundance The relative amount of an element detected in the atmosphere of the Sun, expressed as the ratio of the number of atoms of that element to the number of hydrogen atoms.

solar day The 24-hour period of Earth's axial rotation that brings the Sun back to the same local meridian.

solar eclipse Blocking of all or part of the Sun by the Moon.

solar flare Explosive events on the Sun's surface associated with complex sunspot groups and strong magnetic fields.

solar maximum (pl. **maxima**) The time, occurring about every 11 years, when the Sun is at its peak activity, meaning that sunspot activity and related phenomena (such as prominences, flares, and coronal mass ejections) are at their peak.

solar neutrino problem The historical observation that only about a third as many neutrinos as predicted by theory seemed to be coming from the sun.

Solar System The gravitationally bound system made up of the Sun, planets, dwarf planets, moons, asteroids, comets, and KBOs, and their associated gas and dust.

solar tide The component of Earth's tide due to the Sun's gravity. Also see *tide* and *lunar tide.*

solar time Time based on the apparent rotation period of Earth. A 24-hour solar day is the average interval between two successive passages of the Sun across the local meridian.

solar wind The stream of charged particles emitted by the Sun that flows at high speeds through interplanetary space.

solstice ("sun standing still") (a) One of the two most northerly and southerly points on the ecliptic. (b) The time of year when the Sun is at one of these two points.

south celestial pole The southward projection of Earth's rotation axis onto the celestial sphere.

South Pole The location in the Southern Hemisphere where Earth's rotation axis intersects the surface of Earth.

spacetime The four-dimensional continuum in which we live, and which we experience as three spatial dimensions plus time.

special relativity The consequences of the speed of light being a constant for nonaccelerating frames of reference; discovered by Albert Einstein. Compare with *general theory of relativity.*

spectral type A classification system for stars based on the presence and relative strength of absorption lines in their spectra. Spectral type is related to the surface temperature of a star.

spectrograph A device that spreads out the light from an object into its component wavelengths.

spectrometer A spectrograph in which the spectrum is generally recorded by electronic means. See *spectrograph.*

spectroscopic parallax Use of the spectroscopically determined luminosity and the observed brightness of a star to determine the star's distance.

spectroscopy The study of electromagnetic radiation from an object in terms of its component wavelengths.

spectrum (a) The intensity of electromagnetic radiation as a function of wavelength. (b) Waves sorted by wavelength.

speed The rate of change of an object's position with time without regard to the *direction* of movement. Units: m/s, km/h. See *velocity.*

spherically symmetric Describes an object whose properties depend only on distance from the object's center, so that the object has the same form viewed from any direction.

spin-orbit resonance A relationship between the orbital and rotation periods of an object such that the ratio of their periods can be expressed by simple integers.

spiral density wave A stable spiral-shaped change in the local gravity of a galactic disk that can be produced by periodic gravitational kicks from neighboring galaxies or from nonspherical bulges and bars in spiral galaxies.

spiral galaxy A galaxy of Hubble type "S" class, with a discernible disk in which large spiral patterns exist.

spoke One of several narrow radial features seen occasionally in Saturn's B ring. Spokes appear dark in backscattered light and bright in forward, scattering light, indicating that they are composed of tiny particles. Their origin is not well understood.

sporadic meteor A meteor not associated with a specific meteor shower.

spreading center A zone from which two tectonic plates diverge.

spring tide An especially strong tide that occurs near the time of a new or full Moon, when lunar tides and solar tides reinforce each other.

stable equilibrium An equilibrium condition in which the system returns to its former condition after a small disturbance. Compare with *unstable equilibrium.*

standard candle An object whose luminosity either is known or can be predicted in a distance-independent way, so its brightness can be used to determine its distance via the inverse square law of radiation.

standard model The theory of particle physics that combines electroweak theory with quantum chromodynamics to describe the structure of known forms of matter.

star A luminous ball of gas that is held together by gravity. A normal star is powered by nuclear reactions in its interior.

star cluster A group of stars that all formed at the same time and in the same general location.

static equilibrium A state in which the forces within a system are all in balance so that the system does not change. Compare with *dynamic equilibrium.*

Stefan–Boltzmann constant (σ) The proportionality constant that relates the flux emitted by an object to the fourth power of its absolute temperature.

Stefan, Josef (1835–1893) The Austro-Hungarian physicist best known for his discovery of the power law of radiation from a black body.

Stefan's Law Gives the amount of electromagnetic energy emitted from the surface of a body, summed over the energies of all photons of all wavelengths emitted, which is proportional to the fourth power of the temperature of the body.

stellar mass loss The loss of mass from the outermost parts of a star's atmosphere during the course of its evolution.

stellar occultation An event in which a planet or other Solar System body moves between the observer and a star, eclipsing the light emitted by that star.

stellar population A group of stars with similar ages, chemical compositions, and dynamical properties.

stereoscopic vision The way an animal's brain combines the different information from its two eyes to perceive the distances to objects around it.

stony-iron meteorite A meteorite consisting of a mixture of silicate minerals and iron–nickel alloys.

stony meteorite A meteorite composed primarily of silicate minerals, similar to those found on Earth.

stratosphere The atmospheric layer immediately above the troposphere. On Earth it extends upward to an altitude of 50 km.

strong nuclear force The attractive short-range force between protons and neutrons that holds atomic nuclei together; one of the four fundamental forces of nature, mediated by the exchange of *gluons.*

S-type asteroid An asteroid made of material that has been modified from its original state, likely as the outer part of a larger, differentiated body that has since been broken into pieces.

subduction zone A region where two tectonic plates converge, with one plate sliding under the other and being drawn downward into the interior.

subgiant A giant star smaller and lower in luminosity than normal giant stars of the same spectral type. Subgiants evolve to become giants.

subgiant branch A region of the H-R diagram defined by stars that have left the main sequence but have not yet reached the *red giant branch.*

sublimation The process of a solid becoming a gas without first becoming a liquid.

subsonic Moving within a medium at a speed slower than the speed of sound in that medium.

summer solstice The time of year in the Northern Hemisphere (about June 21) when the Sun is at its most northerly distance from the celestial equator.

sungrazer A comet whose perihelion is within a few solar diameters of the surface of the Sun.

sunspot A cooler, transitory region on the solar surface produced when loops of magnetic flux break through the surface of the Sun.

sunspot cycle The approximate 11-year cycle during which sunspot activity increases and then decreases. This is one-half of a full 22-year cycle in which the magnetic polarity of the Sun first reverses, then returns to its original configuration.

supercluster A large conglomeration of galaxy clusters and galaxy groups; typically more than 100 million light-years in size and containing tens of thousands to hundreds of thousands of galaxies.

superluminal motion The appearance (though not the reality) that a jet is moving faster than the speed of light.

supermassive black hole A black hole of 1,000 solar masses or more that resides in the center of a galaxy, and whose gravity powers active galactic nuclei (AGNs).

supernova A stellar explosion resulting in the release of tremendous amounts of energy, including the high-speed ejection of matter into the interstellar medium. See *Type I supernova* and *Type II supernova.*

supersonic Moving within a medium at a speed faster than the speed of sound in that medium.

superstring theory The theory that conceives of particles as strings in 11 dimensions of space and time; the current contender for a *theory of everything (TOE).*

surface brightness The amount of electromagnetic radiation emitted or reflected per unit area.

surface wave Seismic wave that travels on the surface of a planet or moon.

symmetry (a) The property that an object has if the object is unchanged by rotation or reflection about some point, line, or plane. (b) In theoretical physics, the correspondence of different aspects of physical laws or systems, such as the symmetry between matter and antimatter.

synchronous rotation The case in which the period of rotation of a body on its axis equals its period of revolution in its orbit around another body. A special type of spin-orbit resonance.

synchrotron radiation Radiation from electrons moving at close to the speed of light as they spiral in a strong magnetic field; named because this kind of radiation was first identified on Earth in particle accelerators called synchrotrons.

synodic A period or time related to apparent changes in celestial objects with respect to the Sun as seen from a planet. See also *sidereal.*

S0 galaxy A galaxy with a bulge and a disklike spiral, but smooth in appearance like ellipticals.

T

tail A stream of gas and dust swept away from the coma of a comet by the solar wind and by radiation pressure from the Sun.

tectonism Deformation of the lithosphere of a planet.

telescope The basic tool of astronomers. Working over the entire range from gamma rays to radio, astronomical telescopes collect and concentrate electromagnetic radiation from celestial objects.

temperature A measure of the average kinetic energy of the atoms or molecules in a gas, solid, or liquid.

terrestrial planet An Earth-like planet, made of rock and metal and having a solid surface.

Tertiary Period The period in Earth's history from 65 million until 1.8 million years ago.

theoretical (a) Using mathematics to formalize rules or laws from ideas and/or empirical results. (b) Using rules and laws to explain observed phenomena or predict phenomena not yet seen.

theoretical model A detailed description of the properties of some object or system in terms of known physical laws or theories. Often a computer calculation of predicted properties based on such a description.

theory A well-developed idea or group of ideas that are consistent with known physical laws and make testable predictions about the world. A very well-tested theory may be called a physical law, or simply a fact.

Theory of Everything (TOE) A theory that unifies all four fundamental forces of nature: strong nuclear, weak nuclear, electromagnetic, and gravitational forces.

Theory of Evolution   First published in 1859 by Charles Darwin, the theory maintains that variation or evolution within species occurs randomly, and that the survival or extinction of an organism is determined by its ability to adapt to its environment. Modern theory involves both evolutionary change and the concept of natural selection as the agent of change. See also *natural selection*.

thermal conduction   The transfer of energy in which the thermal energy of particles is transferred to adjacent particles by collisions or other interactions. The transport of energy by thermal conduction is most important in solids.

thermal energy   The energy that resides in the random motion of atoms, molecules, and particles, by which we measure their temperature.

thermal equilibrium   The state in which the rate at which thermal energy emitted by an object is equal to the rate at which thermal energy is absorbed.

thermal motion   The random motion of atoms, molecules, and particles that gives rise to thermal radiation.

thermal radiation   Electromagnetic radiation resulting from the random motion of the charged particles in every substance.

thermosphere   That portion of Earth's atmosphere between the exosphere and the mesosphere.

third quarter Moon   The phase of the Moon in which the eastern half of the Moon as viewed from Earth is illuminated. The third quarter Moon occurs about one week after the full Moon.

tidal bulge   A distortion of a body resulting from tidal stresses.

tidal locking   Synchronous rotation of an object caused by internal friction as the object rotates through its tidal bulge.

tidal stress   Stress due to differences in the gravitational force of one mass on different parts of another mass.

tide   (a) The deformation of a mass due to differential gravitational effects of one mass on another because of the extended size of the masses. (b) On Earth, the rise and fall of the oceans as Earth rotates through a tidal bulge caused by the Moon and Sun.

time dilation   The relativistic "stretching" of time.

TOE   See *Theory of Everything*.

Tombaugh, Clyde William (1906–1997) The American astronomer best known for his discovery of the planet Pluto in 1930.

topographic relief   The differences in elevation from point to point on a planetary surface.

tornado   A violent rotating column of air, typically 75 m across with 200 km/h winds. Some tornadoes can be more than 3 km across, and winds up to 500 km/h have been observed.

total lunar eclipse   A lunar eclipse in which the Moon passes through the umbra of Earth's shadow.

total solar eclipse   The type of eclipse that occurs when Earth passes through the umbra of the Moon's shadow so that the disk of the Sun is completely blocked by the Moon.

transform fault   The actively slipping segment of a fracture zone.

Trans-Neptunian Object (TNO)   Another name for a *Kuiper Belt object*.

transverse wave   A wave in which the oscillations are perpendicular to the direction of the wave's propagation.

triple-alpha process   The nuclear fusion reaction that combines three helium nuclei, or "alpha particles," together into a single nucleus of carbon. See also *CNO cycle*.

Trojan asteroid   One of a group of asteroids orbiting in the L_4 and L_5 Lagrangian points of Jupiter's orbit.

tropical year   The time between one crossing of the vernal equinox and the next. Due to precession of the equinoxes, a tropical year is slightly shorter than the time that it takes for Earth to orbit once about the Sun.

tropics   The region on Earth between 23.5° south latitude and 23.5° north latitude and in which the Sun appears directly overhead twice during the year.

tropopause   The top of a planet's troposphere.

troposphere   The convection-dominated layer of a planet's atmosphere. On Earth the atmospheric region closest to the ground within which most weather phenomena take place.

T Tauri star   A young stellar object that has dispersed enough of the material surrounding it to be seen in visible light.

tuning fork diagram   The two-pronged diagram showing Hubble's classification of galaxies into ellipticals, S0s, spirals, barred S0s and spirals, and irregular galaxies.

turbulence   The random motion of blobs of gas within a larger cloud of gas.

Type I supernova   A supernova explosion in which no trace of hydrogen is seen in

the ejected material. Most supernovae of this type are thought to be the result of runaway carbon burning in a white dwarf star onto which material is being deposited by a binary companion.

Type II supernova   A supernova explosion in which the degenerate core of an evolved massive star suddenly collapses and rebounds.

U

ultraviolet (UV) radiation   Describes electromagnetic radiation with frequencies and photon energies greater than those of visible light but less than those of X-rays, and wavelengths shorter than those of visible light but longer than those of X-rays.

umbra   (a) The darkest part of a shadow, where the source of light is completely blocked. (b) The darkest, innermost part of a sunspot.

unbalanced force   The nonzero net force acting on a body.

unbound orbit   An orbit in which the velocity is greater than the escape velocity.

unified model of AGN   A model in which many different types of activity in the nuclei of galaxies are all explained by accretion of matter around a supermassive black hole.

uniform circular motion   Motion in a circular path at a constant speed.

universal gravitational constant (G)   The constant of proportionality in the universal law of gravity. Value: $G = 6.673 \times 10^{-11}$ Nm^2/kg^2.

universal law of gravitation   The gravitational force between any two objects is proportional to the product of their masses and inversely proportional to the square of the distance between them.

universe   All of space and everything contained therein.

unstable equilibrium   An equilibrium state in which a small disturbance will cause a system to move away from equilibrium.

V

vacuum   A region of space which contains very little matter. However, in quantum mechanics and general relativity, even a perfect vacuum has physical properties.

variable star   A star with varying luminosity. Many periodic variables are found within the instability strip on the H-R diagram.

velocity The rate and direction of change of an object's position with time. Units: m/s, km/h. Compare with *speed*.

vernal equinox (a) The point where the ecliptic crosses the celestial equator, with the Sun moving from south to north. (b) The day on which the Sun appears at this location; the first day of spring in the Northern Hemisphere (about March 21).

virtual particle A particle which, according to quantum mechanics, comes into momentary existence. According to theory, fundamental forces are mediated by the exchange of virtual particles.

visual binary A binary star in which the two stars can be seen individually from Earth.

Vogt-Russell theorem A theorem that states that the structure of a star is completely determined by its mass and chemical composition.

volatile material Material that remains gaseous at moderate temperature, sometimes referred to as an "ice." Compare with *refractory material.*

volcanic dome Steep-sided, dome-shaped volcanic feature, produced by viscous lavas.

volcanism The occurrence of volcanic activity on a planet or moon.

vortex (pl. vortices) Any circulating fluid system: (a) An atmospheric anticyclone or cyclone. (b) A whirlpool or eddy.

W

waning The changing phases of the Moon as it becomes less fully illuminated between full Moon and new Moon as seen from Earth.

watt (W) A measure of *power*. Units: J/s.

wave A disturbance moving along a surface or passing through a space or a medium.

wavefront The imaginary surface of an electromagnetic wave, either plane or spherical, oriented perpendicular to the direction of travel.

wavelength The distance on a wave between two adjacent points having identical characteristics. The distance a wave travels in one period. Unit: m.

waxing The changing phases of the Moon as it becomes more fully illuminated between new Moon and full Moon as seen from Earth.

weak nuclear force The force underlying some forms of radioactivity and certain interactions between subatomic particles. It is responsible for radioactive beta decay and for the initial proton–proton interactions that lead to nuclear fusion in the Sun and other stars. One of the four fundamental forces of nature.

Wegener, Alfred (1880–1930) The multidisciplinary German scientist who in 1915 first proposed the theory of continental drift, for which he was generally ridiculed. His theory finally became widely accepted in the 1950s. See also *plate tectonics.*

weight The force equal to the mass of an object times the local acceleration due to gravity; or (in general relativity) the force equal to the mass of an object times the acceleration of the reference frame in which the object is observed.

Whipple, Fred (1906–2004) The American planetary astronomer who proposed the "dirty snowball" model of comet nuclei.

white dwarf The stellar remnant left at the end of the evolution of a low-mass star. A typical white dwarf has a mass of 0.6 $M_\odot$, and a radius about that of Earth: it is made of nonburning, electron degenerate carbon.

Wien's Law A relationship describing how the peak wavelength, and therefore the color, of electromagnetic radiation from a glowing black body changes with temperature. Also known as Wien's Displacement law. See *lambda peak.*

Wilson, Robert (1936–) The Bell Laboratories physicist who, together with Arno Penzias, in the early 1960s discovered the cosmic background radiation, for which they won the Nobel Prize in physics.

WIMP "Weakly interacting massive particle." A hypothetical massive particle that interacts through gravity but not with electromagnetic radiation and is a candidate for dark matter. See also *MACHO.*

winter solstice The time of year (about December 22) when the Sun is at its most southerly distance from the celestial equator. The first day of Northern Hemisphere winter.

X

X-ray Electromagnetic radiation having frequencies and photon energies greater than those of ultraviolet light but less than those of gamma rays, and wavelengths shorter than those of UV light but longer than those of gamma rays.

X-ray binary A binary system in which mass from an evolving star spills over onto a collapsed companion such as a neutron star or black hole. The material falling in is heated to such high temperatures that it glows brightly in X-rays.

Y

year The time it takes Earth to make one revolution around the Sun. A solar year is measured from equinox to equinox. A sidereal year, Earth's true orbital period, is measured relative to the stars.

Z

zenith The point on the celestial sphere located directly overhead from an observer.

zodiac The 12 constellations lying along the plane of the ecliptic.

zodiacal dust Particles of cometary and asteroidal debris less than 100 μm in size that orbit the inner Solar System close to the plane of the ecliptic. Compare with *meteoroid* and *planetesimal.*

zodiacal light A band of light in the night sky caused by sunlight reflected by zodiacal dust.

zonal wind The planetwide circulation of air that moves in directions parallel to the planet's equator.

Credits

These pages contain credits for all 21 chapters in the complete volume of *21st Century Astronomy,* Second Edition. The Solar System Edition does not include Chapters 13 and 15–20. The Stars and Galaxies Edition does not include Chapters 6–12.

Part I Opener STScI. **Part II Opener** NASA/JPL. **Part III Opener** NASA/ESA/STScI/J. Hester and P. Scowen (Arizona State University). **Part IV Opener** Jason T. Ware/Photo Researchers, Inc.

Chapter 1

Opener Courtesy of Anthony Ayiomamitis (http://www.perseus.gr). **1.1** Craig Lovell/Corbis. **1.2** Dr. Bernard G. Lindsay, Physics and Astronomy Department, Rice University. **1.3a & b** Neil Ryder Hoos; **1.3c** Owen Franken/Corbis; **1.3d** George Hall/Corbis; **1.3e** PhotoDisc; **1.3f & g** American Museum of Natural History; **1.3h** NASA/JPL/Caltech. **1.4** Bettmann/Corbis. **1.5 left** NASA/STScI/Arizona State University/Hester/Ressmeyer/Corbis; **1.5 right** Ron Watts/Corbis. **1.6a** Michael J. Tuttle (NASM/NASA); **1.6b** NSSDC. **1.7** NRAO/AUI, James J. Condon, John J. Broderick, and George A. Seielstad. **1.8** ©1969 by The New York Times Co. Reprinted by permission. **1.9 top left** Bettmann/Corbis; **1.9 top right** Special Relativity: *The M.I.T. Introductory Physics Series* by A. P. French. ©1968, 1966 by Massachusetts Institute of Technology. Used by permission of W. W. Norton & Company, Inc; **1.9 center** Seth Joel/Corbis; **1.9 bottom left** Musée d'Orsay, Paris, France. Photo: Réunion des Musées Nationaux/Art Resource, NY. Photograph by Herve Lewandowski; **1.9 bottom right** Robbie Jack/Corbis. **1.10** Bettmann/Corbis. **1.11** Four seasons photographs, Jim Schwabel/Index Stock Imagery/PictureQuest. **1.12** NON SEQUITUR by Wiley Miller. Dist. by Universal Press Syndicate. Reprinted with permission. All rights reserved.

Chapter 2

Opener Jon Arnold Images/Alamy. **2.1** British Library Maps. **2.7 left** Pekka Parviainen/Photo Researchers, Inc.; **2.7 right** David Nunuk/Photo Researchers, Inc. **2.19 top left** AFP/Getty Images; **2.19 top right** AP Images; **2.19 bottom left** Lynn Goldsmith/Corbis; **2.19 bottom right** Lindsay Hebberd/Corbis. **2.24** Roger Ressmeyer/Corbis. **2.26** Courtesy of Nick Quinn. **2.27** Courtesy of Koen van Gorp, www.koenvangorp.be **2.28a** Reuters New Media Inc./Corbis. **2.28b** Photograph ©2001 by Fred Espenak, courtesy of www.MrEclipse.com.

Chapter 3

Opener Richard Cummins/Corbis. **3.11** Courtesy of Alan Bean.

Chapter 4

Opener Courtesy David Malin. **4.8** Eyewire Collection/Getty Images. **4.28** NASA/NSSDC/GSFC.

Chapter 5

Opener NASA/JPL. **5.1** Junenoire Photography. **5.2** Gianni Tortoli/Photo Researchers, Inc. **5.3** Jim Sugar/Corbis. **5.4a** Yerkes Observatory photographs: Richard Dreiser; **5.4b** W. M. Keck Observatory/CARA. **5.5** ESO Telescope Systems Division. **5.8b** Howard Voss. **5.10b** Howard Voss. **5.11** Howard Voss. **5.13c** Steve Percival/Photo Researchers, Inc. **5.17** Keck Observatory. **5.18 left to right** NASA; NASA/CXC/SAO; Graphic courtesy Orbital Sciences Corp; NOAO/AURA/NSF; NASA/JPL/Caltech; Joint Astronomy Center in Hilo, Hawaii; NRAO VLA Image Gallery; The National Radio Astronomy Observatory, Green Bank; Dave Finley, Courtesy National Radio Astronomy Observatory and Associated Universities, Inc. **5.19a** Dr. Jeremy Burgess/Photo Researchers, Inc.; **5.19b** Hulton-Deutsch Collection/Corbis. **5.20** NASA/IPAC/NED/JPL/Caltech/SSDS. **5.21b** Jean-Charles Cuillandre (CFHT). **5.22a** Courtesy of Ed Grafton; **5.22b** Steve Larson/University of Arizona. **5.24a** Roger Ressmeyer/Corbis; **5.24b** David Parker/Photo Researchers, Inc. **5.25** Image courtesy of NRAO/AUI. **5.26** ESO Telescope Systems Division. **5.27** NASA/NSSDC. **5.28** NASA/JPL. **5.29** Fermilab/Photo Researchers, Inc. **5.30** Cern/Photo Researchers, Inc. **5.31** Frank Summers, Space Telescope Science Institute; Chris Mihos, Case Western Reserve University; Lars Hernquist, Harvard University.

Chapter 6

Opener Caltech/NASA/JPL. **6.1a** STScI photo: Karl Stapelfeldt (JPL); **6.1b** D. Padgett (IPAC/Caltech), W. Brandner (IPAC), K. Stapelfeldt (JPL) and NASA; **6.1c** D. Padgett (IPAC/Caltech), W. Brandner (IPAC), K. Stapelfeldt (JPL) and NASA; **6.1d** D. Padgett (IPAC/Caltech), W. Brandner (IPAC), K. Stapelfeldt (JPL) and NASA. **6.2** Photograph by Pelisson, SaharaMet. **6.4** Reuters/Corbis. **6.11** NASA/JPL/Northwestern University.

Chapter 7

Opener NASA/GRIN. **7.1** Stockli, Nelson, Hasler, Goddard Space Flight Center/NASA. **7.2** USGS Hawaiian Volcano Observatory. **7.3b** NASA/JSC. **7.4** Montes De Oca & Associates. **7.5** Photograph by D.J. Roddy and K.A. Zeller, USGS, Flagstaff, AZ. **7.6** Don Davis. **7.7** NASA/JSC. **7.8** NASA/JPL/Caltech. **7.9** Dennis Flaherty/Photo Researchers, Inc. **7.10** NASA/JPL/Caltech. **7.13b** Grant Heilman Photography, Inc. **7.14** Photo courtesy of Ron Greeley. **7.19** NASA/JSC. **7.20** NSSDC/NASA. **7.21** NASA/Magellan Image/JPL. **7.23** NASA/JSC.

7.24 NASA/JSC. **7.25** NASA/JPL/Caltech. **7.26** NASA/MOLA Science Team. **7.27** NASA/JPL. **7.28** NASA/JPL/Malin Space Science Systems. **7.29a** NASA/JPL/Caltech; **7.29b** NASA//JPL/Caltech. **7.30** NASA/JPL/Caltech, courtesy Alfred McEwen, USGS. **7.31a** NASA/JPL/Malin Space Science Systems; **7.31b** NASA/JPL/Cornell. **7.32** ESA/DLR/FU Berlin (G. Neukum).

Chapter 8

Opener Adam Block/NOAO/AURA/NSF. **8.1 left** NASA/NSSDC; **8.1 center** NASA/NSSDC; **8.1 right** NASA/STScI. **8.9a** Louis A. Frank, The University of Iowa; **8.9b** Raymond Gehman/Corbis. **8.11** Lyndon State College/Nolan Atkins. **8.13** Courtesy Peter Arnold. **8.14** NOAA. **8.15** Eric Nguyen/Jim Reed Photography/Corbis. **8.16** NASA/JPL. **8.18a** Craig Aurness/Corbis; **8.18b** Philip James Corwin/Corbis. **8.19** NASA/JPL/Caltech. **8.20** NASA/JPL/Caltech. **8.21a** NASA/JPL/Caltech; **8.21b** NASA/STScI.

Chapter 9

Opener NASA/JPL/Space Science Institute. **9.1a** NASA/JPL/University of Arizona; **9.1b** NASA and E. Karkoschka (University of Arizona). **9.2a** NASA/JPL; **9.2b** NASA/JPL. **9.3** NASA/JPL/Caltech. **9.6a** NASA/STScI; **9.6b** NASA/JPL. **9.7** CICLOPS/NASA/JPL/University of Arizona. **9.8a** NASA/JPL/Caltech. **9.9a** E. Karkoschka (LPL) and NASA; **9.9b** NASA/JPL; **9.9c** L. Sromovsky (University of Wisconsin) and NASA. **9.10** NASA/JPL/Caltech. **9.18** N.M. Schneider, J.T. Trauger, Catalina Observatory. **9.19a** J. Clarke (University of Michigan), NASA; **9.19b** Courtesy of John Clark, Boston University and NASA/STScI.

Chapter 10

Opener NASA/JPL. **10.7** Christopher Mackay. **10.12** NASA/JPL/Caltech. **10.13a** NASA, ESA, H. Weaver, and T. Smith (STScI); **10.13b** New Mexico State University Astronomy Department. **10.14** NASA/AURA.

Chapter 11

Opener NASA/JPL/Space Science Institute/Photo Researchers, Inc. **11.1** NASA/JPL/Caltech. **11.3** NASA/JPL/Space Science Institute. **11.5** NASA/JPL/Caltech. **11.6** NASA/JPL/Caltech. **11.7**: NASA/JPL/Caltech. **11.8** NASA/JPL/Caltech. **11.9** NASA/JPL/Caltech. **11.10a** NASA/JPL/Caltech; **11.10b** NASA/JPL/Space Science Institute. **11.11a** NASA/JPL/Space Science Institute; **11.11b** NASA/JPL/Caltech. **11.12** NASA/JPL/Space Science Institute. **11.13** NASA/JPL/Caltech. **11.14** NASA/JPL/Caltech. **11.15** NASA/JPL/Caltech. **11.16** NASA/JPL/Caltech. **11.17a** NASA/JPL/Space Science Institute; **11.17b** NASA/JPL/Space Science Institute. **11.18** NASA/JPL/Caltech. **11.19** NASA/JPL/Caltech. **11.20** NASA/JPL/Caltech. **11.21a** NASA/JPL/Space Science Institute; **11.21b** NASA/JPL/Space Science Institute; **11.21c** NASA/JPL/Space Science Institute. **11.22** NASA/JPL/ESA/University of Arizona. **11.23a** NASA/JPL/ESA/University of Arizona; **11.23b** NASA/JPL. **11.25** NASA/JPL/Space Science Institute. **11.26a** NASA/JPL/Caltech; **11.26b** Courtesy of Lucasfilms Ltd, *Star Wars: Episode IV—A New Hope* ©1977 & 1997 Lucasfilm Ltd. & TM, All rights reserved. **11.27** NASA/JPL/Caltech. **11.28** Dr. R. Albrecht, ESA/ESO Space Telescope European Coordinating Facility/NASA. **11.29** M. Brown (Caltech), C. Trujillo (Gemini), D. Rabinowitz (Yale), NSF, NASA. **11.31** NASA, ESA, J. Parker (Southwest Research Institute), P.Thomas (Cornell University), and L. McFadden (University of Maryland, College Park).

Chapter 12

Opener Jerry Lodriguss/Photo Researchers, Inc. **12.1** From *Meteorites and their parent planets*, H.Y. McSween, Cambridge. **12.2** Courtesy of Ron Greeley. **12.3** NASA/JPL/Cornell. **12.6** NASA/JPL/Caltech. **12.7** NASA/Johns Hopkins University Applied Physics Laboratory (JHU/APL). **12.8 left** NASA/JPL/Cornell; **12.8 right** ESA/DLR/FU Berlin (G. Neukum). **12.10b** Courtesy of Terry Acomb. **12.11a** NASA/NSSDC/GSFC; **12.11b** HST/WFPC2 H. Weaver (ARC), NASA. **12.12** NASA/JPL/Caltech. **12.13** NASA/JPL/Caltech/UMD. **12.14** NASA/JPL/Caltech/UMD. **12.16** Roger Lynds/NOAO/AURA/NSF. **12.19a** NASA/JPL/MPIA; **12.19b** NASA/STScI; **12.19c** R. Evans, J. T. Trauger, H. Hammel and the HST Comet Science Team, and NASA. **12.20** Courtesy of the Wolbach Library, Harvard-Smithsonian Center for Astrophysics, Cambridge, MA. **12.21b** Getty © Barrie Rokeach. **12.22** Armagh Observatory. **12.23** Tony Hallas/Science Faction. **12.24** Courtesy of Joe Orman. **12.25** NASA.

Chapter 13

Opener Jon Lomberg/Photo Researchers, Inc.

Chapter 14

Opener Neil deGrasse Tyson, 2001. **14.9a** Brookhaven National Laboratory; **14.9b & c** Courtesy of Tomasz Barszczak, University of California, Irvine, for the Super-Kamiokande Collaboration; **14.9d** Lawrence Berkely National Laboratory/Photo Researchers, Inc. **14.10** NOAO/AURA/NSF. **14.11** SOHO/ESA/NASA. **14.13** Nigel Sharp, NOAO/NSO/Kitt Peak FTS/AURA/NSF. **14.14a** Stefan Seip, www.astromeeting.de; **14.14b & c** 2001 by Fred Espenak, courtesy of www.MrEclipse.com. **14.15** NASA/Photo Researchers, Inc. **14.16** SOHO/ESA/NASA. **14.19** SOHO/ESA/NASA. **14.20** Göran Scharmer/Institute for Solar Physics of the Royal Swedish Academy of Sciences, Sweden. **14.22c & d** SOHO/ESA/NASA. **14.24** SOHO/ESA/NASA. **14.25** SOHO/EIT and SOHO/LASCO (ESA and NASA).

Chapter 15

Opener Canada-France-Hawaii Telescope/J.-C. Cuillandre/Coelum. **15.1 top** Image courtesy of Dr. A. Mellinger; **15.1 bottom** NASA/NSSDC/GSFC Diffuse Infrared Background Experiment (DIRBE) instrument on the Cosmic Background Explorer (COBE). **15.4** NSSDC/GSFC Infrared Astronomical Satellite (IRAS). **15.5** NSSDC/GSFC Roentgen Satellite (ROSAT). **15.6** Courtesy of Ron Reynolds, WHAM project. **15.7a** Courtesy of Stefan Seip; **15.7b** Courtesy of Emmanuel Mallart www.astrophoto-mallart.com & www.axisinstruments.com. **15.8** ESO Telescope Systems Division. **15.9** C. Jones, Smithsonian Astrophysical Observatory. **15.10** Patrick Hartigan, Rice University. **15.15** Jeff Hester and Paul Scowen (Arizona State University), Bradford Smith (University of Hawaii), Roger Thompson (University of Arizona), and NASA. **15.19** Karl Stapelfeldt/JPL/NASA. **15.20** Jeff Hester (Arizona State University), WFPC2 Team, NASA. **15.21** Anglo-Australian Observatory/Royal Observatory, Edinburgh, photograph from UK Schmidt, plates by David Malin.

Chapter 16

Opener NASA, ESA and H.E. Bond (STScI). **16.11a** Minnesota Astronomical Society; **16.11b** NASA and The Hubble Heritage Team (STScI/AURA). **16.12 top left** Dr. Raghvendra Sahai (JPL) and Dr. Arsen R. Hajian (USNO), NASA and The Hubble Heritage Team (STScI/AURA); **16.12 center left** A. Hajian (USNO) et al., Hubble Heritage Team (STScI/AURA), NASA; **16.12 bottom left** J. P. Harrington and K. J. Borkowski (University of Maryland), HST, NASA; **16.12 top right** NASA, A. Fruchter and the ERO team (STScI); **16.12 center right** Anglo-Australian Telescope, photograph by David Malin; **16.12 bottom right** Bruce Balick (University of Washington), Vincent Icke (Leiden University, The Netherlands), Garrelt Mellema (Stockholm University), and NASA.

Chapter 17

Opener NASA/HST/ASU/J. Hester et al. **17.7** STScI/NASA/Arizona State University/Hester/Ressmeyer/Corbis. **17.10** Anglo-Australian Observatory, photographs by David Malin. **17.11a** Joachim Trumper, Max-Planck-Institut für extraterrestrische Physik; **17.11b** Courtesy Nancy Levenson and colleagues, American Astronomical Society; **17.11c** Jeff Hester, Arizona State University, and NASA. **17.15a** European Southern Observatory, ESO PR Photo 40f199; **17.15b & c** Jeff Hester, Arizona State University, and NASA.

Chapter 18

Opener NASA/ESA/STScI/Photo Researchers, Inc. **18.1** NASA, ESA, S. Beckwith (STScI) and the HUDF Team. **18.3** Courtesy David Malin; **18.3 top right** Canada-France-Hawaii Telescope/J.-C. Cuillandre/Coelum. **18.4** The Hubble Heritage Team (AURA/STScI/NASA). **18.6** NOAO/AURA/NSF. **18.7** Courtesy David Malin. **18.8** C. Howk (JHU), B. Savage (University of Wisconsin), N. A. Sharp (NOAO)/WIYN/NOAO/NSF. **18.10 top left** George Jacoby, Bruce Bohannan, Mark Hanna/NOAO/AURA/NSF; **18.10**

top right European Southern Observatory (ESO). **18.11 bottom left** Courtesy David Malin; **18.11 bottom right** Courtesy David Malin. **18.12** R. Windhorst and D. Burstein (Arizona State University). **18.13 left** Todd Boroson/NOAO/AURA/NSF; **18.13 right** Richard Rand, University of New Mexico. **18.15** Courtesy David Malin. **18.16b** Z. Frei, Institute of Physics, Eotvos University, Hungary. **18.17** Images courtesy G. Fabbiano, Harvard-Smithsonian Center for Astrophysics. **18.18** J. Bahcall (Institute for Advanced Study), M. Disney (University of Wales) and NASA. **18.19** NRAO/Alan Bridle et al. **18.22b** Courtesy Bill Keel, University of Alabama, and NASA; **18.22c** Holland Ford, STScI/Johns Hopkins University, and NASA; **18.22d** R. P.van der Marel, STScI, F. C. van den Bosch, University of Washington, and NASA. **18.23** F. Owen, NRAO, with J. Biretta, STScI, and J. Eilek, NMIMT. **18.24** Holland Ford, STScI/Johns Hopkins University; Richard Harms, Applied Research Corp.; Zlatan Tsvetanov, Arthur Davidsen, and Gerard Kriss at Johns Hopkins University; Ralph Bohlin and George Hartig at STScI; Linda Dressel and Ajay K. Kochhar at Applied Research Corp. in Landover, MD; and Bruce Margon from the University of Washington, Seattle/NASA. **18.25** John Biretta, STScI. **18.26** Halton Arp/Caltech.

Chapter 19

Opener Gordon Garradd 1996. **19.1a** Axel Mellinger; **19.1b** C. Howk and B. Savage (University of Wisconsin); N. Sharp (NOAO)/WIYN, Inc. **19.2** U.S. Patent Office Library. **19.3** Hubble Heritage Team (AURA/STScI/NASA). **19.4** NOAO/AURA/NSF. **19.5** Courtesy Jeff Hester. **19.7** Courtesy J. Binney. **19.9** Anglo-Australian Observatory. **19.12** The Electronic Universe Project. **19.16a** Carl Heiles, UC Berkeley; **19.16b** C. Howk and B. Savage (University of Wisconsin), N. Sharp (NOAO). **19.18** NRAO. **19.19** Anglo-Australian Observatory, photograph by David Malin.

Chapter 20

Opener NASA/WMAP Science Team. **20.2** TIMEPIX/Getty Images. **20.3 top** Courtesy

of Jeff Hester; **20.3 bottom** NASA William C. Keel (University of Alabama, Tuscaloosa). **20.5** WIYN/NOAO/NSF, WIYN Consortium, Inc. **20.14** Lucent Technologies' Bell Labs. **20.15** Ann and Rob Simpson. **20.17a & b** NASA COBE Science Team; **20.17c** NASA/WMAP Science Team.

Chapter 21

Opener NASA/JPL/Origins. **21.2a** Anglo-Australian Observatory; **21.2b** Omar Lopez-Cruz and Ian Shelton/NOAO/AURA/NSF. **21.3** Max Tegmark/SDSS Collaboration; **21.6** AAO/David Malin. **21.7** NASA, A. Fruchter and the ERO team (STScI, ST-ECF). **21.8** Images produced by Donna Cox Smithsonian Institution and Motorola Corporation. **21.14** From *The Elegant Universe* by Brian Greene. ©1999 by Brian R. Greene. Used by permission of W. W. Norton & Company, Inc. **21.16** Kenneth EWard/Biografx/Photo Researchers, Inc. **21.17 top left** Mark A. Schneider/Visuals Unlimited; **21.17 middle left** K. Sandved/Visuals Unlimited; **21.17 bottom left** Ken Lucas/Visuals Unlimited; **21.17 top right** Science VU/National Museum of Kenya/Visuals Unlimited; **21.17 middle right** Ken Lucas/Visuals Unlimited; **21.17 bottom right** Ken Lucas/Visuals Unlimited. **21.18** Jeff J. Daly/Visuals Unlimited. **21.19** NASA. **21.20** NASA. **21.21** F. Drake (UCSC) et al., Arecibo Observatory (Cornell, NAIC). **21.22** FOXCLIPS. **21.23 bottom left** Dr. Michael Perfit, University of Florida, Robert Embley/NOAA; **21.23 bottom right** Woods Hole Oceanographic Institution, Deep Submergence Operations Group, Dan Fornari. **21.24** Isaac Gary.

Epilogue NASA/Manned Spacecraft Center/Image # 68-HC-870.

Appendix 8 The International Astronomical Union/Martin Kornmesser.

Every attempt has been made to contact the permission holders of each image. If there is a change or correction, please contact the Photo Permissions Department at W. W. Norton & Company, 500 Fifth Avenue, New York, NY 10110.

Index

This Index contains page references for all 21 chapters in the complete volume of *21st Century Astronomy*, Second Edition. The Solar System Edition does not include pages 374–401 and 430–605. The Stars and Galaxies Edition does not include pages 161–372.